T0179192

MOLECULARLY IMPRINTED MATERIALS

Science and Technology

edited by
MINGDI YAN
OLOF RAMSTRÖM

CRC Press
Taylor & Francis Group
Boca Raton London New York

CRC Press is an imprint of the
Taylor & Francis Group, an **informa** business

CRC Press
Taylor & Francis Group
6000 Broken Sound Parkway NW, Suite 300
Boca Raton, FL 33487-2742

First issued in paperback 2020

© 2005 by Taylor & Francis Group, LLC
CRC Press is an imprint of Taylor & Francis Group, an Informa business

ISBN 13: 978-0-367-57819-0 (pbk)
ISBN 13: 978-0-8247-5353-5 (hbk)

Library of Congress Cataloging-in-Publication Data
A catalog record for this book is available from the Library of Congress.

Visit the Taylor & Francis Web site at
http://www.taylorandfrancis.com

and the CRC Press Web site at
http://www.crcpress.com

Foreword

The science and technology of molecular imprinting lie at the meeting point of several areas of high activity in chemical sciences: polymer chemistry, supra-molecular chemistry, analytical chemistry, and of course organic, inorganic, and bio-chemistry. They present many features of great current interest from both the basic and applied points of view, from the design of molecular recognition processes to their implementation in separation science.

Presenting the panorama of molecular imprinting research and development thus requires a multidisciplinary approach that must nevertheless remain accessible to the many scientists of varied backgrounds who wish to enter the field and take advantage of its achievements.

The present book is an excellent road map toward this goal, by providing expert presentations of the very diverse facets of molecular imprinting science and technology.

The authors and editors are to be complimented for their high quality and timely work. They deserve the gratitude of the chemical community.

Jean-Marie Lehn
Laboratoire de Chimie Supramoléculaire
ISIS/ULP
Strasbourg, France

Preface

Molecular imprinting, a technique of tailor-making network polymers for the recognition of specific analyte molecules, has attracted increasing interest in recent years, as evidenced by the exponential growth in the number of publications during the past decade. Every year, researchers new to this technology enter the field, wishing to learn the synthesis, characterization, and uses of these materials for a variety of applications. Our goal has been to offer in this book a comprehensive tutorial that will serve as a reference book especially for those who wish to learn the basic techniques and make contributions to the field. Readers will find many in-depth discussions and guidelines as well as detailed experimental protocols that are meant to help beginners jump-start into the field of molecular imprinting. On the other hand, skilled researchers will find new information and recent developments.

The book starts with a general overview of molecular imprinting that surveys various aspects of molecular imprinting technology. The issue of how to qualitatively evaluate the imprinting effect is presented and discussed, and guidelines are provided in a flowchart format to show how the effect can be determined. As with advances of many fields of science and technology, the pioneer work of key individuals has shaped and promoted the development of molecular imprinting technology. Three such individuals are profiled in Chapter 2, through interviews discussing their past and present achievements and their views for the future. In Part II, various approaches to the preparation of molecularly imprinted polymers (MIPs) are presented. Strategies for the design and synthesis of MIPs are discussed in Part III. Part IV details the means for the design and analysis of binding characteristics of MIPs. In Part V, methods for the synthesis of uniformly sized and shaped MIPs are described. Part VI surveys the use of MIPs in various applications. The book ends with an appendix that includes many useful addresses and links to assist researchers in initiating projects in molecular imprinting.

The editors are deeply grateful to all the contributors and many other individuals who provided time and effort to review and proofread the manuscripts. We are thankful for the financial support provided by the NSF International Division, which made our collaboration possible.

Mingdi Yan
Olof Ramström

Contents

Contributors

Cameron Alexander   School of Pharmacy and Biomedical Sciences, University of Portsmouth, Portsmouth, United Kingdom

Richard J. Ansell   School of Chemistry, University of Leeds, Leeds, United Kingdom

Claudio Baggiani   Dipartimento di Chimica Analitica, Università di Torino, Torino, Italy

Jennifer J. Brazier   Department of Chemistry, Portland State University, Portland, Oregon, U.S.A.

Oliver Brüggemann   Technische Universität Berlin, Berlin, Germany

Biening Chen   Institute of BioScience and Technology, Cranfield University at Silsoe, Silsoe, Bedfordshire, United Kingdom

Iva Chianella   Institute of BioScience and Technology, Cranfield University at Silsoe, Silsoe, Bedfordshire, United Kingdom

Peter G. Conrad II   Department of Chemistry, University of California, Irvine, California, U.S.A.

David Cunliffe   School of Pharmacy and Biomedical Sciences, University of Portsmouth, Portsmouth, United Kingdom

Sheng Dai   Chemical Sciences Division, Oak Ridge National Laboratory, Oak Ridge, Tennessee, U.S.A.

Jessica L. Defreese   Department of Chemical Engineering, University of California at Berkeley, Berkeley, California, U.S.A.

B. Dirion   Institut für Umweltforschung (INFU), Universität Dortmund, Dortmund, Germany

Shouhai Gao   Toronto Research Chemicals, Inc., North York, Ontario, Canada

Karsten Haupt   Department of Bioengineering, UMR CNRS 6022, Compiègne University of Technology, Compiègne, France

Alexander Katz   Department of Chemical Engineering, University of California at Berkeley, Berkeley, California, U.S.A.

Nicole Kirsch[*]   Institute of Food Research, Norwich, United Kingdom

Takaomi Kobayashi   Department of Chemistry, Nagaoka University of Technology, Nagaoka, Japan

F. Lanza   Institut für Umweltforschung (INFU), Universität Dortmund, Dortmund, Germany

Andrew G. Mayes   School of Chemical Sciences and Pharmacy, University of East Anglia, Norwich, United Kingdom

Klaus Mosbach   Pure and Applied Biochemistry, Chemical Center, Lund University, Lund, Sweden

George M. Murray   Technical Services Department, Johns Hopkins University, Laurel, Maryland, U.S.A.

Ian A. Nicholls   Department of Chemistry and Biomedical Sciences, University of Kalmar, Kalmar, Sweden

Sergey A. Piletsky   Institute of BioScience and Technology, Cranfield University at Silsoe, Silsoe, Bedfordshire, United Kingdom

Olof Ramström   Department of Chemistry, Royal Institute of Technology, Stockholm, Sweden

Ronald H. Schmidt   Pure and Applied Biochemistry, Chemical Center, Lund University, Lund, Sweden

---

[*]*Current affiliation*: Department of Chemistry and Biomedical Sciences, University of Kalmar, Kalmar, Sweden.

B. Sellergren Institut für Umweltforschung (INFU), Universität Dortmund, Dortmund, Germany

Kay Severin Institute of Chemical Sciences and Engineering, Swiss Federal Institute of Technology Lausanne (EPFL), Lausanne, Switzerland

Kenneth J. Shea Department of Chemistry, University of California, Irvine, California, U.S.A.

Ken D. Shimizu Department of Chemistry and Biochemistry, University of South Carolina, Columbia, South Carolina, U.S.A.

Glen E. Southard Johns Hopkins University, Laurel, Maryland, U.S.A.

David A. Spivak Department of Chemistry, Louisiana State University, Baton Rouge, Louisiana, U.S.A.

Anthony P. F. Turner Institute of BioScience and Technology, Cranfield University at Silsoe, Silsoe, Bedfordshire, United Kingdom

Mathias Ulbricht Lehrstuhl für Technische Chemie II, Universität Duisburg-Essen, Essen, Germany

Binghe Wang Department of Chemistry and Center for Biotechnology and Drug Discovery, Georgia State University, Atlanta, Georgia, U.S.A.

Wei Wang Department of Chemistry, University of New Mexico, Albuquerque, New Mexico, U.S.A.

Michael J. Whitcombe Institute of Food Research, Norwich, United Kingdom

Günter Wulff Institute of Organic Chemistry and Macromolecular Chemistry, Heinrich-Heine-University Düsseldorf, Düsseldorf, Germany

Mingdi Yan Department of Chemistry, Portland State University, Portland, Oregon, U.S.A.

Lei Ye Pure and Applied Biochemistry, Chemical Center, Lund University, Lund, Sweden

Ecevit Yilmaz Pure and Applied Biochemistry, Chemical Center, Lund University, Lund, Sweden

1
Molecular Imprinting— An Introduction

Olof Ramström   Royal Institute of Technology, Stockholm, Sweden

Mingdi Yan   Portland State University, Portland, Oregon, U.S.A.

I. INTRODUCTION

All living systems are based on interactions between molecules, and the recognition that takes place in these interactions. The formation of complex structures, such as membranes, DNA duplexes, and whole cells, is essentially a consequence of a multitude of such binding processes. In contrast to the strong (covalent) forces keeping single molecules together in defined species, these complexes are normally maintained by weaker binding forces, leading to dynamics in their formation and breakdown. This dynamic property is a prerequisite for the functioning of many biological processes, and the possibility of rapid organization between different units is the foundation of such diverse reactions as DNA replication, enzymatic catalysis, and protein biosynthesis. Molecular interactions are furthermore a basis for chemical and biological information. All processes comprising molecular interactions, such as hormone responses and cell adhesions, are a consequence of "weak" communications between molecules or groups of molecules. In addition, the identification of self- and non-self in immune systems is based on the molecular recognition of identity markers. This molecular interplay is thus responsible for the expression of life per se.

The concept of molecular interactions is very old, having already been used by the Greek and Roman empires [1]. By the later half of the 19th century, modern ideas about these interactions began to emerge, e.g., through the work of Johannes Diderik van der Waals in his studies of interactions between atoms in the gaseous state, and Alfred Werner's theory on coordination chemistry. In 1894, Emil Fischer presented his famous "lock-and-key" metaphor depicting the way a substrate interacts with an enzyme [2]. In this prophetic statement, an enzyme, which is large compared to the substrate, has clefts and depressions on its surface complementary to the shape of the substrate. Thus, the substrate fits like a key into the lock of the enzyme's active site.

The study of chemistry beyond the molecule, supramolecular chemistry [3], essentially based on weaker but often complex interactions between discrete molecular entities [4], and the ability to mimic natural binding phenomena has

1

intrigued scientists over a long period [5]. These issues have led to the establishment of the field of biomimetic chemistry [6,7], in which imitations of natural binding entities, such as enzymes and antibodies, are being studied. The term "biomimetic" generally refers to any aspect in which a chemical process imitates a biochemical reaction. As the structures and mechanisms of biochemical systems become known, scientists are attempting to transfer this knowledge to synthetic strategies. Often, these synthetic approaches aim at reducing the degree of complexity of biological systems into smaller, "miniaturized" models, e.g., enzyme models that lack macromolecular peptide backbones, but contain catalytically active groups, oriented in the geometry dictated by the active sites of enzymes.

In biological systems, molecular complexes are often formed by a plethora of noncovalent interactions such as hydrogen bonds, ion pairing, and hydrophobic interactions. Although these interactions, when considered individually, are weak in nature compared to covalent bonds, the simultaneous action of several of these weaker bonds often leads to complexes with very high stability. This is the case in biotin–avidin binding, for example, where a dissociation constant in the femtomolar range, corresponding to a binding energy of approximately 90 kJ/mol at 25°C, has been measured in aqueous media [8]. Thus, the concerted action of several weak non-covalent bonds leads to a very stable complex. These features are also employed in artificial systems, where this "chelate effect" may result in highly efficient structures, capable of recognizing their partners with a very high degree of selectivity and strength.

II. WHAT IS MOLECULAR IMPRINTING?

One attractive synthetic approach to mimic nature is *Molecular Imprinting*, which is the concept of preparing substrate-selective recognition sites in a matrix using a molecular template in a casting procedure. The template molecule is first allowed to form solution interactions/bonds with one or several types of functional elements in a prearrangement step, and subsequent locking-in of these interactions/bonds leads to the formation of a matrix which accommodates recognition sites selective for the template. By this procedure, natural binding sites can be mimicked synthetically in a simple but effective way and the resulting molecularly imprinted materials are often characterized by having very high chemical and physical stability.

In the spirit of Fischer's lock-and-key metaphor, Molecular Imprinting can be described as a way of making artificial "locks" for "molecular keys". In the following very simplistic view, a brief outline on the Molecular Imprinting procedure is drawn (Fig. 1).

The "molecular key" may, in principle, be any substance ranging from small molecules such as drug substances, amino acids, steroid hormones, or metal ions to larger molecules such as peptides or proteins. Large molecules and molecular assemblies, such as cells and viruses, may also be perceived. However, the difficulty of making the imprinted materials generally increases with the size of the selected key molecule. When preparing molecularly imprinted polymer against large imprint species, the template could be trapped permanently inside the polymer after polymerization. The rebinding process, which relies on the diffusion of the template molecules to the recognition sites, will also be hindered by this size effect.

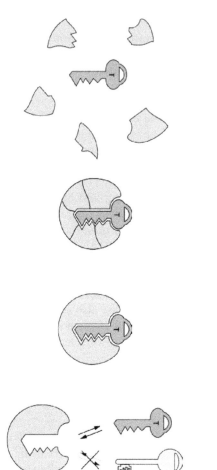

The selected key molecule is in the first step mixed with lock building blocks.

The building blocks and the key are allowed to, either firmly or loosely, associate with each other.

The so formed complexes between the key and the building blocks are subsequently "glued" together in order to fix the building block positions around the key.

Removing the molecular key then leaves a construction which, if everything works properly, is selective for the original key.

Figure 1 Principle of Molecular Imprinting inspired by Fischer's lock-and-key metaphor.

III. IMPRINTING TERMINOLOGY

By definition, "imprinting" is referred to as a process of producing a mark, or an "imprint", onto a surface by pressure, a terminology that may also be used in a transferred sense in, e.g., imposing ideas or certain properties onto an object. This terminology is commonly used in (book-) printing technology, but several other areas have also adopted the wording. In engineering, for example, imprinting is used to describe the process of transferring patterns from an original mold to a surface. When the size of the patterns is in the range of micrometers or nanometers, the procedure is called imprint lithography or nanoimprint lithography [9]. In modern genetics, the term imprinting, often referred to as genomic imprinting or parental imprinting, is the phenomenon in which certain maternal and paternal genes are differentially expressed in the offspring [10].

In "Molecular Imprinting" molecules are instead used to create the marks or imprints, normally within a network polymer. A number of expressions have been used in the past to describe the present technology. These include, for example: "enzyme-analogue built polymers" [11], "host-guest polymerisation" [12], "template synthesis" or "template polymerization" [13,14], creation of "footprints" [15], and preparation of "specific adsorbents" [16].

"Imprints" as a term associated to the concept appeared in the early literature [17], but the complete catch-phrase "Molecular Imprinting" first emanated from a series of publications in the 1980's [18]. Since then, the expression has gained much popularity in the community, and "Molecular Imprinting" together with "Molecularly Imprinted Polymer" (MIP) [19] are currently widely accepted and have become the standard terminology in the field.

In molecular imprinting, the "key molecules" depicted above can be denoted by a variety of expressions such as "templates" (T), "template molecules", "target molecules", "analytes", "imprint molecules", "imprint antigens", or "print molecules", any of which is frequently encountered. The "lock building blocks" are normally called functional monomers (M), although polymers have also been used as imprinting building blocks. The "molecular glue" used to fix the key-building block complexes is almost always perceived as crosslinkers (X) or crosslinking monomers. The entire imprinting procedure is performed in a solvent, which can be denoted "porogen", since one of its functions is to fill out space between the aggregated network polymer so as to induce a porous construction. Since most imprinting protocols make use of a radical polymerization process, a free radical initiating agent—or initiator—is also added.

IV. APPROACHES TO MOLECULAR IMPRINTING

Currently, two fundamental approaches to Molecular Imprinting may be distinguished, presented in detail in Part II. One of these is the covalent approach, mainly developed by Günter Wulff and coworkers, where the template–monomer construct in solution prior to polymerization is maintained by reversible covalent bonds, and the recognition between the polymer is dependent on the formation and cleavage of these bonds (Chapter 4). The other major type is the noncovalent approach, advocated mainly by Klaus Mosbach and coworkers, where the prearrangement between the template and the monomer(s) is formed by noncovalent (Chapter 3), or (weak) metal coordination interactions (Chapter 6), and subsequent recognition is also dependent on these interactions. In parallel with these strategies, another method has been developed that takes advantage of a combination of these approaches, with strong covalent bonds being used in the imprinting step, and noncovalent interactions used in the recognition process after cleavage of the template from the polymer. This so-called semi-covalent approach has recently experienced increased interest, in part due to new and improved systems (Chapter 5).

All of these approaches have their respective advantages and drawbacks, and the choice of system is largely dependent on what template and what application are specifically targeted. Generally, the covalent and semi-covalent approaches can be successful in creating well-defined recognition sites, but both these techniques suffer from limitations in what templates can be used. Essentially, the same situation is true for the use of metal–coordination interactions, and therefore these approaches have enjoyed their principal success in rather specific systems. The noncovalent approach is generally regarded as being of more versatile nature, inasmuch as it can be applied to almost any type of template. On the other hand, the inherent weakness of the interactions makes this approach more difficult to control, often associated with a higher degree of heterogeneity in the binding sites formed. This approach is also highly sensitive to the polarity of the solvent, as a consequence of the noncovalent interactions employed. Nevertheless, this approach has attracted the greatest number of researchers, and the majority of applications currently presented in the literature complies with this technique.

Other variations to the above themes are the techniques of surface imprinting (Chapter 9) and scaffold imprinting (Chapter 10). In these cases, an organizational element capable of holding the interactional elements in the right arrangement is instead used and the "cross-linker" may be excluded from the protocol. This can either be an extended surface of diverse structure, or a molecular scaffold that may be suitably decorated with the functional element. This template-assisted synthesis, leading to an artificial recognition, is thus performed in a very direct way.

V. HOW TO PREPARE A MOLECULARLY IMPRINTED POLYMER

The process of Molecular Imprinting starts with the selection of an appropriate imprinting approach, either covalent or noncovalent. Many factors influence the choice of selection; the most important of which is the template molecule. The type and availability of functional groups on the template molecule play important roles. The structure, size, and solubility of the template, and the end-uses of the imprinted materials also dictate the selection of the imprinting method. After a specific approach is selected, functional monomer(s) need to be chosen so that a good template–monomer complex can guide the formation of the template-specific recognition sites. Other components such as the cross-linker, the polymerization initiator, and the porogen influence the structures and performance of the polymer and therefore should also be carefully selected and studied. A comprehensive discussion on how to select the functional monomers, cross-linkers and the porogen can be found in Chapters 7 and 15. This book contains a multitude of different protocols, useful as guides for initial attempts in the art of Molecular Imprinting.

When free-radical polymerization is employed, polymerization can be initiated either thermally or photochemically, thus starting the cleavage of the initiator to generate free radicals. This radical formation then initiates the polymerization of the functional monomers and the crosslinker, which leads to the formation of a cross-linked polymer network. Note that polymerization is carried out in the presence of the template molecules, and therefore, the template molecules are "trapped" inside the polymer network at the end of this process.

The prepared molecularly imprinted polymer is then subjected to a work-up scheme. If the polymer is made as a solid monolith, a particle fragmentation is normally performed by grinding and sieving, where particles with an average size of 10–25 μm are collected.

Before the molecularly imprinted polymer may be used for rebinding studies, one step remains—extraction of the original template molecules from the polymer matrix. A number of extraction protocols have been developed for different imprinting procedures. In the covalent and metal coordination approaches, an appropriate reagent is needed to break the bonds formed between the template and the functional elements. For the noncovalent approach, a polar solvent, often

containing an acidic or a basic component, can effectively interrupt ionic or hydrogen bonds, thus competing off the template molecules from their binding sites. A final wash with a volatile solvent such as methanol facilitates the removal of the reagents and the drying of the polymer.

VI. BINDING EVALUATION

Once the material has been prepared, the binding characteristics of the material have to be evaluated. This can be achieved in several different ways. For example, the material can be subjected to simple batch binding assays, where the substance of interest is allowed to equilibrate between the polymer phase and a surrounding liquid phase. Measuring the amount of ligand in the liquid by any conventional analytical method such as UV–vis or fluorescence spectroscopy, scintillation counting, etc., will provide information on how much substance is bound to the polymer (Chapter 25). Characteristics such as binding site capacity and homogeneity, as well as binding constants and distribution, can then be evaluated by applying various binding isotherms to the data (Chapter 16). Other techniques, such as HPLC, capillary electrophoresis, etc., may also be used (Chapters 20 and 21). With some of these techniques, the binding evaluation is not as straightforward as equilibrium batch rebinding, since the transport rate of the substance through the material influences the rebinding kinetics. The solvent flow rate obscures binding sites possessing slow rebinding kinetics or low accessibility within the material.

A. The Imprinting Effect

Even if you go through the above steps, or follow the protocols presented in this book or in the imprinting literature, how do you know if your polymer is actually a MIP at all? What if the imprinting protocol failed to produce any specific recognition sites? What are the measures that need to be taken in order to prove that imprinting is actually taking place?

These questions are of constant concern to scientists working in the field, and different solutions have come up over the years. An excellent way to prove the effect is to use one enantiomer of a chiral compound as the template. Since both enantiomers are physically identical in the absence of a chiral environment, chiral discrimination can be used to demonstrate that imprinting has taken place. However, this is true only if no chiral template is left in the polymer after synthesis and workup, and thorough washing is always necessary. Chiral induction may also occur during polymerization that is not associated to specific binding sites, but may result in chiral discrimination. Therefore, it is extremely important to use adequate CONTROLS to support the fact that imprinting has actually taken place.

The "imprinting effect" is a term used within the imprinting community to describe the outcome of a positive (successful) MIP as compared to polymers that prove inefficient. The term is generally used in a slightly nonspecified manner, but can be freely defined as:

The establishment of *template-specific* recognition sites through the locking-in of functional recognition elements around the template.

What does this mean?

1. The template in itself is the source of the binding sites. Any other binding effects are nonspecific and not due to the imprinting process, but rather inherent to the polymer system or the preparation process.
2. The template is essentially the best ligand for the MIP, unless the binding sites produced are changed after imprinting.
3. The recognition elements introduced in the process, together with steric effects produced by structural elements, are responsible for specific binding.

That is, the template should give rise to a structural change in the produced material, complementary to the template. This effect is not as easy to prove as it may sound. Firstly, the template may induce structural changes in the material during imprinting that change its adsorption properties, but no specific binding sites are created. Secondly, even though binding sites appear to be formed, other molecules bind equally good or better to the material than does the original template as a consequence of the combined specific and nonspecific effects, and possibly also to shrinking/swelling in the polymer upon removal of template or change of solvent.

B. Strategy for Demonstrating a Real Imprinting Effect

In order to show any binding effects to the synthesized MIP, analysis of the adsorption data needs to be performed. This is not a trivial task, and guidelines on how to apply various isotherm models to the data are presented in Chapter 16. Nevertheless, even if these analyses are performed, the imprinting effect may still be difficult to demonstrate. Most materials show various degrees of binding, whether they contain imprinted sites or not, and binding data from MIPs as well as various control samples need to be compared to corroborate a true imprinting effect.

The following flow charts can be followed to demonstrate the outcome of likely imprinting effect in a newly prepared MIP. In this strategy, it is supposed that all polymers/materials have been sufficiently washed and worked up, and that control polymers/materials have been prepared and analysed.

The following controls are recommended:

- NP—a nonimprinted polymer prepared exactly the same way as the MIP, but in the absence of the template.
- MIP2—an imprinted polymer prepared exactly the same way as the MIP, but against a compound (T2) that is structurally related to the template.

The rationale is that NP in itself cannot be used as the only comparison, since the morphology and physical characteristics of the NP may be completely different from those of the MIP. For example, the surface area and polarity of the materials may differ substantially, resulting in large differences in nonspecific binding.

As can be deduced from these flowcharts, a likely imprinting effect can be demonstrated following either the enantiomer strategy (Fig. 2) or the general strategy (Fig. 3). In the first case, at least two different materials have to be prepared (although more are always recommended)—the MIP and the NP—and the imprinting is analyzed by a combination of chiral discrimination of the enantiomers (e.g., chiral chromatography) and binding strength of various analytes. Alternatively, a

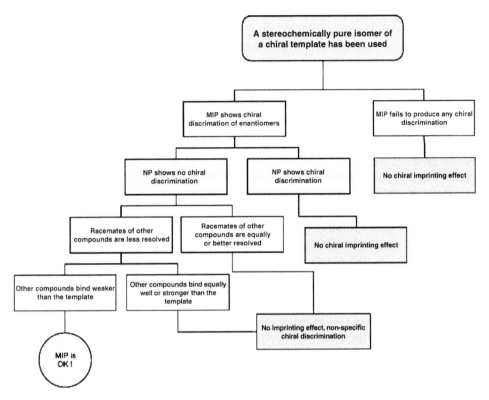

Figure 2 Flowchart for demonstrating the outcome of an imprinting effect following the enantiomer strategy.

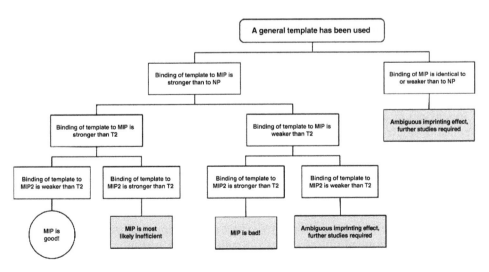

Figure 3 Flowchart for demonstrating the outcome of an imprinting effect following the general strategy.

second MIP2, prepared in the same way against a related enantiomerically pure compound (not the enantiomer of the template) can be analyzed with both compounds similar to the general strategy. In the general strategy, at least three different materials need to be prepared—MIP, NP, and MIP2—and the binding strengths to the various polymers probed for each template. A second MIP2 is necessary since the NP may not be similar enough to be used as a true reference.

Another control that is useful is a polymer/material prepared against fragments of the template. With this polymer, the chemical environment can be better simulated than an NP. For example, methacrylic acid tends to dimerize in the absence of the template. In the presence of the template molecule, however, dimerization is often prevented due to the interactions with the template, and as a result, the MIP and NP may have different physical properties. As a means to prepare a control polymer that better simulates the chemical environment of the MIP, one can choose a set of fragments, each carrying a single functional group present in the template molecule. The hope is that the functional groups in these fragments interact with the monomers in a relatively similar manner as the template, thus generating a control polymer that imitates the MIP with respect to its physical properties. However, in this case, the recognition elements are more randomly organized in the polymer in contrast to the situation in the MIP, where the chelate effect of the functional recognition elements may occur, and therefore the comparison is not necessarily ideal.

In addition, several compounds should always be tested against the polymer/material in order to further evaluate the binding effect, both template-related and nonrelated. It is only by careful cross-reactivity studies that the efficiency of the material may be fully established.

VII. CONFIGURATIONS OF MOLECULARLY IMPRINTED POLYMERS

Molecularly imprinted polymers have been used in several configurations. By far the most widely employed preparation method produces bulk polymer monoliths in the shape of the reaction vessel, for example, a test tube or flask. The polymer is then ground and sieved, and particles of usually about 25 μm are collected for subsequent studies. To make the imprinted polymer suitable for specific applications, other configurations have been developed to produce polymers that are more homogeneous in dimensions and morphology. These include (A) Beads, prepared either through suspension, emulsion, dispersion polymerization, or by grafting/coating of imprinted polymers on preformed silica or polymer microspheres (Chapter 17). In this manner, spherical particles with narrow size distribution can be obtained, providing, for example, good flow performances in chromatography. (B) Films and membranes generated by direct casting on a surface or a device (Chapter 18). This configuration is frequently desirable when the imprinted polymer is used in separation or sensing. Membranes can also be prepared by synthesizing imprinted polymers inside the pores of a porous membrane that acts as a solid support, or gluing the polymer particles together using a particle binding agent obtaining, for example, coated glass plates similar to those used in thin layer chromatography. (C) Micromonoliths by in situ polymerization inside a chromatography column, in a capillary, or on a silicon chip for applications in chromatography and miniaturized devices (Chapter 19). An added advantage with the direct preparation methods is that particle sizing is not necessary and that the recognition sites left in the polymer are undamaged from the fragmentation process.

VIII. CHARACTERISTICS OF MOLECULARLY IMPRINTED POLYMERS

Molecularly imprinted polymers are highly cross-linked thermosets, and therefore porosity has been a necessary feature of their morphology to allow permeability and transport of template molecules to the bulk polymer phase. A high internal surface area ensures that the vast majority of the polymer mass is within several molecular layers of the surface and allows access of the template molecules to the majority of the polymer mass. A broad distribution of pore sizes is desirable for the use of these materials in chromatographic applications. Mesoporosity of amorphous porous materials is most commonly evaluated using a porosimeter by analyzing the N_2 adsorption/desorption isotherms. Parameters that can be obtained from the measurements include surface area, average pore size, and pore size distribution.

Apart from the more obvious recognition properties of molecularly imprinted polymers, their physical and chemical characteristics are highly appealing. These materials exhibit physical and chemical resistance towards various external degrading factors. Thus, molecularly imprinted polymers are remarkably stable against mechanical stress, elevated temperatures, and high pressures, resistant against treatment with acid, base, or metal ions and stable in a wide range of solvents. The long-term endurance of the polymer is also very high: storage for several years at ambient temperature leads to no apparent reduction in performance. Further, the polymers can be used repeatedly, in excess of 100 times during periods of years without loss of the "memory effect". In comparison with natural, biological recognition elements, which are often proteins that are sensitive and fragile, these properties are highly appealing.

IX. MIPs—FOR WHAT PURPOSE?

In addition to studies where the nature of the recognition events per se has been the major issue, a number of application areas have been explored for imprinted matrices viz. (A) Chromatography, where the imprinted polymer is used as the stationary phase for separation and isolation (Chapter 20). This application is based on the fact that the imprinted polymer has a better retention for the template molecules than others. A related area of application is solid-phase extraction, where the imprinted polymer is used as a "sponge" to concentrate the molecule of interest (Chapter 23). (B) Immunoassay-type analyses in which molecularly imprinted polymers are used as antibody and receptor mimics (Chapter 25). (C) Catalysis where the molecularly imprinted polymers are used as enzyme mimics (Chapter 24). (D) Sensors and biosensor-like devices where the molecularly imprinted polymer is the recognition element (Chapters 26 and 27).

X. CONCLUDING REMARKS

The technique of Molecular Imprinting was built upon a simple concept of molding a recognition "lock" around a molecular "key". It is remarkable that antibody-like recognition specificity can be achieved by polymerizing/cross-linking a simple monomer in the presence of a template molecule. However, these synthetic materials are still relatively primitive compared to what nature provides. The affinity and selectivity of molecularly imprinted materials are often orders of magnitude lower than their natural counterparts. Other issues and challenges include low binding capacity and heterogeneity of the binding sites. In designing imprinted materials with improved recognition abilities, lessons can again be learned from the biological systems. While imprinted materials are rigid cross-linked network, enzymes and antibodies are structurally flexible entities. In the absence of the substrates, many enzymes have an open structure, and binding of the substrates often induces conformation changes at the recognition sites. This structural flexibility is known to enhance the specificity of the enzyme–substrate recognition and the catalytic activity of the enzyme. In developing the next generation imprinted materials with improved molecular recognition properties, one possible focus would be the introduction of local conformational flexibility to the structures that allows for "induced-fit" of the template molecules to the recognition sites.

REFERENCES

1. Zimmerman, S.C. Rigid molecular tweezers as hosts for the complexation of neutral guests. Topics Curr. Chem. **1993**, *165*, 71–102.
2. Fischer, E. Einfluss der Configuration auf die Wirkung der Enzyme. Chem. Ber. **1894**, *27*, 2985–2993.
3. Lehn, J-M. *Supramolecular Chemistry. Concepts and Perspectives*. Weinheim, VCH: Verlagsgesellschaft mbH, 1995.
4. Buckingham, A.D.; Legon, A.C.; Roberts, S.M. *Principles of Molecular Recognition* Blackie Academic and Professional: London, UK, 1993.
5. Dugas, H. Bioorganic Chemistry. In: *A Chemical Approach to Enzyme Action*, 3rd ed. Springer-Verlag: New York 1996.
6. Breslow, R. Biomimetic Chemistry. Chem. Soc. Rev. **1972**, *1*, 553–580.
7. Breslow, R. Biomimetic chemistry and artificial enzymes: catalysis by design. Acc. Chem. Res. **1995**, *28*, 146–153.
8. Bayer, E.A.; Wilchek, M. Applications of avidin–biotin technology to affinity-based separations. J. Chromatogr. **1990**, *510*, 3–11.
9. Chou, S.Y.; Krauss, P.R.; Renstrom, P.J. Nanoimprint lithography. J. Vac. Sci. Technol. B **1996**, *14*, 4129–4133.
10. Voet, D.; Voet, J.G.; Pratt, C.W. *Fundamentals of Biochemistry*. Wiley: New York, 1999.
11. Wulff, G.; Sarhan, A. Use of polymers with enzyme-analogous structures for the resolution of racemates. Angew. Chem., Int. Ed. Engl. **1972**, *11*, 341.
12. Arshady, R.; Mosbach, K. Synthesis of substrate-selective polymers by host–guest polymerization. Makromol. Chem. **1981**, *182*, 687–692.
13. Shea, K.J.; Thompson, E.A. Template synthesis of macromolecules. Selective functionalization of an organic polymer. J. Org. Chem. **1978**, *43*, 4253–4255.
14. Shea, K.J.; Sasaki, D.Y. On the control of microenvironment shape of functionalized network polymers prepared by template polymerization. J. Am. Chem. Soc. **1989**, *111*, 3442–3444.

15. Beckett, A.; Anderson, P. "Footprints" in adsorbents. J. Pharm. Pharmacol. **1959**, *11*, 258T–260T.
16. Dickey, F.H. The preparation of specific adsorbents. Proc. Natl. Acad. Sci. USA **1949**, *35*, 227–229.
17. Beckett, A.H.; Andersson, P. A method for the determination of the configuration of organic molecules using 'stereo-selective adsorbents'. Nature **1957**, *179*, 1074–1075.
18. Mosbach, K. Novel affinity techniques. In *Affinity Chromatography and Biological Recognition. Part IV: Affinity Methods—Design and Development*; Chaiken, I.M., Wilchek, M., Parikh, I., Eds.; Academic Press: Orlando, FL, 1983, pp 209–222.
19. Ramström, O.; Andersson, L.I.; Mosbach, K. Recognition sites incorporating both pyridinyl and carboxy functionalities prepared by molecular imprinting. J. Org. Chem. **1993**, *58*, 7562–7564.

2

A Brief History of the "New Era" of Molecular Imprinting

Andrew G. Mayes School of Chemical Sciences and Pharmacy,
University of East Anglia, Norwich, United Kingdom

I. INTRODUCTION

For this book, I have chosen to take a more arbitrary and anecdotal approach to imprinting history, by tracing the development of what I have chosen to call the "New Era" of molecular imprinting. A more detailed historical perspective is the view taken by Andersson and Nicholls [1], who spent a great deal of time researching the early history of imprinting. I recommend their chapter to get a fascinating insight into how some of the ideas and concepts of what we now know as molecular imprinting first arose. I define this "New Era" as the point where research transferred from predominantly silica-based systems to synthetic organic polymers in the early 1970s. Many researchers have contributed, to a greater or lesser extent, to the development of ideas and techniques in this area, but three individuals stand out as having shaped this "New Era" through their pioneering work—Günter Wulff, Ken Shea, and Klaus Mosbach (Fig. 1). Each of these individuals has made a tremendous contribution to the development of the technique. In the following pages, I will explain what they were working on and thinking about at the time, how they got involved in imprinting research and where it led them. Finally, I will present their views on the current problems and limitations of the technique and a vision of where the field might develop in the coming years.

II. HOW DID THEY GET INVOLVED?

Wulff studied chemistry at the University of Hamburg and then went on to study for a Ph.D. in Bonn in the group of Prof. Rudolf Tschesche, working on the isolation and structural elucidation of plant saponins. Following his Ph.D., he continued with this work and published a number of papers, including the discovery of a completely new type of glycoside structure of particular biochemical relevance. At a conference for young lecturers, where he reported this work an elder and highly respected colleague suggested to him that while the work he reported was good, the credit would always be given to his boss (a scenario very familiar to most young researchers!). The

Figure 1 Günter Wulff, Ken Shea, and Klaus Mosbach.

colleague proposed that he should make his own way and find a new area of research that he could establish as his own.

Since he was working in an institute for organic chemistry and biochemistry, he was familiar with the work of the Israeli's Katchalski-Katzir and Patchornik, who used polymers as supports for enzyme immobilization and for performing reactions on functional polymers. It was also the time when the mechanism of chymotrypsin was elucidated. This led Wulff to contemplate the possibility of producing synthetic polymers with catalytic sites that worked in the same way as enzyme active sites. There was already some literature on using synthetic polymers as catalytic models of enzymes but these did not have specific structural binding sites as enzymes did. Clearly a new approach would be required to generate such sites with specific predetermined structure.

Figure 2 Günter Wulff and Ali Sarhan.

Realizing that he needed polymer chemistry skills Wulff spent some time at the University of Mainz working in the macromolecular chemistry group of Prof. R.C. Schulz, where his knowledge increased and his ideas started to take shape. Back in Bonn, different approaches were tried leading to the idea of "molecular imprinting". These ideas were outlined in a research proposal to the Deutsche Forchungsgemeinschaft, which was fortunately funded. Wulff commented "I got a rather generous grant that was prolonged several times, and looking back now I am surprised how long I got this support without being successful with my main idea!" At the same time he developed some preparative organic chemistry work, related to the original application, in glycoside synthesis, in order to have some results to report. This work progressed much more smoothly than the imprinting work, and it was for the glycoside work that he completed his "Habilitation" in 1970.

The first positive imprinting results were obtained in 1971 by his student Ali Sarhan (now a full professor at the University of El Mansourah in Egypt) (Fig. 2). These results were presented at the annual "Freiburger Makromolekulares Kolloquium"—a meeting for academic and industrial polymer scientists— in 1972 [2]. Wulff recalls that his 20-minute lecture was the last one before lunch, but that it provoked a violent and controversial discussion, polarizing the audience of 300 into two halves; those that liked the idea and those that did not believe the results at all! Wulff suggests that some perhaps did not like the fact that they had been generated by an "outsider" to the world of polymer chemistry. The "10-min" discussion lasted for 45 min and made everybody late for lunch, but, as the chairman remarked, it certainly brought the presentation to the attention of the audience!

Wulff's first patent on imprinting was also filed in 1972 (though it was not published until 1974 after academic papers had come out). In this document, and the US filing (as a continuation-in-part) from 1975 that contained some additional material, a range of basic imprinting concepts was outlined. This included polymers made by radical polymerization, polycondensation and polyaddition to form polyurethanes and included examples using both covalent and noncovalent interaction, though the noncovalent interactions were only used in association with covalent interactions in the examples quoted. Many other things are also mentioned, such as the use of UV and gamma radiation for curing, which was later adopted and developed by many groups. The first academic research paper on imprinting in synthetic polymers was published by Wulff et al. [3], and he still regards this as the most important publication from his group, since it set out the principle underpinning the molecular imprinting concept for the first time. It was also to have a significant impact on the research direction of Shea, as indicated below.

Shea studied chemistry at the University of Toledo and went on to complete a Ph.D. in physical organic chemistry with P.S. Skell at The Pennsylvania State University. His thesis research involved mechanistic and kinetic studies of free radical brominations. He moved on to a postdoctoral position at Caltech with Professor R.G. Bergman where he was acquainted with problems associated with the influence of organized assemblies such as micelles on the outcome of chemical reactions. During this time, he was looking for an appropriate topic to research and came across Wulff's early Tetrahedron Letters paper on imprinting of sugars using boronate esters [3]. He liked the simplicity of the general concept and realized that the fixed geometry imparted by imprinting would be a much more promising candidate for influencing the course of a chemical reaction than the dynamic, fluid micelles he had been using during his Ph.D. He developed this idea into an NIH postdoctoral proposal to use imprinting to create "microreactor cavities" that could be used to direct chemical reactions. The project was funded and took him to the University of California at Irvine, where he began his first experiments to imprint chiral molecules with the aim of generating microreactors that could induce asymmetry in subsequent reactions. Shea recalls that his first meeting with Mosbach was at a Gordon Conference on affinity techniques, where he presented some of his preliminary results with George Whitesides and Klaus Mosbach sitting in the front row. At that time, very few people were attempting to do molecular imprinting and most people told about it thought that it was an interesting concept, though probably a little crazy!

Shea's early work on imprinting led on to a number of projects including his work on bis-ketals for which he is probably best known [4,5]. The aim of this project was to define the details of the rebinding of small molecules into imprinted receptor sites using IR and solid-state NMR techniques. Shea recalled that this project was in fact extremely frustrating and that he came close to giving up because, having made the covalent adducts and imprinted them, they could not find a way to cleave them again and remove them from the polymer. Studies of polymers using solvation-sensitive fluorescent probes [6] gave a better understanding of polymer morphology, however, and led to new formulations that allowed the template to be removed successfully. The project was kept alive, says Shea, by the persistence and determination of several very good students.

Through the years since this work, Shea has maintained some involvement in imprinting research (Fig. 3) while pursuing a variety of other research interests. As a result of this, he has been less prolific than the other pioneers, but his publications have always been notable for their chemical rigor and the insight they have brought to understanding the fundamental mechanisms of the imprinting process and the subsequent recognition of template by the imprinted polymers.

Figure 3   Ken Shea and Börje Sellergren.

Mosbach was born in Leipzig, Germany, but moved to Sweden as a young man where he began his university studies in Lund in 1953. He completed his Ph.D. at the same university in 1960 under G. Ehrensward on the subject of "The Biosynthesis of Aromatic Compounds in Fungi and Lichens". He was then awarded the Waksman–Merck postdoctoral fellowship and worked for 1.5 years at the Institute of Microbiology at Rutgers University in New Jersey, working with Professors Waksman and Schaffner. On returning to Sweden, he continued to work on aspects of secondary metabolism and became associate professor and finally full professor in 1970. Throughout the late 1960s and 1970s, Mosbach made many contributions in the fields of cell and enzyme immobilization and biotechnological aspects of enzyme engineering, and edited a number of books in this area.

Throughout the late 1960s and 1970s, he was probably best known for his many contributions to the emerging field of affinity chromatography, in particular methodology for coupling ligands to solid supports and for work on immobilized nucleotides for purifying dehydrogenase and kinase enzymes. It was this work on generating "convex" receptor structures in polymer gels to recognize the "concave" binding sites in enzymes and proteins that started him thinking about the concept of imprinting. If one could do effective enzyme immobilization and affinity chromatography, perhaps one could also generate "concave" receptors in polymer gels to recognize the "convex" structures of small molecules or even whole proteins. Since it was possible to entrap enzymes in cross-linked polymer gels with retention of structure and activity, it seemed reasonable to suppose that if the gel was slightly less cross-linked, an enzyme might be able to diffuse in and out of the material and reversibly occupy the "receptor sites" in the gel. A similar strategy should also be possible for small molecules, leading to concave cavities like enzyme or antibody binding sites. Unlike Wulff, Mosbach recalls that in the early days, it was almost impossible to get any funding for the concept. Reviewers considered it far-fetched and impossible to accomplish and could not imagine what it could be useful for even if it was possible! (If you are a writer of scientific research proposals, this will no doubt sound very familiar to you ...) Being a biochemist by background and working in a totally different area, Mosbach was unaware of Wulff's emerging work, but continued to work away at the problem. It was natural for him to utilize the types of simple noncovalent interactions (ion-pairs, H-bonds, etc.) used in nature to create receptor sites in proteins. It was this work that led eventually to his first publication on imprinting using only noncovalent interactions in 1981 [7]. Interestingly, the approach was not initially called "molecular imprinting", but rather "host–guest polymerisation", although "imprints" was also used in the text (Mosbach remembers heated semantic discussions with his co-author, a more traditional polymer chemist). Mosbach believes that the term was first adopted in an article on novel affinity techniques in 1983 [8]. Surprisingly, the abbreviation of "molecularly imprinted polymer" to "MIP", now in common usage, was not adopted until 1993 [9]. Prior to that Mosbach had rather referred to them as "MIA" (molecularly imprinted adsorbents) until it was pointed out to him that this was the standard abbreviation for "missing in action". This was not a ringing endorsement for the efficacy of the imprinted receptors, given that many researchers still doubted that they really existed!

The noncovalent approach to imprinting remains the area of the technique for which Mosbach is best known, although he has also worked with a variety of other approaches. Since the mid-1980s the work has gone from strength to strength and his group has now published almost 200 papers in the area and has over 20 patents. Of all these contributions, the most important was probably "Drug assay using antibody mimics prepared by molecular imprinting" published in *Nature* in 1993 [10]. It is not only the scientific content of this paper that was pivotal—the realization that imprinted receptors made by the simple and flexible noncovalent approach had truly remarkable affinity and selectivity, but the fact that it was read by such a wide audience from diverse backgrounds. This took the technique from the position of academic curiosity researched by only a few groups into the scientific mainstream for the first time. The exponential growth in interest and activity in the field of imprinting can be traced back to the influence of this publication. I was working in Mosbach's group in Lund at the time when this paper was submitted to *Nature*, and I recollect that some vigorous argument was required to convince the Editor of *Nature* that it was worthy of publication. The impact it made clearly shows that it was indeed a highly significant contribution.

Through the late 1980s and 1990s Mosbach had a number of very good Ph.D. students working on imprinting and was fortunate also to attract a string of talented postdoctoral workers and exchange students who added to a vibrant and international research group (Fig. 4). It also sometimes led to confusion and I remember a number of occasions when Mosbach used English, German and Swedish in a single sentence when having discussions! The international "school of imprinting" that Mosbach effectively created through this activity is what gives him the most pleasure, he claims. In all he worked with some 40 students, postdocs, and guest professors, mostly at the Lund Institute of Technology, but also during his period at the ETH in Zurich. Most of these are listed in the acknowledgements to his keynote lecture to "MIP 2000" in Cardiff [11]. A surprisingly large number of these individuals (including myself) have been infected with his love and fascination with the

Figure 4   Klaus Mosbach with co-authors Ronald Schmidt and Ecevit Yilmaz.

technique of imprinting and have gone on to academic careers where they have built their own groups in this research area.

III. WHAT ARE THE MAIN PROBLEMS AND LIMITATIONS OF IMPRINTING AT THE PRESENT TIME?

Considering the current limitations of our knowledge of the imprinting process, Shea points to the lack of understanding of exactly how and when in the polymerization process the imprinted receptor sites are actually produced. Is it, for instance, simply the case that template/monomer complexes (which have been shown spectroscopically to be present in solution [12,13]) are incorporated into the growing polymer network? This might be possible for covalent imprinting, but it seems less likely for most noncovalent imprinting where the complexes are generally weak and are constantly exchanging in solution. It is true that polymerization is very rapid in the vicinity of a newly formed radical, but many processes of diffusion, exchange, and heat generation are also occurring at this site. Would the weak transient complexes survive this? An alternative view is that the template rather interacts with pendant functional groups that have been randomly incorporated into the growing polymer chains as the secondary and tertiary structure of the polymer network evolves. The template can then influence the "maturation" of the structure towards its final state. This proposal has the advantage that it can exploit the "avidity effect". Simultaneous interaction of the template with two or more functional groups keeps the template in position since it is unlikely that all the interactions will dissociate simultaneously. As the network continues to evolve the template becomes more and more tightly associated, until it finally becomes "trapped" within the receptor it has templated. The latter proposal is consistent with results where imprints have been made by post-cross-linking of preformed functional polymers, though in this process the results are rarely as good as when the polymer is synthesized from monomers in the presence of the template. Careful thought needs to be put into design of experiments to test these ideas.

A related question is "where exactly within the macroporous structure of the imprinted polymer are the receptor sites?". The answer to this question is currently unclear but it has quite profound implications for application of imprinted polymers in analytical science. Shea points to the work of Guyot on the detailed structure of macroporous polymer materials [14–16] as being of particular importance in understanding the general physical properties of imprinted polymers, though further work is still needed to elucidate the fine details of the imprinting process.

For Mosbach, a major current problem is the inability to make effective imprints in aqueous systems (an area also highlighted by Shea). This is no doubt informed by the fact that Mosbach has spent most of his scientific career working at the interface between biology and chemistry (affinity chromatography, enzyme engineering etc.) and hence has a particular interest in biological molecules such as peptides and proteins, sugars, nucleotides etc. Most of these compounds are insoluble in the organic systems generally used in current noncovalent imprinting strategies, and the H-bond-dominated interactions usually relied upon for high quality noncovalent imprinting are very weak in water, so it is clear that some radical new thinking is required to advance this area.

Wulff sees the polydispersity of the imprinted receptor sites, nonspecific binding problems and the poor mass transfer properties of typical imprinted materials as key problems. He has recently published work on imprinting using strong noncovalent interactions to achieve near-stoichiometric association of templates and functional monomers in an attempt to improve the homogeneity of receptor sites and minimize the fraction of functional monomers dispersed non-specifically in the polymer structure. Continuation of this work, along with related work from other groups, will no doubt lead to further advances in this area.

IV. WHAT IMPRINTING PROJECTS ARE THEY WORKING ON AT PRESENT?

Shea's current work involves the synthesis of imprinted polymer films and studies of the transport properties and mechanisms of these materials. There are several reports of transport properties in the current literature but the conclusions are in some cases contradictory [17]. The general consensus seems to be that imprinted polymers provide a facilitated diffusion path for the template molecule that increases the flux of this compound compared with other similar compounds. Mechanistically, however, it is by no means clear why this should be so. Possible explanations include a "hopping" mechanism transporting molecules from site to site, or a simple explanation based on the imprinted sites increasing the concentration of template in the microenvironment of the polymer film and hence increasing the flux. Work on deposition of thin films of imprinted polymer is being applied to sensor research.

The project that is most exciting for Mosbach currently is what he likes to call the next generation of molecular imprinting, involving the generation of "anti-idiotypic" imprints or "imprints of imprints", and direct molding [18]. The former concept here is that if imprints are made to a drug or molecular structure of interest, the imprinted receptors can subsequently be used as templates to assemble new molecules with structural similarity to the original template. In this way, new libraries of small molecules or macromolecules can be created that might be useful for lead discovery of new pharmaceutical compounds. Similarly, if bacteria or viruses could be imprinted, new polymeric nanoparticles could be generated by "nano-molding" in the receptor sites. In this way, it might be possible to produce synthetic particles with similar structure and surface chemistry to the original agent that could be used, for instance, in vaccine development.

It is particularly interesting that Wulff's current research activity in some ways takes him back to the roots of his original idea to produce "enzyme analog built polymers" or enzyme mimics. He has developed methods to prepare highly cross-linked soluble "nanogels" with a diameter around 10 nm, a molecular weight (M_n) of about $2–7 \times 10^5$ and a polydispersity (M_w/M_n) of about 3–4. These particles resemble biological proteins quite closely in terms of their dimensions and basic properties. Recently, he has been able to imprint catalytically active sites in such nanogels, with an average of approximately one active center per particle. This is coming very close to creating true "artificial enzyme mimics" and it can be envisaged that such materials could be treated and processed much more like biological molecules. For instance, if high affinity receptor sites can be created in such nanoparticles, the resulting popula-tion could be purified by affinity chromatography to generate a near-homogeneous population of receptor particles. These imprinted particles would truly mimic monoclonal antibodies for the first time and could be incorporated into assay technology much more easily than current materials.

V. WHERE WILL MIPS BE APPLIED AND WHERE WILL RESEARCH GO IN THE FUTURE?

Molecular imprinting in synthetic polymers has now been around for about 30 years and has enjoyed a decade of vigorous research activity with many new groups becoming involved. Research activity is very diverse, covering the whole range from the highly applied to the truly fundamental. With this in mind, it is interesting to consider what the three pioneers consider to be the most likely application areas during the next few years. Wulff suggests that sensor technology is a likely area for early exploitation, where the acknowledged robustness and stability of imprinted receptors is highly attractive. There has certainly been a great deal of activity in this area, as a review of the recent literature will confirm. A wide range of different transducer technologies has been investigated for suitability. Currently, work on acoustic devices seems to be leading the way, though in the future I expect to see increased activity in the electrochemical and optical sensor areas.

Shea also considers sensors to be a likely area for exploitation, but is clear that many problems remain to be overcome. In particular, the reliable fabrication of thin films of imprinted polymer on transducer surfaces remains a problem (one being addressed by a number of groups including Shea's and my own). This type of development may also lead on to a variety of microfabricated devices containing MIPs as part of "lab-on-a-chip" type devices. Shea was also keen to point out that although imprinted receptors are very good in some cases, they are still far from natural antibodies in terms of their selectivity, nonspecific binding properties and, in particular, affinity. If they are to be used as real antibody substitutes for sensor devices (or assays) the affinity and capacity of the best binding sites needs to be significantly improved (preferably along with the homogeneity of these sites). Another issue for MIP sensor development is the need to find the right market. The market needs to be large enough to justify and generate the investment and commercial pressure that will be required to move imprinted materials from the research lab to commercial reality.

Selective adsorbents for separation and recovery of high value products are another area that Wulff considers ripe for exploitation, though the functional capacity of the materials may need to be further enhanced to make this process cost-effective. The use of stoichiometric noncovalent imprinting procedures should contribute to such an increase in receptor sites. Shea also considers SPE and extraction a possible niche application, though he was rather more cautious. He was very clear, however, that chromatographic separations, such as chiral separations would not be a commercial reality. A huge amount of work in imprinting has relied on chromatographic evaluation of chiral systems. Chromatography is used because it provides a convenient method using readily available equipment, and chiral systems are chosen as models since they provide the ultimate test of the imprinting effect in a system where the two enantiomers have identical physicochemical properties. The large body of work suggests that chiral resolution is a target application for imprinting, but this is, in fact, unlikely to be the case. In practice the low capacity, over-high receptor affinity and poor exchange kinetics result in inferior dynamic performance of imprinted polymers. This leads to bad peak broadening and tailing and very poor separations, despite the often high α-values. As Shea observes, the world already has many highly successful commercial chiral phases, and in the face of this competition from proven products, it is difficult to see what advantages imprinted chiral phases would have.

Mosbach has always been confident that MIPs would one day find applications in a wide variety of areas and that in the coming years it will become commonplace to find MIPs incorporated into many different types of devices and processes. He points out that commercial endeavors have already been accomplished, including solid-phase extraction matrices and materials for separation and recognition. Small companies have thus been formed, and sizeable ones are likely to follow. In large companies, the technique is often used in-house. As for chiral separation, he believes that the materials can prove highly useful especially for "enantiopolishing", that is removal of the "wrong" enantiomer. In addition, Mosbach strongly believes that the technique may witness a great future in drug discovery. Those of us who have been involved in this field for some years, and the many new groups entering for the first time, clearly believe that imprinting has a future and continue to pursue research in many different aspects of the technique. Will MIP research still be as popular and trendy in 10 years time? Shea thinks not, unless there are major new breakthroughs that take to field to new levels. "Ten years is a long time in research", he points out. It is likely that many people will still be developing it and using it as a tool to make receptors (in the same way as monoclonal antibodies are produced today) but it will no longer have so much impact as a research area in its own right. Unless, of course, you will prove otherwise . . .

ACKNOWLEDGMENTS

I would like to extend my thanks to professors Wulff, Shea, and Mosbach for providing me with both factual and anecdotal material to assist with the writing of this chapter.

I would also like to thank Prof. Mosbach, the molecular imprinters there at the time and the many other members of the Department of Pure and Applied Chemistry for two (mostly!) very happy years (1992–1994), doing postdoctoral studies on molecular imprinting as a Royal Society Research Fellow at the University of Lund. This has undoubtedly influenced my thinking, opinions, and current research, though I hope it has not unduly colored this introduction.

Finally I would like to thank the Royal Society for the funding to go and work in Lund at a very exciting time for molecular imprinting research.

REFERENCES

1. Andersson, H.S.; Nicholls, I.A. A Historical perspective of the development of molecular imprinting. In *Molecularly Imprinted Polymers—Mon-Made Mimics of Antibodies and their Applications in Analytical Chemistry*; Sellergren, B., Ed.; Elsevier Science B.V: Amsterdam, 2001; Vol. 23, 1–20.
2. Wulff, G.; Sarhan, A. Use of polymers with enzyme-analogous structures for the resolution of racemates. Angew. Chem. Int. Ed. Engl. **1972**, *11*, 341.

3. Wulff, G.; Sarhan, A.; Zabrocki, K. Enzyme-analogue built polymers and their use for the resolution of racemates. Tetrahedron Lett. **1973**, *14*, 4329–4332.
4. Shea, K.J.; Dougherty, T.K. Molecular recognition on synthetic amorphous surfaces. The influence of functional group positioning on the effectiveness of molecular recognition. J. Am. Chem. Soc. **1986**, *108*, 1091–1093.
5. Shea, K.J.; Sasaki, D.Y. On the control of microenvironment shape of functionalized network polymers prepared by template polymerization. J. Am. Chem. Soc. **1989**, *111*, 3442–3444.
6. Shea, K.J.; Stoddard, G.J.; Sasaki, D.Y. Flourescence probes for the evaluation of diffusion of ionic reagents through network polymers. Chemical quenching of the flourescence emission of the dansyl probe in macroporous styrene-divinylbenzene and styrene-diisopropylbenzene copolymers. Macromolecules **1989**, *22*, 4303–4308.
7. Arshady, R.; Mosbach, K. Synthesis of substrate-selective polymers by host–guest polymerization. Makromol. Chem. **1981**, *182*, 687–692.
8. Mosbach, K. Novel Affinity Techniques. In *Affinity Chromatography and Biological Recognition. Part IV: Affinity Methods—Design and Development*; Chaiken, I.M., Wilchek, M., Parikh, I., Eds.; Academic Press: Orlando, FL, 1983, 209–222.
9. Ramström, O.; Andersson, L.I.; Mosbach, K. Recognition sites incorporating both pyridinyl and carboxy functionalities prepared by molecular imprinting. J. Org. Chem. **1993**, *58*, 7562–7564.
10. Vlatakis, G.; Andersson, L.I.; Müller, R.; Mosbach, K. Drug assay using antibody mimics made by molecular imprinting. Nature **1993**, *361*, 645–647.
11. Mosbach, K. Toward the next generation of molecular imprinting with emphasis on the formation, by direct molding, of compounds with biological activity (biomimetics). Anal. Chim. Acta **2001**, *435*, 3–8.
12. Sellergren, B.; Lepistö, M.; Mosbach, K. Highly enantioselective and substrate-selective polymers obtained by molecular imprinting utilizing non-covalent interactions. NMR and chromatographic studies on the nature of recognition. J. Am. Chem. Soc. **1988**, *110*, 5853–5860.
13. Andersson, H.S.; Nicholls, I.A. Spectroscopic evaluation of molecular imprinting polymerization systems. Bioorg. Chem. **1997**, *25*, 203–211.
14. Guyot, A. Synthesis and structure of polymer supports. In *Synthesis and Separations using Functional Polymers*; Sherrington, D.C., Hodge, P., Eds.; John Wiley and Sons: London; 1988, 1–42.
15. Guyot, A. Polymer supports with high accessibility. Pure Appl. Chem. **1988**, *60*, 365–376.
16. Guyot, A. Some Problems in the physical and chemical characterization of functionalized supports. React. Polym. **1989**, *10*, 113–129.
17. Yoshikawa, M. Molecularly imprinted polymeric membranes. Bioseparation **2002**, *10*, 277–286.
18. Yihua, Y.; Ye, L.; Haupt, K.; Mosbach, K. Formation of a class of enzyme inhibitors (drugs), including a chiral compound, by using imprinted polymers or biomolecules as molecular-scale reaction vessels. Angew. Chem. Int. Ed. **2002**, *41*, 4459–4463.

3
The Noncovalent Approach

Ecevit Yilmaz, Ronald H. Schmidt, and Klaus Mosbach
Chemical Center, Lund University, Lund, Sweden

I. INTRODUCTION

In any cell of an organism, noncovalent interactions account for nearly all communications at the intracellular level (between the components of an individual cell) and intercellular level (between cells or between cells and organs). If we examine how antibodies bind to their antigens, how receptors interact with hormones, and how enzymes catalyze reactions, we find that nearly all steps in these processes are dominated by noncovalent, reversible interactions. Our genetic information is stored in long DNA chains that exist as double helices that are held together by hydrogen bonding between complementary base pairs. Also, the characteristic shape of a protein is held together by noncovalent interactions between the various side groups of the residues that comprise the chain. An antibody possesses a recognition site that complements the molecular shape and chemical functionality of its antigen These recognition sites frequently contain several interaction points to ensure that the antibody forms a strong, yet highly specific, interaction with the antigen. The 20 different amino acids that serve as building blocks in natural proteins display a wide variety of molecular interactions. Very different types of interactions are likely to occur for amino acids that are hydrophobic (e.g., phenylalanine), highly polar (e.g., asparagine), and charged (e.g., glutamic acid or lysine).

Inspired by the rich diversity of noncovalent interactions found in nature, researchers have pursued the preparation of artificial antibodies, receptors and enzymes using molecular imprinting technology. Figure 1 depicts schematically the underlying principle of noncovalent imprinting. A complex is formed in self-assembly mode between a template (or as often called print) molecule (T) and monomers (M) that possess functional groups complementary to those on the template. Cross-linking monomers are added, and the mixture is polymerized in order to permanently and rigidly fix the spatial arrangement of the functional monomers. Following polymerization, the template is extracted from the polymer matrix, leaving behind cavities whose size, shape, and chemical functionality complement that of the template. These empty cavities can selectively and reversibly rebind molecules identical (or very similar) to the original template. The molecularly imprinted materials prepared in this manner are commonly abbreviated as MIPs [1].

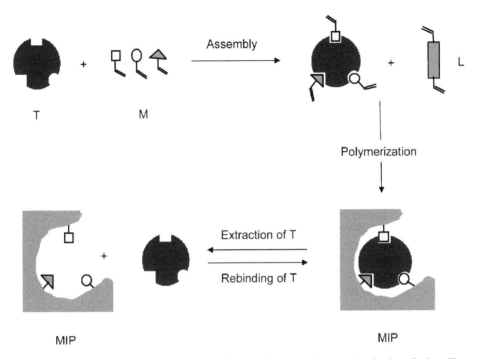

Figure 1 Representation of the general scheme of noncovalent molecular imprinting. For a template molecule (or target or print molecule), T appropriate functional monomers M are chosen and allowed to form a self-assembly construct. By co-polymerization with a cross-linking monomer L, a polymer network is formed in which the self-assembly is set. Thereby, the position and the spatial conformation of the monomers are constructed according to the template. The embedded template T can then be extracted from and rebind to the molecularly imprinted polymer (MIP).

The specific and reversible binding that occurs between a MIP and its target analyte are analogous to those seen in natural antibody–antigen systems. This observation has led us to refer to this class of materials as plastic antibodies or "plastibodies". MIPs have exhibited selectivities and binding strengths similar to those seen with natural antibodies [2–5]. Furthermore, MIPs can mimic enzymes, in that they can catalyze chemical reactions [6,7]. In principle, it should be fairly straightforward to synthesize an MIP for a target or reaction for which no antibody or enzyme is known. The highly cross-linked structure of an MIP renders it relatively resistant to harsh chemicals and temperature extremes. MIPs can be used and recycled many times and stored for long periods without losing their recognition properties. A wide variety of MIP formats have been prepared to suit specific applications, as summarized in Table 1.

At the last section, a demonstration of molecular imprinting and protocols for synthesizing and characterizing MIPs is given.

In a historical perspective, the inspiration for molecular imprinting may be traced to two main sources. The first is Linus Pauling, who proposed an "instructional theory" for the formation of antibodies [8]. Figure 2 depicts Pauling's theory, which suggested that an antibody in some initially nonspecific configuration adopts a new,

Table 1 Configurations of Molecularly Imprinted Polymers (Reviews by Ye and Mosbach [48] Brüggemann et al. [49])

Configuration	Application	Reference
Ground polymer particles from monoliths	Materials for chromatography and assays systems	[3,50]
Polymer beads prepared by suspension, emulsion, or precipitation polymerization	Beads for chromatography and assay systems	[51–53]
Composite polymer beads	Beads for chromatography	[54,55]
In situ-prepared polymer rods	Materials for chromatography	[56]
Polymer particles bound in thin layers	Thin layer chromatography	[57]
Thin polymer membranes or films	Sensor surfaces and selective membranes	[58,59]
Surface imprints	Sensor surfaces, separation materials	[60,61]

highly specific and complementary structure in the presence of an antigen. Thus, the antigen "instructs" the antibody to adopt a configuration that is complementary to its own shape. Pauling proposed this theory long before it was recognized that the three-dimensional configuration and binding specificity of an antibody are ultimately determined by an organism's genetic code [9]. Although Pauling's model was eventually rejected, it suggested a strategy for the synthesis of artificial molecular receptors and accordingly, F.H. Dickey, a student of Pauling, used sol–gel chemistry to prepare silica-based materials that could in part selectively adsorb dyes [10]. It was a series of developments in the 1960s and 1970s, however, that led to the development of molecular imprinting as a viable technology.

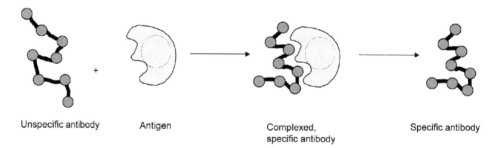

Unspecific antibody   Antigen   Complexed,   Specific antibody
                                                      specific antibody

Figure 2 A cartoon of Pauling's instructive theory for the formation of how antibodies may be formed. An unspecific antibody precursor, represented as an oligomeric entity here, comes in contact with an antigen and engulfs a portion of the antigenic surface. In this contact formation, the antibody precursor would adopt a complementary structure to the antigen. In Pauling's theory, the antibody precursor would then mature to a specific antibody for that particular antigen and would then retain its conformation and specificity even after dissociation. Note: it is now known that the conformation and specificity of antibodies is purely dependent on the genetic code. Antibodies originate from specific cell clones and details about the mechanism of the immune system can be found in biochemistry textbooks [9].

During this period, Mosbach and coworkers wrote a series of publications starting in 1966 in which they characterized enzymes that they had immobilized in soft, hydrophilic polymer gels prepared by the copolymerization of acrylamide with methylenebisacrylamide [11–13]. These immobilized enzymes retained their activity within the synthetic polymer gel, and in some cases, the stability of the enzyme was increased relative to the free enzyme. Such preparations have had a significant impact on the manufacture of biotechnological products by facilitating the handling of enzymes during synthetic steps and simplifying their removal during the final workup. In the context of these studies, we conceived the idea that if the immobilized enzymes were held relatively loosely, they should be able to be extracted from the polymer matrix, leaving behind cavities whose size and shape complemented the original immobilized enzyme. These cavities could later serve as specific binding sites to rebind the enzyme (Fig. 3).

The Mosbach group was also active in the field of affinity chromatography during this period [14,15]. Affinity chromatography relies on specific interactions between complementary molecule pairs like inhibitor–enzyme or antigen–antibody, much like molecular imprinting does. The understanding we gained, in particular the importance of shape and complementarity for the biorecognitions to occur, helped us enormously in the design of molecularly imprinted systems.

Thus, combining the ideas from these two lines of research, led us to the concept of noncovalent imprinting. The first successful MIPs that utilized noncovalent

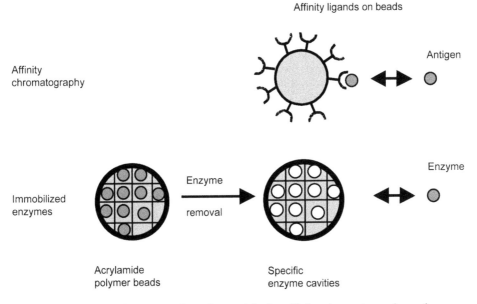

Figure 3 A schematic presentation of materials for affinity chromatography and enzymes immobilized in acrylamide gel beads and how the two fields led to the idea of molecular imprinting. In affinity chromatography, specific affinity ligands on polymeric beads rebind antigens with high selectivity and strength. It was believed that such specific sites could also be obtained from beads with embedded enzymes. Upon extraction and removal of the embedded enzymes cavities that are complementary to the original enzyme would be left behind and these should theoretically rebind the enzyme with high selectivity and strength [17].

interactions however were based on small organic molecules as templates [16]. These were imprinted in cross-linked organic polymer networks utilizing mainly hydrogen bonding interactions between the template and the functional monomers. For a more detailed account of the insight that led to noncovalent molecular imprinting, the reader is referred to the keynote lecture from the first International Workshop on Molecularly Imprinted Polymers [17].

II. CONSIDERATIONS IN THE DESIGN OF NONCOVALENT IMPRINTING SYSTEMS

A. Types of Interactions in Molecular Imprinting

MIPs can be divided into two main categories based on the types of interactions between the template molecule and the functional groups, namely noncovalent and covalent interactions. Covalent imprinting was pioneered by Wulff and coworkers and is discussed in greater detail in Chapter 4 in this volume. In short, the technique involves the formation of covalent bonds between the template and the functional monomer in the synthesis of the polymer. During rebinding, the ligands interact with the polymer via reversible, labile covalent bonds. Whitcombe and coworkers introduced semi-covalent imprinting which can be looked upon as a hybrid approach relying on covalent bonds to first form the template–monomer complex with subsequent rebinding to the polymer occurring via noncovalent interactions (Chapter 5). Potential shortcomings of covalent and semi-covalent imprinting include their narrow range of appropriate functional monomers and the need to perform additional synthetic steps to assemble the template–monomer complex. However, the possible occurrence of heterogeneity of the binding sites with the noncovalent approach is not encountered to the same degree by these two approaches. Covalent imprinting systems display other features being addressed in more detail in the corresponding chapters in this volume.

The introduction of MIPs based on noncovalent interactions [2,16] has led to rapid growth in the imprinting field. The popularity of this technique (in our estimate, currently over 90% of the published work make use of this approach) can be attributed to several factors. First, the preparation of a noncovalent MIPs is relatively straightforward and requires few simple synthetic steps. A wide variety of commercially available functional monomers enable imprinters to take advantage of many different types of intermolecular interactions (Fig. 4, also see Chapter 7).

This in turn leads to a much broader range of applicability compared to covalent and semi-covalent techniques. The diversity of target compounds that have been imprinted using noncovalent techniques is vast and a brief general overview of substance classes is listed in Table 2. Finally, compared to covalent techniques, the noncovalent approach utilizes interactions that are more similar to those seen in natural, biochemical processes. Thus, the noncovalent approach is regarded as the most promising technique for mimicking nature's selectivity.

B. Noncovalent Types of Interactions

Noncovalent interactions are the basis of reversible binding and recognition events in biochemical systems that rely on noncovalent interactions such as salt bridges, hydrogen bonds, and hydrophobic interactions. The binding energies of such types

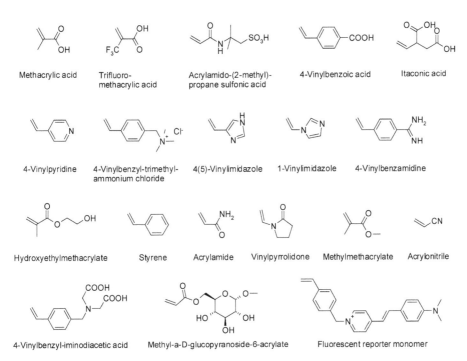

Methacrylic acid   Trifluoro-   Acrylamido-(2-methyl)-   4-Vinylbenzoic acid   Itaconic acid
                    methacrylic acid   propane sulfonic acid

4-Vinylpyridine   4-Vinylbenzyl-trimethyl-   4(5)-Vinylimidazole   1-Vinylimidazole   4-Vinylbenzamidine
                   ammonium chloride

Hydroxyethylmethacrylate   Styrene   Acrylamide   Vinylpyrrolidone   Methylmethacrylate   Acrylonitrile

4-Vinylbenzyl-iminodiacetic acid   Methyl-a-D-glucopyranoside-6-acrylate   Fluorescent reporter monomer

Figure 4   Representation of functional monomers that can serve as building blocks in noncovalent molecular imprinting. The first row represents acidic functional monomers with methacrylic acid as the most widely used functional monomer. In the second row, a selection of basic functional monomers is shown. Neutral monomers are lined up in the third row and the last row contains monomers that were designed and synthesized for a specific purpose. For example, 4-vinylbenzyl-iminodiacetic acid serves as a metal chelating monomer [89], methyl-α-D-glucopyranoside-6-acrylate was used in the imprinting of a protected amino acid [90]. As the last example, a monomer that would fluorescence upon rebinding of the template is depicted [91].

of bond are weak compared to a covalent bond. However, when weak interactions are present as multiple docking points, they lead to strong binding between two entities. An example is the strong binding between the pair biotin–avidin; the dissociation constant within the femtomolar range and the binding energy between the two entities corresponds to 90 kJ/mol [18]. This example shows that the accumulation of a multitude of noncovalent weak forces leads to exceptionally high binding strengths between two entities.

The design and synthesis of successful imprints require a basic understanding of the fundamental forces that lead to self-assembly of the template–monomer complex and to rebinding of ligands to the final MIP. For noncovalent imprinting, the most important interactions include van der Waals (VDW) forces, hydrogen bonding, ionic interactions, and hydrophobic forces. This section is intended to provide a brief overview of these forces; more thorough theoretical treatments can be found elsewhere [19]. Examples of these forces are shown schematically in Table 3.

Interactions between oppositely charged ions or between ions and dipoles are tremendously important in imprinting. The most common functional monomer,

Table 2 A Selection of Substance Classes Used as Templates in Molecular Imprinting Protocols Via Noncovalent Imprinting Techniques (Reviews by Mosbach and Ramström [50] and Sellergren [62])

Substance class	Examples	Reference
Amino acids	Free and derivatized amino acids	[63]
Peptides	Enkephalin, oxytocin	[64,65]
Steroids	Cholesterol, testosterone, estrogen	[66,67]
Carbohydrates	Derivatized sugars, glycosides	[68,69]
Nucleotides	AMP, IMP	[70,71]
Dyes	Safranine O, rhodanil blue	[72]
Pesticides	Atrazine, 2,4 dichlorophenoxyacetic acid	[32,73]
Metal ions	Cu^{2+}, Ca^{2+}	[74,75]
Drugs	Propranolol, morphine, nicotine, penicillin	[31,76,77,78]
Proteins	Transferrin, RNase A, myoglobin, IgG, urokinase	[22,26,72,79,80]
Microorganisms	*Saccharomyces cerevisiae, Staphylococcus aureus, Listeria monocytogenes*	[44,81]
Crystals	Calcite	[82]

methacrylic acid (MAA), can transfer its acidic proton to basic functional groups on the template molecule (e.g., amino groups as exemplified in Tables 3 and 4). This results in a positive electrostatic interaction between the deprotonated carboxylate anion and the protonated amino group. While the strength of the interaction varies depending on the chemical environment, the typical energies may reach up to 60 kJ/mole [20]. Hydrogen bonding is also among the most important mechanisms of interaction between the template and functional monomer. Basically, in a hydrogen bond, a H atom is covalently attached to one electronegative atom (mostly O or N), forming a hydrogen bond donor with a high dipole moment. A second electronegative atom, the hydrogen bond acceptor, is attracted to the positive pole in the donor, forming an exceptionally strong dipole–dipole interaction. The strength of a typical hydrogen bond is between 10 and 40 kJ/mol, which is much stronger than typical VDW interactions ($\sim$1 kJ/mol) but still considerably weaker than covalent bonds ($\sim$ 500 kJ/mol) [20,21]. It is assumed that dispersion or van der Waals forces play a minor role in typical imprinting procedures, as these interaction types do not display high binding energies.

Hydrophobic interactions also play an important role in the self-assembly processes. Highly hydrophobic molecules like long-chain hydrocarbons tend to aggregate in, e.g., water due to entropic effects. When such a hydrophobic molecule comes in contact with water, the hydrogen bonds between individual water molecules are disturbed. It is possible for water molecules to rearrange themselves in a manner that allows them to maintain their normal hydrogen bonding interactions; however, such rearrangements are highly entropically unfavorable. A consequence of this hydrophobic effect is a relatively strong attractive interaction between two

Table 3 Overview Over Some Types of Noncovalent Interactions with Importance to Molecular Imprinting Represented by Schematic Models, Examples, and Approximate Bond Energies

Type of interaction	Model	Example	Approximate bond energy
Charge–charge			Upto ~60 kJ/mol
Hydrogen bond			Upto ~40 kJ/mol
Charge–dipole			Upto ~8 kJ/mol
Dipole–dipole			~1 kJ/mol
van der Waals			0.1–1 kJ/mol

Table 4 Schematic Overview Over Examples of Binding Sites Formed with Some Functional Monomers Given in Fig. 4

Template–monomer system	Schematic presentation of the binding site	Reference
Nicotine + TFMAA or MAA		[83]
2,4-Dichlorophenoxy-actic acid + vinylpyridine		[32]
p-Nitrophenyl-methyl-phosphonate + vinylimidazole		[84]

(*Continued*)

Table 4 (*Continued*)

Template – monomer system	Schematic presentation of the binding site	Reference
Creatinine + acryla\mindo-methyl-propane sulfonic acid		[85]
Bisimidazole + vinylbenzyl-iminodiacetic acid		[86]
Boc-Tryptophane + acrylamide		[23]

hydrophobic moieties in water compared to their interaction in a hydrocarbon-based solvent. As this interaction force is most pronounced in polar and aqueous environments, hydrophobic interactions may be most favorably be exploited in such systems.

C. Monomers and Cross-Linkers

Because the self-assembly process is governed by attractive interactions between the monomers and template, the choice of appropriate functional monomers is critical for achieving good imprints. Generally, one chooses monomers with functional groups that complement those found on the template molecule. For example, if the template contains carboxylic or sulfonic acid groups, one might select a functional monomer that contains an amine group, which can form strong ionic interactions with the template. Conversely, basic groups on the template can form strong ionic interactions with an acidic functional monomer. The number and type of hydrogen bond donors/acceptors on the template are also important factors to take into account. Finally, when choosing the functional monomer one should consider the chemical environment and application in which the MIP is to be used.

Figure 4 depicts a range of various commercially available polymerizable functional building blocks of which many are commonly used monomers for molecular imprinting. In Table 4, examples of templates imprinted with monomers are presented. While by no means exhaustive, this list demonstrates the rich diversity of interactions that can be exploited for making good imprints. The most commonly used functional monomer is methacrylic acid (MAA). In addition to the strong ionic interactions that MAA can form with basic functional groups on the template, the carboxyl group on this monomer is an excellent hydrogen bond donor and acceptor. Frequently, trifluoromethacrylic acid (TFMAA) is used instead of MAA because the former is a stronger acid due to its electronegative fluorines than the latter, so that TFMAA is capable of forming stronger ionic interactions. Basic functional monomers like vinylpyridine or vinylimidazole can participate in the formation of hydrogen bonds and can also form ionic interactions with acidic template molecules. Metal chelating monomers (e.g., vinylbenzyliminodiacetic acid and vinylimidazole) are also frequently used in molecular imprinting. For a discussion of how metal chelating interactions are applied to molecular imprinting, the reader is referred to Chapter 6. For some applications, it is advantageous to use "neutral" monomers, e.g., styrene or methylmethacrylate or acrylamide, the latter having being used in imprinting of proteins in aqueous systems [22] also, acrylamide has been advantageously utilized as a hydrogen bonding functional monomer [23].

The synthesis of custom-made monomers to suit specialized molecular imprinting applications, exemplified here in the last row with, e.g., a fluorescent reporter monomer is discussed in Chapter 7 (see also Chapter 17 for a scintillation reporting monomer).

The main purpose of the cross-linking monomer is to rigidly fix the functional monomers in place to produce stable binding cavities. This generally requires that the cross-linker must be present in very high proportions in the final polymer. Thus, the chemical environment and morphology of the MIP are greatly affected by the choice of cross-linking monomer, so that careful consideration must be given to its choice. Some of the most common cross-linkers are shown in Fig. 5. While the most commonly used cross-linker EDMA has two acrylate groups, TRIM has a third

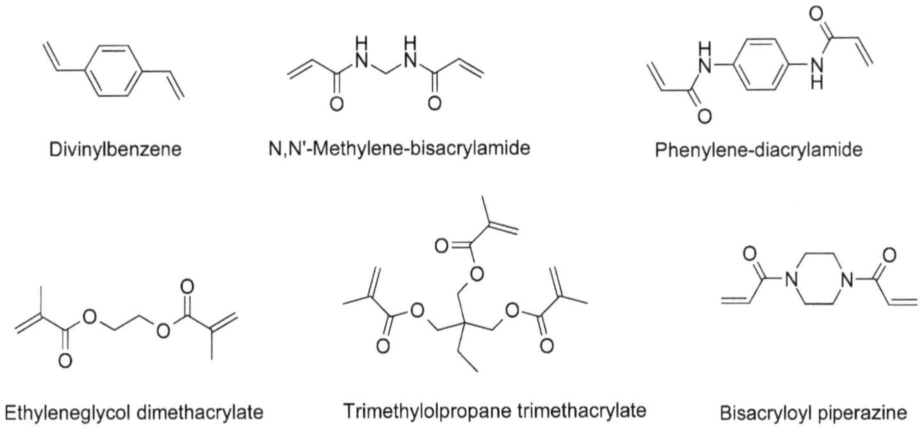

Figure 5 Selection of common cross-linkers used in molecular imprinting protocols. Both ethyleneglycol dimethacrylate (EDMA) and divinylbenzene (DVB) are very common cross-linkers in molecular imprinting. Other acrylate-based cross-linking monomers commonly used include the branched cross-linker trimethylolpropane trimethacrylate (TRIM)-[24]. Among the water-soluble cross-linkers, there are phenylene-diacrylamide, *N,N*-methylene diacrylamide [22], and *bis*-acryloylpiperazine [92], which have been used in aqueous systems for the imprinting of, e.g., enzymes.

acrylate group that enables it to form more rigid polymers. This enhanced rigidity has been shown to lead to MIPs with higher capacities and selectivities than those prepared with EDMA [24]. It is generally the case that capacity and selectivity of MIPs are strongly related to rigidity and thus cross-linking density [25].

The optimal cross-linker is also dependent upon the template and solvent. When imprinting proteins, for example, a water-soluble cross-linker is needed like the water-soluble cross-linker *N,N'*-methylene diacrylamide (methylenebisacrylamide). Divinylbenzene (DVB) is also a popular cross-linker and has been shown to be efficient for certain imprinting systems. An example of this is in the second imprinting protocol in this chapter, where DVB is used as a cross-linker for imprinting the chiral template (*R*) -isoproterenol.

D. Solvents

The choice of solvent is critical for achieving good imprints and for successful rebinding isotherms or chromatography results. One must use a solvent that dissolves all of the components of the prepolymerization mixture, allows for optimal template–monomer interaction, and contributes to good porosity characteristics in the final MIP.

For efficient template–monomer interactions, generally, aprotic organic solvents with low polarities are required. Highly polar aprotic solvents (e.g., DMF) tend to interfere with the ionic and VDW forces between the template and the functional monomer, while protic solvents (methanol, water) interfere with hydrogen bonding interactions. Regarding protein imprinting typically, an aqueous solvent or cosolvent mixture is required to avoid denaturing the template and to provide an appropriate natural environment for the protein. In addition, also the shear size of the template protein has to be taken into account and surface imprinting is a more favorable approach [26]. However, as discussed above, in an aqueous medium, apart from strong ionic interactions, hydrophobic forces may be the sole basis of molecular recognition.

Some typical solvents used for imprinting small molecules include toluene, acetonitrile, and chloroform. These, along with other solvents commonly used in either synthesis or evaluation of MIPs, are listed in Table 5.

Other solvents parameters (which are not discussed in more detail) may give valuable information whether the solvent is appropriate for the synthesis of a particular imprinting system or not. For instance, the solubility parameter may give information about the solubility of compounds, the morphology and the swellability of a polymer in a particular solvent [27–30]. Another important solvent property is the dielectric constant. Especially for electrostatic forces between two charged entities, the dielectric constant of the surrounding media is of great importance. The binding strength between two oppositely charged molecules is high in solvents with low dielectric constants like hydrocarbons, e.g., toluene, and is decreased considerably in polar solvents like water or dimethylformamide. The presence of little amounts of water or other polar solvents that are potent in breaking hydrogen bonds or salt bridges decreases the binding abilities of the MIP. However, although this is a general rule of thumb, there are cases, where the rebinding of MIPs worked very well in aqueous solutions although they were imprinted in nonpolar solvents [31]. Furthermore, there are even cases, where a template was imprinted in a polar solvent mixture with a high dielectric constant (water/methanol, 4:1, v/v) and still led to efficiently imprinted materials where the MIP worked exceptionally well in aqueous solvents [32].

A third parameter that seems to be highly useful in the choice of the optimal solvent system is the hydrogen bond parameter, HBP [30]. It denotes the ability of the solvent to form hydrogen bonds, either with molecules of the same kind or with solutes. A solvent with a high hydrogen bond parameter will easily dissolve polar molecules able to create hydrogen bonds. But it will consequently also interfere with the hydrogen bonds that are formed between functional monomer and template and

Table 5 Physical Properties of Some Solvents Commonly Used in Molecular Imprinting Protocols [30]

Solvent	Dielectric constant	Hydrogen bond type	General effect as imprinting solvent
Toluene	2.40	P	+
Methylene chloride	8.90	P	+
Chloroform	4.80	P	+
Acetonitrile	37.50	P	+
Acetic acid	6.20	S	−
Methanol	32.70	S	−
Dimethylformamide	37	M	−
Water	80.00	S	−

Hydrogen bond type, hydrogen bond donor/acceptor ability: P: poor, M: medium, S: strong; +: favorable as solvent for imprinting, −less favorable as solvent for imprinting.

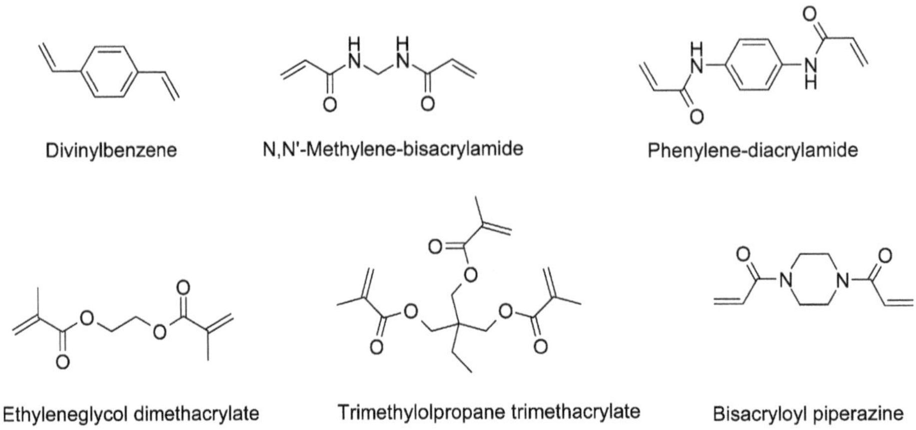

Divinylbenzene   N,N'-Methylene-bisacrylamide   Phenylene-diacrylamide

Ethyleneglycol dimethacrylate   Trimethylolpropane trimethacrylate   Bisacryloyl piperazine

Figure 5 Selection of common cross-linkers used in molecular imprinting protocols. Both ethyleneglycol dimethacrylate (EDMA) and divinylbenzene (DVB) are very common cross-linkers in molecular imprinting. Other acrylate-based cross-linking monomers commonly used include the branched cross-linker trimethylolpropane trimethacrylate (TRIM)-[24]. Among the water-soluble cross-linkers, there are phenylene-diacrylamide, *N,N*-methylene diacrylamide [22], and *bis*-acryloylpiperazine [92], which have been used in aqueous systems for the imprinting of, e.g., enzymes.

acrylate group that enables it to form more rigid polymers. This enhanced rigidity has been shown to lead to MIPs with higher capacities and selectivities than those prepared with EDMA [24]. It is generally the case that capacity and selectivity of MIPs are strongly related to rigidity and thus cross-linking density [25].

The optimal cross-linker is also dependent upon the template and solvent. When imprinting proteins, for example, a water-soluble cross-linker is needed like the water-soluble cross-linker *N,N'*-methylene diacrylamide (methylenebisacrylamide). Divinylbenzene (DVB) is also a popular cross-linker and has been shown to be efficient for certain imprinting systems. An example of this is in the second imprinting protocol in this chapter, where DVB is used as a cross-linker for imprinting the chiral template (*R*)-isoproterenol.

D. Solvents

The choice of solvent is critical for achieving good imprints and for successful rebinding isotherms or chromatography results. One must use a solvent that dissolves all of the components of the prepolymerization mixture, allows for optimal template–monomer interaction, and contributes to good porosity characteristics in the final MIP.

For efficient template–monomer interactions, generally, aprotic organic solvents with low polarities are required. Highly polar aprotic solvents (e.g., DMF) tend to interfere with the ionic and VDW forces between the template and the functional monomer, while protic solvents (methanol, water) interfere with hydrogen bonding interactions. Regarding protein imprinting typically, an aqueous solvent or cosolvent mixture is required to avoid denaturing the template and to provide an appropriate natural environment for the protein. In addition, also the shear size of

may thereby lead to MIPs with a lower selectivity. As these solvents disturb the template–functional monomer complex, the resulting MIPs will have binding sites with lower fidelity and the capacities will be decreased [33]. That solvents with a high HBP have a deleterious influence for imprinted polymers was shown in work of Sellergren and Shea [28] with a model system using a chiral template. An MIP prepared in a low hydrogen bond parameter solvent (e.g., methylene chloride) lead to an MIP that displayed a high separation factor of 8.2, whereas imprinting in dimethylformamide decreased the separation factor down to 2.0. In this study, it was shown that the most influencing parameter on the outcome of the MIP was the HBP, whereas the role of the dielectric constant did not seem to have a significant influence.

At last, it is very desirable to develop water-based imprinting protocols, as this resembles the environment of biochemical processes and might lead to MIPs that approach antibody-like binding abilities [32]. Water is also a more environmental-friendly solvent and cheap compared to common organic solvents.

III. AREAS FOR IMPROVEMENT IN NONCOVALENTLY IMPRINTED POLYMERS

Despite the many advantages of MIPs (Table 6) over natural antibodies (high stability, resistance to harsh chemicals, recyclable, the ease of synthesis, high specificity, and selectivity), several improvements need to be addressed. Whereas monoclonal antibodies or receptors exhibit a well-defined binding energy (monoclonal behavior), the cavities in MIPs exhibit a relatively broad spectrum of binding energies (polyclonal behavior) like natural polyclonal antibodies do. The asymmetric peaks commonly observed in chromatography applications have been widely attributed to the heterogeneous nature of the binding energies in a given MIP [34]. This problem may be solved by stabilizing the template–monomer complex prior to polymerization. Covalent imprinting techniques lead to the most stable monomer–template complexes; however, these strategies remain rather limited in their scope of applicability. Another way to stabilize the template–monomer complex involves immobilizing the template on a sacrificial support, as described in section II.B. Finally, it is thought that by lowering the temperature at which the polymerization takes place, the entropic penalty associated with the self-assembly process is

Table 6 Characteristics of Molecularly Imprinted Polymers [50,87,88]

Feature	Characteristic of the MIP
Physical stability	Resistant against mechanical stress, high pressures, and elevated temperatures
Chemical stability	Resistant against acids, bases, organic solvents, and metal ions
Shelf life	Several years without loss of performance
Capacity	0.1–1 mg template/g MIP
Recovery yield of template	>99%
Yield of polymer	Up to 99%
Binding strength	nM range

minimized, leading to a stronger template–monomer interaction and a more stable complex. Molecular motion is retarded at lower temperatures, which has been shown to lead to improved MIPs [35,36].

Another area to be looked upon of MIPs is their sometimes moderate capacity. While many preparations are sufficient for analytical applications [5], many large-scale processes (e.g., preparatory-scale chromatography) would profit from higher capacities. Several variables, such as the amount of original template used, monomer composition, configuration of the polymer, solvent, etc. may be optimized towards preparations with enhanced capacity. Generally, the capacity of MIPs corresponds often to 1% of the initial amount of template used in the synthesis of the polymer.

The most commonly used functional and cross-linking monomers for molecular imprinting are MAA and EDMA, respectively. Although these monomers often lead to a significant imprinting effect, other monomers may perform better for a given template or for a specific application. Thus, it is frequently necessary to test several cross-linkers, functional monomers, and solvents to determine the optimum recipe for the imprinting of a desired target compound. In combinatorial studies, an assortment of "promising" functional monomers may be combined with a series of appropriate and compatible cross-linkers. Several papers have discussed, on a small scale [25,37], optimization processes where a few functional monomers were combined with several cross-linkers. Optimizing MIP recipes using the approach described thus far requires large amounts of tedious experimental work combined with a little luck in picking effective monomers, cross-linkers, and solvents from the dozens of choices (and thousands of combinations) available. The last few years have seen the developments towards automated systems that can execute the simultaneous synthesis, work-up, and evaluation of large numbers of MIPs. In addition to improved throughput, automation minimizes the amount of random and systematic variation caused by human error. Automation was first applied to MIP synthesis and characterization by Takeuchi and coworkers [38] and by Lanza and Sellergren [39].

Other researchers have attempted to bypass the tedious hit-or-miss experimental approach in favor of computational techniques [40]. The computational approach could turn into a useful tool for the design of MIPs and is covered in Chapter 8.

However, many of the strategies developed for the automated synthesis of MIPs require expensive, specialized equipment that may not be available in many laboratories. For a discussion on combinatorial or computational approaches the reader is referred to Chapters 8 and 14, respectively.

IV. EXAMPLES OF MOLECULARLY IMPRINTED POLYMERS: TYPICAL TECHNIQUES AND RECIPES

The examples presented in this section serve as a starting point for those who are new to the field and are interested in developing rudimentary skills in MIP synthesis and evaluation. Although the functional monomers, cross-linkers, and porogenic solvents are different for the two systems, the relative compositions of the prepolymerization mixtures are the same. In both systems, the molar ratio between the template, functional monomer, and cross-linker is 1:4:20, and the volume ratio between the porogenic solvent and the monomers is 4:3. In general, an imprinting effect can be observed over a fairly large range of compositions, and the optimal composition varies for different templates, monomers, and applications. This process gives irregular MIP granules; a typical batch of such traditional MIPs is shown in the scanning electron micrograph (SEM) in Fig. 10a. For a comparison, spherical beads can also be produced, which are uniform in size and shape (Fig. 10b). The preparation of such beaded imprinted materials is discussed in a wider scope in Chapter 17

In the first example, the cross-linking and functional monomers are EDMA and MAA, respectively, and these are polymerized photochemically using UV lamps whose emission maximum is centered at around 360 nm (note that the 254 nm radiation emitted by low-pressure mercury lamps is not useful since it is strongly absorbed by most laboratory glassware). Thermal initiation is used in the second example to polymerize DVB and TFMAA. Commercially obtained monomers contain inhibitors to prevent their polymerization during transport and storage. In the case of DVB, the inhibitor is typically *tert*-butylcatechol, which can be removed by passing the DVB through a column of basic alumina immediately before use. Removal of the inhibitor from acrylate-based cross-linkers (e.g., EDMA) typically requires cumbersome vacuum distillation procedures, however, and in most cases, complete polymerization of the vinyl groups proceeds when a sufficient amount of initiator is used (typically 0.7 molar percent relative to the total number of vinyl groups). Following the polymerization, a hard monolith is obtained that must be crushed and ground until all of the particles can pass through a sieve with (typically) 25 µm openings. The powder obtained from this procedure must be washed thoroughly in order to extract the template. If the MIP is to be used as a stationary phase in HPLC columns, it is important to remove very small particles ("fines") since they tend to lead to very high back pressures. This is usually achieved using a repetitive sedimentation procedure as described in the second example. The sedimentation procedure can often be avoided (and therefore the effective yield of polymer can be increased) if the MIP is to be used for batch-mode analysis or solid-phase extraction applications

The first example, depicted in Fig. 6, describes the synthesis and evaluation of a polymer imprinted with the bronchodilating drug theophylline, which is used in the treatment of asthma. Originally published in the journal *Nature* [2], this work drew considerable attention to the field of molecular imprinting because it was the first study to show that an MIP could be substituted for a natural antibody in a standard clinical assay. The MIP and antibody-based assays exhibit similar selectivities, and both can discriminate between theophylline and structurally related compounds. An equilibrium binding assay is described which uses radiolabeled theophylline as a marker. Data are presented for which nonradioactive theophylline, caffeine, and theobromine are used in competitive binding assays. These assays provide valuable information about the capacity and selectivity of the MIP.

In the second example, we will imprint a chiral template, (*R*)-isoproterenol (as depicted in Fig. 9), which is another common asthma medication [41]. The synthesis of materials that can discriminate between enantiomers [24] demonstrates the vast potential of the imprinting technique. Because the chemical and physical properties of enantiomers are identical in a nonchiral environment, the observation of enantioselectivity in MIPs can only be attributed to the three-dimensional arrangement of the functional monomers around the template molecule. Chiral selectivity in MIPs can be readily observed in batch-mode binding and competition studies [31] similar to the one described in the theophylline MIP protocol. In this case, however, we will present a procedure for evaluating the MIP as a HPLC packing material in order to demonstrate that a racemic mixture of (*R-S*)-isoproterenol can be resolved with a predetermined elution order.

V. PROTOCOL 1: AN IMPRINTED POLYMER SPECIFIC FOR THE DRUG THEOPHYLLINE

Reagents needed	Equipment needed
Theophylline	Borosilicate test tubes with a screw top
Azoisobutyronitrile (AIBN)	Sonicator (optional)
Ethylene glycol dimethacrylate (EDMA)	Photoreactor (e.g., Lamag UV-lamp)
	4°C cold-room (optional)
Methacrylic acid (MAA)	Mortar and pestle
Toluene (containing ≤0.5% water)	25 μm sieve (e.g., available from Retsch)
Nitrogen or argon source	
Acetone	Centrifuge equipped for 15–50 mL tubes
Methanol	Vacuum source
Acetic acid	Magnetic stirrer and stir bar
Tetrahydrofuran	1.5 mL polypropylene tubes
	(Eppendorf or equivalent)
^{3}H-labeled theophylline	Rocking table
(e.g., available from Sigma)	
Scintillation fluid (e.g., Ecoscint O,	Vortex mixer (optional)
available from National Diagnostics)	
	Centrifuge equipped for Eppendorf tubes
	Scintillation vials
	Beta-counter (e.g., Rackbeta 1219 from
	LKB Wallac, Turku, Finland)

A. Preparation of a Theophylline MIP and the Corresponding Nonimprinted Polymer

An MIP and a nonimprinted polymer (NP) should be prepared in parallel and with identical compositions (except that the template is to be omitted from the NP). Weigh the template (theophylline; 0.6 mmol, 108 mg; *omit from the NP*) and initiator (AIBN; 0.26 mmol, 43 mg) into a glass test tube with a screw-on cap. Add the cross-linker (EDMA; 12 mmol, 2.26 mL), functional monomer (MAA; 2.4 mmol, 0.205 mL), and porogenic solvent (toluene; 3.3 mL). If necessary, sonicate until all the solids have dissolved. Cool the mixtures with an ice water bath (to prevent evaporation) and sparge for 2 min with nitrogen or argon (i.e., to remove dissolved oxygen, which can inhibit free radical polymerizations). Seal the sparged tubes, and place them in the UV photoreactor (which should ideally be located in a cold room at 4°C). For the photoreactor specified above, polymerization should be allowed to proceed for 12–18 hr. More intense UV sources may polymerize the mixture more quickly, while less intense sources may require longer polymerization times.

After the photolysis, a hard polymer monolith should be obtained, and this must be crushed and ground into a fine powder with a mortar and pestle. Suspend the

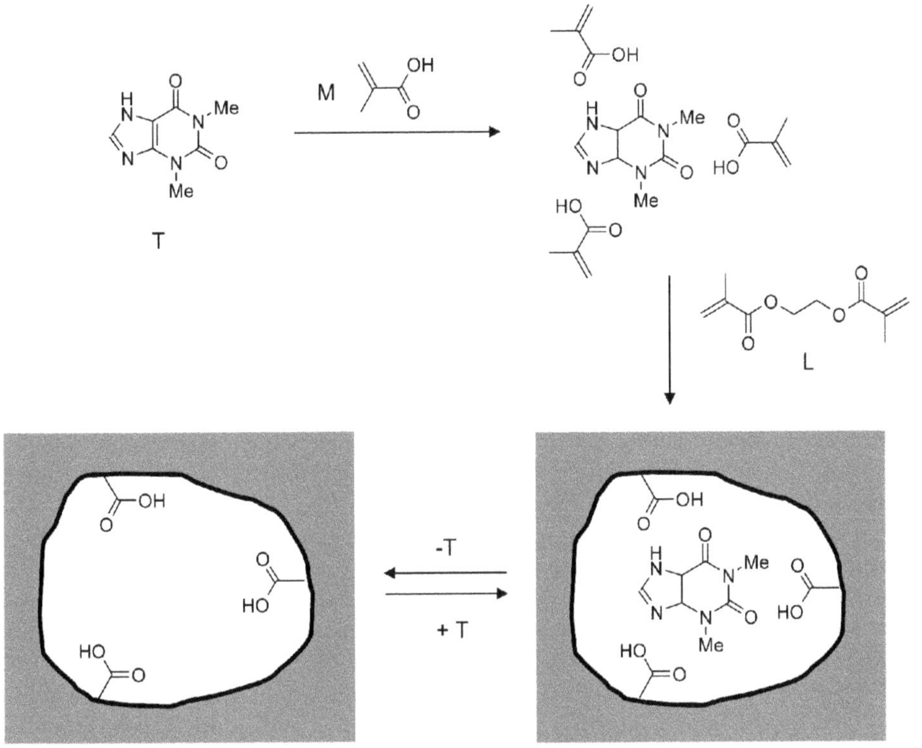

Figure 6 The template theophylline (T) is allowed to interact with the functional monomer methacrylic acid (M) to form a self-assembly. The monomers will interact with theophylline and will form a self-assembly complex mainly based on hydrogen bonding. This self-assembly and the positions of the functional monomers are then frozen and held in place by copolymerization with cross-linker ethyleneglycol dimethacrylate (L). This leads to a rigid polymer scaffold that retains the spatial conformation and thus the specific binding cavity of the original template. After extraction of the template, a molecularly imprinted polymer (MIP) is obtained and the imprinted cavity is able to specifically rebind the template and other, even similar structures are excluded from the binding site cavity.

powder in acetone or methanol and pass as much of the material through a 25 μm sieve as possible. The fraction that does not pass through should be collected for further grinding and sieving, and this process should be repeated until all of the polymer can pass through the sieve. For MIPs that will be used as chromatography packing materials (refer to the preparation of the isoproterenol MIP), it is important to remove the fine particles, because these tend to restrict the flow in HPLC columns, resulting in very high back pressures. However, since the theophylline MIP and NP will be used in batch-mode analyses, removal of the fines is not necessary in the present case.

The sieved polymer must be collected and washed thoroughly with 1:9 acetic acid/methanol (two 15–20 mL washings for at least 2 h followed by a third washing overnight). During each washing period, the mixture should be agitated with a rocking table, and after the washing period has been completed, the tubes should be centrifuged so that the solution can be easily removed by decantation. After the acetic acid/methanol washings, the polymer should be washed several times with pure methanol to extract the acetic acid and facilitate drying. Repeat the methanol washings until the mixture no longer smells strongly of acetic acid. After drying the polymers overnight under vacuum at room temperature, the materials are ready for testing.

B. Evaluating Theophylline MIP and NP with Batch-Mode Binding Assays

Prepare stock suspensions of the MIP and NP by weighing 50 mg of each polymer into separate glass vials and adding 5.0 mL of binding solvent (9:1 toluene/THF, v/v). Stir the mixtures magnetically to obtain homogeneous suspensions. Add 0.4 pmole of ^{3}H-labeled theophylline into each vial. A pipette scheme is listed in Table 7, where appropriate amounts of the stirred polymer suspension, binding solvent, and radioligand solution are mixed in small 1.5 mL Eppendorf microtubes (total volume 1 mL).

This scheme is followed for both the imprinted and the nonimprinted polymer (one may wish to use a different color of ink to distinguish between the MIP and NP samples).

Sample 0 contains 900 μL of binding solvent and 100 μL of radioligand solution (i.e., no polymer) and serves as a standard, which is necessary for determining the number of counts/minute corresponding to 0% binding. Incubate the samples over-night on a rocking table. Centrifuge the samples the next day to separate the polymer from the supernatant. Pipette 200 μL of each supernatant solution into separate scintillation vials containing 10 mL of scintillation fluid. Quantify the amount of radioligand in each sample using a beta-counter. Plot the percent of bound radioligand against the concentration of the suspended polymer, and compare the results to those shown in Fig. 7.

For both polymers, the amount of free radioligand decreases as the concentration of polymer increases. However, the binding curve for the MIP is much steeper than that of the NP. $[P]_{50}$ values (the concentration of polymer necessary to achieve 50% binding) usually correspond to the inflection point of a binding curve (when

Table 7 Pipette Scheme for the Binding Assay

Number of tube	Polymer suspension (μL)	Radioligand (μL)	Binding solvent (μL)
0	0	100	900
1	1	100	899
2	3	100	897
3	10	100	890
4	30	100	870
5	100	100	800
6	300	100	600
7	900	100	0

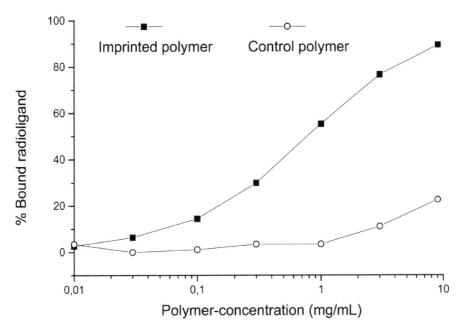

Figure 7 Plot of the binding assay (details in text).

plotted in semi-logarithmic coordinates, as in Fig. 7), and their magnitudes are therefore a convenient measure of the relative capacities of different polymers (e.g., the MIP vs. the NP). While approximately 1 mg/mL of MIP is capable of binding 50% of the radioligand (i.e., for the data in Fig. 7; results may vary), inspection of the plot (and extrapolation to higher polymer concentrations) reveals that the $[P]_{50}$ value of the NP is greater than 10 mg/mL. Thus, we conclude that the MIP has a higher capacity towards the radioligand than does the NP. If, in the system currently under consideration, the $[P]_{50}$ value for the MIP is much greater than 1 mg/mL, this may indicate a weak imprinting effect. Possible causes include the presence of water in the prepolymerization mixture (water disturbs the formation of the hydrogen bonds which are largely responsible for the self-assembly of the template–monomer complex) or incomplete polymerization. One may consider repeating the above procedure, taking care to use anhydrous solvents and to dry the test tubes in an oven prior to use. To increase the degree of polymerization, one may choose to increase the amount of initiator, or alternatively, after an 8 hr period of photolysis at 4°C, the tubes can be placed in an oven at 65–70°C for 12 hr.

The specificity of the MIP will be evaluated from competitive binding assays, as follows. Based on the $[P]_{50}$ (1 mg polymer per mL), a 10 mg/mL stock suspension (5 mL) is prepared in binding solvent. Keep the polymer in suspension by agitating the mixture with a small magnetic stirrer. Prepare competitor solutions containing 100 and 1 μg/mL of each theophylline, caffeine, or theobromine in binding solvent. Pipette these stock solutions into Eppendorf tubes according to the scheme in Table 8 (see pg. 46).

For Eppendorf tubes 1–4 a competitor concentration of 1 μg/mL and for tubes 5–7, a competitor concentration of 100 μg/mL was used.

As in the previous experiment, standards must be prepared that contain 800 µL of binding solvent, 100 µL of radioligand solution, and 100 µL polymer suspension. These standards will give 0% competition and correspond to 50% binding value of the radioligand to the polymer. Incubate the mixtures overnight, and work up the samples in the same manner described above. Calculate the percent binding in each tube, and plot the data as shown in Fig. 8 (see pg. 46).

Inspection of the data reveals that the percentage of bound radioligand decreases with increasing competitor ligand for all three cases. However, theobromine and caffeine are far less efficient competitors for the radioligand than theophylline. In fact, the fraction of bound radioligands remained nearly constant for the samples containing theobromine and caffeine up to concentrations of 1 µg/mL, whereas the addition of only 10^{-3} µg/mL of unlabeled theophylline resulted in a significant decrease in the fraction of bound radioligand. As can be seen from the competition assay, unlabeled theophylline is most potent as a competing agent. Much higher concentrations of the other xanthines, caffeine, and theobromine are necessary to decrease radiolabeled theophylline to bind to the MIP. Approximately 1000 times more caffeine or theobromine is required to compete out the same amount of radioligand than theophylline does. According to antibody terminology, these values correspond to cross-reactivities of 0.1% signifying that the polymer is highly selective. This is a strong indication for the specificity of this particular MIP. The aspects of MIP analysis are covered more deeply by Spivak and Shimizu in this volume in Chapters 15 and 16.

VI. PROTOCOL 2: ENANTIOSELECTIVE MIP FOR (R)-ISOPROTERENOL

Reagents needed	Equipment needed
(R)-isoproterenol	Glass vial with screw-cap
(R-S)-isoproterenol	Sonicator (optional)
(S)-isoproterenol (optional)	Oven
Trifluoromethacrylic acid (TFMAA)	Mortar and pestle
Divinylbenzene (DVB; tech. grade, 80% mixture of isomers)	25 µm sieve (e.g., available from Retsch)
Basic alumina	200 mL Erlenmeyer flask
Acetonitrile (HPLC grade)	Stainless steel column, 250 × 4.6 mm, with endfittings
Nitrogen or argon source	
Azoisobutyronitrile (AIBN)	Air-driven fluid pump and fittings for packing HPLC columns (e.g., available from Haskel)
Acetic acid	
	HPLC equipped with a detector sensitive to 280 nm light

A. Preparation of a (R)-Isoproterenol Imprinted Polymer

The molecular scheme for this polymer is presented in Fig. 9. Mix the template ((R) -isoproterenol; 396 mg, 2.5 mmol), initiator (AIBN; 95 mg, 0.58 mmol),

Table 8 Pipette Scheme for the Competition Assay

Number of tube	Polymer stock suspension (μL)	Radioligand solution (μL)	Competitor solution (μL)	Binding solvent (μL)
0	100	100	0	800
1	100	100	1	799
2	100	100	10	790
3	100	100	100	600
4	100	100	800	0
5	100	100	10	790
6	100	100	100	700
7	100	100	800	0

functional monomer (TFMAA; 1.05 g, 10 mmol), cross-linker (DVB; 6.7 mL, 50 mmol), and porogenic solvent (acetonitrile; 10 mL) in a screw-top glass vial. If necessary, sonicate the mixture until all the solids dissolve. Place the vial in an ice bath, and sparge the mixture with nitrogen or argon for at least 2 min. Seal the vial tightly, and place it in an oven at 65°C for at least 16 hr. After the polymerization has been completed, remove the polymer from the vial. Follow the grinding and sieving procedure described in the theophylline MIP protocol. Note, however, that one should take care to grind the polymer gently and for only a short period of time before attempting to pass the powder through a 25 μm sieve. Extensive grinding will result in an excessive amount of very small particles ("fines"), which must be separated

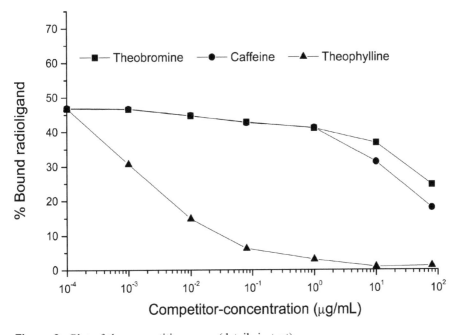

Figure 8 Plot of the competition assay (details in text).

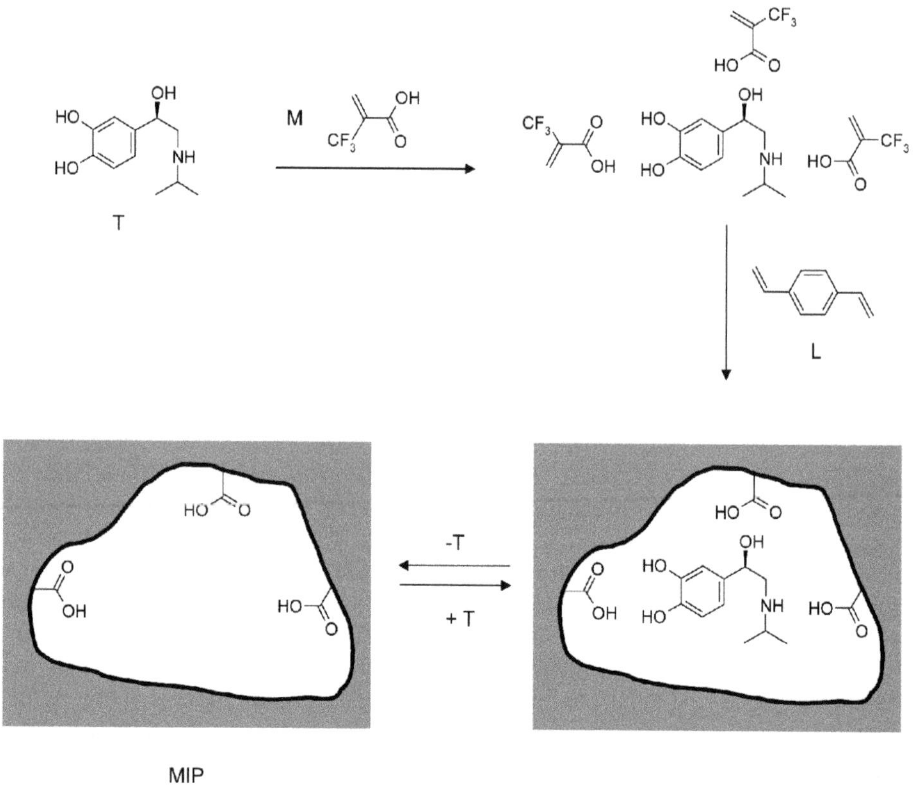

Figure 9 Schematic presentation of the imprinting of (−)-isoproterenol. A self-assembly construct forms between template (−)-isoproterenol (T) and functional monomers (trifluormethacrylic acid, M) in accordance with the interactions. The basic secondary nitrogen group in the template forms an ionic salt bridge with the strongly acidic monomer; the hydroxyl groups form hydrogen bonds with the monomer. This specific assembly is then solidified by copolymerization with a cross-linker (divinylbenzene, L). After polymerization, the template is extracted and the MIP rebinds the imprinted (−)-enantiomer in preference to the (+)-enantiomer.

and discarded prior to packing the column. A larger number of grinding/sieving cycles will therefore result in a greater yield of useful particles.

Removal of the fines is accomplished with the following sedimentation procedure. Suspend the sieved particles in approximately 100 mL acetone in a 200 mL Erlenmeyer flask. Shake the flask briefly, and then allow the mixture to sit undisturbed. After 30 min, decant the acetone (which contains suspended fines), taking care not to disturb the particles that had settled to the bottom of the flask. Repeat this procedure as many times as necessary until the supernatant is clear (this usually requires 3–5 cycles). Figure 10a shows an SEM image of the particles obtained by this procedure. The size of these particles, which ranges from 10 to 25 μm in diameter, is appropriate for chromatography applications. For a comparison, spherical beads are presented in Fig. 10b; these can be easily prepared by various methods, which are discussed in more detail in Chapter 17.

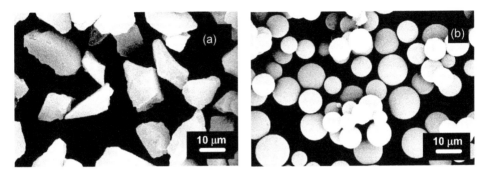

Figure 10 (a) SEM of the MIP granules, approximate size range between 10 and 25 μm, magnification × 1000. (b) As an addition, uniform and spherical beads of MIP can also be produced and an SEM of MIP beads, approximate size range between 10 and 15 μm, magnification × 1000 is shown.

B. Evaluating the (R)-Isoproterenol MIP Using HPLC

Prepare a slurry of the MIP, approximately 3 g polymer in 50 mL 1:3 acetonitrile/ water. Pump the slurry into the column, which is capped on one end with an appropriate endfitting, using an air-driven fluid pump or other suitable device. Continue pumping the slurry until the column has been completely filled with polymer (the backpressure necessary to achieve this is expected to be on the order of 200 bar). Cap the other end of the column with an endfitting, and install the column in the HPLC, taking care to ensure that the direction of flow is the same as it was during packing.

Extract the template by pumping a 1:4 acetic acid/acetonitrile solution through the column using a flow rate of 1 mL/min. Removal of the template can be followed in real-time by monitoring the absorbance of the eluent at 280 nm. After several hours of washing, the baseline will have stabilized, indicating that all the template has been extracted. Change the mobile phase to a 25 mM sodium citrate buffer (pH 3.0; containing 10% acetonitrile) and continue pumping at a flow rate of 1 mL/min until a stable baseline is obtained. Prepare a solution of 2 mM (R-S)-isoproterenol in the mobile phase spiked with 1 ppm acetone, and inject 20 μL onto the column. The chromatogram is monitored at 280 nm (Fig. 11). The first peak, centered around 7 min corresponds to the void marker acetone. The second peak is caused by the (S) -enantiomer and has its maximum around 16.5 min, while the third, highly asymmetric peak around 23 min, corresponds to the (R) -enantiomer. The assignment of the peaks can be confirmed by injecting the pure (R) and (S) enantiomers. As the polymer was imprinted against the R-form of the molecule isoproterenol, the polymer retained R much stronger than the nonimprinted S-form. The elution order shown in Fig. 11 is typical of chiral imprints. An imprinted polymer prepared against the other enantiomers would display the reversed elution order, confirming a chiral imprinting effect. The present column could be used for instance, for the quantification of the amount of each enantiomer in a given sample. Especially in the pharmaceutical industry, where many drugs are chiral compounds, the enantiomeric purity is of great importance and MIPs could be applied here to monitor the amount of each enantiomer. However, it has to be noted that the latter asymmetric peak is not ideal

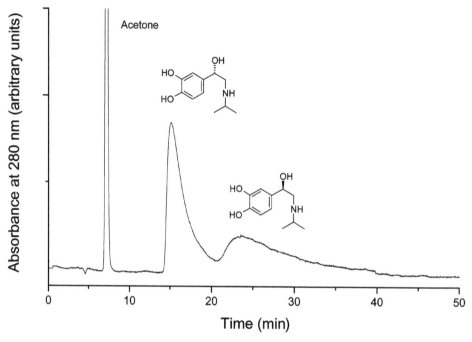

Figure 11 Chromatogram of the imprinted stationary phase. This imprinted polymer does work in aqueous solvents and chromatographic characterizations are performed using an aqueous mobile phase (25 mM sodium citrate, pH 3.0) containing 10% MeCN at a flow rate of 1 mL/min and chromatograms, run in isocratic mode and recorded at 280 nm. Injections of 20 μL of 2 mM racemic isoproterenol HCl (40 μmol) dissolved in the mobile phase are done in order to evaluate the enantioselectivity towards its imprinted print molecule isoproterenol. Eluent used was a sodium citrate buffer (pH 3.0, 25 mM citrate, 10% MeCN) flow rate 1 ml/min, peak detection at 280 nm, injection of 20 μL of a racemic isoproterenol hydrochloride solution (2 mM), acetone was used as void marker. The structures of + and − isoproterenol are given in Fig. 8.

for analysis, whereas the first peak is more symmetrical and more straightforward to analyze.

VII. IMPRINTING WITH IMMOBILIZED TEMPLATES

A recently developed new approach to molecular imprinting is the imprinting of a template that is immobilized to a support material. By immobilization, the molecular motion of the template prior to polymerization is greatly diminished and the stability of the template–monomer complex is thought to be enhanced, leading to greater degeneracy among the binding cavities. Also, the template is fixed in an oriented way so that all templates expose only one side of the molecule. By that, more stable template–monomer complexes may be expected and an increased uniformity of the sites should convey a more homogeneous population of binding cavities. Further, after polymerization, the template with the sacrificial carrier is removed and the cavities that are obtained are predominantly located at the surface of the polymer, which should lead to improved accessibility of the sites even with larger entities. The binding kinetics would be greatly enhanced by this approach. Altogether, it is anticipated that it may lead to imprinted cavities that are more uniform in size, shape, and binding energy.

Researchers have devised a synthesis strategy that entails the use of a sacrificial porous solid support on which the template is immobilized. An example that demonstrates the feasibility of this approach is represented in Fig. 12 [42]. Presently, this new concept is explored further and the imprinting of other immobilized templates has been published [43].

This new approach entails the imprinting of a template that is attached to a solid surface and the monomers are in the liquid prepolymer phase. Imprinting performed on such an interface (solid–liquid) is a rather new concept and the phenomena at the interface governing the self-assembly and recognition are not completely resolved yet. Although the technique is still in its beginnings, it may eventually lead to MIPs with novel and even superior binding and recognition properties.

A disadvantage of this method is that an appropriate support material has to be provided. Either, this material is synthesized or purchased from commercial sources.

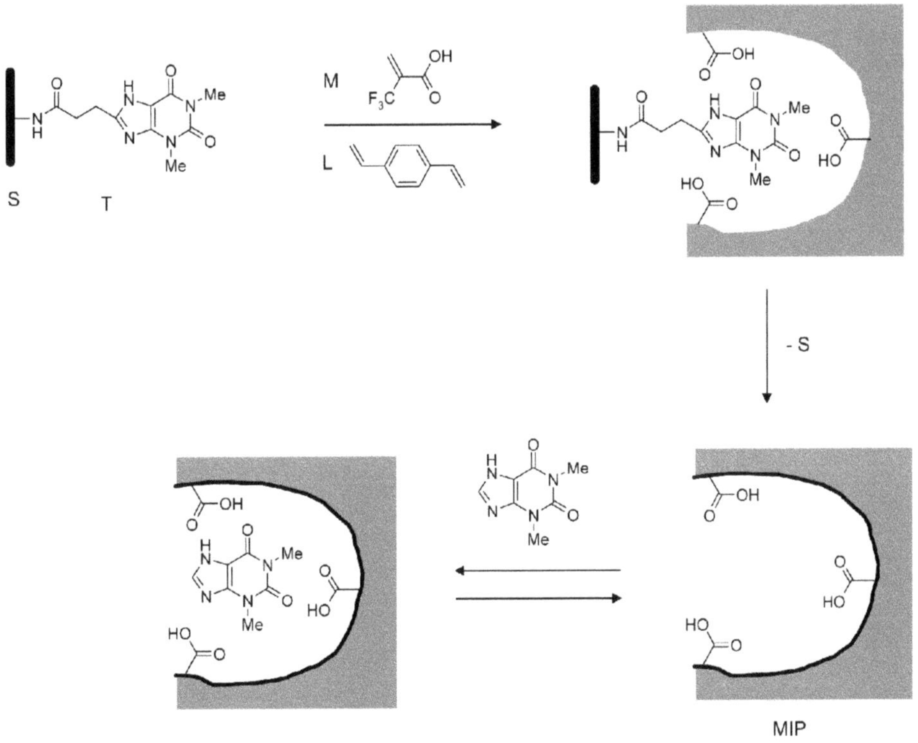

Figure 12 Imprinting of a theophylline derivative immobilized on silica. The template (a carboxypropyl-derivative of theophylline) was immobilized on porous and spherical amino-functionalized silica gel via formation of an amide bond. This construct is then imprinted with traditional imprinting monomers (TFMAA and DVB). Following polymerization, the composite material is treated with HF to dissolve and remove the silica gel, leaving spherical porous MIP particles which mirror the original silica in size, shape, and porosity.

Further, imprinting of immobilized templates involves additional synthetic steps and the support material is often destroyed in the process.

VIII. "ANTI-IDIOTYPIC IMPRINTING" AND "DIRECT MOLDING": THE NEXT GENERATION OF MIPs

Until now, most research activities have focused on generating imprinted materials recognizing different target entities ranging from small molecules to proteins [22], and, even to whole cells [44].

We have recently gone a step further and used preimprinted binding sites to generate bioactive compounds mimicking the original template, which was chosen from a known enzyme inhibitor [45]. This "double imprinting" is analogous to the formation of anti-idiotypic antibodies in the immune response system, where the combining site of the secondary antibody is an "internal image" of the original antigen. We envision that our synthetic anti-idiotypic or double imprinting approach should be useful for finding new drug candidates, especially when the three-dimensional structure of a biological target, a prerequisite for the rational drug design, is unresolved.

In our model anti-idiotypic imprinting, we have chosen the medicinally interesting proteinase kallikrein as a model system. In the first step, a previously identified inhibitor [1] was used as the original template to prepare noncovalently an imprinted polymer. The polymer contains specific binding "cavities" mimicking the enzyme's active site. These "cavities" were then used in a second round of imprinting to

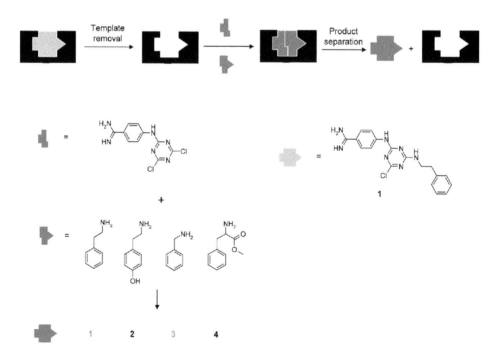

Figure 13 Schematic presentation of the approach of anti-idiotypic imprinting for generating bioactive molecules [45].

(a)

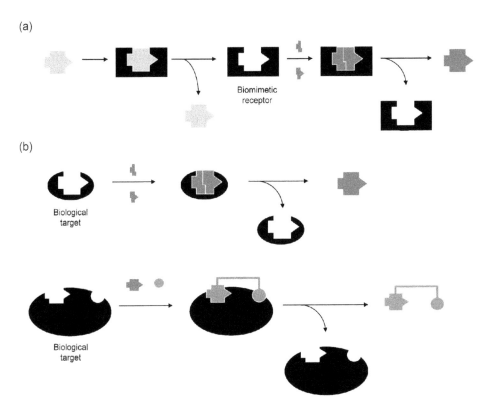

(b)

Figure 14   Schematic representation of the "anti-idiotypic" and "direct molding" methodologies. (a) The "anti-idiotypic imprinting" can lead to new compounds mimicking the original template [45]. (b) "Direct molding" assembles ligands by polymerization/condensation within or between sites of a biological target [47].

synthesize the original template, as well as that of new inhibitors using a small library of building blocks (Fig. 13). In addition to the original template, we have identified a new enzyme inhibitor using this approach. In principle, our concept of using MIPs to screen building blocks instead of product libraries should greatly simplify the drug discovery process by reducing the number of synthetic operations, since only the hit reactions need to be scaled up for further investigation.

The above approach can also be more schematically depicted as shown in Fig. 14. Alternatively, a biographical target can be used as a template and in its cavity polymerization/condensation allowed to take place in a sort of "direct molding" process leading to synthetic molecules such as an inhibitor affecting biological recognition processes. We expect these approaches (for relevant patent literature, see Ref. 47) to turn out to be useful new tools especially in drug discovery and for the formation of specific affinity materials.

ACKNOWLEDGMENTS

Ecevit Yilmaz was financed by the Swedish Center for BioSeparation. Ron Schmidt acknowledges financial support from the Wenner–Gren foundation.

Klaus Mosbach and his co-workers have been supported generously over the years by the Swedish Research Council and it deserves mentioning in this context that almost half of the total number of 27 chapters in this book were written by scientists that have been associated with our Lund group at one time or another, either as PhD students or postdoctoral fellows (R. Ansell (25), O. Brüggemann (21), K. Haupt (26), A. Mayes (2), I.A. Nicholls et al. (14), R. Schmidt (3), B. Sellergren (8), L. Ye (17, 23), E. Yilmaz (3,17) and the co-editor O. Ramström (1,7).

REFERENCES

1. Mosbach, K. Molecular imprinting. Trends Biochem. Sci. **1994**, *19* (1), 9–14.
2. Vlatakis, G.; Andersson, L.I.; Müller, R.; Mosbach, K. Drug assay using antibody mimics made by molecular imprinting. Nature **1993**, *361*, 645–647.
3. Haupt, K.; Mosbach, K. Plastic antibodies: developments and applications. Trends Biotechnol. **1998**, *16*, 468–478.
4. Haupt, K.; Mosbach, K. Molecularly imprinted polymers in chemical and biological sensing. Biochem. Soc. Trans. **1999**, *27*, 344–350.
5. Haupt, K.; Mosbach, K. Molecularly imprinted polymers and their use in biomimetic sensors. Chem. Rev. **2000**, *100* (7), 2495–2504.
6. Ramström, O.; Mosbach, K. Synthesis and catalysis by molecularly imprinted materials. Curr. Opin. Chem. Biol. **1999**, *3*, 759–764.
7. Wulff, G. Enzyme-like catalysis by molecularly imprinted polymers. Chem. Rev. **2002**, *102* (1), 1–27.
8. Pauling, L. A theory of the structure and process of formation of anibodies. J. Am. Chem. Soc. **1940**, *62*, 2643–2657.
9. Mathews, C.K.; Holde, K.Ev. In *Biochemistry*, 2nd Ed.; The Benjamin/Cummings Publishing Company, Inc., 1996.
10. Dickey, F.H. The preparation of specific adsorbents. Proc. Natl. Acad. Sci. USA **1949**, *35* (5), 227–229.
11. Mosbach, K.; Mosbach, R. Entrapment of enzymes and microorganisms in synthetic cross-linked polymers and their application in column techniques. Acta. Chem. Scand. **1966**, *20*, 2807–2810.
12. Mosbach, K. In *Immobilized Enzymes. Methods in Enzymology*; Mosbach, K., Ed.; Academic Press, Inc: New York, 1976; Vol. XLIV.
13. Mosbach, K. In *Immobilized Enzymes and Cells. Methods in Enzymology*; Mosbach, K., Ed.; Academic Press: Orlando, FL, 1987; Vol. 135–137.
14. Guilford, H.; Larsson, P-O.; Mosbach, K. On adenine nucleotides for affinity chromatography. Chem. Scripta **1972**, *2*, 165–170.
15. Mosbach, K.; Guilford, H.; Ohlsson, R.; Scott, M. General ligands in affinity chromatography. Biochem. J. **1972**, *127*, 625–631.
16. Arshady, R.; Mosbach, K. Synthesis of substrate-selective polymers by host–guest polymerization. Makromol. Chem., Rapid Commun. **1981**, *182*, 687–692.
17. Mosbach, K. Keynote lecture: towards the next generation of molecular imprinting with emphasis on the formation, by direct molding, of compounds with biological activity (biomimetics). Anal. Chim. Acta **2001**, *435*, 3–8.
18. Bayer, E.A.; Wilchek, M. Applications of avidin-biotin technology to affinity-based separations. J. Chromatogr. **1990**, *510*, 3–11.
19. Israelachvili, J.N. In *Intermolecular and Surface Forces*; Academic Press Limited: San Diego, 1992.
20. van Holde, K.E.; Johnson, W.C.; Ho, P.S. In *Principles of Physical Biochemistry*; Prentice-Hall, Inc: Upper Saddle River, NJ, 1998.

21. Solomons, T.W.G. In *Organic Chemistry*; John Wiley & Sons, Inc: New York, 1992.
22. Liao, J.L.; Wang, Y.; Hjertén, S. A novel support with artificially created recognition for the selective removal of proteins and for affinity chromatography. Chromatographia **1996**, *42* (5/6), 259–262.
23. Yu, C.; Mosbach, K. Molecular imprinting utilizing an amide functional group for hydrogen bonding leading to highly efficient polymers. J. Org. Chem. **1997**, *62*, 4057–4064.
24. Kempe, M. Antibody-mimicking polymers as chiral stationary phases in hplc. Anal. Chem. **1996**, *68*, 1948–1953.
25. Yilmaz, E.; Haupt, K.; Mosbach, K. Influence of functional and cross-linking monomers and the amount of template on the performance of molecularly imprinted polymers in binding assays. Anal. Comm. **1999**, *36*, 167–170.
26. Kempe, M.; Glad, M.; Mosbach, K. An approach towards surface imprinting using the enzyme ribonuclease a. J. Mol. Recognit. **1995**, *8*, 35–39.
27. Sellergren, B. Imprinted dispersion polymers: a new class of easily accessible affinity stationary phases. J. Chromatogr. A **1994**, *673*, 133–141.
28. Sellergren, B.; Shea, K.J. Influence of polymer morphology on the ability of imprinted network polymers to resolve enantiomers. J. Chromatogr. **1993**, *635*, 31–49.
29. Cowie, J.M.G. In *Polymers: Chemistry & Physics of Modern Materials*, 2nd Ed.; Stanley Thornes Ltd: Cheltenham, UK, 1998.
30. Bandrup, J.; Immergut, E.H. In *Polymer Handbook*, 3rd Ed.; Wiley: New York, 1989.
31. Andersson, L.I. Application of molecular imprinting to the development of aqueous buffer and organic solvent based radioligand binding assays for *s*-propranolol. Anal. Chem. **1996**, *68*, 111–117.
32. Haupt, K.; Dzgoev, A.; Mosbach, K. Assay system for the herbicide 2,4-dichlorophenoxyacetic acid using a molecularly imprinted polymer as an artificial recognition element. Anal. Chem. **1998**, *70*, 628–631.
33. Nicholls, I.A. Towards the rational design of molecularly imprinted polymer. J. Mol. Recognit. **1998**, *11*, 79–82.
34. Sellergren, B.; Shea, K.J. Origin of peak asymmetry and the effect of temperature on solute retention in enantiomer separations on imprinted chiral stationary phases. J. Chromatogr. **1995**, *690*, 29–39.
35. O'Shannessy, D.J.; Ekberg, B.; Mosbach, K. Molecular imprinting of amino acid derivatives at low temperature (0°C) using photolytic homolysis of azobisnitriles. Anal. Biochem. **1989**, *177*, 144–149.
36. Chen, Y.; Kele, M.; Sajonz, P.; Sellergren, B.; Guiochon, G. Influence of thermal annealing on the thermodynamic and mass-transfer kinetic properties of D- and L-phenylalanine anilide on imprinted polymeric stationary phases. Anal. Chem. **1999**, *71*, 928–938.
37. Mayes, A.G.; Lowe, C.R. In *Optimization of Molecularly Imprinted Polymers for Radioligand Binding Assays. Drug Development Assay Approaches, Including Molecular Imprinting and Biomarkers*; Reid, E., Hill, H.M., Wilson, D., Eds.; Guilford Academic Associates: Guilford, UK, 1998; Vol. 25.
38. Takeuchi, T.; Fukuma, D.; Matsui, J. Combinatorial molecular imprinting: An approach to synthetic polymer receptors. Anal. Chem. **1999**, *71*, 285–290.
39. Lanza, F.; Sellergren, B. Method for synthesis and screening of large groups of molecularly imprinted polymers. Anal. Chem. **1999**, *71*, 2092–2096.
40. Piletsky, S.A.; Karim, K.; Piletska, E.V.; Day, D.J.; Freebairn, K.W.; Legge, C.; Turner, A.P.F. Recognition of ephedrine enantiomers by molecularly imprinted polymers designed using a computational approach. Analyst **2001**, *126*, 1826–1830.
41. Sellergren, B.; Ekberg, B.; Mosbach, K. Molecular imprinting of amino acid derivatives in macroporous polymers. J. Chromatogr. **1985**, *347*, 1–10.

42. Yilmaz, E.; Haupt, K.; Mosbach, K. The use of immobilized templates – a new approach in molecular imprinting. Angew Chem. Int. Ed. Engl. **2000**, *12* (39), 2115–2118.
43. Titirici, M.M.; Hall, A.J.; Sellergren, B. Hierarchically imprinted stationary phases: mesoporous polymer beads containing surface-confined binding sites for adenine. Chem. Mater. **2002**, *14* (1), 21–23.
44. Alexander, C.; Vulfson, E.N. Spatially functionalized polymer surfaces produced via cell-mediated lithography. Adv. Mat. **1997**, *9* (9), 751–755.
45. Mosbach, K.; Yu, Y.; Andersch, J.; Ye, L. Generation of new enzyme inhibitors using imprinted binding sites: the anti-idiotypic approach, a step toward the next generation of molecular imprinting. J. Am. Chem. Soc. **2002**, *123* (49), 12420–12421.
46. Ye, L.; Mosbach, K. Molecular imprinted materials: towards the next generation. Mater. Res. Soc. Symp. Proc. **2002**, *723*, 51–59.
47. Yu, Y.; Ye, L.; Haupt, K.; Mosbach, K. Formation of a class of enzyme inhibitors (drugs), including a chiral compound, by using imprinted polymers or biomolecules as molecular-scale reaction vessels. Angew Chem. Int. Ed. Engl. **2002**, *41*, 4460–4463.
48. Ye, L.; Mosbach, K. The technique of molecular imprinting—principle, state of the art, and future aspects. J. Incl. Phenom. Macro. Chem. **2001**, *41*, 107–113.
49. Brüggemann, O.; Haupt, K.; Ye, L.; Yilmaz, E.; Mosbach, K. New configurations and applications of molecularly imprinted polymers. J. Chromatogr. A **2000**, *889*, 15–24.
50. Mosbach, K.; Ramström, O. The emerging technique of molecular imprinting and its future impact on biotechnology. *Bio/Technology* **1996**, *14*, 163–170.
51. Mayes, A.G.; Mosbach, K. Molecularly imprinted polymer beads: suspension polymerization using a liquid perfluorocarbon as the dispersing phase. Anal Chem **1996**, *68* (21), 3769–3774.
52. Mayes, A.G. Polymerisation techniques for the formation of imprinted beads. Tech Instrum. Anal. Chem. **2001**, *23*, 305–324.
53. Ye, L.; Cormack, P.A.G.; Mosbach, K. Molecularly imprinted monodisperse microspheres for competitive assay. Anal. Comm. **1999**, *36*, 35–38.
54. Glad, M.; Reinholdsson, P.; Mosbach, K. Molecularly imprinted composite polymers based on trimethylolpropane trimethacrylate (trim) particles for efficient enantiomeric separations. React. Polym. **1995**, *25*, 47–54.
55. Yilmaz, E.; Ramström, O.; Möller, P.; Sanchez, D.; Mosbach, K. A facile method for preparing spherical imprinted polymer spheres using spherical silica templates. J. Mater. Chem. **2002**, *12*, 1577–1581.
56. Takeuchi, T.; Matsui, J. Miniaturized molecularly imprinted continuous polymer rods. J High Resolut. Chromatogr. **2000**, *23* (1), 44–46.
57. Kriz, D.; Kriz, C.B.; Andersson, L.I.; Mosbach, K. Thin-layer chromatography based on the molecular imprinting technique. Anal. Chem. **1994**, *66*, 2636–2639.
58. Jakusch, M.; Janotta, M.; Mizaikoff, B.; Mosbach, K.; Haupt, K. Molecularly imprinted polymers and infrared evanescent wave spectroscopy. A chemical sensors approach. Anal. Chem. **1999**, *71* (20), 4786–4791.
59. Dzgoev, A.; Haupt, k. Enantioselective molecularly imprinted polymer membranes. Chirality **1999**, *11*, 465–469.
60. Norrlöw, O.; Månsson, M.-O.; Mosbach, K. Improved chromatography: prearranged distance between boronate groups by the molecular imprinting approach. J. Chromatogr. **1987**, *396*, 374–377.
61. Hwang, K.-O.; Sasaki, T. Imprinting for the assembly of artificial receptors on a silica surface. J. Mater. Chem. **1998**, *8* (9), 2153–2156.
62. Sellergren, B. Imprinted polymers with memory for small molecules, proteins, or crystals. Angew Chem. Int. Ed. Engl. **2000**, *39* (6), 1031–1037.
63. Kempe, M.; Mosbach, K. Molecular imprinting used for chiral separations. J Chromatogr. A **1995**, *694*, 3–13.

64. Andersson, L.I.; Mueller, R.; Mosbach, K. Molecular imprinting of the endogenous neuropeptide leu5-enkephalin and some derivatives thereof. Macromol. Rapid Comm. **1996**, *17* (1), 65–71.

65. Rachkov, A.; Minoura, N. Recognition of oxytocin and oxytocin-related peptides in aqueous media using a molecularly imprinted polymer synthesized by the epitope approach. J Chromatogr A **2000**, *889* (1 + 2), 111–118.

66. Sreenivasan, K. Molecularly imprinted polyacrylic acid containing multiple recognition sites for steroids. J. Appl. Polym. Sci. **2001**, *82* (4), 889–893.

67. Ye, L.; Yu, Y.; Mosbach, K. Towards the development of molecularly imprinted artificial receptors for the screening of estrogenic chemicals. Analyst **2001**, *126* (6), 760–765.

68. Striegler, S.; Tewes, E. Investigation of sugar-binding sites in ternary ligand-copper(ii)-carbohydrate complexes. Eur. J. Inorg. Chem. **2002**, *2*, 487–495.

69. Mayes, A.G.; Andersson, L.I.; Mosbach, K. Sugar binding polymers showing high anomeric and epimeric discrimination obtained by noncovalent molecular imprinting. Anal. Biochem. **1994**, *222* (2), 483–488.

70. Sallacan, N.; Zayats, M.; Bourenko, T.; Kharitonov, A.B.; Willner, I. Imprinting of nucleotide and monosaccharide recognition sites in acrylamidephenylboronic acid-acrylamide copolymer membranes associated with electronic transducers. Anal. Chem. **2002**, *74* (3), 702–712.

71. Tsunemori, H.; Araki, K.; Uezu, K.; Goto, M.; Furusaki, S. Surface imprinting polymers for the recognition of nucleotides. Bioseparation **2002**, *10* (6), 315–321.

72. Glad, M.; Norrloew, O.; Sellergren, B.; Siegbahn, N.; Mosbach, K. Use of silane mono-mers for molecular imprinting and enzyme entrapment in polysiloxane-coated porous silica. J. Chromatogr. **1985**, *347* (1), 11–23.

73. Matsui, J.; Miyoshi, Y.; Doblhoff-Dier, O.; Takeuchi, T. A molecularly imprinted synthetic polymer receptor selective for atrazine. Anal. Chem. **1995**, *67* (23), 4404–4408.

74. Dai, S.; Burleigh, M.C.; Shin, Y.; Morrow, C.C.; Barnes, C.E.; Zue, Z. Imprint coating: a novel synthesis of selective functionalized ordered mesoporous sorbents. Angew Chem. Int. Ed. Engl. **1999**, *38* (9), 1235–1239.

75. Rosatzin, T.; Andersson, L.I.; Simon, W.; Mosbach, K. Preparation of ca2+ selective sorbents by molecular imprinting using polymerizable ionophores. J. Chem. Soc. Perkin Trans. **1991**, *28*, 1261–1265.

76. Kriz, D.; Mosbach, K. Competitive amperometric morphine sensor based on an agarose-immobilized molecularly imprinted polymer. Anal. Chim. Acta. **1995**, *300* (1–3), 71–75.

77. Matsui, J.; Takeuchi, T. A molecularly imprinted polymer rod as nicotine-selective affinity media prepared with 2-(trifluoromethyl)acrylic acid. Anal. Comm. **1997**, *34* (7), 199–200.

78. Skudar, K.; Brueggemann, O.; Wittelsberger, A.; Ramström, O. Selective recognition and separation of -lactam antibiotics using molecularly imprinted polymers. Anal. Comm. **1999**, *36* (9/10), 327–331.

79. Shi, H.; Tsai, W.-B.; Garrison, M.D.; Ferrari, S.; Ratner, B.D. Template-imprinted nanostructured surfaces for protein recognition. Nature **1999**, *398*, 593–597.

80. Zhou, M.; Chen, J.-N.; Chen, D.-C.; Hua, Z.-C. Synthesis of human urokinase imprinted polysiloxane polymer. Int. J. Bio-Chromatogr. **2001**, *6* (1), 1–5.

81. Hayden, O.; Dickert, F.L. Selective microorganism detection with cell surface imprinted polymers. Adv. Mat. **2001**, *13* (19), 1480–1483.

82. D'Souza, S.M.; Alexander, C.; Carr, S.W.; Waller, A.M.; Whitcombe, M.J.; Vulfson, E.N. Directed nucleation of calcite at a crystal-imprinted polymer surface. Nature **1999**, *398*, 312–398.

83. Matsui, J.; Doblhoff-Dier, O.; Takeuchi, T. 2-(trifluoromethyl)acrylic acid: a novel functional monomer in non-covalent molecular imprinting. Anal Chim. Acta **1997**, *343* (1–2), 1–4.

84. Robinson, D.K.; Mosbach, K. Molecular imprinting of a transition state analogue leads to a polymer exhibiting esterolytic activity. J. Chem. Soc. Chem. Commun. **1989**, *14*, 969–970.

85. Panasyuk-Delaney, T.; Mirsky, V.M.; Wolfbeis, O.S. Capacitive creatinine sensor based on a photografted molecularly imprinted polymer. Electroanalysis **2002**, *14* (3), 221–224.

86. Dhal, P.K.; Arnold, F.H. Template-mediated synthesis of metal-complexing polymers for molecular recognition. J. Am. Chem. Soc. **1991**, *113*, 7417–7418.

87. Ansell, R.J.; Ramström, O.; Mosbach, K. Towards artificial antibodies prepared by molecular imprinting. Clin. Chem. **1996**, *42* (9), 1506–1512.

88. Svenson, J.; Nicholls, I.A. On the thermal and chemical stability of molecularly imprinted polymers. Anal. Chim. Acta **2001**, *435* (1), 19–24.

89. Dhal, P.K.; Arnold, F.H. Metal-coordination interactions in the template-mediated synthesis of substrate-selective polymers: recognition of bis(imidazole) substrates by copper(ii) iminodiacetate containing polymers. Macromolecules **1992**, *25*, 7051–7059.

90. Liu, X.-C.; Dordick, J.S. Sugar acrylate-based polymers as chiral molecularly imprintable hydrogels. J. Polym. Sci. Polym. Chem. **1999**, *37*, 1665–1671.

91. Turkewitsch, P.; Wandelt, B.; Darling, G.D.; Powell, W.S. Fluorescent functional recognition sites through molecular imprinting. A polymer-based fluorescent chemosensor for aqueous camp. Anal. Chem. **1998**, *70*, 2025–2030.

92. Piletsky, S.A.; Andersson, H.S.; Nicholls, I.A. Combined hydrophobic and electrostatic interaction-based recognition in molecularly imprinted polymers. Macromolecules **1999**, *32*, 633–636.

4
The Covalent and Other Stoichiometric Approaches

Günter Wulff   Heinrich-Heine-University Düsseldorf, Düsseldorf, Germany

I. INTRODUCTION

The starting point of our research on molecular imprinting was the desire to prepare enzyme mimics that resemble the structure of natural enzymes as closely as possible (for recent review, see [1]). When we started our work in 1968, it was just the time when first detailed investigations on the structure of natural enzymes and their catalytic action became known (for a review see [2]). Especially, chymotrypsin was investigated in these early days. In this enzyme, as found for many other examples in later investigations, the shape of the active site is complementary to the chemical structure of the substrate. This was already postulated by Fischer [3] in his "Key and Lock" principle more than 100 years ago. Functional groups are arranged in this cavity acting as binding sites and as catalytic sites. The three-dimensional arrangement of these groups inside the active site is of special importance and has a strong influence on selectivity and catalytic activity of the enzyme.

In order to prepare models of enzymes, we have chosen synthetic polymers as scaffold for the active site. The use of polymeric substances instead of low molecular weight substances would have some advantages since enzymes are polymers as well. Many of their unique features are directly connected to their polymeric nature. This is especially true for the high cooperativity of the functional groups and the dynamic effects such as the *induced fit*, the *allosteric effect*, and the *steric strain* as shown by the enzymes.

In order to prepare synthetic polymers showing enzyme-like catalytic activity, a number of prerequisites have to be fulfilled. First, a cavity or a cleft has to be made with a defined shape corresponding to the shape of the substrate or, even better, to the shape of the transition state of the reaction. At the same time, functional groups have to be introduced that act as binding sites, coenzyme analogues, or catalytic sites within the cavity and are in a defined stereochemistry. Since binding and catalysis in enzymes are rather complex procedures, simplified structures have to be found that can be handled more easily.

To fulfill these prerequisites, we have introduced a novel strategy for obtaining "enzyme-analogue built polymers" [4,5]. For this, a cross-linked polymer is formed around a molecule that acts as a template. The monomer mixture contains functional monomers that can interact with the template through covalent or noncovalent interaction. After removal of the template, an imprint containing functional groups in a certain orientation remains in the highly cross-linked polymer. The shape of the imprint and the arrangement of the functional groups are complementary to the structure of the template (see Scheme 1 [6]).

This imprinting procedure was quite different from earlier imprinting attempts using silica as the scaffold (see, e.g. [7] and Chapter 11). In these experiments, silicic acid is precipitated in the presence of, e.g., dyes. After drying, the dyes were washed out and the silica gels obtained showed an increased affinity for the template dye. In this case, it is impossible to introduce certain functional groups into the cavity since no specific binding occurs during imprinting. Only silanol and siloxane groupings are present, which might show some molecular recognition but a predetermined catalytic arrangement cannot be obtained. Furthermore, the medium cannot be changed inside the cavities unlike in synthetic polymers in which by the proper choice of comonomers the polarity and the hydrophobicity can be controlled to an appreciable extend.

For the preparation of polymeric structures resembling the structure of natural receptors or enzymes, it is therefore preferable to use synthetic polymers. As will be discussed later in some detail, strong interaction between binding site monomers and template during imprinting will control the arrangement of the functional groups

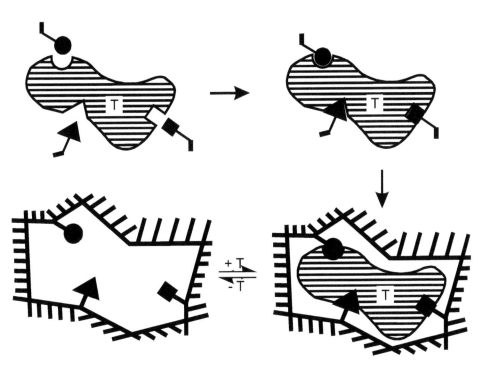

Scheme 1 Schematic representation of the imprinting of specific cavities in cross-linked polymers by a template (T) with three different binding groups (adapted from Ref. 6).

inside the cavity most perfectly. This is the reason why we started with covalent inter-actions during the imprinting and turned later onto stoichiometric noncovalent interaction. Usual noncovalent interactions (electrostatic interaction, hydrogen bonding, etc.) are suitable for molecular recognition, but causes problems in the construction of catalytically active enzyme models with defined arrangement of functional groups.

II. AN EXAMPLE FOR IMPRINTING WITH COVALENT INTERACTIONS

The polymerization of the template monomer **1** is an example of the imprinting method using covalent interaction. Extensive optimization work was carried out on the methodology with this system (see [6] and reviews [8,9]). The template is phenyl-α-D-mannopyranoside (**2**) to which two molecules of 4-vinylphenylboronic acid are bound by esterification with two OH groups of the sugar. Boronic acid was chosen as the binding group (binding site) as it undergoes a rapid and reversible reaction with diols.

Monomer **1** is subjected to radical copolymerization with large amounts of a cross-linking agent such as ethylene dimethacrylate in the presence of an inert sol-vent (which acts as a porogen to give a porous structure). This yields macroporous polymers with a large inner surface area and a permanent pore structure. Up to 95% of the templates can be split off again from polymers of this type by treatment with water or methanol (Scheme 2).

The accuracy of the steric arrangement of the binding sites and the proper shape in the resulting imprinted cavity can be tested by the ability of the polymer to resolve the racemate of the template, namely phenyl-α-D,L-mannopyranoside. The

(a)   (b)

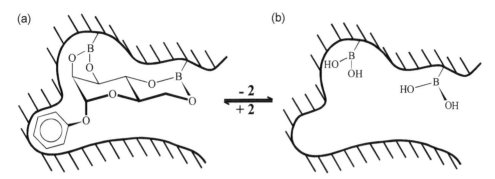

Scheme 2 Schematic representation of a cavity (a) obtained by polymerization of **1**. The template **2** can be removed with water or methanol to give (b). Addition of **2** causes the cavity to be reoccupied, giving (a) again (adapted from Refs. 8,9).

polymer was equilibrated in a batch procedure with a solution of the racemate under conditions that allowed a thermodynamically controlled partition of the enantiomers between polymer and solution. The enrichment of the antipodes in the polymer and in solution was determined and the separation factor α, i.e., the ratio of the distribution coefficients of the D and L enantiomer between polymer and solution was calculated. After extensive optimization of the procedure, α values between 3.5 and 6.0 were obtained [9]. This is an extremely high selectivity for racemic resolution, which cannot be reached by most other methods commonly used for the separation of racemates.

Polymers obtained by this procedure can be used for the chromatographic separation of the racemates of the corresponding template molecules [6,9]. The selectivity of the separation process is fairly high (separation factors up to $\alpha = 4.56$), and under particular conditions (high temperatures, gradient elution) resolution values of $R_s = 4.2$ with baseline separation have been obtained (see Fig. 1).

These sorbents can be prepared conveniently and possess excellent thermomechanical stability. Even when used at 80°C under high pressure for a long time, no leakage of the stationary phase or decrease of selectivity during chromatography was observed.

Meanwhile, a large number of different templates have been investigated by us and many other groups in the world (see books [11,12] and reviews [9,13,14]. An interesting extension of the concept of molecular imprinting was introduced by Mosbach and coworkers [15], who exclusively used noncovalent interactions; this aspect is dealt with in Chapter 3.

III. OPTIMIZATION OF THE POLYMER STRUCTURE

Covalent interactions using boronic ester linkages have been used for the optimization of the polymer structure in the imprinting procedure. This optimization turned out to be rather complicated.

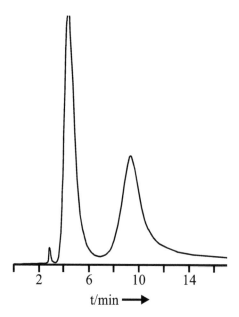

Figure 1 Chromatographic resolution of D,L-**2** on a polymer imprinted with **1** (elution with a solvent gradient at 90°C (adapted from Ref. 10).

On one hand, the polymers should be rather rigid to preserve the structure of the cavity after splitting off the template. On the other hand, a high flexibility of the polymers should be present to facilitate a fast equilibrium between release and re-uptake of the template in the cavity. These two properties are contradictory to each other, and a careful optimization became necessary. Furthermore, good accessibility of as many cavities as possible is required as well as high thermal and mechanical stability. Since the initial experiments on imprinting most examples until now are based on macroporous polymers with a high inner surface area [100–600 m^2/g] that show, after optimization, good accessibility as well as good thermal and mechanical stability.

The selectivity is mainly influenced by the kind and amount of cross-linking agent used in the synthesis of the imprinted polymer [16,17]. Figure 2 shows the selectivity dependence for racemic resolution of the racemate of **2** on the structure of polymers of the type described in Scheme 2.

Below a certain amount of cross-linking in the polymer (around 10%), no selectivity can be observed because the cavities are not sufficiently stabilized. Above 10% cross-linking, selectivity increases steadily. Between 50% and 66%, a surprisingly high increase in selectivity takes place, especially in the case of ethylene dimethacrylate as a cross-linker. This cross-linking agent is now preferred by most groups working in the field. Cross-linking with divinylbenzene (either the commercial mixture or pure regioisomers) results in reduced selectivity, yet this cross-linker has the advantage of higher chemical stability (bonds are not hydrolyzable) and less interaction with functional groups [17].

The optimization and analysis of the polymer structure will be discussed more in detail in Chapters 7 and 8.

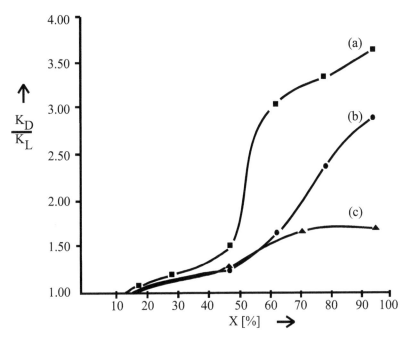

Figure 2 Selectivity of imprinted polymers (see Scheme 2) as a function of the type and amount (X) of the cross-linking agent [16]. Polymers are prepared from **1**. Selectivity for racemic resolution of the racemate of **2** is measured in the batch procedure. Cross-linking agents: (a) ethylene dimethacrylate, (b) tetramethylene dimethacrylate, (c) divinylbenzene (adapted from Ref. 16).

IV. GENERAL ASPECTS OF THE BINDING SITE INTERACTION

Besides the polymer structure, the type of the binding site interaction is the most important characteristic feature of the molecular imprinting procedure. This interaction should optimally show the following properties during the different steps of their use in the imprinting procedure:

1. During polymerization, the interaction between binding site and template needs to be stable. Template and binding site monomer should be present in a stoichiometric ratio. Therefore, stoichiometric binding with stable covalent bonds or with noncovalent interactions and high association constants is desirable. The binding interaction should possess a specific geometric directionality, as present in covalent bonds or hydrogen bonds.
2. The template should be able to split off under mild conditions and as completely as possible.
3. Equilibration with substrates has to be rapid.
4. There must be a favorable binding equilibrium.

The first point implies the need for thermodynamically and kinetically stable interactions, whereas readily reversible interactions are demanded to satisfy the other three points. Mostly covalent (a), noncovalent (b), stoichiometric noncovalent (c), coordinative bonding (d), and semi-covalent bonding (e) have been used during the imprinting procedure.

(a) Covalent interaction during imprinting fulfills the first point of the above-mentioned requirements most perfectly. Unfortunately, there is only a limited number of covalent interactions that can easily be split afterwards to a high percentage, such as boronic diesters or Schiff bases (see, e.g., Ref. 18). Boronic acids, diols, as well as aldehydes (and acids after oxidation) and amines can thus be introduced in the cavity for use in binding or catalysis. With boronic acids, it was evident from chromatographic experiments that a very quick mass transfer with diols is possible at higher temperature in presence of certain bases [18,19]. This type of interaction will be discussed in Section V of this chapter.

(b) The use of noncovalent interactions during imprinting is the easiest way to introduce functional groups acting as binding and/or catalytic sites into a cavity (for a thorough discussion, see Chapter 3). Mostly electrostatic interactions and hydrogen bonding have been employed in this respect. Hydrogen bonding interactions are particularly important in noncovalent polymer imprinting due to their specific geometric directionality.

This type of interaction is relatively weak. Hydrogen bonds show low association constants of, e.g., $K_{ass} = 1.7\ M^{-1}$ even in completely nonpolar solvents like CCl_4 (which cannot be used in molecular imprinting due to its chain-transfer reaction during radical polymerization and its poor solubilizing properties) (see Table 1).

Electrostatic interactions between carboxylic acids and basic nitrogen with additional hydrogen bonds are stronger with association constants of, e.g., $K_{ass} = 3.3\ M^{-1}$ or $K_{ass} = 30\ M^{-1}$, respectively (in acetonitrile), depending on the structure of the compounds. But even this results in only a low degree of association using the partners in stoichiometric ratio. At concentrations of $0.1\ mol\ L^{-1}$ of both partners (typical concentration during imprinting) only about 13% (with hydrogen bonds), 21%, and 57% (with electrostatic interaction) of the templates are attached to the binding site monomer (Table 1). Therefore, an excess of binding site monomers (usually fourfold) is necessary to saturate the template. The situation becomes more problematic since these values decrease considerably at higher temperatures (60°C) and hydrogen bonding of the binding site monomers is also possible to other constituents of the polymerization mixture (like ethylene dimethacrylate).

Due to the strong excess of binding site functional monomers during the imprinting procedure, the binding site functional groups are distributed all over the polymer and not only inside the cavity.

(c) Of special interest are stoichiometric noncovalent interactions [18,26–28]. Multiple hydrogen bonds can have very high binding constants ($K = 10^3–10^7\ M^{-1}$) (see Table 1), so that they behave during the imprinting like a covalent bond. We have used, for example, the amidine group to bind carboxyl groups, phosphonic monoesters and phosphates. The N,N'-diethyl-4-vinylbenzamidine 3 is especially suitable [29,30]. Association constants with carboxylic acids (see 4) are typically well above $10^6\ M^{-1}$ in $CHCl_3$ [18,28]. Templates are easily split off from an imprinted polymer, and equilibration is rapid. Sufficient binding occurs even in aqueous solutions. In contrast to usual noncovalent interactions, a re-uptake of nearly 100% is easily possible [28]. Guanidinium binding sites have also been used to prepare imprints of phosphates and phosphonates in SiO_2 xerogels [31].

Table 1 Association Constants for Noncovalent Interactions (Model Systems) that can be Used in Molecular Imprinting Determined by ^1H NMR if not Indicated Otherwise

Binding site	Template	K_{ass} (M^{-1})	Solvent[a]	T (°C)	Complexation (%)[b]	References
		1.7^c	CCl$_4$	25	12.9	[20]
		3.3^c	ACN	25	20.7	[21]
		3.0×10	ACN	23	56.6	[15]
		5.5	ACN	60	28.3	[15]
		2.0×10^2	CDCl$_3$	22	80.0	[22]

Template	Monomer	K	Solvent[a]	T (°C)	%[b]	Ref.
(5-methylpyrazole trifluoroacetamide)	(N-acetyl-valyl-valine methyl ester)	8.9×10^2	$CDCl_3$	25	90.0	[23,26]
(guanidine, H_2N–C(=NH)–NH_2)	acetic acid (HO–CO–CH_3)	$7.9 \times 10^{3\,d}$	DMSO	25	96.5	[24]
(tetrachloro-OSty quinone)	Ampicillin	$> 3.0 \times 10^4$	DMSO	25	> 98.2	[25]

[a] Deuterated solvents have been used for NMR experiments.
[b] Calculated percentage of formed complex of template and monomer at concentrations of each 0.1 mol L^{-1}.
[c] Determined by IR spectroscopy.
[d] Calorimetric determination.

(d) Coordinative binding is an interesting type of interaction for molecular imprinting (for a review, see Chapter 6), and closely resembles that in ligand exchange chromatography. The advantage of this kind of bond is that its strength can be controlled by experimental conditions. Definite interactions can occur during polymerization under proper conditions, and an excess of binding groups is unnecessary. Especially, Arnold and coworkers [32] have developed this method by using mostly copper complexes of polymerizable iminodiacetic acid. Another promising approach uses triazacyclononane$-$Cu^{2+} chelates for binding [33].

(e) To overcome the problems with noncovalent interactions, a covalent bond is used during the imprinting procedure, whereas after splitting off the template, the newly generated group remains accessible for noncovalent interaction [4] (for reviews see [18,34] and Chapter 5). In this respect, amides have been used to introduce an electrostatic interaction between carboxylic acids and amines [4] or carbonate esters to introduce the possibility of hydrogen bonding (see Chapter 5). When Schiff bases are used during imprinting, removal of the aldehyde template leaves a polymer-bound amine behind that can interact electrostatically with carboxylic acids [18].

V. DIFFERENT POSSIBILITIES OF COVALENT BINDING

A. Boronic Acid-Containing Binding Site Functional Monomers

In the example given in Scheme 2, boronic acids are used as binding sites. This group is very suitable for covalent binding. With diol-containing templates, relatively stable trigonal boronic esters are formed (see Eq. (a)).

Ester formation occurs with complete stoichiometric conversion at relatively low reaction rates ($t_{\frac{1}{2}} \approx 100$–600 s). In the absence of water, no back reaction takes place. There is a very slow ester$-$ester exchange reaction ($t_{\frac{1}{2}} \approx 5000$ s). Templates can easily be released from boronic esters in water/alcohol; this reaction is, in principle, fast and complete ($t_{\frac{1}{2}} \approx 100$ s). The template can be released from the imprinted macroporous polymer in 85–95%.

The uptake of template afterwards is relatively slow, but more than 90% of the free cavities can be reoccupied. Fortunately, in aqueous alkaline solution or in the presence of certain nitrogen bases (for example, NH_3 or piperidine) tetragonal boronic esters are formed (Eq. (b)), which equilibrate extremely rapidly with tetragonal boronic acid and diol [18,19]. In these cases, the rate of equilibration ($t_{\frac{1}{2}} \approx 10^{-4}$ s) is comparable to that of noncovalent interactions. At molar ratios in Eq. (b), conversion amounts to 97%. The extent of binding can be controlled by the addition of water or methanol.

These exchange equilibria can be even more rapid if the interaction with nitrogen bases occurs intramolecularly (see Table 2 entry e) [18,19,35]. In those cases in which a basic nitrogen atom is located at a favorable distance from the boronate ester, this leads to an acceleration of 10^8–10^9 fold compared to the unsubstituted phenylboronate [35].

Mostly templates with diol groups have been used. Templates can be bound by one boronic acid via a boronic ester bond. Imprinted polymers with higher selectivity are obtained if two boronic acid binding sites are attached to the templates as in most saccharide derivatives (see, e.g., Scheme 2).

An important question in molecular imprinting has been addressed using covalent binding by two boronic acids: to what extent can imprinted polymers also bind substances other than the template. For example, are racemates of other substances resolvable? In the first experiments on glyceric acid esters 5 with a certain ester as template, imprinted polymers were shown to resolve a whole series of racemates even when the alcohol group in the racemate is varied (methyl, ethyl, benzyl, or 4-nitrophenyl) [39]. Aromatic amino acids were shown to behave similarly. Here, the aromatic group in the racemates can vary. A racemate resolution is possible provided that the rest of the structure remains the same [40].

5

Table 2 Examples of Molecular Imprinting Using Boronic Acids as Binding Sites

Entry	Template	Binding site	Binding during equilibration	References
a	Diol containing templates (like diols, saccharides, L-DOPA)			[18]
b	Saccharides			[18]
c	Glycoproteins	R = OH −CH₂−CH−CH₂−O−(CH₂)₃−Si(OH)₃	at SiO₂ / at SiO₂	[36]
d	Mandelic acid			[37]
e	Diols, saccharides			[19]
f	Monoalcohols			[19,38]

If the arrangement of functional groups is the same, the size and shape of the rest of the molecule are important factors for the separability, as shown, for example, in various bisketals [41].

To determine to what extent the arrangement of the binding groups in the cavity, rather than its shape, affects the resolving power, we polymerized a series of similar monomers with various arrangements of the functional groups or with different spatial properties [9,42]. For example, imprinted polymers were prepared in the usual way with template monomers 1, 6a–c, and 7. After removal of the template, the resolving power was determined for various sugar racemates (Table 3).

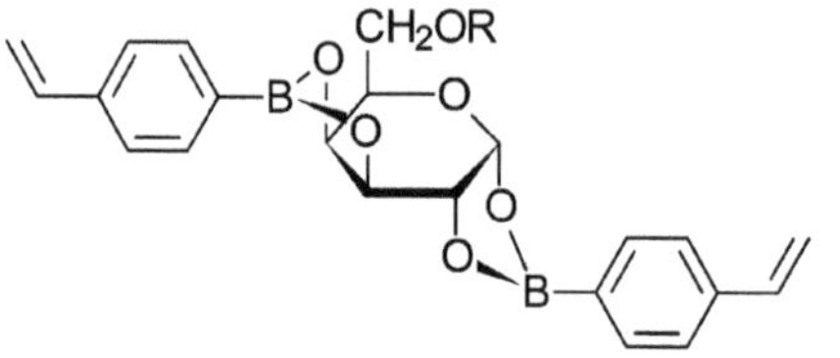

6a   R = H

6b   R = methyl

6c   R = benzyl

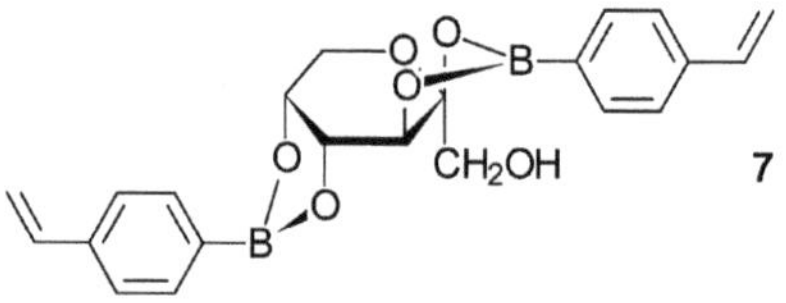

7

All polymers were able, as expected, to resolve racemates of their own template. D,L-mannose can be resolved on a polymer that has been imprinted with phenyl-α-D-mannopyranoside (entry 2, Table 3). This shows that even racemates with significantly smaller spatial requirements than the template can be resolved. The same is true for the resolution of D,L-galactose on a polymer imprinted with 6-o-benzyl-D-galactose (entry 5, Table 3). In these cases, the template and racemate have the same arrangement of functional groups. However, with the exception of D,L-fructose, it was not possible to resolve other sugar racemates on a polymer imprinted with 1. The cross-equilibration of polymers that had been imprinted with D-galactose (P-6a) or D-fructose (P-7) is interesting. Whereas on equilibration with its own

Table 3 Selectivity of Various Polymers for Racemate Resolution [9,42]

Entry	Template monomer	Racemate	α
1	1	Phenyl-α-D,L-mannoside	5.0
2	1	D,L-mannose	1.60
3	1	D,L-fructose	1.34
4	6a	D,L-galactose	1.38
5	6c	D,L-galactose	1.58
6	7	D,L-fructose	1.63
7	7	D,L-galactose	0.85 (1.17)[a]
8	6a	D,L-fructose	0.80 (1.25)[a]
9	6b	D,L-fructose	0.71 (1.41)[a]

[a]To give a better comparison of the selectivities, reciprocal values are given in parantheses (K_L/K_D).

racemate, the template is bound preferentially, the polymer imprinted with D-fructose preferentially binds L-galactose (structure **7**), and the polymer imprinted with D-galactose preferentially binds L-fructose (structures **8,9**) out of the racemic mixture. This inverse selectivity on cross-equilibration can be explained if one considers models of the preferred conformations of the molecules [42]. As an example, Fig. 3 shows the structures of β-L-fructopyranose and α-D-galactopyranose, which are both preferentially bound by polymer **P-6a**.

Both compounds have the same arrangement of hydroxyl groups that react with the boronic acids (in black). Only the hydroxymethyl groups (crosshatched) are at different positions in the molecule, which causes significantly different shapes for the molecules. Similar considerations apply to β-L-galactopyranose and α-D-fructopyranose and are also valid for phenyl- α-D-mannopyranoside and β-D-fructopyranose [42]. From these results, it follows that the arrangement of the functional groups in the cavity is the decisive factor for the selectivity, while the shape of the cavity is somewhat less important. Classes of substances with the same arrangement of interacting groups therefore have a group selectivity.

One might expect that three covalent interactions should be especially favorable for imprinting and for the subsequent molecular recognition. However, it has been shown that three boronic acid interactions [4,43], three different covalent interactions [44], or three or more noncovalent interactions [45,46] all lead to poorer resolving power (that is, a smaller number of theoretical plates N_{th}) for the polymers. Here, the main factor must be the low rate of formation of polymer–substrate complexes.

The effect of the flexibility of the binding group on the resolving power of the polymers was also investigated exemplarily for boronic acids [10,47]. For this, nine different polymerizable boronic acids with different degrees of mobility between the boron and the polymerizable double bond were used, with phenyl-α-D-mannopyranoside as the template. As the mobility of the groups in the cavity increased, the selectivity in racemate resolution decreased.

α-Hydroxy carboxylic acids, such as mandelic acid (Table 2, entry d), also bind to boronic acids, and imprinted polymers of this type possess remarkably good resolving power for the racemate of mandelic acid [37].

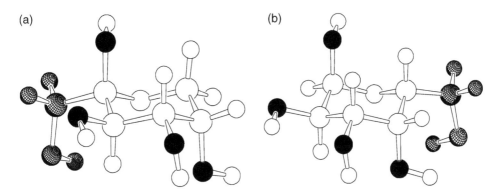

Figure 3   Ball-and-stick model of the preferred conformation of β-L-fructopyranose (a) and α-D-galactopyranose (b). The OH groups that react with the boronic acid are represented in black. The hydroxy methylene groups which influence the shape are hatched (adapted from Ref. 42).

Boronic acid binding sites can also be used in combination with other covalent or noncovalent interactions. Glyceric acid and derivatives were thus bound as boronic esters aided by electrostatic, hydrophobic, charge-transfer interactions, or hydrogen bonding [39]. A boronic ester bond and a Schiff base have been used for the imprinting of D-glyceraldehyde [4] and of L-DOPA in 8 [48].

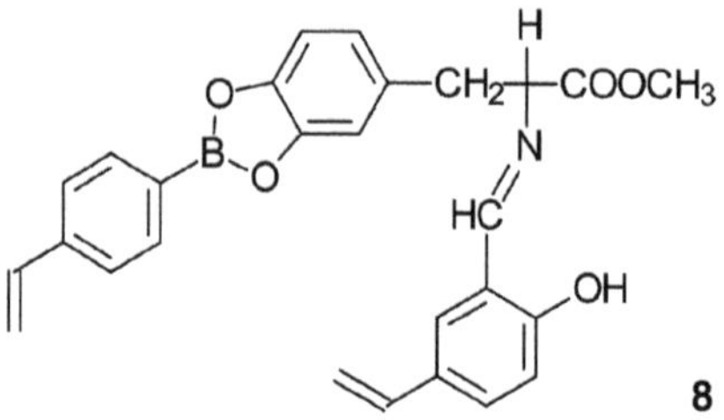

8

As shown in entries b and c in Table 2, it is also possible to couple silane-containing boronic acids to the surface of silica.

The boronic acid binding site can also be used for binding monoalcohols. In the case of an ortho hydroxymethylene group (see entry f), an intramolecular cyclic monoester is formed (boronophthalide). This compound has still one hydroxyl group left for the esterification of monoalcohols, a reaction that can be used in the imprinting procedure [19]. This group has recently been used by Whitcombe and coworkers [38] for binding steroid alcohols.

B. Using Azomethines, Ketals, and Sulfides for Covalent Binding

The suitability of Schiff bases for covalent binding was also extensively investigated (see Table 4, entries a–e). Schiff bases are, in principle, suitable for imprinting and subsequent binding, since the position of the equilibrium is favorable. The rate of equilibration is, however, too low for rapid chromatography. By using catalysts or suitable intramolecular neighboring groups, the rate of equilibration can be considerably increased [49] (see entry d). Schiff base binding allows the binding site to be either an amine (entries a, b) or an aldehyde (entries c–e). Similarly, aldehyde-containing binding sites have been successfully applied for the imprinting with amino acid derivatives (see 8 and 9) [48,49]. In this case, the application of amino acid templates in the form of anilides was especially suitable as shown for phenyl alanine anilide 9 derivative.

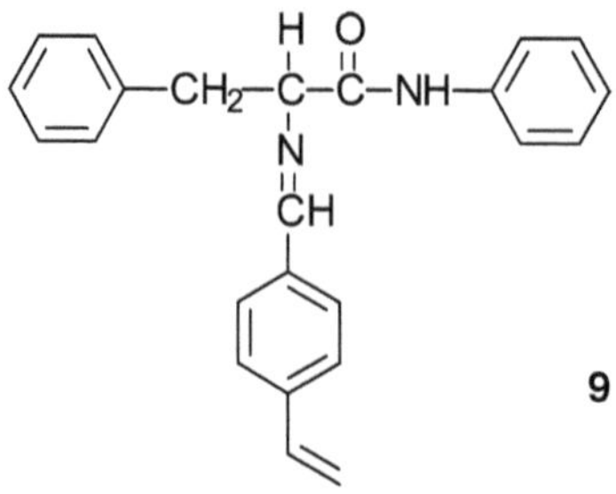

9

Table 4 Examples of Covalent Molecular Imprinting Using Azomethines, Ketals, and Sulfides

Entry	Template	Binding site	Binding during equilibation	References
a	Aldehydes	(vinyl-phenyl)—CH$_2$NH$_2$	—Ar—CH$_2$—N=CHR	[50]
b	Aldehydes	CH$_3$O—Si(CH$_3$)(CH$_3$)—(phenyl)—NH$_2$	—O—Si(CH$_3$)(CH$_3$)—Ar—N=CHR at SiO$_2$	[50]
c	Amines	(vinyl-phenyl)—CHO	(phenyl)—CH=NR	[48,49]
d	Amines	(vinyl-phenyl, OH)—CHO	(phenyl, OH)—CH=NR	[48,49]
e	Amines	CH$_3$O—Si(CH$_3$)(CH$_3$)—(phenyl)—CHO	—O—Si(CH$_3$)(CH$_3$)—Ar—CH=NR at SiO$_2$	[44]
f	Ketones	(vinyl-phenyl)—CH(CH$_2$OH)(CH$_2$OH)	—Ar—CH(CH$_2$O—C(R^1)(R^2)—OCH$_2$)	[13,41,51]
g	Alcohols	R''—NH, R', OH, CH$_2$—O—CH (fused ring); R' = H, OCH$_3$; R'' = —CO—C(CH$_3$)=CH$_2$	—NH, R', OR, CH$_2$—O—CH (fused ring)	[52]
h	Disulfide	(vinyl-phenyl)—CH$_2$SH	(phenyl)—CH$_2$—S—S—R	[53]

Binding as Schiff bases often involved amine-containing binding sites and double fixation as Schiff bases (entries a, b). Thus, with the aid of bisaldehydes of differing structures, two amino groups could be introduced in microcavities of different shapes. Remarkable selectivity for rebinding the template was obtained (see entry a) [50]. In connection with these experiments, the question arose whether or not the distance between two binding groups on a surface could also lead to substrate selectivity [50]. In the typical imprinting of polymers or silica gels, the selectivity of the products so obtained depends both on the shape of the cavity and the arrangement of the functional groups within it.

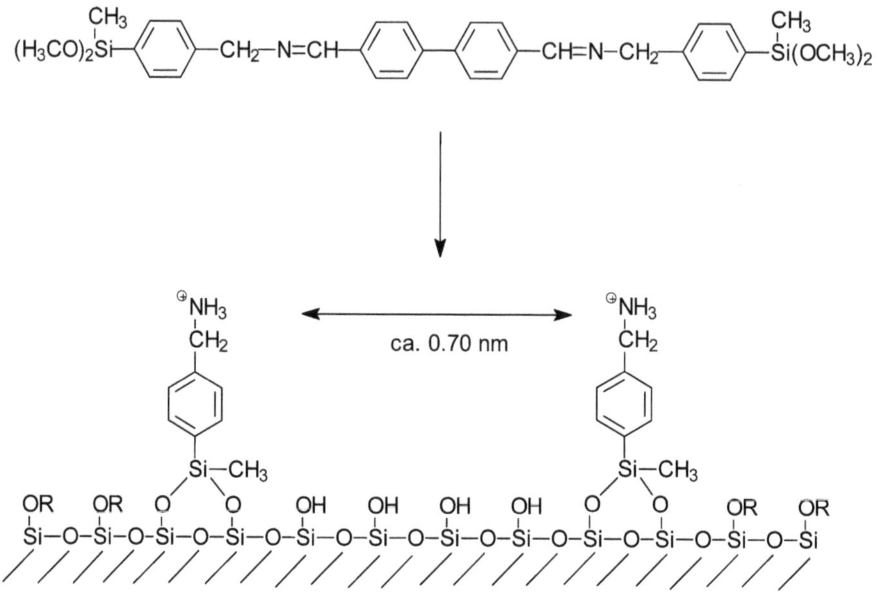

Scheme 3 Preparation of a silica gel surface functionalized with pairs of amino groups at a definite distance apart. The distance between the groups (about 0.7 nm) is fixed by using biphenyl-4,4′-dialdehyde as the template [9,18,50] (adapted from Ref. 50).

Bisazomethines were condensed on the surface of wide-pore silica gels for a test of distance selectivity (see Scheme 3). The free silanol groups on the surface of silica were then blocked by hexamethyldisilazane to prevent nonspecific adsorption. After cleavage of the template, two amino groups were left behind at a definite distance apart (Scheme 3).

If good selectivity is to be obtained, the number of condensed templates must not be too high, as otherwise, after removal of the templates and subsequent equilibration with dialdehydes, bonding takes place through adjacent amino groups that did not originally form a pair. If dialdehydes in which the aldehyde groups are at varying distances apart are exposed to such modified silica gels, the dialdehyde that was used to carry out the imprinting is bound preferentially [9,50]. Dicarboxylic acids with different distances between the carboxyl groups can also be very effectively separated on these silica gels.

Using a trialdehyde as template, it was also possible to introduce three amino groups as a TRIPOD at the silica surface [54].

Shea [13] and Shea and Sasaki [41,51] used ketals (entry f) for imprinting synthetic polymers. Diketones with various distances between the keto groups were treated with polymerizable diols, to obtain template monomers. After polymerization and removal of the template, these polymers showed considerable selectivity for the original template when the imprinted polymers were reloaded with a mixture of diketones. Unlike the reloading experiments mentioned earlier, this bond formation takes place slowly, and therefore it seems to be kinetically controlled in contrast to the bonding reactions described so far, which are thermodynamically controlled. If the arrangement of functional groups is the same, the size and shape of the rest of the molecule are important factors for the separability as shown, for example, in various bisketals.

With these template monomers, very careful investigations in the mechanism of the imprinting procedure have been performed by Shea's group. The influence of the cooperativity of binding sites, of the shape of the cavity, and of the occurrence of one-or two-point binding has been taken into consideration.

With suitable cyclic half-acetals (see Table 4, entry g), it is possible to form full acetals with monoalcohols and use this reaction for reversible binding of alcohols. The kinetics of the exchange reaction can be enhanced by neighboring group effects (entry g, $R = OCH_3$) [52].

(A)

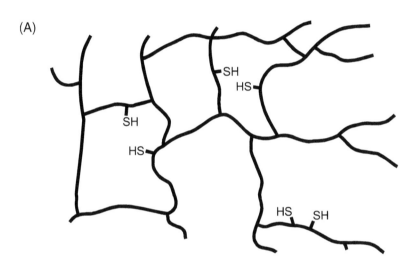

(B)

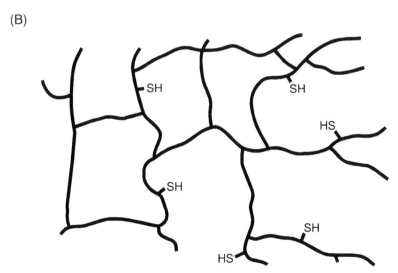

Scheme 4 Mercapto groups as nearest neighbors by copolymerization of **10** and subsequent reduction (A) or distributed at random by copolymerization of (4-vinylbenzyl) thioacetate and subsequent reduction (B) (adapted from Ref. 53).

Another possibility to introduce two functional groups in a certain neighborhood was investigated by polymerizing disulfide **10** under imprinting conditions (see entry h). Subsequent reduction of the disulfide bond with diborane provides polymers with two closely positioned mercapto groups (see Scheme 4). Under comparable conditions, polymers with randomly distributed mercapto groups were obtained from (4-vinylbenzyl)-thioacetate. The degree of cooperativity in each set of polymers with mercapto groups was determined quantitatively by oxidation with I_2. The percentage of reoxidation by I_2 to disulfide depends on the exact position of each pair of mercapto groups, which in turn depends on the degree of cross-linking of the polymer, the concentration of the mercapto groups, the degree of swelling and the reoxidation temperature. With proper conditions, reoxidation of polymers from **10** was achieved in 99% yield. Randomly distributed mercapto groups resulted, under certain conditions, in complete site separation and no reoxidation [53].

VI. EXAMPLES OF STOICHIOMETRIC NONCOVALENT INTERACTION

A. General

As already outlined in Section IV, stoichiometric noncovalent interactions combine the advantages of covalency and noncovalency without suffering from their disadvantages. Stoichiometric interaction in our sense means, if a 1:1 equimolar mixture of template and binding site monomer is used, more than 90% (or even better 95%) of the template should be bound to the binding site monomer.

It is, therefore, not necessary to use any excess of binding site monomers in order to saturate the template nearly completely. In order to reach in equimolar concentrations (0.1 molar) during imprinting 90% or better 95% degree of association, association constants of 900 (or better 3800) M^{-1} are necessary. Usual noncovalent interactions show much lower values. Due to this stoichiometric interaction, we have called this type "stoichiometric noncovalent interaction" [18,26,27].

Since no excess of binding site monomer is needed to shift the formation of the template assembly to completion, after polymerization the functional groups are exclusively located within the imprinted cavities. In this way, the high splitting yields and fast rebinding kinetics usually observed with noncovalent, nonstoichiometric interaction are combined with the high accessibility of the free cavities for molecular recognition that is typical for covalent imprinting.

Reviews on stoichiometric noncovalent interactions can be found in Refs. 18,28. In Ref. 28 also methods for the determination of the association constants are described in some detail.

B. Benzamidines and Guanidines as Binding Sites

The amidine group is a very efficient binding site able to form multiple hydrogen bonds as well as electrostatic interactions. It strongly interacts with oxyanions, such as carboxylates and phosphonates, and represents an excellent mimic of the guanidinium group of the natural amino acid arginine.

Since some years, the application of N,N'-diethyl-4-vinyl-benzamidine (3) has been investigated in our group [29,30,55]. Synthesis of the monomer is straightforward starting from 4-vinylbenzoyl chloride exhibiting high yield over three steps (for experimental details see Refs. 55,56 and the protocol in Section VIII). The monomer forms 1:1 complexes with various oxyanions (Table 5), e.g., carboxylates, phosphonates, and phosphates. Even in polar solvents like acetonirile and at higher temperatures, strong interaction occurs with association constants between 10^3 and 10^6. Thus, the stated criteria for stoichiometric noncovalent interaction are fulfilled.

The full structure of an amidinium–carboxylate complex has been recently resolved by X-ray analysis of 4-bromobenzamidinium benzoate by Kraft and coworkers (Fig. 4) [57]. Two strong H-bonds (with NH···O angles of 160° and 168°, NH···O distances of 1.97 and 1.71 Å, and N···O distances of 2.75 and 2.70 Å) are found between the carboxylate O-atoms and the amidinium N-atoms in which the hydrogen atoms are located nearer to the amidinium nitrogen atoms. A torsion angle of 81° between the benzene ring and the NCN plane of the base is caused by the space demanding zigzag conformation of the diethylamidinium group.

Sellergren [58] has used the same type of binding partners in a reverse manner, preparing pentamidine imprinted polymers for solid-phase extraction with methacrylic acid as functional monomer. Addition of an excess of acid to an isopropanolic, template containing polymerization mixture led to formation of a precipitate which was dissolved by adding water.

While complexes of the unsubstituted 4-vinyl-benzamidine also require polar solvents as DMSO and DMF for dissolution, salts of N,N'-diethyl-4-vinyl-benzamidine (3) show ideal solution behavior [30]. Even nonpolar solvents such as chloroform and benzene can be used which has been attributed to the introduction of the alkyl chains at the nitrogen atoms that render formation of *inter*molecular hydrogen bonds between different complexed templates impossible.

The copolymerization reactivity ratios of 3 with other monomers correspond to a $Q = 0.82$ and an $e = 0.44$ in the Alfrey-Price Q–e scheme, where Q describes the "general monomer reactivity" and e takes account of polar factors influencing copolymeri-

Table 5 Association Constants of Different Acids With Amidine 3 (Data from Ref. 28)

	Solvent	K_{ass} (M^{-1})	Complexation (%)
Carboxylic acid[a]	Chloroform	3.4×10^6	99.9
Carboxylic acid[a]	Acetonitrile	1.2×10^4	97.2
Phosphonate[b]	Acetonitrile	8.7×10^3 (25°C)	96.7
		7.6×10^3 (60°C)	96.4
Phosphate[c]	Acetonitrile	4.6×10^3	95.4

Association constants determined by ^{1}H-n.m.r. spectroscopy at 25°C. Complexation percentage at equimolar concentrations (0.1 m) of acid and amidine 3.
[a] 3,5-Dimethyl-benzoic acid.
[b] 3,5-Dimethyl-benzyl-phosphonic-mono(3,5-dimethyl-phenyl) ester.
[c] Di(3,5-dimethyl-phenyl) phosphate.

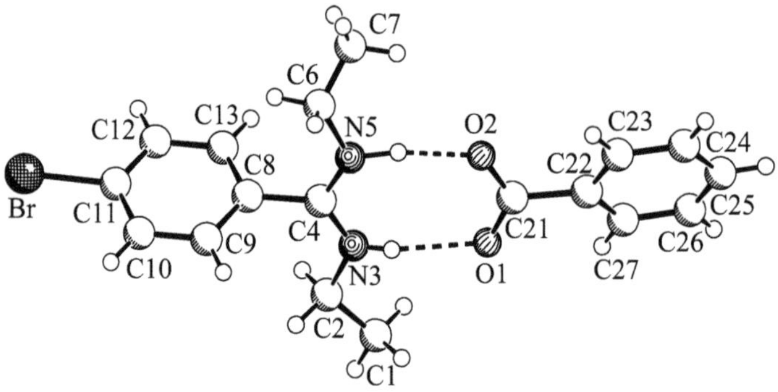

Figure 4 Crystal structure of 4-bromo-N,N'diethylbenzamidinium benzoate (adapted from Ref. 57).

zation [55]. This means that N,N'-diethyl-4-vinyl-benzamidine (3) is an electron deficient monomer, which is almost as reactive as styrene ($Q_{styrene} = 1.0$, $e_{styrene} = -0.8$).

Therefore, 3 has been employed as a binding site functional monomer for imprinted polymers in several application areas like enantioselective separation, sensing layers, and catalytically active polymers mimicking natural enzyme action.

Enantioselectively imprinted polymers have been prepared using a number of chiral dicarboxylic acids (11–13). Thus, N-terephthaloyl-D-phenylglycine 11 and two equivalents of amidine 3 as binding site monomers are used to prepare imprinted polymers (Scheme 5). The conformationally restricted dicarboxylic acid 11 represents a good template for enantioselective polymers due to its structural diversity of the four ligands at the asymmetric carbon atom. Polymers were synthesized using ethylene dimethyacrylate and THF as porogen. Splitting off of the template was carried out by stirring with 1 N HCl/acetonitrile 1:1 (v/v) followed by washing with 0.1 N NaOH/acetonitrile 1:1 (v/v). Following this procedure, up to 95% of N-terephthaloyl-D-phenylglycine 11 could be recovered.

Binding studies revealed the high quality of the obtained cavities. Figure 5 shows the binding isotherm of the polymer in methanol. Various amounts of template have been offered to the polymer particles with free cavities and the amount of substance taken up was determined. The plot shows two distinct sections: at ratios of template offered to free cavities below one, almost all of the template is bound to the polymer. At higher ratios, the isotherm directly levels off to a re-uptake of about

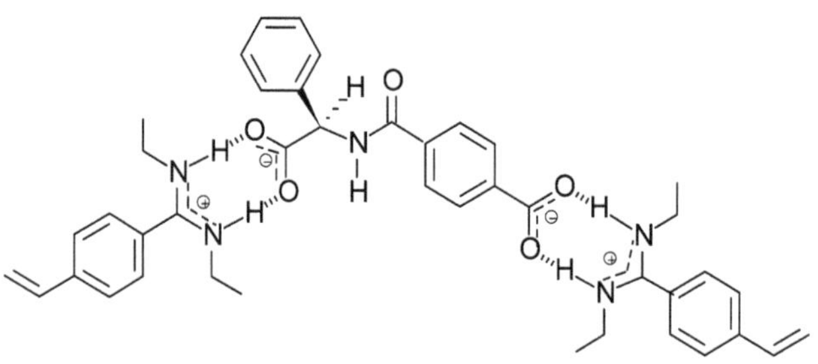

Scheme 5 Structure of bisamidinium complex of 3 formed from a conformationally stable dicarboxylic acid 11 for the preparation of enantioselective imprinted polymers [18,28,30] (adapted from Ref. 28).

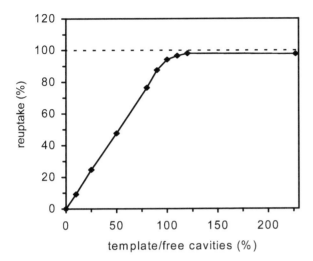

Figure 5   Re-uptake curve for a polymer prepared from template **11** (see Scheme 5) (adapted from Ref. 28).

98%. The polymer is saturated. This is a favorable example of an imprinted polymer using stoichiometric noncovalent interactions: re-uptake is quantitative, but nevertheless no unspecific binding is observed. This stands in contrast to conventional noncovalent imprinted polymers which only show poor re-uptake.

The enantioselectivity of this polymer has been investigated by equilibration with the racemate of template **11**. In dependence with the degree of rebinding, α-values up to $\alpha = 2.8$ at a re-uptake of 2.8% have been obtained.

Enantioselective separations in thin layers have been performed by using surface-enhanced Raman scattering (SERS) for the detection of the compounds in the layer. Thus polymer layers imprinted with (L) -aspartic acid derivative **12** or (2*S*,3*S*) -tartaric acid derivative **13** have been investigated [59]. The advantage of this procedure is the direct quantitative substance-selective detection of substances taken up or released from the layer.

The amidine group can act as a binding site or a catalytic site at the same time. As will be described more in detail in Chapter 21, stable transition-state analogues for the basic ester hydrolysis like **14** or **16** can be used as bis-amidinium salts to get efficient enzyme mimics for the ester hydrolysis of **15** [29] and **17**. In case of **17**, even enantioselective reactions are possible [60].

Similarly, diphenyl phosphate as amidinium salt **18** acts as a stable transition-state analogue for carbonate **19** and carbamate hydrolysis [27] (for a detailed review, see [1]).

Association constants of guanidinium ions are similar to those of amidinium ions as mentioned before (Table 1). Shea and coworkers have made use of the guanidinium group for stoichiometric noncovalent imprinting. They prepared synthetic receptors for phosphates and phosphonates in SiO_2 xerogels via a surface imprinting procedure [31]. As binding site monomer, N-(3-(trimethoxy-silyl)-propyl)-guanidinium chloride was used. The imprinted xerogels showed an increased K_{ass} for phosphates and phosphonates compared to a randomly functionalized xerogel as well as to analogous small molecule receptors.

More recently, the same group used guanidinium ions for imprinting in polymers for binding of carboxylic acids [61]. In this case, the binding site monomers were N-(2-methyl-acryloyl)-guanidine or N-(3-guanidino-propyl)-2-methyl-acrylamide. Imprinted polymers with good binding for carboxylic acids could be obtained. Polymers imprinted with optically active templates, though, did not show enantioselective recognition.

C. Imprinted Polymers with Multiple Hydrogen Bonded Templates

It is also possible to use for molecular imprinting multiple hydrogen bonding. In general, when two or more hydrogen bonds are formed between the template molecule and a binding site monomer, very stable assemblies with large association constants can be formed in aprotic media, which in turn are easily cleaved in solvents such as water or alcohols. Karube, Takeuchi and coworkers [62] introduced the polymerizable 2,6-diaminopyridine derivative 20 as novel binding site monomer that is able to interact with barbiturates through multiple hydrogen bonds.

Though quantitative determinations are lacking for the monomer, from other model substances like 2,6-bis(acetamido)pyridine, association constants with barbiturates like hexobarbital 21 are known [22]. In this case, the association constant is $K_{ass} = 200$ (Table 1). This shows that this group represents an intermediate between nonstoichiometric and stoichiometric noncovalent interactions. At 0.1 M concentration for both partners in $CHCl_3$, 80% complex formation is expected to occur.

Polymers have been prepared with various templates. Cyclobarbital 22 imprinted polymers show in HPLC experiments preferred retention of the template compared to other barbiturates like hexobarbital **21**. Compounds that do not possess two $-CO-NH-CO-$motifs—as hexobarbital **21**—are only poorly bound.

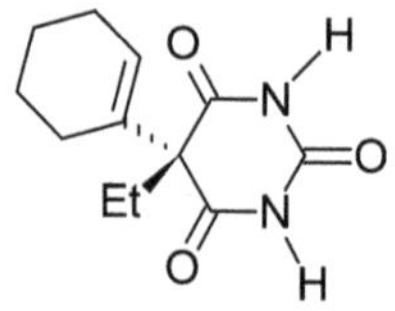

Another binding site with multiple hydrogen bonds was investigated in our institute [23,26]. First, the binding of amido-pyrazoles **23** with dipeptides was carefully investigated with low-molecular-weight model substances. In this case one amido-pyrazole interacts with the top face of the dipeptide by a three-point binding process, whereas a second amido-pyrazole is bound to the bottom face by a two-point binding. Complexation stabilizes the β-sheet conformation in the dipeptide (see Fig. 6).

The effect of substituents at the amino group and in 5 position is demonstrated in Table 6. Electron-withdrawing residues in the acyl group enhance binding by increasing the hydrogen bond donor property of the amide proton, whereas electron-donating substituents in the 5 position of the pyrazole seem to enforce the electron density of the heterocycle which leads to better hydrogen acceptor property. Therefore, 3-trifluoroacetylamino-5-methyl pyrazole (**23c**) with an association constant of 890 M^{-1} in chloroform is the strongest receptor. The binding constant of the corresponding polymerizable 3-trifluoroacetylamino-5-(4-styryl) pyrazole **23f** with 570 M^{-1} is somewhat lower.

This type of interaction has now been used to prepare a variety of imprinted polymers where N-toluoyl-glycyl-L-valine methyl ester acted as chiral template and 3-trifluoroacetylamino-5-(4-styryl)pyrazole **23f** as the binding site monomer. In batch rebinding experiments, some of them showed a pronounced enantioselectivity towards rebinding of template (up to $\alpha = 2.7$). The re-uptake of substrate to the polymers was generally only low.

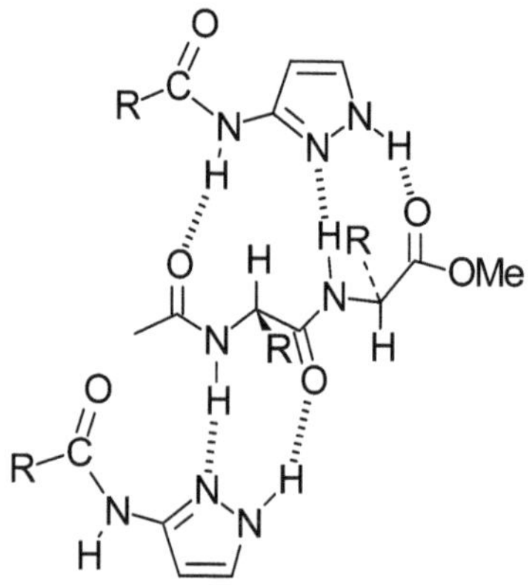

Figure 6 Side view of the computer-calculated dipeptide ester amido pyrazol 2:1 complex (adapted from Ref. 23).

Table 6 Association Constants K_{ass} of 3-Aminopyrazoles **23** With the N/C-Protected Dipeptide Ac-L-Val-L-Val-OMe in Dependence of Substituents. Determined by ^{1}H NMR Titrations in CDCl$_3$ at 25°C. Adapted from Ref. 23

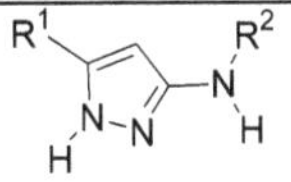

Compound 23	R^1	R^2	K_{ass} (M^{-1})
a	CH$_3$-	H-	10
b	CH$_3$-	Methacroyl	80
c	CH$_3$-	Trifluoroacetyl	890
d	4-vinylphenyl-	Pivaloyl-	81
e	4-vinylphenyl-	Acetyl-	370
f	4-vinylphenyl-	Trifluoroacetyl-	570

To overcome this problem, the amidopyrazol binding site has been combined with a guanidinium group. This binding site shows a high association constant in acetonitrile with, e.g., *N*-toluoyl-glycyl-L-valine as shown in Fig. 7 [63].

Thus, the advantages of the guanidinium resp. amidinium group are combined with the selective interaction of the amidopyrazol. Polymers with these binding site groups are now under investigation.

D. Imprinting with Polymerizable Cyclodextrins Employing Host–Guest Inclusion

Imprinted polymers have been prepared using polymerizable cyclodextrins as the binding site monomer. Mostly steroids were used as the templates. In general, cyclodextrins are of use as binding sites in molecular imprinting if the template is a rather extended molecule, so that, independent of the binding inside the cyclodextrin cavity, an imprinting effect in the polymer is to be expected. In cases of complexation of the template by two cyclodextrin binding site groups, an introduction of two cyclodextrin moieties into the polymer can lead to a certain cooperativity of the binding site groups. Imprinting with cyclodextrins is only possible in polar solvents (preferably aqueous mixtures), since in this case the association constants are sufficiently high.

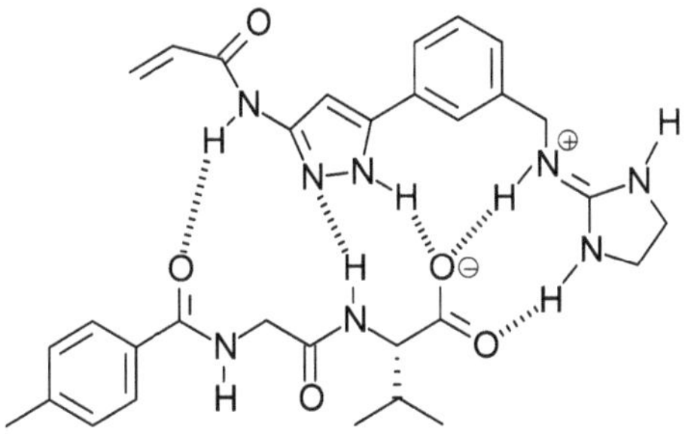

Figure 7 Interaction of a guanidine-amido pyrazol with a dipeptide acid (adapted from Ref. 63).

In case of estradiol, the association constant with β-cyclodextrin in MeOH/H$_2$O (45:55 v/v) of $K = 4.7 \times 10^2$ M^{-1} has been measured in solution. So, in principle, this is a further possibility of a stoichiometric noncovalent interaction.

Quite some examples have been published in recent years (see review [28] and, e.g. [64,65]). Despite the high association constants of the cyclodextrins, there are still some problems in their application as stoichiometric noncovalent binding site. They are mainly connected with the necessity to use very polar solvents and to get the steroids soluble. Komiyama and coworkers [64] used cholesterol as the template and prepared inclusion complexes with β-cyclodextrin which are cross-linked by toluene 2,4-diisocyanate in DMSO. It is shown that first a 1:1 complex with cholesterol is formed and it is assumed that during the imprinting the stoichiometry changes to 1:2 or even 1:3. The imprinted polymer shows double the uptake of cholesterol compared to a control polymer.

L-Phenylalanine as template was used by Nicholls and coworkers and a diacryloyl-β-cyclodextrin as the binding site monomer. In addition, a polymerizable sulfonic acid was used in the conventional noncovalent fashion. An enantioselectivity, though relatively low, was observed independent of the inherent enantioselectivity of the cyclodextrin moiety [65].

E. Stoichiometric Noncovalent Interactions for the Imprinting of Ampicillin

Whitcombe and coworkers [25] reported an interesting example of two independent types of stoichiometric noncovalent interaction for the template ampicillin (a penicillin derivative).

With ampicillin **24** possessing both an amino group and a carboxyl group, functional monomers are needed which interact with both groups of the template but not with each other. This problem was solved by introducing two different electroneutral binding site monomers for molecular imprinting.

As binding site monomer for carboxylates, a polymerizable derivative **26** of a boron-containing receptor has been prepared (Scheme 6). The interaction of carboxylates with these type of compounds has been investigated before in solution [66]. The association constant of the polymerizable receptor **26** with tetrabutylammonium acetate has now been determined to be 1.4×10^2 M^{-1} in DMSO-d_6 at 25°C [25]. This is one order of magnitude lower than the association constant of the nonpolymerizable unsubstituted receptor [66], which was attributed to the inductive effect of the oxygen of the ether bridge. Binding to the ampicillin salt was somewhat higher with $K_{ass} = 2.8 \times 10^2$ M^{-1} (in DMSO-d_6 at 25°C).

$n-\pi$ Interaction was used for binding of the amino group in ampicillin. For that purpose, a polymerizable electron-deficient quinone **27** was synthesized in one step from p-vinylphenol and chloranil (Scheme 7). As the planar chinone **27** offers two

a)

b)

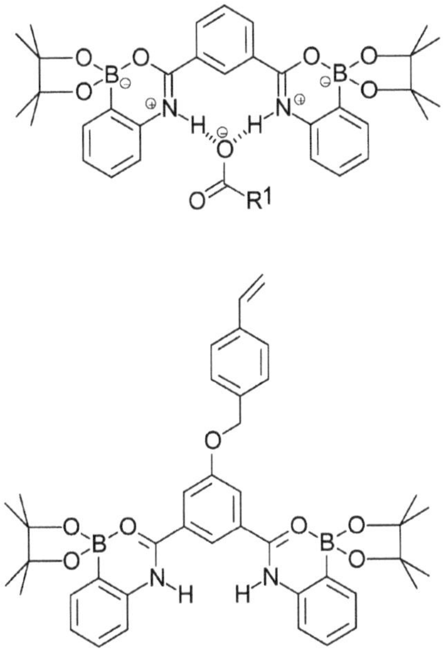

26

Scheme 6 (a) Low-molecular-weight boron-containing receptor for carboxylates [66]. (b) Polymerizable derivative **26** as synthesized by Whitcombe et al. (adapted from Ref. 25).

coordination sites for the amine, formation of a 2:1 complex of functional monomer and ampicillin **24** in DMSO-d_6 has been observed. Due to the strong binding, the association constant could only be estimated to be well above $3 \times 10^4 \, M^{-1}$ at 25°C.

With those two new binding site monomers, imprinted polymers have been prepared for ampicillin **24** as the template using all three complex partners in stoichiometric amounts.

These polymers revealed higher uptake of ampillicin **24** compared to the nonimprinted polymer which has been prepared with both functional monomers but no template at all. Batch rebinding studies were undertaken in which around 20% of the template has been offered to the polymer (in relation to the number of free cavities).

a)

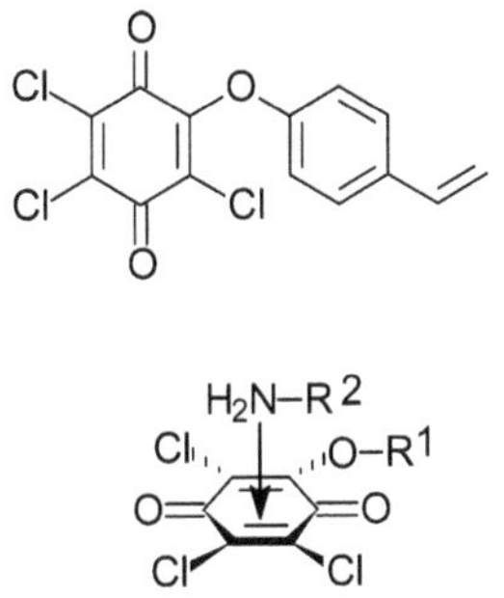

27

b)

Scheme 7 (a) Polymerizable, electron-deficient chinone **27** as synthesized by Whitcombe et al. (b) **27** as binding site for amines via n–π interaction (adapted from Ref. 25).

Binding of ampicillin **24** was favored by the imprinted polymer compared to structurally related compounds like cephalexin **25**. This is a cephalosporin antibiotic which differs from ampicillin **24** in having a six-membered instead of a five-membered sulfur-containing ring. Offering the same concentrations of both antibiotics, only approximately half of the amount of cephalosporin **25** was bound compared to the template. Careful rebinding studies have been performed.

Evidence was given that both functional monomers contribute to this strong binding in water by the preparation of two imprinted polymers, in which only one of the functional monomers has been used. However, the main part could be ascribed to quinone **27**, as expected due to its higher association constant that it already revealed in the low-molecular-weight investigations. Furthermore, the Scatchard plot shows that all cavities can be reoccupied, in contrast to nonstoichiometric binding.

These two new functional monomers demonstrate the high potential of molecular imprinting employing stoichiometric noncovalent interactions.

VII. CONCLUSION

Covalent interactions were the first interactions to be used in the molecular imprinting procedure. Due to the strong interaction during the polymerization, the binding site functional groups are only situated in the imprinted cavities after removal of the template. The number of possible covalent interactions for molecular imprinting is limited. Most covalent interactions are too stable for splitting and too slow in the reversible interaction. Boronic acids esters and Schiff bases present very suitable covalent interactions. Very promising interactions are noncovalent ones with high association constants between template and binding site (stoichiometric noncovalent interaction). In this case, the binding site functional groups can again be introduced exclusively in the cavities. Binding equilibria are fast and the position of the equilibrium can easily be controlled. These imprinted polymers show complete re-upstake of the template after removal of the template. They are especially useful for the preparation of catalytically active polymers and for preparative separations since their capacity is much higher compared to polymers prepared with conventional noncovalent interaction. A disadvantage of this procedure is the sometimes complicated synthesis of the binding monomers.

VIII. PROTOCOLS OF THE PREPARATION OF TWO TYPICAL EXAMPLES OF IMPRINTED POLYMERS

A. Polymer from Scheme 2

1. Preparation of Template Monomer 1 [17]

Phenyl-α-D-mannopyranoside (**2**)(6 g, 23.4 mmol) and tris(4-vinylphenyl)-boroxin (6.075 g, 15.6 mmol) were heated in benzene with removal of the water by azeotropic distillation. After complete removal of water, the solvent was evaporated, and the residue was crystallized from diethyl ether. Yield: 9.7 g (86%), m.p. 139°C.

2. Preparation of the Polymer [10]

Template monomer **1** (0.75 g), initiator azobis(isobutyronitrile)(120 mg), cross-linker ethylene dimethacrylate (15.0 g) in tetrahydrofuran (15.0 g) were filled in a tube, carefully degassed by three freeze–thaw cycles, sealed under argon, and polymerized for 4 days at 65°C. The tube was then cooled and broken, and the polymer was milled with an Alpine Contraplex 63 C, and sieved to a grain size of 125–63 μm. Alternatively, the polymer could be milled to a finer powder and separated to a particle diameter fraction of 8–15 μm by a wind-sieving machine (Alpine Multiplex 100 MRZ). This material was first extracted with dry diethyl ether before being dried in vacuum at 40°C. The template was removed from the polymer by a continuous extraction with methanol–water (250 mL per g of polymer). After the solvent was evaporated, the residue was dissolved in a defined volume of methanol and the content of 2 was determined polarimetrically.

This polymer possessed an inner surface area of 322 m^2/g, a splitting percentage of the template of 82%, a swelling ability in methanol of 1.20, and an α-value for the separation of D, L-2 of 4.52.

B. Polymer from Template 14 and Binding Site 3 [55,56]

1. N-Ethyl-4-Vinyl Benzamide

4-Vinylbenzoic acid chloride [67] (81.60 g, 0.49 mol) in dry CH_2Cl_2 (100 mL) was dropped to a solution of ethyl amine (45.09 g, 1.00 mol) in dry CH_2Cl_2 (250 mL) at −20 °C. After warming up to ambient temperature, phenothiazine (0.20 g) was added. The solution was stirred for 15 h. The precipitated ethyl ammonium chloride was filtered off and washed with a small amount of dry CH_2Cl_2. The solvent was removed in vacuo and the residue recrystallized from EtOAc to give colorless crystals (172.4 g, 98 %).

2. N-Ethyl-4-Vinylbenzocarboximide Acid Ethyl Ester

N-Ethyl-4-vinyl benzamide (39.95 g, 0.228 mol) was added to a solution of triethyloxonium tetrafluoroborate (58.5 g, 0.308 mol) in dry CH_2Cl_2 (150 mL) under argon. The solution was stirred for 2 h at ambient temperature, phenothiazine (0.20 g) was added, the stirring was continued for 36 h, and the solvent removed in vacuo. Remaining oil was treated with 3 M NaOH (80 mL) and directly extracted with ice cooled Et_2O (250 mL). The aqueous layer was extracted another four times with Et_2O, the combined organic layers were dried over Na_2SO_4, filtered, concentrated in vacuo and distilled to give a colorless liquid [46.19 g, ~100%, b.p. 53°C, 10^{-3} mbar].

3. N,N'-Diethyl-4-Vinylbenzamidine (3)

Dry ethylammonium chloride (63.61 g, 0.78 mol) was added to a solution of N-ethyl-4-vinylbenzocarboximide acid ethyl ester (121.97 g, 0.60 mol) in dry EtOH (270 mL) under argon. After stirring at 15°C for 5 h, 4-tert.-butylbrenzcatechol (0.5g) was added, the stirring was continued for 5 days at ambient temperature, and the solvent was removed in vacuo. The residue was treated with ice cooled 6 M NaOH (500 mL) and directly extracted with an ice cooled mixture of EtOAc/Et_2O (1:1). The aqueous layer was extracted another four times, the combined organic layers were dried with Na_2SO_4. The solvent was removed in vacuo, and the residue was sublimated twice at 0.01 mbar to give white crystals (105.24 g, 87%, m.p. 76°C).

4. Preparation of the polymer [29,56]

The polymer is prepared similarly to the first example of this protocol from a mixture of ethylene dimethacrylate (4.67 g), methyl methacrylate (0.22 g), N,N'-diethyl-4-

vinylbenzamidine (**3**) (0.442 g, 2.188 mmol), template **14** (0.350 g, 1.094 mmol), azobis(isobutyronitrile)(56.8 mg), and tetrahydrofuran (5.64 mL) as the porogen. Polymerization in a sealed ampule was carried out for 70 hr at 60°C. After the usual work-up procedure, the template was removed from the particles by extraction with methanol, followed by washing twice with 0.1 N NaOH/acetonitrile (1:1), water, and acetonitrile. The recovery of the template from the washings was determined by HPLC [RP 8-column, eluent 0.2% trifluoroacetic acid in water/acetonitrile (55:45)]. Recovery of template was 85%. The inner surface area of the polymer amounted to 228 m^2/g, the swelling ratio in methanol was 1.45.

This polymer was used for enzyme-like catalysis [29].

ACKNOWLEDGMENT

These investigations were supported by financial grants from the Deutsche Forschungsgemeinschaft, Minister für Wissenschaft und Forschung des Landes Nordrhein-Westfalen, and Fonds der Chemischen Industrie.

REFERENCES

1. Wulff, G. Enzyme-like catalysis by molecularly imprinted polymers. Chem. Rev. **2002**, *102*, 1–28.
2. Fersht, A. *Enzyme Structure and Mechanism.* W. H. Freeman: New York, **1985**.
3. Fischer, E. Einfluss der Konfiguration auf die Wirkung der Enzyme. Ber. Dtsch. Chem. Ges. **1894**, *27*, 2985–2993.
4. Wulff, G.; Sarhan, A.; Zabrocki, K. Enzyme-analogue built polymers and their use for the resolution of racemates. Tetrahedron Lett. **1973**, *44*, 4329–4332.
5. Wulff, G.; Sarhan, A. Enzym-analoge Polymere. German Patent Application, (Offenlegungsschrift) DE-A 2242796, 1974. Chem. Abstr. **1975**, *83*, P60300. Greatly enlarged US Patent, continuation in part US-A 4127730, 1978.
6. Wulff, G.; Vesper, W.; Grobe-Einsler, R.; Sarhan, A. Enzyme-analogue built polymers IV. On the synthesis of polymers containing chiral cavities and their use for the resolution of racemates. Makromol. Chem. **1977**, *178*, 2799–2819.
7. Dickey, F.H. Preparation of specific adsorbents. Proc. Natl. Acad. Sci. USA **1949**, *35*, 227–229.
8. Wulff, G. Molecular recognition in polymers prepared by imprinting with templates. In *Polymeric Reagents and Catalysts*; Ford, W.T. Ed.; American Chemical Society; ACS Symposium Series 308: Washington, 1986; 186–230 pp.
9. Wulff, G. Molecular imprinting in cross-linked materials with the aid of molecular templates—a way towards artificial antibodies. Angew Chem. Int. Ed. Engl. **1995**, *34*, 1812–1832.
10. Wulff, G.; Minarik, M. Template imprinted polymers for h.p.l.c. separation of racemates. J. Liq. Chromatogr. **1990**, *13*, 2987–3000.
11. Bartsch, R.A., Maeda, M. Eds.; *Molecular and Ionic Recognition with Imprinted Polymers*; American Chemical Society; ACS-Symposium Series 703: Washington, 1998.
12. Sellergren, B., Ed.; *Molecularly Imprinted Polymers—Man-Made Mimics of Antibodies and their Application in Analytical Chemistry.* Elsevier: Amsterdam, 2001.
13. Shea, K.J. Molecular imprinting of synthetic network polymers: the de novo synthesis of macromolecular binding and catalytic sites. Trends Polym. Sci. **1994**, *2*, 166–173.
14. Mosbach, K.; Ramström, O. The emerging technique of molecular imprinting and its future impact on biotechnology. Biotechnology **1996**, *14*, 163–170.

15. Sellergren, B.; Lepistö, M.; Mosbach, K. Highly enantioselective and substrate-selective polymers obtained by molecular imprinting utilizing noncovalent interactions. J. Am. Chem. Soc. **1988**, *110*, 5853–5860.

16. Wulff, G.; Kemmerer, R.; Vietmeier, J.; Poll, H.-G. Chirality of vinyl polymers. The preparation of chiral cavities in synthetic polymers. Nouv. J. Chim. **1982**, *6*, 681–687.

17. Wulff, G.; Vietmeier, J.; Poll, H.-G. Enzyme-analogue built polymers, 22. Influence of the nature of the crosslinking agent on the performance of imprinted polymers in racemic resolution. Makromol. Chem. **1987**, *188*, 731–740.

18. Wulff, G.; Biffis, A. Molecular imprinting with covalent or stoichiometric non-covalent interactions. In *Molecularly Imprinted Polymers—Man-Made Mimics of Antibodies and their Application in Analytical Chemistry; Sellergren, B.,* Ed.; Elsevier: Amsterdam, **2001**, 71–111 pp.

19. Wulff, G.; Dederichs, W.; Grotstollen, R.; Jupe, C. On the chemistry of binding sites, Part II. Specific binding of substances to polymers by fast and reversible covalent interactions. In *Affinity Chromatography and Related Techniques*; Gribnau, T.C.J., Visser, J., Nivard, R.J.F., Eds.; Elsevier: Amsterdam, **1982**, 207–216 pp.

20. Grunwald, E.; Coburn, W.C. Calculation of association constants for complex formation from spectral data. Infrared measurements of hydrogen bonding between ethanol and ethyl acetate, and ethanol and acetic anhydride. J. Am. Chem. Soc. **1958**, *80*, 1322–1325.

21. Albrecht, G.; Zundel, G. Carboxylic acid–nitrogen base hydrogen bonds with large proton polarizability in acetonitrile as a function of the basicity of the hydrogen bond acceptors. Z. Naturforsch. **1984**, *39*, 986–992.

22. Schneider, H.-J.; Juneja, R.K.; Simova, S. Solvent and structural effects on hydrogen bonds in some amides and barbiturates. An additive scheme for the stability of corresponding host–guest complexes. Chem. Ber. **1989**, *122*, 1211–1213.

23. Kirsten, C.; Schrader, T. Intermolecular ß-sheet stabilization with aminopyrazoles. J. Am. Chem. Soc. **1997**, *119*, 12061–12068.

24. Hamilton, A.D.; Linton, B. Calorimetric investigation of guanidinium–carboxylate interactions. Tetrahedron **1999**, *55*, 6027–6038.

25. Lübke, C.; Lübke, M.; Whitcombe, M.J.; Vulfson, E.N. Imprinted polymers prepared with stoichiometric template–monomer complexes: efficient binding of ampicillin from aqueous solutions. Macromolecules **2000**, *33*, 5098–5105.

26. Wulff, G.; Groß, T.; Schönfeld, R.; Schrader, T.; Kirsten, C. Molecular imprinting for the preparation of enzyme-analogous polymers. In *Molecular and Ionic Recognition with Imprinted Polymers*; Bartsch, R.A., Maeda, M., Eds.; American Chemical Society; ACS-Symposium Series 703: Washington, **1998**; 10–28 pp.

27. Strikovsky, A.G.; Kasper, D.; Grün, M.; Green, B.S.; Hradil, J.; Wulff, G. Catalytic molecularly imprinted polymers using conventional bulk polymerization or suspension polymerization: selective hydrolysis of diphenyl carbonate and diphenyl carbamate. J. Am. Chem. Soc. **2000**, *122*, 6295–6296. See also: Liu, J.-Q.; Wulff, G. Molecularly imprinted polymers with strong carboxypeptidase A-like activity: combination of an amidinium function with a zinc-ion-binding site in transition-state imprinted cavities. Angew. Chem. Int. Ed. Engl. **2004**, *43*, 1287–1290.

28. Wulff, G. Knorr, K. Stoichiometric noncovalent interaction in molecular imprinting. Bioseparation **2002**, *10*, 257–276.

29. Wulff, G.; Groß, T.; Schönfeld, R. Enzyme models based on molecularly imprinted polymers with strong esterase activity. Angew Chem. Int. Ed. Engl. **1997**, *36*, 1961–1964.

30. Wulff, G.; Schönfeld, R. Polymerizable amidines—adhesion mediators and binding sites for molecular imprinting. Adv. Mater. **1998**, *10*, 957–959.

31. Sasaki, D.Y.; Rush, D.J.; Daitch, C.E.; Alam, T.M.; Assink, R.A.; Ashley, C.S.; Brinker, C.J.; Shea, K.J. In *Molecular and Ionic Recognition with Imprinted Polymers*;

Bartsch, R.A., Maeda, M., Eds.; American Chemical Society; ACS-Symposium Series 703: Washington, **1998**; 314–324 pp.

32. Mallik, S.; Plunkett, S.D.; Dhal, P.K.; Johnson, R.D.; Pack, D.; Shnek, D.; Arnold, F.H. Towards materials for the specific recognition and separation of proteins. New J. Chem. **1994**, *18*, 299–304.
33. Chen, G.; Guan, Z.; Chen, C.-T.; Fu, L.; Sundaresah, V.; Arnold, F.H. A glucose-sensing polymer. Nat. Biotechnol. **1997**, *15*, 354–357.
34. Whitcombe, M.J.; Vulfson, E.N. Covalent imprinting using sacrificial spacers. In *Molecularly Imprinted Polymers—Man-Made Mimics of Antibodies and their Application in Analytical Chemistry*; Sellergren, B., Ed.; Elsevier: Amsterdam, **2001**, 203–212 pp.
35. Wulff, G.; Lauer, M.; Böhnke, H. Rapid proton transfer as cause of an unusually large neighboring group effect. Angew Chem. Int. Ed. Engl. **1984**, *23*, 741–742.
36. Glad, M.; Norrlöw, D.; Sellergren, B.; Siegbahn, N.; Mosbach, K. Use of silane mono-mers for molecular imprinting and enzyme entrapment in polysiloxane-coated porous silica. J. Chromatogr. **1985**, *347*, 11–23.
37. Sarhan, A.; Ali, M.M.; Abdelaal, M.Y. Racemic resolution of mandelic acid on polymers with chiral cavities, 3. Cooperative binding over phenylboronic acid groups and N-bases. React. Polym. **1989**, *11*, 57–70.
38. Alexander, C.; Smith, C.R.; Whitcombe, M.J.; Vulfson, E.N. Imprinted polymers as protecting groups for regioselective modification of polyfunctional substrates. J. Am. Chem. Soc. **1999**, *121*, 6640–6651.
39. Wulff, G.; Lohmar, E. Enzyme-analogue built polymers. Specific binding effects in chiral microcavities of cross linked polymers. Isr. J. Chem. **1979**, *18*, 279–284.
40. O'Shannessy, D.J.; Andersson, L.I.; Mosbach, K. Molecular recognition in synthetic polymers. Enantiomeric resolution of amide derivatives of amino acids on molecularly imprinted polymers. J. Mol. Recognit. **1989**, *2*, 1–5.
41. Shea, K.J.; Sasaki, D.Y. On the control of microenvironment shape of functionalized network polymers prepared by template polymerization. J. Am. Chem. Soc. **1989**, *111*, 3442–3444.
42. Wulff, G.; Schauhoff, S. Racemic resolution of free sugars with macroporous polymers prepared by molecular imprinting. Selectivity dependence on the arrangement of func-tional groups versus spatial requirements. J. Org. Chem. **1991**, *56*, 395–400.
43. Wulff, G.; Schulze, I.; Zabrocki, K.; Vesper, W. Bindungsstellen im Polymer mit unterschiedlicher Zahl der Haftgruppen. Makromol. Chem. **1980**, *181*, 531–544.
44. Wulff, G.; Oberkobusch, D.; Minárik, M. Chiral cavities in polymer layers coated on wide-pore silica. React. Polym. Ion Exch. Sorbents **1985**, *3*, 261–275.
45. Moradian, A.; Mosbach, K. Preparation of a functional, highly selective polymer by molecular imprinting. A demonstration with L-*p*-aminophenylalanine anilide as a template molecule allowing multiple points of attachment. J. Mol. Recognit. **1989**, *2*, 167–169.
46. Sellergren, B. Molecular imprinting by noncovalent interactions: tailor-made chiral stationary phases of high selectivity and sample load capacity. Chirality **1989**, *1*, 63–68.
47. Wulff, G.; Poll, H.G. Enzyme-analogue built polymers, 23. Influence of the structure of the binding sites on the selectivity for racemic resolution. Makromol. Chem. **1987**, *188*, 741–748.
48. Wulff, G.; Vietmeier, J. Enzyme-analogue built polymers, 26. Enantioselective synthesis of amino acids using polymers possessing chiral cavities obtained by an imprinting procedure with template molecules. Makromol. Chem. **1989**, *190*, 1727–1735.
49. Wulff, G.; Best, W.; Akelah, A. Enzyme-analogue built polymers, 17. Investigations on the racemic resolution of amino acids. React. Polym. Ion Exch. Sorbents **1984**, *2*, 167–174.

50. Wulff, G.; Heide, B.; Helfmeier, G. Molecular recognition through the exact placement of functional groups on rigid matrices via a template approach. J. Am. Chem. Soc. **1986**, *108*, 1089–1091.

51. Shea, K.J.; Sasaki, D.Y. Ananalysis of small-molecule binding to functionalized synthetic polymers by ^{13}C CP/MAS NMR and FT-IR spectroscopy. J. Am. Chem. Soc. **1991**, *113*, 4109–4120.

52. Wulff, G.; Wolf, G. Über die Eignung verschiedener Aldehyde und Ketone als Haftgruppen für Monoalkohole. Chem. Ber. **1986**, *119*, 1876–1889.

53. Wulff, G.; Schulze, I. Directed cooperativity and site separation of mercapto groups in synthetic polymers. Angew Chem. Int. Ed. Engl. **1978**, *17*, 537–538.

54. Tahmassebi, D.C.; Sasaki, T. Synthesis of a new trialdehyde template for molecular imprinting. J. Org. Chem. **1994**, *59*, 679–681.

55. Wulff, G.; Schönfeld, R.; Grün, M.; Baumstark, R.; Wildburg, G.; Häußling, L.B.A. Polymerizable amidines. German Patent Application (Offenlegungsschrift) DE-A 19720345 A1, 1998; Chem. Abstracts **1998**, *128*, 49155.

56. Wulff, G. Templated synthesis of polymers—molecularly imprinted materials for recognition and catalysis. In *Templated Organic Synthesis*; Diederich, F., Stang, P.J., Eds.; Wiley-VCH: Weinheim, **1999**, 39–73 pp.

57. Peters, L.; Fröhlich, R.; Boyd, A.S.F.; Kraft, A. Noncovalent interactions between tetrazole and an *N,N′*-diethyl-substitued benzamidine. J. Org. Chem. **2001**, *66*, 3291–3298. See also: Peters, L., Ph.D. thesis, Heinrich-Heine University of Düsseldorf, 2001.

58. Sellergren, B. Direct drug determination by selective sample enrichment on an imprinted polymer. Anal. Chem. **1994**, *66*, 1578–1582.

59. Kostrewa, S.; Emgenbroich, M.; Klockow, D.; Wulff, G. Surface-enhanced Raman scattering on molecularly imprinted polymers in water. Macromol. Chem. Phys. **2003**, *204*, 481–487.

60. Wulff, G.; Emgenbroich, M. Chem. Eur. J. **2003**, *9*, 4106–4117.

61. Shea, K.J.; Spivak, D. Molecular imprinting of carboxylic acids employing novel functional macroporous polymers. J. Org. Chem. **1999**, *64*, 4627–4634.

62. Tanabe, K.; Takeuchi, T.; Matsui, J.; Ikebukuro, K.; Yano, K.; Karube, I. Recognition of barbiturates in molecularly imprinted copolymers using multiple hydrogen bonding. J. Chem. Soc. Chem. Commun. **1995**, *18*, 2303–2304.

63. Wulff, G.; Knorr, K. Unpublished results. See: Knorr, K., Ph.D. thesis, Heinrich-Heine University of Düsseldorf, 2003.

64. Asanuma, H.; Kakazu, M.; Shibata, M.; Hishiya, T.; Komiyama, M. Molecularly imprinted polymer of beta-cyclodextrin for the efficient recognition of cholesterol. Chem. Comm. **1997**, *20*, 1971–1972.

65. Piletsky, S.A.; Andersson, H.S.; Nicholls, I.A. The rational use of hydrophobic effect-based recognition in molecularly imprinted polymers. J. Mol. Recogn. **1998**, *11*, 94–97.

66. Hughes, M.P.; Smith, B.D. Enhanced carboxylate binding using urea and amide-based receptors with internal Lewis acid coordination: a cooperative polarization effect. J. Org. Chem. **1997**, *62*, 4492–4499.

67. Ishizone, T.; Hirao, A.; Nakahama, N. Protection and polymerization of functional monomers. Macromolecules **1989**, *22*, 2895–2901.

5

The Semi-Covalent Approach

Nicole Kirsch* and **Michael J. Whitcombe** Institute of Food
Research, Norwich, United Kingdom

I. WHY USE A SEMI-COVALENT APPROACH?

In the previous two chapters you learned about two different approaches to the
creation of recognition sites in synthetic polymers by molecular imprinting, i.e.,
via the noncovalent and the covalent approach. Both methods have inherent
advantages and disadvantages.

The noncovalent approach has been widely adopted due to its simplicity and
broad applicability to a range of template structures. In its simplest form, the starting
materials are relatively cheap and little specialist knowledge of polymer chemistry
is required to prepare materials with good selectivity and high affinity for the template. The fact that chiral stationary phases may be prepared in this way relatively
routinely certainly vindicates the method, however, the need to use an excess of functional monomer for efficient formation of the prepolymerization complex leads to a
large number of statistically distributed functional monomer units, not associated
with template, which can contribute to high nonspecific binding. In addition, the
equilibrium (and therefore dynamic) nature of the template–monomer interaction
leads to the creation of a distribution of binding sites with different affinities for the
template monomer. The development of new approaches to noncovalent imprinting
through the design of functional monomers with high affinity for specific structural
motifs of the template can do away with the need for an excess of functional monomer,
but at present there are few such general purpose high affinity monomers.

The covalent approach does not suffer from any of the problems caused by using
an excess of functional monomer, since the template is covalently bound to an appropriate stoichiometric amount of functional monomer in the polymerization mixture.
The result is that all functional groups in the resultant imprinted polymer are present
only in the imprint sites and in the precise spatial arrangement for rebinding. This
would appear to represent an ideal situation for the creation of an imprint, however,
the range of template functionality for which efficient reversible complex formation is

---

*Current affiliation: University of Kalmar, Kalmar, Sweden.

possible is rather limited. Most, if not all, covalent imprinting strategies involve condensation reactions with the loss or addition of one or more molecules of water in the bond breaking (template removal) or bond making (rebinding) processes, respectively. The consequence is slow interchange of template with the sites, which is aggravated by the increased spatial requirement for the condensation process, for example, boronate ester formation or hydrolysis proceeds via a tetrahedral intermediate at boron, whereas the trigonal boronate ester is imprinted.

The "semi-covalent" approach attempts to combine the advantages of both methods by the polymerization of a template, covalently bound to a functional monomer by a cleavable linkage. Template removal, typically by hydrolysis, leaves imprint bearing functional groups, which are capable of interacting with the template in a noncovalent sense in the rebinding step. In other words, the imprinting is covalent, but the rebinding is noncovalent in nature. The consequences are that: (i) all of the functional groups introduced in the imprinting step are associated with introduced template; (ii) there are no randomly distributed functional groups because no excess monomer has been used; (iii) the sites are more uniform in nature; and (iv) template rebinding is not subject to any kinetic restrictions except diffusion. In addition, template that is not removed in the hydrolysis step remains covalently bound to the polymer and is unlikely to contribute to template "bleeding" which can occur in noncovalently imprinted polymers. For further discussions on the problem of template bleeding from noncovalently imprinted polymers and methods to overcome it, see the chapters by Sellergren (Chapter 8) and Ye (Chapter 23).

There have been two main methodologies for the semi-covalent approach: (a) by direct connection of template and monomer by an ester or amide linkage or (b) using a spacer between the template and polymerizable recognition element in the imprinting step. Examples of both of these approaches will be given in the following sections.

II. SEMI-COVALENT APPROACH USING CARBOXYLIC ESTER TEMPLATES

In this approach, the target molecule and the polymerizable groups are directly connected via an ester linkage. After the polymer has been obtained, hydrolysis of the ester group releases the template and reveals the complementary function in the binding site. This functionality is then used to establish noncovalent interactions with the target analyte. Damen and Neckers [1] and Shea and Thompson [2] employed ester chemistry for the imprinting of carboxylic acids in DVB-based polymers. Both of these examples were fully covalent in their approach (binding was also covalent, reforming stable carboxylic esters in the template sites) and designed to show that the polymers retain some form of stereochemical memory for the template by influencing the course of reactions carried out in the polymer. Moderate stereochemical preferences were seen in both cases.

Other groups have sought to use similar ester chemistry, but in the semi-covalent sense, where rebinding of the target analyte is based on hydrogen bonding to the products of ester hydrolysis which remain as polymer-bound residues in the imprint sites after template removal. This can best be illustrated by an example such as that shown in Fig. 1, which shows the schematic representation of this strategy using testosterone methacrylate as the template monomer, as used by Cheong et al. [3]. The

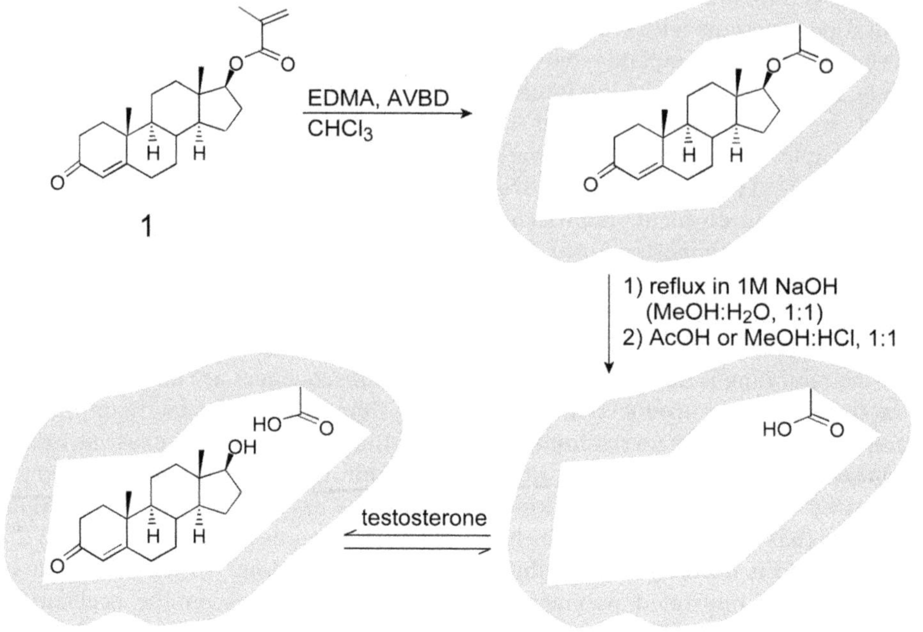

Figure 1 Scheme for the imprinting of testosterone, starting from testosterone methacrylate **1** (adapted from Ref. 3).

template monomer was prepared by esterification of testosterone with methacryloyl chloride in the presence of triethylamine, which was then copolymerized with EDMA in chloroform. The resultant polymer was hydrolyzed with 1 M NaOH in 50:50 aqueous methanol followed by acidification to yield a polymer bearing methacrylic acid residues capable of binding testosterone via hydrogen bonding (unesterified methacrylic acid was also added in order to noncovalently complex the testosterone keto function). This approach was hampered by severe difficulties in hydrolyzing the link between the polymer and the template monomer; furthermore, the imprinted polymer prepared by this route was unable to separate testosterone and β-estradiol when used as an HPLC stationary phase. The authors concluded that in this case noncovalent-imprinted polymers were superior to those prepared by their semi-covalent method.

A similar approach was used by Joshi et al. [4], who prepared imprinted polymers for the removal of phenol from bisphenol A. In this case, phenyl methacrylate **2** was used as template monomer to generate MIPs to bind phenol, and similarly cumylphenyl methacrylate **3** was used as the template monomer for cumylphenol rebinding (Fig. 2).

Previously, the same group had shown [5] that phenyl methacrylates were superior to the corresponding phenyl-4-vinylbenzoate esters as template monomers. The phenol-imprinted polymer showed selective binding of phenol over bisphenol A, whereas the opposite selectivity was observed with the cumylphenol-imprinted polymer. These results show that size and shape selectivity can be achieved with this simple imprinting method. Furthermore, it was shown that the choice of solvent is

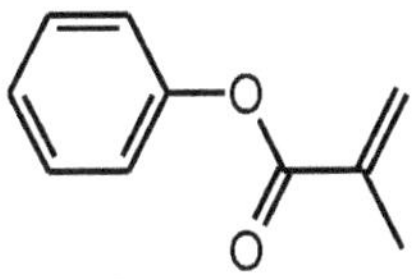

Figure 2 The structures of phenyl methacrylate **2** and cumylphenyl methacrylate **3**, used as templates by Joshi et al. (Ref. 4).

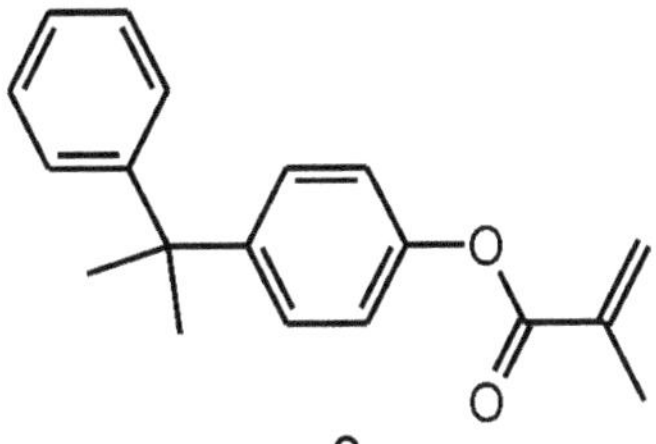

important for the selectivity: rebinding in ethyl acetate did not result in discrimination between phenol and bisphenol A (BPA), whereas in 1% acetic acid in methanol, the phenol/BPA adsorption ratio was greater than 19. An increase in the imprinting efficiency β (α MIP/α NP, where α = amount of phenol adsorbed/amount of BPA adsorbed) was noticed with increasing solvent polarity. This effect was due to competition between the analyte and the solvent for nonspecific binding sites. (The hydrolysis conditions used to remove the template may have led to partial hydrolysis of the EDMA backbone and hence the generation of excess methacrylic acids capable of establishing hydrogen bonding with the phenols in nonimprinted regions of the polymer [6]).

The use of a similar approach to the preparation of solid phase extraction (SPE) materials for the extraction of phenolics from environmental water samples has recently been reported [7]. Polymers were imprinted with 4-nitrophenol, either in the conventional noncovalent sense, using methacrylic acid as the functional monomer, or in a semi-covalent protocol using the 4-nitrophenyl methacrylate ester as template monomer. The latter is relatively easily cleaved by basic hydrolysis due to the electron-withdrawing effect of the nitro-group. The polymers were investigated in SPE mode for their ability to extract the template and a range of other simple phenols from aqueous samples, including spiked river water. The conclusions were that the noncovalently imprinted material was the most selective, but the semi-covalent imprinting route led to higher recoveries.

The semi-covalent ester approach can also be used to generate binding sites for amines as shown by Sellergren and Andersson [8]. A structural analogue was used for p-aminophenylalanine ethyl ester **4**, namely N^2-propionyl-O^1-acryloyl-2-amino-3-(O^4-acryloyl-4-hydroxyphenyl) 1-propanol **5** (Fig. 3) which was copolymerized with DVB.

The carboxyl groups introduced into the polymer after basic hydrolysis were capable of forming hydrogen bonds and ionic interactions with the amino groups of the amino acid. However the polymer showed preferential binding of the D-form of **4**, despite imprinting the L-configuration of the template.

Byström et al. [9] used MIPs generated by the ester methodology for the selective reduction of steroid ketones. The sterol was converted at the 3- or 17-position into the acrylate esters (17-steroid acrylate ester **6** shown in Fig. 4).

The polymers were generated by suspension polymerization using DVB as comonomer and cross-linker. The ester groups were reduced with LiAlH$_4$ to remove

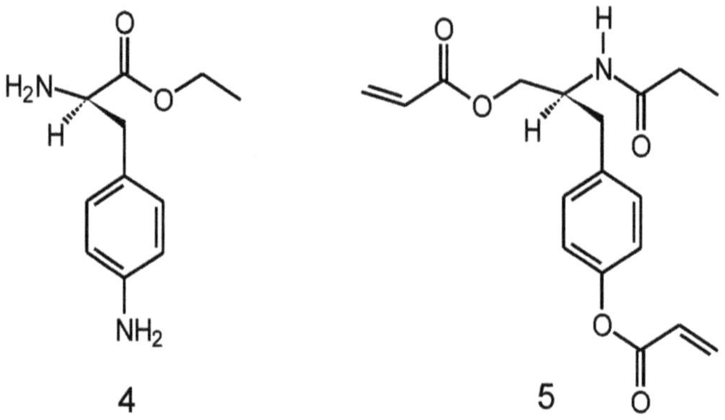

Figure 3 The target compound, *p*-aminophenylalanine ethyl ester **4** and the corresponding covalent template analogue **5**, used in the preparation of imprinted polymers by Sellergren and Andersson (Ref. 8). Despite imprinting with the L-form of the template, the polymer preferentially bound the D-form of **4**.

the template, as basic hydrolysis did not result in significant cleavage of the template. Treatment with additional hydride reagent created an imprinted polymeric reducing agent by reaction of $LiAlH_4$ with the hydroxyl groups left by template cleavage. The steric requirements of the imprint site meant that approach of the diketone **7** to the hydride group was restricted if the steroid approached in an orientation which did not correspond to that of the template. This led to exclusive reduction at the 17-position. This is stark contrast to reduction of **7** in free solution, where both keto groups are reduced. In addition, a shift in the stereoisomer ratio was also seen: in solution, hydride is almost exclusively delivered at the 17-position from the less hindered α-face, leading to the β-isomer of the alcohol; at the polymer surface, a significant amount of hydride deliver takes place on the β-face, leading to an enhanced yield of the α-isomer (Fig. 5).

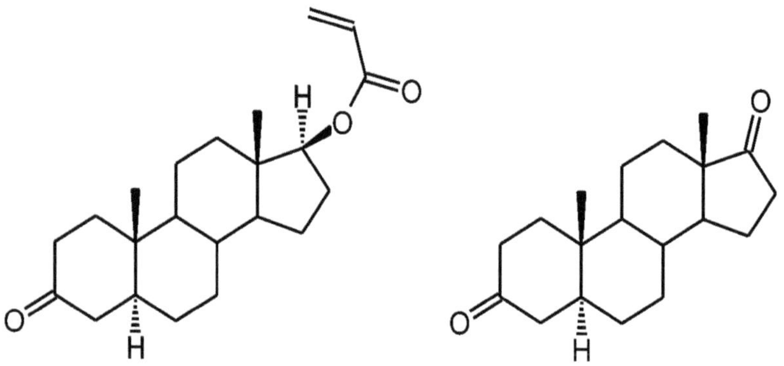

Figure 4 The 17-β-sterol acrylate ester template **6**, used by ByStöm et al. (Ref. 9) to prepare regio- and stereo-selective polymeric reducing agents for the diketone **7**.

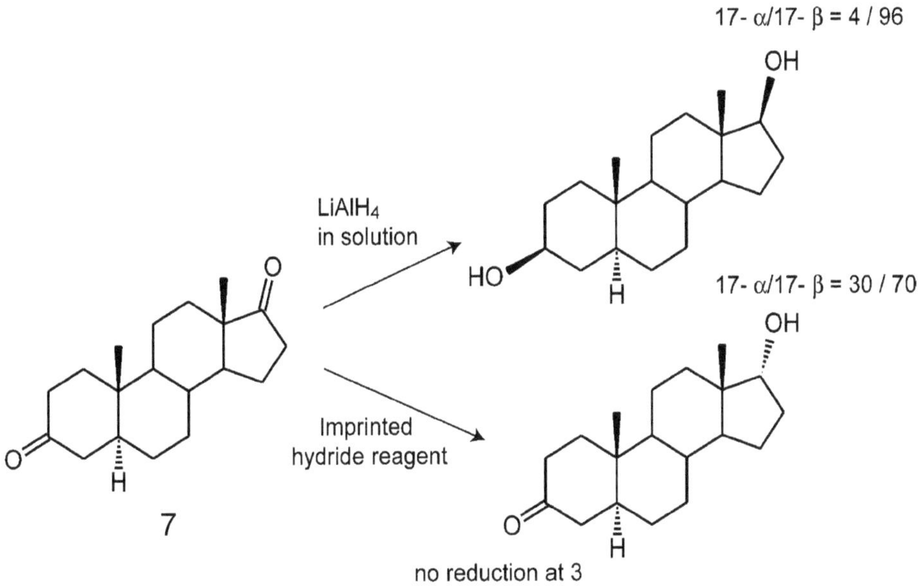

17-α/17-β = 4 / 96

OH

LiAlH₄ in solution

HO

H

17-α/17-β = 30 / 70

OH

O

Imprinted hydride reagent

H

O

H

7

O

no reduction at 3

Figure 5 Stereochemical consequence of the reduction of diketone **7** by the imprinted hydride reagent of Byström et al. (Ref. 9) or with a soluble reducing agent.

In this section we have shown that a simple semi-covalent approach, based on the polymerization of carboxylic esters of hydroxyl-functionalized templates, can be used to create imprint sites bearing carboxylic acid groups, in direct analogy to noncovalent imprinting with methacrylic acid as the functional monomer. These imprints are capable of binding analytes, via hydrogen bonding or ionic interactions, in a noncovalent sense and in some cases were highly selective towards the imprint molecules [4,9]. Nevertheless, the ester approach does suffer from some severe drawbacks, in some cases the template species were difficult [3] or impossible [9] to hydrolyse or only a small percentage (~20%) of the theoretical binding sites could be re-occupied [4]. In addition, the binding sites generated with this method cannot rebind the template in the exact orientation in which it was imprinted, as this would entail elements of both the monomer residues in the binding site and of the template occupying the same space within the binding site (Fig. 6). Methodologies that attempt to address this issue of steric crowding will be discussed in the following sections.

III. SEMI-COVALENT APPROACH USING SPACERS

To allow the imprinted cavity and the functional group interactions to work together in the binding site in a similar way to that evident in noncovalent imprinting, some kind of spacer group needs to be introduced between the template and functional group precursor, such that the rebound target molecule can both fit the "hole" left by the template and form a hydrogen bond with imprint site functionality without severely compromising either aspect of recognition. The key to this approach is the development of chemistries by which a spacer of suitable geometry can be introduced

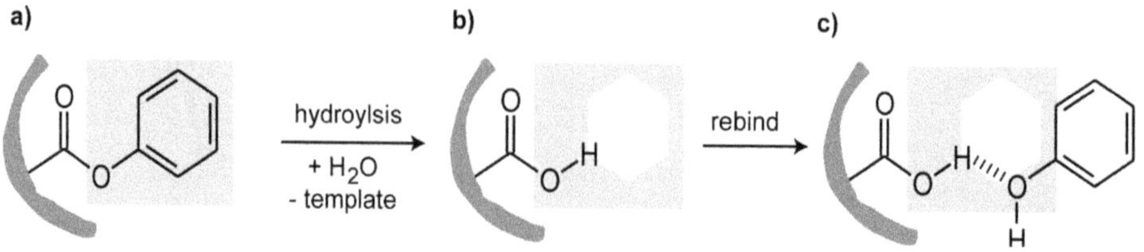

Figure 6 Steric crowding in the rebinding event when imprinting has been carried out with an ester template: (a) the ester group holds the template close to the wall of the imprint cavity; (b) cleavage, by acid or base, results in the addition of the elements of water to form acid and alcohol or phenol; (c) rebinding in the same position within the recognition cavity would require the –OH groups of the acid and alcohol to occupy the same space, therefore, hydrogen bonding can only occur when the template is displaced from its original position.

between the template and functional group precursors. This spacer will be lost (sacrificed) in the template cleavage step and hence this method is also known as the sacrificial spacer methodology. To date, there have been a growing number of reports describing the use of diverse spacer chemistries to introduce various functional groups into polymers to recognize a range of different compound classes.

A.   Carbonate Esters

The requirement of a sacrificial spacer can be met by the introduction of a carbonyl group between the template alcohol and a monomer residue, in the case where this polymerizable group is also an alcohol (or phenol), the template monomer would therefore be a carbonate ester. Base hydrolysis of the carbonate ester results in loss of the carbonyl group as CO_2 and cleavage of the template, as shown by the example of Fig. 7, depicting the imprinting of cholesterol [10]. This example, from our group, was the first reported use of a sacrificial spacer in the imprinting of an organic template, combining the advantages of covalent imprinting with a noncovalent recognition step.

Cholesterol (as cholesteryl chloroformate) was attached to 4-vinylphenol in a simple coupling reaction to give cholesteryl, (4-vinyl) phenyl carbonate, **8**. This was copolymerized with EDMA in a thermal polymerization with AIBN as the initiator. The resultant polymer was collected, broken-up and ground, in the same way as other imprinted polymer monoliths. At this stage, the template group was still covalently bound to the polymer. While the initial experiments were carried out on monolithic materials, the stable covalent nature of the template assembly means that, unlike most conventional noncovalent, or indeed covalent methodologies, where the presence of water should be avoided, the polymerization of carbonate ester templates can be carried out in aqueous dispersions by suspension [11] and emulsion [12] polymerization. Cholesterol was cleaved from these polymers by hydrolysis with methanolic NaOH, followed by an acidic work-up. The cholesterol liberated from the imprint sites was removed by thorough washing of the polymer particles.

Rebinding of cholesterol to the polymers occurs in nonpolar solvents such as hexane, the major interaction being the formation of a hydrogen bond between the polymer-bound phenol residue and the hydroxyl of cholesterol. This hydrogen bond can still be formed when the cholesterol is precisely reoccupying the space vacated

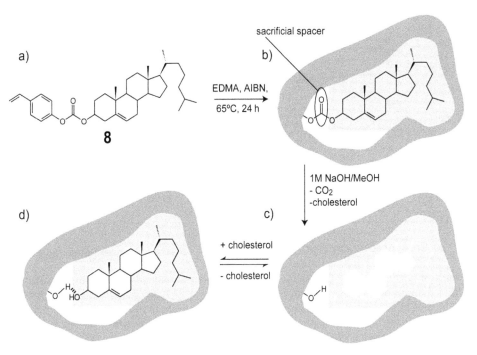

Figure 7 Sacrificial spacer method, as exemplified by the imprinting of cholesterol: (a) cholesteryl (4-vinyl)phenyl carbonate **8** is used as the template monomer to form a covalently imprinted polymer; (b) in the polymerization step, the carbonyl group of the carbonate ester holds the functional monomer and template oxygen atoms apart by two bond distances; (c) hydrolysis results in loss of the template and loss of the spacer as CO_2; (d) rebinding can now occur with the cholesterol ligand occupying essentially the same space as the template cholesteryl group (adapted from Ref. 10).

by the cholesteryl residue of the template (Fig. 7), which illustrates the advantage of the spacer approach (compare with Fig. 6). Although the hydrogen bond geometry is not the most energetically favorable linear $(O-H-O)$ arrangement of atoms, which might be expected in the "best" sites created by noncovalent imprinting, hydrogen bonding can occur over a wide range of angles and the "bent" bond shown here will still lead to binding. Hydrogen bonding by the phenol can be "switched-off" by modification of the polymer with acylating agents (such as acetyl chloride), converting the phenol to acetate ester groups with a corresponding reduction in ability of the polymer to bind cholesterol [10].

In order to demonstrate the imprinting effect, a control polymer, using phenyl (4-vinyl) phenyl carbonate as the template, was prepared under the same condition as the MIP. This was chemically identical to the cholesterol-imprinted polymer after template cleavage, but would have a much smaller imprint cavity associated with the phenol in the recognition site, too small to accommodate the cholesterol molecule. An additional control, prepared in the absence of either template was also made, to test the effect of the hydrolysis conditions on the polymer matrix. Both of these polymers in their hydrolyzed state and the cholesterol-imprinted polymer in its unhydrolyzed state (template still bound in the polymer) were unable to bind cholesterol,

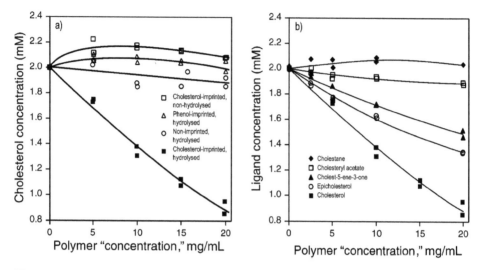

Figure 8 (a) Binding of cholesterol to cholesterol-imprinted and control polymers, from a 2 mM solution of cholesterol in hexane, as a function of polymer concentration. (b) Binding of cholesterol and various cholesterol analogues (2 mM) to the cholesterol-imprinted polymer, prepared by the sacrificial spacer method. Reprinted with permission from Journal of the American Chemical Society. Copyright 1995 American Chemical Society (Ref. 10).

while the hydrolyzed-imprinted polymer did (Fig. 8a). Furthermore the cholesterol MIP showed selective rebinding of cholesterol over close structural analogues (cholestane **9**, cholesteryl acetate **10**, epicholesterol **11** and cholest-5-en-3-one **12**, Figs. 8b and 9).

A prediction that one can make about the ideal semi-covalent approach is that the distribution of binding site affinities seen in noncovalent imprinting ought not to occur, as the distance between template and functional monomer is fixed during imprinting. The result should be a uniform binding affinity for the imprint sites. Furthermore, since there is no excess functional monomer, nonspecific binding should be relatively low, only being due to general adsorption to the surface of the polymer, or due to incidental functional group interactions such as with initiator residues or with methacrylic acid residues created by hydrolysis of a small fraction of the EDMA matrix. The binding site distribution for the cholesterol polymer was probed by determination of the isotherm for cholesterol binding (Fig. 10) and indeed shows a relatively uniform binding constant of around $1700 \, M^{-1}$ ($K_d = 0.59 \pm 0.12 \, mM$) and a capacity of $114 \pm 6 \, \mu mol \, g^{-1}$[10]. For a more detailed comparison of the isotherm and binding site distribution for this polymer and other typical noncovalently imprinted polymers, the reader is suggested to consult the papers by Umpleby et al. [13,14] and the chapter by Shimizu (Chapter 16).

One consequence of the uniform binding site distribution and high capacity of these polymers is that they may be rather better suited to the preparation of chromatographic (HPLC) stationary phases than noncovalently imprinted materials. In fact this was the conclusion arrived at by Hwang and Lee [15], who compared the chromatographic performance of cholesterol-imprinted polymers prepared by the semi-covalent (carbonate spacer) method with noncovalent MIPs incorporating

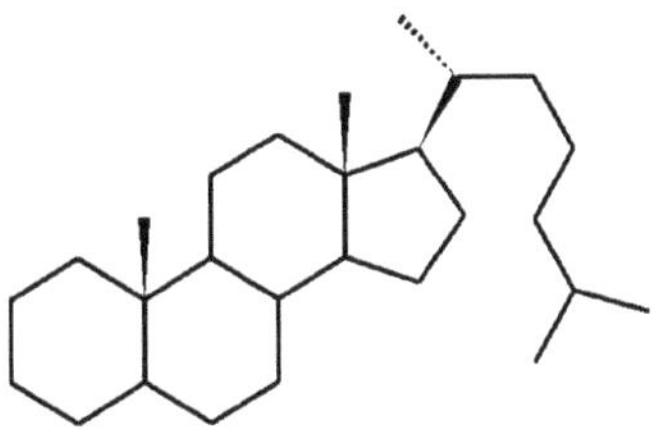

Figure 9 The structures of cholestane **9**, cholesteryl acetate **10**, epicholesterol **11** and cholest-5-en-3-one **12**.

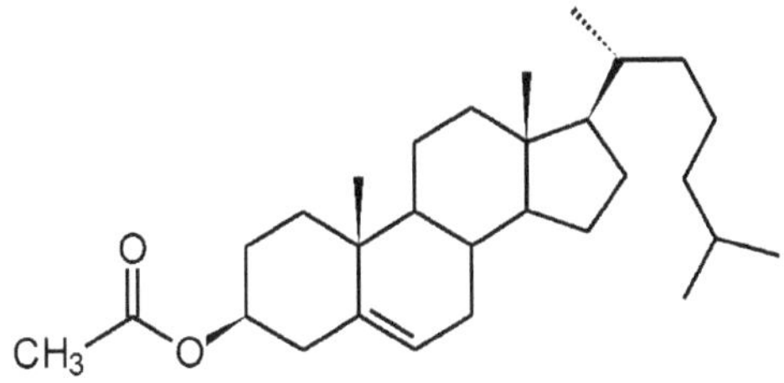

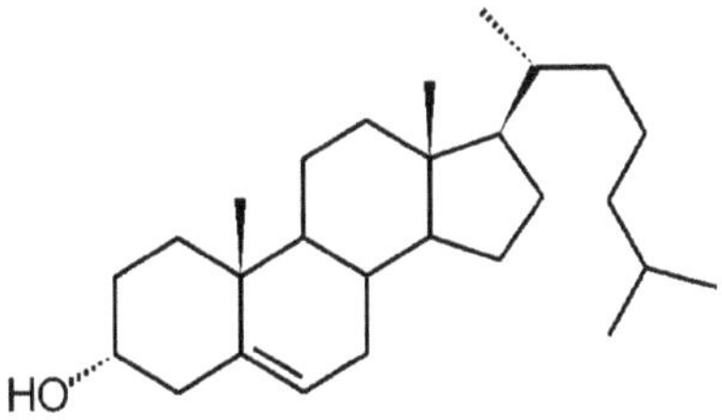

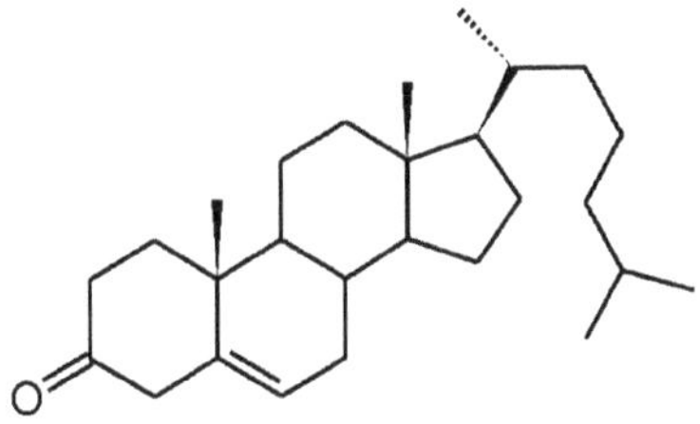

either 4-vinylpyridine (4-VP) or methacrylic acid (MAA) as functional monomer. All of the polymer stationary phases were able to resolve estradiol and cholesterol, but the peak shape with the semi-covalent MIP was much better (narrower peaks with little tailing) than with either of the noncovalently prepared materials. As well as an improved peak shape, this material also showed the highest selectivity factor and the greatest adsorption capacity for cholesterol and the highest chromatographic efficiency, with a theoretical plate number N of 1240, compared to 220 and 240 for the MAA and 4-VP-based materials, respectively. The authors concluded that the sacrificial spacer material has the potential to be used in quantitative analysis.

Highly selective MIP-coated quartz crystal microbalance (QCM) sensors have been prepared for nandrolone [16], a performance-enhancing substance that has been banned in international sports. The authors hope to address the need for a sensor to rapidly detect this substance of abuse. MIP was prepared with a 4-vinylphenol carbonate ester of the template, but also in the presence of methacrylic acid to target the ketone functionality. The sensor was able to detect nandrolone down to 0.2 ppm and was highly selective over closely related structures such as testosterone and epitestosterone.

Another example of this chemistry, which was also aimed at using imprinted polymers prepared by the sacrificial spacer method as the basis of a rapid quantitative assay, is the imprinting of propofol [17], the active constituent of an intravenous anesthetic. Propofol, **13**, Fig. 11 (2,6-diisopropylphenol), like cholesterol, has only a single hydroxyl group for recognition by polar functional monomers, the presence of the two bulky isopropyl groups however means that it is a poor candidate for noncovalent imprinting. Imprinting with the carbonate ester **14** was successful in creating polymers with good binding for propofol from solutions in methanol; moreover, grinding

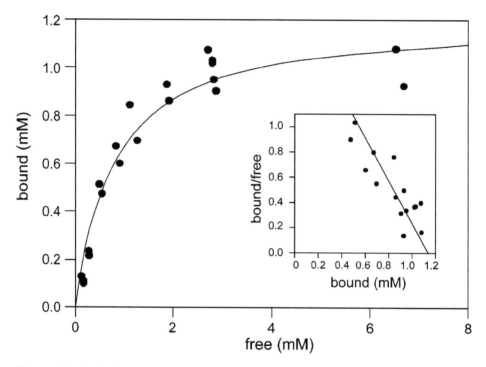

Figure 10 Binding isotherm and (inset) Scatchard plot for cholesterol adsorption by the cholesterol-imprinted polymer ($10\,mg\,mL^{-1}$), prepared by the sacrificial spacer method. Reprinted with permission from Journal of the American Chemical Society. Copyright 1995 American Chemical Society (Ref. 10).

to low particles sizes led to polymers with very low nonspecific binding ($<2\%$) and a rapid specific uptake of the analyte. The polymers were also reported to exhibit very low cross-reactivity with other phenols and require only a short incubation time.

Not only can phenols be introduced into the recognition site by this method but so can other hydroxy-monomers. For example, 2-hydroxyethyl methacrylate (HEMA) residues can be introduced [5] as their corresponding carbonate esters, **15** as can allyl alcohol residues [18], via allyl carbonates **16** (Fig. 12). Both these

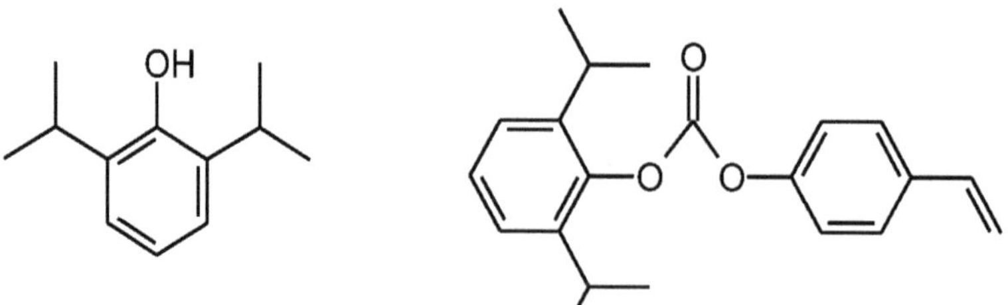

Figure 11 Structures of the intravenous anesthetic propofol **13** and the carbonate ester template of propofol **14** (Ref. 17).

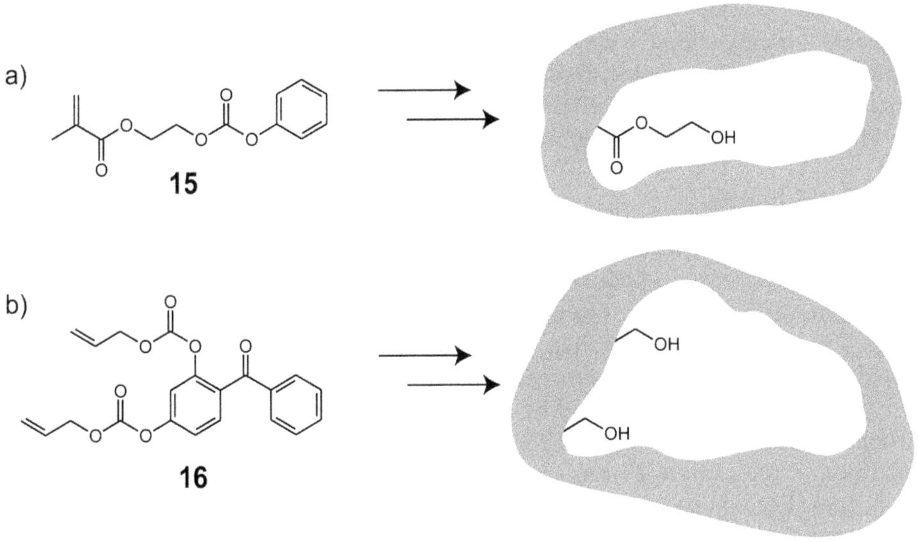

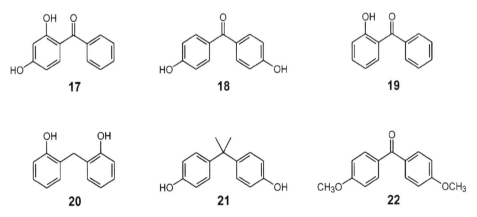

Figure 12 Carbonate ester templates: (a) based on 2-hydroxyethyl methacrylate **15** (adapted from Ref. 5); (b) based on allyl alcohol **16** (adapted from Ref. 18).

approaches produced MIPs that showed selectivity for their target molecule over structurally related compounds. The importance of the solvent in the rebinding step can be crucial. Whereas 2,4-dihydroxybenzophenone **17** was rebound only to the extent of ~40% to the MIP prepared with **16** in 1% acetic acid in methanol, 95% binding was observed in THF. Nevertheless, the selectivity of the MIP towards **17** over structural analogues **18–22** (Fig. 13) increased with increasing polarity of the solvent. Furthermore, the selectivity increased with structural differences no matter which solvent was used.

Figure 13 Structures of: 2,4-dihydroxybenzophenone **17**; 4,4′-dihydroxybenzophenone **18**; 2-hydroxybenzophenone **19**; 2,2′-dihydroxybenzophenone **20**; bisphenol A **21,** and 4, 4′-dimethoxybenzophenone **22**.

These examples show that the use of the carbonate carbonyl group as a sacrificial spacer is a highly effective way of preparing selective-imprinted polymers for both batch and chromatographic applications. The requirement for some chemistry in the template preparation is offset by a high capacity in the resultant polymer and a uniform binding affinity throughout the polymer.

B. Carbamates

Carbonyl spacers can also be introduced in the form of a carbamate (urethane) group. In this variation, either a hydroxyl [19] or an amino [20,21] functionality can be introduced into the binding site. The efficiency of this method was shown by Khasawneh et al. [19], who prepared an imprinted polymer for nortriptyline 23 (Fig. 14) by copolymerization of the template monomer 24 with TRIM.

The template was removed by basic hydrolysis leaving the phenolic hydroxyl group in the binding site. The MIP was then packed into a capillary column and the retention of the analyte 23 was compared with that of a broad range of structural analogues. For comparative reasons, noncovalently imprinted polymers were also synthesized with either MAA or 4-vinylphenol as the functional monomer. In all cases, 23 was more strongly retained than the structural analogues, no matter which imprinting approach was used. The control polymers (prepared either with no functional monomer or with functional monomer but no template present) did not show any retention of the compounds. However, as was seen (above) in the MIP-HPLC of cholesterol, the peak shape for the template molecule 23 was much improved on the semi-covalent MIP compared with the noncovalent MIPs (asymmetry factors of 1.72, 5.43, and 4.85 for the semi-covalent, MAA and vinylphenol MIPs, respectively). In common with Hwang and Lee [15], the authors concluded that the semi-covalent MIP stationary phase, prepared by a carbonyl spacer method, was suitable for quantitative determination of the template.

The carbamate approach is not limited to just organic polymeric materials, as was shown by Katz and Davis [20], Graham et al. [21] and Ki et al. [22] who used it

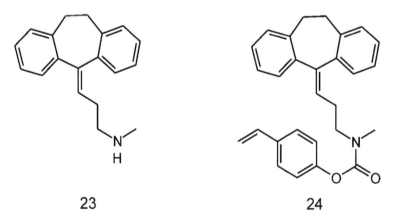

Figure 14 Structures of the tricyclic antidepressant nortriptyline 23 and the carbamate template monomer 24 used to prepare an imprinted capillary HPLC stationary phase (adapted from Ref. 19).

in sol–gel materials. In the first of these examples [20], a series of templates were used to introduce up to three amino groups, using templates such as **25**, in an imprinted porous silica gel (Fig. 15). The materials were prepared as shape-selective basic catalysts; for example, imprints of the two-point template (not shown) catalyzed the Knoevenagel condensation of malononitrile with isophthalaldehyde in a 1:1 stoichiometry, the addition of the second malononitrile being inhibited by the steric requirements of the imprinted site. The authors probed the structure of the imprinted sites by a number of techniques including ^{29}Si NMR and Ar porisimetry, in the latter experiment the additional pore volume created by the loss of the template was seen. While in this example the object was not to create a binding site for the template, as it was in the following examples, the study demonstrates a very thorough characterization of an imprinted material. For a more detailed description of imprinting in sol–gel materials, the reader should consult the chapter by Katz (Chapter 11).

In the study by Graham et al. [21], an analogue for DDT was used to prepare the imprinted structure **26** (Fig. 16). Treatment with LiAlH$_4$ removed the template to leave amino groups in the recognition site imprinted for DDT, **27**.

A slightly different approach was used by Ki et al. [22], who used the same isocyanate approach to prepare a template monomer based on estrone, **28**. This was used to prepare imprinted silica spheres of 1.5–3 μm in diameter, however, instead of cleaving the template by hydrolysis or reduction, the carbamate bond was decomposed to the starting materials, isocyanate and sterol, by heating to 180°C in DMSO, except the isocyanate was now bound in the imprinted site. The isocyanate was then either treated with water, in which case the amine was formed, or with ethylene glycol to form the hydroxyethyl carbamate (Fig. 17). Interestingly, the second strategy

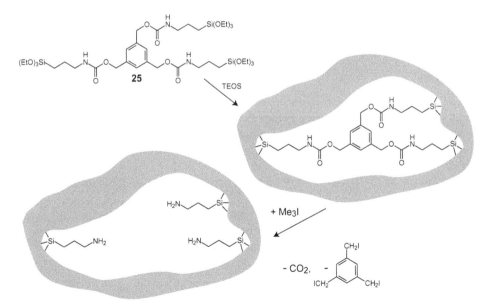

Figure 15 Schematic representation of the preparation of imprinted silica gel catalyst from the reactive silane monomer **25** (adapted from Ref. 20).

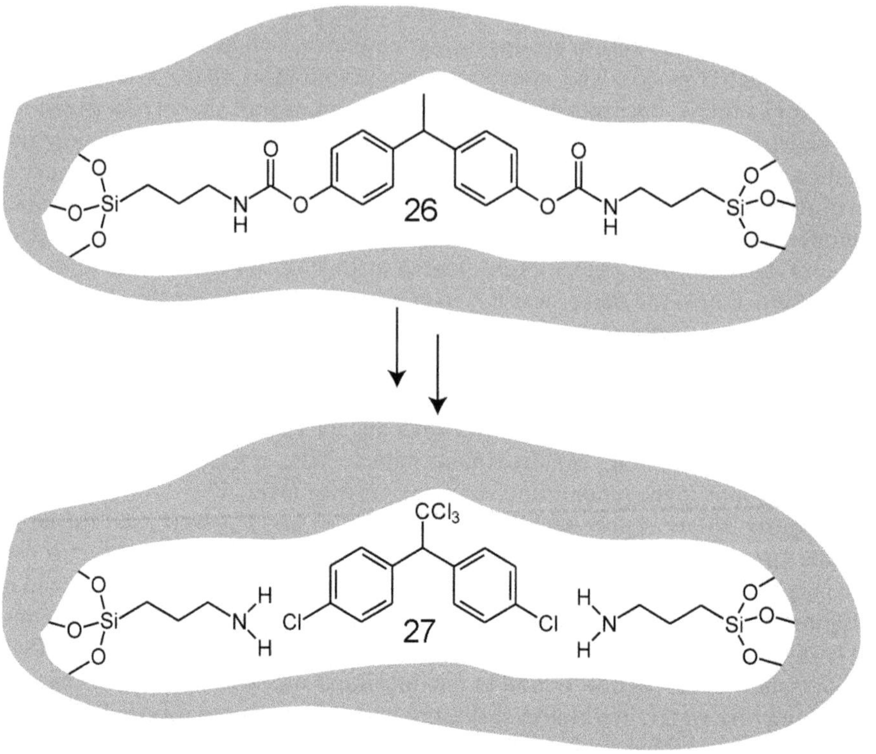

Figure 16 Schematic representation of the preparation of an imprinted silica **26** for the pesticide DDT **27** (adapted from Ref. 21).

gave a material which showed a stronger interaction with the template than the amine-functionalized binding site.

C. Urea

Urea linkages, where two nitrogen atoms are bound to the carbonyl group, can also be used, with the C=O group acting as sacrificial spacer in the template monomer. This is another method to introduce amine groups into the imprinted material. In the imprinting of dioxin derivatives, these groups were intended to form weak hydrogen bonds with aromatic chlorine [23].

Lübke et al. [23] used this method to prepare the template monomer **29** (Fig. 18), which was copolymerized with an additional monomer capable of forming a charge transfer complex with the electron-deficient aromatic dioxin analogue. This study showed that it was possible to use a combination of noncovalent and sacrificial spacer methodologies to prepare an imprinted material.

D. Salicylamide Spacer

The role of sacrificial spacer can also be filled by larger species, for example, salicylic acid. This approach is particularly useful in the imprinting of amine templates, as demonstrated by the model compound **30** (Fig. 19). This was prepared by the reaction

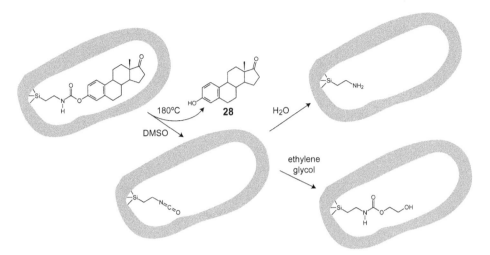

Figure 17 Scheme for the imprinting of estrone **28** in silica microspheres: the template is removed by thermal treatment (at 180°C), which regenerates the isocyanate. The isocyanate-derivatized imprinted silica can then either be treated with water, to generate the amine, or with ethylene glycol to form the hydroxyethyl carbamate in the imprinted site (adapted from Ref. 22).

of 2-methacryloyloxybenzoyl chloride with the appropriate amine (in this case, 4-hexadecylaniline). Infra-red and NMR spectroscopy were used to demonstrate the presence of the intramolecular hydrogen bond between the amide hydrogen atom and the methacylate oxygen atom [23]. This helps to control the relative conformation of template monomer and ensures that the methacrylic acid residue is placed to interact with functionality of the analyte after template cleavage. An added advantage is that the phenyl methacrylate ester bond is fairly labile and can be cleaved with base to effect template removal.

With this method, template monomers for dioxin [23] and a tripeptide [24] have been synthesized (Fig. 19). The template monomer **31** and **32** were both copolymerized with DVB. In the case of the tripeptide, vinylpyridine was added to form non-covalent interactions with the carboxylic acid groups of the peptide. The imprinted polymers bound the target molecule better than the control polymers in both cases.

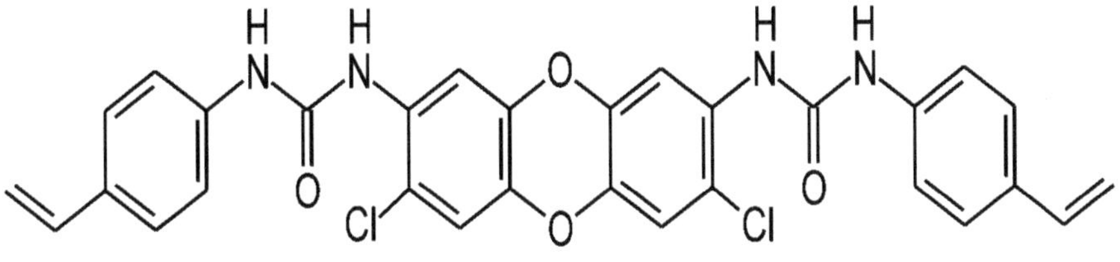

29

Figure 18 Structure of the diurea template **29** used to imprint 2,3,7,8-tetrachlorodibenzo-dioxin (TCDD) (Ref. 23).

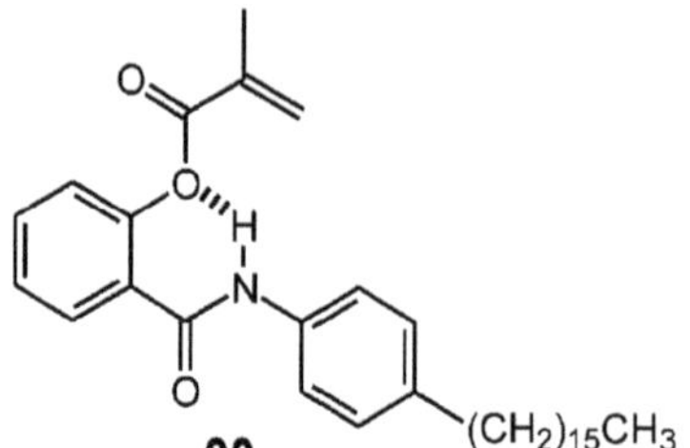

Figure 19 Templates prepared using salicylamide spacers and the model compound **30**: template for the imprinting of TCDD **31** (Ref. 23) and the tripeptide, Lys-Trp-Asp **32** (Ref. 24).

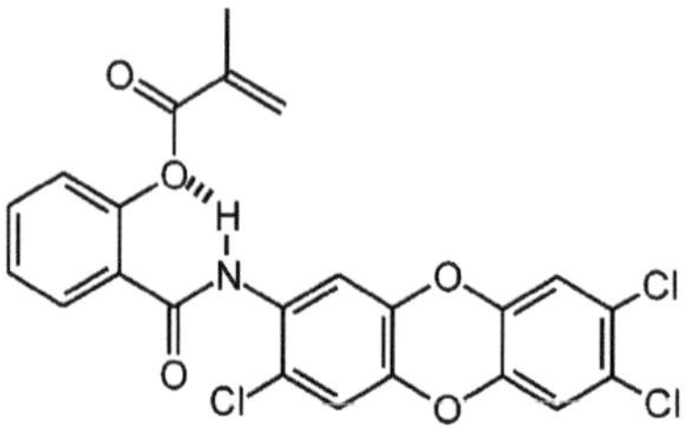

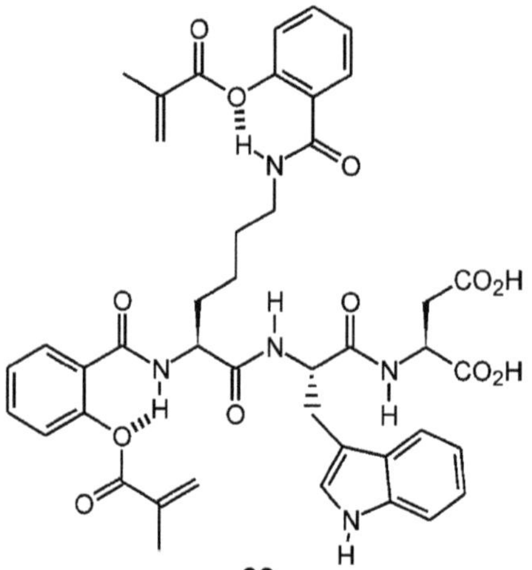

Furthermore, the peptide-imprinted polymer bound its target selectively over other tri-peptides, in addition, the polymer also showed selectivity for the *N*-terminal dipeptide subunit over other similar dipeptides (Table 1).

E. Silicon-Based Spacers

The sacrificial spacer methodology can also be used for molecules where it has proven difficult to generate MIPs by the conventional noncovalent and covalent

Table 1 Binding of Peptides to a Lys-Trp-Asp-Imprinted Polymer (MIP) and Control Polymer (NP), (1mM Peptide in Acetonitrile/Water 4:1 v/v, 30 mg mL^{-1} Polymer) (adapted from Ref. 24)

Peptide	Percentage bound	
	MIP	NP
Lys-Trp-Arg	43	11
Arg-Trp-Asp	24	9
Leu-Trp-Asp	< 2	6
Gln-Trp-Asp	< 2	< 2
Lys-Phe-Asp	4	< 2
Lys-Trp-Glu	5	3
Lys-Trp	35	17
Lys-Phe	9	5
Lys-Val	15	4

methods. Noncovalent imprinting of aromatic *N*-heterocycles was successful when two or more basic nitrogen atoms were present in the target analyte. However, for simple *N*-heterocycles, such as pyridine, no imprinting effect was observed with this method [25] due to high nonspecific binding to the control polymer. As mentioned earlier, the advantage of the sacrificial spacer method is a reduction in the nonspecific binding, since the functional monomers are only located in the imprint sites. Since *N*-heterocycles cannot readily be modified to form labile covalent bonds, a template analogue approach was developed. This analogue had to fulfill certain criteria: (i) the covalent bond should be strong enough to withstand the polymerization conditions but also be easy to cleave; (ii) the template should possess a group capable of acting as a hydrogen bond donor after cleavage; and (iii) the geometry had to be such that donor functionality would be correctly positioned for rebinding when the analyte occupies the space vacated by the template in the imprint sites. Our solution to this problem was to use silyl ethers **33** or silyl esters **34**, which appear to fulfill all the above criteria (Fig. 20).

Silyl ethers are widely used in organic chemistry as protecting groups for alcohols in organic synthesis, therefore, efficient methods for their synthesis and cleavage are known. As a template, structures such as **33** can be used to introduce a phenolic hydroxyl group, after polymerization and acid hydrolysis, capable of interacting with pyridine via a hydrogen bond [26]. In this way, the dimethylsilyl group is acting as a sacrificial spacer in much the same way as the carbonyl group in the carbonate ester templates.

While silyl ethers introduce an alcohol (or phenol) in the recognition site, silyl ester analogues such as **34** can be used to introduce a carboxylic acid group into the binding site [27], with similar spatial and hydrogen bonding implications for the heterocyclic analyte (Fig. 20). Both approaches were successful in creating affinity in imprinted polymers for the target pyridine after template removal. In the case of the silyl ether-imprinted materials, it was possible to increase the selectivity for pyridine

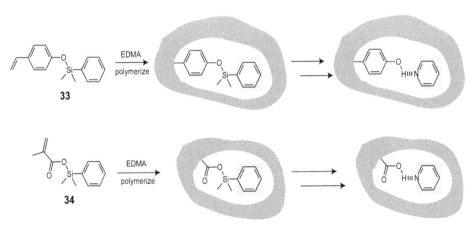

Figure 20 Scheme showing the imprinting of *N*-heterocycles using silyl ether **33** (Ref. 27) and silyl ester **34** (Ref. 29) templates. The dimethyl silyl groups acts as a sacrificial spacer.

over the larger homologues, quinoline and acridine, by selective modification of the binding site distribution with reagents of different steric bulk [26].

IV. PROTOCOL 1: PREPARATION OF 4-VINYLPHENOL [10,28]

A. Reagents and Equipment

4-Acetoxystyrene	CO_2 cylinder or some solid CO_2 (dry ice) in a stoppered Büchner flask as a CO_2 source
Potassium hydroxide	Rubber tubing and gas trap
Water	Filtration equipment
THF	Desiccator
250 mL conical flask	Glass tube or Pasteur pipette
Ice-bath	
Magnetic stirrer and follower	

B. Method

The following should be carried out in a fume hood: Weigh out 10 g of 4-acetoxystyrene into a conical flask and add a cooled solution of KOH (8.6 g) in water (85 mL) and 1 mL of THF. Add a magnetic follower, clamp the flask in an ice-bath and stir until the oily 4-acetoxystyrene has dissolved (up to 2 hr). If there are any gummy residues at this stage, it is best to filter the aqueous solution through a fluted filter paper into a clean flask before proceeding. Stand the flask in a fresh ice-bath and attach the source of CO_2 to a glass tube or Pasteur pipette and blow a steady stream of carbon dioxide through the solution. (Caution: when using a cylinder, always ensure that there is a trap between the cylinder valve and the reaction mixture, so that the flask contents cannot be sucked back into the cylinder if there is a drop in pressure.) The product, 4-vinylphenol, should crystallize out of the mixture as the gas dissolves in the solution and the pH falls. The phenol can be collected as shiny colorless plates by filtration on a Büchner funnel and washed with water. The product can be dried in a vacuum desiccator but should either be used within a day or two or stored in a freezer for longer periods of up to two or three weeks.

Optional. The crude phenol obtained by filtration can be purified for immediate use by dissolving in diethyl ether, drying the solution over $MgSO_4$, filtering and evaporating and recrystallization of the residue in hexane.

C. Comments

4-Vinylphenol is relatively unstable on storage and should be prepared before use.

V. PROTOCOL 2: PREPARATION OF CHOLESTERYL (4-VINYL) PHENYL CARBONATE [10]

A. Reagents and Equipment

4-Vinylphenol (see Protocol 1, above)	Magnetic stirrer and follower
Cholesteryl chloroformate	Ice-salt bath
Anhydrous THF	Dichloromethane
Anhydrous triethylamine	500 mL separating funnel
2,6-di-*tert*-butyl-4-methyl-phenol (BHT)	Magnesium sulfate (dried)
250 mL 3-necked ROUND-BOTTOMED FLASK fitted with CaCl$_2$ guard tubes, dropping funnel and thermometer	Rotary evaporator Propan-2-ol Filtration equipment Desiccator

B. Method

The following should be carried out in a fume hood: set up the round-bottomed flask, with dropping funnel, thermometer, and drying tube. Charge the flask with 4-vinylphenol (2 g) dissolved in a mixture of THF (60 mL) and triethylamine (4 mL). Add a few crystals of BHT as a precaution against premature polymerization. Cool the flask and contents in ice-salt bath, taking care not to introduce moisture by condensation by the use of guard tubes on the flask and dropping funnel. Charge the dropping funnel with a solution of cholesteryl chloroformate (7.5 g) in THF (40 mL) and add dropwise with stirring, such that the temperature in the flask does not exceed +5°C. During the addition, a thick white precipitate of triethylamine hydrochloride will form. Once the addition of the chloroformate is complete, the ice-bath can be removed. Stir for 2–3 h or overnight, whichever is convenient. Transfer the contents of the flask to a single-neck ROUND-BOT-TOMED FLASK and rotary evaporate to remove the THF and excess triethyla-mine. The temperature of the water-bath should be kept below 40°C. Dissolve the residue in dichloromethane and transfer to a separating funnel. Wash the lower organic layer with several portions of water and finally with saturated sodium chloride (brine). Carefully separate the organic layer after the final wash into a clean, dry flask and dry over MgSO$_4$. Filter the dried organics into a clean ROUND-BOTTOMED FLASK and evaporate. The residue can now be purified by recrystallization from aqueous acetone or propan-2-ol to give the template monomer.

C. Comments

We have repeated this preparation many times and have always found it reliable and the product is sufficiently pure to not require chromatographic purification. It is important to ensure that the product is free from 4-vinylphenol, as this may interfere in the polymerization. The molecular weight of the product is 532.88 g mol^{-1}.

VI. PROTOCOL 3: PREPARATION OF CHOLESTEROL-IMPRINTED POLYMER [10]

A. Reagent and Equipment

Cholesteryl (4-vinyl) phenyl carbonate (see Protocol 2, above)	Filtration equipment
	Sinter funnel
Ethyleneglycol dimethacrylate (EDMA)	50 mL tube with glass joint
Azo-bis-isobutyronitrile (AIBN)	Stopcock adaptor
n-Hexane (redistilled)	Vacuum/nitrogen line
Toluene (redistilled)	Dewar filled with liquid nitrogen
1 M NaOH solution	Temperature-controlled water-bath
Saturated aqueous sodium chloride solution (brine)	Grinding mill
Magnesium sulfate (dried)	Methanol
Activated neutral alumina (for chromatography)	Soxhlet apparatus
250 mL separating funnel	Vacuum oven

B. Method

The following should be carried out in a fume hood: recrystallize the initiator AIBN from methanol and store the obtained crystals at 4°C. Remove the inhibitor from the cross-linker as follows: add EDMA into a separation funnel and extract with three portions of NaOH solution (add saturated sodium chloride solution (brine) as necessary to encourage separation, the density of EDMA is close to that of water) and finally wash the monomer with pure brine. Carefully separate the monomer after the final wash into a clean, dry flask and dry over $MgSO_4$. Filter the dried monomer into a clean glass vial with screw top. The monomer can be stored in the fridge for several weeks. Prior to polymerization allow the EDMA to warm to room temperature, fill a short (3–4 × 1.5 cm) column with activated neutral alumina and pass through about 12–15 mL EDMA.

   To a 50 mL capacity quickfit test tube, add cholesteryl (4-vinyl)phenyl carbonate (1.24 g, 5 mol%), EDMA (8.76 g, 8.34 mL, 95 mol%), AIBN (0.15 g, 1 mol% wrt double bonds), toluene (2 mL), and n-hexane (18 mL). Close the tube with a stopcock adapter and connect to the vacuum line, leave the stopcock closed at this point. Carefully freeze the polymerization mixture in liquid nitrogen. Once the mixture is frozen, open the stopcock to evacuate the headspace while the tube is still standing in liquid nitrogen. After a few minutes, close the stopcock, lower away the liquid nitrogen Dewar and allow the mixture to thaw. To speed up the thawing process, a beaker of lukewarm water can be used (be careful as the mixture can bump in the tube). When the mixture has warmed to room temperature, gently release the vacuum by admitting nitrogen to the tube. Now close the stopcock and repeat the freeze–pump–thaw process for two more times. After the last thawing, leave the tube contents under reduced pressure and place the tube into a water-bath at 65°C to carry out the polymerization. Shake the tube frequently during the first few minutes, before the contents gel, in order to dissolve the monomer and ensure that the prepolymerization mixture is homogeneous. Clamp the tube, so that the liquid level is just below the level of the water-bath and make arrangements to compensate for evaporation of the water-bath contents, so that the water level does not drop appreciably during the polymerization period. (One method is to clamp an inverted 2 L flask, filled with water, with its mouth just touching the surface of the water-bath, in combination with floating plastic balls to reduce evaporation.)

After 24 h, take the tube out of the water-bath, allow it to cool, release the vacuum and remove the stopcock. In a fume hood, break up the brittle white solid with a spatula and wash with methanol onto a sintered glass funnel. Let the polymer air-dry and grind the lumps of polymer using a grinding mill. Transfer the ground polymer into a suitably sized cellulose thimble, place into a Soxhlet apparatus and extract with methanol to remove any unreacted monomers, initiator fragments and any other soluble material. Remove the thimble after 12–18 h and allow the remaining methanol to evaporate in a fume cupboard before drying in a vacuum oven at 70°C.

C. Comments

The quantities are for the preparation of 10 g of polymer. When working with a vacuum line always inspect the glassware for star-cracks and replace any suspect items before evacuating to high vacuum. Isohexane may be used as a direct replacement for n-hexane as it has a lower toxicity, but its use has not been tested in the above protocol.

D. Note

The mass of each monomer m_1, m_2 ..., etc., used in a polymer preparation of m_T g total mass, using n monomers of molecular weight M_1, M_2, ... and molar fractions X_1, X_2 ... etc., is easily calculated for each monomer using an equation of the following form, expressed here for monomer 1 (Formula 1):

$$m_1 = m_T \times \frac{M_1 \times X_1}{\sum_{i=1}^{i=n} M_i \times X_i} \qquad \text{(Formula 1)}$$

Consequently the mass of initiator, m_{In} of molecular weight M_{In}, used at a molar percentage of MP_{In} with respect to polymerizable double bonds is given by Formula 2, where d_i is the number of polymerizable double bonds per molecule of the ith monomer:

$$m_{In} = M_{In} \times \frac{MP_{In}}{100} \times \sum_{i=1}^{i=n} \frac{d_i \times m_i}{M_i} \qquad \text{(Formula 2)}$$

VII. PROTOCOL 4: TEMPLATE REMOVAL [10]

A. Reagents and Equipment

Cholesterol-imprinted polymer (see Protocol 3, above)	Condenser
Sodium hydroxide pellets	Magnetic stirrer-hotplate
Dilute HCl	Silicone oil bath
Deionized water	Sinter funnel
Methanol	Soxhlet apparatus
Hexane	Vacuum oven
Diethyl ether	Separating funnel
Magnesium sulfate (dried)	
50 mL round-bottomed flask	

B. Method

The following should be carried out in a fume hood: to 2 g of cholesterol-imprinted polymer (Protocol 3, above) in a round-bottomed flask equipped with a stirrer bar, add 1 g of sodium hydroxide pellets and 25 mL of methanol. Fit the condenser to the flask and lower into a preheated (90–100°C) oil bath on a stirrer-hotplate. Ensure that the stirrer is adequately dispersing the suspension and heat under reflux for 6 hr. Remove from the oil bath, separate the flask from the condenser and add the contents to an excess of dilute HCl to quench the reaction. Wash any remaining solid from the flask with a little extra methanol. Allow the mixture to stand for 15–20 min before collecting the polymer on a sinter funnel by filtration. Wash the collected solids carefully with water, followed by methanol, then with ether. Extract the polymer in a Soxhlet apparatus first with methanol and then with hexane. Dry the polymers as described above (Protocol 3). To determine the amount of cholesterol removed gravimetrically, transfer the hydrolysate and washings (filtrate) into a separation funnel. Add sufficient brine to ensure separation of the layers and extract the aqueous mixtures with three portions of diethyl ether. Combine the ether extracts, dry ($MgSO_4$) and filter into a preweighed flask, remove the solvent by rotary evaporation and dry the solid to a constant weight. The extracted solid can be verified as cholesterol by spectroscopic (NMR, IR) and chromatographic means (HPLC, TLC).

C. Comments

Since the template is covalently bound to the polymer at the outset, the polymer will lose mass as the template is removed, corresponding to the loss of one molar equivalent of cholesterol, the same of CO_2 and the addition of one molar equivalent of water. This must be taken into account when calculating the yield and theoretical capacity of the polymer.

VIII. PROTOCOL 5: PREPARATION OF 2-METHACRYLOYLOXYBENZOIC ACID [23]

A. Reagents and Equipment

Salicylic acid (2-hydroxybenzoic acid)	Ice-bath
Anhydrous pyridine	Dilute HCl
Methacrylic anhydride	Diethyl ether
250 mL ROUND-BOTTOMED FLASK and $CaCl_2$ guard tube	Separating funnel
	Anhydrous $MgSO_4$ (or Na_2SO_4) for drying
Magnetic stirrer and follower	Hexane

B. Method

The following should be carried out in a fume hood. Place the salicylic acid (25 g) in a round-bottomed flask with pyridine (90 mL) and the magnetic follower. It may be necessary to warm the flask, to dissolve the acid if it forms a solid lump at the bottom of the flask. Once the acid has been dissolved or adequately dispersed, cool the flask in an ice-bath with rapid stirring so that the reprecipitated acid crystals remain in suspension. When cold, add methacrylic anhydride (33.5 g), in a single addition, stir the mixture overnight, allowing the flask and contents to warm to room temperature as the ice melts. After stirring overnight, all traces of solid should have disappeared. Pour the contents onto an excess of dilute HCl (approx. 3 M) and crushed ice. Transfer the mixture to a separating funnel and extract the oily product with several portions of diethyl ether. The combined ether extracts should be washed with water (consider using $CuSO_4$ solution to aid in the removal of pyridine residues), dried ($MgSO_4$), filtered and evaporated. The resultant oil can be rather difficult to crystallize. Repeated recrystallization from hexane should eventually yield a pure product.

C. Comments

Seeding the mother liquor is often much more successful than waiting for spontaneous crystallization to occur in the recrystallization of this compound as it is very prone to forming an oil. Keeping a few crystals aside to seed the next batch is recommended.

IX. PROTOCOL 6: PREPARATION OF 2-METHACRYLOYLOXYBENZOYL CHLORIDE [23]

A. Reagents and Equipment

2-Methacryloyloxybenzoic acid (see above)	Gas scrubber (or an inverted funnel, just in contact with a large bowl of water,
Oxalyl chloride	connected by flexible tubing to the
N,N-dimethylformamide (DMF)	neck of the reaction flask)
50 mL ROUND-BOTTOMED FLASK	Water pump (aspirator)
	High-vacuum pump and trap
Magnetic stirrer and follower	Vacuum distillation apparatus

B. Method

The following should be carried out in a fume hood. Place a portion of the acid (between 2 and 10 g) in the flask with sufficient oxalyl chloride to react with the acid (at 1:1 stoichiometry) and an excess amount to enable efficient stirring of the flask. Add the magnetic follower and fit the gas trap. While stirring the mixture, briefly open the top of the flask and admit only one or two drops of DMF, at which point the reaction will start and gases (HCl, CO_2, and CO) will be generated. Ensure that the water in the gas trap cannot be sucked back into the flask by ensuring that the funnel just touches the water surface. The acid will dissolve through the course of the reaction. When gas production has ceased, remove the excess reagent by applying vacuum from a water pump (aspirator), this will ensure that the volatile and corrosive oxalyl chloride is hydrolyzed in a large excess of water. Transfer the oily residue to a distillation setup and distil at high vacuum. The pure compound is a mobile colorless oil.

X. NEWER APPROACHES IN SEMI-COVALENT IMPRINTING

A. Disulphide-Linked Templates

A recent development in semi-covalent imprinting involves the use of disulphide template monomers, such as allyl phenyl disulfide, **35**, Fig. 21. This was reported by Mukawa et al. [29], as an effective method for the imprinting of phenol templates. The polymer-bound mercaptan (thiol) residue is revealed in the site by reduction of the disulphide with sodium borohydride, with the loss of thiophenol. A comparison of retention factors for phenol, thiophenol, aniline, and pyridine showed that the polymer possessed the highest imprinting effect for phenol, in agreement with the calculated bonding energies for the formation of a hydrogen bond between the analytes and methyl mercaptan. The conclusion was that this method would be generally applicable to the imprinting of phenols. This method is somewhat in between the true "spacer" approach and the "direct connection" semi-covalent approach, as exemplified by the use of an ester template. The phenolic oxygen of the ligand and polymer-bound sulfur are separated by the distance of the S–S bond in the template design, but no allowance is made for the hydrogen atoms of the –OH or –SH groups. This is obviously not too severe a restriction on the rebinding and a slight conformational change to achieve binding can take place in the recognition site.

An extension of this methodology has also been reported by the same group [30]. By oxidation of the polymer-bound mercaptan to a sulfonic acid with hydrogen peroxide in acetic acid, a polymer of very different binding characteristics can be formed from the same type of precursor. In contrast, the polymers bearing –SH groups, the charged sulfonic acid imprints are suitable for the imprinting of amines.

B. Dendrimer-Based Monomolecular Imprints

Wendland and Zimmerman [31] had remarked in their 1999 paper that what they referred to as "cored dendrimers" represented a system closely analogous to polymer imprinting, but within the framework of a discrete single molecule. In their approach, a dendrimer is prepared by the assembly of branched polyether "dendrons", around a central core by esterification. The periphery of the dendrimer is fully substituted

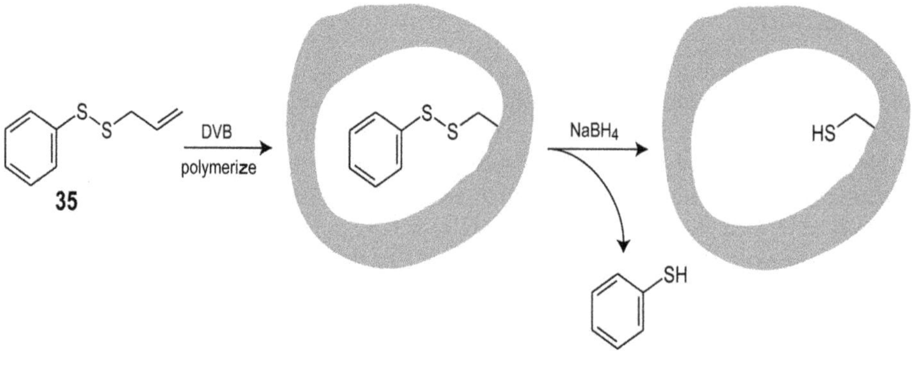

Figure 21 The preparation of thiol-functionalized imprinted polymer selective for phenols by the polymerization of a disulfide template **35** (adapted from Ref. 29).

with olefin groups, in the form of but-4-enyl ethers. If these structures are treated with an olefin metathesis catalyst (Grubb's catalyst), cross-linking of the outer core occurs. This cross-linking is reversible in the presence of the catalyst and so continual bond making and breaking allow the lowest energy forms of the cross-linked dendrimer to form. By control of the conditions of the reactions, mainly intramolecular (rather than intermolecular), cross-linking occurs, resulting in the formation of individual cross-linked dendrimer structures. Cleavage of the ester bonds, by which the outer shell is attached to the core, allows the core to be removed leaving a specific arrangement of carboxyl groups surrounding the central cavity of the dendrimer.

The implications for imprinting were set out in a more recent example [32] (Fig. 22) where a porphyrin molecule bearing four dihydroxyphenyl substituents was used as the core. The complete dendrimer was prepared by reaction with eight "wedges" of generation 3 dendrons, such that the final molecule 36 bore 64 olefin residues at the surface. Cross-linking, 37 followed by removal of the core (template), gave a single, soluble polymer molecule with a central cavity bearing eight benzoic acid residues 38. The authors admitted that the core structure was not likely to rebind to the cavity, on account of severe steric crowding. They concluded however that other porphyrins, bearing isomeric phenols, pyridine or pyrimidine residues in place of the 3,5-dihydroxy substituents of the template, would bind. An apparent binding constant K_{app} was measured for a range of

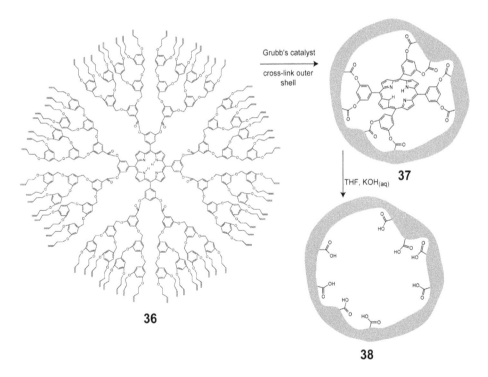

Figure 22 Dendrimer **36** used in the preparation of "cored-dendrimer" unimolecular imprints, by olefin metathesis to the covalently linked template cavity within a cross-linked shell **37** and removal of the template "core" to yield the binding cavity **38** (adapted from Ref. 32).

porphyrin analogues of the central core molecule. The highest binding affinity was seen when 3-pyridyl or 4-pyridyl groups decorated the porphyrin core. A surprising result was that it was necessary to have functionality on all four aromatic groups for binding to occur, as porphyrins bearing one, two, or three pyridyl groups (the remainder being phenyl) were not bound, whereas the compound bearing four pyridines was strongly bound.

C. Ring-Opening Metathesis Polymerization with a Semi-Covalent Imprinting Strategy

As was shown in the previous section, methathesis polymerization is very effective at optimizing the structure of the polymer matrix by the fact that it is a reversible, and therefore thermodynamically controlled process. This has the potential to allow a much better "fit" for the template to be created than is likely in a kinetically controlled free radical polymerization. New developments in this field could lead the way to high affinity MIPs with narrower binding site populations than are currently achieved. In a first step in this direction, Patel et al. [33] have prepared bulk MIPs by ring-opening metathesis polymerization (ROMP), using dicyclopentadiene 39 as the matrix-forming monomer. The authors elected to use a covalently bound template in these initial studies for simplicity (avoiding the use of functional monomers) and

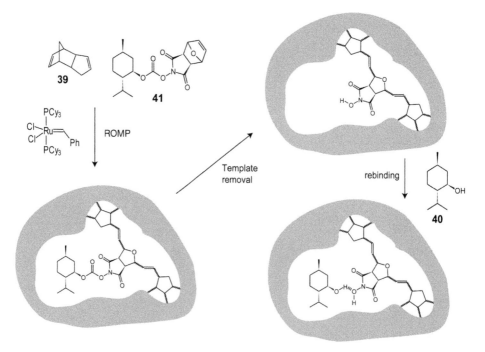

Figure 23   The preparation of enantioselective MIPs, imprinted with L-menthol **40** by ring-opening metathesis polymerization (ROMP). A copolymer of dicyclopentadiene **39** and the template monomer **41** was prepared by treatment with Grubb's catalyst. Template removal by treatment with an amine resulted in polymers capable of enantioselective recognition of **40** (adapted from Ref. 33).

consequently chose to use the carbonate spacer approach to imprint L-menthol **40** as the template monomer **41** (Fig. 23). Polymerization gave the expected ROMP-MIP in quantitative yield and the template was removed by treatment with an amine after grinding, also in high yield.

Batch binding with the L-menthol-imprinted ROMP-MIP and an equimolar mixture of D- and L-menthol in hexane:chloroform (1:1 v/v), resulted in an uptake of around 75% of the available ligand in the ratio 60% L-menthol to 40% D-menthol. This result was confirmed both by analysis of the supernatant and of the menthol desorbed from the polymer, which showed equal and opposite enrichment of the two enantiomers. In addition, D- and DL-imprinted polymers were prepared, which showed the expected behavior, both polymers bound the same amount of template, the D-ROMP-MIP showing preferential binding of D-menthol and the DL-ROMP-MIP exhibiting no enantioselectivity. The enrichment of the imprinted isomer in these simple uptakes experiments corresponded to an enantiomeric excess (ee) of almost 20% and a separation factor (α) of 2.1 was calculated, which is superior to many chiral stationary phases. The enantioselectivity of this polymer is remarkable when one considers that binding occurs only through hydrogen bonding to a single alcohol functionality within the recognition sites. Chiral recognition is therefore entirely determined by the polymer matrix and is totally steric in nature. These results certainly show the potential of ROMP to be a powerful technique in MIP synthesis, particularly in combination with the semi-covalent (sacrificial spacer) template approach.

XI. CONCLUSIONS

The semi-covalent method in its modern guise offers a versatile approach to the synthesis of polymers selective to a large number of template types. Obvious advantages such as tolerance to more severe polymerization conditions and the presence of water, the avoidance of an excess of functional monomer and the higher capacity and uniform binding constant, have to be offset against the need for some organic chemistry to design and synthesize a suitable template. This naturally imposes a barrier to its wider acceptance as a method of choice for the imprinting of certain templates. If the chemical companies could be persuaded to list some of the key intermediates (such as 4-vinylphenyl chloroformate) in their catalogues, the uptake of this method would be much greater. What has become clear to us in researching this chapter is that polymers based on the carbonyl "sacrificial spacer" approach, initially developed in our laboratory, may be ideally suited to chromatographic applications such as HPLC and capillary electrochromatography. While semi-covalent imprinting is never likely to displace noncovalent imprinting as the method of choice, its unique characteristics will guarantee it a place in the future of MIP technology.

REFERENCES

1. Damen, J.; Neckers, D.C. Memory of synthesized vinyl polymers for their origins. J. Org. Chem. **1980**, *45*, 1382–1387.
2. Shea, K.J.; Thompson, E.A. Template synthesis of macromolecules. Selective functionalization of an organic polymer. J. Org. Chem. **1978**, *43*, 4253–4255.

3. Cheong, S.H.; McNiven, S.; Rachkov, A.E.; Levi, R.; Yano, K.; Karube, I. Testosterone receptor binding mimic constructed using molecular imprinting. Macromolecules **1997**, *30*, 1317–1322.

4. Joshi, V.P.; Karmalkar, R.N.; Kulkarni, M.G.; Mashelkar, R.A. Effect of solvents on selectivity in separation using molecularly imprinted adsorbents: separation of phenol and bisphenol A. Ind. Eng. Chem. Res. **1999**, *38*, 4417–4423.

5. Joshi, V.P.; Karode, S.K.; Kulkarni, M.G.; Mashelkar, R.A. Novel separation strategies based on molecularly imprinted adsorbents. Chem. Eng. Sci. **1998**, *53*, 2271–2284.

6. Sellergren, B.; Shea, K.J. Influence of polymer morphology on the ability of imprinted network polymers to resolve enantiomers. J. Chromatogr. A **1993**, *635*, 31–49.

7. Caro, E.; Masqué, M.; Marcé, R.M.; Borrull, F.; Cormack, P.A.G.; Sherrington, D.C. Non-covalent and semi-covalent molecularly imprinted polymers for selective on-line solid-phase extraction of 4-nitrophenol from water samples. J. Chromatogr. A **2002**, *963*, 169–178.

8. Sellergren, B.; Andersson, L. Molecular recognition in macroporous polymers prepared by a substrate-analog imprinting strategy. J. Org. Chem. **1990**, *55*, 3381–3383.

9. Byström, S.E.; Börje, A.; Akermark, B. Selective reduction of steroid 3-ketones and 17-ketones using LiAlH$_4$ activated template polymers. J. Am. Chem. Soc. **1993**, *115*, 2081–2083.

10. Whitcombe, M.J.; Rodriguez, M.E.; Villar, P.; Vulfson, E.N. A new method for the introduction of recognition site functionality into polymers prepared by molecular imprinting—synthesis and characterization of polymeric receptors for cholesterol. J. Am. Chem. Soc. **1995**, *117*, 7105–7111.

11. Flores, A.; Cunliffe, D.; Whitcombe, M.J.; Vulfson, E.N. Imprinted polymers prepared by aqueous suspension polymerization. J. Appl. Polym. Sci. **2000**, *77*, 1841–1850.

12. Pérez, N.; Whitcombe, M.J.; Vulfson, E.N. Molecularly imprinted nanoparticles prepared by core-shell emulsion polymerization. J. Appl. Polym. Sci. **2000**, *77*, 1851–1859.

13. Umpleby, R.J.; Bode, M.; Shimizu, K.D. Measurement of the continuous distribution of binding sites in molecularly imprinted polymers. Analyst **2000**, *125*, 1261–1265.

14. Umpleby, R.J.; Baxter, S.C.; Chen, Y.Z.; Shah, R.N.; Shimizu, K.D. Characterization of molecularly imprinted polymers with the Langmuir–Freundlich isotherm. Anal. Chem. **2001**, *73*, 4584–4591.

15. Hwang, C.C.; Lee, W.C. Chromatographic characteristics of cholesterol-imprinted polymers prepared by covalent and non-covalent imprinting methods. J. Chromatogr. A **2002**, *962*, 69–78.

16. Percival, C.J.; Stanley, S.; Braithwaite, A.; Newton, M.I.; McHale, G. Molecular imprinted polymer coated QCM for the detection of nandrolone. Analyst **2002**, *127*, 1024–1026.

17. Petcu, M.; Cooney, J.; Cook, C.; Lauren, D.; Schaare, P.; Holland, P. Molecular imprinting of a small substituted phenol of biological importance. Anal. Chim. Acta. **2001**, *435*, 49–55.

18. Joshi, V.P.; Kulkarni, M.G.; Mashelkar, R.A. Molecularly imprinted adsorbents for positional isomer separation. J. Chromatogr. A **1999**, *849*, 319–330.

19. Khasawneh, M.A.; Vallano, P.T.; Remcho, V.T. Affinity screening by packed capillary high performance liquid chromatography using molecular imprinted sorbents II. Covalent imprinted polymers. J. Chromatogr. A **2001**, *922*, 87–97.

20. Katz, A.; Davis, M.E. Molecular imprinting of bulk, microporous silica. Nature **2000**, *403*, 286–289.

21. Graham, A.L.; Carlson, C.A.; Edmiston, P.L. Development and characterization of molecularly imprinted sol–gel materials for the selective detection of DDT. Anal. Chem. **2002**, *74*, 458–467.

22. Ki, C.D.; Oh, C.; Chang, J.Y. The use of a thermally reversible bond for molecular imprinting of silica spheres. J. Am. Chem. Soc. **2002**, *124*, 14838–14839.

23. Lübke, M.; Whitcombe, M.J.; Vulfson, E.N. A novel approach to the molecular imprinting of polychlorinated aromatic compounds. J. Am. Chem. Soc. **1998**, *120*, 13342–13348.

24. Klein, J.U.; Whitcombe, M.J.; Mulholland, F.; Vulfson, E.N. Template-mediated synthesis of a polymeric receptor specific to amino acid sequences. Angew Chem. Intl. Ed. **1999**, *38*, 2057–2060.

25. Andersson, H.S.; Koch-Schmidt, A.-C.; Ohlson, S.; Mosbach, K. Study of the nature of recognition in molecularly imprinted polymers. J. Mol. Recogn. **1996**, *9*, 675–682.

26. Kirsch, N.; Alexander, C.; Lübke, M.; Whitcombe, M.J.; Vulfson, E.N. Enhancement of selectivity of imprinted polymers via post-imprinting modification of recognition sites. Polymer **2000**, *41*, 5583–5590.

27. Kirsch, N.; Alexander, C.; Davies, S.; Whitcombe, M.J. Sacrificial spacer and non-covalent routes towards the molecular imprinting of "poorly functional" *N*-heterocycles. Anal. Chim. Acta. **2004**, *504*, 63–71.

28. Corson, B.B.; Heintzelman, W.J.; Schwartzman, L.H.; Tiefenthal, H.E.; Lokken, R.J.; Nickels, J.E.; Atwood, G.R.; Pavlik, F.J. Preparation of vinylphenols and isopropenyl-phenols. J. Org. Chem. **1958**, *23*, 544–549.

29. Mukawa, T.; Goto, T.; Nariai, H.; Aoki, Y.; Imamura, A.; Takeuchi, T. Novel strategy for molecular imprinting of phenolic compounds utilizing disulfide templates. J. Pharm. Biomed. Anal. **2003**, *30*, 1943–1947.

30. Mukawa, T.; Goto, T.; Takeuchi, T. Post-oxidative conversion of thiol residue to sulfonic acid in the binding sites of molecularly imprinted polymers: disulfide based covalent molecular imprinting for basic compounds. Analyst **2002**, *127*, 1407–1409.

31. Wendland, M.S.; Zimmerman, S.C. Synthesis of cored dendrimers. J. Am. Chem. Soc. **1999**, *121*, 1389–1390.

32. Zimmerman, S.C.; Wendland, M.S.; Rakow, N.A.; Zharov, I.; Suslick, K.S. Synthetic hosts by monomolecular imprinting inside dendrimers. Nature **2002**, *418*, 399–403.

33. Patel, A.; Fouace, S.; Steinke, J.H.G. Enantioselective molecularly imprinted polymers via ring-opening metathesis polymerisation. Chem. Commun. **2003**, 88–89.

6

The Use of Metal Coordination for Controlling the Microenvironment of Imprinted Polymers

Peter G. Conrad II and Kenneth J. Shea   University of California, Irvine, California, U.S.A.

I. INTRODUCTION

Biological receptors and enzymes possess sites for molecular recognition and catalysis [1]. Many of these biological macromolecules contain metal ions, which serve both structural and catalytic functions by orchestrating the three-dimensional arrangement of ligands and by providing the focus for reactivity [2]. The use of metal ion coordination is an extremely powerful method for organizing functional groups to create recognition sites and selective catalysts. In this chapter, we examine the application of metal–ligand interactions as an organizational motif for controlling the microenvironment of imprinted polymers. This strategy has been employed to control both reactivity and selectivity by modulating the coordination sphere of the catalytic center, and as a method to achieve an ordered three-dimensional array of functional groups in a binding site.

Synthetic polymers have often been used to organize arrays of functional groups in the hope of achieving enzyme-like activity [3]. Refinements in our understanding of the mechanism of enzyme catalysis allowed the design of more sophisticated polymer enzyme mimics. One strategy that allows polymers to be fabricated with selective active sites that closely resemble those found in enzymes and proteins is molecular imprinting. Recognition sites in imprinted polymers are created during the polymerization reaction due to favorable interactions between polymerizable functional monomers and a template molecule. In noncovalent imprinting this is due to the low K_a between template and functional monomer. To overcome the low association constants, a large stoichiometric excess of functional monomer relative to the template is typically employed. This results in the formation of small populations of high fidelity binding sites, but this also increases the number of nonspecific binding sites. This effectively broadens the distribution of binding sites leading to a poorer performance. An alternative is to use stronger binding interactions, such as metal–ligand coordination, which can achieve an ordered three-dimensional orientation of functional groups in the molecularly imprinted polymer matrix with fewer nonspecific binding sites.

The well-defined coordination sphere of many transition metal complexes allows control of the orientation of functionality in imprinted polymers. This organizational motif can be used to position ligands around the binding site to create a defined second coordination sphere with ligands that may not be directly involved in binding at the metal center. In addition, metal–ligand binding interactions are typically stronger than noncovalent interactions utilized in molecular imprinting. Therefore, the stronger interactions between the functional monomer and template ligand will reduce the number of randomly oriented functional groups in the polymer, effectively reducing nonspecific interactions. These stronger interactions could allow the use of stoichiometric amounts of template and functional monomer.

There are several guidelines that should be considered for generating metal-coordinated sites in MIPs:

- Compatibility with polymerization conditions: polymerization is often carried out under free radical conditions; therefore, the metal center should not inhibit the initiation or propagation steps.
- Metal centers should possess a well-defined coordination sphere. Several transition metals have very labile coordination geometries that can result in binding sites with arrays of functionality that are not well defined.
- Metal–template binding should be stable under polymerization conditions yet labile enough to allow removal of substrates/templates.

This chapter examines studies that employ transition metal containing imprinted polymers for selective transformations, recognition, and separation processes. This chapter also attempts to summarize the current state of the imprinting field and identifies possible future directions.

II. SELECTIVE ORGANIC TRANSFORMATIONS

The number of useful synthetic transformations that employ transition metal complexes as a reagent has grown significantly in the last decade in large part due to increased selectivity and reactivity of these in new catalysts. Successful transition metal catalysts often employ complex ligand structures to maintain a highly ordered first coordination sphere, which creates a well-defined reaction site. Highly ordered ligands are also utilized in metalloenzymes to create active sites with high selectivity and activity. Metalloenzymes possess an advantage over synthetic transition metal catalysts, however. In addition to a well-defined first coordination sphere, they also possess a second coordination sphere that surrounds the active site. The second coordination sphere is composed of peptide residues that are not directly connected to the metal center. These contribute to the activity and selectivity of the metalloenzyme.

There have been attempts to improve synthetic catalysts by creating an artificial second coordination sphere. These efforts have included covalently attaching a metal center to a macromolecule or by incorporating the transition metals into hosts, such as zeolites [4]. Not surprisingly, there has also been considerable activity focused on the development of catalytically active metal complexes with imprinted polymers. The polymerization process develops a matrix around the active site that can influence the recognition and selectivity. The following discussion examines examples in which molecularly imprinted polymers have been used to define a second coordination sphere for transition metal catalysts.

A. Hydride Transfer Reductions

Recent refinements in hydride transfer reductions have enhanced the utility of oxazaborolidine- and BINAP-Ru (II) complex-catalyzed reductions. A review by Wills describes the development of catalysts for the syntheses of chiral nonracemic secondary alcohols from aryl ketones [5]. Among the more interesting catalysts discussed were η^6-arene ruthenium complexes, which utilize diamine and monotosylated diamine ligands.

A proposed mechanism, shown in Scheme 1, involves a six-member cyclic transition state between the aryl ketone and the active form of the catalyst, 2 [6]. The stable catalyst precursor 1 is transformed to the active catalyst, 2, through the loss of HCl. Treatment with 2-propanol forms ruthenium hydride 3 as a single diastereomer. Complexation of an aryl ketone precedes the hydride transfer step, which results in the reduced product. The mild reaction conditions make this catalyst an excellent candidate for incorporation in an imprinted network. The reported enantiometric excesses (ee's, +90%) serve as a useful benchmark to evaluate the influence of the imprinted polymer on the reduction. To the extent that the ruthenium center is situated in an imprinted cavity, the MIP can influence the approach of the ketone to the metal ion or better accommodate a specific reduction product.

Initial attempts were made by ter Halle et al.[7] to convert the homogeneous ruthenium catalyst into a heterogeneous catalyst by incorporating the metal center in a solid support using polymerizable diamine ligands. Unfortunately, both the conversions and % ee's for the transfer hydrogenations were low compared to the

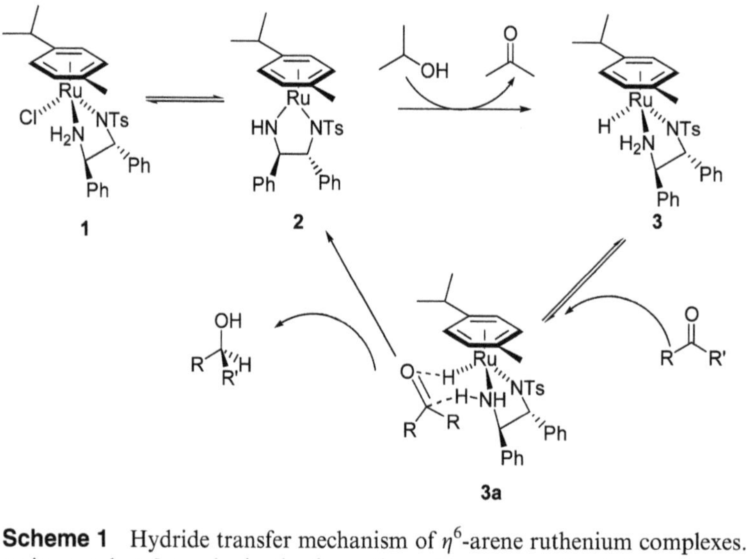

Scheme 1 Hydride transfer mechanism of η^6-arene ruthenium complexes. Reduction of the active catalyst **2** results in the formation of **3**. Approach of the incoming ketone proceeds through a six-membered transition state to form the secondary alcohol and regenerate **2**.

homogeneous catalysts. Polborn and Severin [8] proposed that the performance of the heterogeneous version of the ruthenium catalyst would be improved by incorporating the reactive catalyst in a selective binding site using imprinting techniques. The overall effect would be an increased ability to control the microenvironment of the catalyst. To this end, they reported the development of a catalytically active metal complex with a spectator ligand with one or more polymerizable side chains coordinated to a pseudosubstrate as a template molecule.

The template should be similar to the actual substrate to achieve selective recognition of the substrate. Imprinting a metal complex that resembles the transition state of the catalytic transformation can lead to stereo control of the reaction process. The phosphinato complex 7 was chosen as a template to mimic the transition state of the hydride transfer (Scheme 2). Phosphate analogs are good candidates for

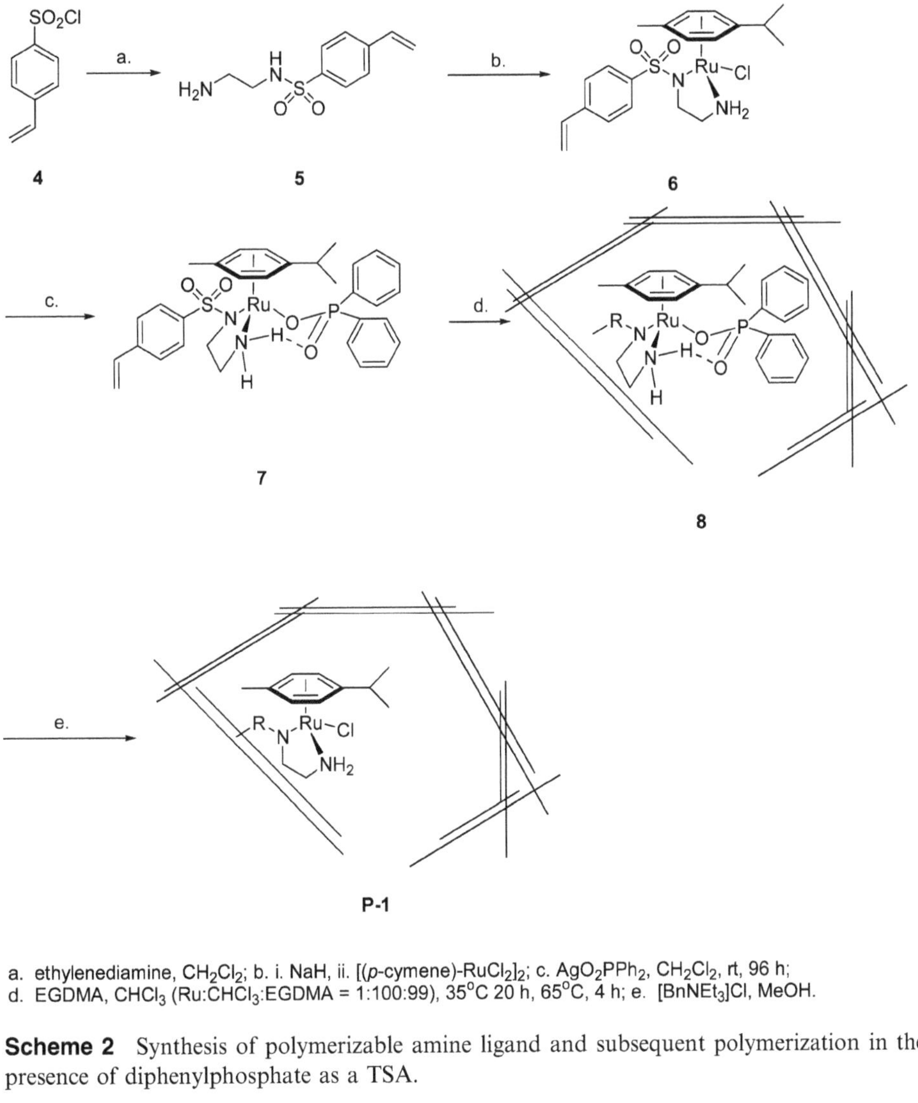

a. ethylenediamine, CH$_2$Cl$_2$; b. i. NaH, ii. [(p-cymene)-RuCl$_2$]$_2$; c. AgO$_2$PPh$_2$, CH$_2$Cl$_2$, rt, 96 h;
d. EGDMA, CHCl$_3$ (Ru:CHCl$_3$:EGDMA = 1:100:99), 35°C 20 h, 65°C, 4 h; e. [BnNEt$_3$]Cl, MeOH.

Scheme 2 Synthesis of polymerizable amine ligand and subsequent polymerization in the presence of diphenylphosphate as a TSA.

transition state mimics because the phosphorous center exists in a tetrahedral geometry similar to the carbonyl carbon geometry expected in the transition state of hydride transfer to ketones.

The chloro-ruthenium complex was synthesized according to Scheme 2. A single crystal X-ray analysis confirmed the structure of 7. Complex 7 was copolymerized with ethylene glycol dimethacrylate (EGDMA) using chloroform as porogen to yield 8 as a highly cross-linked porous polymer. The phosphinato ligand was selectively cleaved with a solution of BnNEt$_3$Cl in methanol to yield polymer P-1.

A control polymer, P-2, was prepared using ruthenium complex 6 as the functional monomer to assess the influence of the imprinted cavity on the reduction. The activity of each polymer was quantified by determining the initial turnover frequencies (TOF) of the catalysts. The MIP P-1 was shown to be three times more active for benzophenone reduction than the corresponding control polymer, P-2, displaying a TOF of 51.4 h^{-1} compared to 16.5 h^{-1} for the reference polymer.

In addition to measuring the activity of the imprinted polymer, Polburn et al. also examined the selectivity of the catalytic site. Use of the diphenylphosphinate ligand creates cavities that resemble the transition state for the reduction of benzophenone. It should follow that the MIP should be most effective towards benzophenone reduction. A series of competition reductions with equimolar amounts of benzophenone and a cosubstrate was conducted using polymers P-1 and P-2 as catalysts (Table 1). It is interesting to note that both the imprinted catalyst and nonimprinted catalyst show a preference towards benzophenone reduction. The MIP P-1 shows approximately 1.3–2.0 fold enhancement of selectivity. As indicated by entries 4–6, the aromatic ketones were not as easily differentiated by the MIP P-1, however, P-1 showed a preference for benzophenone over other aromatic ketones demonstrating the ability to differentiate structurally similar substrates.

Polborn and Severin [9] developed a second-generation catalyst that incorporated a diamino ligand that was modified to include a second polymerizable group (Scheme 3). The additional attachment point to the polymer support was incorporated to reduce the flexibility around the metal center following polymerization in an effort to improve the stereochemical control of the transfer hydrogenation. Temperature-dependent ^{31}P NMR experiments of compound 11 revealed two

Table 1 Substrate-selectivity of the MIP-catalyst P-1 and the Control Polymer P-2

Entry	Cosubstrate[a]	Selectivity[b]	
		P-1	P-2
1	2-Norbornanone	4.3	2.2
2	2-Adamantanone	4.9	2.5
3	Cyclohexyl methyl ketone	14.2	7.4
4	Acetophenone	2.8	1.8
5	α-Tetralone	3.8	2.2
6	Phenyl isopropyl ketone	20.6	13.8

[a]The reactions were carried out with 50 μmol of benzophenone, 50 μmol of the cosubstrate and 1 μmol of the respective catalyst.
[b]The selectivity was calculated by dividing the initial rate of diphenylmethanol formation by the initial rate of product formation for the reduction of the cosubstrate. For norborneol the sum of isomers was used.

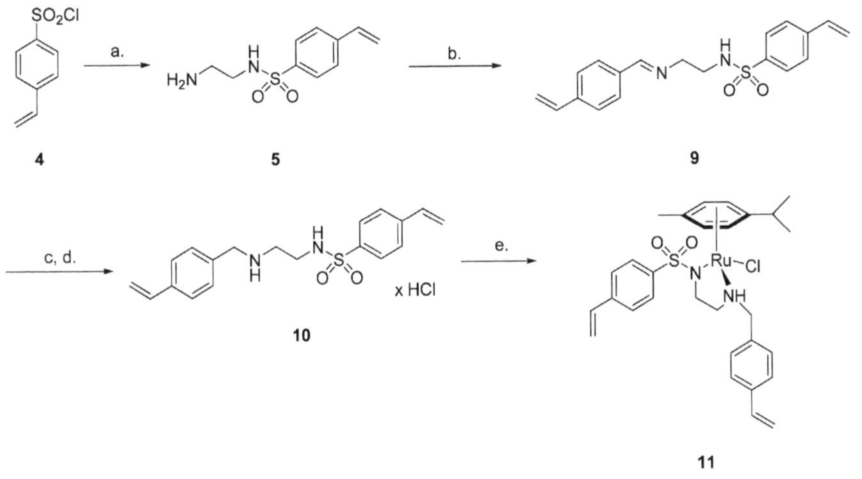

a. ethylenediamine, CH₂Cl₂; b. 4-vinylbenzaldehyde, 4A molecular sieves, CH₂Cl₂; c. NaBH₄, MeOH; d. HCl/Et₂O; e. i. NaH, ii. [(p-cymene)-RuCl₂]₂.

Scheme 3 Synthesis of second-generation polymerizable ruthenium complexes for selective hydride transfer reactions.

isomers present below 50°C (8:1 ratio). It was concluded that **11** forms with high dia-stereoselectivity and that epimerization of one of the stereogenic centers occurs on the NMR timescale. **P-3** was fabricated using a complex between **11** and diphenylpho-sphinate as a functional monomer–template and conditions previously described. The control polymer, **P-4**, was developed under identical conditions without the template diphenylphosphinate.

Transfer hydrogenations were carried out using formic acid as a reducing agent. Formic acid was found to be a better hydrogen donor by making the process quasi-irreversible and allowing conversions to reach 100%. The activity of the poly-mers was high: 1 mol% catalyst at 70°C produced initial turnover frequencies of more than $10\,h^{-1}$ for the reduction of benzophenone. The difference in activity between the second-generation MIP imprinted with diphenylphosphonate and the control polymer was more than a factor of five (Fig. 1). An approximately three-fold increase in relative activity (second generation MIP vs. first generation MIP) was attributed to the more rigid cavity that surrounds the metal center in the second generation MIP.

The second-generation catalysts were also selective towards the imprinted analyte benzophenone. Figure 2 shows the transfer hydrogenation of a mixture of ketones in the presence of catalysts **P-3** or **P-4**. The product distribution after 20 min is shown in the bar graph. The control catalyst **P-4** (light bars) failed to show signifi-cant preference for either benzophenone or the corresponding competing ketone, and gave very low reaction yields. In contrast, the MIP **P-3** (dark bars) was more active by a factor of 6.5 and showed a high preference for benzophenone over alternative substrates. The relative increase in yields for benzophenone over aceto-phenone and tetralone is indicative of the polymer's ability to differentiate between structurally similar compounds.

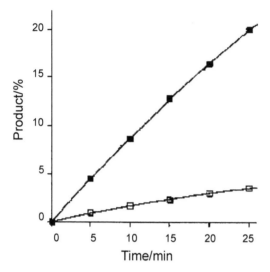

Figure 1   Product formation as a function of time for the transfer hydrogenation of benzo-phenone in the presence of catalyst **P-3** (filled squares) or control polymer **P-4** (open squares).

The MIP **P-3** also exhibited a modest ability to selectively reduce a diaryl ketone in a multifunctional substrate. Transfer hydrogenation of 4-acetylbenzophe-none, **12**, with catalysts **P-3** or **P-4** resulted in a mixture of three products, shown in Scheme 4. The diaryl ketone moiety in **12** is preferentially reduced by **P-3** after 1 h exposure to the catalyst. It is interesting to note the reversal of selectivity by the con-trol polymer (percent yields shown in parentheses), where the methyl ketone in **12** is preferentially reduced. This particular study highlights the effects of imprinting on catalytic activity. Imprinting allows rational changes in the microenvironment that influence the reactivity and selectivity of the catalytic center.

This study did not investigate the transfer hydrogenation of **12** using a homo-geneous catalyst. These studies would have been a useful comparison between a homogeneous catalyst and an MIP for the selective reduction in structurally similar ketones.

Polborn and Severin [10] also utilized a polymerizable CpRh(III) catalyst in which a methylphenylphosinato ligand was used as a template during their investiga-tion of the asymmetric hydrogenation of aryl ketones. Synthesis of the racemic complex **17** is shown in Scheme 5. Association with a phosphinato ligand results in complex **18**, which was shown to be a single isomer by NMR. The free methyl phenyl-phosphinate is achiral and, therefore, the configuration of the metal atom, as well as that of the phosphorous atom, is controlled by the diamino ligand. This indicates that the chiral information is transferred with high efficiency from the ligand to the metal and the pseudosubstrate. Copolymerization of complex **18** with EGDMA was antici-pated to result in the formation of a catalytic site that is selective towards compounds complimentary in shape and size to the transition state analog.

Copolymerization of **18** or **17** with EGDMA using a mixture of chloroform and methanol as a porogen resulted in polymers **P-5** and **P-6**, respectively. The reaction conditions were identical to previously reported polymerization conditions [9]. The resulting polymers were stirred in a solution of benzyltriethylammonium chloride

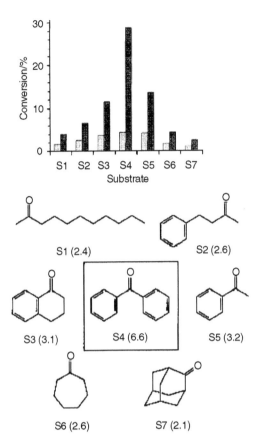

Figure 2 Transfer hydrogenation of a mixture of seven different ketones (**S1**–**S7**) in the presence of catalyst **P-3** (dark bars) or **P-4** (light bars). The conversion percent shows the product distribution after 20 min of reaction time. The numbers in parentheses correspond to the ratio of percentage conversion in the presence of **P-3** to **P-4**.

in methanol (0.1 M) to cleave the phosphinato ligand. Splitting yields were reported as quantitative as measured by in situ ^{31}P and ^{1}H NMR experiments. The polymers **P-5** and **P-6** were then evaluated for the reduction of acetophenone derivatives.

Substrates were exposed to 1 mol% of polymer in the presence of KOH with 2-propanol serving as the solvent. Reaction mixtures were analyzed using GC after the appropriate reaction times. The results are shown in Table 2, which indicate that the cavity generated by molecular imprinting has a pronounced effect on the activity of the immobilized catalyst. Perhaps even more impressive is the ability of the heterogeneous catalyst to compete with the homogeneous analog. Previous reports have shown CpRh(III) catalysts reduce acetophenone with 90% ee [11,12]. Polymer **P-5** demonstrated comparable activity to the homogeneous analog, presumably through ordered access to the metal centers.

Kinetic experiments were designed to determine the imprinted catalyst's substrate selectivity. Acetophenone and an equimolar amount of a second ketone were subjected to reducing conditions in the presence of either **P-5** or **P-6**. Table 3

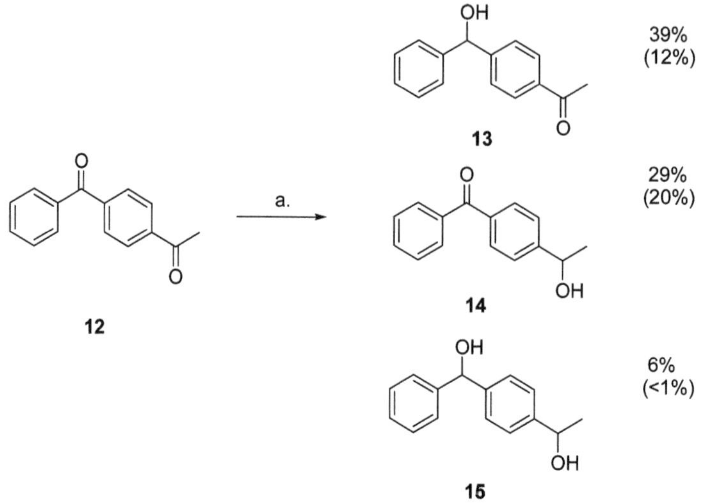

13   39%
                                                                      (12%)

12   a.   29%
                                                                      (20%)

14

15   6%
                                                                      (<1%)

a. azeotropic HCO_2H/NEt_3, CH_3CN, 70°C, either **P-3** or **P-4**.

Scheme 4 Transfer hydrogenation of 4-acetylbenzophenone in the presence of catalysts **P-3** and **P-4**. The product distribution after 1 h is reported (the values in brackets correspond to the reaction with the control catalyst **P-4**).

shows the comparison of the initial reaction rates, and the authors claim these results show that in all cases the imprinted polymer **P-5** displays a higher selectivity for acetophenone than the control polymer, **P-6**. Clearly, the results are less dramatic compared to the ruthenium complexes previously described (vide supra).

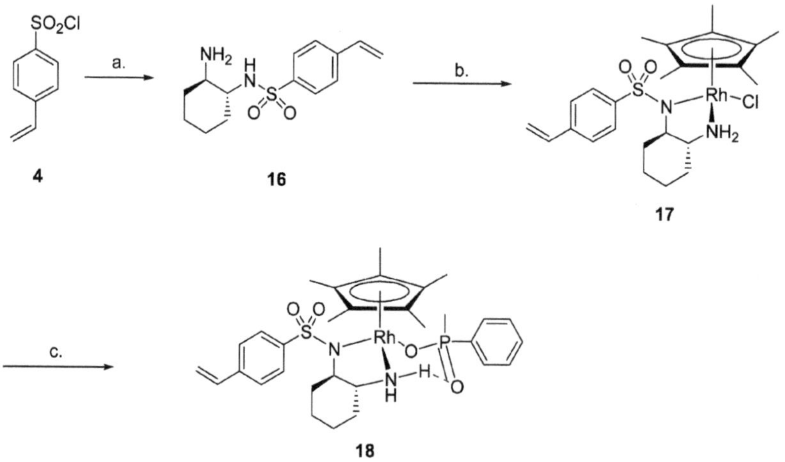

4   a.   16   b.   17

c.

18

a. (R,R)-1,2-diaminocyclohexane, CH_2Cl_2, 0°C; b. NaOMe/MeOH, 1/2 [(Cp)RhCl$_2$]$_2$
c. CH_2Cl_2, AgO$_2$PPhMe

Scheme 5 Synthesis of complex **17** and subsequent association with phosphonate TSA resulting in **18**.

Table 2 Transfer Hydrogenation of Aromatic Ketones in the presence of the Imprinted Catalyst **P-5** and the Control Catalyst **P-6**

Catalyst	Ketone	Time (h)	Yield (%)	ee (%)
P-5 (P-6)	Acetophenone	6	81 (41)	95 (93)
P-5 (P-6)	p-Fluoroacetophenone	6	71 (36)	93 (90)
P-5 (P-6)	o-Fluoroacetophenone	6	85 (52)	84 (76)
P-5 (P-6)	m-Fluoroacetophenone	3	98 (75)	94 (92)
P-5 (P-6)	m-Chloroacetophenone	3	98 (69)	94 (92)
P-5 (P-6)	1-Acetonaphthone	6	52 (24)	83 (74)

It is proposed that methylphenylphosphate serves as a transition state analog (TSA) for acetophenone resulting in a complimentary cavity that is selective towards acetophenone. The authors demonstrated that the selectivity was not merely a size exclusion issue with the following experiment. Polymers imprinted for benzophenone show a 2.8-fold preference to reduce benzophenone over acetophenone while polymers imprinted for acetophenone selectively reduce acetophenone over benzophenone by a ratio of 14:1.

The homogeneous examples cited indicate that the MIP does not appear to play a large role in the catalytic enantioselectivity since there was no significant increase in % ee's for the MIPs compared to the corresponding control polymers. A homogeneous comparison was not reported for the reduction of acetophenone. However, the role of the imprinted binding cavity influences substrate selectivity.

Locatelli et al. [13] investigated the hydride transfer reduction of prochiral ketones using a rhodium based catalyst on a polyurea support. The homogeneous reduction of acetophenone using a rhodium catalyst with two equivalents of (1 S, 2 S)-N,N'-dimethyl-1,2-diphenylethane diamine was conducted to establish an appropriate comparison for the imprinting studies. This control reaction resulted in formation of 1-(R)-phenyl ethanol with 67% ee (Scheme 6). The low enantioselectivity was attributed to a poor coordination sphere surrounding the metal center. The selectivity from the hydride transfer is proposed to arise from the approach of the substrate to the metal center, as shown in Scheme 7. The metal

Table 3 Substrate Selectivity of the Polymeric catalysts **P-5** and **P-6**

Cosubstrate[a]	Selectivity[b]	
	P-5	**P-6**
Cyclohexanone	0.38	0.31
2-Adamantanone	1.4	1.1
Benzophenone	14.0	8.9
α-Tetralone	4.2	3.7
2-Acetylnaphthalene	0.89	0.77

[a]Acetophenone and cosubstrate were exposed to either **P-5** or **P-6**.
[b]The selectivity is calculated by dividing the initial rate of 1-phenylethanol formation by the initial rate of product formation for the reduction of the cosubstrate.

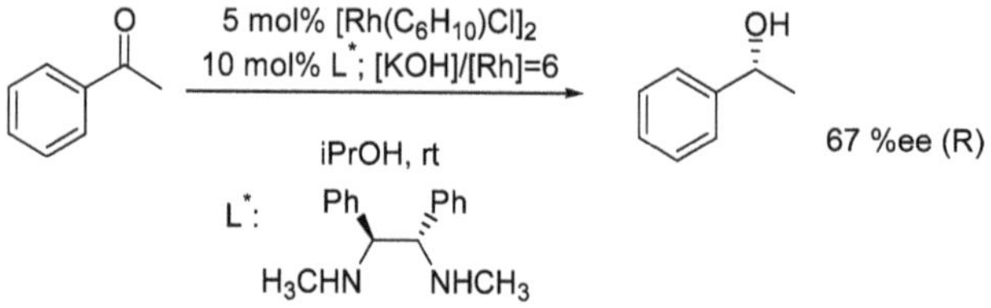

Scheme 6 Homogeneous catalytic hydride transfer reduction of acetophenone. Two equivalents of a diamine ligand (L*) were added to the homogeneous reaction to serve as coordinating ligands on the rhodium center.

center can attack from either the re or si face yielding (S) - or (R) -phenyl propanol, respectively. A modified catalyst with imprinted cavities around the metal center was envisioned to influence approach of the ketone to the metal center.

To enhance the enantioselective reduction of acetophenone, several heterogeneous variations of the chiral rhodium complex were investigated. Included in the study was a polyurea-supported complex in which the rhodium was deposited on a polyurea polymer and a polymerized diamine–rhodium complex that incorporates the rhodium centers within a cross-linked polyurea support. In addition, a rhodium MIP in which 1-(S) -phenylethoxide is used as a template was also prepared. Examination of these heterogeneous examples was aimed to provide insight into the role of secondary ligand structure around the rhodium center on the reduction selectivity of aryl ketones.

The shift from homogeneous to heterogeneous catalysis with a polyurea-supported rhodium complex resulted in an inversion of stereochemistry of the product, 1-phenylethanol, indicating that the microenvironment around the active metal is significantly different for the two systems. Reported % ee's were highest for the polyurea-based supported rhodium complexes when formulated with a cross-linker. The structure of these polymers was more rigid than the noncrosslinked polymers and it was proposed that they contained a more rigid, well-defined active site. The polyurea-supported complex lowered the reaction time from seven days to one, compared to the homogeneous catalysis. However, the % ee's for the polyurea supported rhodium catalysts were similar to the % ee's for the homogeneous analogs. It is likely that the ligands are randomly dispersed around the rhodium center, which results in poorly defined rhodium complexes.

Locatelli and coworkers carried out the design and fabrication of a polymer in which the diamine–rhodium complex was formed prior to polymerization. The formulation of **P-7a,b** is shown in Scheme 8. The advantage of this system over

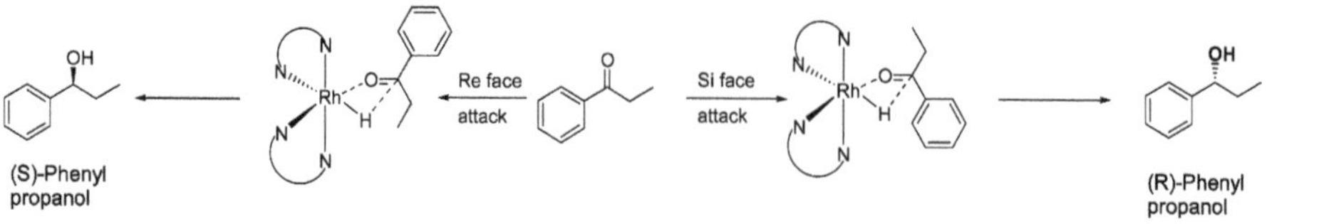

Scheme 7 Proposed mechanism for stereochemical control using the homogeneous rhodium catalyst.

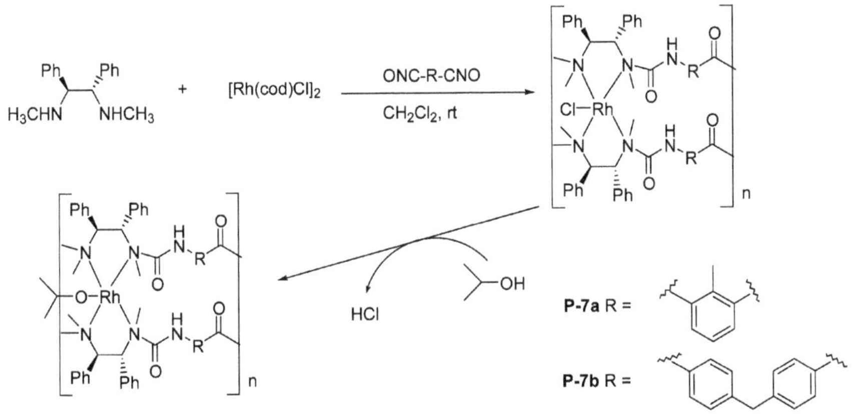

Scheme 8 Polymerization of a preformed diamine–rhodium complex without a template molecule.

the previous polymer is twofold. First, the metal centers in the polymer have a better-defined first coordination sphere. The prepolymerization complex ensures that the nitrogen atoms are coordinated to the ruthenium center, allowing the catalytic center to resemble the homogeneous catalyst more closely. Second, the metal centers are expected to be distributed more evenly throughout the polymer matrix, an improvement over a system that has the active sites predominantly on the surface of the polymer.

The reduction of aryl ketones using **P-7a** and **P-7b** as a catalyst resulted in low % ee's for product formation (Table 4). It is noteworthy that the same enantiomer is preferentially formed when **P-7a,b** is used as the catalyst compared to reductions using the homogeneous rhodium catalyst. The fact that the same enantiomer is formed during the homogeneous reactions suggests that the active sites in the heterogeneous catalyst are similar to those in the homogeneous version. The lower enantioselectivity in the heterogeneous catalyts may indicate that the more selective,

Table 4 Molecular Imprinting Effect in the Reduction of Acetophenone derivatives[a] by Using Rh Catalyst

Entry	Polymer	R[a]	Template Config.	Inductor Config.[b]	Conversion (%)	Time (day)	ee (%) (Config.)
1	**P-7a**	CH_3	—	(R,R)	44	1	33 (S)
2	**P-8a**	CH_3	(S)	(R,R)	42	1	43 (S)
3	**P-8a**	CH_3	(R)	(R,R)	40	1	30 (S)
4	**P-7b**	CH_3	—	(S,S)	98	1	25 (R)
5	**P-8b**	CH_3	(R)	(S,S)	98	1	43 (R)
6	**P-7b**	C_2H_5	—	(S,S)	96	6	47 (R)
7	**P-8b**	C_2H_5	(R)	(S,S)	91	9	66 (R)

[a]The substrate reduced is an acetophenone derivative (PhCOR), where $R = CH_3$ for acetophenone and $R = C_2H_5$ for propiophenone.
[b]The configuration of the diamino ligand.

less reactive centers are inaccessible, or at least the second coordination sphere for the rhodium centers is poorly defined. The authors speculate that the metal centers near the surface of the polymer, which have a poorly defined reactive center, may be the only ones capable of reacting with the substrate.

Locatelli and coworkers then employed molecular imprinting to improve the coordination sphere of the rhodium centers. To achieve this they utilized sodium 1-(S)-phenylethanolate as a template molecule (Scheme 9). The deprotonated template is capable of coordinating to the rhodium center by displacing chloride ion. The bidentate diamine ligand occupies the remaining coordination sites on the metal. Polymerization was carried out by condensation with bisisocyanates to form a polyurea matrix surrounding the metal–template complex. The rhodium microenvironment incorporates a cavity complementary in shape to 1-(S)-phenyl ethanolate. The template was exchanged with isopropanol following polymerization to evaluate the stereoselectivity of the reduction of acetophenone. The authors did not report a splitting yield for the template, although they mention that the elimination process was monitored by chiral GC with 1-(R)-phenylethanolate as an internal reference.

The effects of both template and amine ligands were examined for the reduction of acetophenone derivatives with the rhodium catalysts (Table 4). The reduction of propiophenone using P-8a, b was also examined to eliminate the possibility of the template leaching to distort the product analysis (entry 7). The enantioselectivity was the highest reported of the series, although the results were still unremarkable when compared to the homogeneous examples.

This series of experiments represents a thorough study employing control experiments to evaluate the role of the imprinted polymer. The homogeneous catalyst was used to establish a baseline comparison to the heterogeneous systems. A comparison between deposited rhodium and polymerized rhodium catalyst demonstrated that the catalyst with a poorly defined coordination sphere was not capable of

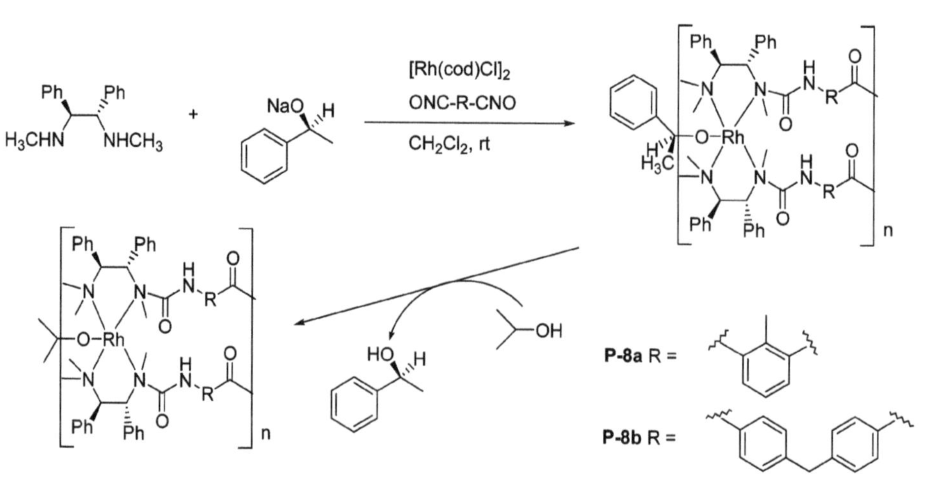

Scheme 9 Use of sodium 1-(S)-phenylethanolate as a template in the fabrication of rhodium MIP catalysts.

enantioselective reduction. Additional analyses to determine the polymer morphology would have been welcome. Elemental analyses on the polymers were conducted to determine Rh% of the polymer. Also, elemental analysis was conducted on rinse solvents, which showed no leaching of metal.

Unfortunately, the imprinted catalyst did not perform as well as the homogeneous system. The causes of the selectivity falloff were not examined systematically. Investigations designed to incorporate different polymer backbones and template molecules might provide insight into the roles that each play during the reduction process.

B. Hydrolysis Reactions

Recently, communications by Mashelkar and coworkers [14] described efforts to develop materials that mimic chymotrypsin-like activity. These proteins hydrolyze ester and amide linkages at the carbonyl group of phenylalanine or tyrosine. The active site in the enzyme is comprised of a hydroxyl, carboxyl, and imidazole side chains, respectively. The rate of hydrolysis displayed by chymotrypsin is attributed in part to cooperative effects brought on by the close proximity of the aforementioned functional groups within the active site of the enzyme.

Mashelkar and coworkers [14] utilized ligands capable of metal ion coordination to ensure the ligands remained in close proximity during the polymer

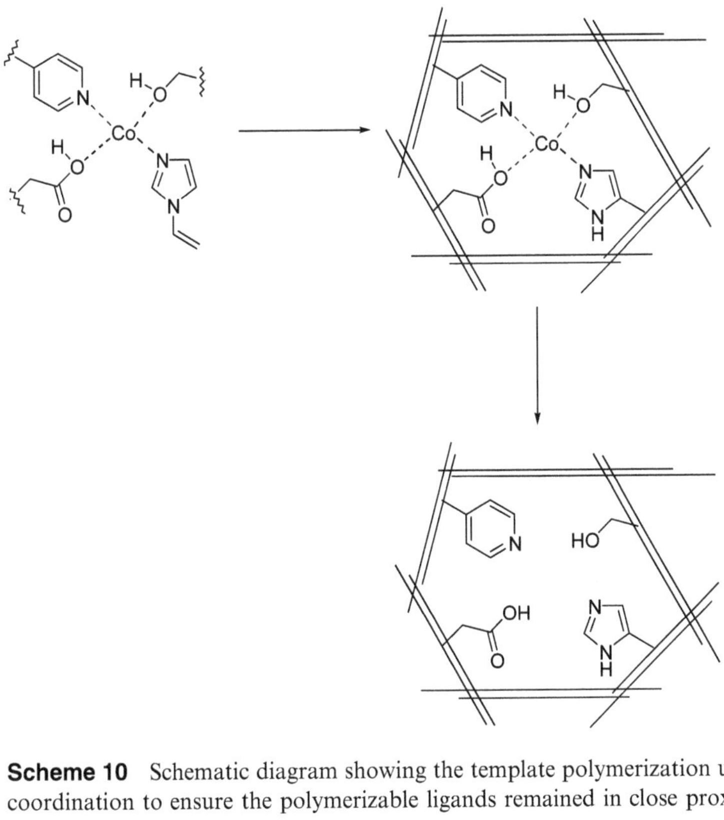

Scheme 10 Schematic diagram showing the template polymerization utilizing metal–ligand coordination to ensure the polymerizable ligands remained in close proximity to each other.

formulation process (Scheme 10). The expectation was that following metal ion extraction, the polymer bound ligands would remain in close proximity, and the resulting polymer would possess enhanced activity due to cooperative interactions between the ligands. The use of metal–ion coordination is especially interesting here because the metal center is not present in the final functional polymer. Rather, it is used to coordinate the active ligands prior to polymerization, so they remain in close proximity. This can lead to polymers with enhanced efficiency of hydrolysis through cooperative effects.

Preliminary investigations were conducted to examine the effects of imidazole on the hydrolysis of 2-methacryloyl ethyl *p*-amino benzoate, **19**. Polymers were formulated with an equimolar mixture of *N*-vinyl imidazole (NVIm), **19** and 2-hydroxyethyl methacrylate (**20**) using EGDMA as a cross-linker [P(**20**-**19**-NVIm)] (Scheme 11A). A second polymer was formulated without *N*-vinyl imidazole [P(**20**-**19**)]. The rate constant for release of *p*-aminobenzoic acid from P(**20**-**19**) was 6.07×10^{-3} day^{-1}, compared to 8.07×10^{-3} day^{-1} for P(**20**-**19**-NVIm). These experiments establish that the presence of imidazole in a network polymer with

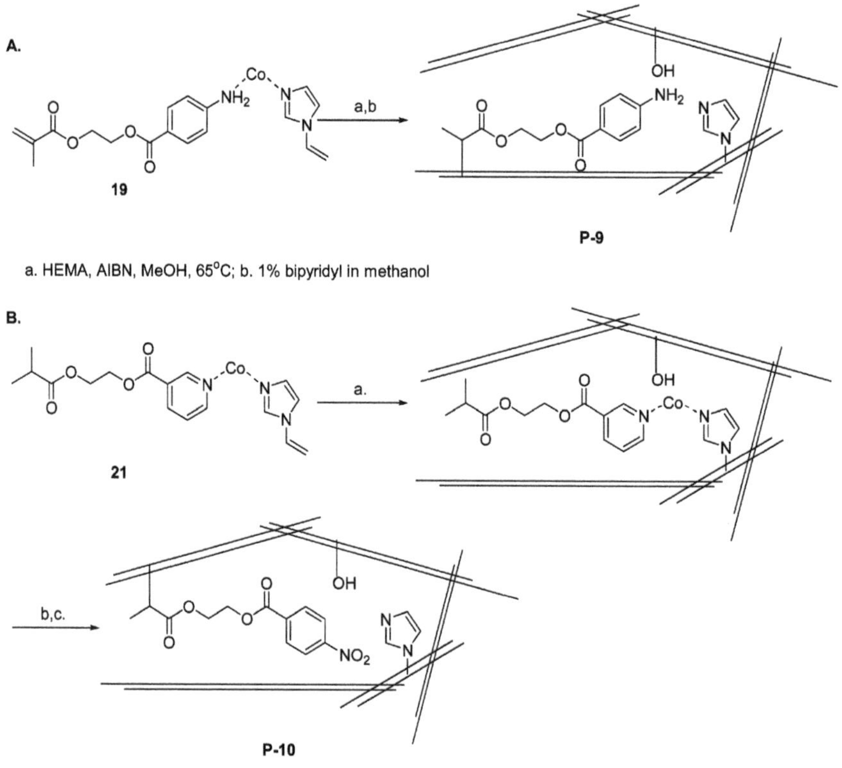

a. HEMA, AIBN, MeOH, 65°C; b. 1% bipyridyl in methanol

a. **20**, AIBN, MeOH, 65°C; b. 0.01 N HCl, 10°C, 24 h; c. i. **22**, MeOH, 12 h; ii. Co^{60}, MeOH, 6 h.

Scheme 11 (A) Coordination of polymerizable ligands to cobalt to promote rate enhanced hydrolysis reactions within polymer matrices. (B) Template polymerization of **21** to develop selective binding sites for the hydrolysis of **22**. Subsequent incorporation of **22** results in **P-10**.

randomly dispersed functional monomers is not sufficient enough to raise the rate of hydrolysis.

Polymers were developed to bring 19 in proximity to the imidazole group within the polymer network by allowing a complex between NVIm, 19 and CoCl$_2$·6H$_2$O to form prior to polymerization in the presence of 20 and EGDMA, where the cobalt would serve as a coordinating center for the functional monomers. The metal–ion was removed from the resulting polymer P-9 by washing with acidic aqueous media. The measured rate for the hydrolysis of 19 was 3.98×10^{-2} day^{-1}, a fivefold rate enhancement over polymers fabricated in the absence of cobalt. Karmalkar et al. infer that the enhanced rate constant is due to the prepolymerization coordination complex that occurs between the cobalt center, imidazole group, and the p-aminobenzoate group.

In their schematic representation (Scheme 11A), only two coordination sites are accounted for in the cobalt binding and there are no structural investigations presented to support the notion of well-ordered active sites. In addition, square planar Co^{II} complexes are substitution labile, which is an inherently bad trait for template complexes. Therefore, there is a high degree of uncertainty regarding the composition of the reactive cavitites.

An important control reaction necessary for this series would be to study the hydrolysis of 19 from a polymer fabricated with 20 and 19 in the presence of cobalt. Possible ligand combinations for the cobalt centers include various combinations of 19 and HEMA. The close proximity of HEMA to 19 could provide an increase in the observed rate of hydrolysis. This type of control would show the role of imidazole more clearly.

A similar coordination strategy was used with a structural analog of 19, a non-polymerizable complex with the metal ion [15]. NVIm and isobutyryl ethyl nicotinate [21] were coordinated to Co(II) in a prepolymer complex, which was polymerized using 20 and EGDMA (Scheme 11B). The template molecule is not a polymerizable ligand and therefore can easily be removed from the resulting network. The metal ion and imprint molecule were washed from the polymer matrix. The hydrogel was then bathed in a solution of 2-methacryloylethyl p-nitrobenzoate, 22. The mixture was exposed to a Co^{60} source of 0.25 Mrad/h for 6 h to polymerize 22 into the hydrogel. All unreacted monomer was extracted from the disc through copious washing, resulting in P-10.

It is interesting to note the authors' claim that the amount of ester incorporated onto the hydrogel was equal to the amount of template used during the formulation step. They also assumed that the ester occupied all of the imprinted cavities. The authors have used UV spectroscopy and ESR to characterize the structure of the complex prior to the removal of the metal center. Although the spectra were not shown, they reported studies which showed that cobalt formed a complex with 20, methacrylic acid (MAA), methacryloyl histidine (MA–His), and the template in the molar ratio 1:1:1:1:1. The authors' interpretation of the spectra and explanation of the cobalt complexes with the polymer are confusing at best. The schematic diagrams that they report incorporate a hydroxyl group in the active site, which is not consistent with their spectral interpretations.

A control polymer (P-10a) was formulated without the cobalt metal to serve as a coordination center. A comparison of the rates of hydrolysis of 22 in the P-10 and P-10a is shown in Fig. 3. The release of p-nitrophenol from polymers imprinted in

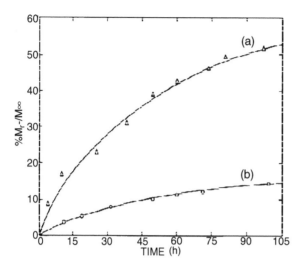

Figure 3 The release of *p*-nitro phenol from **P-10** (triangles) and **P-10a** (circles) in phosphate buffer (0.01 M, pH 8.0).

the presence of cobalt was 5.5 times as fast as the control polymers. These results imply metal coordination of the functional monomers and template places the imprint cavity in close proximity to the functional monomers.

To further enhance the catalytic activity of the polymers, methacrylic acid was included in the polymer formulation. It is known that hydrolytic activity of imidazole polymers is enhanced by the presence of hydroxyl and carboxylic acid groups [15,16]. Although the authors do not offer any spectral evidence, they do not believe that methacrylic acid serves as a ligand to the cobalt complex. The release rate constant for the polymer containing the methacrylic acid group is 1.26×10^{-2} h^{-1}, which is 1.75 times higher than the polymer containing only the imidazole group.

The results presented in these communications are interesting examples of utilizing metal coordination to obtain reactive sites that rely on cooperative interactions between functional groups in close proximity to each other. One should note that the metal center used in the imprinting process is not involved in the resulting polymer's function. Also, the active sites possess an enhanced affinity for template derivatives. Analytes could reassociate with the polymer after the removal of template and metal ion.

Unfortunately, Karmalkar and coworkers did not address the issue of substrate specificity or characterize the cobalt templates in a satisfactory manner. A useful investigation could have involved exposing **P-10** and **P-10a** to structurally diverse esters following the removal of template and metal. The selective recognition of the template molecule would validate the ability of the imprinted cavity to play a crucial role in the hydrolysis of specific substrates.

Following these results, Kulkarni and coworkers [17] examined the role of molecular imprinting and metal coordination in a second generation of polymeric mimics of chymotrypsin. The monomers previously utilized for cooperative rate enhancing effects were replaced by polymerizable amino acids that are present in the active site of the enzyme. A series of polymers were synthesized in which one, two, or three functional monomers were grafted onto a polymer support. Metal coordination using a cobalt(II) center was used to coordinate the functional monomers and the template in a proximal orientation during polymerization. N-Cbz-Tyr-p-nitrophenol [23] was used as a substrate to evaluate the polymer activity towards hydrolysis. However, since 23 is unable to form a complex with the cobalt center, N-nicotinoyltyrosylbenzyl ester 24 was used during the imprinting process to formulate an imprinted cavity around the reactive center (Scheme 12). The imprinting template 24 is similar in size and shape to the analyte 23.

In a typical experiment to prepare the surface-grafted mimics shown in Scheme 12, Co(II) -coordinated monomers–template assembly of MA-serine 25, MA-aspartic acid 26, MA-histidine 27, and 24 in methanol is added to a solution of EGDMA and AIBN. The solution is adsorbed onto poly(glycidyl methacrylate/EGDMA) beads and vacuum is applied to ensure the enhanced sorption of the assembly onto the surface of the polymer matrix. Following polymerization, the beads are washed with acidified methanol to elute unreacted monomer, cobalt and template yielding P-11. The beads are then washed with water, phosphate buffer, and dried under vacuum. Cobalt extraction was quantified by UV absorption. In addition, amino acid grafting was quantified through HCl hydrolysis of the support beads and subsequent Ninhydrin tests. Control polymers were fabricated to examine the source of enhanced reactivity. Table 5 shows the composition of each control polymer support fabricated, in addition to the activity of each polymer.

Evaluation of the polymer's catalytic activity involved monitoring hydrolysis reactions using 50-fold molar excess of 23 with respect to the theoretical amount of functional groups present in each polymer. Figure 4 shows a plot of percent hydrolysis vs. time. Curves a, b, and c correspond to polymers P-12, P-13, and P-11, respectively. It is clear that curves a and b do not display a typical pseudo first-order kinetics. Only curve c displays Michaelis–Menton kinetics, indicating P-11 contains an active site that may bear some features of saturation kinetics (Table 5) wherein cooperative effects from juxtaposed ligands help to enhance the nucleophilicity of the serine-hydroxyl group (Scheme 12).

Polymers P-14–P-16 were formulated to examine the role of the metal ion coordination and template role in the coordination step. P-14 was formulated by grafting all three monomers onto the support beads in the absence of a cobalt center and template. The overall effect of this is a random distribution of monomers on the support. P-14 is catalytically inactive with respect to the hydrolysis of 23 (Table 5).

In contrast, P-15 was formulated using all three monomers in addition to $CoCl_2 \cdot 6H_2O$ to obtain coordination with the Co(II) center. However, 24 is not part of the Co(II)-coordinated assembly. Polymer P-16 was formulated under the identical conditions as P-15 except the template molecule was incorporated in the prepolymerization complex. The hydrolytic activity of P-15 and P-16 was evaluated for Michaelis–Menton kinetic behavior. Results are reported in Table 5. Surprisingly, the rate constant of the control polymer, P-15, was calculated to be 0.22 s^{-1}, while that of P-16 was found to be 0.18 s^{-1}. The authors could not explain the faster rate constant associated with P-15 and this requires additional investigation. However, the value of K_m for P-15 is almost twice as high as that of P-16 ($K_m = 1.17 \times 10^{-3}$ M vs. 5.29×10^{-4} M, respectively), reflecting a higher affinity for 23 in P-16. The polymer formed in the absence of template, P-15, displays a higher catalytic activity despite having a lower affinity for the substrate than the imprinted polymer, P-16.

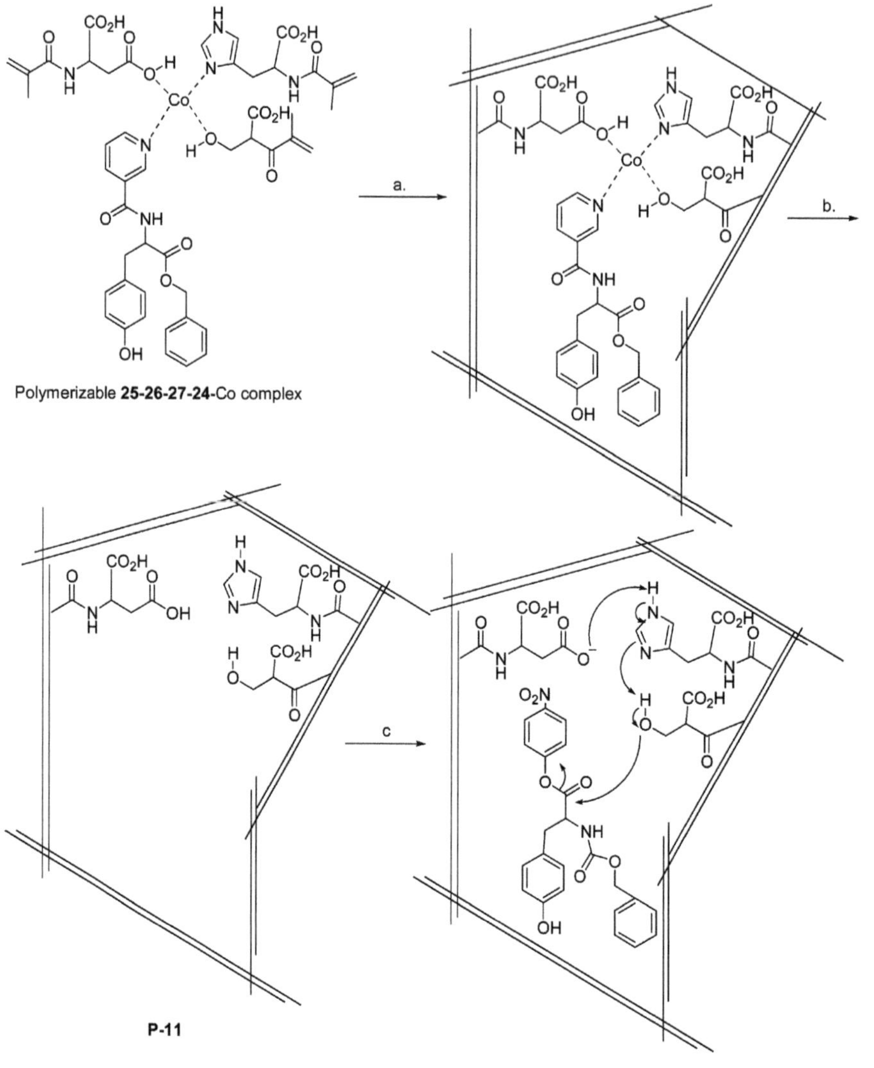

Polymerizable **25-26-27-24**-Co complex

P-11

a. MeOH, EGDMA, AIBN, polymer-support, 75°C, 24 h; b. MeOH, 35%HCl; c. **23**, CH₃CN:phospatebuffer (40:60), 37°C, 60 min.

Scheme 12 Coordination of polymerizable amino acid derivatives and **24** to cobalt followed by polymerization and removal of template and metal center resulting in **P-11**. Hydrolysis of **23** is brought about through cooperative interactions in the selective binding sites within **P-11**.

The increase in affinity for **23** displayed by **P-16** may be attributed to the imprinted cavities formed in close proximity to the functional monomers. The imprinted cavities align the substrate in the proper orientation for hydrolysis. In addition, calculations were made based on the initial velocity of the hydrolysis to determine substrate specificity for **23** within each polymer. The authors report that **P-16** exhibits a three-fold higher substrate specificity than that of

Table 5   Polymer Composition and Activity of Cymotrypsin Mimics

	No.						
	1	2	3	4	5	6	7
Polymer	Blank	P-12	P-13	P-11	P-14	P-15	P-16
MA-serine (M)	—	—	0.00150	0.00150	0.00150	0.00150	0.00150
MA-aspartic acid (M)	—	—	—	0.00150	0.00150	0.00150	0.00150
MA-histidine (M)	—	0.00150	0.00150	0.00150	0.00150	0.00150	0.00150
Template (M)	—	0.00150	0.00150	0.00150	—	—	0.00150
$CoCl_2 \cdot 6H_2O$ (M)	—	0.00150	0.00150	0.00150	—	0.00150	0.00150
Poly (GMA-EGDMA)(g)	0.98	0.36	0.65	0.98	0.98	0.98	0.98
Functional groups per 50 mg (M)	—	1.68×10^{-6}	3.01×10^{-6}	1.53×10^{-6}	nd	1.02×10^{-6}	1.8×10^{-6}
Surface area (m²/g)	49.63	nd	nd	nd	17.42	29.22	16.01
Pore volume (cm³/g)	0.14	nd	nd	nd	0.07	0.09	0.04
$k_{cat}(s^{-1})$[a]	—	nd	nd	nd	N/A[b]	0.22	0.18
K_m(M)[a]	—	nd	nd	nd	N/A[b]	1.17×10^{-3}	5.29×10^{-4}
k_{cat}/K_m (s⁻¹M⁻¹)[a]	—	nd	nd	nd	N/A[b]	196	347

[a]Kinetic measurements for the hydrolysis of N-cbz-tyr-p-nitrophenyl ester.
[b]No activity.

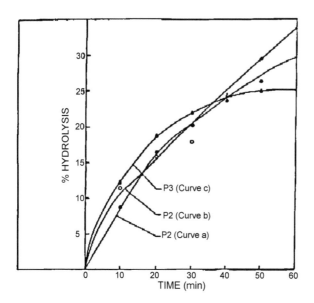

Figure 4 Plot of percent hydrolysis vs. time for the hydrolysis of **23** in the presence of **P-12** (curve a), **P-13** (curve b), and **P-11** (curve c).

P-15 ($V_{\text{max}}/K_{\text{m}} = 6.0 \times 10^{-4}\,\text{s}^{-1}\,\text{m}^{-1}$ vs. $1.8 \times 10^{-4}\,\text{s}^{-1}\,\text{m}^{-1}$ for **P-16** and **P-15**, respectively). These results indicate that the mimic synthesized by surface grafting of Co(II) -coordinated monomers–template assembly exhibit a modest molecular imprinting effect with a slight ability to discriminate between different substrate molecules.

Characterization of the cobalt complexes was conducted with electron-spin resonance (ESR) spectroscopy of $CoCl_2 \cdot 6H_2O$ in methanol solutions with various concentrations of coordinating ligands (Table 6). The ESR spectra of these assemblies showed a decrease in g values as a sequential addition of monomers and template to the $CoCl_2 \cdot 6H_2O$ solution. While this confirms that all ligands coordinate, in part, to the cobalt center in solution, ESR spectroscopy is not sufficient to determine the ratio of ligands bonded to the Co(II) ion. Nonselective binding of ligands should produce a statistical mixture of coordinated species. This is a

Table 6 Coordination of Monomeric Ligands and Template to $CoCl_2 \cdot 6H_2O$ for HEMA, MAA, MA-histidine

Entry	Description of complex	$g_{\parallel}$
1	(0.0005 M) $CoCl_2 \cdot 6H_2O$	2.199
2	(0.0005 M) $CoCl_2 \cdot 6H_2O$ + (0.0005 M) HEMA	2.1390
3	(0.0005 M) $CoCl_2 \cdot 6H_2O$ + (0.0005 M) HEMA + (0.0005 M) MAA	2.0140
4	(0.0005 M) $CoCl_2 \cdot 6H_2O$ + (0.0005 M) HEMA + (0.0005 M) MAA + (0.0005 M) MA − histidine	1.9445
5	(0.0005 M) $CoCl_2 \cdot 6H_2O$ + (0.0005 M) HEMA + (0.0005 M) MAA + (0.0005 M)MA − histidine + (0.0005 M) template	1.9984

distinct possibility since Co^{II} square planar complexes are substitution labile. It is unfortunate that this issue was not addressed, since this could explain some of the results presented.

The ability of the materials to differentiate between similar peptides was not assessed in these studies. This control experiment would have provided more information regarding the active site formed by metal–ligand binding interactions. For example, it is not clear that the active sites in P-16 contained one imidazole, one hydroxyl, and one carboxyl unit. Also, it was not determined what additional ligand was in place of the template during P-15 formation. This could have satisfied the rate dilemma that was observed.

While these studies are interesting, the choice of a substitution labile metal ion and poorly characterized metal complexes created a number of problems that made it difficult to understand the origins of the small effects observed.

C. Carbon–Carbon Bond Forming Reactions

The catalysis of carbon–carbon bond formation remains as one of the most active areas of research in catalytic antibody technology and organic synthesis [18]. Mosbach and coworkers reported the preparation and evaluation of a vinylpyridine–styrene–divinylbenzene copolymer imprinted with an aldol condensation reactive intermediate analog (Scheme 13). Dibenzoylmethane (DMB, 28) was chosen as the imprinted template based on molecular modeling studies for the cobalt (II) ion-mediated aldol condensation of acetophenone, 29, and benzaldehyde, 30, to produce chalcone, 31.

Coordination of 28 to cobalt was expected to consume two coordination sites on the square planar cobalt center. Vinyl pyridine groups occupied the two remaining coordination sites. The strategy for incorporation of vinyl pyridine was threefold: it provided strong binding interactions to coordinate cobalt to the polymer, it would provide the base necessary for generation of the enolate of acetophenone, and interactions with 28 through π–π stacking and van der Waals interactions may provide secondary binding interactions within the polymer backbone. Unfortunately, the authors did not include any spectroscopic data to support the proposed structure of the metal complex.

Imprint polymers were fabricated using the Co^{2+}–28 complex in addition to a series of reference polymers imprinted with either Co^{2+}, 28, or nothing (P-17, P-18, P-19, and P-20, respectively). The role of the 28–Co^{2+} occupied recognition sites was evaluated through aldol reactions carried out in the presence of the 28–Co^{2+} and Co^{2+} MIPs and in solution in the presence of pyridine and cobalt (II) ion (Fig. 5). The rate of the 28–Co^{2+} MIP mediated reaction was eightfold higher than the solution reaction and twice that of the Co^{2+} MIP reaction. Up to 80% conversion of the starting materials was observed when allowed to react in the presence of MIP for three weeks, which corresponds to approximately 138 turnovers per theoretical active site.

In addition to the catalytic activity of the MIP, Matsui et al.[19] investigated the degree of substrate selectivity by substituting adamantal methyl ketone, 32, and 9-acetylanthracene, 33, for acetophenone in condensation reactions with benzaldehyde (Table 7). Lower relative rate enhancements were observed for reaction of the bulkier substrates 32 and 33 with benzaldehyde to form 34 and 35 as products,

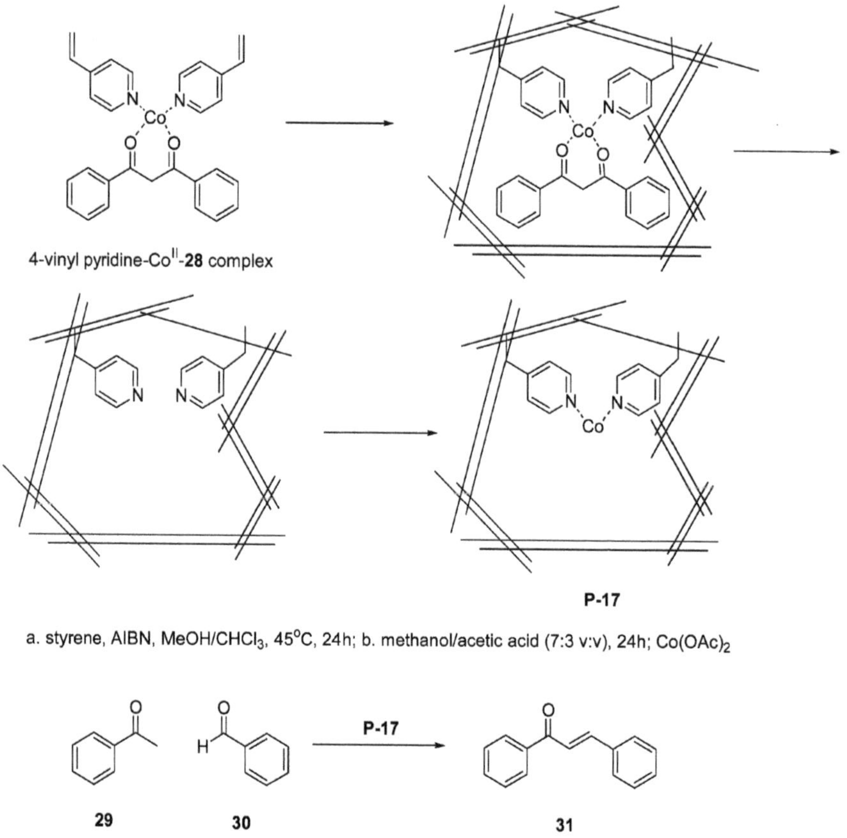

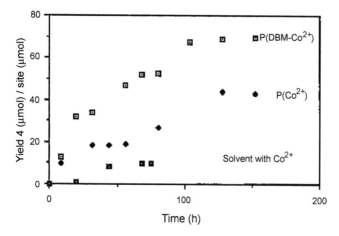

4-vinyl pyridine-Co^{II}-**28** complex
P-17
a. styrene, AIBN, MeOH/CHCl$_3$, 45°C, 24h; b. methanol/acetic acid (7:3 v:v), 24h; Co(OAc)$_2$
P-17
29
30
31

Scheme 13 Schematic representation of molecularly imprinted polymer preparation for **P-17** and subsequent use to catalyze the aldol condensation between acetophenone, **29**, and benzaldehyde, **30**.

Yield 4 (μmol) / site (μmol)
Time (h)
P(DBM-Co^{2+})
P(Co^{2+})
Solvent with Co^{2+}

Figure 5 Production of Chalcone using DBM-Co^{2+} and Co^{2+} MIPs and solvent containing cobalt (II) acetate. Yields were calculated from the average of triplicate determinations of duplicate sets of reactions.

respectively. The planer aromatic structure of **33** was thought to contribute to the increased reactivity relative to **32**. The lower selectivity for the alternate substrates lend support to the shape recognition effects previously observed for recognition in styrene–divinylbenzene MIPs [20].

To confirm that the imprinted recognition site was indeed the reactive center, reactions were conducted in the presence of the imprinting template, **28**, to determine its ability to inhibit the polymer-catalyzed reaction. A series of aldol reactions were conducted with increasing concentrations of **28**. Figure 6 shows a Lineweaver–Burk plot (a) and a Dixon plot (b) illustrating the increase in concentration of **28** leads to the decrease in efficiency of the MIP **P-17** for the catalysis of chalcone formation. The concentration-dependent inhibition of chalcone production by **28** implies the presence of a specific reaction center in the polymer matrix.

These results show MIPs are capable of catalytic turnover, as well as substrate selectivity and rate enhancement for the catalysis of C–C bond formation. Furthermore, Matsui and coworkers presented a thorough evaluation of the fabricated materials, which allowed the authors to state with a high degree of confidence the mode of operation of these molecularly imprinted polymers.

So far, the use of MIPs in catalysis has been limited to transformations involving specifically imprinted compounds. From a practical standpoint, it is unrealistic to envision the current technology utilized in natural product syntheses or method development procedures. Useful catalysts are reactive and selective towards entire families of substrates, and imprinting technology is not currently suited for this.

Table 7 Substrate Selectivity for Co^{2+} Mediated Aldol Condensation using Benzaldehyde, **30**, and either Acetophenone (**29**), Adamantal Methyl Ketone (**32**), or 9-Acetylanthracene (**33**). MIP **P-17** was used as the source of Cobalt

Ketone substrate	Product	Reactivity ratio
29	31	2.0
32	34	1.3
33	35	1.4

(a)

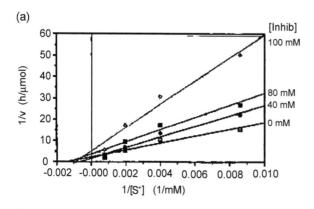

(b)

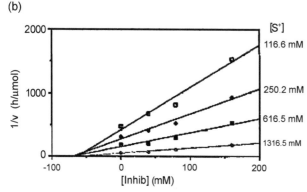

Figure 6 (a) Lineweaver–Burk plot for the production of chalcone (**31**) over a range of inhibitor concentrations (dibenzoylmethane, **28**) [active site concentration 5.76 μmol g (polymer)$^{-1}$; maximum velocity $V_{max} = 0.61 \pm 0.06$ μmol h^{-1}; Michaelis constant $K_m = 1.23 \pm 0.04$ μM]. (b) Dixon plot ($K_i = 60 \pm 10$ μM].

However, the advances in MIP technology over the last five years have been steady. Continued investigations of catalytic organic transformations using imprinted polymers may lead to a better understanding of the active sites in MIPs and better methods for controlling reaction efficiency and selectivity.

III. MOLECULARLY IMPRINTED POLYMERS DESIGNED FOR RECOGNITION AND SEPARATION

Several publications have focused on developing metal imprinted materials for selective recognition. Metal–ligand complexed monomers offer an alternative for creating a receptor site over "traditional" imprinting approaches. The stronger and more specific interactions of the metal–ligand bonds have potential to provide higher selectivity over noncovalent interactions. Traditional hydrogen bonding between functional monomers and template can suffer from low selectivity due to the relatively weak interactions between the template and functional monomer. The following examples describe recent advances in molecular imprinting using metal coordinating ligands as a means to achieve selective recognition and chromatographic separation.

A. Imprinted Microenvironments for Controlling Metal Coordination

Borovik utilized a molecular imprinting technique to develop sensors capable of reversible absorption of gaseous dioxygen and nitric oxide. The systems are porous, network polymers that incorporate metal centers. Molecular imprinting is used to shape the microenvironment around the metal center. However, in this case, imprinting with a template molecule does not improve selective binding of the metal to the ligands, rather, the resulting polymer matrix enhances the stability of the immobilized metal–gas complex.

Borovik rationalized that the use of a kinetically inert metallo-template during polymerization would allow for manipulation of the microenvironment of the metal sites in the network polymer host. The use of inert metal centers prevents undesired reactions, such as ligand substitution, from occurring prior to, during and immediately following polymerization reactions. Substitution inert metal centers would serve as a useful template by complexing to polymerizable ligands during imprinting procedures. Following polymerization, the metal center is converted to a kinetically labile oxidation state, and the space filling ligands removed. The active metal sites are capable of absorbing dioxygen without reacting further to dimers and oxidized products, as is the case in solution. The rigid framework provided by the network polymer provides the increased stability.

To demonstrate this concept, Krebs and Borovik [21] examined dioxygen binding to immobilized cobalt Schiff base complexes. The kinetically inert six-coordinate **36** was immobilized into a methacrylate host according to Scheme 14. Complex **36**

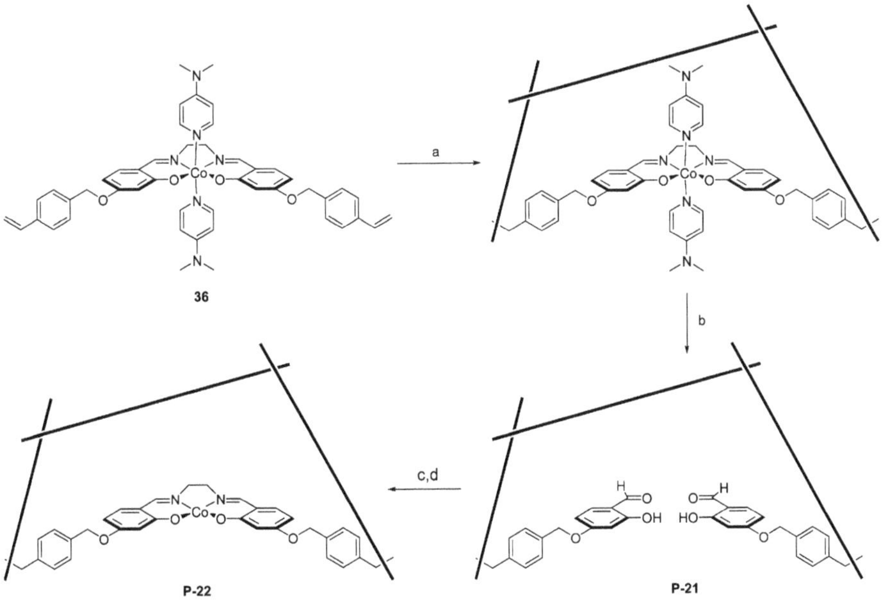

Conditions: a. EDGMA, AIBN, DMF, Ar, 60°C; b. EDTA, H₂O, heat; c. C₂H₈N₂, MeOH; d. Co((OAc)₂, MeOH, Ar

Scheme 14 Synthesis of **P-22**.

was polymerized using EGDMA as a cross-linking reagent and AIBN as an initiator. Following grinding and sieving, the removal of Co^{III} ions and DMAP ligands resulted in **P-21**, a functional polymer containing two salen groups in close proximity. The salen ligand was regenerated by treatment with ethylene diamine, and bathed in $Co(OAc)_2$ and methanol, resulting in a four-coordinate complex, **P-22**. The EPR spectra for **P-22** were consistent with immobilized complexes having the expected square-planar arrangement of donors around the metal ions.

The use of DMAP as axial ligands in the octahedral cobalt^{III} complex during the polymerization resulted in axial space above and below the metal centers. It was anticipated that this would allow for coordination of external ligands, such as nitrogeneous bases and dioxygen to the cobalt sites. Over 80% of the immobilized Co^{II} sites were converted to five-coordinate complexes when **P-22** was treated with 15 equivalents of either pyridine or DMAP under an argon atmosphere (Scheme 14). The EPR spectrum of **P-22(DMAP)** measured at 77 K exhibited the characteristics of a low spin Co^{II} complex incorporating the axial DMAP to afford a square-pyramidal coordination geometry around the Co^{II} ion. The five-coordinate geometry is necessary for Co^{II} complexes to obtain the correct electronic configuration for O_2 binding. These complexes can form 1:1 $Co-O_2$ adducts at room temperature under 1 atm of O_2, as indicated by the change in the EPR spectra of the complex.

It is interesting to note that analysis of the EPR signal showed that 70% of the immobilized Co^{II} sites form stable $Co-O_2$ adducts in **P-22(DMAP)(O$_2$)**, while 52% of the sites in **P-22(py)(O$_2$)** form $Co-O_2$ complexes. This is in contrast to monomeric **Co(37)(DMAP)** and **Co(37)(py)** complexes dissolved in 1:1 methylene chloride: toluene mixture that yield $<10\%$ of the $Co-O_2$ adducts after 1 h exposure to dioxygen at room temperature.

An improved scaffold system was employed in a second-generation polymer by taking advantage of a three-point immobilized ligand to link the template to the network polymer [22]. Noting cobalt's enhanced binding to dioxygen in the presence of pyridine (Scheme 15),[23], Borovik and coworkers designed a substitution-inert cobalt complex in which two polymerizable ligands would occupy five coordination sites around a cobalt center. Two polymerizable sites are available from N,N'-bis [2-hydroxy-4(4-vinylbenzyloxy)-benzaldehyde]ethylenediamine, **37**, and a third from a vinyl pyridine ligand. The remaining axial coordination site would be occupied by a DMAP ligand, which would serve, after polymerization and removal, to create an open cavity for binding of guest molecules within the polymer (Scheme 17). It is worth mentioning that the complex is diamagnetic and substitution-inert under the polymerization conditions.

The synthesis of the template complex is illustrated in Scheme 16. Compound **37** was treated with $Co(OAc)_2$ under argon atmosphere. The complex was allowed to react with DMAP to form $[Co^{II}37(DMAP)]$. EPR studies confirmed the formation of this complex. Following oxidation with ferricenium, treatment with vinyl pyridine yielded **38**. ^1H NMR and ^1H COSY analysis confirmed the final structure of the Co-template monomer.

Complex **38** was immobilized in a porous polymethacrylate matrix according to Scheme 17. Activation of the polymer was achieved through a three-step procedure, where the cobalt center was chemically reduced using $S_2O_4^{2-}$. It was interesting to note that only 10% of the DMAP was removed during this procedure.

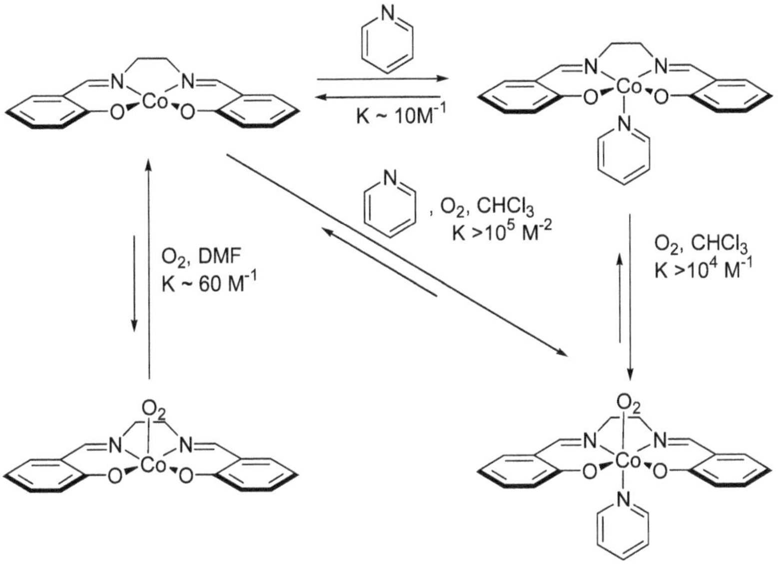

Scheme 15   Rebinding equilibrium for Co^{II}-salen complexes to dioxygen in the presence and absence of pyridine.

Treatment with a methanolic solution of (NMe$_4$)$_2$EDTA forms the apo polymer **P-23** in which 85% of the Co^{II} ions were removed along with an additional 70% of the DMAP ligand. **P-23** readily rebinds to Co^{II} ions to form **P-24**.

The physical properties including the surface area [270 m^2/g] and average pore diameter (22 Å) were reported. EPR spectroscopy was used to characterize the structure of the Co^{II} complex. Depending on the solvent used, 60–70% of the Co^{II} sites exist in a square-planar geometry, while only 30–40% have five coordinate square-pyramidal geometries. Borovik proposed an equilibrium between bound and free pyridine in which the axial pyridine ligand is only loosely bound to the cobalt center. Similar equilibria have been reported for Co^{II} complexes in solution [24].

Exposure of **P-24** to dioxygen in CH$_3$NO$_2$ for 2 min results in the complete conversion to **P-24(O$_2$)**. A well-resolved anisotropic EPR signal is observed for **P-24(O$_2$)** in which approximately 90% of the immobilized cobalt sites bind dioxygen.

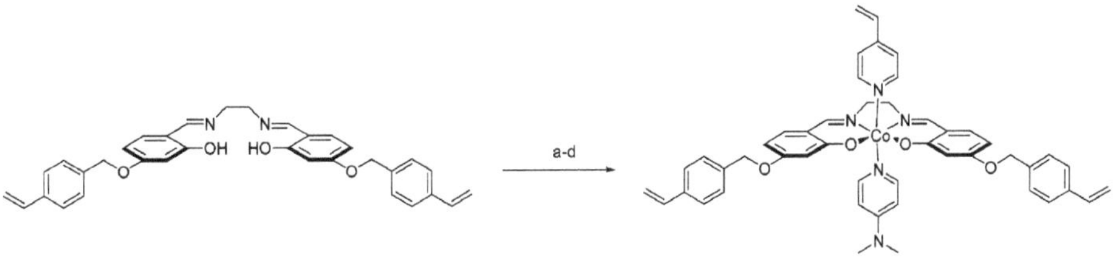

Conditions: a. Co(OAc)$_2$, DCE/MeOH, Ar; b. dmap, DCE; c. [FeCp$_2$]FP$_6$, CH$_3$CN; d. 4-vinylpyridine, DCE.

Scheme 16   Synthesis of cobalt monomer complex **38**.

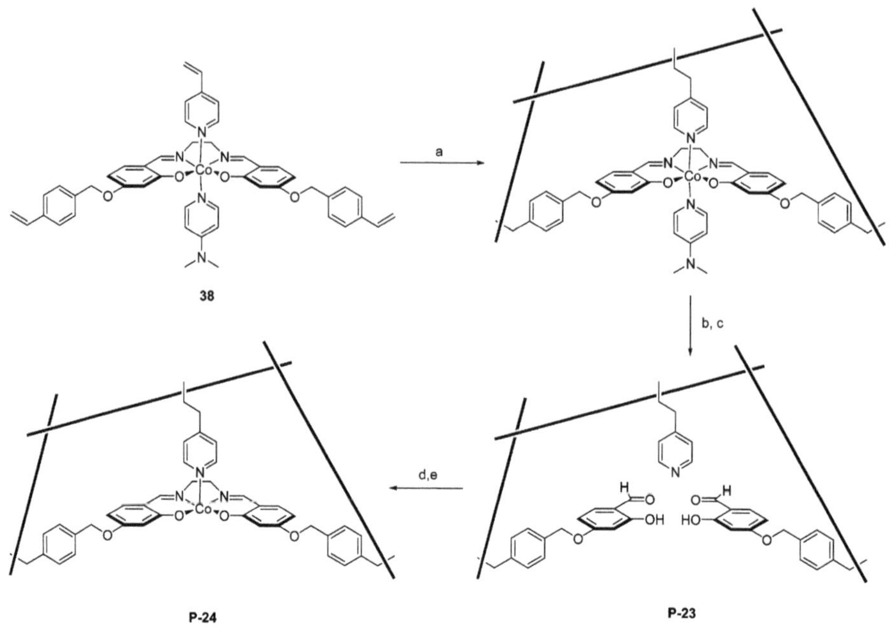

Conditions: a. EDGMA, AIBN, CH$_3$NO$_2$, Ar, 60°C; b. Na$_2$S$_2$O$_4$, MeOH, Ar, 2 h; c. (NMe$_4$)$_2$ EDTA, MeOH, Ar, 24 h; d. Co(OAc)$_2$, MeOH, Ar, 6 h; e. O$_2$, CH$_3$NO$_2$.

Scheme 17 Synthesis of **P-24**.

The signal is consistent with dioxygen binding directly to the immobilized Co^{II} centers, and was a marked improvement over the previously reported MIP Co^{II} complexes. The binding was shown to be reversible by flushing the polymer with N$_2$ for 3 min, which resulted in the loss of signal. Subsequent rebinding to O$_2$ regenerates the EPR spectra.

Interestingly, the EPR signals lacks any contribution from four-coordinate Co^{II} of the **P-24**(O$_2$), indicating that the endogenous pyridine ligands are within binding distance to the cobalt center. Borovik demonstrated that the lack of four-coordinate cobalt complexes is not simply due to dioxygen binding to the four-coordinate sites by measuring dioxygen bonding to **P-22** in a control experiment. They reported that **P-22** has limited dioxygen bonding ability, and discovered < 5% of the sites in **P-22** bound dioxygen. This clearly demonstrates that four-coordinated cobalt is not capable of dioxygen binding, and therefore cobalt must be a five-coordinate species in the polymer, which implies pyridine is directly involved in the cobalt activation process.

A reversible NO binding system was developed in the Borovik lab using a porous network polymer to host Co^{II} centers [25]. The synthesis of the cobalt polymers was carried out according to Scheme 14. The cobalt concentration in **P-22** ranged from 180 to 230 mmol of Co/g of polymer. The resulting macroporous polymers had an average pore diameter of 25 Å.

Absorbance spectroscopy was utilized to characterize NO binding in **P-22** (Fig. 7). The absorbance spectrum of a **P-22** suspension has two bands at 350 and 410 nm with a shoulder centered at 495 nm. In contrast, a suspension of **P-22**(NO) is characterized by two prominent bands at 310 and 400 nm, and a weaker band at

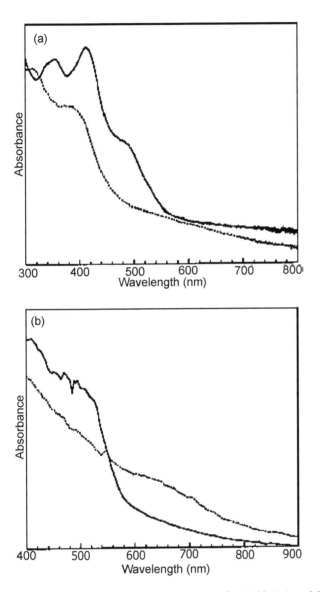

Figure 7 Electronic absorbance spectra for **P-20** (—) and **P-20**(NO)(----). (a) suspended in toluene and (b) solid-state diffuse reflectance spectra.

600 nm. The solid-state absorbance spectra for **P-22** and **P-22**(NO) are shown in Fig. 7b and are qualitatively similar to those obtained as suspensions. The differences in the absorbance properties of **P-22** and **P-22**(NO) lead to dramatic color changes in these materials. In the presence of NO, the orange **P-22** changes to brown–green characteristics of **P-22**(NO). The original orange color returns upon heating at 120°C under vacuum, indicating the release of NO. A control experiment demonstrated that the color change is due to NO binding to cobalt by establishing that poly(EGDMA) remains colorless in the presence of NO.

The proposed square-planar geometry of the cobalt centers is supported by X-ray absorption spectroscopy. The spectra for the immobilized cobalt sites in **P-22** are consistent with cobalt centers that are centrosymmetric, which is indicative of a square-planar coordination geometry. The lose of centrosymmetry in **P-22(NO)** is caused by the formation of a five-coordinate complex in the presence of NO.

The reversible binding of NO was also monitored using EPR spectroscopy. EPR signals for **P-22** are consistent with the cobalt (II) center having a square-planar geometry. Upon NO binding, the EPR signal is quenched, which is consistent with a diamagnetic metal center. This could result from antiferromagnetic coupling of the unpaired electrons on NO and the Co center. The EPR signal is regenerated upon removal of the NO ligand. At room temperature and atmospheric pressure, 40% conversion of **P-22(NO)** to **P-22** is observed over a period of 14 days as determined by comparison of the relative areas of the corresponding EPR signals.

The spectral properties of **P-22** and **P-22(NO)** are nearly identical with those reported for monomeric $Co^{II}(37)$ and $Co^{II}(37)(NO)$ complexes in solution, indicating that the coordination geometries of the immobilized complexes are similar to their solution analogs. Despite these similarities, the immobilized complexes behave quite differently from their solution counterparts. The reversible binding of NO observed for **P-22** contrasts the irreversible binding reported for monomeric $Co^{II}(37)$ in solution.

In addition, $Co^{II}(37)(NO)$ complexes in solution are not stable and oxidize to $Co^{III}-NO_2$ complexes through a dimeric cobalt intermediate species. The immobilized cobalt centers in the polymethacrylate network are not capable of dimerization, and thus do not undergo oxidative processes due primarily to the rigid polymer matrix, which effectively isolates the cobalt centers.

Borovik and coworkers conducted gas uptake measurements for NO, O_2, CO_2, and CO within **P-22** and **P-21** (Fig. 8). Solid samples of **P-22** were exposed to 10.3 Torr (30.5 μmol) of gas and uptake was monitored as a change in pressure with respect to time. According to Fig. 8a, **P-22** absorbed 21 μmol of NO within 2 min. This corresponds to 70% uptake of the available NO. **P-22** failed to show any detectable affinity for O_2, CO_2, or CO under identical conditions. A control polymer, which did not contain any cobalt sites failed to absorb any gas, establishing the cobalt sites as a necessary component for gas absorption.

The source of increased affinity for NO compared to other gases resides primarily with the geometry of the metal sites. It is known that square-planar Co^{II}-salen complexes are very weak binders of O_2, compared to the square-pyramidal Co^{II} complexes. CO_2 and CO binding to square-planar Co^{II} complexes is also known to be weak. Although Borovik and coworkers claim **P-22** binds NO selectively, they failed to conduct competitive binding experiments. It would have been interesting to see if the conversion of the immobilized sites in **P-22** to a square-pyramidal geometry during NO binding would have any affect on binding to O_2, CO_2, or CO.

The gas sensors developed by Borovik utilizes the techniques of template imprinting to provide void space in proximity to immobilized metal centers. This allows binding to guest molecules at the immobilized metal centers. Although the polymers are not traditional molecularly imprinted polymers, in that the template molecule is not used to develop binding sites with enhanced affinity and selectivity towards the template molecule, Borovik and coworkers utilize the imprinting techniques to control the ligand field and void space in the vicinity of the metal binding site.

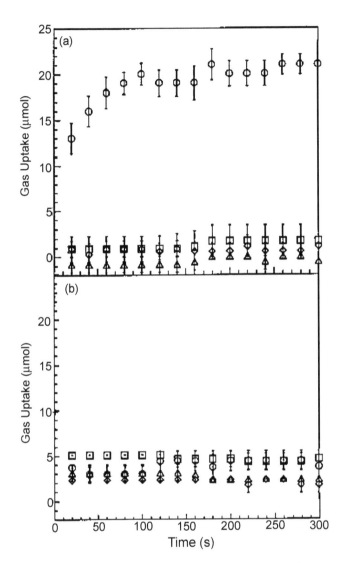

Figure 8 Gas binding vs. time for **P-22** (a) and **P-21** (b). Legend of analyte gases: NO (O), $O_2(\triangle)$, $CO_2(\square)$, and CO ($\diamondsuit$). The gas uptake plots in part A have been corrected for physical adsorption by subtracting the values of **P-21** binding from those of **P-22** for each gas, respectively.

Their work is notable because of the imprinting effects on the reactivity and stability of the cobalt centers. The investigation on the polymer reactivity included a detailed characterization of the reactive sites.

B. Selective Recognition of Imprinted Templates

Fugii et al. [26] reported one of the earliest examples of selective recognition using metal coordinated molecular imprinting. They utilized a polymerizable cobalt (III) Schiff base complex as the recognition element in an MIP designed for the selective recognition of enantiomers of amino acids. Synthesis of the cobalt complex **42** is shown in Scheme 18. Schiff base ligand **41** was prepared from imine condensation of (1*R*, 2*R*)-1,2-diaminocyclohexane with 4-(*p*-vinylbenzyloxy)salicylaldehyde **40**. Introduction of Co(OAc)₂ followed by D-phenylalanine under air oxidation conditions resulted in **42**. The Δ-β₂-structure of **42** was confirmed from electronic, CD, and ¹H NMR spectra. Copolymerization of **42** with styrene and divinylbenzene (mol fraction of **42**:styrene:divinylbenzene = 1:20:4) in THF for 24 h at 63°C, using AIBN as an initiator resulted in the formation of the polymer complex **P-25**.

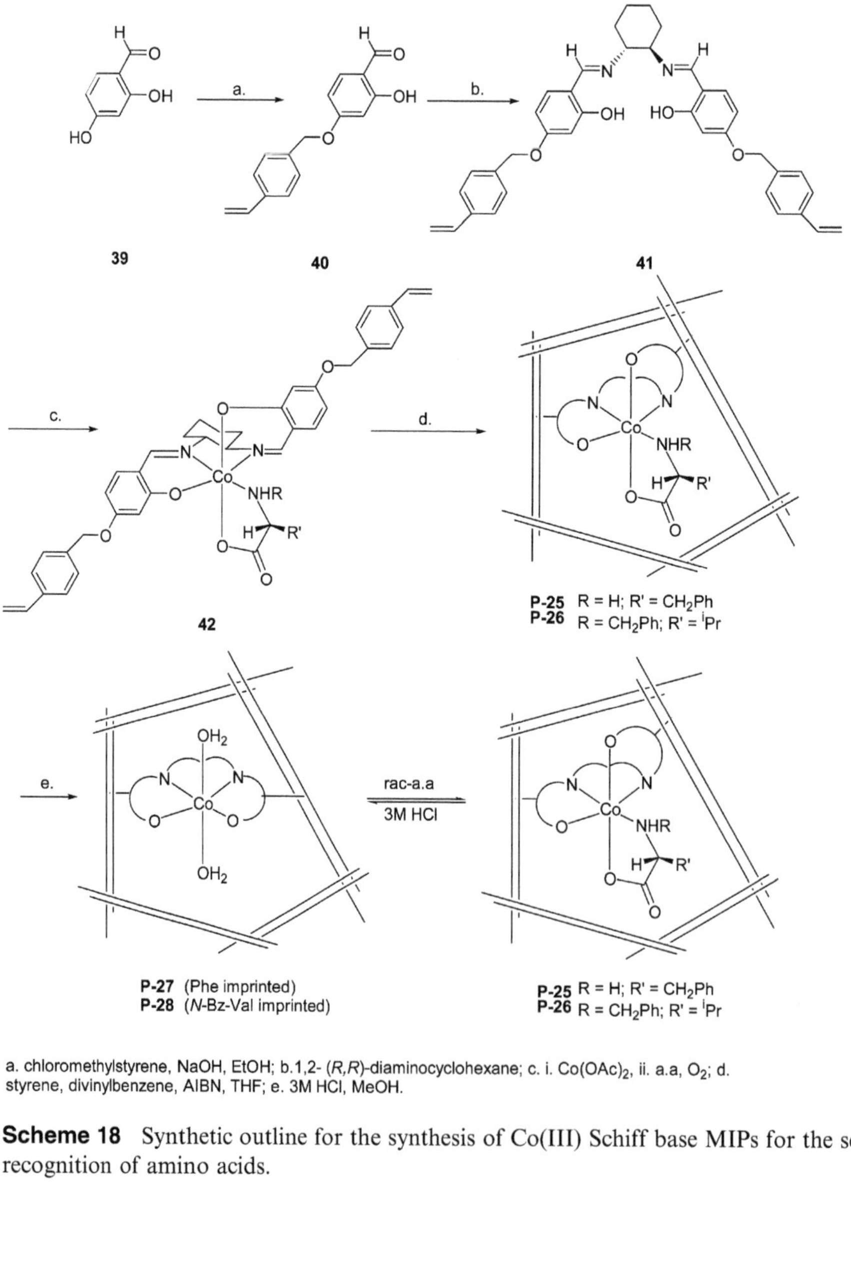

a. chloromethylstyrene, NaOH, EtOH; b.1,2- (*R,R*)-diaminocyclohexane; c. i. Co(OAc)₂, ii. a.a, O₂; d. styrene, divinylbenzene, AIBN, THF; e. 3M HCl, MeOH.

Scheme 18 Synthetic outline for the synthesis of Co(III) Schiff base MIPs for the selective recognition of amino acids.

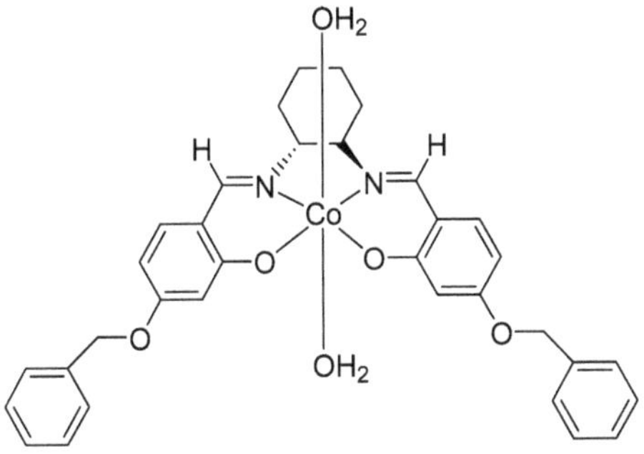

Figure 9   Cobalt (III) *N-N'*-bis(salicylidene)-1,2-(diaminocyclohexane)dihydrate, **43**.

Exposure of the polymer to aqueous acid resulted in a color change from green to brown with concomitant release of the phenylalanine template. Optical rotation confirmed that the released template did not racemize during the polymerization process. The splitting yield was reported as quantitative. Similar results were reported for *N*-benzylvaline imprinting [27].

The intrinsic ability of homogeneous solutions of chiral cobalt complex **43** (Fig. 9) to discriminate between the enantiomers of amino acids was established for a baseline for the imprinting effect [28]. The results are shown in Table 8. The optical purity of bound phenylalanine was 50%, while the optical puritiy for bound *N*-benzylvaline recovered from **43** was found to be 88%.

Rebinding studies were conducted with **P-27** and **P-28** using racemic phenyl-alanine and racemic *N*-benzylvaline, respectively. A suspension of imprinted polymer (**P-27** or **P-28**) in a 1:1 methanol: chloroform solution was stirred for two days in the presence of the corresponding amino acid. The green color returned to the polymer, indicating that the cobalt binding sites reabsorbed the amino acid. The polymer was washed with acidic media to remove the bound amino acid.

The optical purity of the amino acid recovered from the polymer is shown in Table 8. The imprinting effects on phenylalanine recognition results in ~15% improvement in diastereoselectivity over the homogeneous solution, while *N*-benzylvaline recognition is improved by more than 10%.

A cobalt-containing control polymer **P-29** was fabricated in the absence of an amino acid template in order to separate the effects of imprinting from those of immobilization. It was noted that the rate of uptake of **P-29** with racemic phenyl-alanine was slower than that of **P-27**. Likewise, the MIP **P-28** took up racemic

Table 8   Optical Purity % of Recovered Phenylalanine and *N*-Benzylvaline

Complex	Optical Purity % of Phe		Optical purity % of *N*-Bz-Valine	
	From Complex (D-a.a)	From Reaction (L-a.a)	From Complex (D-a.a)	From Reaction (L-a.a)
P-27 or **P-28**	65	16	99.5	26.2
P-29	41	9	86.5	14.5
43	50	10	88.2	21.8

N-benzylvaline faster compared to *N*-benzylvaline binding to **P-29**. The results indicate that the template effect augment the discrimination between the enantiomers of phenylalanine and *N*-benzylvaline from 10% to 15%.

Gagné utilized polymerizable platinum complexes to create imprinted polymer sites. The metal centers were used to probe the ligand substitution reactions on the microenvironment of the binding site [29]. The chiral template (*R*) -1,1'-dibutyl-2-naphtal ((*R*)-Bu$_2$BINOL, **44**) was chosen because the chirality associated with the compound is manifested in a sterically obvious manner and the resulting imprinted cavity should accommodate a ligand with the same absolute configuration as the template. The metallomonomer (vinyl dppe)Pt[(*R*)-Bu$_2$BINOL] **45** was polymerized using EGDMA in toluene following association with **44** (Scheme 18). Removal of the BINOL template from **P-30** leaves a spatially defined cavity selective for substituted (*R*)-BINOL derivatives.

Template **44** was removed from **P-30** using a variety of conditions that are summarized in Scheme 19. Treatment with HCl in CH$_2$Cl$_2$ over 6 h liberated 83 ± 1% of Bu$_2$BINOL resulting in the formation of **P-31**. Bu$_2$BINOL can also be displaced from the MIP by exposure to acidic phenols or BINOL analogs. Exposure

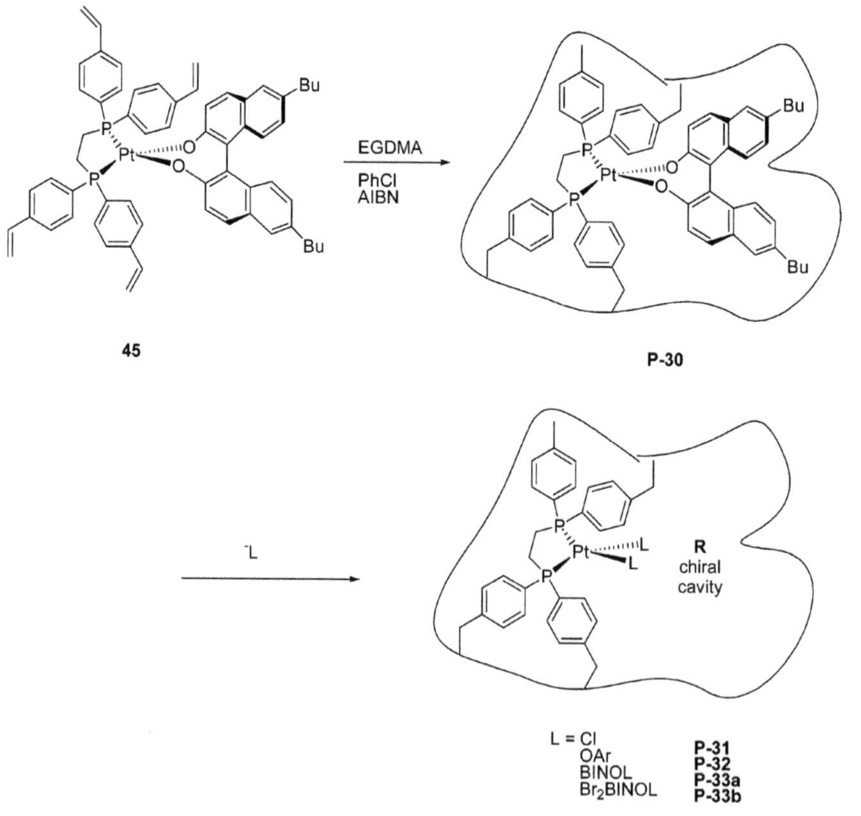

L = Cl P-31
  OAr P-32
  BINOL P-33a
  Br$_2$BINOL P-33b

Scheme 19 Coordination of polymerizable phosphorous ligands and (*R*)-Bu$_2$BINOL to platinum (**45**) and polymerization with EGDMA creates polymers (**P-30** through **P-33b**) with binding sites selective towards (*R*)-BINOL derivatives following removal of the template.

to α,α,α,-trifluoro-*m*-cresol displaced $67 \pm 2\%$ (R)-Bu$_2$BINOL while forming **P-32**. Alternatively, using either BINOL or Br$_2$BINOL displaced $27 \pm 1\%$ and $31 \pm 1\%$, respectively to generate **P-33a** and **P-33b**.

The trend for (R)-Bu$_2$BINOL displacement from **P-30** is dependant on the steric bulk of the incoming ligand, and implies that a distribution of platinum sites with different accessibilities exists throughout the MIP. The most accessible and least restrictive sites are available to the most sterically hindered ligands for (R)-Bu$_2$BINOL displacement. Similarly, these bulky ligands are unable to exchange with (R)-Bu$_2$BINOL in more confined sites. Chloride ion, a much smaller ligand, was able to reach binding sites that are not available to BINOL, and hence more (R)-Bu$_2$BINOL is displaced from **P-30**.

The different accessibilities associated with the binding sites in the MIP prompted Gangé to investigate the selectivity of the binding sites. **P-32** was exposed to excess *rac*-BINOL to assess the effect of imprinted chiral cavities on the selectivity of ligand-exchange reactions at the Pt centers (Scheme 20). The aryl oxide ligands coordinated to the Pt sites in **P-32** are labile and can be easily exchanged using conditions identical to those used to remove the (R)-Bu$_2$BINOL template from the MIP.

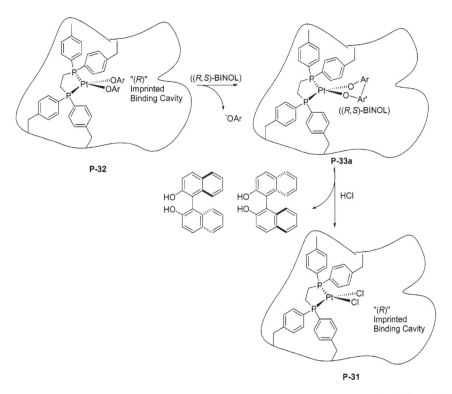

Scheme 20 Imprinted polymer **P-32** is bathed in a racemic solution of BINOL, which displaces the aryl oxide ligands of the platinum center. Removal of BINOL from the polymer matrix and determination of the % ee of BINOL is used to gauge the polymer's selective rebinding ability.

When HCl was used to displace the rebound BINOL for uptake analysis, the % ee BINOL recovered from the MIP is an indication of the selectivity of the most reactive sites (Scheme 20). Figure 10 shows the percentage of total Pt sites in **P-32** rebound by *rac*-BINOL and the % ee of the rebound BINOL as a function of time. It is interesting to note that even after the maximum number of sites rebound is reached (45% after 2 h) the % ee continues to rise. After 8 h, the % ee levels off at approximately 68% ee (not shown on graph). These observations indicate that the selectivity for ligand exchange increases as more Pt sites participate in ligand exchange. This implies that the accessibility of a Pt site plays a role in determining the ability of an imprinted chiral cavity to distinguish between enantiomers of the imprinted ligand. Gangé and coworkers claim that the accessibility of a given Pt site depends on the level of outer-sphere definition of its associated chiral cavity. The highly accessible sites have a poorly defined second coordination sphere and are, therefore, not very selective. In contrast, the Pt sites that have a well-defined coordination sphere due to a highly ordered microenvironment are capable of highly selective rebinding; yet these sites are not as accessible.

Gagné also examined the effects of temperature on the reactivity and selectivity of rebinding reactions. The results showed that increasing the temperature resulted in an increase in the number of Pt sites rebound by *rac*-BINOL and an increase in the % ee of rebound BINOL. These results are interesting and surprising in that the selectivity of the rebinding increases with increasing temperature. To explain this phenomenon, Gagné suggested that the elevated temperature is necessary for the polymer bound Pt[(*R*)-BINOL] and Pt[(*S*)-BINOL] sites to reach equilibrium. The increased temperature allows more sites to become accessible, which effectively enhances the rebinding selectivity. Overall, the behavior of the active sites is dependent on the conditions of the rebinding experiments.

Gagné postulated that sites could be differentiated based on whether the binding sites' selectivity was thermodynamically or kinetically controlled. **P-32** was subjected to rebinding conditions with *rac*-Br$_2$BINOL following removal of excess

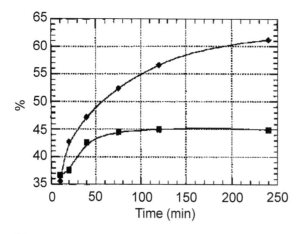

Figure 10   Percent of total Pt sites in **P-32** rebound by *rac*-BINOL (square) and percent ee of rebound BINOL (diamond) as a function of rebinding time for rebinding reactions performed at 40°C.

Scheme 21 Schematic outline used to investigate the kinetic and thermodynamic binding sites within the platinum MIPs.

BINOL left from *rac*-BINOL rebinding (Scheme 21). Br_2BINOL was allowed to exchange with BINOL in the sites that could participate in such an exchange, however BINOL was left bound to Pt in sites that are less accessible. The % ee for Br_2BINOL, therefore, represents the average thermodynamic selectivity of the $n\%$ most reactive Pt sites, while the BINOL % ee reflects the kinetic selectivity of the $m\%$ least reactive sites accessible in the polymer. According to Table 9, an increase in temperature results in sites becoming more accessible and falling into thermodynamic control. The table also emphasizes that the less reactive Pt sites are far more selective.

The work by Gagné and coworkers represents a thorough study with appropriate controls. Trends were dissected and presented in a thoughtful manner. This study of an excellent analysis of binding interactions has made a significant contribution to the imprinting community by demonstrating exhaustive efforts to completely characterize the binding cavity.

A recent publication by Sreenivasan [30] describes an MIP selective for cholesterol, and has several potential applications. In this study, molecular imprinting is utilized to develop materials for the selective absorption of cholesterol. Copper (II) acrylate is used to improve the absorption of cholesterol over other steroids, such as testosterone.

Four polymers were prepared to investigate the absorption of cholesterol using the metal-coordinated approach (Table 10). Two polymers were fabricated using cholesterol as a template with copper(II) acrylate and acrylic acid, MIPCu (**P-34**) and MIPAA (**P-35**), respectively. Control polymers fabricated in the absence of a template using the functional monomers copper (II) acrylate and acrylic acid (AACu (**P-36**) and AAC (**P-37**), respectively) were also investigated. Following template removal, all polymers were soaked in mixed solutions of equimolar amounts of cholesterol and testosterone, and evaluated for the uptake of steroid.

Absorption of substrates by polymers was conducted in methylene chloride. The MIPs selectively absorbed the imprinted cholesterol by a factor of 9:1, regardless of the functional monomer. Also, incorporation of the copper metal resulted in significant increase in the quantity of cholesterol absorbed; **P-34** was shown to

Table 9   Sequential Rebinding of *rac*-BINOL and *rac*-Br_2BINOL to **P-32**

Br_2BINOL rebinding temp (°C)	% Pt sites bound by Br_2BINOL	Br_2BINOL % ee	% Pt sites bound by BINOL	BINOL % ee	% Pt sites rebound (total)
60	41	46	20	89	61
80	46	58	13	94	59
100	50	61	8	94	58

Table 10 Uptake of Testosterone and Cholesterol by **P-34**, **P-35**, **P-36**, **P-37**

Polymer	Testosterone[a]	Cholesterol[a]
P-34	122 ± 4	1217 ± 6
P-35	47 ± 2	440 ± 5
P-36	102 ± 3	64 ± 4
P-37	34 ± 4	39 ± 3

[a]Quantity reported in µg as absorbed by 100 mg of polymer.

adsorb roughly three times more cholesterol than **P-35**. Furthermore, the control polymer binds testosterone preferentially over cholesterol by a ratio of 5:3, whereas **P-34** selectively binds cholesterol over testosterone by a 9:1 ratio. The fact that both the control polymer and the imprinted polymer adsorb roughly equivalent amounts of testosterone is attributed to the nonselective binding associated with the polymers.

Sreenivasan also made qualitative investigations into the extent of hydrogen bonding associated with the template–monomer binding. Carbonyl infrared absorption shifts were measured in the absence and presence of cholesterol. There is a 15 cm^{-1} shift upon cholesterol binding to acrylic acid [1721–1706 cm^{-1}], whereas a more pronounced 26 cm^{-1} shift is observed for the binding of cholesterol to copper acrylate [1729–1703 cm^{-1}]. The larger shift is attributed to an enhanced interaction in the Cu acrylate–cholesterol system.

Unfortunately, no further analyses were conducted on this system. Considering the overall importance and potential for cholesterol binding polymers, it would have been interesting to have more rigorously evaluated the binding isotherm.

Hart and Shea [31] described a peptide–metal interaction in molecularly imprinted polymers for the recognition of N-terminal histidine peptides. The strategy had an advantage over traditional imprinting formulations in that both polymer synthesis and recognition are carried out in aqueous media. Traditional hydrogen bonding interactions between complimentary binding groups are replaced with metal–ligand binding interactions between nickel centers and peptides. The strategy for the formation of selective binding sites takes advantage of the affinity of Ni(II) for N-terminal histidine groups. The nickel center is incorporated into the polymer through coordination to a modified nitrilotriacetic acid complex (NTA, Scheme 22). The uptake and release of amino acids can be controlled with pH, as the imidazole–Ni complex becomes labile under acidic conditions.

The polymerizable methacrylamide–NTA–Ni^{2+} mixed-complex was prepared by combining aqueous solutions of NTA monomer with NiSO$_4$. The prepolymerization complex was then formed by addition of the N-terminal histidine peptide his-ala. The octahedral geometry of **46** was confirmed by comparing the electronic spectra for **46** with known absorbance bands of octahedral complexes of Ni(II) [32]. In addition, negative-ion electrospray mass spectroscopy of **46** showed peaks corresponding to the lithium salt of **46** ($m/z = 546$).

Examination of the his–Ni–NTA complex (**46**) shows six possible diastereomers. However, several constraints lead to a reduction in the actual number of isomers that can be formed. First, the two nitrogen donors of histidine must be *cis*. Second, the three oxygen donors of the NTA ligand must be meridional. The remaining isomers that conform to these constraints can be grouped into two

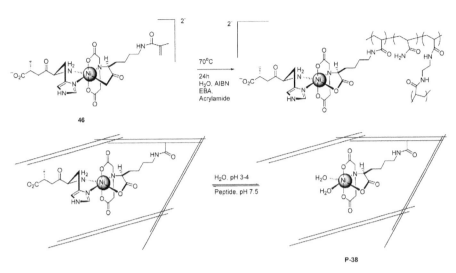

Scheme 22 Schematic representation of the peptide imprinting process using Ni–NTA technology. Copolymerization of the (His-Ala)–Ni–NTA complex (**46**) with mono- and bisacrylamides followed by extraction of the peptide at pH 3-4 provides a polymer (**P-38**) containing Ni–NTA complexes capable of rebinding the template peptide.

categories: those with the imidazole nitrogen of histidine *cis* to the NTA nitrogen and those with a trans orientation. For each of these two groups, there are three possible orientations of the chiral center bearing acetate of the NTA ligand with respect to the chiral center on histidine. Therefore, there are a total of six possible diastereomers for **46**.

Potentiometric studies were conducted on **46**, NTA and his-ala to determine the dominant species in solution at various pH values. Species distribution curves are shown in Fig. 11. The log of the formation constant for the 1:1:1 complex was calculated from these data, and the value ($K_f = 5.03$) compares well with the literature value for the association of histidine with $[Ni(NTA)]^-$ of log. These results indicate that his-ala and histidine have similar association constants for complexation to $[Ni(NTA)]^-$. In addition, the modifications made to the NTA ligand do not appear to have a detrimental effect on the formation of the 1:1:1 complex.

Copolymerization of **46** (5 mol%) with N-N'-ethylenebisacrylamide cross-lining monomer (82 mol%) and acrylamide (13 mol%) provided a pale blue monolith. Polymerization was initiated using the redox initiator system of 0.02 mol% N-N'-tetramethylethylenediamine (TMEDA) and 0.10 mol% ammonium persulfate (APS).

Following removal of the his-ala template by lowering the pH, binding isotherms for the uptake of template his-ala and his-phe to **P-38-P-43** were conducted. Values for B_{max} for these peptides are shown in Table 11. The ratio of the his-ala B_{max} over the his-phe B_{max} is expressed as α, and gives some measure of the selectivity of these polymers in terms of capacity.

Binding studies were performed to evaluate uptake of the template and non-template peptides. As evident in Fig. 12 the (his-ala)-imprinted polymer has a significantly higher capacity for the template peptide over other sequences. The primary binding interaction is the his–Ni interaction, however there are significant

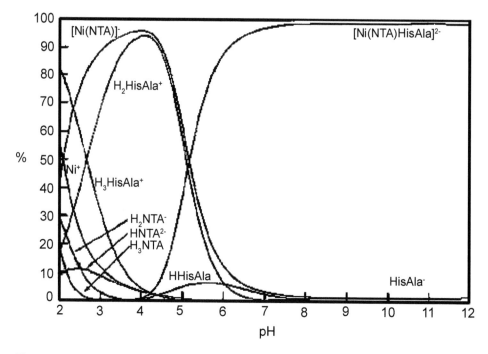

Figure 11 Sepcies distributions of a 0.01 M solution of His-Ala, Ni(II), and NTA ligand for pH values ranging from 2 to 12.

interactions between the peptide and the cavity walls. The binding capacity is reduced to a third for his-phe binding in a his-ala imprinted polymer indicating that there is a size exclusion phenomenon at work for the larger peptides. The tripeptide his-ala-phe has the same binding capacity as his-phe, although there is a distinctly lower uptake at low concentrations of his-ala-phe. The reduced uptake at lower concentrations could be due to a reduced number of strong binding sites available for the tripeptide. This is supported by Scatchard plots for the three peptides (Fig. 13).

There is a distinctly bimodal Scatchard indicating two populations of binding sites with differing affinities for the tri-peptide. In contrast, the data presented for

Table 11 Maximum Binding Capacities for His-Ala and His-Phe Binding, Uptake Selectivity, and Ni(II) Content for **P-33-P-38**

| Polymer | Cross-link (mol %) | B_{max}(μmol bound/μmol Ni) | | α | Ni content (wt%) | |
		His-Ala	His-Phe		Measured	Theoretical
P-38	80	0.44	0.13	3.3	0.51	0.76
P-39	70	0.69	0.32	2.2	0.45	0.81
P-40	60	0.67	0.31	2.2	0.43	0.87
P-41	50	0.83	0.41	2.0	0.42	0.93
P-42	40	0.65	0.36	1.8	0.46	1.01
P-43	30	0.29	0.30	1.0	0.48	1.10

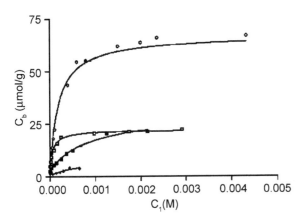

Figure 12 Binding isotherms for the rebinding peptides in aqueous solution to a polymer prepared using His-Ala as the template peptide. His-Ala($\bigcirc$), His-Phe($\square$), His-Ala-Phe($\blacksquare$), Al-Phe ($\Diamond$). A large increase in capacity is found for the template peptide (His-Ala) over other N-terminal histidine di- and tripeptides. Binding of a nonhistidine containing peptide (Ala-Phe) is minimal.

his-ala and his-phe show monomodal binding sites, indicating the binding sites are homogeneous with respect to binding affinity.

Nonspecific interactions were evaluated by fabricating polymers lacking metal centers. Binding isotherms of his-ala to these polymers indicated a B_{max} of 10–20 µmol/g, which is similar to nonimprinted polymers. The low level binding was attributed to the electrostatic interactions between the triacid NTA and the basic sites on the peptide.

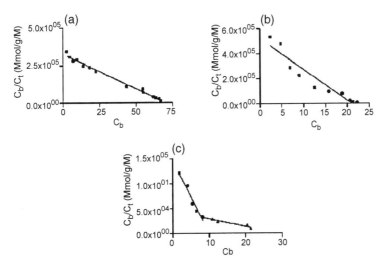

Figure 13 Scatchard plots of data obtained by rebinding peptides to a polymer prepared using His-Ala as the template peptide: (a) template dipeptide His-Ala, (b) dipeptide His-Phe, and (c) tripeptide His-Ala-Phe.

Size exclusion in Ni–NTA MIP's was investigated by using his-phe as the template peptide. The mole fraction of cross-linker used was 50% since this formulation provided the highest selectivity. Results from the binding isotherm studies showed that the values of B_{max} for his-ala, his-phe, and his-ala-phe were nearly identical ($B_{max} \sim 0.7$). These results are in agreement with the theory that the size of the template plays a large role in determining the rebinding capacity. The larger phenylalanine residue allows the formation of larger rebinding cavities, leading to high rebinding capacities for the smaller template.

A bait-and-switch technique was implemented with this system to investigate the role of the metal and its coordination sphere. Nickel was removed from the polymer and replaced by copper, and the binding capacities were measured. Polymers were not prepared in the presence of the copper II complex due to the low polymer yields. The origin of the low yields is thought to occur by inhibition of the free radical polymerization by copper II [33]. Polymers in which nickel is replaced by copper show very similar uptake for imprinted peptides; however, nonimprinted sequences showed a higher uptake than corresponding nickel control polymers. This result indicates that the metal bound to its coordinating ligands must influence the steric environment around the complex.

The effects of increasing the percentage of cross-linker on recognition and rebinding properties was investigated with a series of Ni–NTA polymers fabricated in the presence of his-ala as a template. The ratios for the copolymerization of N-N'-ethylenebis(acrylamide) (EBA) and acrylamide are listed in Table 12. As the percent cross-linking is decreased in the Ni–NTA polymer, there is an increase in binding capacity for both his-ala and his-phe peptides, displayed in the B_{max} values. However, the selectivity of the polymers is decreased as the percent cross-linking is decreased, as is evident in the decreasing α values. The relationship between cross-link density and both selectivity and absolute capacity is shown in Fig. 14. The capacity, as represented by B_{max} for his-ala, is maximized at 50 mol% cross-linker. The decrease in capacity at higher cross-linking is attributed to a larger fraction of sites becoming inaccessible. Alternatively, the decrease in uptake at lower levels of cross-linking could be due to a hydrophobic collapse of the binding cavities. Shea and coworkers state that the origin of the inaccessibility remains unclear.

Table 12 Polymerization Formulation Composition for the Synthesis of EBA Cross-Linked Polymer **P-38-P-44**

Polymer	Template peptide	Mol % cross-linker	NTA complex (mmol)	Acrylamide (mmol)	EBA (mmol)
P-38	His-Ala	80	0.08	0.72	3.20
P-39	His-Ala	70	0.08	1.12	2.80
P-40	His-Ala	60	0.08	1.57	2.35
P-41	His-Ala	50	0.08	1.92	2.00
P-42	His-Ala	40	0.08	2.35	1.57
P-43	His-Ala	30	0.08	2.72	1.20
P-44	His-Ala	10	0.08	3.52	0.40

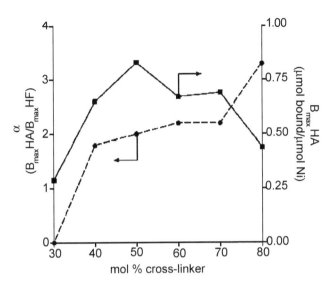

Figure 14 Graph relating the change in α (left Y-axis) and B_{max} HA (right Y-axis) to the mol % cross-linker in the polymer.

There is also a significant dependence of selectivity (α) on cross-link density. The highest degree of selectivity was observed with the highest degree of cross-linking. The observed selectivity may originate from size exclusion principles. This is supported by studies that show the least amount of swelling is observed in polymers with the highest degree of cross-linking, which leads to an overall reduction in accessibility with larger analytes being restricted more than smaller ones, resulting in increased selectivity. These results indicate that there is an optimum cross-link density for achieving selectivity in terms of capacity for these polymers. In addition, cross-link density affects the accessibility of the polymer bound nickel sites in a predictable way.

In this study, the metal–ligand complexes were characterized prior to and after polymerization. Control experiments were also conducted to adequately investigate the possible interactions responsible for the observed selectivity for the imprinted polymer.

C. MIPs as HPLC Stationary Phases

Arnold and coworkers utilized the selective recognition properties of MIPs to develop separation media for imidazole derivatives. They reported the use of a polymerizable copper (II) iminodiacetate (Cu^{II} IDA, **47**) in rigid macroporous polymers. The materials were designed to separate bisimidazole "protein analogs" that cannot be separated by reverse phase HPLC [34]. UV titration experiments conducted on the complex between **48** (Fig. 15) and Cu^{II} IDA-derivatized vinyl monomer **47** revealed a shift in an absorption maximum from 778 to 665 nm, suggesting the presence of a ligand to metal charge transfer between the copper and coordinated imidazole groups. The data also reveal a 2:1 stoichiometry for Cu IDA:**48** with an association constant of 8800 M^{-1}. The complex was polymerized in EGDMA using AIBN as a thermal initiator (Scheme 23). The polymer yields were not reported;

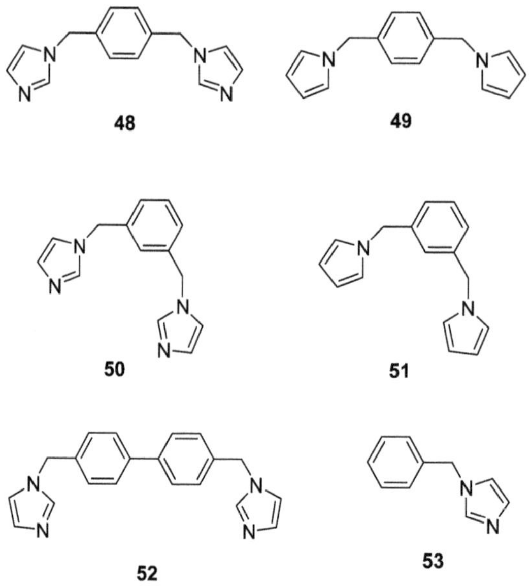

Figure 15 Mono- and bis-imidazole substrates used for molecular imprinting and re-binding studies.

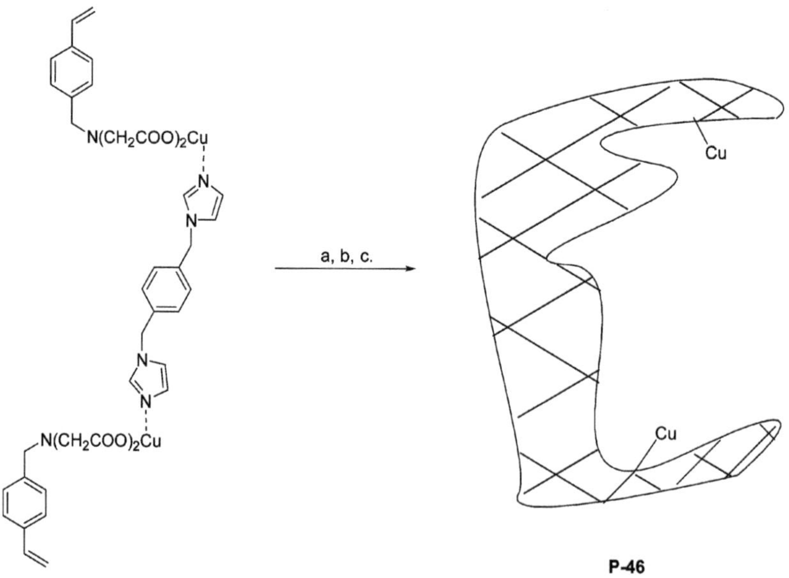

a. EGDMA, MeOH, AIBN, 65°C; b. i. Aq. MeOH, pH 2.5, ii. Triazacyclononane, iii. EDTA; c. CuCl$_2$.

Scheme 23 Template polymerization of Cu(II)-iminodiacetate complexes. Strategically distributed copper binding sites are formed through coordination to bis-imidazoles in prepolymerization mixtures followed by template removal.

however, the resulting polymer was washed in methanol for 24 h to remove an unpolymerized monomer, which was reported as less than 1% indicating a near quantitative polymerization yield. An additional wash with acidified 50% aqueous methanol was carried out to remove the template, followed by washing with EDTA to remove the copper ions. The polymer was reconstituted with copper by bathing in $CuCl_2$ solution. The reloading process was declared quantitative, although no spectral investigation or other analysis was conducted.

Rebinding studies using **48–51** were conducted on bulk polymers imprinted with **48, 50**, or without template (**P-46, P-47**, or **P-45**, respectively). Control polymers were also formulated in the absence of Cu but with template (Table 13). The competitive binding experiments showed very modest results. All of the control polymers failed to distinguish the bis-imidazole isomers, while the imprinted polymers showed slight enhanced selectivity for the imprinted isomer. The average relative substrate selectivity ($\alpha_{i/j}$) measured during competitive binding was 1.16 in favor of the imprinted substrate. It is unclear whether this small enhancement of selectivity could be considered to be beyond statistical error.

Despite the poor resolution, Arnold and coworkers reported further investigations with the Cu^{II} IDA system because of the strong binding associated with copper–imidazole complexes [35]. In fact, systems developed for HPLC chromatography exhibit very long retention times (no peaks were observed using isocratic elution). Imprinted polymers were grafted onto modified silica gel particles to alleviate the problems associated with using ground, bulk MIPs in chromatographic separations. In this technique, silica particles were silanized with

Table 13 Rebinding Studies of Templated Copper-Complexing Copolymers

Entry	Polymer	Original template	Rebinding substrate[a]	Substrate bound [b,c] (mmol/g polymer)	Relative substrate selectivity in competitive binding [b,c]
1	P-45		48	0.42	
2	P-45		50	0.43	
3	P-45		48 + 50		$\alpha_{50/48} = 1.02$
4	P-46	48	48	0.33	
5	P-46	48	50	0.22	
6	P-46	48	49	0.016	
7	P-46	48	48 + 50		$\alpha_{48/50} = 1.17$
8	P-46 (Cu free)	48	48 + 50		$\alpha_{48/50} = 1.04$
9	P-47	50	50	0.24	
10	P-47	50	48	0.17	
11	P-47	50	51	0.014	
12	P-47	50	48 + 50		$\alpha_{50/48} = 1.15$
13	P-47 (Cu free)	50	48 + 50		$\alpha_{50/48} = 1.04$

[a]Saturation binding studies: > five-fold molar excess of substrate over theoretical binding sites.
[b]Amounts bound determined from analysis of unbound substrates after equilibration, using 500 MHz ^{1}H NMR and an internal reverence of known concentration (single substrate binding) or relative peak intensities (for competitive binding).
[c]Competitive binding: equimolar mixuters of substrates used. α_{ij} is the ratio of i to j in the bound state.

3-(trimethoxysilyl)-propylmethacrylate to provide anchoring sites for the MIP. The modified silica particles were combined with EDGMA, $Cu^{II}IDA$, and template and polymerized using AIBN as a thermal initiator. Approximately 97% of the copper was incorporated onto the silica; however, the polymer coating was stripped from the silica particles during chromatographic traces.

Alternatively, the polymerization was carried out using a lower initiation temperature and ammonium peroxysulfate as an initiator. Only 73% of the polymer remained associated with the silica and 71% of the available copper was included in the polymer. However, the coating on the silica beads remained intact after continuous use in chromatographic analyses.

Silica beads coated with polymer imprinted for **48** were slurry packed and evaluated for the separation of **48**–**53** (Fig. 15). All compounds failed to elute through the column until the pH was adjusted to 4.1 after which all compounds eluted simultaneously. Exchanging zinc II ions for copper ions in the coated polymer resulted in beads with lower binding constants for imidazoles. The exchange of ions allows the imidazoles to elute through the column; however, the retention of the substrates was still substantial. According to Table 14 only the mono-imidazole compounds eluded through the column in less than 10 column volumes. The bis-imidazole **48** was retained for over 200 column volumes.

The lower retention times for the mono-imidazole compounds provide evidence for the binding differences between mono- and bis-imidazole compounds on the imprinted columns. During the imprinting process, metal centers are positioned such that two metals can bind to the bis-imidazole template simultaneously. There is only one coordination site involved in the binding to the mono-imidazole compounds, which results in a much lower binding affinity towards the polymer.

Materials were prepared using different templates to validate the concept of two-coordination imprinting (Table 15). Materials prepared using the mono-imidazole **53** as a template were unable to separate the bis-imidazoles. Imprinting for imidazole distributes the binding sites in a random manner; at low loading this prevents the bis-imidazole derivatives from simultaneously coordinating to two binding sites. However, since there is essentially double the opportunity for the substrate to interact with a binding site, all of the bis-imidazoles have longer retention times than the imprinted mono-imidazole. This is in contrast to the polymers imprinted with compounds **48** or **50**, where the template was always the most strongly retained. The authors claimed that there is an inherent cooperative interaction

Table 14 Elution Volumes (V_c) and Chromatographic Separation Factors ($\alpha_{i,j}$) for Substrates on Polymer-coated Silica Prepared using **48** as a Template

Substrate	V_c(column volumes)	$\alpha_{43,j}$
49	0.7	$\rightarrow \infty$
Imidazole	6.8	>33
53	9.5	> 23
48	> 200	—

Copper (II) has been replaced with zinc (II) ions. The 50×4.6 mm I C column was operated at 65°C, 0.5 mL/min 100% methanol, with a sample size of 10 µl of 0.4 mM solution. $\alpha_{i,j} = (V_c - V_0)_j/(V_c - V_0)_i$, where V_0 is the column void volume.

Table 15 Capacity Factors (k') and Chromatographic Separation Factors ($\alpha_{i,j}$) for Imidazole and Substrates **48**, **50**, **52**, and **53** on Polymer-coated Silicas Prepared using **48**, **50**, **52**, or **53** as Templates

| | Material prepared using: | | | | | | | |
| | 53 | | 48 | | 50 | | 52 | |
Substrate	k'	($\alpha_{53,j}$)	k'	($\alpha_{48,j}$)	k'	($\alpha_{50,j}$)	k'	($\alpha_{52,j}$)
Imidazole	0.75	1.16	0.69	8.1	1.5	5.8	1.7	6.6
53	0.87	—	0.69	8.1	1.7	5.1	1.8	6.3
48	2.9	3.0	5.6	—	6.6	1.4	7.7	1.5
50	2.6	0.33	3.5	1.6	8.7	—	6.3	1.8
52	2.7	0.32	1.5	3.9	2.9	3.0	11.3	—

Elution volumes (V_c) were measured on 50×4.6 mm ID columns, 0.5 mL/min 100% methanol, 65°C, with a sample size of 10 μl of 0.4 mM solution. The mobile phase contained zinc acetate in the following concentrations: experiments on **53**-templated material, 50 mM; **8**-templated, 50 mM; **50**-templated, 40 mM; **52**-templated, 30 mM; $k' = (V_c - V_0)_j/V_0$; $\alpha_{i,j} = (V_c - V_0)_j/(V_c - V_0)_i$, where V_0 is the column void volume.

between the multiple binding sites, which allows two binding sites to be closely oriented to one another. Nonetheless, the separation of the bis-imidazoles is modest, at best.

The nature of the binding interactions between substrate and polymer was investigated with equilibrium rebinding studies using ServaGel particles coated with molecularly imprinted polymer. Scatchard plots from these studies are shown in Fig. 16. Polymer imprinted with bis-imidazole **48** exhibits a significant heterogeneity. The binding affinity is highest at low loading levels while a dramatic falloff occurs as more sites become occupied. In contrast, the binding of **53** in the same polymer does not show the same heterogeneity. For the **53**-imprinted polymer the binding is homogeneous to both mono- and bis-imidazole substrates.

These results suggest that the polymers fabricated in the presence of bis-imidazole templates recognize the template by a different mechanism from **53**. The data also suggest that the template influences the organization of metal ions in the polymer, although the extent of the template influence is still unclear.

In order to lessen the retention times, the copper metal sites were exchanged with Zn^{2+} by washing the columns with 10 mM diethylenetriamine in methanol. Zinc acetate was also added to the mobile phase to further reduce the retention times of the substrate. The additional metal ions present in the eluent result in the imidazole having a lower affinity for the binding sites within the polymer network. Polymer-coated silica prepared using template **48** can separate bis-imidazoles **48** and **52** to baseline resolution; however, only a partial separation of **48** and **50** is achieved. This bait-and-switch technique allows a metal with strong binding interactions, like copper, to be used during imprinting in order to create highly defined binding sites while incorporating a weaker chelating agent during separation schemes.

In a related study, Arnold and coworkers [36] applied $Cu^{II}IDA$ ligands to ligand exchange chromatography (LEC) suitable for chiral separations of underivatized amino acids. The use of an achiral ligand (IDA) shows the enantioselectivity of the adsorbent arises from the chirality of the recognition sites created during

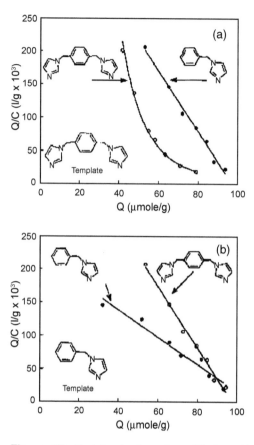

Figure 16 Scatchard plot of equilibrium binding data for substrates **53** and **48** on MIP-coated silicas (ServaGel). Polymerization reaction included monomer at 50 wt% to silica, composition: 5% CuIDA, 30% methylmethacrylate, 65% EGDMA. Q = concentration of substrate bound (µmol/gram). C = concentration of substrate in solution (mM); (a) **53** and **48** on polymer templated with **48**; (b) **43** and **48** on polymer templated with **53**.

polymerization. The supports were fabricated by grafting the imprinted polymer to modified silica particles. Bulk polymers were also prepared for rebinding studies. Following polymerization, the imprinted material was removed followed by removal of the copper center. The copper was reconstituted within the polymer using 100 mM solution of CuSO$_4$.

Several amino acids were utilized as a template during polymerization (Phe, Tyr, Ala, Val, Leu, Ile), by complexation to the copper structure (Fig. 17). The association constants for the amino acids are of the order of 10^3–10^4 M^{-1}, whereas K_{IDA} is of the order of 10^{11} M^{-1}. This assures that the amino acids are not capable of displacing the IDA ligand from the copper center. Experimental conditions called for greater than twofold excess amino acid during polymerization procedures in order to ensure a 1:1 complex between the amino acid and the copper center. Isothermal titration calorimetry (ITC) experiments indicate that the binding between Cu–IDA and L-phenylalanine reaches saturation after two equivalents of amino acid.

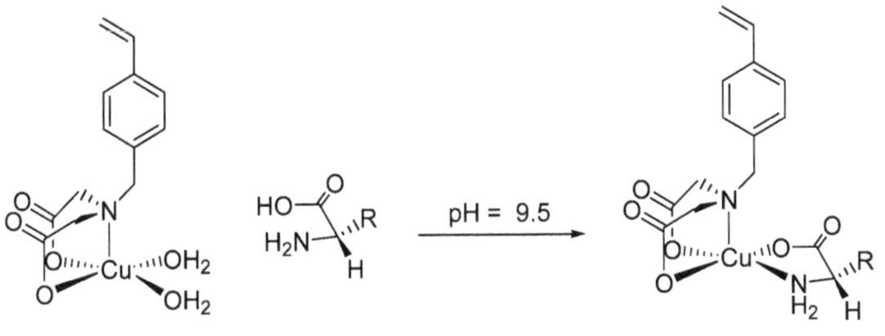

Figure 17 The achiral functional monomer, Cu(II)-[*N*-(4-vinylbenzyl)]-iminodiacetate and complexation to phenylalanine.

Bulk polymers were used for rebinding studies using racemic amino acid solutions. Polymers were bathed in solutions for 24 h and analyzed for the selective uptake of the amino acid. Polymers lacking copper ions showed negligible uptake of amino acid indicating that the metal–amino acid complex is necessary for the recognition process and indicates nonspecific binding is not a major component for amino acid rebinding in the MIPs. These results call for a more careful assessment of the binding interaction of the metal–amino acid complex. While Arnold and coworkers claim a 1:1 complex is involved in the prepolymerization mixture, there is a twofold excess of amino acid.

Polymers imprinted for amino acids with larger side groups generally displayed a higher, yet relatively modest, enantioselectivity. Also, polymers are able to separate amino acids with functional groups that are similar to the imprint amino acid. For example, polymers imprinted for L-phenylalanine displayed a selectivity of 1.45 for L-phenylalanine; 1.36 for L-tyrosine; and 0.99–1.08 for alanine, valine, leucine, and isoleucine. This indicates that the larger cavity created by phenylalanine does not have the ability to differentiate the smaller side chains of the smaller amino acids. Likewise, polymers imprinted with smaller amino acid side chains showed moderate selectivity for similar amino acids, but could not differentiate the larger amino acids.

Acetate buffer (pH = 8) could not effectively elute amino acids during chromatographic separation; therefore, it was necessary to use 1.5 mM glycine solution. Imprinted amino acids eluted between 5 and 8 column volumes, whereas the corresponding enantiomers eluted in 3–4 column volumes. Unfortunately, the imprinted polymers could not achieve baseline separations. The nonimprinted enantiomer was eluted as a sharp peak followed by a broad, overlapping peak, which corresponded to the imprinted enantiomer. Nonimprinted polymers did not show any ability to separate racemic solutions.

Interestingly, polymers imprinted for L-phenylalanine were able to separate tyrosine to the same extent as phenylalanine. However, the column was unable to separate phenylalanine–tyrosine mixtures. Polymers imprinted for other amino acids such as Trp, Ala, Val, Leu, Ile were not able to separate the corresponding racemic mixtures. It appears that the polymers are unable to differentiate small variations in the smaller amino acid side chains. Arnold and coworkers attributed the ability of a polymer to separate materials to a three-point interaction system between substrate and polymer. This type of mechanism was investigated with phenylalanine derivatives.

It was suggested that the formation of the polymer network around the monomer–template complex stabilizes the binding of template yet destabilizes the enantiomer by preventing at least one binding point interaction (Fig. 18). For example, binding of phenylalanine is accomplished through metal–ligand interactions between the amino and carboxyl groups of the amino acid and the metal center. It is claimed that the third interaction is the steric interaction between the amino acid and the polymer walls.

To test the "three-point interactions" requirement, racemic phenylalanine derivatives α-methylphenethylamine, 54, and α-methylhydrocinnamic acid, 55, were separated using a column imprinted for D-phenylalanine. Chromatographic separations were carried out in 1.5 mM acetate at pH 8. The D-phenylalanine

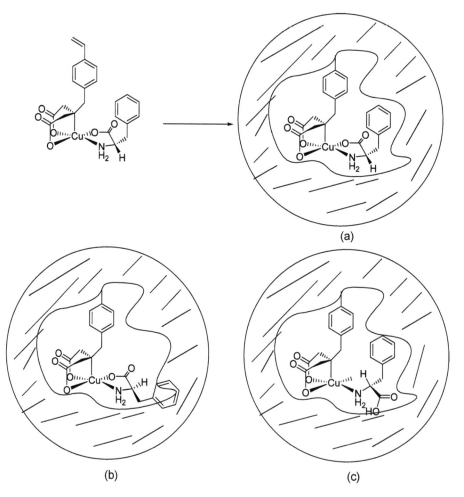

Figure 18 (a) Polymerization of Cu–IDA complexed to L-phenylalanine. Uptake of D-phenylalanine results in either (b) unfavorable interactions with the substrate and polymer wall or (c) one less binding interaction between the substrate and Cu center.

imprinted column was able to resolve the chiral amine **54**; however, it was unable to resolve **55**. The authors claim that the acetate is a stronger binding ligand than the carboxylic acid and therefore at least one binding site is lost. They conclude that the steric interactions between the side chain and the binding cavity involve more than one point of contact. While the carboxyl group must be discounted as a viable binding interaction, there is no alternative structure offered for the amino acid–metal complex. These experiments raise further questions into the true interactions between Cu–IDA and the amino acids.

Matsui and coworkers reported the use of cobalt ion MIPs for chromatography based recognition studies on imprinted compounds. The authors chose to utilize an imprinting system described previously for the catalysis of aldol condensations (vide supra). This system was shown to be amenable to the study of MIP–metal ion mediated recognition. Preliminary studies were conducted to provide evidence for the complex formation between cobalt, polymerizable ligands, and dibenzoyl-methane, **28**. Compleximetric titration of **28** in a model prepolymerization reaction mixture containing cobalt (II) acetate and pyridine in chloroform/methanol (5:1) showed formation of a complex with 1:1 stoichiometry between **28** and Co(II) (Fig. 19).

In addition to the MIP (**P-48**) designed to selectively retain **28**, reference polymers were prepared including polymers containing only cobalt (II) acetate (**P-49**), only **28** (**P-50**), acetylacetonate–cobalt (II) acetate complex (**P-51**), and no imprint species (**P-52**). All polymers contained divinylbenzene and styrene as cross-linking agents, 4-vinylpyridine as the functional monomer, and AIBN as the initiator. Polymers were then ground, sieved, loaded into PTFE-lined liquid chromatography columns, and washed with EDTA buffer to ensure the removal of solvent-accessible metal ions.

Optimal recognition of **28** by the imprinted polymer **P-48** was achieved at 1.0 mM concentration of cobalt ion in the eluent (Fig. 20). In addition, the capacity factors for **P-48** are greater than those for the other reference polymers, indicating that the selective coordination sites have a higher affinity for **28** than the control

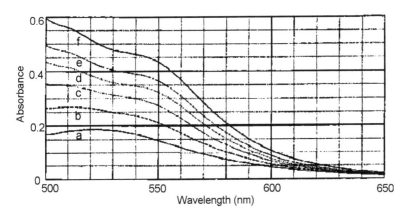

Figure 19   (a) Visible region spectra of cobalt (II) acetate (10 mM) and pyridine (2 eq) in chloroform/methanol (5:1, w/w) with 1:0 eq, (b) 0.125 eq, (c) 0.25 eq, (d) 0.375 eq, (e) 0.5 eq, (f) 1.0 eq dibenzoylmethane.

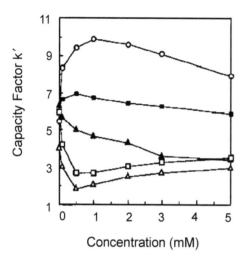

Figure 20   Capacity factors (k') for dibenzoyl methane recognition as a function of eluent cobalt (II) acetate concentration. Open circle-**P-43**, closed square-**P-44**, open triangle-**P-45**, closed triangle-**P-46**, open square-**P-47**.

polymers. Interestingly, **P-49** capacity factors are greater than those for **P-50**, indicating that shape induced recognition alone is not sufficient for stabilizing the binding of **28**. Taken together, the data which support the notion of enhanced binding selectivity are developed through cobalt-**28** interactions, which aide in the recognition of the template.

Alternative analytes were selected to investigate the binding selectivity of the imprinted polymers. Table 15 lists the capacity factors of each polymer for the retention of acetylacetone, **54**, benzoylacetone, **55**, or **28**. The difference between recognition in the presence and absence of cobalt ion provides a measure for the extent of selective binding. Entries under the **P-48** polymer in column (a) for each analyte seem to indicate specific interactions between the phenyl rings of the analyte and the polymer contributes to the recognition of the imprinted template. The number of $\pi-\pi$ stacking interaction between the polymer and substrate increases from **54** to **55** to **28**.

Table 16   Ligand-polymer Selectivity Factors

Polymer	56		57		28	
	a	b	a	b	a	b
P-48	0.8	0.4	3.1	1.9	9.3	5.4
P-50	0.1	0.5	0.8	1.5	2.5	4.7
P-49	1.0	0.5	3.1	1.9	7.0	5.6
P-51	0.5	0.3	1.0	1.4	5.2	5.2
P-52	0.3	0.3	0.3	1.5	3.6	5.2

Elution buffers: (a) methanolacetic acid (3% v/v) + 1.0 mM Co(II) acetate, (b) methanol/acetic acid (3% v/v).

Table 17 Cation Effects on Recognition of **28** on **P-48** and **P-52**

Eluent[a]	Capacity factor k'^{b}		a/b
	a	b	
Co(II)	9.9	2.7	3.7
Ni(II)	13.9	7.1	2.0
Zn(II)	6.6	5.2	1.3

[a]Methanolic Metal (II) acetate (1.0 mM) was used as the eluent to determine the capacity factor for **28**.
[b]Column a refers to k' for **P-48**, while column b refers to k' for **P-52**.

Matsui and coworkers examined the role of metal ions in the eluent by substituting nickel (II) and zinc (II) acetate in the elution buffer. The data shown in Table 16 reveal superior ligand selectivity in the case of cobalt (II) acetate. It was suggested that the distinct coordination geometries of each metal ion have caused the large differences. Cobalt (II) and zinc (II) prefer to exist in a tetrahedral geometry, while nickel (II) centers are known to reside in square-planar orientations. Although the nickel (II) mediated binding is stronger than that of Co(II), as evident by the higher capacity factor, it is not as selective. The authors suggest that strong nonspecific binding modes are responsible for the large capacity factors (Table 17).

Recently, Takeuchi and coworkers [37] reported the use of molecular imprinting for constructing a highly specific porphyrin-based receptor site. 9-Ethyladenine [37] was chosen as the imprint molecule. Two different functional monomers were utilized to bind **58** during the polymerization process, methacrylic acid (MAA) and a polymerizable zinc–porphorin derivative, **59** (**P-53**), as shown in Fig. 20. Reference polymers imprinted with **56** were fabricated using either **59** or MAA (**P-54** and **P-55**, respectively) and corresponding nonimprinted, blank polymers were prepared using MAA and **59**, MAA, or **57** as functional monomers to form polymers **P-56**, **P-57**, and **P-58**, respectively (Fig. 21).

Polymers were used as stationary phases for chromatography in order to evaluate binding characteristics. All nonimprinted polymers allowed rapid elution of **58**,

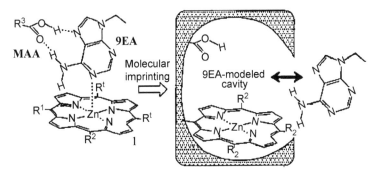

Figure 21 Schematic representation of molecular imprinting of 9-ethyl adenine (9EA) using 5,10,15-tris(4-isopropylphenyl)-20-(4-methacyloloxyphenyl)porphyrin zinc (II) complex and methacrylic acid as a functional monomer.

Table 18   Retention Property of **P-53–P-58**

Polymer	Retention factor			
	58	**62**	**60**	**61**
P-53	28.9	0.0	2.23	0.79
P-54	0.85	2.25	0.52	0.04
P-55	9.57	0.0	1.41	0.48
P-56	0.14	0.58	0.39	0.28
P-57	0.15	0.47	0.86	0.15
P-58	0.09	0.19	0.37	0.14

A mixture of dichloromethane, methanol, and acetic acid (97:2:1, v/v/v) was used as the eluent at a flow rate of $0.5\,L\,min^{-1}$. The column size was $100\,mm \times 4.6\,mm$ i.d. and the sample size was $20\,mL$ ($1.0\,mM$). The elution was monitored by UV absorption at 260 nm. 4-Aminopyridine (4AP) and 2-aminopyridine (2AP) were tested because of their structural analogy to 9EA.

while the imprinted polymers showed significant retention (Table 18). **P-53** displayed the longest retention for the imprint compound indicating a cooperative binding effect. Also noteworthy is the fact that **P-54** does not discriminate between **57** and similar derivatives of 4-aminopyridine, **60**, and 2-aminopyridine, **61**, or its precursor adenine, **62**. However, **P-53** shows significant ability to differentiate between all substrates. A nonlinear Scatchard plot was observed for **57** binding in **P-53**, which suggests the presence of binding sites with various affinities. Only the concentration range exhibiting strong binding site interactions was examined further.

Fluorescence spectra of the polymers were used to investigate the porphyrin-binding sites (Fig. 22). The nonlinear profile suggests that the quenching is due to the binding of **57** to the porphyrin-based recognition site center. This is also

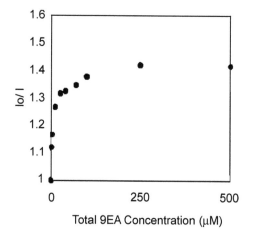

Figure 22   Fluorescence intensity at emission maximum around 605 nm (ex: 423 nm) of **P-56** suspension (2.0 mg) incubated with **55** in dichloromethane (2.0 mL). Dichloromethane was chosen due to suitability for the metal coordination and hydrogen bond formation expected between the polymers and **55**, and the density close to that of the polymer tested for stable suspension.

indicative of porphyrin residues involved in the high affinity binding sites observed in the Scatchard plot at the low concentration range. Competition experiments were conducted using 4-aminopyridine as a reference ligand between **P-53** and **P-54**. **P-53** displayed a superior selectivity for **57** to **P-54** in fluorescent extinction sensitivity. These results show that porphyrin-based binding sites formed in **P-53** uptake ligands based on a different mechanism from that of **P-54**.

IV. CONCLUSIONS AND FUTURE OUTLOOK

The role of metal ions in molecular imprinting has become an active area of research over the last five years, and has provided important steps in advancing the technology of molecular imprinting. To date, studies have shown that metal–ligand interactions can provide strong interactions that can lead to a decrease in nonspecific binding typically found in imprinted polymers. To date, the most successful examples have utilized substitution inert metal centers. The use of labile metal complexes often leads to systems with increased nonspecific binding and lower specificity for analyte recognition.

Studies have shown that MIPs with metal-coordinating binding sites are capable of recognizing and separating structurally similar compounds. In addition, the development of MIP usage in aqueous systems provides several advantages over organic systems. Particularly for imprinting molecules of biological interest, further research in these areas can provide valuable information regarding the nature and mechanisms related to recognition and separation by molecularly imprinted polymers provided proper control experiments are conducted and the phenomena found in systems are properly and thoroughly investigated.

REFERENCES

1. (a) Bertini, I.; Gray, H.B.; Lippard, S.J.; Valentine, J.S. *Bioinorganic Chemistry*; University Science Books: Mill Valley, CA, 1994; (b) Lippard, S.J.; Berg, J.M. *Principles of Bioinorganic Chemistry*; University Science Books: Mill Valley, CA, 1994; (c) Kaim, W.; Schwederski, B. *Bioinorganic Chemistry: Inorganic Elements in the Chemistry of Life*; John Wiley & Sons: New York, 1994.
2. Siegel, H. In *Metal Ions in Biological Systems*; Siegel, A.; Siegel, H. Eds.; Dekker: New York, 1976.
3. (a) Dugas, H. *A Chemical Approach to Enzyme Action*; 2nd Ed; Springer-Verlag: New York, 1989; (b) Fersht, A. *Enzymer Structure and Mechanism*; 2nd Ed; W. H. Freeman: New York, 1985; (c) Wulff, G. Molecular imprinting in cross-linked materials with the aid of molecular templates—a way towards artificial antibodies. Angew. Chem. Int. Ed. Engl. **1995**, *34*, 1812–1832; (d) Wulff, G. Enyzme-like catalysis by molecularly imprinted polymers. Chem. Rev. **2002**, *102*, 1–27.
4. He, X.; Antonelli, D. Recent advances in synthesis and applications of transition metal containing mesoporous molecular sieves. Angw. Chem. Int. Ed. Engl. **2002**, *41*, 214–229.
5. Palmera, M.J.; Wills, M. Asymmetric transfer hydrogenation of C=O and C=N bonds. Tetrahedron: Asymmetry **1999**, *10*, 2045–2061.
6. Noyori, R.; Hashiguchi, S. Asymmetric transfer hydrogenation catalyzed by chiral ruthenium complexes. Acc. Chem. Res. **1997**, *30*, 97–102.
7. ter Halle, R.; Schulz, E.; Lemaire, M. Heterogeneous enantioselective catalytic reduction of ketones. Synlett **1997**, 1257–1258.

8. Polborn, K.; Severin, K. Molecular imprinting with an organometallic transition state analogue. Chem. Commun. **1999**, 2481–2482.

9. Polborn, K.; Severin, K. Biomimetic catalysis with immobilised organometallic ruthenium complexes: substrate- and regioselective transfer hydrogenation of ketones. Chem. Eur. J. **2000**, *6*, 4604–4611.

10. Polborn, K.; Severin, K. Biomimetic catalysis with an immobilised chiral rhodium(III) complex. Eur. J. Inorg. Chem. **2000**, 1687–1692.

11. Mashima, K.; Abe, T.; Tani, K. Asymmetric transfer hydrogenation of ketonic substrates catalyzed by (eta(5)-C5Me5)MCl complexes (M=Rh and Ir) of (1*S*,2*S*)-*N*-(*p*-toluenesulfonyl)-1,2-diphenylethylenediamine. Chem. Lett. **1998**, 1199–1200.

12. Carmona, D.; Lahoz, F.J.; Atencio, R.; Oro, L.A.; Lamata, M.P.; Viguri, F.; San Jose, E.; Vega, C.; Reyes, J.; Joo, F.; Katho, A. Trimerisation of the cationic fragments [(eta-ring)M(Aa)](+)((eta-ring) M=(eta(5)-C5Me5)Rh, (eta(5)-C5Me5)Ir, (eta(6)-p-MeC(6)H(4)iPr)Ru; Aa=alpha-amino acidate) with chiral self-recognition: Synthesis, characterization, solution studies and catalytic reactions of the trimers [{(eta-ring) M(Aa)}(3)](BF4)(3). Chem. Eur. J. **1999**, *5*, 1544–1564.

13. Locatelli, F.; Gamez, P.; Lemaire, M. Molecular imprinting of polymerised catalytic complexes in asymmetric catalysis. J. Mol. Cat. **1998**, *135*, 89–98.

14. Karmalkar, R.N.; Kulkarni, M.G.; Mashelkar, R.A. Pendent chain linked delivery systems: II. Facile hydrolysis through molecular imprinting effects. J. Control. Release **1997**, *43*, 235–243.

15. Karmalkar, R.N.; Kulkarni, M.G.; Mashelkar, R.A. Molecularly imprinted hydrogels exhibit chymotrypsin-like activity. Macromolecules **1996**, *29*, 1366–1368.

16. Kunitake, T.; Okahata, Y. Multifunctional hydrolytic catalyses. 7. Cooperative catalysis of the hydrolysis of phenyl esters by a copolymer of *N*-methacryloyl hydroxamic acid and 4-vinyl imidazole. J. Am. Chem. Soc. **1976**, *98*, 7793–7799.

17. Lele, B.S.; Kulkarni, M.G.; Mashelkar, R.A. Molecularly imprinted polymer mimics of chymotrysin 1. Cooperative effects and substrate specificity. React. Funct. Polym. **1999**, *39*, 37–52.

18. Danishefsky, S. Catalytic antibodies and disfavored reactions. Science **1993**, *259*, 469–470.

19. Matusi, J.; Nicholis, I.A.; Karube, I.; Mosbach, K. Carbon-carbon bond formation using substrate selective catalytic polymers prepared by molecular imprinting: an artificail class II aldolase. J. Org. Chem. **1996**, *61*, 5414–5417.

20. Shea, K.J.; Sasaki, D.Y. An analysis of small-molecule binding to functionalized sythetic polymers by 13C CP/MAS NMR and FT-IR spectroscopy. J. Am. Chem. Soc. **1991**, *113*, 4109–4120.

21. Krebs, J.F.; Borovik, A.S. Dioxygen binding to immobilized Co^{II} complexes in porous organic hosts: evidence for site isolation. Chem. Commun. **1998**, 553–554.

22. Sharma, A.C.; Borovik, A.S. Design, synthesis, and characterization of templated metal sites in porous organic hosts: application to reversible dioxygen binding. J. Am. Chem. Soc. **2000**, *122*, 8946–8955.

23. Cesarotti, M.; Gullotti, A.; Pasini, A.; Ugo, R. Optically active complexes of Schiff bases. Part 5. An investigation of some solvent and conformational effects on the equilibriums between cobalt(II) Schiff-base complexes and dioxygen. J. Chem. Soc. Dalton Trans. **1977**, *8*, 757–763.

24. Walker, F.A.; Bowen, J. EPR Evidence for hydrogen bond donation to the terminal oxygen of $Co-O_2$ model compounds and cobalt oxymyoglobin. J. Am. Chem. Soc. **1985**, *107*, 7632–7635.

25. Padden, K.M.; Krebs, J.F.; MacBeth, C.E.; Scarrow, R.C.; Borovik, A.S. Immobilized metal complexes in porous organic hosts: development of a material for the selective and reversible binding of nitric oxide. J. Am. Chem. Soc. **2001**, *123*, 1072–1079.

26. Fujii, Y.; Kikuchi, K.; Matsutani, K.; Ota, K.; Adachi, M.; Syoji, M.; Haneishi, I.; Kuwana, Y. Template synthesis of polymer schiff base Cobalt (III) complex and formation of specific cavity for chiral amino acid. Chem. Lett. **1984**, *13*, 1487–1490.

27. Fujii, Y.; Matsutani, K.; Kikuchi, K. Formation of a specific co-ordination cavity for a chiral amino acid by template synthesis of a polymer schiff base cobalt(III) complex. J. Chem. Soc. Chem. Commun. **1985**, 415–417.

28. Fujii, Y.; Sano, M.; Nakano, Y. Bull. Chem. Soc. Jpn. **1977**, *50*, 2609.

29. Brunkan, N.M.; Gagne, M.R. Effect of chiral cavities associated with molecularly imprinted platinum centers on the selectivity of ligand-exchange reactions at platinum. J. Am. Chem. Soc. **2000**, *122*, 6217–6225.

30. Sreenivasan, K. The use of metal-containing monomer in the preparation of molecularly imprinted polymer to increase the adsorption capacity. J. Appl. Polym. Sci. **2001**, *80*, 2795–2799.

31. Hart, B.R.; Shea, K.J. Synthetic peptide receptors: molecularly imprinted polymers for the recognition of peptides using peptide–metal interactions. J. Am. Chem. Soc. **2001**, *123*, 2072–2073.

32. Geenwood, N.N.; Earnshaw, A. In *Chemistry of the Elements*, 2nd ed., Butterworth-Heinmann: Oxford, 1999, 1144–1172.

33. Dhal, P.K.; Arnold, F.H. Substrate selectivity of molecularly imprinted polymers incorporating a rigid chelating monomer, bis-methacrylato(4-methyl-4′-vinyl)-2,2′-bipyridine Cu(II). New J. Chem. **1996**, *20*, 695–698.

34. Dhal, P.K.; Arnold, F.H. Template-mediated synthesis of metal-complexing polymers for molecular recognition. J. Am. Chem. Soc. **1991**, *113*, 7417–7418.

35. Plunkett, S.D.; Arnold, F.H. Molecularly imprinted polymers on silica: selective supports for high-performance ligand-exchange chromatography. J. Chromatogr. A **1995**, *708*, 19–29.

36. Vidyasankar, S.; Ru, M.; Arnold, F.H. Molecularly imprinted ligand-exchange adsorbents for the chiral separation of underivatized amino acids. J. Chromatogr. A **1997**, *775*, 51–63.

37. Matsui, J.; Higashi, M.; Takeuchi, T. Molecularly imprinted polymer as 9-ethyladenine receptor having a porphyrin-based recognition center. J. Am. Chem. Soc. **2000**, *122*, 5218–5219.

7

Synthesis and Selection of Functional and Structural Monomers

Olof Ramström   Royal Institute of Technology, Stockholm, Sweden

I. INTRODUCTION

Essential to all successful preparations of molecularly imprinted materials is a careful choice of the imprinting components, well emphasized in most other chapters in this book. These components, or building blocks, are in most instances reactive monomers, capable of forming network polymers or gels that are sufficiently stable to maintain a memory for the template. Not only are the building blocks essential to the molecular recognition of the ligand, but also to the various techniques of preparing the matrices and the physical and chemical properties of the final materials. All imprinting protocols are therefore highly dependent on the components used.

Two major types of materials have until now been applied to molecular imprinting, either organic polystyrenes/polyacrylates, or inorganic polysiloxanes. Variations are abundant but these have nevertheless been the most popular, sometimes also used in combination. In this chapter, a survey of building blocks is given and basic synthetic schemes presented. Since organic matrices are by far the more important, organic polymeric systems based on free-radical polymerization of vinyl monomers are exclusively covered. An account on polysiloxane chemistry is covered by Chapter 11 in this volume.

II. BUILDING BLOCKS IN MOLECULAR IMPRINTING

In most cases, the structure of the target molecule is the starting point to molecularly imprinted materials, and the basic design of how to achieve template recognition has to be sufficiently deduced. Equally important is the choice of the chemistry used (organic, inorganic), as well as the basic format of the imprinted material which requires special considerations. Finally, application of the imprinted matrices to certain functions may involve restrictions on the recognition as well as the chemistry and the format. Any of these prerequisites may set the limits of the others and, depending on situation, govern the system design.

Central to all imprinting protocols is the actual recognition event taking place between the matrix and the imprinted molecule and its analogues. Although the imprinting technique to some extent relies on empirical strategies, many experimental parameters remain to be fully mastered, it is mainly through a (semi)rational design of the recognition sites that the desired effect can be accomplished. This is usually done following one or more strategies. Most straightforward is to design the system by judicious choices from the imprinting literature (cf. the database of the Society for Molecular Imprinting: www.smi.tu-berlin.de). Other methods involve the study of nonpolymeric model systems using common analytical and spectroscopic techniques, and/or by in silico methods (cf. Chapter 14). When these fundamental recognition issues have been determined, the experimenter has to adapt these to the suitable/chosen material structure and format. Normally, these involve the use of monomers for (free) radical polymerization, or for silica gel formation, applied to various preparation techniques depending on structure and function. As mentioned in other chapters in this volume (e.g., Parts II–III, Chapter 15), many other systems have been developed as well, including different radical, and ionic polymerization techniques, as well as hybrid organic–inorganic materials. Obviously, different choices of building blocks have to be made in each case.

Once the design of these recognition systems has been sufficiently sorted out, the next step is to acquire the necessary templates and building blocks and to prepare the preorganized/prearranged structures. In many cases, the acquisition of these building elements is very straightforward since many building blocks are available from general chemical suppliers, or by more specialized companies (see list of suppliers in Appendix). However, often a special building block is necessary for acquiring the desired effect and in this case a certain degree of synthetic chemistry needs to be undertaken. This chapter gives an introduction to basic reaction protocols and techniques that can be used to prepare the desired building blocks from chosen recognition motifs—or so called *Functional Elements*. Several other chapters in this book also contains protocols on building blocks that have been specially prepared for certain imprinting systems. For example, Chapters 3–6 all contain protocols for preparing monomers suitable for covalent, noncovalent, and metal-coordination systems.

Finally, when all elements have been accessed, the imprinted materials can be prepared and the desired effects probed. The actual imprinting (polymerization) step always needs to be more or less empirically adjusted and optimized, the morphology of the material being dependent on experimental setup (choice of porogenic solvent, molar ratios of incorporated building blocks, temperature, etc.). Nevertheless, a recognition system that clearly works in nonpolymeric model systems often functions to some extent in the polymeric state as well. A secondary imprinting effect, as created by structural implications during polymerization, is nontheless more difficult to control, relying on parameters that are more obscure. The imprinting literature has however become very rich, and many of these parameters have been largely sorted out.

III. THE FUNCTIONAL ELEMENT (FE)

As stated, perhaps most often the molecular recognition of the target species constitutes the starting point in MIP design and preparation. Thus, the functionality matching between the target and possible *Functional Elements* (FEs) are of particular

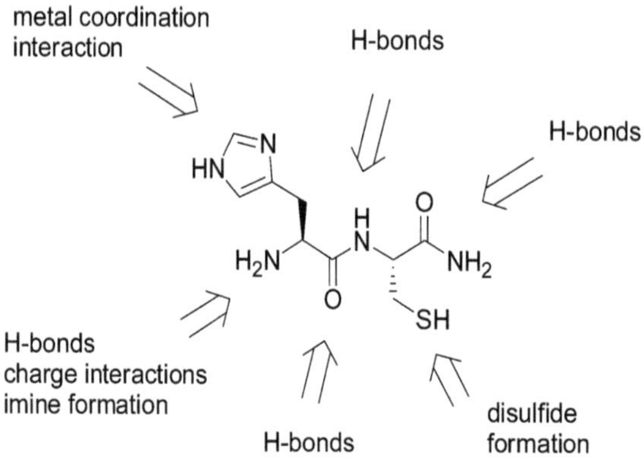

Figure 1 Example of target molecule (L-histidinyl-L-cysteine-amide) containing a variety of functional elements. The molecule may potentially form several noncovalent bonds (hydrogen bonds, ionic interactions, metal coordinations), as well as reversible covalent bonds (imines, disulfides).

importance in order to optimize recognition efficiency. In this context, an FE is any partial structure that can interact with or bind to the target species. Rules for identifying such FEs for certain functionalities of the target species can be found in several other chapters in this volume, and especially Part II and Chapters 8, 14 and 16.

For example, consider the structure in Fig. 1. The target species carries specific functionalities that can be matched by complementary functional elements, and often several matching options can be distinguished in each case. The matching partners may, for example, be hydrogen donors/acceptors that are complementary to hydrogen bonding motifs in the target molecule; charged or polar moieties for electrostatic interactions between permanent charges or dipoles; metal-coordination ligands for metal ions, or (transition) metal ions for coordinating ligand motifs; functional groups that may form reversible covalent bonds to specific counterparts in the target species; etc. These functionalities, either alone or in combinations, may then constitute the FE (Table 1).

IV. FUNCTIONAL AND STRUCTURAL MONOMERS

Basically, there are two types of building blocks used in imprinting protocols—so-called functional building blocks and structural building blocks (Table 2). In the first group lie all components that are directly interacting with, or binding to, the template molecule during polymerization and subsequent recognition. The second group is composed of organizational components that are mainly creating the polymeric network, building up the shape of the recognition site and supporting the functional components in the right three-dimensional structure in this site. This distinction is however not very strict, and many (if not all) building blocks are as well functional and structural, but the classification helps to distinguish fundamental strategies.

Both functional and structural building blocks used in free-radical polymerization protocols can contain a single polymerizable group——simple monomers——, or two or more polymerizable groups——cross-linking monomers or

Table 1 Examples of Complementary Recognition Elements

Target functionality	Functional element (FE)	FE example	Monomer example
Charge	Complementary charge	R–S(=O)(=O)–O⊖	(monomer structure)
H-bond Donor/acceptor	H-bond acceptor/donor	R–COOH	COOH monomer
π-Donor/acceptor	π-Acceptor/donor	adenine (NH₂ purine)	adenine methacrylate monomer
Metal ion	Metal-coordination ligand	triamine ligand	vinylbenzyl triamine monomer
Metal-coordination ligand	Immobilized metal ion	Cu²⁺ complex	Cu²⁺ complex monomer
Amine/aldehyde	Aldehyde/amine	R–CHO	salicylaldehyde monomer
Thiol	Thiol	R–SH	allyl thiol (SH)
Hydroxyl/diol	Boronic acid	phenyl B(OH)₂	vinylphenyl B(OH)₂
Carbonyl	Diol	propanediol (OH, OH)	vinylphenyl diol

Table 2 Types of Building Blocks

Type	Function	Example
FUNCTIONAL (F)	Interactions with the target molecule	Methacrylic acid (M9) 4-Vinylphenylboronic acid (M37)
STRUCTURAL (S)	Organization, shape and structure	Divinylbenzene (X1) Ethylene glycol dimethacrylate (X3)

cross-linkers. Traditionally, cross-linkers are more of structural nature, used to make up the bulk mass of the imprinted material, whereas the simple monomers carry more of the functional role in contact with the template. However, since the role of the monomer does not depend on the number of polymerizable groups in the molecule, this description is rather unclear. For example, cross-linking functional monomers can be chosen to increase the "rigidity" or fidelity of the site when necessary, especially in noncovalent protocols. The template–monomer adduct in covalent protocols normally contains two or more vinyl groups and this structure acts as cross-linker per se.

It is important to note, however, that over the years the imprinting literature has adopted a somewhat standardized terminology when describing and comparing building blocks. Thus, a "Functional monomer" is according to this rule any simple monomer interacting with or binding to the template. Also, a "Cross-linker" is any cross-linking monomer being of structural nature in building up the network material. Traditionally, no further distinction was ever needed, but with more complex systems composed of several different monomers, this terminology is slightly limiting. Many examples exist where cross-linking monomers are highly functional. In this text, however, this classical terminology will be used throughout, and only in cases where further distinction is needed will clarifications be made.

The nature of the functional monomer also depends on the imprinting strategy used. In noncovalent imprinting protocols, the general distinction is fairly clear, and the functional building block is a discrete entity. This is on the other hand not the case with other imprinting techniques, where the functional building block is used to form bonds to the template prior to the imprinting preparation process. For example, in protocols making use of covalent bonds during polymer preparation and noncovalent bonds in the recognition step, the functionality obviously differs. In particular, in the so-called sacrificial-spacer protocols (semi-covalent imprinting Chapter 5), the element looses a fragment upon removal of the template.

V. OVERVIEW OF ORGANIC BUILDING BLOCKS USED IN MOLECULAR IMPRINTING

Tables 3 and 4 display an overview of the majority of organic styrene- or acryl-based monomers used in molecular imprinting to date. As can be seen, many of those are commercially available, whereas most others can be accessed following the experimental procedures in the literature, or via the general guidelines outlined in this chapter. The building blocks have been divided into structural- and functional

Table 3 Structural and Functional Building Blocks Used in Molecular Imprinting. Monomers Carrying One Polymerizable Group. C: Commercially Available, MC: Metal Ion Coordination, NC: Noncovalent, RC: Reversible Covalent, MS: Mainly Structural

Entry	Type	Structure	Name [CAS registry number]	Reference synthesis	Reference imprinting
M1	MC		N-(4-vinylbenzyl)iminodi-acetic acid [46917-20-8]	[21]	[22–25]
M2	MC		N2,N2-bis(carboxymethyl)-N6-(methacryloyl)-L-lysine, [331662-47-6]	[26]	[26]
M3	MC		1-[(4-vinylphenyl)methyl]-1,4,7-triazacyclononane [182306-47-4]	[27–29]	[27–29]
M4	MC		4-vinyl-4′-methyl-2,2′-bipyridine [74173-48-1]	[13,14]	[30]
M5	MC		Diethyl vinylphosphonate [682-30-4]	C	[31]

M6	MC	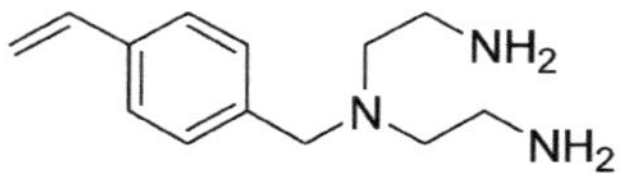	4-(N-vinylbenzyl)diethylenetriamine [46734-05-8]	[4]	[4]
M7	MC	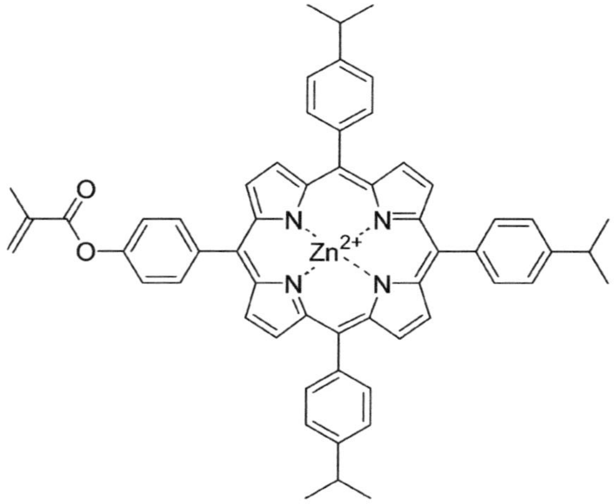	5,10,15-Tris(4-isopropylphenyl)-20-(4-methacryloyloxyphenyl)porphyrin [287402-11-3] (Zn(II)-complex)	[18]	[18,32]
M8	NC		Acrylic acid [79-10-7]	C	[33–45]
M9	NC	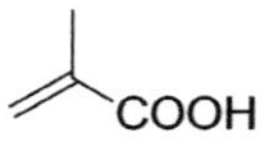	Methacrylic acid [79-41-4]	C	Numerous
M10	NC	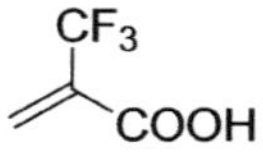	2-(Trifluoromethyl) acrylic acid [381-98-6]	C	[46–55]

Table 3 (*Continued*)

(*Continued*)

Entry	Type	Structure	Name [CAS registry number]	Reference synthesis	Reference imprinting
M11	NC		Itaconic acid [97-65-4]	C	[56]
M12	NC		6-Methacrylamidohexanoic acid [59178-92-6]	[57]	[58]
M13	NC		4-Vinylbenzoicacid [1075-49-6]	C	[59–62]
M14	NC		2-Acrylamido-2,2-dimethyl-ethanesulfonic acid [15214-89-8]	C	[35,42,63–68]
M15	NC		2-Sulfoethyl methacrylate [10595-80-9]	C	[69]
M16	NC		2-(Methacryloyloxy)ethyl phosphate [24599-21-1]	C	[70]
M17	MC NC		4-Vinylpyridine [100-43-6]	C	[53,71–73]

		Structure	Name [CAS]		
M18	MC NC		2-Vinylpyridine [100-69-6]	C	[53,74–80]
M19	MC NC		4(5)-Vinylimidazole [3718-04-5]	[15,81,82]	[83–88]
M20	NC		1-Vinylimidazole [1072-63-5]	C	[89]
M21	NC RC		4-Aminostyrene [1520-21-4]	C	[6,90–96]
M22	NC		Allylamine [107-11-9]	C	[97–101]
M23	NC		*N,N*-diethylaminoethyl methacrylate [105-16-8]	C	[54,94,97,100–103]

(Continued)

Table 3 (*Continued*)

Entry	Type	Structure	Name [CAS registry number]	Reference synthesis	Reference imprinting
M24	NC		N-(2-aminoethyl)metha-crylamide [6298-57-7]	[20]	[20,104]
M25	NC		N,N'-diethyl-4-vinyl-benzenecarboximidamide [200063-00-9]	[105] Chapter 4	[106–109]
M26	NC		N-(diaminoethylene)-2-methylprop-2-enamide [65658-72-2]	[20]	[20]
M27	NC		N-(3-guanidinopropyl) methacrylamide [231934-29-5]	[20]	[20]
M28	NC		N-(3-aminopropyl) methacrylamide [86742-39-4]	[20]	[20]

		Name [CAS]		
M29	NC	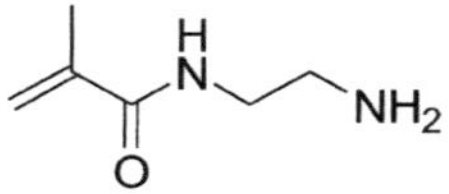 N-(2-aminopropyl)metha-crylamide [63298-57-7]	[20]	[20]
M30	MC NC	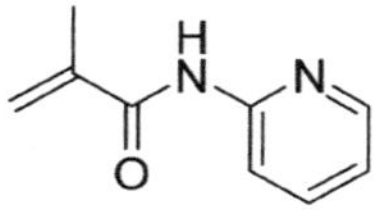 N-(2-aminopyridine) methacrylamide [40000-70-2]	[20]	[20]
M31	NC	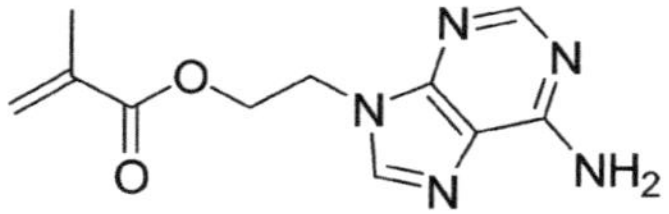 9-(β-Methacryloyloxyethyl) adenine [20245-91-4]	[19]	[20]
M32	NC	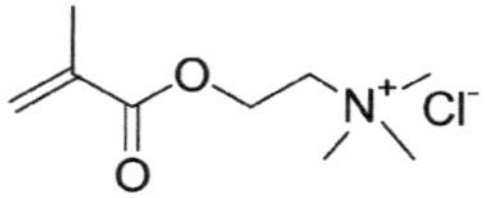 N,N,N-trimethyl-N-(2-methacryloxyethyl) ammonium chloride [5039-78-1]	C	[110,111]
M33	NC	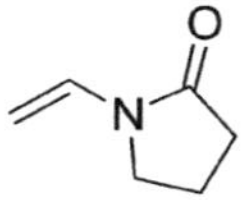 N-vinyl-2-pyrrolidinone [88-12-0]	C	[48,53,112–115]
M34	NC	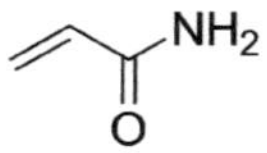 Acrylamide [79-06-1]	C	[26,116–119]
M35	NC	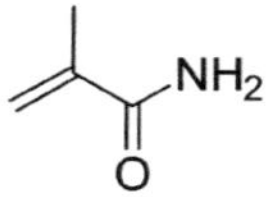 Methacrylamide [79-39-0]	C	[53]

Table 3 (*Continued*)

(*Continued*)

Entry	Type	Structure	Name [CAS registry number]	Reference synthesis	Reference imprinting
M36	NC		Ethyl urocanate [27538-35-8]	[120]	[121]
M37	RC		4-Vinylphenylboronic acid [2156-04-9]	C	Numerous
M38	RC		5-Vinylsalicylaldehyde [68860-34-4]	[122]	[122]
M39	NC RC		4-Vinylbenzylamine [50325-49-0]	[123]	[124,125]
M40	RC		4-Vinylbenzaldehyde [1791-26-0]	[126]	[127,128]
M41	MS		4-Vinylanisole [637-69-4]	C	[94]

(Continued)

		Structure	Name [CAS]		
M42	MS	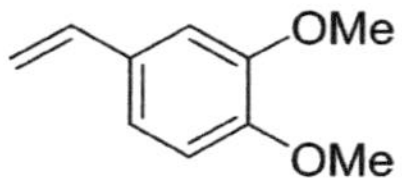	4-Vinylveratrole [6380-23-0]	C	[94]
M43	RC	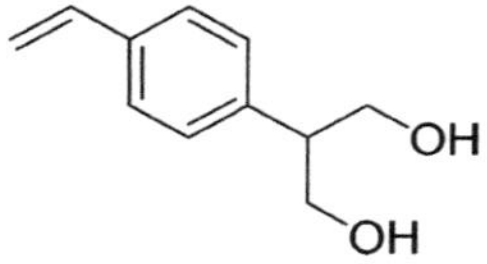	2-(4-Vinylphenyl)-1,3-propanediol [144174-36-7]	[129]	[129]
M44	NC	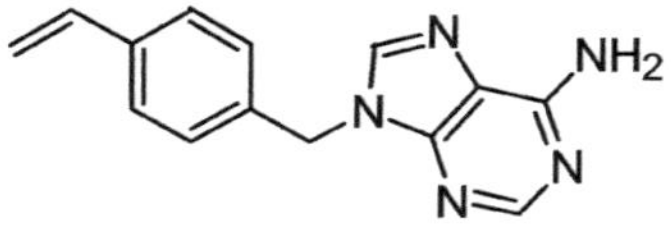	9-(4-Vinylbenzyl)adenine [464181-96-2]	[165]	[94]
M45	MS	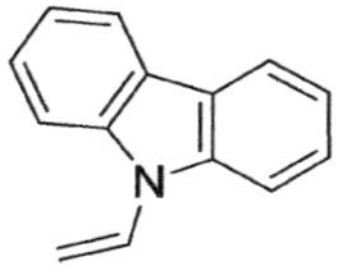	9-Vinylcarbazole [1484-13-5]	C	[94]
M46	MS	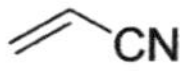	Acrylonitrile [107-13-1]	C	[94,130]
M47	NC	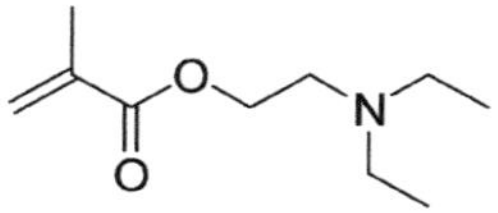	N,N-dimethylaminoethyl methacrylate [2867-47-2]	C	[131]

Table 3 *(Continued)*

Entry	Type	Structure	Name [CAS registry number]	Reference synthesis	Reference imprinting
M48	MS		Methyl methacrylate [80-62-6]	C	[53]
M49	NC		Hydroxyethyl methacrylate [868-77-9]	C	[53,94,97,132]
M50	NC		2-Naphthyl methacrylate [10475-46-4]	C	[94]
M51	NC		9-Anthracenylmethyl methacrylate [31645-35-9]	C	[94]
M52	NC		N-methylmethacrylamide [3887-02-3]	C	[133]

M53	RC		5-(Methacryloylamino) boronphthalide [244178-19-6]	[134]	[134]
M54	RC		Allyl thiol [870-23-5]	C	[135,136]
M55	NC		Methyl-6-O-acrylate-α-D-glucopyranoside [144790-40-9]	NC	[137]

Table 4 Structural and Functional Building Blocks Used in Molecular Imprinting. Cross-Linkers—Monomers Carrying Two or More Polymerizable groups

Entry	Structure	Name [CAS registry number]	Reference synthesis	Reference imprinting
X1		Divinylbenzene (DVB) Ortho [91–14-5] Meta [108-57-6] Para [105–06-6] Mixture [1321-74-0]	Ortho [122] Meta [122] Para C Mixture C	Ortho [138] Meta [138] Para [44,138–143] Mix numerous
X2		m-Diisopropenylbenzene (DIP) [3748-13-8]	C	[144]
X3		Ethylene glycol dimethacrylate (EDMA) [97-90-5]	C	Numerous
X4		Tetramethylene glycol dimethacrylate (TDMA) [2082-81-7]	C	[138,145,146]
X5		Trimethylolpropane trimethacrylate (TRIM) [3290-92-4]	C	Numerous

	Structure	Name		Reference
X6		Pentaerythritol triacrylate (PETRA) [3524-68-3]	C	[147,148]
X7		Pentaerythritol tetraacrylate (PETEA) [4986-89-4]	C	[4,147]
X8		Bisphenol A dimethacrylate [3253-39-2]	C	[138]
X9		1,4:3,6-dianhydro-D-sorbitol 2,5-dimethacrylate [108646-62-4]	[138]	[138,149]
X10		Tetrahydrofuran-3,4-diyl dimethacrylate [108580-23-0]	[138]	[138]

(Continued)

Table 4 (*Continued*)

Entry	Structure	Name [CAS registry number]	Reference synthesis	Reference imprinting
X11		*N*,*O*-diacryloyl-L-phenyl-alaninol [100665-22-3]	[150]	[150]
X12		*N*,*N′*-methylenebisacrylamide [110-26-9]	C	[117,119,151]
X13		*N*,*N′*-ethylenebisacrylamide [2956-58-3]	C	[152,153]
X14		*N*,*N′*-butylenebisacrylamide [10405-38-6]	[154]	[152,155]
X15		*N*,*N′*-hexamethylenebisacryl-amide [7150-41-6]	C	[152,155]
X16		*N*,*N′*-m-phenylenebis(methacrylamide) [25256-12-6]	C	[144]

	Structure	Name / CAS		
X17		*N,N'-m*-phenylenebis-acrylamide [23435-66-7]	[156]	[155,157]
X18		2,6-Bis(acrylamido)pyridine [147075-91-0]	[156,158]	[158–161]
X19		3,5-Bis(acrylamido)benzoic acid [76961-93-8]	[162]	[163]
X20		1,4-Bis(acryloyl)piperazine [6342-17-2]	C	[67]
X21		1,4-Bis(methacryloyl) piperazine [17308-56-4] Not used in imprinting	[144]	[144]
X22		2,2'-Oxybis[*N*-(4-vinylphe-nyl)-*N*-methyl-acetamide] [136821-00-6]	[96]	[96]

(Continued)

Table 4 (*Continued*)

Entry	Structure	Name [CAS registry number]	Reference synthesis	Reference imprinting
X23		*N*,*N*′-bis(4-ethenylphenyl)-5-(octyloxy)-1,3-Benzenedi-carboxamide, [352524-75-5]	[6]	[6]
X24		*N*,*N*′-Bisacrylylcystamine [60984-57-8]	C	[164]
X25		*N*,*N*′-dihydroxyethylene-bis-acrylamide [868-63-3]	C	[117]

monomers, respectively, largely following their basic functionalities. In cases where the monomers are not commercially available, reference to their synthesis is provided. In addition, representative references regarding their use to produce molecularly imprinted materials have been stated. When the monomers have been extensively used, such as methacrylic acid (M9), or 4-vinylphenylboronic acid (M37), the number of references is very large and the reader may also consult Part II or the database of the Society for Molecular Imprinting: www.smi.tu-berlin.de.

A. Structural Building Blocks

A variety of structural building blocks have been used over the years in molecular imprinting. Most popular in this respect is ethylene glycol dimethacrylate (EDMA or EGDMA, X3), used as structural cross-linker in many protocols involving acrylate or acrylamide functional monomers. In recent years, trimethylolpropane trimethacrylate (TRIM, X5) has increasingly been tested and proven superior in several cases. For styrene-based protocols, divinylbenzene (X1) is the most common element. This compound is usually sold as a technical grade mixture of isomers, also containing considerable amounts of ethylbenzene, and reproducible purification is rather difficult to accomplish. Nevertheless, this monomer mixture has been successfully used on a number of occasions without extensive purification. Often, vacuum distillation or passage through a column of activated basic alumina has been regarded as enough purification, mainly resulting in removal of radical quenchers and traces of water. For more reliable results, the pure isomers are recommended. 1,4-divinylbenzene is now available from some commercial sources and may also be rather straightforwardly synthesized in few steps from available starting materials (Experimental Protocol X1).

Neither EDMA/TRIM nor DVB is soluble in water and for aqueous phase imprinting other cross-linking structural monomers have to be used. Acrylamides are generally more soluble than acrylate esters and the most common in this respect is 1,4-bis(acryloyl)piperazine (X20), easily soluble in water and lower alcohols at high concentrations. Another possibility is to make use of N,N'-dihydroxyethylene-bis-acrylamide (X25), or N,N'-ethylene-bisacrylamide (X13), although at least the latter considerably less soluble. Attachment of charged groups may potentially help to solubilize the cross-linking monomers, provided such functionalities do not interfere with the recognition.

Except for its structural and polar properties, the polymer backbone has been little subjected to any additional effects. Attempts to introduce inherent main chain chirality into the structural part of the molecularly imprinted polymer have, however, been made. For example, cross-linking monomers X9–X11 have all been evaluated for their effects with variable results. Another example is the introduction of cleavable groups into the polymer backbone, to tune the binding properties and the accessibility of the polymeric binding sites further. The disulfide bond in bis-acrylyl-cystamine (X24) can, for example, be reversibly cleaved into the corresponding thiols.

Simple structural monomers (containing a single vinyl group) are sometimes also employed. For example, styrene and ethylstyrene are commonly used to decrease the degree of cross-linking in protocols involving divinylbenzene as cross-linking monomer. Similarly, methyl methacrylate (M48) has been used to tune the degree of cross-linking when methacrylate cross-linkers (EDMA, TRIM, etc.) are employed.

B. Functional Building Blocks

The large majority of building blocks belong to the functional class—directly inter-acting with the template molecule, and a great deal of creativity has been invested into this area over the years. The monomers in Table 3 have been divided into three major groups, depending on whether the protocol makes use of general noncovalent inter-actions, metal-coordination, or reversible-covalent protocols, respectively. In some cases, monomers may belong to more than one group. The noncovalent group can furthermore be divided into subgroups, depending on the major interaction type within this group, e.g., mainly acidic, basic, hydrogen bonding, or potentially π–π-stacking, obviously with possibilities of overlapping. Monomers/building blocks suitable for the semi-covalent approach are extensively covered in Chapter 6.

Given the higher versatility and possibly higher simplicity of noncovalent/metal coordinating protocols, most functional monomers belong to this group. The apparently more demanding covalent protocols hampering the generation and evaluation of a similar variety of monomers.

As can be seen from Table 3, the monomers can also be separated into broad-spectrum monomers and specific monomers, depending on their specific matching to a certain functional element or if the monomer expresses recognition to a wider range of functionalities. For example, methacrylic acid can form charge–charge interactions to any positively charged species, as well as fairly strong hydrogen bonds to complementary H-bonding motifs, and is a true broad-spectrum functional monomer. On the other hand, 2,6-bis(acrylamido) pyridine (X18), expressing a donor–acceptor–donor (DAD) H-bond array, is more specific in binding certain acceptor–donor–acceptor (ADA) hydrogen bonding motifs. In general, the more functionally complex the monomer, the more specific it becomes. This is a conse-quence of using the "chelate" (or multifunctional) effect already before the poly-merization step. Monomers capable of forming several interactions to the template in the prepolymerization solution will lead to stronger complexation and potentially higher binding strength to the final polymeric binding site. Nevertheless, most functional monomers contain single functional groups, and imprinters make use of several of these to accomplish the desired chelate effect in the formed imprinted recognition site.

In addition to specificity, the monomers also differ considerably in binding strength potency, ranging from very low (weak noncovalent) to very high (covalent) binding energies. Obviously, these features depend on what major protocol is used (noncovalent, covalent), but also within the noncovalent group can differences be distinguished. Thus, acrylic acid (M8) is less efficient in binding than trifluoro-methacrylic acid (M10), the latter capable of forming stronger hydrogen-bonded reinforced ionic bonds, than the former. Likewise, an amidino- or a guanidino group forms significantly stronger complexes with several acids, than does a simple amino functionality. This increased binding strength is likely to result in an increased binding strength also in the polymeric sites.

Solubility can obviously play a role also when choosing functional monomers, and many monomers express low solubility in common organic imprinting porogens, such as acetonitrile, chloroform, and toluene. Especially when ionic interactions are formed, this may present a problem, and the main way to circumvent this is to add extra lipophilicity to the monomer. This can, for example, be done by adding alkyl chains to amino functionalities, as in monomer N,N'-diethyl-4-vinylbenzene-carboximidamide (**M25**). When water is used as polymerization solvent, the problem is of course the opposite, and in this case hydrophilic groups need to be included if the monomer is poorly soluble in itself. As mentioned above, this can potentially be done by increasing the amount of charged or polar groups in the molecules.

VI. MONOMERIZATION—HOW TO ADD POLYMERIZABLE GROUPS TO YOUR FUNCTIONAL ELEMENT

When the functional elements corresponding to the print molecule have been judiciously chosen, and potential access from commercial sources have failed, the building blocks need to be prepared in the laboratory. In most cases, organic MIPs are based on styrene-type monomers or acryl-type monomers being used together with radical polymerization mechanisms. Variations to these themes exist, for example, polycondensations and substitutive cross-linking, but on large these are the systems that have proven to work best so far. The general synthetic protocols for accessing these monomers from a functional element is rather straightforward and does in many cases not need a vast synthetic experience. If the basic FE chosen can be generated with simple building blocks that are commercially available, then often a simple route to a polymerizable element can be found. In the following, procedures for synthesizing simple styrene-and acrylate-based building blocks are given. Obviously, there are many ways to cook a bouillabaisse, and available synthetic routes are certainly as multifarious as there are chemists. Therefore, only the most common strategies are discussed.

A. General Considerations

Basically, three different approaches to functional and structural monomers have been practiced over the years (Table 5) [1]. The first and foremost is to decorate

Table 5 Approaches to Monomerization. FE Denotes Functional Element. See Text for Further Explanation

the FE with an easily accessible or commercially available monomer fragment, such as a styrene group or an acryl group. Often this can be made in few synthetic steps, and yields are normally good to excellent, thus explaining the popularity. Especially for acryl-type monomers this is the common strategy. The second approach involves the direct attachment of a vinyl group to the FE, normally to generate a styrene building block. Finally, the third approach is to generate a vinyl group by elimination.

B. General Synthesis of Styrene-type Monomers

1. Addition of a Styrene Fragment

In this case, most synthetic protocols used in molecular imprinting make use of 4-vinylbenzyl chloride (4VBC), commercially available from all major sources. This reactant is then applied to the precursor FE, chosen to carry a nucleophilic site, e.g., an amino-, a hydroxy,- or a thiol group, able to displace the chloride in a substitution reaction. The reaction is usually carried out in a polar aprotic solvent (DMF, acetonitrile, etc.), under slightly basic conditions (triethylamine, potassium carbonate, etc.), but other systems may as well be used.

4VBC

R = alkyl,aryl, etc.
Nu = NH$_2$, OH, SH, etc.

Examples of this strategy are monomer **M1** (Protocol M1) and **M6**, both prepared in few steps from 4VBC.

In Experimental protocol M1, 4VBC is allowed to react with one equivalent of iminodiacetic acid in alkaline methanol solution, and the product 4-vinylbenzyliminodiacetic acid (**M1**) is recovered after acidic work-up. This monomer can subsequently be metallated with, e.g., Cu^{2+}, Co^{2+}, or Zn^{2+} to yield a functional monomer that is efficient in metal-coordination interactions. In particular, imidazole may form strong interactions with the metallated form of the monomer.

EXPERIMENTAL PROTOCOL M1 (ADAPTED FROM REFS. 2,3)

Synthesis of 4-Vinylbenzyliminodiacetic Acid (M1)

Iminodiacetic acid (4.4 g, 33 mmol) and NaOH (2.2 g, 55 mmol) were dissolved in MeOH/H$_2$O 1:1, v/v (70 mL), by heating to 60°C. To the stirred solution, half of the vinylbenzyl chloride (5.0 g, 33 mmol) was added dropwise over 30 min. Following addition of another portion of NaOH (2.2 g, 55 mmol), the second half of vinylbenzyl chloride (5.0 g, 33 mmol) was added dropwise over an additional 30 min. The solution was heated for another 30 min, after which period it was concentrated to ~2/3 of the volume. The remaining solution was washed with ether (3 x 15 mL), and the residual aqueous phase was acidified to pH ~2.5 by concentrated HCl. The resulting white precipitate was filtered off, washed and recrystallized from MeOH/H$_2$O 9:1, v/v.

Yield: 40–60%
TLC: $R_r = 0.7$, silica, MeOH/H$_2$O 2/1, (v/v)
1H-NMR (300 MHz, CD$_3$OD): δ 7.4–7.7 (4H, m, ArH), 6.7–6.9 (1H, m, vinyl), 5.8–6.0 (2H, m, vinyl), 4.4 (2H, s, CH_2), 3.9 (4H, s, CH_2).

A similar metal-coordination monomer can be prepared as outlined below [4]. In this case, the primary amino groups in diethylenetriamine are first protected as their phthalimides by heating in phthalic anhydride. The secondary amine is subsequently coupled to 4VBC in the presence of potassium carbonate in acetonitrile. After deprotection of the amines with hydrazine, the functional monomer is isolated, which can be further metallated with, e.g., Cu^{2+} to yield a potent metal coordinating monomer.

M6

An alternative to this route involves 4-vinylaniline or 4-vinylbenzoic acid, relatively easily coupled to various acid counterparts to produce amide linkages. A multitude of amide coupling protocols may in this case in principle be used, for example, by initial activation of the acid by thionyl- or oxalyl chloride and subsequent addition of the corresponding amine in the presence of base (pyridine, triethylamine, etc.) [5,6], or by use of coupling reagents such as dicyclohexylcarbodiimide (DCC).

Although the use of these simple one- or two-step protocols has been the most popular in the imprinting community, and indeed successful ones, other routes to couple a styrene fragment to the functional element may certainly be considered. Although not tested to any larger extent, other commercially available precursors include, for example: 4-vinylaniline that may be used to produce, e.g., amino linkages by nucleophilic substitution or reductive amination, and 4-hydroxystyrene that can be used to introduce a vinylphenyl ether linkage in the structure. The latter has, however, been used on several occasions to produce chloroformate and carbonate esters for use in sacrificial-spacer protocols (cf. Chapter 6). Another possibility is to use 4-vinylphenylboronic acid, often used in itself in reversible covalent protocols to vicinal diols (cf. Chapter 4), which may be used in cross-coupling protocols (e.g., Suzuki couplings) to in principle access a range of building elements [7]. Other cross-coupling reactions may also be envisaged.

Relatively little is known of aryl Grignard reagents or aryllithium reagents, for example, produced from bromostyrene, but the former seems to have a tendency to polymerize during coupling [8]. On the other hand, the corresponding reagents prepared from vinylbenzyl chloride may in principle be used to couple a variety of functionalities to the styrene moiety.

2. Introducing a Vinyl Group to a Derivatized Aryl Group

The vinyl group may in principle also be introduced after the benzene ring has been properly derivatized, however, not applied in many imprinting protocols. In general polymer chemistry, three main routes are normally employed: direct attachment of the vinyl group to a halogenated aromatic ring, e.g., by a Heck reaction; formation of the vinyl group by a Wittig-type reaction starting either from the benzaldehyde or the benzyl halide; and formation of the vinyl group by an elimination reaction. Of all monomers in Tables 3 and 4, the only examples of this strategy in molecular imprinting are monomers X1 (DVB, protocol 1) and M4 (5-methyl-5'-vinyl-2,2'-bipyridine). Decarboxylation of α,β-unsaturated carboxylic acids is another potential route as exemplified by monomer M19.

Experimental protocol X1 describes two alternative routes to 1,4-divinylbenzene (X1) using the Wittig reaction. In the first strategy, 1,4-bis(chloromethyl) benzene is allowed to from the phosphonium chloride salt with triphenyl phosphine by refluxing in dimethylformamide (DMF). The divinylbenzene is subsequently formed from reacting the phosphonium salt with formaldehyde in aqueous alkaline solution, after which X1 can be recovered in good yield.

EXPERIMENTAL PROTOCOL X1 (ADAPTED FROM REFS. 9,10)

Synthesis of Divinylbenzene (X1)

A solution of 1,4-bis(chloromethyl) benzene (88 g, 0.5 mol) and triphenylphosphine (285 g, 1.1 mol) in DMF (1 L) was refluxed under stirring for 3 h. The produced mixture was allowed to cool to room temperature and the colorless crystalline product was filtered off, washed with DMF and ether, and dried in vacuo. Yield: 335 g (93 %).

The phosphonium salt (21 g, 30 mmol) was suspended in 40% aqueous formaldehyde (100 mL), and 12.5 M NaOH (32 mL) was added dropwise under vigorous stirring at a suitable rate to keep the reaction temperature below 40°C. The reaction was allowed to proceed an additional 2 hr, after which time the product was extracted from the reaction mixture with petroleum ether (40–60 °C). Following drying and concentration, the residue was purified by column filtration over neutral alumina (100 g) using petroleum ether (40–60 °C) as eluent.

Yield: 3.6 g (92 %) crystalline X1.
m.p. 27–28°C

o-, and *m*-divinylbenzene can be prepared in the same manner from 1,2-bis (bromomethyl) benzene and 1,3-bis(bromomethyl) benzene, respectively.

The second protocol follows a single-step strategy. Terephthalaldehyde is allowed to react with methyl triphenylphosphonium bromide (MTPP$^+$Br$^-$) by refluxing in basic dioxane solution. After simple work-up, **X1** is produced in good yield. Both methods can be used to prepare either isomer of divinylbenzene.

EXPERIMENTAL PROTOCOL X1 (ADAPTED FROM REFS. 11,12)

Synthesis of Divinylbenzene (X1) Alternative Route

terephthalaldehyde   **S1**

Methyl triphenylphosphonium bromide (MTPP$^+$Br$^-$)(20 mmol), potassium carbonate (3.5 g), 1,4-dioxane (20 mL) containing water (0.3 mL), and terephthalaldehyde (10 mmol) were successively mixed. The reaction mixture was refluxed for approx. 6 hr and the product mixture filtered. After concentration of the filtrate, product **X1** was purified by chromatography (short and broad column) using hexane as eluant.

Yield: 78 %, colorless liquid.
^{1}H-NMR (CDCl$_3$): δ 7.35 (4H, s, Ar*H*), 6.75 (2H, dd, α-vinyl), 5.77 (2H, dd, *cis*-β-vinyl), 5.27 (2H, dd, *trans*-β-vinyl)
13C-NMR (CDCl$_3$): δ 137.17, 136.54, 126.46, 113.83.

The 2,2'-bipyridine functional monomer **M4** can be synthesized using an elimination reaction [13,14]. 5,5'-Dimethyl-2,2'-bipyridine is first lithiated with lithiumdiisopropyl amide (LDA) at one of the methyl groups. This lithium derivative is subsequently reacted with chloromethoxymethane to form the methoxy-ethyl derivative. Finally, the vinyl group is produced by elimination of the methoxy group in basic tetrahydrofuran (THF) at low temperature.

Decarboxylation is used to generate 4(5)-vinylimidazole (**M19**), a useful monomer for metal coordination and catalysis [15]. In this case, neat urocanic acid is simply heated, and the monomer can be directly collected by distillation in a short-necked distillation setup (e.g., a Kugelrohr-apparatus).

urocanic acid   **M19**

C. General Synthesis of Acryl-type Monomers

The majority of imprinting protocols make use of acryl-type monomers and a range of monomers have been specially developed for these applications. One reason to why this particular strategy has been so widely adopted is because acrylates copolymerizes rather well with other acrylates, and the same goes for acrylamides. Thus, the chances of obtaining a MIP where the distribution of the monomers is not affected by the polymerization per se are substantially greater.

1. Addition of an Acryl Fragment

Similar to 4VBC couplings, acrylates and acrylamides are fairly easily accessible from an activated acryl donor such as acryloyl chloride or methacryloyl chloride and either an alcohol or an amine as the nucleophilic element.

R = alkyl, aryl, etc.

R' = H, CH3

A wide range of functional and structural monomers has been prepared by this method (Tables 3 and 4), exemplifying its usefulness. A typical example of this type is presented in Experimental protocol X11. The acid chloride is added portionwise to a solution of the nucleophile in the presence of a base (triethylamine, NaH, etc.) in a suitable (polar) aprotic solvent. Other examples include monomers X16, M7 and M31.

In Experimental protocol X11, *N,O*-bisacryloyl-L-phenylalaninol (X11) is synthesized in a simple step from L-phenylalaninol. The starting material is dissolved in dimethyl formamide (DMF), and acryloyl chloride together with triethylamine are added dropwise/portionwise while cooling to 0°C. The product can subsequently be recovered after washing and recrystallization.

EXPERIMENTAL PROTOCOL X11 (ADAPTED FROM REF. 16)

Synthesis of *N,O*-Bisacryloyl-L-Phenylalaninol

L-phenylalaninol   X11

L-Phenylalaninol (16.6 mmol) was dissolved in ice-cooled DMF (15 mL). Acryloyl chloride (37 mmol, 2.2 eq) and triethylamine (37 mmol, 2.2 eq) were added in portions and the mixture allowed to react overnight at room temperature. The resulting precipitate was filtered off and the solution diluted with 20-fold excess of ethyl acetate. This solution was washed with 0.5 M NaHCO$_3$ and 0.5 M citrate, and after concentration the product **X11** was recrystallized from methanol/water (1/10 v/v).

Yield : 20%
m.p. : 102°C
$[\alpha]_D^{22}$: −28.1° ($c = 0.43$, methanol)
TLC : $R_f = 0.73$, Silica, CHCl$_3$/MeOH 1/1, (v/v)
1H-NMR (CDCl$_3$): δ 2.9 (2H, m, -C$\underline{H}_2$-X6H$_5$), 4.2 (2H, d, $J = 4.5$, -OC$\underline{H}_2$-), 4.5 (1H, m, NH-C$\underline{H}$(CH$_2$)-CH$_2$-, 5.5–6.6 (7H, m, C$\underline{H}_2$=C$\underline{H}$-, -N$\underline{H}$-), 7.3 (5H, m, -X6$\underline{H}_5$)

A similar strategy can be used to synthesize N,N'-bisacryloyl-1,3-diaminobenzene **X16** (Experimental Protocol X16). In this case, 1,3-diaminobenzene is simply added dropwise to a solution of methacryloyl chloride in acetonitrile at 0°C. After filtration, the product is precipitated and recrystallized from acetonitrile in good yield.

EXPERIMENTAL PROTOCOL X16 (ADAPTED FROM REF. 17)

Synthesis of *N,N*-Bisacryloyl-1,3-Diaminobenzene (X16)

1,3-diaminobenzene   **X16**

To a solution of methacryloyl chloride (20.0 mL, 205 mmol) in acetonitrile (400 mL) at 0°C was added dropwise a solution of 1,3-diaminobenzene (24.0 g, 222 mmol) in acetonitrile (200 mL). The mixture was allowed to react for 3 hr at room temperature. The salts were filtered off and washed thoroughly with hot acetonitrile. After concentration to approx. 200 mL, the remaining solution was placed at −20°C for 24 hr, during which time the product **X16** precipitated out. Recrystallization from acetonitrile.

Yield: 19.2 g (77%)
m.p: 148–149°C
1H-NMR [300 MHz, CDCl$_3$]: δ 7.94 (1H, t, $J = 2.0$ Hz, ArH), 7.66 (2H, s, CONH), 7.32 (3H, m, ArH), 5.80 (2H, d, $J = 0.6$ ^{Hz,} vinyl), 5.47 (2H, dd, $J = 1.3$ and 2.0 Hz, vinyl), 2.06 [6H, dd, $J = 1.0$ and 1.4 Hz, CH_3]
^{13}C-NMR [125.8 MHz, CDCl$_3$]: δ 167.4,−141.3, 139.0, 130.2, 120.8, 116.5, 112.2, 19.4
IR (KBr): 3250, 1656, 1605, 1547, 1481, 1421, 1221, 948, 936, 925, 861, 801, 774 cm^{-1}

An example of this strategy to produce porphyrin monomer **M7** is outlined below [18]. The porphyrin is first produced by heating two different aldehydes, one of which carrying a phenol functionality, with pyrrole in isopropanol. After purification, the phenol-derivatized porphyrin is metallated by treating with Zn(II) acetate in chloroform to yield the Zn^{2+}-porphyrin. Finally, this compound is reacted with methacryloyl chloride in basic ether solution to yield the mono-functionalized Zn–porphyrin monomer (**M7**). This monomer can subsequently be employed in metal–coordination protocols to bind adenine.

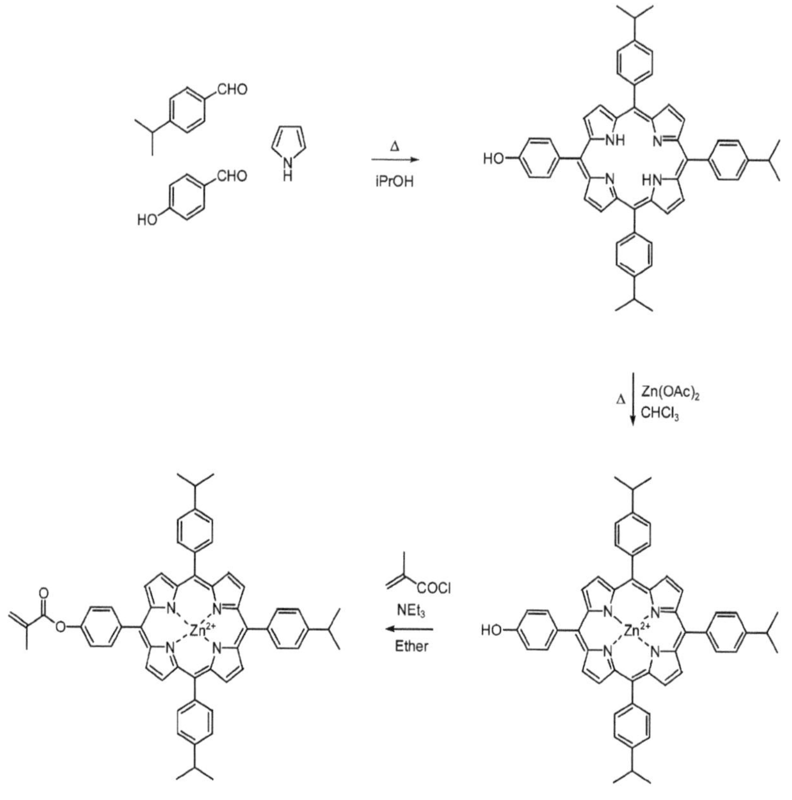

M7

Adenine can also be methacrylated as outlined below [19]. In this case, adenine is first reacted with ethylene carbonate in dimethylformamide (DMF) containing catalytic amounts of NaOH. The resulting 9-hydroxyethyl-adenine is then deprotonated with sodium hydride in DMF, and allowed to react with methacryloyl chloride to yield 9-methacryloyloxyethyl-adenine (**M31**).

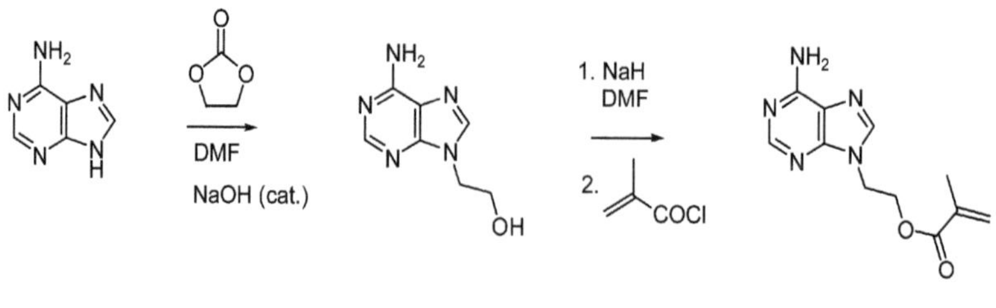

M31

Acryl-type monomers can also be prepared from reactions involving functional groups attached to an acryl (head) group. For example, aminopropylmethacrylamide has been used to generate acrylated guanidine building block, the **M27** [20]. The starting aminopropyl monomer was treated with an *S*-ethylisothiourea salt at moderately basic conditions, and the guanidinium monomer was yielded after treatment by a basic anion exchange resin.

D. Side Reactions

Believe it or not, side reactions are always a threat to successful monomer synthesis. Both for styrene-type monomers and acryl-type monomers, special attention should be taken in order to avoid addition reactions to the double bond, and this type of side reaction is fairly common. Especially for acrylates/acrylamides conjugate addition occurs quite readily, and may interfere with the desired carbonyl addition. However, conjugate addition/elimination is in principle reversible, and the preferred product will be formed unless other side reactions (polymerization, etc.) come into play. Styrenes are less prone to this type of side reaction than are acrylates/acrylamides but caution should nevertheless be taken, especially if the vinyl group is electron deficient from a strongly electron withdrawing aryl group. In the acrylate/acrylamide case, the side reaction can however be minimized by adding stoichiometric amounts of nucleophile to the acyl chloride since this is likely to react more readily. Starting from a less activated donor such as methyl esters may result in increased competing conjugate addition, but normally conjugate addition is less rapid than carbonyl addition once the acyl chlorides are used under normal conditions.

Another source of "side" reaction is polymerization per se, either radical or ionic. Some monomers are very sensitive to polymerization, and if this is the case all reactions and subsequent purification schemes should be performed at lower temperature and protection from light under "neutral" conditions. Another possibility to prevent radical polymerization is to add traces of radical quenchers (e.g., 4-*tert*-butylcatechol, hydroquinone, hydroquinone monomethyl ether) to the reaction mixture. Around 0.01–0.1% of quencher is normally sufficient. A small amount of quencher is also recommended for long-time storage of the monomer. Preferably, the quencher should be removed immediately prior to polymerization, or an increased amount of radical initiator needs to be added.

Cationic/anionic polymerization may occur as a side reaction if (strong) acid or base is formed or added during reaction. When this becomes a problem, it is therefore essential to quench the acid or base to re-establish sufficiently neutral conditions.

VII. SELECTION OF FUNCTIONAL MONOMERS

The rules for selection of functional monomers have been covered in other chapters in this book (cf. Chapters 8, 14 and 15), describing the prerequisites for acquiring strong binding to the chosen target molecule and/or template species. These features have been discussed mainly from a thermodynamic angle, optimizing the complementarity and binding strength between the template and the functional element, especially regarding noncovalent protocols. However, a few other issues are also important in choosing monomers to produce efficient molecularly imprinted polymers. These include the solubility of all included components, to a certain extent discussed above, the polymerization compatibility between monomers, and possible side reactions before and during polymerization. In this section, some very general considerations on monomer selection will be given from this point of view.

A. Solubility

Solubility is one of the most troublesome problems for the practising imprinter, since the concentrations of most included components are very high. Not only may the components in themselves be poorly insoluble, but also the complexes formed between the functional monomers and the templates may precipitate from solution prior to or during polymerization. Obviously, there is no general rule to solve this problem, and it can mainly be solved by trial and error following the common device "like dissolves like". Reversible- and semi-covalent protocols are here less problematic since the adducts are more stable and normally more soluble in organic phase, whereas noncovalent protocols are more sensitive to solvent choice. The most popular organic solvents/porogens in imprinting include acetonitrile, chloroform, dichloromethane, and toluene, all chosen to provide good porogenic effects at the same time as they have good solubilizing properites towards a wide range of compounds. Acetone, tetrahydrofuran (THF), (lower) alcohols, and even dimethylformamide (DMF) have been used in cases where solubility has been particularly severe. In noncovalent protocols, special attention needs to be taken to ensure that complexation is produced, and polar, protic solvents should be used with caution. Also, all solvents should be dried before use to avoid undesired competition from water. If hydrophobic/solvatophobic effects are wanted, however, water or alchols are sometimes used. As a general rule of thumb, the least soluble component should be dissolved first in the best solvent, then the others are added carefully not to cause phase separation. Sometimes this method can be used to prevent precipitation/oiling out of components from solution. If, however, various solvents or solvent mixtures fail to produce homogeneous solutions, a more elaborate method need to be used. The only "safe" way to circumvent severe solubility problems is to suitably decorate the components to enhance solubility.

B. Monomer Compatibility

The compatibility between various monomers is normally described as their "reactivity ratios"—r-values. The reactivity of a certain monomer (M_1) is measured in relation to a second monomer (M_2), expressed as a function of the reactivity of either of the monomer radicals ($M_1\bullet$ and $M_2\bullet$) towards either of the monomers (M_1 and M_2).

The representative rate equations are presented below. Thus, r_1 is the reactivity ratio for M_1, representing the rate ratio for reacting with itself or with the comonomer.

$$M_1^{\bullet} + M_1 \xrightarrow{k_{11}} M_1^{\bullet} \qquad \text{rate} = k_{11}\left[M_1^{\bullet}\right]\left[M_1\right]$$

$$M_1^{\bullet} + M_2 \xrightarrow{k_{12}} M_2^{\bullet} \qquad \text{rate} = k_{12}\left[M_1^{\bullet}\right]\left[M_2\right] \qquad r_1 = k_{11}/k_{12}$$

$$M_2^{\bullet} + M_2 \xrightarrow{k_{22}} M_2^{\bullet} \qquad \text{rate} = k_{22}\left[M_2^{\bullet}\right]\left[M_2\right] \qquad r_2 = k_{22}/k_{21}$$

$$M_2^{\bullet} + M_1 \xrightarrow{k_{21}} M_1^{\bullet} \qquad \text{rate} = k_{21}\left[M_2^{\bullet}\right]\left[M_1\right]$$

In order to acquire random polymerization, which is normally desired in molecular imprinting protocols, the product of the reactivity ratios ($r_1 r_2$) should be close to unity. If $r_1 r_2$ deviates significantly from this ideal value, then either homopolymeric regions will be formed from either M_1 or M_2 (block copolymers), or alternating polymers will be produced where the monomer at the end of the growing chain preferably reacts with the opposite monomer. Neither of these two cases is advantageous in preparing molecularly imprinted polymers. For example, the reactivity ratio in a system with methyl methacrylate and styrene is 0.5 for both monomers, resulting in a product of 0.25. This implies that copolymerization of methyl methacrylate and styrene would result in a more alternating copolymer, thus disadvantageous for most imprinting purposes.

Reactivity ratios are, however, not easily determined, and especially not for three-dimensional network polymers formed at high monomer concentrations. Thus, no representative values for imprinting systems exist and the only data that can be used/generated refer to dilute solutions of linear copolymers. Also, the r-value for a certain monomer will be different when forming complexes with the template. Nevertheless, such values can be used as guides in determining the potential compatibility between monomers, but should be treated with appropriate judgment. As a rule of thumb, it is generally advantageous to choose functional monomers of the same structure, as these are more likely to copolymerize well. For example, mixtures of methacrylates are known to copolymerize more or less randomly, and their respective reactivity ratio products are close to unity. Similarly, different styrenes copolymerizes relatively well, although these systems are rather sensitive to substitution pattern. Thus, if the common structure is kept the same, the more likely it is to achieve random copolymerization.

C. Side Reactions

Undesired reactions may also be caused from the choice of monomers and special attention needs to be taken to avoid these. As stated in the synthesis section, side reactions may occur from certain functionalities and byproducts. Thus, amino- and especially mercapto-functionalities may add to the vinyl groups in general and acryl groups in particular before and during polymerization, and unless these features are controlled, it is safer to use monomers devoid of reactive amines and thiols. Amino groups may also be made less reactive by alkylation.

VIII. CONCLUSIONS

This chapter has presented an overview on functional and structural building blocks for use in molecular imprinting, as well as common ways to their selection and synthesis. Most of these ways are relatively straightforward and do not require extensive synthetic skills. Obviously, sophisticated protocols require a lot more synthetic work, but as long as an efficient functionally interacting entity can be prepared to possess a handle, where the monomerizing unit can be attached without side reactions, then synthesis should progress relatively smoothly. Also evident from the overview is the fact that a multitude of building blocks are available from main chemical manufacturers, making (initial) attempts in molecular imprinting even less demanding. Together with general considerations regarding recognition properties (Chapters 14 and 16) and format design (Part V), this chapter could provide a starting point for a basic imprinting project, and possibly inspiration to optimize a present imprinting system further.

REFERENCES

1. Arshady, R. Functional monomers. J.M.S.—Rev. Macromol. Chem. Phys. **1992**, *C32*, 101–132.
2. Morris, L.R.; Mock, R.A.; Marshall, C.A.; Howe, J.H. Synthesis of some amino acid derivatives of styrene. J. Am. Chem. Soc. **1959**, *81*, 377–382.
3. Kempe, M.; Glad, M.; Mosbach, K. An approach towards surface imprinting using the enzyme ribonuclease A. J. Mol. Recognit. **1995**, *8*, 35–39.
4. Striegler, S. Selective discrimination of closely related monosaccharides at physiological pH by a polymeric receptor. Tetrahedron **2001**, *57*, 2349–2354.
5. Rosatzin, T.; Andersson, L.I.; Simon, W.; Mosbach, K. Preparation of Ca^{2+} selective sorbents by molecular imprinting using polymerisable ionophores. J. Chem. Soc., Perkin Trans. **1991**, *2*, 1261–1265.
6. Castro, B.; Whitcombe, M.J.; Vulfson, E.N.; Vazquez-Duhalt, R.; Barzana, E. Molecular imprinting for the selective adsorption of organosulphur compounds present in fuels. Anal. Chim. Acta. **2001**, *435*, 83–90.
7. De, D.; Krogstad, D.J. Pd(DIPHOS)$_2$-catalyzed cross-coupling reactions of organoborons with free or polymer-bound aryl halides. Org. Lett. **2000**, *2*, 879–882.
8. Mirviss, S.B. Synthesis of ω-unsaturated acids. J. Org. Chem. **1989**, *54*, 1948–1951.
9. Campbell, T.D.; McDonald, N.R. Synthesis of hydrocarbon derivatives by the Wittig synthesis. I. Distyrylbenzenes. J. Org. Chem. **1959**, *24*, 1246–1251.
10. Wulff, G.; Akelah, A. Enzyme-analogue built polymers. 6. Synthesis of 5-vinylsalicylaldehyde and a simplified synthesis of some divinyl derivatives. Makromol. Chem. **1979**, *179*, 2647–2651.
11. Le Bigot, Y.; Delmas, M.; Gaset, A. A simplified Wittig synthesis using solid/liquid transfer processes. Synth. Commun. **1982**, *12*, 107–112.
12. Schlick, H.; Stelzer, F.; Tasch, S.; Leising, G. poly[(*m*-phenylenevinylene)-*co*-(*p*-phenylenevinylene)] derivatives synthesized via metathesis condensation (ADMET). J. Mol. Catal. A: Chem. **2000**, *160*, 71–84.
13. Abruna, H.D.; Breikss, A.I.; Collum, B.D. Improved synthesis of 4-vinyl-4'-methyl-2, 2'-bipyridine. Inorg. Chem. **1985**, *24*, 987–988.
14. Williams, C.E.; Lowry, R.B.; Braven, J.; Belt, S.T. Novel synthesis and characterization of ruthenium tris(4-methyl-4'-vinyl-2,2'-bipyridine) complexes. Inorg. Chim. Acta. **2001**, *315*, 112–119.

15. Overberger, C.G.; Vorchheimer, N. Imidazole-containing polymers. Synthesis and poly-merization of the monomer 4(5)-vinylimidazole. J. Am. Chem. Soc. **1963**, *85*, 951–955.
16. Andersson, L.; Ekberg, B.; Mosbach, K. Synthesis of a new amino acid based cross-linker for preparation of substrate selective acrylic polymers. Tetrahedron Lett. **1985**, *26*, 3623–3624.
17. Shea, K.; Stoddard, G.; Shavelle, D.; Wakui, F.; Choate, R. Synthesis and characteriza-tion of highly cross-linked polyacrylamides and polymethacrylamides. A new class of macroporous polyamides. Macromolecules **1990**, *23*, 4497–4507.
18. Matsui, J.; Higashi, M.; Takeuchi, T. Molecularly imprinted polymer as 9-ethyladenine receptor having a porphyrin-based recognition center. J. Am. Chem. Soc. **2000**, *122*, 5218–5219.
19. Kondo, K.; Iwasaki, H.; Ueda, N.; Takemoto, K.; Imoto, M. Vinyl polymerization CCXIX. Vinyl compounds of nucleic acid bases. 2. Synthesis and polymerization of N-β-methacryloyloxyethyl derivatives of uracil, thymine, theophylline, 2-chloro-6-methylpurine, and adenine. Makromol. Chem. **1968**, *120*, 21–26.
20. Spivak, D.; Shea, K.J. Molecular imprinting of carboxylic acids employing novel functional macroporous polymers. J. Org. Chem. **1999**, *64*, 4627–4634.
21. Morris, L.R.; Mock, R.A.; Marshall, C.A.; Howe, J.H. Synthesis of some amino acid derivatives of styrene. J. Am. Chem. Soc. **1959**, *81*, 377–382.
22. Arnold, F.H.; Striegler, S.; Sundaresan, V. Molecular and Ionic Recognition with Imprinted Polymers. In *Chiral Ligand Exchange Adsorbents for Amines and Underiva-tized Amino Acids*: *"Bait-and-Switch" Molecular Imprinting*; ACS Symposium Series; Bartsch, R., Maeda, M., Eds.; 1998; Vol. 703, 109–118.
23. Vidyasankar, S.; Ru, M.; Arnold, F.H. Molecularly imprinted ligand-exchange adsor-bents for the chiral separation of underivatized amino acids. J. Chromatogr. A **1997**, *775*, 51–63.
24. Kempe, M.; Glad, M.; Mosbach, K. An approach towards surface imprinting using the enzyme ribonuclease. A. J. Mol. Recognit. **1995**, *8*, 35–39.
25. Dhal, P.K.; Arnold, F.H. Metal-coordination interactions in the template-mediated synthesis of substrate-selective polymers: recognition of bis(imidazole) substrates by copper(II) iminodiacetate containing polymers. Macromolecules **1992**, *25*, 7051–7059.
26. Hart, B.R.; Shea, K.J. Synthetic peptide receptors: molecularly imprinted polymers for the recognition of peptides using peptide–metal interactions. J. Am. Chem. Soc. **2001**, *123*, 2072–2073.
27. Lo, H.C.; Chen, H.; Fish, R.H. Metal–ion-templated polymers, 3. Synthesis of a [{mono-N-(4-vinylbenzyl)-1,4,7-triazacyclononane}$_2$Hg](OTf)$_2$ sandwich complex, polymerization of this monomer with divinylbenzene, and Hg^{2+} ion selectivity studies with the demetallated resin. Eur. J. Inorg. Chem. **2001**, 2217–2220.
28. Chen, H.; Olmstead, M.M.; Albright, R.L.; Devenyi, J.; Fish, R.H. Metal–ion-templated polymers: synthesis and structure of N-(4-vinylbenzyl)-1,4,7-triazacyclo-nonanezinc(II) complexes, their copolymerization with divinylbenzene, and metal–ion selectivity studies of the demetalated resins—evidence for a sandwich complex in the polymer matrix. Angew. Chem., Int. Ed. Engl. **1997**, *36*, 642–645.
29. Chen, G.; Guan, Z.; Chen, C.-T.; Fu, L.; Sundaresan, V.; Arnold, F.H. A glucose-sensing polymer. Nat. Biotechnol. **1997**, *15*, 354–357.
30. Dhal, P.K.; Arnold, F.H. Substrate selectivity of molecularly imprinted polymers incor-porating a rigid chelating monomer, bis-methacrylato(4-methyl-4′-vinyl)-2,2′-bipyridine Cu(II). New J. Chem. **1996**, *20*, 695–698.
31. Efendiev, A.A.; Kabanov, V.A. Selective polymer complexons prearranged for metal ions sorption. Pure Appl. Chem. **1982**, *54*, 2077–2092.

32. Takeuchi, T.; Mukawa, T.; Matsui, J.; Higashi, M.; Shimizu, K.D. Molecularly imprinted polymers with metalloporphyrin-based molecular recognition sites coassembled with methacrylic acid. Anal. Chem. **2001**, *73*, 3869–3874.
33. Kobayashi, T.; Fukaya, T.; Abe, M.; Fujii, N. Phase inversion molecular imprinting by using template copolymers for high substrate recognition. Langmuir **2002**, *18*, 2866–2872.
34. Nakano, T.; Satoh, Y.; Okamoto, Y. Synthesis and chiral recognition ability of a cross-linked polymer gel prepared by a molecular imprint method using chiral helical polymers as templates. Macromolecules **2001**, *34*, 2405–2407.
35. Sergeyeva, T.A.; Matuschewski, H.; Piletsky, S.A.; Bendig, J.; Schedler, U.; Ulbricht, M. Molecularly imprinted polymer membranes for substance-selective solid-phase extraction from water by surface photo-grafting polymerization. J. Chromatogr. A **2001**, *907*, 89–99.
36. Perez, N.; Whitcombe, M.J.; Vulfson, E.N. Molecularly imprinted nanoparticles prepared by core-shell emulsion polymerization. J. Appl. Polym. Sci. **2000**, *77*, 1851–1859.
37. Haginaka, J.; Sakai, Y. Uniform-sized molecularly imprinted polymer material for (S)-propranolol. J. Pharm. Biomed. Anal. **2000**, *22*, 899–907.
38. Asanuma, H.; Kajiya, K.; Hishiya, T.; Komiyama, M. Molecular imprinting of cyclodextrin in water for the recognition of peptides. Chem. Lett. **1999**, *7*, 665–666.
39. Kanekiyo, Y.; Sano, M.; Ono, Y.; Inoue, K.; Shinkai, S. Facile construction of a novel metal-imprinted polymer surface without a polymerization process. J. Chem. Soc., Perkin Trans. **1998**, *2*, 2005–2008.
40. Wang, H.Y.; Kobayashi, T.; Fujii, N. Preparation of molecular imprint membranes by the phase inversion precipitation technique. Langmuir **1996**, *12*, 4850–4856.
41. Marx-Tibbon, S.; Wilner, I. Photostimulated imprinted polymers: a light-regulated medium for transport of amino acids. J. Chem. Soc., Chem. Commun. **1994**, *10*, 1261–1262.
42. Dunkin, I.R.; Lenfeld, J.; Sherrington, C.D. Molecular imprinting of flat polycondensed aromatic molecules in macroporous polymers. Polymer **1993**, *34*, 77–84.
43. Sellergren, B.; Ekberg, B.; Mosbach, K. Molecular imprinting of amino acid derivatives in macroporous polymers. Demonstration of substrate- and enantio-selectivity by chromatographic resolution of racemic mixtures of amino acid derivatives. J. Chromatogr. **1985**, *347*, 1–10.
44. Garcia, R.; Pinel, C.; Madic, C.; Lemaire, M. Ionic imprinting effect in gadolinium/lanthanum separation. Tetrahedron Lett. **1998**, *39*, 8651–8654.
45. Kobayashi, T.; Wang, H.Y.; Fujii, N. Molecular imprinting of theophylline in acrylonitrile–acrylic acid copolymer membrane. Chem. Lett. **1995**, *10*, 927–928.
46. Matsui, J.; Doblhoff-Dier, O.; Kugimiya, A.; Takeuchi, T. 2-(Trifluoromethyl)acrylic acid: a novel functional monomer in non-covalent molecular imprinting. Anal. Chim. Acta. **1997**, *343*, 1–4.
47. Zander, A.; Findlay, P.; Renner, T.; Sellergren, B.; Swietlow, A. Analysis of nicotine and its oxidation products in nicotine chewing gum by a molecularly imprinted solid-phase extraction. Anal. Chem. **1998**, *70*, 3304–3314.
48. Lanza, F.; Sellergren, B. Method for synthesis and screening of large groups of molecularly imprinted polymers. Anal. Chem. **1999**, *71*, 2092–2096.
49. Takeuchi, T.; Fukuma, D.; Matsui, J. Combinatorial molecular imprinting: an approach to synthetic polymer receptors. Anal. Chem. **1999**, *71*, 285–290.
50. Matsui, J.; Kubo, H.; Takeuchi, T. Molecularly imprinted fluorescent-shift receptors prepared with 2-(Trifluoromethyl)acrylic acid. Anal. Chem. **2000**, *72*, 3286–3290.
51. Takeuchi, T.; Dobashi, A.; Kimura, K. Molecular imprinting of biotin derivatives and its application to competitive binding assay using nonisotopic labeled ligands. Anal. Chem. **2000**, *72*, 2418–2422.

52. Yilmaz, E.; Haupt, K.; Mosbach, K. The use of immobilized templates—a new approach in molecular imprinting. Angew. Chem. Int. Ed. **2000**, *39*, 2115–2118.

53. Quaglia, M.; Chenon, K.; Hall, A.J.; De Lorenzi, E.; Sellergren, B. Target analogue imprinted polymers with affinity for folic acid and related compounds. J. Am. Chem. Soc. **2001**, *123*, 2146–2154.

54. Tarbin, J.A.; Sharman, M. Development of molecularly imprinted phase for the selective retention of stilbene-type estrogenic compounds. Anal. Chim. Acta. **2001**, *433*, 71–79.

55. Mosbach, K.; Yu, Y.; Andersch, J.; Ye, L. Generation of new enzyme inhibitors using imprinted binding sites: the anti-idiotypic approach, a step toward the next generation of molecular imprinting. J. Am. Chem. Soc. **2001**, *123*, 12420–12421.

56. Fischer, L.; Müller, R.; Ekberg, B.; Mosbach, K. Direct enantioseparation of β-adrenergic blockers using a chiral stationary phase prepared by molecular imprinting. J. Am. Chem. Soc. **1991**, *113*, 9358–9360.

57. Batz, H.G.; Koldehoff, J. Monomeric and polymeric succinimido esters of ω-methacryloylamino acids; preparation and reaction with amines. Makromol. Chem. **1976**, *177*, 683–689.

58. D'Souza, S.M.; Alexander, C.; Carr, S.W.; Waller, A.M.; Whitcombe, M.J.; Vulfson, E.N. Directed nucleation of calcite at a crystal-imprinted polymer surface. Nature **1999**, *398*, 312–316.

59. Fukusaki, E.-I.; Saigo, A.; Kajiyama, S.-I.; Kobayashi, A. An artificial plastic receptor that discriminates axial asymmetry. J. Biosci. Bioeng. **2000**, *90*, 665–668.

60. Joshi, V.P.; Karode, S.K.; Kulkarni, M.G.; Mashelkar, R.A. Novel separation strategies based on molecularly imprinted adsorbents. Chem. Eng. Sci. **1998**, *53*, 2271–2284.

61. Andersson, L. Preparation of amino acid ester-selective cavities formed by non-covalent imprinting with a substrate in highly cross-linked polymers. React. Polym., Ion Exch., Sorbents **1988**, *9*, 29–41.

62. Bae, S.Y.; Southard, G.L.; Murray, G.M. Molecularly imprinted ion exchange resin for purification, preconcentration and determination of UO_2^{2+} by spectrophotometry and plasma spectrometry. Anal. Chim. Acta. **1999**, *397*, 173–181.

63. Panasyuk-Delaney, T.; Mirsky, V.M.; Wolfbeis, O.S. Capacitive creatinine sensor based on a photografted molecularly imprinted polymer. Electroanalysis **2002**, *14*, 221–224.

64. D'Oleo, R.; Alvarez-Lorenzo, C.; Sun, G. A new approach to design imprinted polymer gels without using a template. Macromolecules **2001**, *34*, 4965–4971.

65. Panasyuk-Delaney, T.; Mirsky, V.M.; Ulbricht, M.; Wolfbeis, O.S. Impedometric herbicide chemosensors based on molecularly imprinted polymers. Anal. Chim. Acta **2001**, *435*, 157–162.

66. Piletsky, S.A.; Matuschewski, H.; Schedler, U.; Wilpert, A.; Piletska, E.V.; Thiele, T.A.; Ulbricht, M. Surface functionalization of porous polypropylene membranes with molecularly imprinted polymers by photograft copolymerization in water. Macromolecules **2000**, *33*, 3092–3098.

67. Piletsky, S.A.; Andersson, H.S.; Nicholls, I.A. Combined hydrophobic and electrostatic interaction-based recognition in molecularly imprinted polymers. Macromolecules **1999**, *32*, 633–636.

68. Steinke, J.H.G.; Dunkin, I.R.; Sherrington, D.C. Molecularly imprinted anisotropic polymer monoliths. Macromolecules **1996**, *29* (1), 407–415.

69. Takeuchi, T.; Seko, A.; Matsui, J.; Mukawa, T. Molecularly imprinted polymer library on a microtiter plate. High-throughput synthesis and assessment of cinchona alkaloid-imprinted polymers. Instrum. Sci. Technol. **2001**, *29*, 1–9.

70. Kugimiya, A.; Kuwada, Y.; Takeuchi, T. Preparation of sterol-imprinted polymers with the use of 2-(methacryloyloxy)ethyl phosphate. J. Chromatogr. A **2001**, *938*, 131–135.

71. Kempe, M.; Fischer, L.; Mosbach, K. Chiral separation using molecularly imprinted heteroaromatic polymers. J. Mol. Recognit. **1993**, *6*, 25–29.
72. Baggiani, C.; Giovannoli, C.; Anfossi, L.; Tozzi, C. Molecularly imprinted solid-phase extraction sorbent for the clean-up of chlorinated phenoxyacids from aqueous samples. J. Chromatogr. A **2001**, *938*, 35–44.
73. Suedee, R.; Srichana, T.; Saelim, J.; Thavonpibulbut, T. Thin-layer chromatographic separation of chiral drugs on molecularly imprinted chiral stationary phases. J. Planar Chromatogr.—Modern TLC **2001**, *14*, 194–198.
74. Moller, K.; Nilsson, U.; Crescenzi, C. Synthesis and evaluation of molecularly imprinted polymers for extracting hydrolysis products of organophosphate flame retardants. J. Chromatogr. A **2001**, *938*, 121–130.
75. Lin, J.-M.; Yamada, M. Chemiluminescent reaction of fluorescent organic compounds with $KHSO_5$ using cobalt(II) as catalyst and its first application to molecular imprinting. Anal. Chem. **2000**, *72*, 1148–1155.
76. Klein, J.U.; Whitcombe, M.J.; Mulholland, F.; Vulfson, E.N. Template-mediated synthesis of a polymeric receptor specific to amino acid sequences. Angew. Chem. Int. Ed. **1999**, *38*, 2057–2060.
77. Ramstrom, O.; Ye, L.; Gustavsson, P.E. Chiral recognition by molecularly imprinted polymers in aqueous media. Chromatographia **1998**, *48*, 197–202.
78. Lin, J.M.; Uchiyama, K.; Hobo, T. Enantiomeric resolution of dansyl amino acids by capillary electrochromatography based on molecular imprinting method. Chromatographia **1998**, *47*, 625–629.
79. Ramstrom, O.; Andersson, L.I.; Mosbach, K. Recognition sites incorporating both pyridinyl and carboxy functionalities prepared by molecular imprinting. J. Org. Chem. **1993**, *58*, 7562–7564.
80. Tan, Z.J.; Remcho, V.T. Molecular imprint polymers as highly selective stationary phases for open tubular liquid chromatography and capillary electrochromatography. Electrophoresis **1998**, *19*, 2055–2060.
81. Overberger, C.G.; Glowaky, R.C.; Pacansky, T.J.; Sannes, K.N. Poly[4(5)-vinylimidazole]. Macromol. Synth. **1974**, *5*, 43–49.
82. Chang, J.Y.; Do, S.K.; Han, M.J. A sol–gel reaction of vinyl polymers based on thermally reversible urea linkages. Polymer **2001**, *42*, 7589–7594.
83. Yamazaki, T.; Yilmaz, E.; Mosbach, K.; Sode, K. Towards the use of molecularly imprinted polymers containing imidazoles and bivalent metal complexes for the detection and degradation of organophosphotriester pesticides. Anal. Chim. Acta. **2001**, *435*, 209–214.
84. Mathew, J.; Buchardt, O. Molecular imprinting approach for the recognition of adenine in aqueous medium and hydrolysis of adenosine 5′-triphosphate. Bioconjugate Chem. **1995**, *6*, 524–528.
85. Ohkubo, K.; Urata, Y.; Hirota, S.; Honda, Y.; Fujishita, Y.-I.; Sagawa, T. Homogeneous esterolytic catalysis of a polymer prepared by molecular imprinting of a transition state analog. J. Mol. Catal. **1994**, *93*, 189–193.
86. Robinson, D.K.; Mosbach, K. Molecular imprinting of a transition state analog leads to a polymer exhibiting esterolytic activity. J. Chem. Soc., Chem. Commun. **1989**, *70*, 969–970.
87. Leonhardt, A.; Mosbach, K. Enzyme-mimicking polymers exhibiting specific substrate binding and catalytic functions. React. Polym., Ion Exch., Sorbents **1987**, *6*, 285–290.
88. Kawanami, Y.; Yunoki, T.; Nakamura, A.; Fujii, K.; Umano, K.; Yamauchi, H.; Masuda, K. Imprinted polymer catalysts for the hydrolysis of *p*-nitrophenyl acetate. J. Mol. Catal. A: Chem. **1999**, *145*, 107–110.
89. Kempe, M.; Fischer, L.; Mosbach, K. Chiral separation using molecularly imprinted heteroaromatic polymers. J. Mol. Recognit. **1993**, *6*, 25–29.

90. Vigneau, O.; Pinel, C.; Lemaire, M. Ionic imprinted resins based on EDTA and DTPA derivatives for lanthanides(III) separation. Anal. Chim. Acta. **2001**, *435*, 75–82.

91. Mohr, G.J.; Citterio, D.; Demuth, C.; Fehlmann, M.; Jenny, L.; Lohse, C.; Moradian, A.; Nezel, T.; Rothmaier, M.; Spichiger, U.E. Reversible chemical reactions as the basis for optical sensors used to detect amines, alcohols and humidity. J. Mater. Chem. **1999**, *9*, 2259–2264.

92. Luo, L.; Britt, P.F.; Buchanan, A.C., III; Desai, D.O.; Makote, R.D. Chromatographic characterization of molecularly imprinted polymers for nitroaromatics. Polym. Mater. Sci. Eng. **1999**, *81*, 546.

93. Luebke, M.; Whitcombe, M.J.; Vulfson, E.N. A novel approach to the molecular imprinting of polychlorinated aromatic compounds. J. Am. Chem. Soc. **1998**, *120*, 13342–13348.

94. Lanza, F.; Hall, A.J.; Sellergren, B.; Bereczki, A.; Horvai, G.; Bayoudh, S.; Cormack, P.A.G. Development of a semiautomated procedure for the synthesis and evaluation of molecularly imprinted polymers applied to the search for functional monomers for phenytoin and nifedipine. Anal. Chim. Acta. **2001**, *435*, 91–106.

95. Ju, J.Y.; Shin, C.S.; Whitcombe, M.J.; Vulfson, E.N. Imprinted polymers as tools for the recovery of secondary metabolites produced by fermentation. Biotechnol. Bioeng. **1999**, *64*, 232–239.

96. Rosatzin, T.; Andersson, L.I.; Simon, W.; Mosbach, K. Preparation of Ca^{2+} selective sorbents by molecular imprinting using polymerizable ionophores. J. Chem. Soc., Perkin Trans. **1991**, *2*, 1261–1265.

97. Suarez-Rodriguez, J.L.; Diaz-Garcia, M.E. Fluorescent competitive flow-through assay for chloramphenicol using molecularly imprinted polymers. Biosens. Bioelectron. **2001**, *16*, 955–961.

98. Wizeman, W.J.; Kofinas, P. Molecularly imprinted polymer hydrogels displaying isomerically resolved glucose binding. Biomaterials **2001**, *22*, 1485–1491.

99. Huval, C.C.; Bailey, M.J.; Braunlin, W.H.; Holmes-Farley, S.R.; Mandeville, W.H.; Petersen, J.S.; Polomoscanik, S.C.; Sacchiro, R.J.; Chen, X.; Dhal, P.K. Novel cholesterol lowering polymeric drugs obtained by molecular imprinting. Macromolecules **2001**, *34*, 1548–1550.

100. McNiven, S.; Kato, M.; Levi, R.; Yano, K.; Karube, I. Chloramphenicol sensor based on an in situ imprinted polymer. Anal. Chim. Acta **1998**, *365*, 69–74.

101. Piletsky, S.A.; Piletskaya, E.V.; Panasyuk, T.L.; El'skaya, A.V.; Levi, R.; Karube, I.; Wulff, G. Imprinted membranes for sensor technology: opposite behavior of covalently and noncovalently imprinted membranes. Macromolecules **1998**, *31*, 2137–2140.

102. Cheong, S.-H.; Rachkov, A.E.; Park, J.-K.; Yano, K.; Karube, I. Synthesis and binding properties of a noncovalent molecularly imprinted testosterone-specific polymer. J. Polym. Chem., A. Polym. Chem. **1998**, *36*, 1725–1732.

103. Piletsky, S.A.; Piletskaya, E.V.; Elgersma, A.V.; Yano, K.; Karube, I.; Parhometz, Y.P.; Elskaya, A.V. Atrazine sensing by molecularly imprinted membranes. Biosens. Bioelectron. **1995**, *10*, 959–964.

104. Beach, J.V.; Shea, K.J. Designed catalysts. A synthetic network polymer that catalyzes the dehydrofluorination of 4-fluoro-4-(*p*-nitrophenyl)butan-2-one. J. Am. Chem. Soc. **1994**, *116*, 379–380.

105. Baumstark, R.; Wildburg, G.; Haeussling, L.; Guenther, W.; Schoenfeld, R.; Gruen, M. *Polymers from Unsaturated Amidines for Use in Adhesives*; Ger. Offen. DE, (BASF A.-G., Germany), 1997; 12 pp.

106. Kim, J.-M.; Ahn, K.-D.; Wulff, G. Cholesterol esterase activity of a molecularly imprinted polymer. Macromol. Chem. Phys. **2001**, *202*, 1105–1108.

107. Strikovsky, A.G.; Kasper, D.; Gruen, M.; Green, B.S.; Hradil, J.; Wulff, G. Catalytic molecularly imprinted polymers using conventional bulk polymerization or suspension

polymerization: selective hydrolysis of diphenyl carbonate and diphenyl carbamate. J. Am. Chem. Soc. **2000**, *122*, 6295–6296.

108. Wulff, G.; Schoenfeld, R. Polymerizable amidines. Adhesion mediators and binding sites for molecular imprinting. Adv. Mater. **1998**, *10*, 957–959.

109. Wulff, G.; Gross, T.; Schonfeld, R. Enzyme models based on molecularly imprinted polymers with strong esterase activity. Angew. Chem. Int. Ed. **1997**, *36*, 1962–1964.

110. Kugimiya, A.; Takeuchi, T. Surface plasmon resonance sensor using molecularly imprinted polymer for detection of sialic acid. Biosens. Bioelectron. **2001**, *16*, 1059–1062.

111. Kugimiya, A.; Takeuchi, T.; Matsui, J.; Ikebukuro, K.; Yano, K.; Karube, I. Recognition in novel molecularly imprinted polymer sialic acid receptors in aqueous media. Anal. Lett. **1996**, *29*, 1099–1107.

112. Chen, Z.; Takei, Y.; Deore, B.A.; Nagaoka, T. Enantioselective uptake of amino acid with overoxidized polypyrrole colloid templated with L-lactate. Analyst **2000**, *125*, 2249–2254.

113. Lele, B.S.; Kulkarni, M.G.; Mashelkar, R.A. Molecularly imprinted polymer mimics of chymotrypsin. 1. Cooperative effects and substrate specificity. Reac. Func. Polym. **1999**, *39*, 37–52.

114. Sreenivasan, K. Effect of the type of monomers of molecularly imprinted polymers on the interaction with steroids. J. Appl. Polym. Sci. **1998**, *68*, 1863–1866.

115. Sreenivasan, K.; Sivakumar, R. Interaction of molecularly imprinted polymers with creatinine. J. Appl. Polym. Sci. **1997**, *66*, 2539–2542.

116. Ohkubo, K.; Urata, Y.; Hirota, S.; Funakoshi, Y.; Sagawa, T.; Usui, S.; Yoshinaga, K. Catalytic properties of novel L-histidyl group-introduced polymers imprinted by a transition state analogue in the hydrolysis of amino acid esters. J. Mol. Catal. A **1995**, *101*, L111–L114.

117. Burow, M.; Minoura, N. Molecular imprinting: synthesis of polymer particles with antibody like binding characteristics for glucose oxidase. Biochem. Biophys. Res. Commun. **1996**, *227* (2), 419–422.

118. Yu, C.; Mosbach, K. Molecular imprinting utilizing an amide functional group for hydrogen bonding leading to highly efficient polymers. J. Org. Chem. **1997**, *62* (12), 4057–4064.

119. Zayats, M.; Lahav, M.; Kharitonov, A.B.; Willner, I. Imprinting of specific molecular recognition sites in inorganic and organic thin layer membranes associated with ion-sensitive field-effect transistors. Tetrahedron **2002**, *58*, 815–824.

120. D'Auria, M.; Racioppi, R. Photochemical dimerization of esters of urocanic acid. J. Photochem. Photobiol., A: Chem. **1998**, *112*, 145–148.

121. Chianella, I.; Lotierzo, M.; Piletsky, S.A.; Tothill, I.E.; Chen, B.; Karim, K.; Turner, A.P.F. Rational design of a polymer specific for microcystin-LR using a computational approach. Anal. Chem. **2002**, *74*, 1288–1293.

122. Wulff, G.; Akelah, A. Enzyme-analog built polymers, 6. Synthesis of 5-vinylsalicylaldehyde and a simplified synthesis of some divinyl derivatives. Makromol. Chem. **1978**, *179*, 2647–2651.

123. Zhou, W.-J.; Kurth, M.J.; Hsieh, Y.-L.; Krochta, J.M. Synthesis and characterization of new styrene main-chain polymer with pendant lactose moiety through urea linkage. Macromolecules **1999**, *32*, 5507–5513.

124. Shea, K.J.; Sasaki, D.Y.; Stoddard, G.J. Fluorescence probes for evaluating chain solvation in network polymers. An analysis of the solvatochromic shift of the dansyl probe in macroporous styrene-divinylbenzene and styrene–diisopropenylbenzene copolymers. Macromolecules **1989**, *22*, 1722–1730.

125. Wulff, G.; Dhal, P.K. Chirality of polyvinyl compounds. 6. Unusual influences of the comonomer structures on the chiroptical properties of optically active vinyl copolymers

with chirality arising from configurational relationships in the main chain. Macromolecules **1988**, *21*, 571–578.

126. Ren, J.; Sakakibara, K.; Hirota, M. The synthesis of poly[*N*-(*p*-vinylbenzylidene)-tert-butylamine *N*-oxide] and characterizations of reaction with nitrogen dioxide. Bull. Chem. Soc. Jpn. **1993**, *66*, 1897–1902.

127. Polborn, K.; Severin, K. Biomimetic catalysis with immobilized organometallic ruthenium complexes: substrate- and regioselective transfer hydrogenation of ketones. Chem. Eur. J. **2000**, *6*, 4604–4611.

128. Wulff, G.; Vietmeier, J. Enzyme-analogue built polymers. 25. Synthesis of macroporous copolymers from alpha.-amino acid-based vinyl compounds. Makromol. Chem. **1989**, *190*, 1717–1726.

129. Shea, K.J.; O'Dell, R.; Sasaki, D.Y. Free radical induced macrocyclizations. Tetrahedron Lett. **1992**, *33*, 4699–4702.

130. Kobayashi, T.; Murawaki, Y.; Reddy, P.S.; Abe, M.; Fujii, N. Molecular imprinting of caffeine and its recognition assay by quartz–crystal microbalance. Anal. Chim. Acta. **2001**, *435*, 141–149.

131. Kugimiya, A.; Takeuchi, T. Application of indoleacetic acid-imprinted polymer to solid phase extraction. Anal. Chim. Acta. **1999**, *395*, 251–255.

132. Sreenivasan, K. Imparting cholesterol recognition sites in radiation polymerized poly(2-hydroxyethyl methacrylate) by molecular imprinting. Polym. Int. **1997**, *42*, 169–172.

133. Arnold, F.; Dhal, P.; Shnek, D.; Plnkett, S. In *Template Polymerization Using Metal Chelates and Fluid Imprint Matrices for Use in Protein Purification*. PCT International, 1992 pp.

134. Alexander, C.; Smith, C.R.; Whitcombe, M.J.; Vulfson, E.N. Imprinted polymers as protecting groups for regioselective modification of polyfunctional substrates. J. Am. Chem. Soc. **1999**, *121*, 6640–6651.

135. Mukawa, T.; Goto, T.; Takeuchi, T. Post-oxidative conversion of thiol residue to sulfonic acid in the binding sites of molecularly imprinted polymers: disulfide based covalent molecular imprinting for basic compounds. Analyst **2002**, *127*, 1407–1409.

136. Subrahmanyam, S.; Piletsky, S.A.; Piletska, E.V.; Chen, B.; Day, R.; Turner, A.P.F. "Bite-and-switch" approach to creatine recognition by use of molecularly imprinted polymers. Adv. Mater. **2000**, *12*, 722–724.

137. Liu, X.-C.; Dordick, J.S. Sugar acrylate-based polymers as chiral molecularly imprintable hydrogels. J. Polym. Sci. Part A: Polym. Chem. **1999**, *37*, 1665–1671.

138. Wulff, G.; Vietmeier, J.; Poll, G.H. Enzyme-analog built polymers. 22. Influence of the nature of the crosslinking agent on the performance of imprinted polymers in racemic resolution. Makromol. Chem. **1987**, *188*, 731–740.

139. Biffis, A.; Wulff, G. Molecular design of novel transition state analogues for molecular imprinting. New J. Chem. **2001**, *25*, 1537–1542.

140. Alexander, C.; Smith, C.R.; Whitcombe, M.J.; Vulfson, E.N. Imprinted polymers as protecting groups for regioselective modification of polyfunctional substrates. J. Am. Chem. Soc. **1999**, *121*, 6640–6651.

141. Sellergren, B.; Andersson, L. Molecular recognition in macroporous polymers prepared by a substrate analog imprinting strategy. J. Org. Chem. **1990**, *55*, 3381–3383.

142. Cammidge, A.N.; Baines, N.J.; Bellingham, R.K. Synthesis of heterogeneous palladium catalyst assemblies by molecular imprinting. Chem. Commun. **2001**, *24*, 2588–2589.

143. Brueggemann, O.; Freitag, R.; Whitcombe, M.J.; Vulfson, E.N. Comparison of polymer coatings of capillaries for capillary electrophoresis with respect to their applicability to molecular imprinting and electrochromatography. J. Chromatogr. A **1997**, *781*, 43–53.

144. Shea, K.J.; Stoddard, G.J.; Shavelle, D.M.; Wakui, F.; Choate, R.M. Synthesis and characterization of highly crosslinked poly(acrylamides) and poly(methacrylamides). A new class of macroporous polyamides. Macromolecules **1990**, *23*, 4497–4507.

145. Olsen, J.; Martin, P.; Wilson, I.D.; Jones, G.R. Methodology for assessing the properties of molecular imprinted polymers for solid phase extraction. Analyst **1999**, *124*, 467–471.

146. Yoshizako, K.; Hosoya, K.; Iwakoshi, Y.; Kimata, K.; Tanaka, N. Porogen imprinting effects. Anal. Chem. **1998**, *70*, 386–389.

147. Kempe, M. Antibody-mimicking polymers as chiral stationary phases in HPLC. Anal. Chem. **1996**, *68*, 1948–1953.

148. Schweitz, L.; Andersson, L.I.; Nilsson, S. Capillary electrochromatography with molecular imprint-based selectivity for enantiomer separation of local anesthetics. J. Chromatogr. A **1997**, *792*, 401–409.

149. Biffis, A.; Graham, N.B.; Siedlaczek, G.; Stalberg, S.; Wulff, G. The synthesis, characterization and molecular recognition properties of imprinted microgels. Macromol. Chem. Phys. **2001**, *202*, 163–171.

150. Andersson, L.; Ekberg, B.; Mosbach, K. Synthesis of a new amino acid base crosslinker for preparation of substrate selective acrylic polymers. Tetrahedron Lett. **1985**, *26*, 3623–3624.

151. Chen, G.H.; Guan, Z.B.; Chen, C.T.; Fu, L.T.; Sundaresan, V.; Arnold, F.H. A glucose sensing polymer. Nature Biotechnol. **1997**, *15*, 354–357.

152. Ohkubo, K.; Sawakuma, K.; Sagawa, T. Shape-and stereo-selective esterase activities of cross-linked polymers imprinted with a transition-state analogue for the hydrolysis of amino acid esters. J. Mol. Catal. A: Chem. **2001**, *165*, 1–7.

153. Hart, B.R.; Shea, K.H. Molecular imprinting for the recognition of N-terminal histidine peptides in aqueous solution. Macromolecules **2002**, *35*, 6192–6201.

154. Lee, S.K.; Seo, K.H.; Kim, W.S. Temperature dependence of the binding of methyl orange by crosslinked poly(4-vinylpyridine). Polymer **1993**, *34*, 2392–2396.

155. Sellergren, B.; Shea, K.J. Influence of polymer morphology on the ability of imprinted network polymers to resolve enantiomers. J. Chromatogr. **1993**, *635*, 31–49.

156. Oikawa, E.; Motomi, K.; Aoki, T. Synthesis and properties of poly(thioether amides) from 2,6-bis(acrylamido)pyridine and dithiols. J. Polym. Sci., Part A: Polym. Chem. **1993**, *31*, 457–465.

157. Green, B.S.; Priwler, M. Molecularly imprinted polymers for the treatment and diagnosis of medical conditions. U.S. Pat. Appl. Publ. US, (Israel). 2002; 15 pp.

158. Yano, K.; Tanabe, K.; Takeuchi, T.; Matsui, J.; Ikebukuro, K.; Karube, I. Molecularly imprinted polymers which mimic multiple hydrogen bonds between nucleotide bases. Anal. Chim. Acta. **1998**, *363*, 111–117.

159. Kugimiya, A.; Mukawa, T.; Takeuchi, T. Synthesis of 5-fluorouracil-imprinted polymers with multiple hydrogen bonding interactions. Analyst **2001**, *126*, 772–774.

160. Das, K.; Duffy, D.J.; Hsu, S.L.; Penelle, J.; Rotello, V.M. Experimental study of release and uptake in well defined imprinted polymer films. Polym. Prepr. **2000**, *41*, 1173–1174.

161. Duffy, D.J.; Das, K.; Hsu, S.L.; Penelle, J.; Rotello, V.M.; Stidham, D.H. Molecularly imprinted polymer systems for selective recognition via hydrogen bonding interactions. Polym. Mater. Sci. Eng. **2000**, *82*, 69–70.

162. Morikawa, H.; Kono, S.; Fusaoka, Y. *Diacrylamidobenzoic Acid Derivatives as Polyfunctional Monomers and Their Manufacture and Purification*: Jpn. Kokai Tokkyo Koho. Jp (Toray Industries, Inc., Japan), 1999; 5 pp.

163. Arshady, R.; Mosbach, K. Synthesis of substrate-selective polymers by host–guest polymerization. Macromol. Chem. Phys. **1981**, *182*, 687–692.

164. Hiratani, H.; Alvarez-Lorenzo, C.; Chuang, J.; Guney, O.; Grosberg, A.Y.; Tanaka, T. Effect of reversible cross-linker, N,N'-bis(acryloyl)cystamine, on calcium ion adsorption by imprinted gels. Langmuir **2001**, *17*, 4431–4436.

165. Srivatsan, S.G.; Parvez, M.; Verma, S. Modeling of prebiotic catalysis with adenylated polymeric templates: Crystal structure studies and kinetic characterization of template-assisted phosphate ester hydrolysis. Chem. Eur. J. **2002**, *8*, 5184–5191.

8
Combinatorial Approaches to Molecular Imprinting

F. Lanza, B. Dirion, and B. Sellergren Universität Dortmund, Dortmund, Germany

I. WHY DO WE NEED TO ADOPT A COMBINATORIAL APPROACH TO THE SYNTHESIS OF MIPs?

Molecular Imprinting is nowadays an established technique for the production of polymeric materials having affinities and selectivities which in some cases are comparable to those of biological systems (e.g. receptors or antibodies) [1]. As far as noncovalent imprinting is concerned (see Chapter 3 in this book), the success of the technique relies mainly on its intriguing simplicity and on the use of easily available, relatively low cost reagents. In particular, the procedure of preparation of molecularly imprinted polymers (MIPs) to be used for chromatographic/SPE applications is straightforward (mix and bake) and the postpolymerization processing (crush, sieve, and pack) does not require particular skills (see Part VI of this volume).

The fact that a large number of templates has been successfully imprinted using basic recipes has led to the proliferation of MIPs of relatively low complexity, most of which have been based on a restricted number of synthetic trials (even though in some cases supported by spectroscopic or computational hints). The majority of basic templates have been imprinted using methacrylic acid (MAA) as monomer (see Table 5.6A in Ref. 1), the acidic ones using vinylpyridines (VPY) (see Table 5.6B in Ref. 1). Ethylene glycol dimethacrylate (EDMA) is the most used crosslinker (see Table 2.5 in Ref. 1), acetonitrile, dichloromethane, chloroform, toluene the most used solvents.

Most of these imprinting protocols were never optimized and led to materials showing high nonspecific binding, low water compatibility, template leaching, slow kinetics, and insufficient affinity. Furthermore, for many templates, it has been difficult to generate specific binding sites at all. In order to obtain materials with superior performances which can successfully cope with real separation or detection problems (as chromatographic phases, solid-phase extraction materials, membranes, catalysts, sensors, enzyme mimics) more advanced synthetic approaches need to be applied.

We distinguish between design and combinatorial approaches. In the former the area of host–guest chemistry has inspired the design of functional monomers targeted towards structural motifs in "difficult templates." In the latter a large collection of MIPs (based on different functional monomers, comonomers, crosslinkers, solvents, initiators) are generated in a format facilitating the selection of individual members possessing interesting properties (as in combinatorial screening). In both cases, the result will be a starting point for optimization. Due to the many variables that control the imprinting process, the optimization then needs to be achieved by a multivariate strategy such as Experimental Design or Modelling (see Chapter 14).

The development and implementation of high-throughput screening techniques is expected to significantly accelerate the discovery of new, high-performance MIPs.

II. THE DISCOVERY OF NEW IMPRINTING PROTOCOLS: FROM WHERE DO WE START?

Thus far, the development or discovery of a suitable imprinting protocol (or imprinting recipe), consisting of functional monomers, crosslinker, solvent and initiator for a certain template has often involved a few trial-and-error experiments based mainly on intuition. This is not negative if we consider that a great many important discoveries in chemistry originated from astute observations of unintentional experiments.

However, if we want to produce higher performance materials in a short time, we will need to perform a large number of trials accompanied by the smallest percentage of errors. Reducing the number of errors will reduce the cost of the experiments which will, in turn, contribute to a more efficient MIP development. After all, if we are able to find a good starting point we will probably have greater chances of success. This sounds obvious. But from where do we start?

In the noncovalent approach (Chapter 3 of this book), which is by far the most common technique of imprinting, many studies have demonstrated that the solution structure of the monomer–template assemblies defines the subsequently formed binding sites. In other words, the amount and quality of recognition sites in the MIP depends on the number and strength of specific interactions occurring between the template and the monomers in the prepolymerization mixture. These are in turn influenced by the quality of the solvent, crosslinking monomer, temperature and pressure used in the polymerization.

If we then wish to get the highest number of "hits" in our trials, a general criterion is to maximize the stability of these interactions. This is easily obtained by increasing the amount of monomers in the polymerization solution, which, however, leads to an increase in undesired nonspecific binding sites. A more promising approach is based on the use of stoichiometric, noncovalent interactions like for instance multiple hydrogen bonding. In this case, very stable assemblies with large associations constants can form in aprotic media and the number of nonspecific binding sites will be minimized since there will be a reduction in the amount of free non-associated functional monomer. Finally, the position of the equilibrium between free template–monomer(s) and their corresponding complexes can also be affected by temperature and pressure as well as by the choice of the porogen. However, varying these synthetic–related factors will also affect the porous properties of the materials, which in turn determine their kinetic properties, e.g., diffusional mass transfer limitations, size exclusion effects, bleeding. This shows the complexity of the system and highlights the need for high-throughput synthesis and screening techniques.

A. Some Hints on the Choice of System Components

Before getting started, we should take some time to carefully choose our monomers and template for the initial experiment (so called building blocks in Chapter 7 of this book). Here are some general hints. The readers should also consult Chapters 14 and 16 for discussions on the topic.

1. Template

In conventional imprinting protocols, in order to avoid contaminating leaching, the template should be different from the target if the materials are to be employed for its detection or separation at trace levels. Nevertheless, the sites obtained should be selective for both the target and the template, which implies that the target and template should be very closely related analogues. In addition to being a closely related analogue to the target, the template should possess functionalities that interact with the functional monomer. Obviously these functionalities should be complementary to those of the functional monomer.

Examples? Terbutylazine and simazine were used as structural analogue of the target atrazine [2], pentycaine for bupivacaine [3], compound E for the lead drug F [4] Check Table 1 for the structures.

2. Monomers

General considerations for choosing monomers are found in Chapters 7 and 16 in this book. For templates containing acidic groups, basic functional monomers are

Table 1 Examples of Structural Analogues of the Target Chosen as Templates

Reference	Analyte	Chosen Template
[2]		
[3]		
[4]		

preferably chosen and for carboxylic acids and amides high selectivities have been observed with acrylamide and methacrylamide. In the case of poorly polar to apolar templates, with few polar functional groups that can provide interaction sites for polar functional monomers, it may be instead beneficial to use amphiphilic monomers, stabilizing the monomer–template assemblies by hydrophobic and Van der Waals forces (see Table 5.6A–C in Ref. 1).

The search for the optimal structural motif which can complement the template functionality may be guided by results from the areas of host–guest and ligand–receptor chemistry. Similarly to biological systems, a large number of complementary interactions are expected to increase the strength and fidelity of recognition. Thus, the use of comonomers which may target different subunits of a complex template is also a successful strategy [5].

A starting point is to search the imprinting literature for previously published examples of polymers imprinted using similar templates. A comprehensive literature database is available at the website of the Society for Molecular Imprinting (http://www.smi.tu-berlin.de). Otherwise scan through the Chemicals Catalogues for lists of commercially available monomers!

A primary indication on how well the monomers have been chosen is to simply see whether they are capable of assisting solubilization of the template in the prepolymerization mixture. A small-scale solubility test may thus be a good way to initially screen the monomers for strong monomer–template interactions. Weak interactions may be revealed by complexation induced spectral changes (in NMR, UV or fluorescence spectra). The complexation induced shifts of the characteristic ^1H–NMR signals of the template upon increasing monomer concentrations are often used to estimate the monomer–template association constants. Prior to this, however, knowledge about the stoichiometry of the monomer–template complexation and the tendency of the monomer and template to self–associate are required. The former can be obtained by means of a so-called Job's plot whereas the latter by a dilution experiment.

One possible way to increase the number of hits in the process of discovery of new MIPs is the use of results of monomer scoring, based on molecular modeling and computational methods (see Chapter 14 of this book). Given a number of possible monomers for a given template, these methods will provide you with a scored list of possible monomer combinations and their ratios. It is always good to insert these combinations into your screening array.

3. Solvents

First check which solvents and crosslinking monomers are able to individually solubilize your monomers and template. For monomer–template interactions stabilized by polar forces you should select nonprotic solvents of low polarity: they have less likelihood to compete with your monomers for the template. The functional monomer–template complexes are often based on hydrogen bond interactions. If the solvent is a good hydrogen bond donor or acceptor, it will compete with the monomers and destabilize the complexes. Chlorinated solvents like dichloromethane and chloroform are good solvents for these systems, but acetonitrile and toluene may be preferred for their lower health hazard and for environmental reasons. For monomer–template systems stabilized by apolar forces, more polar solvents and higher temperatures are favorable.

4. Temperature and Initiators

In the case of electrostatic interactions, lower temperatures of polymerization are known to increase the stability of monomer–template assemblies. Choose the initiators that allow you to work at lower temperatures (for example 2,2′–azobis(4–methoxy–2,4–dimethylvaleronitrile) and 2,2′–azobis(2,4–dimethylvaleronitrile) by Wako (V–70 and V–65) or photo–polymerize with AIBN (azobisisobutyronitrile) at 15°C with UV light). An environmentally friendly alternative to AIBN is dimethyl 2,2′–azobis(2–methylpropionate)(V-601 manufactured by Wako). Thermally initiated polymerization has a more general scope and may be preferable in certain monomer–template systems. These systems have been critically compared for the imprinting of terbutylazine [6].

Once the possible candidate monomers, crosslinkers, solvent, and initiators for a given template have been selected, we still have a decision to take: how many trials are we going to make? Which parameters will we favor in our screening? How many MIPs can we generate in parallel?

III. TOOLS FOR COMBINATORIAL APPROACHES: MINIMIPS, 96–WELL MIPS

Combinatorial synthesis is based on the preparation of a large number of compounds or materials, if possible in a form that is easy to screen for properties of interest. To screen a very large number of chemicals (monomers, solvents, crosslinker, templates) each experiment needs to be small. This allows one to reduce the amount of reagents used and to speed up the screening process.

Small scale procedures for the rapid synthesis and screening of large groups of MIPs have been recently developed [7,8], based on the preparation of ~50 mg of materials (MiniMIPs) on the bottom of small vials or of 96-well plates [9] and their in situ testing by equilibrium batch rebinding. The preparation of the materials, which represents a scaled down version of the traditional monolith approach (Chapter 15), is outlined in Fig. 1. For each MiniMIP, a corresponding

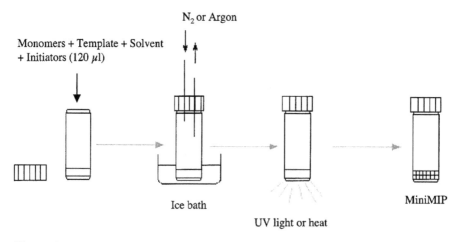

Figure 1 Preparation of MIPs on the bottom of small glass vials (MiniMIPs).

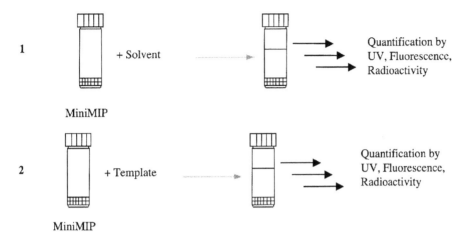

Figure 2 Screening of MiniMIPs: (1) in situ template extraction; (2) equilibrium batch rebinding.

nonimprinted polymer (NP) is prepared on the same scale (MiniNP) by polymerizing all the components in the absence of the template. After polymerization, MiniMIPs and MiniNPs are first incubated with the porogen in a release step. It is here assumed that the rate of release and the amount of released template might give a rough indication of the presence of binding sites. One hundred percent release in this step can be taken as a criterion for discarding a particular MiniMIP: if the template is completely released, it is highly probable that no binding sites were formed in the imprinting process. Following the release step, the template is exhaustively extracted from the polymer with a series of washing steps. After template removal all the polymers are exposed to a solution of the template in the porogen or in another solvent (rebinding step) (Fig. 2). The amount of the nonbound template left in solution (C_{free}) is quantified at definite time intervals by any analytical method (UV, MS, fluorescence, radioactivity measurements). In the rebinding step, the imprinting factor K_{MIP}/K_{NP} as defined in Ref. 8 (where $K = n/C_{free}$ and n is the amount of template rebound to 1 g of dry polymer) is calculated based on the weight of the polymer and the original concentration of the template in the rebinding solution.

The thus calculated partition coefficients K (K_{MIP} and K_{NP}) and imprinting factors are the most important responses used to evaluate the polymers. Alternatively, the fluorescence intensity of each polymer upon binding of the template has been used as response (for an example see Refs. 9 and 10).

Owing to the low consumption of reagents and the absence of any manipulation of the polymers, the small scale protocol can be partly automated (Fig. 3). In particular the use of modern plate technology (plate readers, pipetting robots) can be exploited for the preparation and screening of 96-well MIPs. Thus, we can easily prepare 96 MiniMIPs and test them in equilibrium batch rebinding. The quantification of the nonbound fraction will take a different time depending on if it is performed simultaneously for all the MiniMIPs using a plate reader or sequentially by HPLC-UV. Even though we are able to quickly prepare and screen a large number of MIPs,

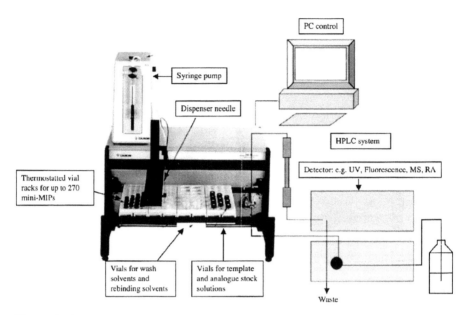

Figure 3 Semi–automated system for the preparation and screening of MiniMIPs.

it is, however, important to assign priorities to the many variables we are going to screen.

IV. EXAMPLES OF COMBINATORIAL SCREENING APPLIED TO THE SEARCH FOR MONOMERS

The functional monomer is a very important parameter of any imprinting protocol. Therefore the first MiniMIP screening should focus on the search for a reasonably good combination of monomers to target the main functionalities of the given template. A low polar, nonprotic solvent capable of dissolving all the components can be used as polymerization diluent and EDMA or DVB are usually the first choice of crosslinker for this initial group of MiniMIPs. The most successful monomer combinations will be taken as starting points for other screening experiments (i.e., optimization of crosslinker and/or porogen) and finally for the fine-tuning which will involve the variation of the ratios of all the components of the imprinting protocol.

Here follows a protocol and some examples of the search for functional monomers for different types of templates. Most examples concern monomer–template systems where association is driven by electrostatic interactions.

V. PROTOCOL 1: GENERAL MONOMER OPTIMIZATION

A. Reagents and Equipment

Crosslinker (ethylenglycol dimethacrylate (EDMA) or divinylbenzene (DVB) or ethylenglycol trimetacrylate (TRIM)), initiators (e.g. 2,2′–azobisisobutyronitrile

(AIBN) or 2,2′-azobis (2,4–dimethyl) valeronitrile (ABDV)), solvent (for instance acetonitrile, dichloromethane), monomers chosen for the screening. The monomers and solvents should be purified and dried prior to use. The following equipment may be used for the preparation of MiniMIPs: glass vials (1.5 mL) with silicone/teflon septa, 96-well plates, high-pressure mercury vapour lamp (e.g., Philips HPK 125W), oven or thermostatted bath, programmed liquid handler (e.g., Gilson 233 XL or 222XL) equipped with microplate holders, HPLC system equipped with a variable wavelength detector (or diode array detector) or a fluorescence detector, microplate reader (for instance Spectra Fluor Plus by TECAN, Austria).

B. Method

1. Synthesis

Prepare two identical solutions by mixing crosslinker, initiator and the solvent you have selected. The volume of the solutions depends on the number of MiniMIPs you are planning to prepare. If you are planning to prepare 100 MiniMIPs you will need approx. a 12-ml solution. Mix for example 4.75 mL of EDMA (25 mmol) with 7 mL of solvent and 50 mg (0.3 mmol) of AIBN. Add 1.25 mmol of template to one of the solutions and label it as mother solution with template (MT). Label the other solution as mother solution blank (MB). The solutions should be prepared shortly before the screening experiment or otherwise stored frozen. Transfer them into the reservoirs of the dispensing robot just before use.

Prepare 2 M solutions of each monomer you wish to screen in the solvent that you have selected. It will be enough to prepare 1 mL of each solution (25 μL will be dispensed). Keep the solutions refrigerated before use. Transfer them into the containers of the pipetting robot just before use.

Sparge all the glass vials with argon and close them with sealed caps. Dispense 95 μL of the mother solution MB into half of the vials and 95 μL of the mother solution MT into each remaining vial. Cool all the vials at 5°C and degass them with argon for 2 min. Dispense 25 μL of a monomer solution (or of a mixture of two lmonomer solutions) in each vial depending on the library you want to create. Sparge again all the vials for 2 min with argon and then initiate the polymerization either thermally in the oven at 60°C (if you use AIBN) or 40°C (with Wako V-65) or photochemically at 15°C using the UV lamp.

To prepare 96-well MIPs, reduce the dispensed volumes by half, then sparge the plate with argon and place it in a heating (40°C) or photopolymerization chamber under a nitrogen atmosphere. In the former case use Wako V-65 as initiator instead of AIBN.

2. Removal of Template

After the polymerization incubate the MiniMIPs for 24 hr with 1 mL of the solvent you have used for the polymerization (under shaking or stirring preferably). Then measure the concentration of the template in the supernatant at definite time intervals (for example over a 10 hr period). You may use any analytical method to determine the concentration of the template. If you are using HPLC–UV, you can directly inject the supernatant into your system without separating it from the polymers.

Compare the amount of template found in each supernatant and identify the MIPs whose supernatants show the lowest template content. These will probably be among the best "hits"of the screening. On the other hand, 100% release indicates that the polymer will probably perform poorly in the rebinding step.

Remove the supernatant from each vial and expose the polymers to harsh wash conditions in order to remove the template from the polymers. This might require a rather large series of successive washing steps based on solvents of different polarity and hydrogen bond capacity, acidity, basicity, and so on. The washing procedure depends on the nature of the template, on the strength of the template–polymer inter-actions and on the nature of the polymer backbone. Methanol/water/acetic acid mixtures were found to be effective in washing out basic templates, such as triazines, from MAA/EDMA-based MIPs. Each washing step should consist of the addition of the solvent and then in a few hours of incubation. Afterwards you can check the washing solution for the template by using the same analytical method previously established for the analysis of the supernatants. Then move to another washing step. The miniNPs should be washed in the same way even though they do not contain any template.

Once you believe to have exhaustively removed the template, incubate miniMIPs and MiniNPs for the last time with the solvent to be used for the rebinding step. Allow equilibration for 24 hr and then check the supernatants. If no trace of template can be detected you can proceed to the batch rebinding test.

3. Batch Rebinding Test

Prepare a solution of the template in the solvent used for the polymerization. The concentration of the solution depends on the amount of template that you have used in the polymerization and on the sensitivity of your detection technique. If you have used a 1/4/20 molar ratio of template/functional monomer/crosslinker you should prepare a 1 mM solution. A 1 mM solution contains an amount of template that would cover *ca.* 10% of the theoretical sites, a typical yield of imprinted sites based on the conventional protocol. Dispense 1 mL of this solution into each vial containing either miniMIPs or miniNPs. Let the solution incubate for 24 hr possibly under stirring or shaking, then measure the concentration of the template in the supernatant and cal-culate the amount of template (in $\mu mol/g$) rebound to 1 g of the polymer (either MIP or NP). Divide the amount of template rebound to 1 g of polymer by the concentration of the template present in solution (in $\mu mol/mL$). Indicate this number with K (K_{MIP} or K_{NP}). Divide K_{MIP} by K_{NP} to obtain the imprinting factor (IF= K_{MIP}/K_{NP}). This value reflects the enhanced uptake of the template to the MIP compared to the NP and most often correlates with the rebinding strength and selectivity exhibited by the upscaled batch. Identify the MIPs with the higher imprinting factors. These correspond to the best combinations of functional monomers to be used for the imprinting of your given template. These combinations should be selected for further optimization.

If K_{MIP} and K_{NP} are similar and pretty high, i.e., the template strongly binds both to MIP and NPs in the porogen, no conclusive answer can be drawn from the batch rebinding test in the porogen, and the MIPs should not be discarded at this stage. In this particular case, conditions should be sought where the nonspecific binding is minimized, for example upon addition of acids or bases (see also Chapters 7–13, 16, 20, and 22).

C. Monomer Combinations

The use of combinations of monomers has often led to enhancement in selectivity and affinity. This occurs especially when the functional monomers target different functional groups in the template. Examples include the combination of two different acidic monomers (acid monomer + basic monomer, acid + amide, base + amide), acidic or basic monomers, π-donor−acceptor monomers.

D. Comments

You can adopt the same protocol for screening factors other than the functional monomers (i.e., solvents or crosslinkers). Remember in this case to modify the components of the mother solutions (they should contain everything except the component you want to screen for) and the composition of the dispensed solutions (they should contain the component you want to screen for).

It is always advisable to check the concentration of the template in the supernatant both by HPLC-UV and spectrophotometry (using a reader) when a new screening experiment is being performed. This is because the template might undergo oxidation or degradation processes during the polymerization and the equilibrium batch rebinding which might change its spectra. In addition HPLC-UV only allows the template to be separated from the interferents copresent in the supernatant (particularly abundant in the release and washing fractions) and thus accurately determined. Once the stability of the template has been assessed and the screening procedure has been established, the spectrophotometric analysis of the equilibrium batch rebinding fractions will be enough to identify the best candidates.

VI. EXAMPLE 1: TERBUTYLAZINE

A. Screening the Functional Monomers for the Imprinting of Terbutylazine

Template: terbutylazine
Monomers used: MAA, TFM, HEMA, MMA, VPY, NVP (Fig. 4)
Polymerization mixture: EDMA, CH$_2$Cl$_2$, AIBN (Fig. 4)

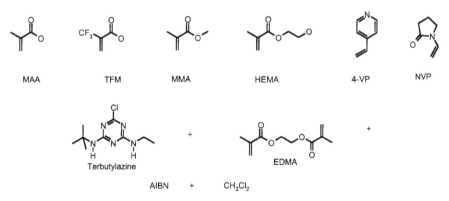

Figure 4 Components of the polymerization mixture selected for the search for functional monomers of Example 1 (imprinting of terbutylazine).

Table 2 Recovery of Template from MiniMIPs (Terbutylazine/EDMA/AIBN/CH$_2$Cl$_2$) Incubated Overnight with Dichloromethane

Polymer (functional monomer)	C_{free} (mM)	Recovery yield (%)
p-MAA	3.56	32
p-TFM	5.53	49
p-MMA	10.4	92
p-HEMA	8.38	75
p-VPY	9.92	88
p-NVP	10.9	97

The screening experiment was performed by manually pipetting the solutions into the vials and photo-polymerizing the mixtures at 15°C. The MiniMIPs were then incubated with the solvent (CH$_2$Cl$_2$) overnight and the supernatants were analyzed by HPLC-UV. The concentration of the template in the supernatant solutions and the percentage of recovered template are shown in Table 2. The MiniMIPs prepared using MAA and TFM showed the lowest concentration of template in the supernatant solutions. This agrees with predictions based on observed solution complexation between substituted pyridines and carboxylic acids and is in agreement with the results of Takeuchi et al. [7]. The MiniMIPs prepared using TFM and MAA were selected for the rebinding test.

After complete extraction of the template, the polymers were submitted to equilibrium batch rebinding with a 1 mM solution of terbutylazine in CH$_2$Cl$_2$. The imprinting factors (K_{MIP}/K_{NP}) of MIPs prepared using MAA and TFM as functional monomers were 11 and 6, respectively.

This screening experiment was repeated by automatically dispensing the solution with a pipetting robot (Gilson 233 XL). Figure 5 shows the percentage of

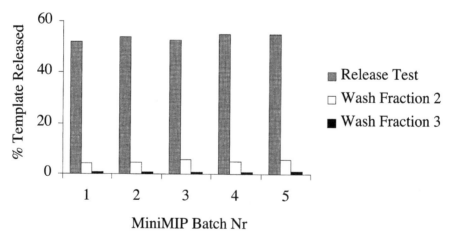

Figure 5 Percentage of ametryn detected in: (i) the supernatants of five parallel MiniMIPs after 24 hr incubation with 1 mL dichloromethane (black bars); (ii) the second and third wash fractions (1 mL MeOH/H$_2$O/CH$_3$COOH 60/10/30, incubated for 2 hr). The MiniMIPs were prepared and tested using the semi-automated system shown in Fig. 3.

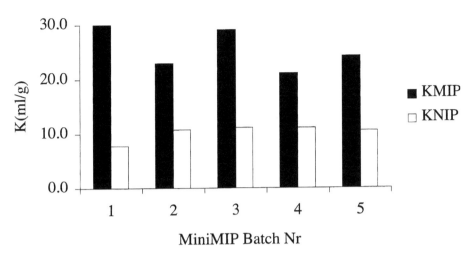

Figure 6 Partition coefficients K_{MIP} and K_{NP} of five parallel MiniMIPs measured after 24 hr incubation with 1 mM ametryn in dichloromethane.

template found in the supernatant solution for five replicates of the same MIP (MAA/ametryn/EDMA/CH_2Cl_2). The percentage of template detected in two wash fractions (methanol/water/acetic acid 60/10/30) is also shown. Figure 6 shows the partition coefficients K_{MIP} and K_{NP} for the five MIP and NP replicates.

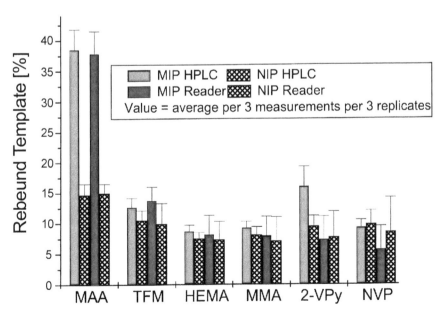

Figure 7 Percentage of template rebound to MIPs and NPs prepared using different functional monomers after 24 hr incubation with 1 mM terbutylazine in acetonitrile. MiniMIPs and MiniNPs were prepared in a 96-well plate and evaluated in parallel by HPLC-UV and plate reader.

B. Screening the Functional Monomers (for Terbutylazine) in a 96–well Format (WellMIPs)

The search for monomers was here performed in a 96-well format. The components of the polymerization mixture were the same. The polymerization was thermally initiated. The solutions were dispensed using a pipetting robot and the supernatants analyzed in series by HPLC-UV and in parallel with a plate reader (see Fig. 7). Six WellMIPs were prepared and screened for each monomer. The reproducibility of the automated procedure is shown in Fig. 8 over six replicates for three different monomers (MAA, TFM, HEMA).

VII. EXAMPLE 2: SCREENING THE FUNCTIONAL MONOMERS FOR THE IMPRINTING OF PHENYTOIN

Template: phenytoin (one of the most commonly prescribed anticonvulsant drugs in the treatment of epilepsy).

Monomers used: MAAM, AAM, DAA, DAP, VBA (see Fig. 9 for the chemical structures).

MAAM, AAM: commercial monomers, hydrogen bond donor/acceptor; DAA: commercial monomer, capable of forming multiple hydrogen bonds, meant to target the hydantoin ring; DAP: synthetic monomer already used for the imprinting of barbiturates, also capable of forming multiple hydrogen bonds; VBA: adenine-based synthetic monomer (in a study reported in the literature 9-ethyladenine was reported to co-crystallize into a 2:1 complex with phenytoin).

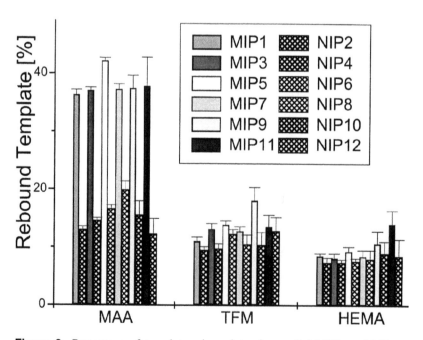

Figure 8 Percentage of template rebound to six parallel MIPs and NPs prepared using different functional monomers, after 24 hr incubation with 1 mM terbutylazine in acetonitrile (the samples were analyzed with the plate reader).

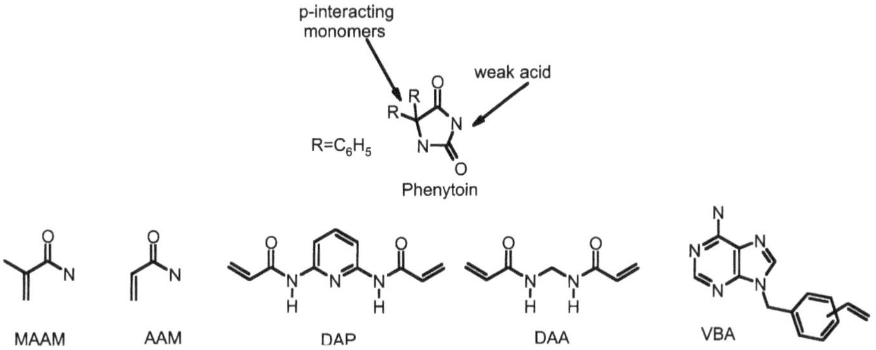

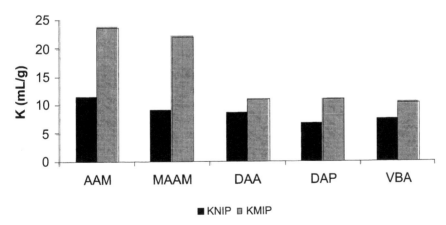

Figure 9 Functional monomers selected for the screening of Example 3 (imprinting of phenytoin).

Polymerization mixture: EDMA, AIBN, THF/CH$_3$CN 12/58 (for pAAM and pMAAM), THF/CH$_3$CN 18/40 (for pDAA) DMF/CH$_3$CN 70/30 (for pDAP), Toluene/DMF 27/53 (for pVBA)). The choice of the solvent mixtures is related to the solubility of the template in combination with the particular functional monomer (scarce in acetonitrile, very good in polar solvents such as THF and DMF).

Three commercial and two synthesized monomers were used for the screening. The results of the rebinding tests are shown in Fig. 10. The commercial hydrogen bonding monomers MAAM and AAM turned out to be the most successful in this case (larger difference in partition factors K between MIPs and NPs). The MAAM-based MIPs were also prepared on a larger scale and evaluated as chromatographic stationary phases (see Fig. 11). The retention behavior of phenytoin is markedly different between NPs and MIPs, especially at low sample loads, as expected for materials showing high affinity towards the template.

Figure 10 Partition coefficients K_{MIP} and K_{NP} of MiniMIPs and MiniNPs after 24 hr incubation with 0.5 mM phenytoin in the porogens (for the composition, see Example 3).

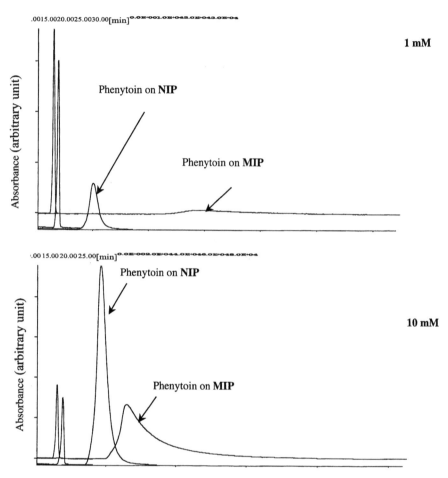

Figure 11 Elution profiles of 1 and 10 nmol injections of phenytoin onto MIP and NP columns (mobile phase acetonitrile 100%, injection volume 10 μL, flow 1 mL/min, the polymers were prepared using MAAM as functional monomer).

VIII. EXAMPLE 3: SCREENING THE FUNCTIONAL MONOMERS FOR THE IMPRINTING OF NIFEDIPINE

Template: nifedipine (Fig. 12) (archetype of the dihydropyridine calcium entry blockers, increasingly used as a probe drug for the assessment of cytochrome P-450 IIIA4 enzyme activity in humans).

Monomers used: MAA, TFM, AcrN, HEMA, DEAEMA, NaphtMA, AnthraMMA, AS, 4-VP, 2-VP, PMS, DMS, NVC, (see Fig. 12 for the chemical structures).

Polymerization mixture: EDMA, AIBN, CH_2Cl_2, V–65 (azobisdivaleronitrile).

Here several commercial monomers were screened in order to target the few functionalities of the molecule: acidic monomers such as MAA, TFM to target the ring nitrogen or the ester groups, basic monomers such as AS, 4-VP, 2-VP, DEAEMA

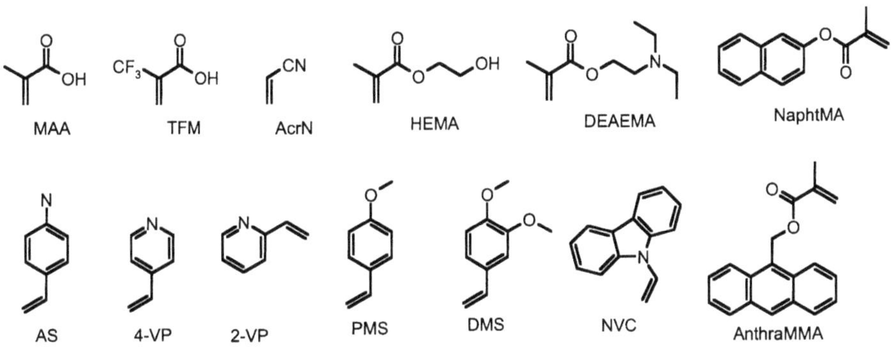

Monomer library

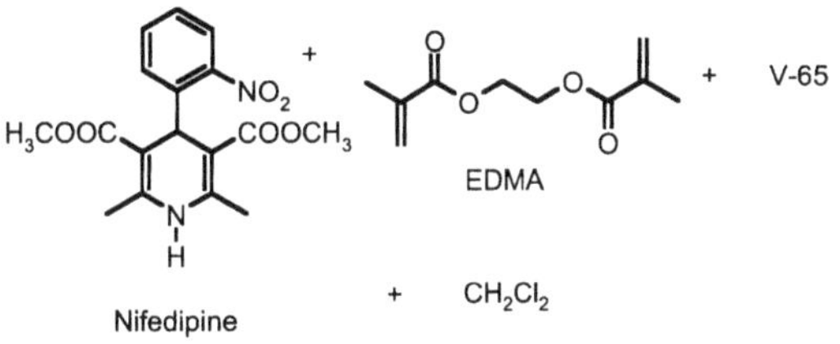

Figure 12 Components of the polymerization mixture selected for the screening of Example 4 (imprinting of nifedipine).

to target the slightly acidic exocyclic proton, hydrogen bonding monomers such as HEMA, AS, MAA, TFM to target the ester groups and finally π-donors/acceptors such as AnthraMMA, NaphtMA or NVC to target the aromatic ring.

From the results of the incubation with the porogen (CH_2Cl_2), the basic monomers 4-VP, 2-VP, the acidic monomer TFM and the π−donor AnthraMMA could be identified as the most promising monomers (Fig. 13). The results of the equilibrium batch rebinding test partially confirmed the first findings. AS, MAA, HEMA, and AcrN could also be identified as good monomers (see Fig. 14). In this context, note that the π-donor monomers are compatible with the acidic ones and thus they could be combined in a combinatorial approach. An important issue concerns the choice between a batch equilibrium or dynamic mode assessment of the rebinding selectivity. In these studies, the promising selectivity observed in the batch mode did not translate into a selective chromatographic phase. This highlights the need to evaluate the polymers in a dynamic situation, particularly if the final material is to be implemented in solid-phase extraction or chromatography.

IX. EXAMPLE 4: SCREENING THE FUNCTIONAL MONOMERS FOR THE IMPRINTING OF METHOTREXATE

Template: methotrexate (MTX) (inhibitor of the enzyme dihydrofolate reductase, it is used in cancer therapy since it preferentially slows down the cell growth of rapidly growing cancerous cells).

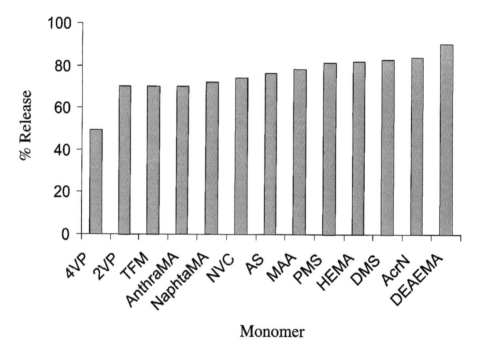

Figure 13 Example 4 (imprinting of nifedipine). Percentage of template detected in the supernatant after 30 min sonication in the presence of 1 mL CH_2Cl_2.

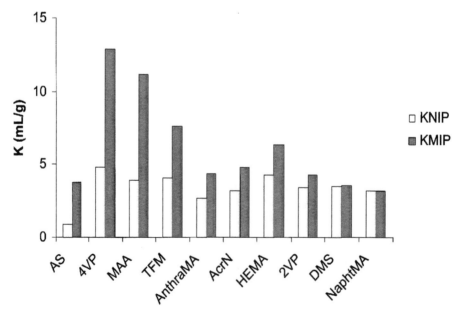

Figure 14 Example 4 (imprinting of nifedipine). Partition coefficients K_{MIP} and K_{NP} of MiniMIPs after overnight incubation with 1 mM nifedipine in dichloromethane.

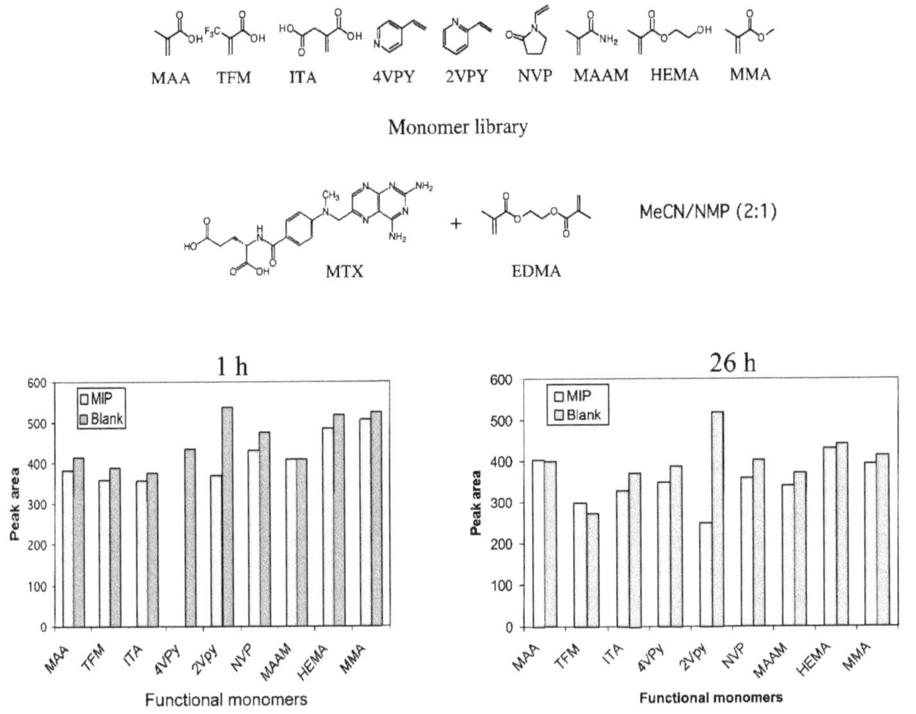

Monomer library

MeCN/NMP (2:1)

Figure 15 Example 5 (imprinting of methotrexate). Components of the polymerization mixture selected for the screening of methotrexate and peak areas measured in the supernatants after 1 and 26 h incubation with 0.16 mM of methotrexate.

Monomers used: MAA, TFM, ITA, 4-VP, 2-VP, NVP, MAAM, HEMA, and MMA (see Fig. 15 for the chemical structures).

Polymerization mixture: EDMA, AIBN, CH_3CN/NMP 2/1.

The problem here was that the target molecule is very complex and unstable. The results of the rebinding test are shown after 1 and 26 h of equilibration (expressed as peak area values). The absolute absorbance decreased during this time due to template decomposition. 2-Vinyl pyridine appeared to be the most successful monomer based on the equilibrium batch rebinding tests (Fig. 15). The 2-VPY materials prepared on a larger scale and tested as chromatographic stationary phases also exhibited a certain selectivity towards the template (methotrexate, MTX) and its closely related analogues (leucovorin and folic acid) (Fig. 16).

A. Feasibility Test for the Imprinting of a Lead Drug

Template: F (structural analogue of lead compound E, see Fig. 17)

Monomer used: MAA

Polymerization mixture: EDMA, AIBN, CH_3CN

In this case, the experiment was aimed at proving that an MIP selective towards the lead drug E could be obtained using the structural analogue F. Four polymers were prepared: two MIPs (P_E and P_F) using the structural analogue E (P_E) or the drug lead

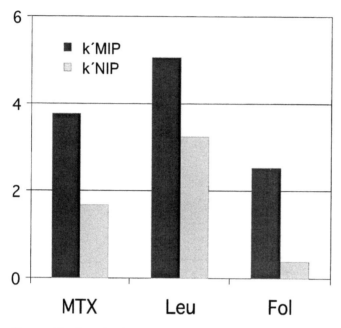

Figure 16 Capacity factors of methotrexate (MTX), leucovorin (Leu), and folic acid (Fol) injected on columns packed with p-2-VPY MIP and NPs (mobile phase $CH_3CN/CH_3COOH/H_2O$ 92.5/5/2.5, 10 nmol injection).

compound F (P_F), as templates; two NIPs (P_{B1} and P_{B2}). The results of the rebinding experiments (see Table 3) showed that it was possible to use F as template instead of E and the HPLC evaluation of the up-scaled materials confirmed that P_F is also selective towards E, as shown in Fig. 18.

X. OPTIMIZATION OF IMPRINTING PROTOCOLS

A. Statistics and Chemometrics

Due to the complexity of the imprinting process, effective optimization is unlikely to be obtained by Changing One Separate factor at a Time (COST strategy). Indeed if the factors (components and parameters) are not independent, a COST strategy will not lead to the real optimum and will not allow the real effects to be separated from "noise."

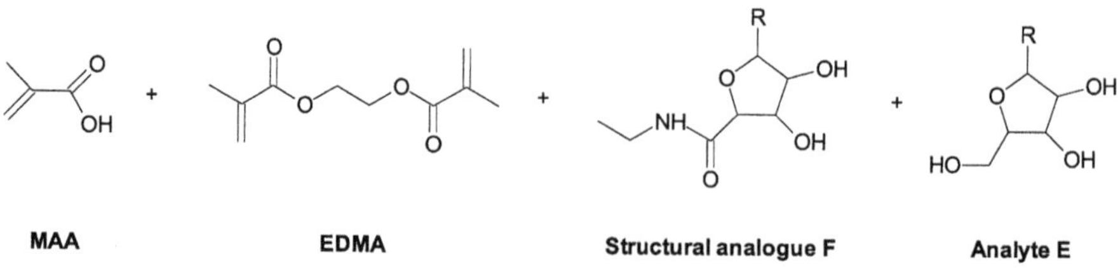

| MAA | EDMA | Structural analogue F | Analyte E |

Figure 17 Components of the polymerization mixture selected for the experiment 6 (feasibility test for the imprinting of the lead drug F).

Table 3 Small-scale Batch Rebinding Experiments for MiniMIPs Prepared using the Drug Lead E (P_E) and Its Structural Analogue F (P_F) as Template

Polymer	Rebinding of E (%)	Rebinding of F (%)
P_F	74	82
P_E	65	n.d.
P_{B1}	61	47
P_{B2}	60	n.d.

The table gives the rebinding percentage of (i) the drug lead compound E to P_E, P_F, P_{B1} and P_{B2}; (ii) its structural analogue F to P_F, P_{B1} and P_{B2}. The rebinding percentage was calculated for NP and MIPs as the relative difference between the concentration of template used in the rebinding solution (C_{rebind}) and the concentration of free template (C_{free}) detected in the supernatant. (Rebinding $\% = 100\ (C_{rebind} - C_{free})/ C_{rebind}$)).

What we need to do is to change all relevant factors together over a set of experimental runs and to chart and connect the results by means of a semi-empirical mathematical model. This strategy is called statistical design and modeling [11]. In this way the "noise" is decreased by averaging, the functional space is efficiently mapped for interactions and synergies between factors (components of the mixture and parameters of the polymerization) and a model is gained which correlates the

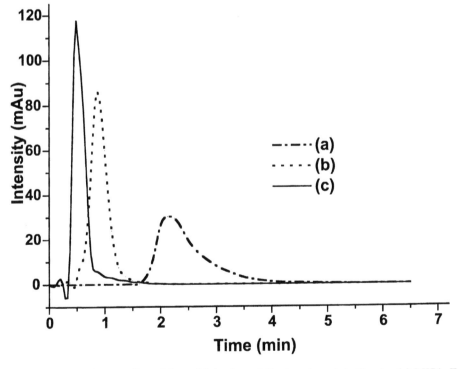

Figure 18 Elution profiles of 2 nmol injection of the target analyte E onto: (a) MIP1, F as template, (b) MIP2, E as template, and (c) NP. Mobile Phase: ACN/AcOH/H_2O 92.5/5/2.5 Flow 1 mL/min Wavelength of detection: 260 nm.

performances of the MIPs and factors. Statistical design can also be applied to material optimization. For example, Viklund et al. [12] applied design to the optimization of organic monoliths (co-GDMA-TRIM) with controlled porous properties. Similar protocols could be implemented to the optimization of MIP preparation. First of all we have to identify: (1) the factors, i.e., the components of the mixture and the parameters of the polymerization which affect the performance of the resulting materials; (2) the limits within which it makes sense to vary these factors; (3) the responses, i.e., the physical or chemical variables which can be used to monitor the performance of the materials. In the case of the organic monoliths (GDMA-TRIM copolymer), the chosen factors were: porogen to monomer ratio, crosslinker ratio (TRIM/GMA) and poor solvent/good solvent (isooctane/toluene) ratio.

The responses were selected among the physical variables used to monitor the porosity of the materials, i.e., maximum of the pore size distribution curve (D_p mode), mean pore diameter (D_p mean), total pore volume (V_p), specific surface area (S_p). The slope of the backpressure vs. flow rate curve was also chosen in order to monitor the chromatographic performance of the materials.

In the case of MIPs, the choice of the responses strictly depends on the field of application which is foreseen for the materials. If, for instance, the MIPs are to be used as chromatographic stationary phases, capacity and imprinting factors as well as separation and resolution factors should be used as responses. If the MIPs will be used as sensing elements, the response will be dependent on the nature of the transducer (intensity of absorption, fluorescence, voltage, conductivity, resistance, and so on) and can be expressed as the equilibrium value of the signal or as the rate of signal generation. If the MIPs will be used in solid-phase extraction, the recovery of the target analyte, as well as that of closely related analogues, will be the natural variable to be chosen as response.

Thus, as an alternative to expressing the imprinting effect as an imprinting factor, the NPs can be neglected and binding selectivity directly assessed in a competitive mode by adding a mixture of the target compound and a structurally related analogue. If no immediate application is foreseen for the materials, the variables will be chosen based on the methods selected for the evaluation of the MIPs (chromatography, batch rebinding, radio-labeling). In most cases, the performance of the MIPs is compared with those of the corresponding NPs and the relative response (MIP over NP) gives the imprinting factor IF.

At first, it is generally better to select a larger number of responses because it may be unclear which one is relevant to the goal. A preliminary experiment (screening) should also be planned to finally establish the main factors and factor ranges.

This screening is normally based on a reduced number of trials (i.e., only few MIP should be prepared and evaluated) and a large number of responses. The number of MIPs to be prepared for both screening and optimization depends on the number of factors. In general, a two-level factorial design is adopted where the number of experiments is given by 2^p for p factors. This means that each response is measured at the extremes of each factor range (p ranges for p factors).

In the case of the organic monoliths [12] a 2^3 full factorial design was applied, i.e., eight polymers were prepared (corresponding to eight corners of a cube) plus three replicates of the centre point (12 polymers in total). The porous properties of each polymer were measured and the responses pertaining to each point were thus obtained. The responses (Y matrix) were correlated with the factors (X matrix) using partial least square regression (PLS) which yields information on the importance and effect of each factor included and on their interactions. The outcome of the regression analysis is a series of so-called "effect plots" (one for each response) showing the change in response due to the variation of one factor from one extreme of the interval to the other one. By using the effect plots, it is possible to estimate the effect of each factor on the responses and hence on the properties of the materials. If the predictive capacity of the model is high (and this can be judged from the goodness of prediction Q^2) the model can be used for direct prediction of the outcome of new experiments.

B. Checklist

Let us write down a checklist of the operations you need to take into account to plan an experiment of statistical design applied to the synthesis of MIPs.

1. Which Factors Should I Select?

Having performed the initial selection of components, the factors of choice can be ratios between components of the polymerization mixture (e.g., the ratio between two functional monomers, the ratio between monomer and template, the ratio between functional monomers and crosslinker, the ratio between monomers and porogen, etc.). Physical parameters, such as temperature and pressure of polymerization, can also be screened. Remember to define the interval of variation for each factor (maximum and minimum).

2. Which Responses?

Think about the final application of the materials you want to prepare or about the way you want to characterize them. If you evaluate the MIPs by batch rebinding, choose partition coefficients K or the imprinting factors IF. If you also evaluate them as stationary phases, choose the capacity factors.

3. How Many Experiments Should I Perform?

The choice of the experiments depends on the design you select: either screening designs such as full factorial, fractional factorial, Plackett–Burmann [11] and D-optimal [11], or Response Surface Modeling and Optimization designs such as Central Composite [11], full factorial design at three levels [11]. Start first with a screening design then move to optimization. It might happen that you find a very good MIP with a simple screening design. If you work with two factors and a single response and choose a two-level full factorial design, the experiments correspond to the corners of a square whose sides have a length equal to the interval of variation of each factor. For three factors, the points are the corners of a cube. Remember to insert also three replicates of the center point to check the reproducibility of your experiments (i.e., the center of the square or of the cube). If you work with more than three factors or with more than two levels, design of the choice of points is more elaborate and you need dedicated software.

4. How can I Connect Responses and Factors?

You do not know the nature of the function which correlates the composition of the polymerization mixture and the properties of the MIPs, but you can fit the experimental points. Use linear or interaction models and least squares regression

(Multiple Linear Regression or Partial Linear Regression). You need statistical design software to perform the regression, e.g., Modde 6.0 by Umetrics AB, Umea, Sweden.

The software will suggest to you the number of experiments to perform for a given number of factors and responses and it will compute the goodness of prediction, Q^2, and goodness of fit, R^2, as well as the effect plots for each response. From the fitting function, the software can elaborate contour plots which will give you an idea of how to vary the factors in order to improve the performance of the MIPs you have in hand. You might decide to use gradient methods and graphical tools to look for the direction of deepest ascent or descent or, alternatively, to complement the design with additional experimental points or, finally, to adopt a more complex design to elaborate a response surface.

C. Optimizing an Imprinting Protocol by Varying the Monomer Ratios

Thus far, no example can be found in the literature where the techniques of statistical design have been applied to the synthesis of MIPs. Some tentative steps in this direction can be seen in the work done by Takeuchi et al. [13], where the ratio of functional monomers (MAA and 2-sulfoethylmethacrylate, SEM) was varied for a large number of polymers prepared in a combinatorial fashion. The responses chosen were the selectivity factor (amount of atrazine bound divided by that of metribuzin) and the amount of atraton produced by catalytic decomposition of atrazine in the specific sites of the polymer. However, no model was worked out from the experiments and the experimental points were arbitrarily selected. In the same way, the authors had previously studied the dependence of the amount of atrazine bound and of the selectivity factor on the MAA/TFM ratio [7]. With a smaller number of experiments, the techniques of statistical design would provide the authors with a model accounting for the variation of many more factors, in addition to the possibility of optimizing their formulation.

XI. CONCLUSIONS

As a conclusion to this tutorial on the combinatorial techniques in molecular imprinting, it might be useful to recall the different steps and operations which need to be taken into account to design and optimize MIPs selective for a given template.

1. Selection of the possible components of the MIP protocol (functional monomer(s), crosslinker, solvent, initiator, temperature of polymerization, use a template analogue?). Get inspiration from the literature. Use a previously described protocol if it was successful.
2. Select possible alternative formulations and prepare the corresponding MIPs on a small scale (MiniMIP, WellMIPs).
3. Evaluate those MIPs and select the best materials. These will be the starting point for the statistical design.
4. Define factors and responses for the MIPs you want to optimize.
5. Define the interval of variation of each factor.
6. Choose a screening design and find out which are the factors of importance (with the help of the effect plots).

7. Identify the direction to go in order to optimize the performance of your MIPs (with the help of contour plots).
8. Use gradient techniques or complement your design in order to optimize the performances of MIPs.
9. Apply RSM or optimization design to fine tune your formulation.

It might happen that you are lucky and that you get a very good MIP from the start (without applying the techniques of statistical design) or by applying a simple screening design. However, in most cases, hard work is required and we hope that this tutorial at least gives you an idea of how to proceed. Good luck!

REFERENCES

1. Sellergren, B. . In *Molecularly Imprinted Polymers, Man-made Mimics of Antibodies and Their Application in Analytical Chemistry*; Elsevier Publishers: Amsterdam, 2001.
2. Ferrer, I.; Lanza, F.; Tolokan, A.; Horvath,V.; Sellergren, B.; Horvai, G.; Barceló, D. Selective trace enrichment of chlorotriazine pesticides from natural waters and sediment samples using terbutylazine molecularly imprinted polymers. Anal. Chem. **2000**, *72*, 3934–3941.
3. Andersson, L. Efficient sample pre-concentration of bupivacaine from human plasma by solid-phase extraction on molecularly imprinted polymers. Analyst **2000**, *125*, 1515–1517.
4. Dirion, B.; Lanza, F.; Sellergren, B.; Chassaing, C.; Venn, R.; Berggren, C. Selective solid phase extraction of a drug lead compound using molecularly imprinted polymers prepared by the target analogue approach. Chromatographia **2002**, *56*, 237.
5. Quaglia, M.; Chenon, K.; Hall, A.J.; de Lorenzi, E.; Sellergren, B. Target analogue imprinted polymers with affinity for folic acid and related compounds. J. Am. Chem. Soc. **2001**, *123*, 2146–2154.
6. Lanza, F.; Hall, A.J.; Sellergren, B.; Berezcki, A.; Horvai, G.; Bayoudh, S.; Cormack, P.A.G.; Sherrington, D.C. Development of a semiautomated procedure for the synthesis and evaluation of molecularly imprinted polymers applied to the search for functional monomers for phenytoin and nifedipine. Anal. Chim. Acta **2001**, *435*, 91–106.
7. Takeuchi, T.; Fukuma, D.; Matsui, J. Combinatorial molecular imprinting: an approach to synthetic polymer receptors. Anal. Chem. **1999**, *71*, 285–290.
8. Lanza, F.; Sellergren, B. Method for synthesis and screening of large groups of molecularly imprinted polymers. Anal. Chem. **1999**, *71*, 2092–2096.
9. Takeuchi, T.; Seko, A.; Matsui, J.; Mukawa, T. Molecularly imprinted polymer library on a microtiter plate. High-throughput synthesis and assessment of Cinchona alkaloid-imprinted polymers. Instr. Science Techn. **2001**, *29*, 1–9.
10. Rathbone, D.; Su, D.; Wang, Y.; Billington, D. Molecular recognition by fluorescent imprinted polymers. Tetrahedron Lett. **2000**, *41*, 123–126.
11. Erikkson, L.; Johansson, E.; Kettaneh-Wold, N.; Wiström, C.; Wold, S. . In *Design of Experiments: Principles and Applications*; Umea Learnways AB: Stockolm, 2000.
12. Viklund, C.; Pontén, E.; Glad, B.; Irgum, K.; Hörstedt, P.; Svec, F. "Molded" Macro-porous poly(glycidyl methacrylate-co-trimethylolpropane trimethacrylate) materials with fine controlled porous properties: preparation of monoliths using photoinitiated polymerization. Chem. Mater. **1997**, *9*, 463–471.
13. Takeuchi, T.; Fukuma, D.; Matsui, J.; Mukawa, T. Combinatorial molecular imprinting for formation of atrazine decomposing polymers. Chem. Lett. **2001**, *30*, 530–531.

9
Surface Imprinting

David Cunliffe and Cameron Alexander University of Portsmouth, Portsmouth, United Kingdom

I. INTRODUCTION

The generation of molecular imprints at surfaces is one of the most active areas of research in this rapidly growing scientific field. This is partly because surfaces and interfaces are crucially important in industrial and biomedical applications such as separations, catalysis, and membrane filtration, but also through a growing realization that molecular recognition phenomena in general are strongly dependent on the surface chemistry of the host–guest interface. There are many different techniques being devised to imprint surfaces as a result of the great variety of formats and application for these materials, but in most cases there are distinct advantages of an imprinted surface compared to that of a conventional–imprinted polymer targeted towards the same template.

The standard method of preparing conventional molecularly imprinted polymers involves bulk polymerization of template, functional monomers, and cross-linking reagents in the presence of a solvent to act as a porogen. Upon completion of polymerization, the imprinted polymer is usually washed, dried, and ground to a fine powder. This is a time consuming and energy costly process, yielding polymer particles of irregular shape and size that often need to be sieved in order to obtain particles of the required diameter. In addition, the binding sites in imprinted polymers manufactured by this method are dispersed throughout the polymer matrix, and this generally results in a distribution of binding site quality with regard to, shape volume and accessibility. The use of such materials in analysis, separation of reaction products, catalysis, and in the resolution of racemic mixtures consequently suffers from the rate limitations of mass transfer of the target molecules through the polymer matrix in order to encounter the binding sites.

The generation of molecular imprints at surfaces is an obvious solution to the problems of mass transfer and accessibility, and potentially affords control over more subtle parameters such as binding site orientation and local solvation state. While a standard general protocol does not as yet exist for surface imprinting, a number of strategies that create the required recognition cavities at a surface or interface have now been developed and selected examples of these are considered in this chapter.

II. IMPRINTING OF POLYMER PARTICLES

The synthesis of uniform polymer particles by heterogeneous polymerization methods such as suspension [1,2], dispersion [3,4], precipitation [5], and emulsion polymerization [6] is one method by which imprint sites can be rendered accessible (cf. Chapter 17). The production of very small polymer particles by mini, and micro-emulsion polymerization [7,8] can greatly improve the kinetics of mass transfer, as a very high surface area is produced for a given mass of polymer. Molecularly imprinted polymers which are obtained in a colloidal form by heterogeneous poly-merization methodology can be used as a dispersion or can be further processed to form films or coatings and can be deposited on supports of any geometry.

The incorporation of monomers possessing functionality, which will bind to the template molecule and can behave as a surfactant offers the potential of producing very small polymer particles with a thin imprinted surface layer. A two-stage emul-sion polymerization method was utilized by Whitcombe and coworkers [9,10] to obtain colloidal polymer particles with a core–shell morphology, in which the shell is imprinted with cholesterol. In this work, the authors used the combination of cova-lent imprinting and noncovalent rebinding developed in their previous study of bulk polymers imprinted with cholesterol [11], using cholesterol vinyl phenyl carbonate (Fig. 1) as a functional monomer to incorporate cholesterol within the imprinted polymers. The binding sites of the imprinted polymers were exposed after cross-link-ing by alkaline hydrolysis of the carbonate ester linkage, to leave a phenolic group within a "cholesterol-shaped" cavity, capable of hydrogen bonding with the steroidal hydroxyl group.

The seed latexes used as the cores of the imprinted particles were prepared from hydrophilic or hydrophobic polymers. The hydrophilic seeds were prepared from methyl methacrylate and methyl methacrylate/ethyleneglycol dimethacrylate copo-lymers, while the hydrophobic seeds were composed of polystyrene or styre-ne/divinyl benzene copolymers. Hydrophilic- and hydrophobic-imprinted shells were then laid over these cores. It was found that the best cholesterol recognition was obtained with a hydrophilic-imprinted shell and a poly(methyl methacrylate) core. However, the performance deteriorated when the core was lightly cross-linked with ethyleneglycol dimethacrylate. In a second paper [10], imprinted polymers were prepared by the noncovalent approach with cholesterol rebinding relying upon hydrophobic interactions between cholesterol and the imprinted shell. To achieve this, the template was modified to give it the characteristics of a surfactant. The structure of the template surfactant is illustrated in Fig. 2.

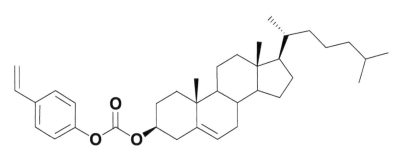

Figure 1 Cholesterol vinyl phenyl carbonate.

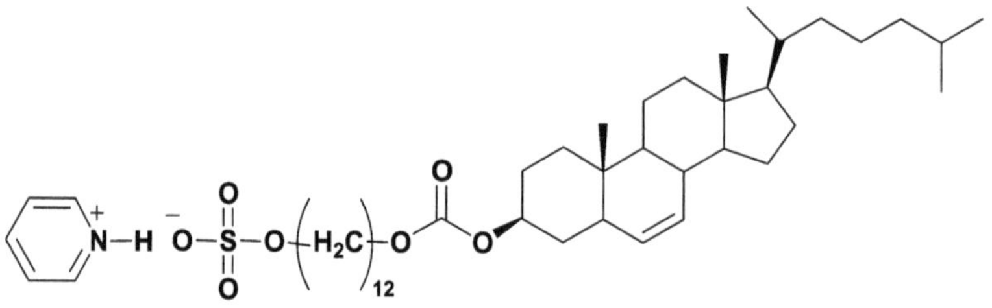

Figure 2 Template surfactant derived from cholesterol.

Here cholesterol forms the hydrophobic tail of the surfactant and is attached via a spacer to a hydrophilic pyridinium sulfate ester head. The effect of molecular imprinting in these particles was elegantly demonstrated by an "immunoprecipitation" experiment, in which α,ω-cholesterol-functionalized poly(ethylene oxide) was mixed with a suspension of the imprinted particles which led to bridging flocculation and precipitation.

The use of surfactants that are also able to bind molecules of interest has long been of interest in the preparation of molecularly imprinted polymers, prepared by emulsion polymerization processes [12–21]. Surfactants containing functionality capable of binding to metal ions have been used in emulsion polymerization. The bifunctional surfactant 1,12-dodecane-diol-O,O'-diphenyl phosphonic acid (Fig. 3) was used in the synthesis of polymers imprinted with zinc ions in a water in oil emulsion [22]. This α,ω-functional surfactant sits at the interface between the organic and aqueous phases during polymerization.

After polymerization and grinding, a molecularly imprinted polymer with a high internal surface area is obtained with metal ion recognition sites at the surface. The uptake of Zn^{2+} and Cu^{2+} by these imprinted polymers from aqueous solutions over the pH range of 1–6 was compared. It was found at pH < 2 that the imprinted polymers showed no tendency to remove zinc or copper ions from solution, whereas in less acidic conditions (pH > 2) the polymers displayed a preference towards zinc ion binding. No copper uptake was observed at pH < 3 and this increased to only 50% at pH 6. The high preference of zinc ions over copper was considered to be due to the presence of cavities that retained a shape- and size specificity for Zn^{2+} ions in the polymer matrix. In the design of the phosphonic acid surfactant, it was found that the structure of the phosphonic acid-functionalized molecule greatly influenced the selectivity of the imprinted polymers towards metal ions. Polymers containing a bifunctional phosphonic acid molecule exhibit a much better metal ion recognition

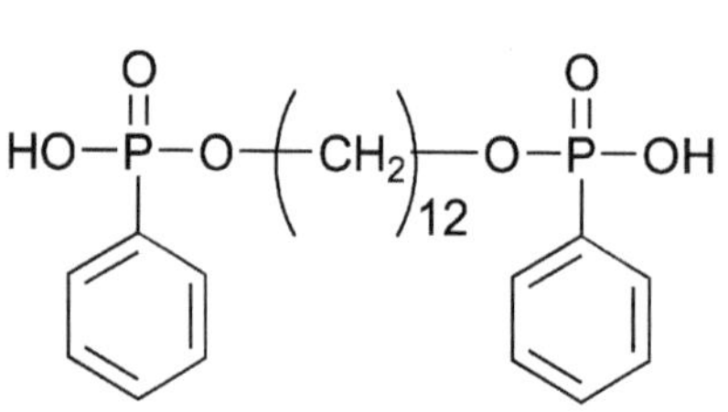

Figure 3 Diphosphonic acid surfactant.

effect than those with a monofunctional surfactant, also the presence of an aromatic ring adjacent to the phosphonic acid moiety was found to enhance metal ion recognition. This was probably due to the increased rigidity imparted to the host species as it was incorporated within the polymer matrix helping to maintain the phosphonate groups in the desired conformation for metal binding.

Surfactants containing phosphonic acid head groups have also been used to prepare surface-imprinted polymers for organic molecules. Yoshida et al. [22,23] synthesized polymers with imprinted surfaces capable of distinguishing amino acid enantiomers. The monophosphonic acid *n*-DDP (Fig. 4) was used as the host molecule in water in oil emulsion polymers imprinted with the methyl ester of tryptophan.

Batch binding was used to determine the binding constants of D and L isomers. For those polymers imprinted with the L isomer, the binding constants were $K_L = 2.9 \times 10^3$ and $K_D = 2.0 \times 10^3 \, M^{-1}$, while for polymers imprinted with the D isomer, the corresponding values were $K_D = 4.8 \times 10^3$ and $K_L = 3.4 \times 10^3 \, M^{-1}$. These binding constants gave separation factors of 1.5 and 1.4 for the L- and D-imprinted polymers, respectively, indicating that the surface-imprinted polymers were capable of differentiating between the different chiral guest molecules.

The binding mechanism was investigated by spectroscopic examination of the host/guest complexes prior to polymerization. In the infrared spectrum of L-Trp-OMe, the indole N–H stretching was assigned to the resonance seen at $3287 \, cm^{-1}$, after complexation with *n*-DDP this was shifted to $3217 \, cm^{-1}$. The carbonyl stretching of L-Trp-OMe was observed at $1747 \, cm^{-1}$, while in the complexed form the absorption occurred at $1751 \, cm^{-1}$, indicative of hydrogen bonding between the indole N–H and the P=O of the phosphonic acid group. The sharp carbonyl stretching observed at 1747 and $1751 \, cm^{-1}$ in the free amino acid and in the complex, respectively, indicated that there was no hydrogen bonding between the C=O group and P–OH. NMR spectra of L-Trp-OMe showed the proton attached to the indole

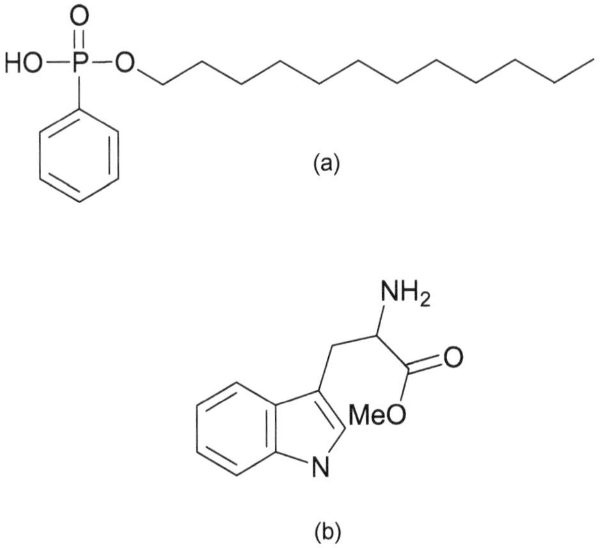

(a)

(b)

Figure 4 Structures of (a) *n*-DDP and (b) Trp-OMe.

nitrogen with a chemical shift of 11.45 ppm which shifted to 11.28 ppm when complexed with n-DDP. This upfield shift was a further indication of $-NH-O=P$ interaction in the complex. The resonance of the protons attached to the cationic ammonium group displayed a dramatic downfield shift from 6.43 to 8.6 ppm, attributed to $-NH_3^+-{}^-O-P$ in the complex. These data clearly indicate that specific intermolecular interactions were present in the preimprinting complexes, however, as with most investigations of imprinted polymers, the nature of these recognition interactions after polymerization could not be probed in the same level of detail and therefore must be inferred rather than considered proven.

The same types of surface-imprinted polymers were also used to discriminate between the enantiomers of bifunctional amino acids [24]. In this study, the functional host molecule was the quaternary ammonium salt and the templates were the N-protected amino acids shown in Fig. 5.

The value of the separation factor for Z-Glu was 1.2, which is lower than the value obtained for the L-Trp-OMe-imprinted polymers (1.5), while the Z-Asp- and Z-Gln-imprinted polymers displayed separation factors close to one (0.94 and 0.97, respectively), illustrating that the imprinted polymers showed enantioselectivity only for the imprinted amino acid.

Phosphonic acid functional molecules have also recently been used to prepare core–shell-imprinted polymers for the recognition of caffeine and theophylline [25]. These polymers were composed of a cross-linked polystyrene core with a poly(ethyleneglycol dimethacrylate) and oleyl-, phenyl-hydrogen phosphate shell. The ability of the imprinted polymers to discriminate between the imprinted molecule and a competing analogue was investigated. Those polymers imprinted with caffeine showed a preference for caffeine in the presence of theophylline with a separation factor of 1.88, however, those polymers imprinted with theophylline were unable to selectively adsorb theophylline from solution in the presence of caffeine, showing a greater tendency towards caffeine uptake, these polymers displayed a separation factor of 0.85. This is a significant result in terms of recognition specificity, as some degree of hydrophobic interaction must have been retained through the core–shell imprinting procedure to effect caffeine selectivity. Previous methods relying on H-bonding interactions have demonstrated better specificity for theophylline compared to caffeine most probably because of the presence of a possible extra binding position at the unsubstituted imidazole nitrogen of theophylline.

III. IMPRINTED POLYMER COMPOSITE PARTICLES

Many of the molecularly imprinted polymer particles reported to date as packaging materials in HPLC columns are limited in their efficiency in analyte separation due to the limitations of mass transfer kinetics and the inaccessibility of many of the binding sites. These can be overcome by synthesizing imprinted composite particles, where an imprinted polymer is attached to the surface of a porous inorganic substrate (see also Chapters 11 and 13).

There are essentially two ways of generating an imprinted surface on these kinds of substrate. The imprinted polymer can be grafted onto the surface by attaching polymerizable functional groups onto the substrate. These are then incorporated into the imprinted polymer during the cross-linking matrix formation

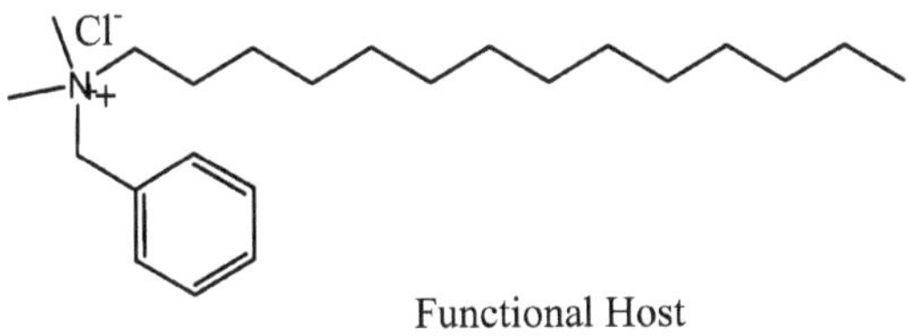

Functional Host

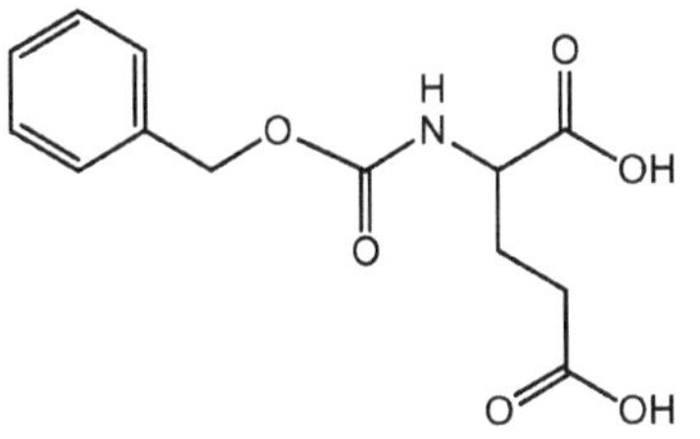

Z-Glu

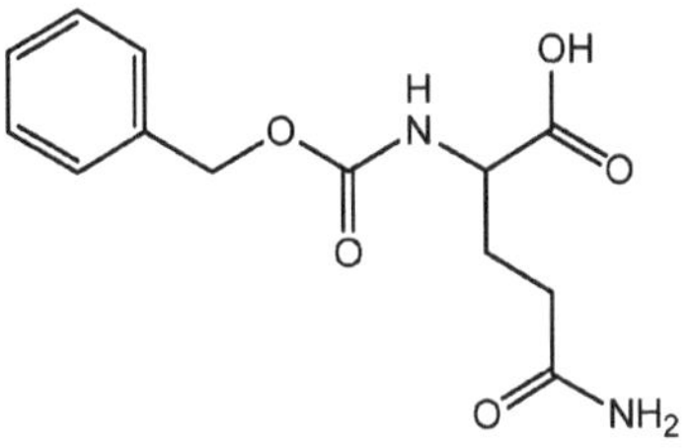

Z-Gln

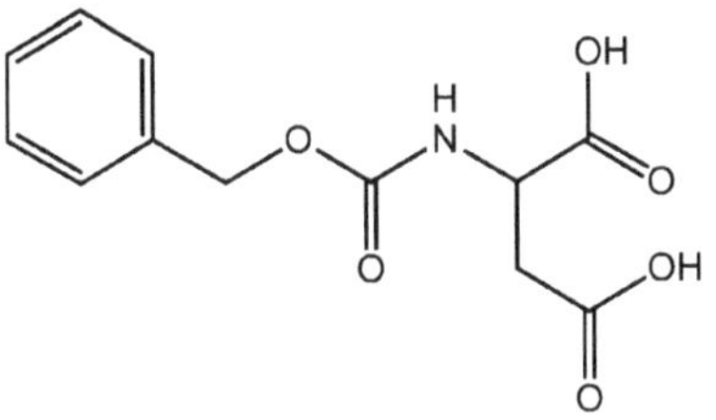

Z-Asp

Figure 5 Structure of functional host and N-protected amino acids used as templates.

stage. Alternatively, an imprinted polymer can be grown directly from a surface that has been modified with an initiator.

Imprinted polymer grafts onto inorganic substrates have included methods involving coating the wall of capillary electrophoresis columns and onto silica particles. Prior to polymerization a vinylic group is attached to the surface, which can be incorporated into the growing polymer. Hirayama et al. [26] has prepared an imprinted polymer layer using lysozyme as the template on silica particles and a quartz crystal microbalance sub-layer. In this process amine functional silica was treated with acryloyl chloride to afford acrylamido groups on the surface. These were then added to a solution of two different monomer mixtures, either acrylamide and N,N-dimethylaminopropylacrylamide, or acrylamide and acrylic acid. Both types of monomer-functionalized surfaces were cross-linked with methylenebisacrylamide or N,N'-(1,2 dihydroxyethylene)bisacrylamide prior to template removal. The uptake of lysozyme was investigated by HPLC, zeta potential measurements of the composite particles, and by frequency shifts of the quartz microbalance. Competitive binding of lysozyme in the presence of hemoglobin was investigated. It was found that the extent to which the imprinted polymers could discriminate between the two proteins was highly dependent upon the specific composition of the imprinted materials. Those imprinted polymers that showed the greatest selectivity for lysozyme had a zeta potential closest to that of this particular enzyme.

The efficiencies of imprinted polymers used as stationary phases in chromatography are often greatly improved when used in capillary electrochromatography columns. These polymers are usually produced in the column, by attaching the imprinted polymer to the fused silica wall [27–30]. This is often done by derivatizing the column wall with a silylated monomer such as [3-(methacryloyloxy)propyl]-trimethoxysilane, or adsorbing an initiator to the surface [31–34] as illustrated in Fig. 6.

The vinyl groups on the surface can then be incorporated within the growing polymer matrix.

The presence of initiator in the monomer/template/solvent mixture and the influence of polymerization conditions can result in a wide variety of imprinted polymer morphologies and imprinting efficiencies as demonstrated by Brüggemann [28].

A better control of the morphology of imprinted composite materials is obtained if the imprinted polymer is grown from a substrate that has been modified with an initiator. A free radical initiator can be attached to a surface either chemically or by adsorption. Schweitz [34] attached 4,4′-azobis(4-cyanopentanoic acid) **(6a)** using carbodiimide coupling to a fused capillary wall which had been treated with 3-aminopropyl triethoxysilane.

An imprinted polymer for *S*-propanolol was then grown from the surface, using methacrylic acid as the functional monomers and 1,1,1-tris-(hydroxymethyl)propane trimethacrylate as the cross-linker, using toluene, dichloromethane, or acetonitrile as the porogenic solvent. The ability of the polymer-coated column to separate the *R*- and *S*-enantiomers of propanolol was investigated. Those polymers prepared in

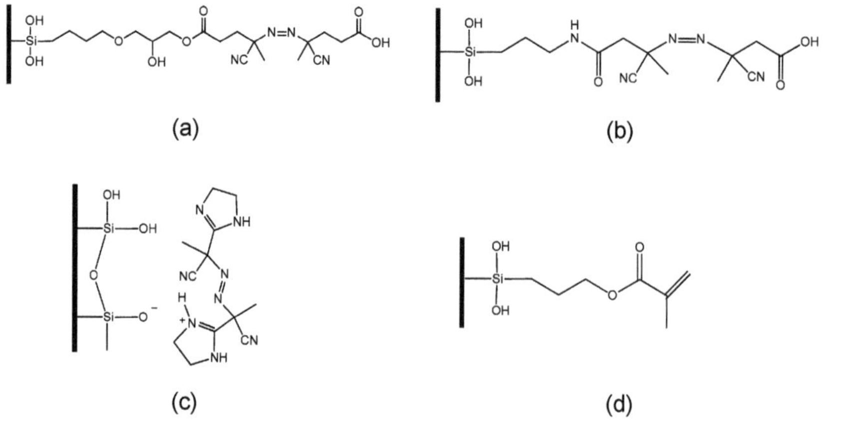

(a)   (b)

(c)   (d)

Figure 6 Modified silica particles; (a) and (b) chemically attached initiators, (c) physically adsorbed initiator, (d) chemically attached polymerizable group.

toluene and dichloromethane generally performed better than the polymer prepared
in acetonitrile.

Sellergren and coworkers [32] investigated the influence of the morphology of
imprinted surfaces grown from silica particles upon the efficiency of the resultant
imprinted composite material. In this study, silica supports were modified by chemically or physically adsorbing a free radical initiator to the surface or a polymerizable
group was attached to the surface. Imprinted polymers were then synthesized with
L-phenylalanine anilide as the template and ethyleneglycol dimethacrylate cross-
linker with methacrylic acid as the functional monomer. The ability of the imprinted
polymer films to separate the D and L isomers of the template was investigated by
HPLC analysis. It was found that the column efficiency was dependant upon the
thickness of the imprinted film: particles with thin films (3.8 nm) exhibited the greatest column efficiencies with elution times of 5 min or less, whereas thicker polymer
films (7–12 nm) displayed much higher column capacity. These results show that it is
possible to vary polymerization conditions such that imprinted composite particles
can be prepared with high separation efficiency for analytical uses, or with a high
capacity for preparative purposes. When the particles with film thickness of 0.9,
2.2, and 3.8 nm, respectively, were used as stationary phases in capillary electrochromatography [31], the ability of the particles to resolve the enantiomers of
phenylalanine anilide followed the trend 0.9 < 2.2 < 3.8 nm. Living/controlled polymerization techniques have also been applied to growing imprinted polymers from
surfaces [33] as in, for example, the modification of surfaces with a photosensitive
dithiocarbamate (Fig. 7) that can act as an iniferter.

Irradiation of these surfaces with ultraviolet light generates a benzyl radical and
a stable dithiocarbamoyl radical: subsequent addition of a cross-linking monomer
together with a functional monomer and template produces an imprinted polymer
film. When the light source is removed, the dithiocarbamoyl radical terminates polymer chain growth by recombination. The controlled nature of this methodology has
been demonstrated by Sellergen et al. [33], who prepared polymer films imprinted with
D- and L-phenylanaline on the surface of silica particles. The initial imprinted films
showed considerable preference for their imprinted targets. These preferences were
then inverted by reinitiation of polymerization in the presence of the opposite isomer.

Porous silica particles can be prepared by the sol/gel process with a wide range
of pore volumes and radii, which can be chosen to be close to the approximate size of
template molecules of interest. The inner surface of the silica can be modified such

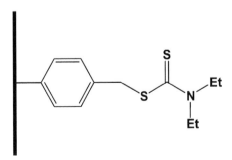

Figure 7 Photosensitive dithiocarbamate attached to surface.

that chemical groups can be introduced, which will have some affinity for the template molecule, this mimics the covalent imprinting/noncovalent binding approach. This method has been used by Ki et al. [35], who reacted (3-isocyanatopropyl)-triethoxysilane with estrone, one of the natural estrogens believed to have an adverse environmental effect when present in large quantities. The phenolic functionality on the estrone A-ring was reacted with the isocyanate group to form a thermally labile urethane bond, and then silica particles were prepared by a sol/gel reaction of the estrone-modified silane with triethoxysilane. The template was removed by heating and the liberated isocyanate was converted to free amine by reacting with water, or a hydoxyl group was introduced into the binding site by reacting with ethyleneglycol. The rebinding capacity of the amine and hydroxyl-functionalized cavities was determined by batch binding experiments. Those imprinted particles with amine functionality in the binding site had a higher binding capacity for estrone than the hydroxyl. When the imprinted particles were challenged with testosterone propionate, which has a similar size and shape to estrone, but has the A-ring hydroxyl group masked as a propionate ester, both imprinted particles had an almost identical but greatly reduced binding capacity.

IV. SURFACE IMPRINTING FOR SENSOR APPLICATIONS

Sensor devices used in chemical assays and bioassays are of increasing relevance in a diverse number of fields, such as environmental monitoring, forensic science, clinical, agricultural, and food analysis. Molecularly imprinted polymers are highly suited to these sorts of uses as they are relatively cheap to make and are more robust than antibody-based sensors and enzyme-linked immunosorbent assay (ELISA) devices. One particular advantage of MIP-based sensors is the possibility for use in nonaqueous conditions; therefore, the incorporation of an imprinted polymer within or on the surface of a sensor device is an attractive proposition (see also Chapter 25). A robust and very sensitive analytical tool that has found increasing use in recent years is the oscillating microbalance, which employs a piezoelectric crystal that shows a change in resonance frequency when an analyte is adsorbed to it. There are two types of oscillating microbalance used as mass sensitive sensors, the quartz crystal microbalance (QCM) that is used at frequencies at the low MHz range and the surface acoustic wave (SAW), which operates at frequencies up to 2.5 GHz. The QCM is the "all-round" device, which can be used in liquids and gases, while the SAW is mainly used in gaseous media due to the heavy damping experienced in liquids. As all analytes have some mass, these devices can be used for the detection of any kind of compound. Molecularly imprinted polymers can either be prefabricated and attached to the balance surface, or grown on the surface by the "grafting onto" or "growing from" method.

Molecularly imprinted polymers in conjunction with QCM have been used in the rapid detection of hormones. Percival et al. [36] has modified an QCM by spin coating a covalently imprinted polymer for the detection of the anabolic steroid, nandrolone. The 4-vinyl phenol carbonate ester of nandrolone (Fig. 8) was copolymerized with ethylenegylcol dimethacrylate and methacrylic acid.

After hydrolysis and washing to remove the template, the finely ground polymer was suspended in a solution of poly-(vinyl chloride) in THF. This was then spin coated on the surface of the QCM. The frequency response of an QCM is also sensitive to

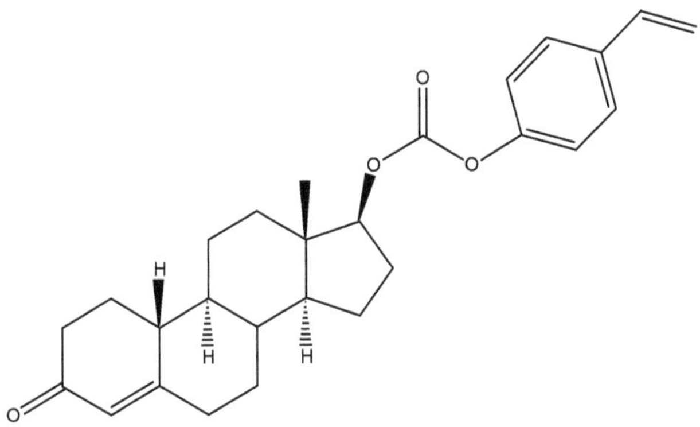

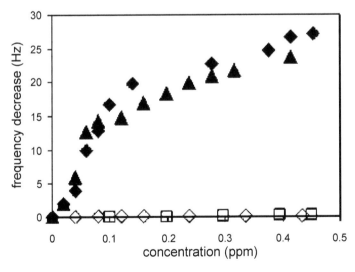

Figure 8 Polymerizable nandrolone derivative.

the operating temperature, to overcome this problem the resonant frequencies of microbalances coated with a nonimprinted polymer were measured at the same time as the imprinted polymer. The rebinding of nandrolone was described with a single point Langmuir type isotherm in Eq. (1) and the results are illustrated in Fig. 9

$$\frac{Kc}{\Delta f/\Delta f_\infty} + \frac{\Delta f}{\Delta f_\infty} = Kc \tag{1}$$

where K is the binding constant, c is the free concentration of nandrolone at equilibrium, Δf is the frequency change, and Δf_∞ is the frequency change at complete coverage.

Although the adsorption isotherm is only an approximation of how the template interacts with the polymer, the authors obtained a value of $123.2 \pm 10.9\ \mu M^{-1}$ which is similar to the value reported by Vlatakis et al. [37] When the imprinted

Figure 9 Frequency response of QCM in the presence of nandrolone (solid symbols) and testosterone (open diamonds), and epitestosterone (open squares). (From Ref. 36.)

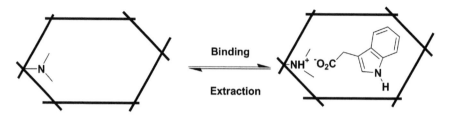

Figure 10 Recognition of indoleacetic acid by imprinted polymer.

polymer was exposed to testosterone and epitestosterone, no significant change in the resonant frequency of the microbalance was observed up to steroid concentrations of 2 ppm. However, upon repeated exposure of the imprinted polymer to a solution of nandrolone, the response of the QCM was seen to become less reproducible.

A noncovalent-imprinted polymer (Fig. 10) for the plant growth factor indoleacetic acid has been assembled on an QCM by Kugimiya and Takeuchi [38].

The microbalance was found to give a linear response to template at concentrations between 10 and 200 nM. The selectivity of the imprinted and nonimprinted polymers was tested with the analogous compounds indolebutyric acid and indole ethanol. The results obtained are given in Table 1.

The greater sensitivity of the imprinted polymer on the QCM for indoleacetic acid over indolebutyric acid is consistent with the authors' previous study using HPLC [39]. In comparison to the QCM imprinted with nandrolone they found that the device was stable enough for up to 50 repeated uses. Similarly, a QCM coated with an imprinted polymer has been used to determine the concentration of caffeine in human serum and urine [40]. A novel and highly ingenious way of coating a QCM with an imprinted polymer surface has been reported by Friggeri et al. [41] In this case rather than preparing the imprinted polymer as a solid from a solution of monomers and template, the imprinted polymer was prepared in solution and then attached to the surface. Poly(L-lysine) with boronic acid pendant groups (Fig. 11) is known to undergo higher order conformational changes in the presence of sugars.

At neutral pH poly(L-lysine) adopts a β-sheet conformation, as the pH is raised the protein takes on a α-helix structure and finally a random coil at high pH. When this polymer has pendent boronic acid groups, in the presence of monosacharides the α-helix content of the peptide chain increases and the pH at which the maximum helix content is observed is reduced. The molecular imprinting was carried out at pH 11 in the presence of a large excess of D-glucose or D-fructose. After equilibration, a gold-coated QCM crystal was placed in the solution. This allowed the polypeptide to adsorb onto the gold surface either as a random coil or β-turn,

Table 1 Relative Frequency Shift for Indoleacetic Acid Imprinted and Nonimprinted Polymers

	Indoleacetic acid	Indolebutyric acid	Indole	Indole-3-ethanol
Imprinted	100	8	4	5
Nonimprinted	13	15	1	3

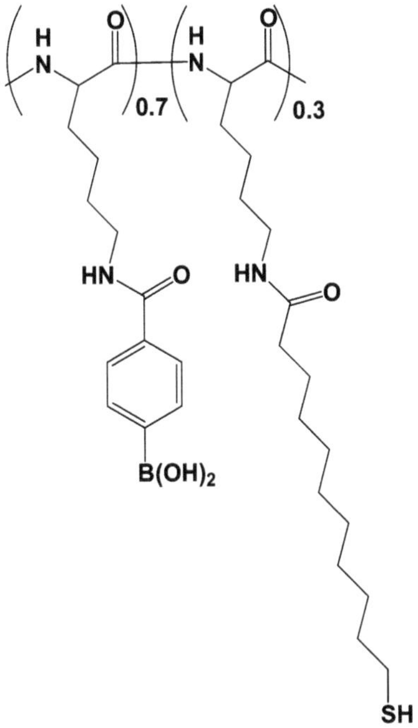

Figure 11 Poly-(L-lysine) with pendant boronic acid groups and thiol moiety for chemisorption to a gold-coated QCM.

depending on the monosaccharide present. After removal of the template, sugar rebinding was monitored. The response of the QCM to added sugar is given in Table 2.

The greater response of the nonimprinted polymer towards D-fructose than D-glucose is a consequence of the higher binding constant for the phenyl boronic acid D-fructose complex (log K_a =3.64) compared to the binding constant for D-glucose (log K_a = 2.04). When D-glucose was exposed to the D-fructose-imprinted polymer, the response of the QCM was similar to that of the nonimprinted polymer. Sugar selectivity was only observed with the D-glucose-imprinted polymer, whereas the QCM coated with this polymer gave a greater response in the presence of D-glucose than in the presence of D-fructose.

A recent method of preparing a very thin molecularly imprinted film on a QCM is the electropolymerization of phenols and aromatic amines [42]. The preparation

Table 2 Influence of Sugar Rebinding upon QCM Response with Adsorbed Imprinted Poly-(L-lysine). In Each Case, the Imprinted Surface Was Exposed to a 1 mmol Solution of Sugar at pH 8

Sugar	Nonimprinted	D-glucose imprinted	D-fructose imprinted
		$-\Delta F$ (Hz)	
D-glucose	42 ± 7	64 ± 5	33 ± 8
D-fructose	72 ± 11	41 ± 10	69 ± 9

and characterization of imprinted films prepared by the electropolymerization of
o-aminophenol have recently been discussed [43]. Polymerization was carried out
on a gold-coated QCM employing 2,4-dichlorophenoxyacetic acid as the template.
Although the specificity of the imprinted polymer was not as great as that obtained
for imprinted polymers prepared by the more traditional free radical polymerization
process, a linear response was obtained between Δf and the log analyte concentration
in the concentration range 4.0×10^{-5}–2.0×10^{-3} M. The selectivity of the QCM
coated with the electropolymerized-imprinted polymer was tested against the com-
pounds listed in Table 3. Here K is equal to $(\Delta f_t / \Delta f_i) \times 100$, where Δf_t is the frequency
shift of the imprinted film in the presence of the template, and Δf_i is the frequency
shift of the imprinted film in the presence of one of the structurally similar
compounds.

A broader range of sensor devices has been developed, where the sensor
response is modulated by the presence of an imprinted polymer, or an imprinted
surface. Perhaps the simplest way of preparing an imprinted surface is in the use
of self-assembled monolayer (SAM) techniques. A cholesterol-specific SAM was
prepared by Piletsky et al. [44], in which a gold electrode was treated with a
solution of cholesterol and hexadecanethiol in methanol. After the SAM had been
established, the template was removed, leaving cholesterol-shaped cavities in the
imprinted surface. Amperometric measurements were then used to investigate the
sensitivity of the electrode towards cholesterol. It was found that the unmodified
electrode, and the imprinted electrode displayed the same peak current. However,
exposure of the imprinted electrode to a solution of template caused a decrease in

Table 3 Selectivity Factors of Polymerized *o*-Aminophenol

Compound	$K_{(imprinted)}$ (%)	$K_{(nonimprinted)}$ (%)
	100	100
	37	113
	68	109
	49	76
	29	85

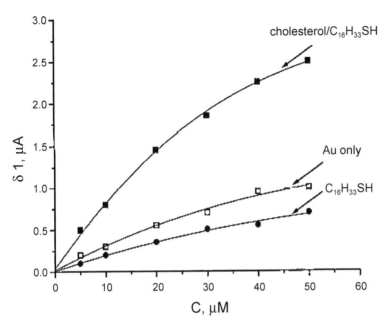

Figure 12 Electrode response to increasing concentration of cholesterol. (From Ref. 44.)

the peak current, dependent upon the concentration of analyte (Fig. 12). The imprinted electrode exhibited a much greater response to increasing cholesterol concentration than the bare gold electrode and the thiol-modified nonimprinted electrode. A template effect was illustrated by probing the electrode with either cholic acid or deoxycholic acid solutions (Fig. 13): the greatest response was obtained with cholesterol, while the two bile acid steroids gave a much smaller change in peak current.

Cyclic molecules that have cavities of the right size and internal functionality are often used as an alternative to assembling the cavity around the target molecule of interest, an example of this are the cyclodextrins (CDs) which are macrocyclic compounds composed of sugar units. Kitano and Taira [45] have prepared SAMs of thiolated α-cyclodextrins on gold electrodes for the analysis of bisphenols (Fig. 14) in aqueous solutions. The selectivity of cyclodextrin was initially determined spectroscopically using methyl orange (MO) as the probe.

The addition of bisphenols to the MO/α-CD mixtures caused an increase in absorption at 465 nm, as the MO was displaced from the CD cavity. The association constants for the cyclodextrins in solution and as SAMs are shown in Table 4.

The order of the binding constants of the bisphenols with the free cyclodextrins is: bisphenol S > bisphenol B > bisphenol F > bisphenol A. This does not follow the expected order of bisphenol hydrophobicity which is bisphenol B > bisphenol A > bisphenol F > bisphenol S; the large K_{assoc} of bisphenol S being attributed to the hydrogen bonding between the sulfonyl group at the center of the molecule and the hydroxyl groups on the rim of the cyclodextrin. The electrochemical behavior of the SAM-modified gold electrodes was examined by cyclic voltammetry using the redox behavior of hydroquinone in the presence of bisphenols.

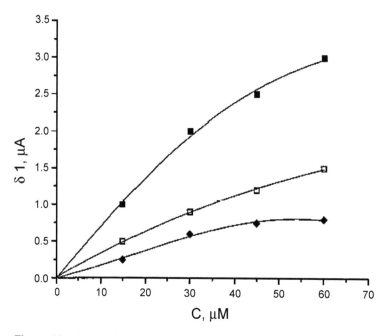

Figure 13 Comparison of imprinted electrode response to cholesterol (-■-), cholic acid (-□-), and deoxycholic acid (-◆-). (From Ref. 44.)

Electrochemical sensors using surfaces modified with imprinted polymers are finding increasing use in analytical chemistry [46]. Electrode surfaces can be modified in a similar way to the crystal surface of a QCM. The application of imprinted polymer particles as an ink to the surface of an electrode is a convenient way of preparing an imprinted surface and can provide a cheap and relatively simple route to prepare a sensing device. Imprinted electrode surfaces have also been prepared

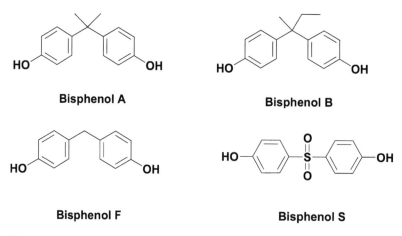

Figure 14 Structure of bisphenols.

Table 4 Association Constants for Bisphenols with α-Cyclodextrin in Solution and as SAMs

Template	K_{assoc} (SAM) ($10^{-3}\,M^{-1}$)	K_{assoc} (free) ($10^{-3}\,M^{-1}$)
Bisphenol A	1.8	6.5
Bisphenol B	1.6	84
Bisphenol F	1.8	21
Bisphenol S	8.3	110

directly, on indium tin oxide electrodes to detect theophylline [47] and on titanium oxide by a sol/gel process for the detection of dicarboxylic acids [48].

A screen-printed carbon electrode MIP sensor for I-hydroxypyrene, a metabolite of polycyclic aromatic hydrocarbons, that can accumulate in fish livers following pollution episodes has been illustrated by Kirsch et al. [49]. The MIP sensor was prepared by conventional bulk polymerization and particles were generated by grinding the polymer to a fine powder before screen printing onto the electrode. Scatchard analysis of batch binding experiments with suspended particles revealed two types of binding sites with binding constants of 1.39×10^{-4} and 1.68×10^{-3} mol dm^{-3}, respectively. The imprinted polymers were then mixed with a carbon-based ink and printed onto the surface of an electrode for cyclic voltammetry measurements. The response of the electrode showed the same trend as obtained as the batch binding experiments (Fig. 15).

A recent development in electroanalysis by imprinted polymers is the use of the changes in capacitance of the imprinted film attached to electrode surface. The earliest example of this was demonstrated by Mirsky et al. [50], who prepared SAMs of alkyl thiolates on a gold electrode in the presence and absence of barbiturate molecules as templates; the different surface types are given in Table 5. It can be seen from Table 5 that SAMs composed purely of the alkyl thiol or thiobarbituric acid did not bind barbituric acid at the interface, suggesting that hydrogen bonding was not the primary interaction in template recognition for these surfaces. Binding was also absent at the interface when 1-butanethiol or 2-methyl-2-propanethiol was used to prepare the monolayers, implying that a minimum layer thickness is required in order to achieve recognition at the electrode surface – i.e., to generate a binding "cavity". There are, however, two problems that must be overcome for capacitive detection to be effective in imprinted polymer films. The first is that it is only applicable to ultrathin polymer films, although this can be obviated by applying the technique of electropolymerization, which affords polymer films that are much thinner than those generally obtained by more conventional polymerization methods. The second problem for capacitative detection is that most imprinted films lack the structural regularity of SAMs, although this too this can be alleviated by filling in the structural defects after polymerization with an alkane thiol.

Combinations of detection mechanisms have also been employed in MIP sensors. For example, an electropolymerized film of polyphenol imprinted with phenylalanine was prepared on a gold electrode by Panasyuk et al. [51], who employed surface plasmon resonance and reduction in electrode capacitance to effect recognition sensing. The change in capacitance upon exposure to a mixture of amino acids

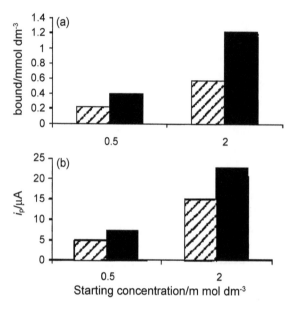

Figure 15 (a) Binding of 1-hydroxypyrene to imprinted (solid) and nonimprinted (hatched). (b) Peak current for 1-hydroxypyrene bound to imprinted (solid) and nonimprinted (hatched). Both with initial concentrations of 0.5 and 2 mmol dm^{-3}. (From Ref. 49.)

was seen to be additive, indicating that nonspecific adsorption takes place in areas other than the imprint sites.

Imprinted polymer membranes have become a useful tool and can be used to modulate the response of sensor devices and in the separation of mixtures of chemical species by the selective retention of the compounds of interest, and can be used as efficient preconcentration devices prior to analysis. The presence of trace quantities of pesticides and herbicides in foodstuffs is of increasing concern in terms of human and animal health, hence a rapid means of detection and removal of these compounds is highly desirable: imprinted surfaces and

Table 5 Binding Sites for Barbituric Acid Formed in Monolayers

| | | Capacitance change upon addition of analyte | | |
SAM	Template	Barbituric acid	5,5-Diethylbar-bituricacid	Pyridine
1-Dodecanthiol	—	None	—	None
—	2-Thiolbarbituric acid	None	None	None
1-Dodecanthiol	2-Thiolbarbituric acid	High	None	Low
Hexanthiol	2-Thiolbarbituric acid	High	None	Low
Butanthiol	2-Thiolbarbituric acid	None	None	—
2-Methyl-2-propanthiol	2-Thiolbarbituric acid	None	None	—

membranes are particularly suitable for such applications. Atrazine, a key member of the common aminotriazine group of herbicides, has been successfully imprinted into membranes for electroconductivity detection by Piletsky et al. [52,53]. Noncovalent and covalently imprinted membranes for atrazine, L-phenylalanine, 6-amino-1-propyluracil, and sialic acid were prepared on glass filter surfaces using methacrylic acid or (N,N'-diethylaminoethyl)methacrylate as the functional monomer for the noncovalent-imprinted membranes and tris(4-vinylphenyl) boroxine as the functional monomer for the covalently imprinted membrane, with ethyleneglycol dimethacrylate as the cross-linker. This membrane was then used to separate the two electrodes of an electrochemical cell, and the conductance of an electrolyte solution in the presence of the template measured. In the aqueous environment, the membranes became swollen causing a reduction in the pore diameter. This restricted the diffusion of the background electrolyte through the membrane. When atrazine was added to the solution, rebinding of the template to the recognition sites caused the membrane to shrink. This opened up the pores and channels through the membrane and increased the conductivity of the membrane. With the covalently imprinted membranes for sialic acid, the presence of template in the recognition sites caused a decrease in the conductivity of the solution, due to the presence of boronic acid groups in the polymer membrane. In the absence of template, the membrane was highly swollen allowing ion transport through the membrane. Upon addition of sialic acid the membranes reduced in swelling, the domains compacted and the diffusion of ions through the membrane became much more restricted.

As previously mentioned, imprinted membranes can also be used in the concentration and purification of particular compounds. Perhaps the simplest way of preparing an imprinted membrane is to mix the template with a preformed polymer and to cast a thin film. This was demonstrated by Kobayashi et al. [54], who used a method known as phase inversion, where a soluble copolymer of acrylonitrile and acrylic acid was mixed with theophylline in dimethyl sulfoxide. This solution was cast onto a glass slide and washed with water to coagulate the polymer. The acrylic acid residues in the polymer provided the recognition sites in the polymer, while the acrylonitrile groups allowed membrane formation. The resulting polymers were then used as ultrafiltration membranes for the separation of theophylline from caffeine. However, this method of preparing imprinted polymers suffered the major drawback of low substrate flux through the membranes.

The molecular recognition capabilities of polyelectrolyte multilayers have also been investigated by Laschewesky [55], while imprinted films have been grown on membrane surfaces in approaches similar to the phase inversion method for preparing a membrane imprinted with theophylline. Wang et al. [56] adopted an acrylonitrile/dithiocarbamoyl-methylstyrene copolymer (Fig. 16) to effect separation of caffeine from the theophylline-imprinted membrane.

Once the polymer membrane was prepared, the photosensitive dithiocarbamate group was used to initiate the polymerization of methylenebis(acrylamide) and acrylic acid in the presence of the template. These membranes selectively permeated caffeine in the filtration of a theophylline/caffeine solution, although the membrane substrates typically required up to twenty 24 h to prepare.

Composite membranes have been generated by modifying commercially available polymeric membranes, significantly reducing the preparation time. Molecularly

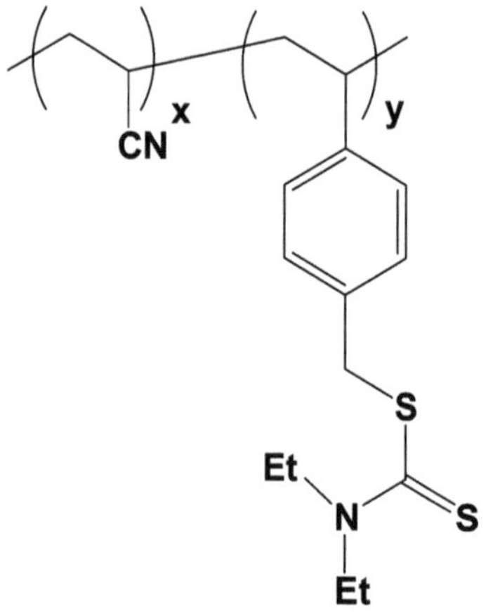

Figure 16 Poly-(AN-co-DTCS).

imprinted ultrafiltration membranes have been prepared in this way from porous polypropylene [57] and poly(vinylidene fluoride) [58–61] for the removal of desmetryn and other triazine herbicides from aqueous solutions. These membranes were treated with a photochemical inititiator, such as benzophenone or benzoin ethylether (Fig. 17). The radicals generated by benzophenone are capable of hydrogen abstraction from the membrane surface allowing initiation by surface bound radicals. As a consequence, the imprinted film is chemically attached to the membrane. However, radicals generated by benzoin ethylether are incapable of reacting with the membrane and initiation takes place exclusively in solution. In this case, the imprinted polymer is deposited on the membrane surface.

The monomers used to prepare the imprinted polymers for the triazine herbicides were acrylamidopropyl sulfonic acid (AMPS) and methylenebisacrylamide (BIS). This methodology has also been used in the synthesis of imprinted membranes used in the solid phase extraction of terbumeton, another member of the group of triazine herbicides form water [59].

A nonconventional imprinting method has been used to prepare membranes capable of separating the enantiomers of amino acids. Recognition moieties for

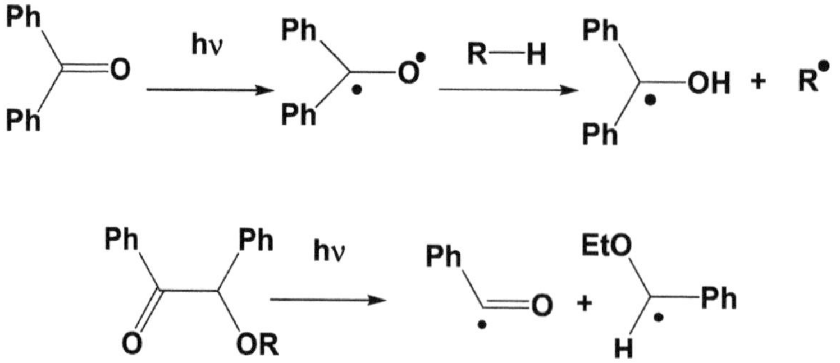

Figure 17 Photo dissociation of benzophenone and benzoin ethylether.

D- or L-amino acids were attached to polystyrene beads using standard Merrifield solid phase pepetide synthesis. These modified beads were subsequently mixed with a solution of a copolymer of styrene and acrylonitrile and the print molecule before casting as a thin film [62–66] (Fig. 18). Polystyrene beads were modified with the tetra peptide H-Asp(OcHex) -Ile-As (OcHex) -Glu(OBz) -CH$_2$- (DIDE-resin), H-Glu(OBz) -Glu(OBz) -Glu(OBz) -(EEE resin) , and H-Glu(OBz) -Phe-Phe (EFF resin, Fig. 19).

The target molecules for these imprinted membranes were the *N*-protected D and L-tryptophanes. The selectivity of the membranes was examined by static binding experiments and electrodialysis. For the DIDE-resin membranes imprinted with Boc-D-Trp, there was no difference in the permeation of the D and L isomers through the membrane, while for the Boc-L-Trp-imprinted membrane the permeation ratio (P_D/P_L) was 1.4. This value was increased to 5 with the imprinted EEE resins and was further increased to 6.8 with the EFF resins.

V. IMPRINTED SURFACES FOR RECOGNITION OF BIOPOLYMERS

A major challenge to molecular recognition research in general is the synthesis of artificial materials capable of recognizing biological macromolecules such as proteins. The large size of these molecules makes diffusion through a matrix very difficult; consequently, these macromolecules can only encounter recognition elements at the polymer surface. The incompatibility of these molecules with organic solvents

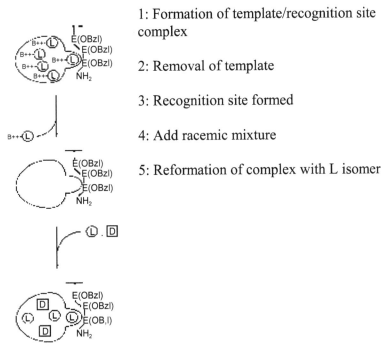

1: Formation of template/recognition site complex

2: Removal of template

3: Recognition site formed

4: Add racemic mixture

5: Reformation of complex with L isomer

Figure 18 Schematic of amino acid recognition by EEE resin. (From Ref. 64.)

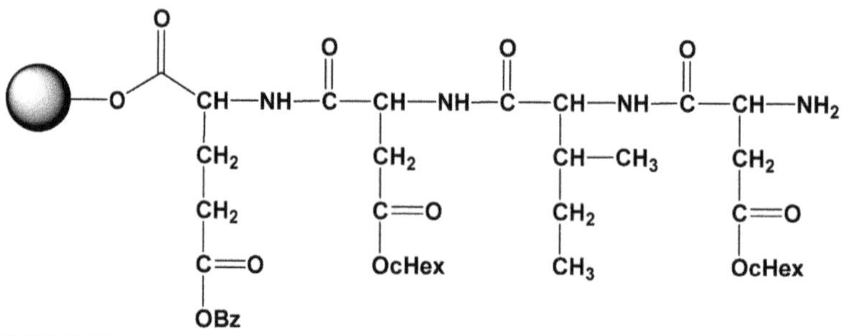

DIDE-Resin

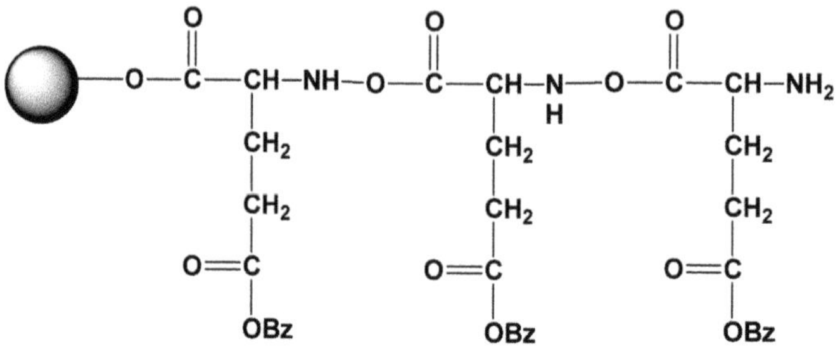

EEE-Resin

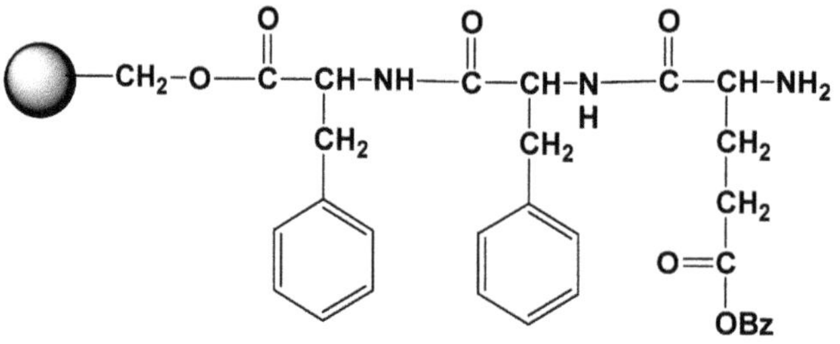

EFF-Resin

Figure 19 Structure of imprinted resins.

also presents a considerable challenge for molecular imprinting. The assembly of a large number of complementary recognition sites at a polymer surface would appear to be a suitable means of producing an imprinted structure capable of recognizing the shape and arrangement of functional groups on the surface of a protein molecule. The electropolymerization of 3-aminophenylboronic acid has been shown [51] to be capable of producing very thin imprinted films. This has been extended by Bossi et al. [65], to the preparation of protein-imprinted polymer films for rapid detection in micro titer plates. Polymers were prepared in the presence of microperoxidaes, horseradish peroxidase, lactoperoxidase cytochrome c, and hemoglobin. The amount of protein bound to the polymers was determined by the peroxidase activity using 2-2′-azino-bis(3-ethylbenzthiazoline-6-sulfonic acid). The imprinted surfaces showed a marked increase in their capacity to bind proteins, as the pH was raised from 5 to 8. The binding specificity of these polymers is shown in Table 6.

Table 6 Binding Specificity of Imprinted Polymers Grafted to the Surface of Microtiter Plates

	$10^6 \times K_D$ (M)				
Polymer	Microperoxidase	Horsradish peroxidase	Hemoglobin HbAo	Lactoperoxidase	Cytochrome c
Blank	10.8	0.24	> 10	0.54	> 10,000
MIP	1.5	0.082	0.056	0.036	> 10,000

VI. CATALYTIC-IMPRINTED SURFACES

The catalytic activity of imprinted materials would also be improved if the active sites were located primarily at the solid liquid interface, owing to better mass transfer of reactants and products into and out of these sites. Emulsion polymerization is, as already mentioned, a means by which imprinted polymers can be prepared with a high surface area and accessible sites. This methodology has been adapted to prepare an artificial biocatalyst [67], employing oleoyl imidazole as the functional "host" monomer, N-α-t-Boc-L-histidine as template and N-α-Boc-L-alanine p-nitrophenyl ester as the substrate for hydrolysis (Fig. 20).

The imprinted polymer was produced by oil in water emulsion polymerization and, after washing and drying, the imprinted material displayed considerably greater activity than the nonimprinted polymer, as well as the functional host monomer assayed as a catalyst in solution. The preparation of catalytic cavities at the surface of silica particles and other inorganic oxides has also been demonstrated by Markowitz et al. [68,69]. Silica particles were grown in a microemulsion, where the

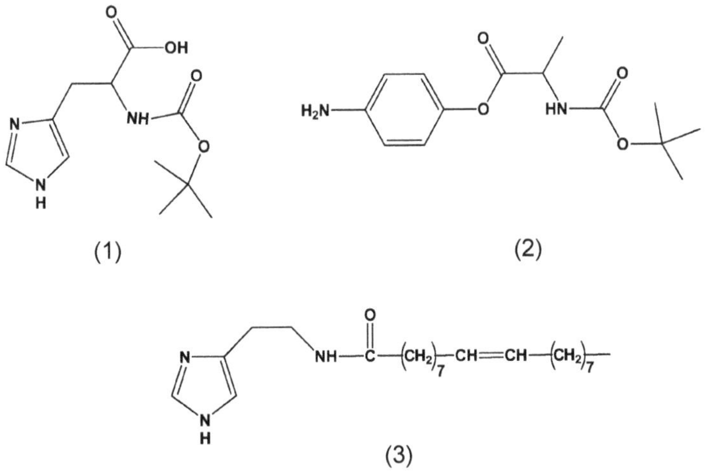

Figure 20 Molecular structures of: (1) print molecule N α-tBoc-L-Histidine, (2) subtrate N α-tBoc-L-alanine p-nitrophenyl ester, (3) functional host oleoyl imidazole.

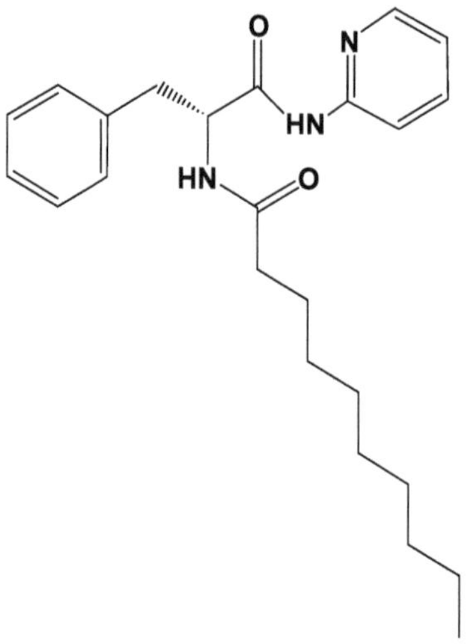

Figure 21 Surfactant-modified TSA of α-chymotrypsin.

surface of the particles was imprinted with a surfactant derivative of a transition state analogue of α-chymoytrypsin (*N*-α-Decyl-D-phenylalanine-2-aminopyridine) (Fig. 21).

Three different silanes were used in the gel process to obtain silica particles, the structures of which are given in Fig. 22.

The catalytic activity of these particles was determined for the hydrolysis of the D- and L-isomers of benzoyl-*p*-nitroanilide (D-BAPNA and L-BAPNA). Particles

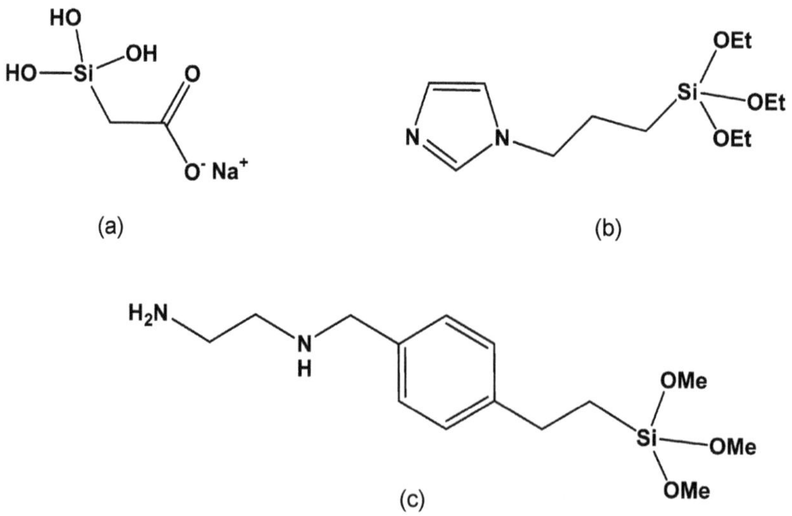

Figure 22 Structure of silanes used to prepare imprinted silica microparticles; (a) CTES, (b) IPTES, (c) PEDA.

imprinted with N-α-decyl-D-phenylalanine-2-aminopyridine were found to selectively hydrolyze L-BAPNA and have no effect upon D-BAPNA. The organosilanes used in the preparation of the silica particles were also found to influence the catalytic activity of the imprinted materials. Particles prepared with the ethylenediamine terminated silane (PEDA) exhibited a higher initial reaction rate than those particles made with IPTES (N-(3-triethoxy-silylpropyl)-4,5-dihydroimidazole), although those particles manufactured with IPTES would be expected to be more basic than those containing PEDA. It was speculated that this effect was due to cooperativity between the two amino groups in PEDA in establishing catalysis. When particles were made containing an equal amount of PEDA, IPTES, and CTES, the initial rate of hydrolysis was nearly as high as for those particles containing only PEDA as the functional silane. The amount of PEDA in these particles was reported to be only 2 wt% of the total silica content compared with 10 wt% for those particles with just PEDA as the functional silane, thus indicating a strong cooperative effect between the dihydroimidazole, the ethylenediamine, and the carboxylate groups in these particles.

Catalysis by imprinted surfaces has been extended to transition metal catalyst hydrolysis and hydrogenation [70–73]. For the catalytic hydrogenation of alkenes, the dimeric and monomeric rhodium complexes were attached to silica surfaces as shown in Fig. 23.

After attachment to the surface, chemical vapor deposition was used to grow a silicon oxide film around the catalytic site. The trimethoxyphosphine ligands were displaced by an alkene during hydrogenation and thus the size of the ligands dictated the accessibility of the catalytic sites in the imprinted polymer. In this example, the P(OMe)$_3$ ligand was acting as a model for the partially hydrogenated intermediate of 3-ethylpent-2-ene, and served to limit the ability of larger alkenes to gain access to the catalytic sites. However, smaller alkenes were able to migrate through the surface matrix and encounter the rhodium species. This was demonstrated by the large turnover frequency for the hydrogenation of 2-pentene, which was an order of magnitude higher for the imprinted catalyst than for the supported catalyst. As the ease at which alkenes can encounter the catalytic centre was dictated by the size of the template ligand, it was expected that alkenes which were smaller than P(OMe)$_3$ would show the greatest rate enhancement. This was indeed found as reactions involving linear and branched alkenes smaller than 3-ethyl-2-pentene exhibited the greatest rate enhancements. A similar influence of ligand size and structure was found upon the rates of ester hydrolysis by imprinted niobium and rhodium complexes [71].

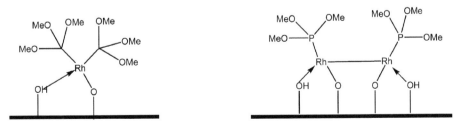

Figure 23 Monomeric and dimeric rhodium complexes.

VII. IMMOBILIZED TEMPLATES: IMPRINTING OF BIOMOLECULES, INORGANIC CRYSTALS AND CELLS

A recent development in the synthesis of imprinted materials has been to construct the imprinted polymer around a fixed surface of the substrate of interest rather than build the matrix around individual randomly oriented templates. This can be considered as a conceptual change from the conventional imprinting approach wherein multiple polymerization loci and templates are present: this methodology should in theory involve multiple directional interactions at single template sites. There are several potential advantages in immobilizing the templates used for imprinting; one is that templates which are otherwise insoluble can be imprinted by attachment to a soluble carrier, and secondly the positioning and orientation of the template with respect to monomers can be controlled by the chemistry, solubility and phase behavior of the immobilizing species.

This approach has been used to imprint theophylline [74], which was attached to the surface of silica gel particles and an imprinted polymer assembled around the inorganic particles. When polymerization was complete, the silica was dissolved away with hydrofluoric acid. This yielded imprinted polymer particles with a narrow pore size distribution with diameters between 254 and 257 Å compared with conventionally prepared imprinted polymers which have pore diameters in the range 30–1000 Å. The imprinted polymer showed a high selectivity for theophylline over the related compounds caffeine and theobromine.

A conceptually similar, if practically rather different, method has also been applied for the recognition of proteins via template-imprinted nanostructured surfaces [75,76]. In this work, atomically flat mica surfaces were used to adsorb protein molecules from aqueous buffer. Disaccharides were then spin-cast from solution to form a 10–50Å coating on the protein, and the whole sample was then exposed to a radio-frequency glow-discharge in the presence of hexafluoropropene. The resultant plasma polymer formed above and around the sugar-coated protein, yielding a smooth and stable film. This film was then mounted onto a second support via an epoxy resin layer, and the mica surface was removed. Treatment of the sample with aqueous base removed the protein molecules, exposing imprinted nanocavities in the sugar-coated surface of the plasma polymer (Fig. 24).

The resultant cavities were shown by AFM to correspond closely in size and shape to the template proteins, and ESCA combined with TOF-SIMS indicated a sharp reduction in nitrogen content and protein fragment peaks at the surface following elution with aqueous base. Concomitant increases in signals due to oxygen and saccharide groups were also observed. In addition, advancing contact angle at the surfaces following protein removal (θ_a 35 $\pm$ 9°) was consistent with exposed hydrophilic sugar residues. Thus, not only was shape complementarity generated at the imprinted surfaces, but also functionality, with saccharide groups positioned and oriented correctly to re-bind the template proteins. Competitive adsorption experiments from aqueous solutions showed that the imprinted surfaces showed a higher selectivity for the templates from which they were produced. This selectivity was expressed by a higher competition ratio required to cause a 50% reduction in the maximum protein adsorption, and in the case of bovine serum albumin (BSA) imprints this increase was 5–10-fold, whereas for immunoglobulin G (IgG) the increase was 4–7-fold. Interestingly, for the proteins lysozyme and ribonuclease A,

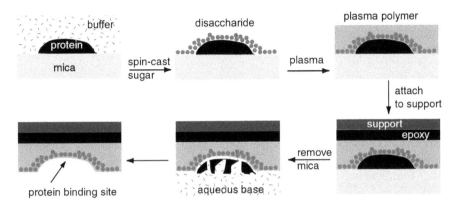

Figure 24 Schematic of protein imprinting at surfaces via multistage adsorption, spin-coating and plasma polymerization routes. The protein recognition sites are exposed by removal of the mica base layer and alkaline hydrolysis of template.

which possess very similar structural rigidity and charge but different shapes, the competition selectivities for the imprinted surfaces were 20–26-fold in preference for their own templates. Further experiments with streptavidin, imprinted onto micropatterned surfaces, indicated that the proteins were able to retain their activity (in this case by binding of biotinylated colloidal gold) suggesting that the imprinting process can generate recognition sites for these complex macromolecules without causing structural breakdown and loss of surface binding domains. The fact that protein recognition was observed in competition experiments strongly suggests that cooperative interactions were involved in binding, due to specific positioning of sugar residues as a result of imprinting. Thus, a template protein, on reaching the imprinted cavity, is able to form large numbers of hydrogen bonds to complementary hydroxyl groups at the surface, leading to a strong overall binding energy. In addition, the functional group positioning and shape complementarity leads to favorable van der Waals interaction and minimizes steric repulsions between template and surface.

Another demonstration of surface imprinting was reported by D'Souza et al. [77,78], who adopted a variation of the "immobilized" template concept to imprint inorganic crystals with the specific aim of using the imprint sites to direct subsequent crystal growth. The ability to control the morphology and/or habit of crystals is of particular interest in the pharmaceutical industry and in the development of novel composite materials, however, hitherto it has only been possible to influence crystal growth by addition of surfactants or soluble polymers to solutions of the crystallizing species. By adsorbing a functional monomer (6-methacryloyl) aminohexanoic acid, to template calcium carbonate crystals in organic solvents and subsequently cross-linking a polymer matrix around the monomer–crystal aggregates, it was possible to generate functional imprints of the crystal surfaces. The imprinting of two different polymorphs of calcium carbonate, calcite, and aragonite, produced, after removal of the calcium carbonate with acidic methanol, "isomeric" polymers with recognition sites complementary to the template crystal surfaces (Fig. 25).

Crystallization experiments were then carried out in supersaturated solutions of calcium carbonate. Under these conditions, rhombohedral calcite crystals were observed in the recognition sites of the calcite-imprinted polymer. Very few crystals were observed at the surface of the nonimprinted polymers or for those polymers prepared in the absence of the functional monomer. This was to be expected if the imprinted sites promoted crystal growth via nucleation on the complementary functional surfaces at a faster rate than the growth of calcite in solution. The imprinted polymer also produced calcite at the imprint sites under conditions where aragonite would normally be formed in solution. However, polymers imprinted with aragonite were less effective at producing aragonite under conditions that would favor calcite formation: this indicated that the recognition sites in these imprinted polymers were exhibiting kinetic rather than thermodynamic control over crystal growth, as calcite is the more thermodynamically stable polymorph.

The technique of imprinting against a surface has also been applied to the templating of cells and microorganisms. In a process that can be considered analogous to lithography, Aherne et al. [79] imprinted bacterial cells into a cross-linked polyamide matrix. In this multistep synthesis, bacterial cells (*Listeria monocytogenes* or *Staphylococcus aureus*) were suspended in an aqueous medium containing a dispersed organic phase. The nonaqueous phase contained a solution of adipoyl chloride, 1,6-hexandioldiacrylate, and azobis (isobutyronitrile) in dibutyl ether. The aqueous phase was buffered with MOPS and stirred at room temperature for 30 min. During this time, the bacterial cells migrated to the interface between the water and the organic droplets. To this suspension, an aqueous solution of poly(allylamine) was added allowing the free amine groups on this prepolymer to react with the diacid chloride at the interphase and thus embed the bacteria in the polyamide shell.

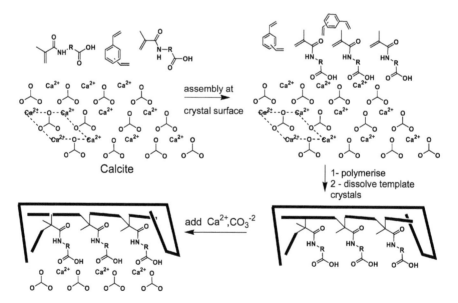

Figure 25 Schematic of imprinting calcium carbonate. Functional monomer $HC = C(CH_3)NH–CO–R–CO_2H$ ($R = (CH_2)_5$) in organic solvent is allowed to assemble at crystal surface, divinylbenzene is added and polymerization carried out. Following dissolution of the crystals with acidic methanol, carboxyl functionality is exposed at the imprint sites. Resuspension of the polymer surfaces in aqueous solutions containing Ca^{2+} and CO_3^{2-} ions results in nucleation at the sites and crystal growth.

The core of the microspheres was then photochemically cured to produce robust polymer beads with cells imprinted at the surface. The beads, which contained exposed residual amine functionality on the parts of the surface not covered by bacteria, were then reacted with a diisocyanatoperfluoro polyether to produce a low energy surface outside the imprint sites. After treatment with acidified methanol, bacteria-shaped depressions were left in the polyamide shell, which contained chemical functionality complementary in space to those found on the cell walls of the bacteria. In the last synthetic step, the recognition cavities in the polyamide shell were modified with a generic binding ligand, the lectin Concanavalin A to target sugar residues on bacterial cell walls. The resultant polymer beads were shown to exhibit adsorption selectivity for their imprint species, although the degree of bacterial discrimination was within one order of magnitude. This bacterial imprinting procedure was further modified to enable the generation of bacterial-imprinted planar surfaces with epoxide and hydroxyl functionality: by allowing cells to position at an unstirred interface between a viscous prepolymer solution and the aqueous layer, surfaces with very high imprinted cell density were produced [80].

Dickert et al. [81–83] have demonstrated an alternative approach to the imprinting of living organisms. In this approach, polyurethane prepolymers or a

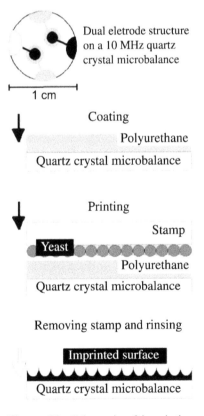

Figure 26   Schematic of imprinting yeast cells onto the surface of a QCM. The cells are pressed onto the surface until polymerization is complete: any cells remaining on the polymer surface are then easily washed away. (From Ref. 81.)

sol/gel layer were deposited on the surface of a QCM and yeast cells or bacteria were pressed into the mixture. After curing and washing to remove the cells, the imprinted surface was left behind. A schematic of surface preparation is outlined in Fig. 26.

As can be seen in Fig. 26, only half of the QCM surface was imprinted thus eliminating the effects of nonspecific adsorption upon the sensor output. The microbalance was imprinted with yeast cells and *Escherichia coli.* The microbalances were able to measure yeast cell concentrations in the range 10^4–10^9 per mL under flow conditions of 10 mL min^{-1} and were able to distinguish between the template cells and other microorganisms in complex mixtures. However, the microbalances proved to be less sensitive towards imprinted bacterial cells.

VIII. CONCLUDING REMARKS

Although significant progress has been made in molecular imprinting in general, and surface imprinting in particular, in recent years, a number of challenges remain. These principally relate to the degree of control that is required to generate a well-defined, specifically shaped and optimally oriented recognition site at an accessible surface. For those who wish to pursue research in this area, perhaps the key questions are:

1. Can an imprinted surface be synthesized in a step-wise fashion, in a manner analogous to solid phase peptide synthesis, such that each "functional building brick" of the imprint site is positioned in exactly the right place for the chemical task (e.g., binding, catalysis, controlled release) it is designed to do?
2. Can this be done in three dimensions so that the matrix and binding cavity are fully defined stereochemically?
3. What are the chemistries needed to adopt successfully a step-wise assembly approach? For example, should a bio-inspired combination of covalent and noncovalent bonds be used where covalently linked linear chains are folded into the desired architecture by noncovalent interactions? Or should covalent chemistries be employed throughout to ensure the resultant materials are as robust as conventional MIPs?
4. If true step-wise assembly is not possible, what other macromolecule-forming chemistries can be used to generate imprinted surfaces?
5. How can the structures and functions of imprint sites at a surface be studied, chemically analyzed and properly characterized?

In addition to these fundamental questions, a number of "real-world" issues need to be addressed. These include:

1. Can sufficient numbers of binding sites be imprinted into surfaces such that the materials are of practical value?
2. At what stage does the need for high surface density of binding cavities begin to impinge on site isolation and adversely affect recognition specificity?
3. Are the chemistries already developed for surface imprinting portable for derivatization of existing material surfaces such as chromatographic separation media or catalytic supports?

4. Can existing functional monomers and cross-linkers be used in step-wise assembly processes or if not, can the more specialized reagents and chemistries be employed on sufficient scales to be commercially useful?
5. Are imprinted surfaces competitive with bulk-imprinted polymers? For example, can small particles of crushed and sieved bulk MIPs embedded in permeable polymer thin films perform as well as surface-initiated or grafted MIPs?

Answers to many of these questions, especially the fundamentals of synthesizing and charactezing surface imprints are being gathered. For example, controlled radical polymerizations are being used to address step-wise assembly of MIP binding sites, while methods of growing polymer brushes of known functionality and film thickness from surfaces are being developed. Combinations of these techniques are allowing the preparation of functional materials with very specific structures and architectures to be produced [84–86]. In addition, reversibly assembling MIPs are being prepared that offer the combination of covalently joined polymer chains and noncovalent cross-links to form materials with many features analogous to biological macromolecules but which are much more robust [87,88].

Ultimately, it is the ingenuity of MIP researchers in addressing these issues that will decide whether imprinted surfaces become the basis of practical technologies or remain lab curiosities, albeit fascinating ones. The innovative work in preparing, for example, protein-specific imprints [75,76] and QCM biosensors [81–83] suggests that practical and useful materials will soon be adopted commercially while the next generation of imprinted surfaces with yet more advanced properties are being developed in the laboratory.

REFERENCES

1. Flores, A.; Cunliffe, D.; Whitcombe, M.J.; Vulfson, E.N. Imprinted polymers prepared by aqueous suspension polymerization. J. Appl. Polym. Sci. **2000**, *78*, 1841–1850.
2. Mayes, A.G.; Mosbach, K. Molecularly imprinted polymer beads: Suspension polymerisation using a liquid perfluorocarbon as the dispersing phase. Anal. Chem. **1996**, *68*, 3769–3774.
3. Sellergren, B. Imprinted dispersion polymers–a new class of easily accessible affinity stationary phases. J. Chromatogr. A. **1994**, *673*, 133–141.
4. Cooper, A.I.; Hems, W.P.; Holmes, A.B. Synthesis of highly cross-linked polymers in supercritical carbon dioxide by heterogeneous polymerization. Macromolecules **1999**, *32*, 2156–2166.
5. Ye, L.; Weiss, R.; Mosbach, K. Synthesis and characterization of molecularly imprinted microspheres. Macromolecules **2000**, *33*, 8239–8245.
6. Masci, G.; Aulenta, F.; Crescenzi, V. Uniform-sized clembuterol molecularly imprinted polymers prepared with methacrylic acid or acrylamide as an interacting monomer. J. Appl. Polym. Sci. **2002**, *83*, 2660–2668.
7. *Polymer Latexes*, Daniels, E.S., Sudol, E.D., El-Asser, M.S., Eds.; American Chemical Society: Washington, DC, USA, 1992.
8. Emulsion Polymerization and Emulsion Polymers, Lovell, P.A., El-Asser, M.S., Eds.; John Wiley and Sons: Chichester, UK, 1997.
9. Pérez, N.; Whitcombe, M.J.; Vulfson, E.N. Molecularly imprinted nanoparticles prepared by core-shell emulsion polymerization. J. Appl. Polym. Sci. **2000**, *77*, 1851–1859.
10. Pérez, N.; Whitcombe, M.J.; Vulfson, E.N. Surface imprinting of cholesterol on sub-micrometer core-shell emulsion particles. Macromolecules **2001**, *34*, 830–836.

11. Whitcombe, M.J.; Rodriguez, M.E.; Villar, P.; Vulfson, E.N. A new method for the introduction of recognition site functionality into polymers prepared by molecular imprinting-synthesis and characterization of polymeric receptors for cholesterol. J. Am. Chem. Soc. **1995**, *117*, 7105–7111.

12. Uezu, K.; Tazume, N.; Yoshida, M.; Goto, M.; Furusaki, S. Characterization and control of matrix for surface molecular-imprinted polymer. Kagaku Ronbunshi **2001**, *27*, 753–755.

13. Araki, K.; Yoshida, M.; Uezu, K.; Goto, M.; Furusaki, S. Lanthanide-imprinted resins prepared by surface template polymerization. J. Chem. Eng. Jpn. **2000**, *33*, 665–668.

14. Yoshida, M.; Hatate, Y.; Uezu, K.; Goto, M.; Furusaki, S. Metal-imprinted microsphere prepared by surface template polymerization and its application to chromatography. J. Polym. Sci. Part A: Polym. Chem. **2000**, *38*, 689–696.

15. Uezu, K.; Nakamura, H.; Goto, M.; Nakashio, F.; Furusaki, F. Metal-Imprinted microsphere prepared by surface template polymerization with W/O/W emulsions. J. Chem. Eng. Jpn. **1999**, *32*, 262–267.

16. Yoshida, M.; Uezu, K.; Goto, M.; Furusaki, S. Metal ion imprinted microspheres prepared by surface molecular imprinting technique using water-in-oil-in-water emulsion. J. Appl. Polym. Sci. **1999**, *73*, 1223–1230.

17. Uezu, K.; Yoshida, M.; Goto, M.; Furusaki, S. Molecular recognition using surface template polymerization. Chemtech. **1999**, *29*, 12–18.

18. Yoshida, M.; Uezu, K.; Goto, M.; Furusaki, S. Required properties for functional monomers to produce metal template effect by a surface molecular imprinting. Macromolecules **1999**, *32*, 1237–1243.

19. Yoshida, M.; Uezu, K.; Nakashio, F.; Goto, M. Spacer effect of novel bifunctional organophosphorus monomers in metal-imprinted polymers prepared by surface template polymerization. J. Polym. Sci. Part A: Polym. Chem. **1998**, *36*, 2727–2734.

20. Uezu, K.; Nakamura, H.; Kanno, J.; Sugo, T.; Goto, M.; Nakashio, F. Metal ion-imprinted polymer prepared by the combination of surface template polymerization with postirradiation by gamma-rays. Macromolecules **1997**, *30*, 3888–3891.

21. Yoshida, M.; Uezu, K.; Goto, M.; Nakashio, F. Metal ion-imprinted resins with novel bifunctional monomers by surface template polymerization. J. Chem. Eng. Jpn. **1996**, *29*, 174–176.

22. Yoshida, M.; Uezu, K.; Goto, M.; Furusaki, S. Surface imprinted polymers recognizing amino acid chirality. J. Appl. Polym. Sci. **2000**, *78*, 659–703.

23. Yoshida, M.; Hatate, Y.; Uezu, K.; Goto, M.; Furusaki, S. Chiral-recognition polymer prepared by surface molecular imprinting technique. Colloids and Surfaces A: Physicochem. Eng. Aspects **2000**, *169*, 259–269.

24. Araki, K.; Goto, M.; Furusaki, S. Enantioselective polymer prepared by surface imprinting technique using a bifunctional molecule. Anal. Chim. Acta. **2002**, *469*, 173–181.

25. Carter, S.R.; Rimmer, S. Molecular Recognition of Caffeine by shell molecular imprinted core-shell polymer particles in aqueous media. Adv. Mater. **2002**, *14*, 667–670.

26. Hirayama, K.; Sakai, Y.; Kameoka, K. Synthesis of polymer particles with specific lysozyme recognition sites by a molecular imprinting technique. J. Appl. Polym. Sci. **2001**, *81*, 3378–3387.

27. Schweitz, L.; Andersson, L.I.; Nilsson, S. Capillary electrochromatography with predetermined selectivity obtained through molecular imprinting. Anal. Chem. **1997**, *69*, 1179–1183.

28. Brüggermann, O.; Freitag, F.; Whitcombe, M.J.; Vulfson, E.N. Comparison of polymer coatings of capillaries for capillary electrophoresis with respect to their applicability to molecular imprinting and electrochromatography. J. Chromatogr. A **1997**, *781*, 43–53.

29. Lin, J.M.; Nakagama, T.; Uchiyama, K.; Hobo, T. Capillary electrochromatographic separation of amino acid enantiomers using on-column prepared molecularly imprinted polymer. J. Pharmaceut. Biomed. Anal. **1997**, *15*, 1351–1358.

30. Schweitz, L.; Andersson, L.L.; Nilsson, S. Molecularly imprinted CEC sorbents: investigations into polymer preparation and electrolyte composition. Analyst **2002**, *127*, 22–28.

31. Quaglia, M.; De Lorenzi, E.; Sulitzky, C.; Massolini, G.; Sellergren, B. Surface initiated molecularly imprinted polymer films: a new approach in chiral capillary chromatography. Analyst. **2001**, *126*, 1495–1498.

32. Sulitzky, C.; Rückert, B.; Hall, A.J.; Lanza, F.; Unger, K.; Sellergren, B. Grafting of molecularly imprinted polymer films on silica supports containing surface-bound free radical initiators. Macromolecules **2002**, *35*, 79–91.

33. Sellergren, B.; Rückert, B.; Hall, A.J. Layer-by-layer grafting of molecularly imprinted polymers via iniferter modified supports. Adv. Mater. **2002**, *14*, 1204–1208.

34. Schweitz, L. Molecularly imprinted polymer coatings for open-tubular capillary electro-chromatography prepared by surface initiation. Anal. Chem. **2002**, *74*, 1192–1196.

35. Do Ki, C.; Oh, C.; Oh, S.G.; Chang, J.Y. The use of a thermally reversible bond for molecular imprinting of silica spheres. J. Am. Chem. Soc. **2002**, *124*, 14838–14839.

36. Percival, C.J.; Stanley, S.; Braithwaite, A.; Newton, M.I.; McHale, G. Molecular imprinted polymer coated QCM for the detection of nandrolone. Analyst. **2002**, *127*, 1024–1026.

37. Vlatakis, G.; Andersson, L.I.; Muller, R.; Mosbach, K. Drug assay using mimics made by molecular imprinting. Nature. **1993**, *361*, 645–647.

38. Kugimiya, A.; Takeuchi, T. Molecularly imprinted polymer-coated quartz microbalance for detection of biological hormone. Electroanalysis **1999**, *11*, 1158–1160.

39. Kugimiya, A.; Takeuchi, T. Effect of 2-hydroxyethyl methacrylate on polymer network and interaction in hydrophilic molecularly imprinted polymers. Anal. Sci. **1999**, *15*, 29–33.

40. Liang, C.; Peng, H.; Bao, X.; Nie, L.; Yao, S. Study of a molecular imprinted polymer coated BAW bio-mimetic sensor and its application to the determination of caffeine in human serum and urine. Analyst **1999**, *124*, 1781–1785.

41. Frigeri, A.; Kobayashi, H.; Shinkai, S.; Reinhoudt, D.N. From solutions to surfaces: A novel molecularly imprinting method based on the conformational changes of boronic-acid-appended poly (L-lysine). Angew. Chem. Int. Ed. **2001**, *40*, 4729–4731.

42. Malitesta, C.; Losito, I.; Zambonin, P.G. Molecularly imprinted electropolymerized polymers: New materials for biomimetic sensors. Anal. Chem. **1999**, *71*, 1366–1370.

43. Peng, H.; Yin, F.; Zhou, A.H.; Yao, S.Z. Characterization of electrosynthesized poly (o-aminophenol) as a molecular imprinting material for sensor preparation by means of quartz crystal impedance analysis. Anal. Lett. **2002**, *35*, 435–450.

44. Piletsky, S.A.; Piletskaya, E.V.; Sergeyeva, T.A.; Panyasuk, T.L.; El'skya, A.V. Molecularly imprinted self-assembled films with specificity to cholesterol. Sens. Actuat. B-Chem. **1999**, *60*, 216–220.

45. Kitano, H.; Taira, Y. Inclusion of bisphenols by a self-assembled monolayer of thiolated cyclodextrins on gold electrodes. Langmuir **2002**, *18*, 5835–5840.

46. Merkoçi, A.; Alegret, S. New materials for electrochemical sensing IV. molecular imprinted polymers. TrAC. Trends Anal. Chem. **2002**, *21*, 717–725.

47. Yoshima, Y.; Ohdaira, R.; Liyama, C.; Sakai, K. "Gate effect" of thin layer of molecularly-imprinted poly (methacrylic acid-co-ethyleneglycoldimethacrylate). Sens. Actuat B: Chem. **2001**, *73*, 49–53.

48. Zayats, M.; Lahav, M.; Kharitonov, A.B.; Willner, I. Imprinting of specific molecular recognition sites in inorganic and organic thin layer membranes associated with ion-sensitive field-effect transistors. Tetrahedron **2002**, *58*, 815–824.

49. Kirsch, N.; Hart, J.P.; Bird, D.J.; Luxton, R.W.; McCalley, D.V. Towards the development of molecularly imprinted polymer based screen-printed sensors for metabolites of PAHs. Analyst **2001**, *126*, 1936–1941.

50. Mirsky, V.M.; Hirsch, T.; Piletsky, S.A.; Wolfbeis, O.S. A spreader bar approach to molecular architecture: Formation of stable artificial chemoreceptors. Angew. Chem. Int. Ed. **1999**, *38*, 1108–1118.

51. Panasyuk, T.L.; Mirsky, V.M.; Piletsky, S.A.; Wolfbeis, O.S. Electropolymerized molecularly imprinted polymers as receptor layers in capacitive chemical sensors. Anal. Chem. **1999**, *71*, 4609–4613.

52. Piletsky, S.A.; Pileyskaya, E.V.; Elgersma, A.V.; Yano, K.; Karube, I.; Parhometz, Y.P.; El'skaya, E.V. Atrazine sensing molecularly imprinted membranes. Biosens. Bioelectron. **1995**, *10*, 959–964.

53. Piletsky, S.A.; Piletskya, E.V.; Panasyuk, T.L.; El'skya, A.V.; Levi, R.; Karube, I.; Wulff, G. Imprinted Membranes for sensor technology: Opposite behavious of covalently and noncovalently imprinted membranes. Macromolecules **1998**, *31*, 2137–2140.

54. Kobayashi, T.; Wang, H.Y.; Fuji, N. Molecular imprinting of theophylline in acrylonitrile-acrylic acid copolymer membrane. Chem. Lett. **1995**, *10*, 927–928.

55. Laschewsky, A.; Wischerhoff, E.; Denziger, S.; Ringsdorf, H.; Delcorte, A.; Bertrand, P. Molecular recognition by hydrogen bonding in polyelectrolyte multilayers. Chem. Eur. J. **1997**, *3*, 34–38.

56. Wang, H.Y.; Kobayashi, T.; Fujii, N. Surface molecular imprinting on photosensitive dithiocarbamoyl polyacrylonitrile membranes using photograft polymerization. J. Chem. Tech. Biotechnol. **1997**, *70*, 355–362.

57. Piletsky, S.A.; Matuschewski, H.; Schedler, U.; Wilpert, A.; Piletska, E.V.; Thiele, T.A.; Ulbricht, M. Surface Functionalization of porous polypropylene membranes with molecularly imprinted polymers by photograft copolymerization in water. Macromolecules **2000**, *33*, 3092–3098.

58. Kochkodan, V.; Weigel, W.; Ulbricht, M. Molecularly imprinted composite membranes for selective binding of desmetryn from aqueous solutions. Desalination **2002**, *149*, 323–328.

59. Sergeyeva, T.A.; Matuschewski, H.; Piletsky, S.A.; Bendig, J.; Schedler, U.; Ulbricht, M. Molecularly imprinted polymer membranes for substance-selective solid-phase extraction from water by surface photo-grafting polymerization. J. Chromatogr. A **2001**, *907*, 89–99.

60. Kochkodan, V.; Weigel, W.; Ulbricht, M. Thin layer molecularly imprinted microfiltration membranes by photofunctionalization using a coated α-cleavage photoinitiator. Analyst **2001**, *126*, 803–809.

61. Ulbricht, M.; Belter, M.; Langenhangen, U.; Schneider, F.; Weigel, W. Novel molecularly imprinted polymer (MIP) composite membranes via controlled surface and pore functionalizations. Desalination **2002**, *149*, 293–295.

62. Yoshikawa, M.; Izumi, J.; Kitao, T.; Koya, S.; Sakamoto, S. Molecularly imprinted polymeric membranes for optical resolution. J. Memb. Sci. **1995**, *108*, 171–175.

63. Yoshikawa, M.; Izumi, J.; Kitao, T.; Sakamoto, S. Molecularly imprinted polymeric membranes containing DIDE derivatives for optical resoultion of amino acids. Macromolecules **1993**, *29*, 8197–8203.

64. Yoshikawa, M.; Ooi, T.; Izumi, J. Novel membrane materials having EEE derivatives as a chiral recognition site. Eur. Polym. J. **2001**, *37*, 335–342.

65. Bossi, A.; Piletsky, S.A.; Piletska, E.V.; Righetti, P.G.; Turner, A.P.F. Surface-grafted molecularly imprinted polymers for protein recognition. Anal. Chem. **2001**, *73*, 5281–5286.

66. Yoshikawa, M.; Fujisawa, T.; Izumi, J. Molecularly imprinted polymeric membranes having EFF derivatives as a chiral recognition. Macromolec. Chem. Phys. **1999**, *200*, 1458–1465.

67. Toorisaka, E.; Yoshida, M.; Uezu, K.; Goto, M.; Furusaki, S. Artificial biocatalysts prepared by the surface molecular imprinting technique. Chem. Lett. **1999**, 387–388.

68. Markowitz, M.A.; Kust, P.R.; Deng, G.; Schoen, P.E.; Dordick, J.S.; Clark, D.S.; Gaber, B.P. Catalytic silica particles via template-directed molecular imprinting. Langmuir **2000**, *16*, 1759–1765.

69. Markowitz, M.A.; Kust, P.R.; Klaehn, J.; Deng, G.; Gaber, B.P. Surface-imprinted silica particles: the effects of added organosilanes on catalytic activity. Anal. Chim. Acta. **2001**, *435*, 177–185.

70. Tada, M.; Sasaki, T.; Iwasawa, Y. Performance and Kinetic Behavior of a new SiO_2-attached molecular-imprinting Rh-dimer catalyst in size- and shape-selective hydrogenation of alkenes. J. Catal. **2002**, *211*, 496–510.

71. Suzuki, A.; Tada, M.; Sasaki, T.; Shido, T.; Iwasawa, Y. Design of catalytic sites at oxide surfaces by metal-complex attaching and molecular imprinting techniques. J. Molec. Catal. A-Chem. **2002**, *182*, 125–136.

72. Tada, M.; Sasaki, T.; Iwasawa, Y. Novel SiO_2-attached molecular-imprinting Rh-monomer catalysts for shape-selective hydrogenation of alkenes; preparation, characterization and performance. Phys. Chem. Chem. Phys. **2002**, *4*, 4561–4574.

73. Tada, M.; Sasaki, T.; Shido, T.; Iwasawa, Y. Design, characterization and performance of a molecular imprinting Rh-dimer hydrogenation catalyst on a SiO_2 surface. Phys. Chem. Chem. Phys. **2002**, *4*, 5899–5909.

74. Yilmaz, E.; Haupt, K.; Mosbach, K. The use of immobilized templates – a new approach in molecular imprinting. Angew. Chem. Int. Ed. **2000**, *39*, 2115–2118.

75. H.Q., Shi; W.B., Tsai; Garrison, M.D.; Ferrari, S. Ratner, B.D. Template-imprinted nanostructured surfaces for protein recognition. Nature **1999**, *398*, 595.

76. H.Q., Shi; Ratner, B.D. Template recognition of protein-imprinted polymer surfaces. J. Biomed. Mater. Res. Part A **2000**, *49*, 1.

77. D'Souza, S.M.; Alexander, C.; Carr, S.W.; Waller, A.M.; Whitcombe, M.J.; Vulfson, E.N. Directed nucleation of calcite at a crystal-imprinted polymer surface. Nature **1999**, *398*, 312–316.

78. D'Souza, S.M.; Alexander, C.; Whitcombe, M.J.; Waller, A.M.; Vulfson, E.N. Control of crystal morphology *via* molecular imprinting. Polym. Int. **2001**, *50*, 429–432.

79. Aherne, A.; Alexander, C.; Payne, M.J.; Perez, N.; Vulfson, E.N. Bacteria-mediated lithography of polymer surfaces. J. Am. Chem. Soc. **1993**, *116*, 8771–8772.

80. Alexander, C.; Vulfson, E.N. Spatially functionalized polymer surfaces produced *via* cell-mediated lithography. Adv. Mater. **1997**, *9*, 751–755.

81. Dickert, F.L.; Hayden, O.; Halikias, K.P. Synthetic receptors as sensor coatings for molecules and living cells. Analyst **2001**, *126*, 766–771.

82. Dickert, F.L.; Hayden, O. Bioimprinting of polymers and sol-gel phases. Selective detection of yeasts with imprinted polymers. Anal. Chem. **2002**, *74*, 1302–1306.

83. Hayden, O.; Dickert, F.L. Selective microorganism detection with cell surface imprinted polymers. Adv. Mater. **2001**, *13*, 1480–1483.

84. Shah, R.R.; Merreceyes, D.; Husemann, M.; Rees, I.; Abbott, N.L.; Hawker, C.J.; Hedrick, J.L. Using atom transfer radical polymerization to amplify monolayers of initiators patterned by microcontact printing into polymer brushes for pattern transfer. Macromolecules **2000**, *33*, 597–605.

85. Huang, W; Baker, G.L.; Bruening, M.L. Controlled synthesis of cross-linked ultrathin polymer films by using surface-initiated atom transfer radical polymerization. Angew. Chem. Int. Ed. **2001**, *40*, 1510–1512.

86. Jones, D.M.; Brown, A.A.; Huck, W.T.S. Surface-initiated polymerizations in aqueous media: effect of initiator density. Langmuir **2002**, *18*, 1265–1273.

87. Alvarez-Lorenzo, C.; Guney, O.; Oya, T.; Sakai, Y.; Kobayashi, M.; Enoki, T.; Takeoka, Y.; Ishibashi, T.; Kuroda, K.; Tanaka, K.; Wang, G.; Grosberg, A.Y.; Masamune, S.; Tanaka, T. Polymer gels that memorize elements of molecular conformation. Macromolecules **2000**, *33*, 8693–8697.

88. Enoki, T.; Tanaka, K.; Watanabe, T.; Oya, T.; Sakiyama, T.; Takeoka, Y.; Ito, K.; Wang, G.Q.; Annaka, M.; Hara, K.; Du, R.; Chuang, J.; Wasserman, K.; Grosberg, A.Y.; Masamune, S.; Tanaka, T. Frustrations in polymer conformation in gels and their minimization through molecular imprinting. Phys. Rev. Lett. **2000**, *85*, 5000–5003.

10
Scaffold Imprinting

Takaomi Kobayashi Nagaoka University of Technology, Nagaoka, Japan

I. INTRODUCTION

Molecular recognition of enzymes, biological antibodies, and receptors can be achieved by collective weak forces between the guest molecule and host sites [1–4]. Because such bio-polymers can precisely recognize their counterparts to achieve the desired function, mimicking the recognition property by a chemical procedure is a great achievement for an artificial system involving imprinted polymers. To design such a system, molecular imprinting has been an important strategy to try to mimic biological molecular recognition systems with synthetic polymers. As mentioned in Sec. III, imprinting techniques have focused on designing recognition sites with polymeric materials [5–11]. These polymers are mainly prepared based on two important principles to imprint molecular shapes into polymeric matrix. First, a target molecule shape must be fixed by enveloping it with several functional monomers. Secondly, polymerization using radical and condensation reactions was carried out for solidification of the functional monomer/template complex in the matrix. This is commonly carried out via polymerization in the presence of a template-functional vinyl monomer complex with a cross-linker. The resultant solid polymer matrix gives rise to the known polymer effect on imprinting. In these processes, monomer elements play an important role in the imprinting process since each element causes interactive binding to the target molecule before imprinting polymerization.

Therefore, the availability of scaffolding elements that can bind to a template molecule becomes a key element in the inability to selectively encode recognition of the target molecule into the synthetic polymer. To allow more sophisticated recognition interactions by imprinted materials, the monomer–template interaction needs to be a cooperative interaction of multiple associating functionality. For that purpose, binding sites that are lined with multiple interacting functional groups are desired that can hold the target molecule via multiple interactions. However, conventional molecular imprinting is not generally predisposed to generate binding sites with multiple interactions.

Intermolecular forces of hydrogen bonds, ionic bonds, and hydrophobic forces are considerably weaker than primary covalent bonds. However, such weak intermolecular forces can produce strong binding interactions when operating in concert. For example, the cooperative effect of covalent and noncovalent imprinting can realize imprinting shapes, which can mimic large organic molecules such as peptides [12,13]. The key feature of this approach is the presence of both bifunctional covalent bonding sites and noncovalent interaction sites by two carboxylic acid groups with the template. After cross-linking polymerization, the imprinted matrix constructs multi-binding sites via hydrogen bonds, resulting in successful imprinting of peptides.

$$(10.1)$$

Although positioning recognition elements in imprinting materials can allow creation of a three-dimensional scaffolding space with specific binding ability, the noncovalent imprinting of functional monomer and template produces the binding sites. For the purposes of sophisticated imprinting, one approach is to use a scaffold monomer instead of functional monomers. Here, the term "scaffold monomer" can mean multi-mer, which is defined as cross-linkable monomers that can interact with the template via multiple binding interactions. Namely, such multi-mer contains both polymerization groups and multi-complexable groups with template. Molecule-assembling matrix of large size molecules having multi-complexable ability classifies as imprinting scaffold. In addition, polymers having multi-complexable groups are another strategical way for scaffolding imprinting. The review presented here will concentrate upon the design of scaffold imprint matrices. Also, a representative technique for creating recognition sites by scaffold molecular imprinting is described.

II. MOLECULAR DESIGN AND CREATION OF SCAFFOLD IMPRINTING SITES

Design approaches for scaffolding imprinting polymers have to use molecule scaffold media in imprinted matrices. To generate binding sites complementary to the target molecule, molecule design is obtained with polymer–template multiple interactions. Therefore, polymerizable monomers having multi-complexable sites are important approaches in scaffolding imprinting. Because of the multi-mer nature, this is achieved by forming a highly cross-linked matrix of monomer-performing components. Relative to multi-mer, the use of a scaffolding polymer, which envelops the template, is also available for scaffold imprinting. As illustrated in Fig. 1, scaffold

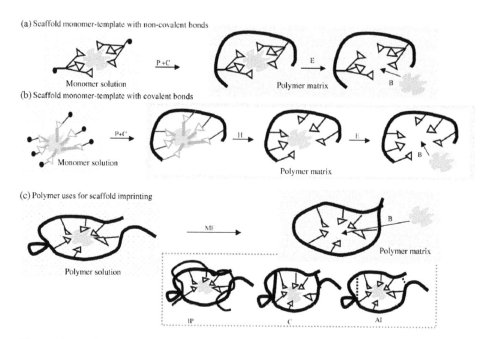

(a) Scaffold monomer-template with non-covalent bonds

Monomer solution

P+C

Polymer matrix

E

B

(b) Scaffold monomer-template with covalent bonds

Monomer solution

P+C

Polymer matrix

H

E

B

(c) Polymer uses for scaffold imprinting

Polymer solution

MF

Polymer matrix

B

IP

C

AI

Figure 1 Molecular imprinting of scaffolding monomer having: (a) noncovalent bond and (b) covalent bonds between monomer and template, and (c) scaffolding polymers with polymer matrix formation (MF) using interpenetration (IP), cross-linking (C) and aggregation interaction (AI) for matrix formation. P + C is for polymerization and cross-linking processes and H is for hydrolysis of covalent bonds. E and B stand for extraction and binding of substrate, respectively.

elements can be either monomers (a and b) or polymers (c) which have multi-interaction sites. In scaffold monomers, the complex between the multi-functional scaffold and the template can be either covalent or noncovalent. Similar to the conventional imprinting [14–16], cross-linking, then result in an insoluble polymer including the template-assembly scaffolding sites. Because of the assembling manner, the imprinted element can form highly ordered components. As a result, removal of the molecule from the polymer matrix can leave memory sites. On the other hand, polymer used for molecular imprinting is an advantage on the introduction of multi-complexable groups on polymer chains that behave as holding sites for imprint molecules. That is, in the process, polymer chains having the functional groups can envelop the template molecule and then fix the three-dimensional space of the template in the polymer microenvironment.

III. TEMPLATE CLASSIFICATION OF SCAFFOLDING MONOMERS HAVING MULTI-COMPLEXABLE TEMPLATE

A scaffolding monomer having multi-interaction sites for the target molecule is a general approach that can utilize the principles of both covalent and noncovalent imprinting. Many cases have been described using scaffold multi-mers, which interact with the template via multiple binding interactions and subsequently cross-linked via radical polymerization and condensation polymerization. For multi-binding interactions, many approaches to noncovalent imprinting, as described in Chapter 3 can be applied, for instance, using mainly hydrogen bonds and Coulomb interaction between the template and matrix [17–29]. In addition to the noncovalent multi-mer, such imprinting containing covalent linkage between polymerable group and template reported using boronic ester linkages [14], carboxylic ester [15,16], amide [17,18], Schiff bases [19], and ketal [20,21] bonds, would be applicable for scaffolding imprinting. The following cross-linking process for conventional imprinting is used as mentioned in the literature [6,26,30,34]. Table 1 lists representative examples of multifunctional scaffold monomers. The cooperative interactions of these multimers with the template are predicted to generate imprinted with higher affinity and capacity for the target molecule. There is a possibility for great diversity among scaffold elements both at covalent and noncovalent imprinting, presenting continuing challenges and attractive research targets. For example, the monomers can contain multiple amide bonds to form hydrogen bonds, or be a covalent monomer with multi-carbamate groups. Large molecular monomers containing such groups as cyclodextrin, porphyrin, and fullerene can also form novel scaffolding imprint media.

A. Uses of Scaffold Monomers

One main difficulty in noncovalent imprinting is heterogeneity in imprinted sites (Chapter 12). By virtue of its directionality, specificity and biological relevance [32], hydrogen bonding is the favorite intermolecular force to form self-ordered assembling system [33]. In imprinted polymers, scaffold elements bearing the ability of multi-hydrogen pairs were developed by Takeuchi et al. The chemical structure is shown in Table 1. These systems have demonstrated that it is favorable to have both vinyl-monomer groups containing multiple hydrogen bonding sites. They reported that highly selective recognition for target molecules of barbiturates [34,35] and uracil derivatives [36] using 2,6-bis(acrylamide)pyridine (BAP) [37] and ethyleneglycol dimethacrylate (EGDMA) copolymers. This afforded high affinity and selectivity when compared to a functional monomer with only a single hydrogen bonding interaction for cyclobarbiturate.

(10.2)

AQ2   **Table 1**   Examples of Functional Monomer Having Scaffold Elements for Molecular Imprinting

Scaffolding functional monomer	Template	Type of interaction	Ref.
BAP BAP	Cyclobarbiters, Fluorouracile	Multi-hydrogen bonds	33,36
Cyclodextrin 	Cholesterol	Guest inclusion by hydrophobic/van der Waals	46
Metal chelator 	Metal ion	Multi-metal coordinates in a monomer	40
	Lanthanide (III)	Diethylenetriamine pentaacetic acid (DTPA) and Ethylenediaminetetraacetic acid (EDTA) derivatives	39

(Continued)

Table 1 (*Continued*)

Scaffolding functional monomer	Template	Type of interaction	Ref.
Carbamate sol–gel monomer	Sol–gel imprinted pore formation	Covalent silica-oxygen bonding as site anchor Introduction of ordered-amino groups in imprinted matrix	51
Boroester–fullerene	Saccharide	Covalent bonding on fullerene	52

The BAP functional monomer was prepared by 2,6-diamino pyridine and acryloyl chloride in triethylamine–chloroform solution (54% yield) [37]. Cyclobarbital-imprinted copolymers were prepared by the following procedure: BAP (1 mmol), template cyclobarbital (0.5 mmol) and EGDA cross-linker (20 mmol) were dissolved in $CHCl_3$. In the presence of a radical initiator, 2,2-azo (2,4-dimethylvaleronitrile), radical copolymerization was performed in N_2 atmosphere with a two-step process for 6 h at 40°C and then for 3 h at 90°C. The obtained polymer was ground and sieved in the range of 26–63 μm. Selectivity of the cyclobarbital-imprinted polymer was assessed by chromatographic analysis for 5-barbiturates. They also used BAP monomer to imprint antitumor active compound of 5-fluorouracil (5-FU) [36]. The 5-FU imprinted polymers showed a higher affinity for 5-FU than for 5-FU derivatives. The BAP imprinted polymer was prepared in the presence of a co-monomer, 2-(trifluoromethyl)-acrylic acid.

Other examples of multi-mer having a cooperative complex with template were achieved by metal–ligand coordination [38]. An imprinted polymer containing scaffolding sites that was selective for metal ions was synthesized from ethylene diaminetetraacetic acid (EDTA) derivatives [39] and N-vinylbenzyl derivative [40]. To achieve high metal ion capture, scaffolding imprinting was carried out using preorganized-metal ion binding monomer. Triethylentetraamine (TETA) monomers can be prepared from a mixture of vinylbenzyl chloride and TETA hydrate. The chelate monomer has four-interacting groups that can coordinate metal ions in the prepolymerization complex and in the resultant imprinted polymers. Polymerization of these monomers with a cross-linker produced polymers capable of achieving high selectivity in metal ion binding studies.

B. Large Molecule Elements as Imprinting Scaffold

Macrocyclic molecules such as Cyclodextrin (CyD) have an ability to be scaffold. Namely, molecule inclusion of small molecule inside CyD can be applicable as scaffold imprinting platform. CyD, which is a representative inclusion supra-molecule, has been proposed as an imprinting element for a large hydrophobic guest molecule [41,42]. CyDs are cyclic oligosaccharides comprising six to eight D-glucopyranoside units linked by a 1,4-glycosidic bond [43]. The cavity of CyD was selected as a scaffolding site for imprinting cholesterol molecules because CyDs bind angstrom-sized guests through cooperative solvaphobic interactions. Komiyama et al. [44] cross-linked CyD in the presence of steroids with di-isocyanates. The imprinted CyD polymers were obtained as depicted in 10.10.3 B for a cholesterol template. The 3:1 β-CyD-cholesterol complex (5.5 g containing 4.36 mmol of CyD) was dissolved in 50 mL of dry DMSO. Then, the diisocyanate cross-linker agent was added with the molar ratio of diisocyanate/CyD = 4.2. The mixtures were kept at 65°C for a few hours. After condensation polymerization, the resulting cross-linked polymer was precipitated in acetone. The polymer was ground with a mortar and washed with sufficient hot water, THF, and hot ethanol to remove template, unreacted monomer and free CyD.

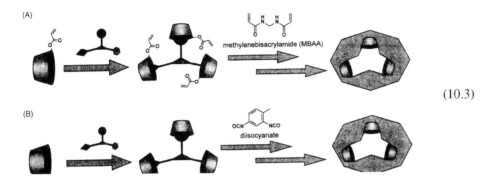

(10.3)

For adsorption experiments of the substrate, the imprinted polymer containing 0.5 mmol of β-CyD segments was incubated in 11 mL of THF/water solution with 0.05 mmol of guest. Cholesterol selectivity was compared with stegmasterol and phenol. Relative to phenol, the imprinted polymer cross-linked with hexamethylene diisocyanate showed 4.3-fold higher selectivity. However, the polymers showed a stegmasterol/cholesterol selectivity of only 1.4. When epichlorohydrin was the cross-linker, only weak binding to cholesterol was observed. They also demonstrated that solvent effect of selectivity indicated high recognition in DMSO solution.

The CyD imprinted polymer consisted of two or three CyD molecules that cooperatively organized to bind large steroid molecules. Use of DMSO was necessary for efficient imprinting since CyD was shown to cooperatively include the template molecule in this solvent.

(10.4)

The molecularly imprinted copolymer of 2-hydroxyethyl methacrylate (HEMA) and β-CyD-coupled HEMA was synthesized in chloroform to study its interaction with a pair of steroids: cholesterol and testosterone [42]. The molecularly imprinted copolymer was found to absorb the print molecule several times better than an imprinted poly HEMA. Asanuma et al. [44] concluded that cholesterol binding by the HEMA polymer was not due to CyD inclusion because no inclusion of CyD to cholesterol can form in chloroform. Another strategy developed for imprinting using CyD elements is the vinyl CyD (Table 1 and 10.3(A)), which has been successfully applied for imprinting antibodies, oligo-peptides [45], and vancomycin [46].

Works concerning molecule assemblies for molecular imprinting were carried out using organogel [47]. The functionalized gluconamides were used for the organogel formation used in scaffolding molecular imprinting. Functionalized gluco-namides (10.5) having phenyl, pyridine, imidazole, and cyclohexyl groups were synthesized and shown to form hierarchal supramolecular assemblies. The phenyl and cyclohexyl gluconamides were synthesized via reaction of excess octylamine with benzoylchloride and cyclohexylchloride, respectively. Imidazole gluconamide was synthesized via process ii (10.5).

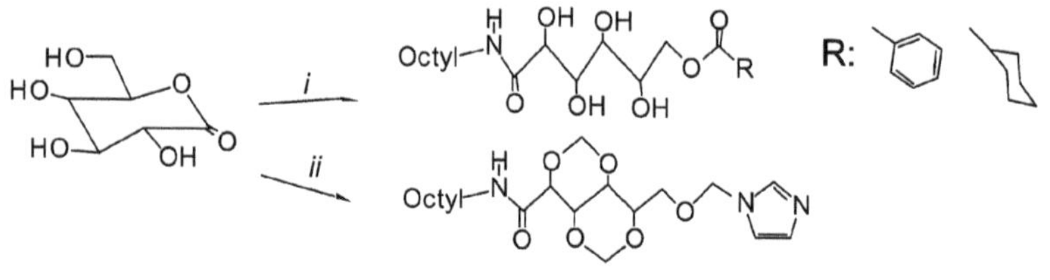

(10.5)

It was reported that ordered aggregation of phenyl and cyclohexyl types of glucona-mides can form fibers in chloroform and ethylacetate, respectively. The latter supra-molecule fiber showing double and multiple standard helices was provided by scaffolding imprinting to induce molecular chirality. In order to obtain imprinted polymer, they dissolved the benzoate or cylohexanoate derivative in methyl metha-crylate and *n*-butyl methacrylate (1:4 v/v) with benzoin ethyl ether (polymerization initiator). Although recognition details were not shown for the imprinted matrix memorizing the supramolecule fibers, it is possible to prepare imprints of glucona-mide assemblies into the polymer matrix for cyclohexyl gluconamide.

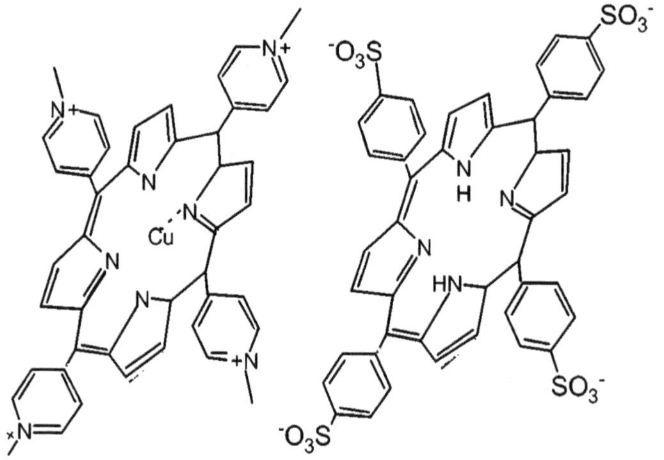

(10.6)

Porphyrin is another representative large heterocyclic molecule [48,49]. Anionic tetraphenylporphyrin and copper cationic porphyrin having four binding sites in a molecule ring (10.6) lead to chiral complex in water in the presence of chiral L- or D-aromatic amino acids [50] and are stable enough to maintain memory of the templating amino acid chirality. As well as CyD and gluconamide organogel, this kind of large molecule will be a candidate of imprint scaffold in the near future, since the functional sites of porphyrin phenyl groups can be modified to proper multi-complexable groups with target molecule.

C. Covalent-Bondable Scaffold Monomers

Table 1 also lists examples of covalently imprinted scaffold [51,52]. Since covalent imprinting can produce more homogeneous imprinted sites in the resultant polymer bulk, Katz and Davis developed a sol–gel multi-mer, which contains three carbamate sites for introducing three convergent amino groups into the sol–gel silca matrix. Using nitrogen adsorption studies, the micro-porosity generated from the imprinting process could be directly observed. See Chapter 11 for details on this approach.

Use of fullerene C_{60} as a covalent imprintable recognition platform for saccharide recognition was recently reported [52]. Ishii et al. chose saccharide both as a template and guest molecule because saccharides have been demonstrated to be capable of introducing arranged two-boronic acid into a platform such as C_{60} and for reversible re-binding to the saccharide template (10.7). For preliminary experiments of saccharide binding on fullerene matrix, 3-O-methyl-D-glucose was selected. The imprinted matrix quantitatively bound the template saccharide.

(10.7)

IV. CLASSIFICATION OF POLYMER MATRIX AS SCAFFOLDING IMPRINTING MEDIA

In contrast to the necessity of utilizing multiple complexable monomers to generate cooperative binding sites, polymers contain multi-interactive groups within a single macromolecule chain that can readily form more stable prepolymerization complexes and therefore stronger multi-functional binding sites. Table 2 lists examples of using polymers in imprinting. There are two strategies for scaffolding imprinting polymers, which are with and without covalent cross-linking. The conventional imprinting process polymerizes the monomer–template prepolymerization complex in the presence of a cross-linking agent to structurally stabilize the resulting binding sites. Alternative methods for imprinting linear polymers containing multiple interacting groups are to fix the orientation and distance of these groups by noncovalent cross-linking. For example, macromolecular binding sites from a linear polypeptide can hold positions by scaffolding sites [8] (10.8).

AQ2 **Table 2** Examples of Polymer uses for Scaffold Molecular Imprinting

Polymers having scaffolding sites	Template	Type of interaction	Ref.
Proteins Albumin		Hydrogen bonding	55,56
Polyacrylonitrile copolymers	Theophylline (THO)	Hydrogen bonding	63–65
—(CH₂-CH)—(CH₂-CH)— CN COOH	Naringin		69
—(CH₂-CH)—(CH₂-CH)— CN [pyridine]	Caffein		68
—(CH₂-CH)—(CH₂-C)— CN			
		Hydrogen bonding	70
Polysulfone	Dibenzofuran (DBF)	Charge transfer interaction	74
Nylon 6	Amino acid	Multi-hydrogen bonding networks	71,72

(Continued)

Table 2 (*Continued*)

Polymers having scaffolding sites	Template	Type of interaction	Ref.
Polyallylamine crosslinked	Glucose phosphate mono-sodium salt	Crosslinking with epichlorohydrin, ethyleneglycol, diglycidyl ester and grycerol diglycidyl ester	75
Polyallylamine crosslinked	Bile acid	Epichlorohydrin crosslinker was used	76
Hydroxyethylmethacrylate (HEMA) + polyurethane Cholesterol		Radical cross-link polymerization of EGDMA was performed for semi-interpenetrating membrane	77
Polyacrylic acid prepolymer bearing vinyl benzoate	Cinchonidine	Radical cross-link polymerization of EGDMA was performed as the purpose of chromathographic media	78

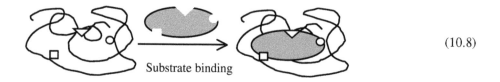

$$(10.8)$$

Substrate binding

In this case, weak cohesive inter-chain forces [3,53] and physical entanglement [54] maintain the structure of the imprinted binding sites. The imprinted structure in this second strategy is inherently less robust because it relies on a multitude of noncovalent interactions. However, this strategy also has the synthetic advantage that the matrix can be imprinted using covalent cross-linking as shown in Fig. 1(c).

A. Molecular Imprinting Using Proteins

A good example of using polymeric scaffold elements is imprinting proteins. Production of abiotic receptors was developed using bovine serum albumin (BSA) [55]. Kilvanov et al. utilized molecule aggregation of the protein during lyophilization in the presence of an L-malic acid template molecule (10.9).

$$(10.9)$$

L-malic acid   BSA   Complex of L-malic acid with BSA   Imprinted BSA

Process (a) involves a solution containing 0.5 M L-malic acid and 0.75 mM BSA; then, pH was adjusted to 2. Finally, the solution was lyophilized. Process (b) is a washing process by anhydrous tetrahydrofuran to remove the template from the BSA powder. The resultant solid powder exhibits selective binding with high capacity for the target molecule. However, the protein in this example actually lost its memory for the template molecule in water because the protein would refold into its natural state. The binding experiments could only be carried out in organic solvents with the strict exclusion of water. They also extended this method to dextran and other polymers and reached a similar conclusion [56]. On a mechanism of imprinting, the protein ligand forms hydrogen bonds with two or more residues, thereby folding a segment of polymeric chain around the target, which is frozen when lyophilized. Therefore, this template directed refolding leads to weakly stable recognition sites. Also, a similar procedure using albumin was reported on enzymatic hydrolysis of (4R, 4S)-4 hydroxy-4-(4-nitrophenyl) butan-2-one [57]. The imprint process using glutaraldehyde was used for cross-linking albumin for glutathione (GSH) derivatives. The imprinted protein with GSH sites shows chemical mutation resulting in synthetic enzymatic activity [58].

B. Polymer Aggregates as a Scaffold Imprinting Media

Small molecule aggregates such as micelles, bilayers, vesicles and biological membranes, form spontaneously by self-aggregation in aqueous solution [59,60]. As

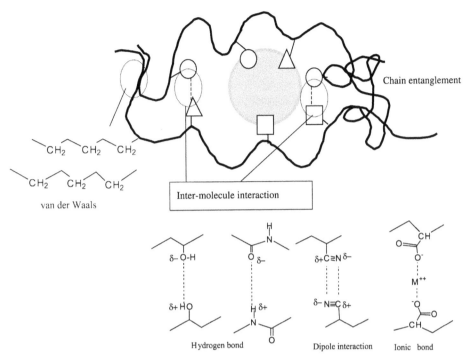

Figure 2   Schematic representation of the concept of interpolymer interactions for molecular imprinting.

shown in Fig. 2, these same weak noncovalent forces between polymer chains via hydrogen bonds, electrostatic bonds, and hydrophobic forces can be sufficient to enable stable polymer structure and therefore the preparation of an imprinted polymer matrix [53,54].

    Utilization of synthetic polymers greatly expands the potential of molecular imprinting because there is a great range of synthetic and many commercially available polymers. Most polymers are soluble in organic solvent and a polymer phase inversion process can be applicable to polymer assembly formation [54,61,62] when the solution is immersed into a nonsolvent, resulting in polymer aggregation. This phenomenon can be utilized for imprinting without covalent cross-linking [63,64]. In general, weak cohesive forces behave as noncovalent cross-links, which force polymer to aggregate [53,54,61]. Therefore, this kind of technique leads to an entirely different approach to imprinting from that using a cross-linker. In addition, molecular weight of the polymer can considerably influence chemical, physical, and mechanical properties of polymer membranes and films because the number of chain entanglements increases with increasing chain length [62]. This leads, in turn, to strong aggregation of polymeric chain entanglement. Therefore, if the polymer contains interaction sites to the template molecule, solidified polymers can memorize template shape in imprinted matrices.

    For instance, polyacrylonitrile (PAN) is capable of having strong dipole inter-action with nitrile groups on every other carbon atom along the chains. Using this strategy, phase inversion imprinting of PAN copolymers with carboxylic acid groups first appeared in 1995 for THO recognition [64]. Here, the highly cohesive acrylonitrile segments in the copolymer aggregate to form a solid polymer matrix that can hold the COOH segments in the proper geometry interacting with the target molecule (10.10).

$$\text{(10.10)}$$

To memorize volumetric space of THO and caffeine in the polymer, COOH segments [65–67] and pyridine segments [68] are necessarily incorporated into the polyacrylonitrile segments. Typically, for an acrylonitrile–acrylic acid copolymer (PAN-co-AA)–theophylline (THO) system, preparation procedure involves phase inversion process of polymer (Fig. 3); preparation of P(AN-co-AA) by phase inversion method was carried out by the following procedure [63,64]. The THO template was dissolved in dimethylsulfoxide (DMSO) in the range of 0–4.7 wt% concentration. At higher concentrations than 4.7 wt%, solubility of the template molecule in DMSO solution is not good. DMSO cast solution containing 10wt% P(AN-co-AA) copolymer and various weight percentages of the template were well mixed at 50°C for 20 h. The solutions were cast on a glass plate warmed at 50°C and coagulated in water at different temperatures [66]. Then, resultant polymers were rinsed with a

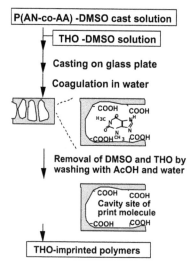

Figure 3 Schematic illustration of imprint process of THO in copolymer membrane P(AN-co-AA) by phase inversion method [63,64]. From Ref. 46.

large excess of water to remove DMSO and template. Binding capacity of THO imprinted P(AN-co-AA) depends on polymer–template composition and the conditions of phase inversion imprinting [65,67]. P(AN-co-AA) was also imprinted using a similar procedure for naringin recognition with 16.6 mol% AA segments [69]. In the extraction process of the template, ethanolic acetic acid was used without any loss of binding capacity.

Moreover, efficiency of phase inversion imprinting can be improved with pre-forming complex of monomer–template (Table 2) in copolymerization [70]. The THO–acrylic acid or methacrylic acid precomplex monomer was copolymerized with acrylonitrile in DMSO. The resultant viscous solution contents were used for phase inversion in water after template copolymerization. Template copolymers can improve binding capacity of THO. From ^{1}H-NMR analysis, this is due to tailor-made modification of a copolymer backbone for the template molecule. Also, comparison was made between copolymers of acrylic acid and methacrylic acid in THO selectivity of the imprinted polymers. Presence of the methacryl methyl group is more efficient in the tailor-made structure of the THO template.

In addition to proteins, synthetic polyamides can form hydrogen-bonding networks that can be used as imprinting media for noncovalent bonding [71,72]. In these works, hydrogen bond networks between nylon chains (10.11a) were used as imprinting media for chiral recognition of amino acids. Furthermore, polyethersulfone has the ability for donor–acceptor complex formation between the segment and template. The charge transfer interaction between template and polymeric segments (10.11b) is available for imprinting scaffold forces [73,74]. The resultant porous membrane, with dibenzofuran recognition, showed excellent membrane-adsorbent properties.

(a) Hydrogen bond network   (b) Charge transfer complex   (10.11)

C. Prepolymer-Crosslinker Systems as Scaffolding Imprint Matrix

Prepolymers containing multiple binding functionality have been utilized in creating scaffolding sites, which were covalently cross-linked in the presence of template via side chain-linking and semi-interpenetrating polymer networks. Imprinted networks composed of poly(allylamine hydrochloride) and glucosephosphate monosodium salt (GPS) are capable of specific binding of glucose [75].

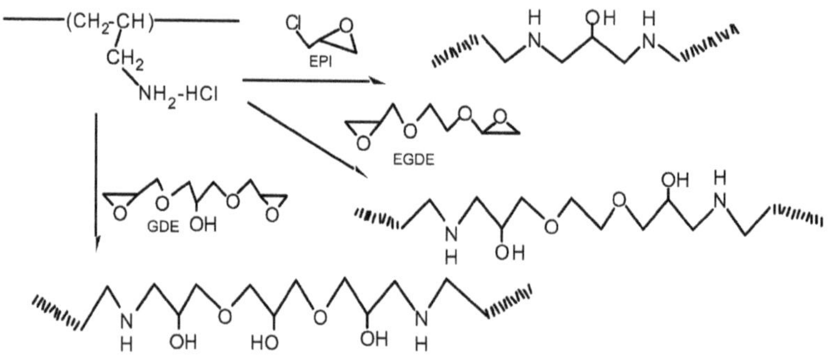

(10.12)

Batch equilibrium studies were performed to determine binding capacity of glucose and the results demonstrate isomeric specificity for glucose. In this work, epichlorohydrin (EPI), ethylene glycol diglycidyl ether (EGDE), and glycerol diglycidyl ether (GDE) were used as cross-linkers (Fig. 10.12). Ionic association of the template to positively charged cross-linked networks can encode the molecule shape into the hydrogel networks. The imprinted hydrogel networks have capacities of 0.58 g of dried gel for glucose and zero for fructose when epichlorohydrin was used at 0.5% GPS concentration. The imprinted hydrogels prepared from another two glycidyl ethers have a low selectivity for glucose over fructose. Cholesterol-lowering polymer drugs have also been obtained by molecular imprinting using polyallylamine [76]. The polymer is a polyelectrolyte and has good water solubility. Also, a high density of amino groups can be used for subsequent chemical modification. Sodium salt of cholic acid, which is a bile acid (Table 2), was used as template in the range of 0–11.5 mmol when amino group and EPI were 53.42 and 37.5 mmol. The maximum binding capacity was about 2 mmol of ligand/g of polymer for the target bile acid.

Semi-interpenetrating networks of EGDMA were formed with pre-polymers, which can hold template shape by scaffolding sites. Polyurethane and 2-hydroxyethyl methacrylate (HEMA) were mixed in the presence of EGDMA cross-linker for cholesterol imprinting [77]. The semi-interpenetrating network-imprinted polymer having membrane shape has a strong mechanical resistance with 35 MPa stress and 620% strain. Affinity of the imprinted polymer for cholesterol and testosterone was 5.5 and 0.67 mg/100 mg polymer, respectively. Prepolymers of methacrylic acid bearing 4-vinylbenzyl pendants were utilized for cinchonidine imprinting [78]. To induce imprint effect, EGDMA cross-linking was performed (10.13).

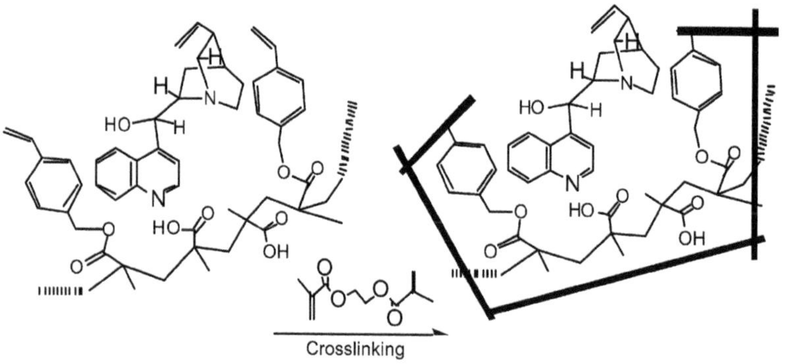

Crosslinking

(10.13)

After cross-linking polymerization, the grounded power was packed in chromatography columns using an acetonitrile–acetic acid solution; then, assessment of affinity was examined for cinchonidine target molecule. Preliminary results proved that the prepolymer method presents potential utilization as an artificial receptor.

V. APPLICATIONS AND FUTURE SCOPE OF SCAFFOLDING IMPRINTING

At this stage, a few kinds of multi-interaction sites for scaffold monomers have been used to encode target molecule shape into cross-linked imprinted polymers. This is in contrast to most reported monomers used for molecular imprinting such as methacrylic acid, which only contains a single binding functionality. Notwithstanding, such an imprinting strategy using scaffolding imprinting will be a very attractive field in the future because technical developments in the world require more sophisticated imprinting in several fields. One advantage of scaffolding imprinting is in application to responsive receptors [44,55,78]. Stronger recognition ability enables new uses of adsorbents [56,74], sensors [68], and chromatographic materials [34–36]. The syntheses of polymeric drugs are also available using this technique [76]. In addition, polymer assembling architectures are possible for scaffolding imprinting enabling formation of imprinted membranes and films [63–67,69,77]. Imprinted polymers containing cyclodextrins, porphyrins, fullerene, and molecule assembling aggregates that form specific and strong binding interactions are important scaffold imprinting elements. For polymer imprinting, scaffolding imprinting methods are promising because polymeric materials with imprinting activities are an important strategy for imprint architecture, which contains micro or nano size-ordered structure. Therefore, polymeric structural units will become more complicated through application of several polymerization procedures and processing. Concerning scaffold imprinting elements using supra-molecules and ordered-aggregate polymers, there will be a very wide diversity of techniques for the purpose of highly ordered homogeneous imprinted architectures.

VI. CONCLUSION

In imprinting processes involving multi-binding sites and assembling units, several functional groups play scaffold roles. Scaffolding elements can be classified into two categories of molecules and macromolecules. Molecules having multi-interaction sites are capable of imprinting when held together by a cross-linker. However, there are not many scaffolding monomers because of the difficulty in their synthesis. Larger supra-molecular functional groups such as cyclodextrins and fullerene were used to molecularly imprinted large guest molecules in cross-linking copolymers. Another strategy for scaffolding imprinting is to use polymers. In these cases, covalently bound scaffold elements on polymer chains can be organized to behave cooperatively to form high affinity sites. However, systematic comparison between polymer imprinting and conventional imprinting has not been done yet. Also, there is an advantage in polymer imprinting as they can be synthesized without cross-linking for formation of film and membranes. This is in contrast to most imprinting materials, which are hard monoliths that can only be grounded for use in chromatography.

In conclusion, imprinting a target molecule into the polymeric matrix will entail a variety of processes, each of which contains several scaffold ideas. Diversity, therefore, increasingly lends itself to future appearance of novel imprinted materials.

REFERENCES

1. Hobza, P.; Zahradnik, R. In *Intermolecular Complexes: The Role of van der Walls Systems in Physical Chemistry and in the Biodisciplines*; Elsevier: Amsterdam, 1988, 185–254.
2. Mariuzza, R.A.; Phillips, S.E.V.; Poljak, R.J. The structural basis of antigen–antibody recognition. Ann. Rev. Biophys. Biophys. Chem. **1987**, *16*, 139–159.
3. Stite, W.E. Protein–protein interactions: interface structure, binding thermodynamics, and mutational analysis. Chem. Rev. **1997**, *97*, 1233–1250.
4. Pauling, L. A theory of the structure and process of formation of antibodies. J. Am. Chem. Sci. **1940**, *62*, 2643–2657.
5. Bartsch, R.A.; Maeda, M. In *Molecular and Ionic Recognition with Imprinted Polymers*. Oxford University Press, Washington DC, 1998, 1–8.
6. Wulff, G. Molecular imprinting in cross-linked materials with aid of molecular templates. A way toward artificial antibodies. Angew. Chem. Int. Ed. Engl. **1995**, *34*, 1812–1832.
7. Mosbach, K. Molecular imprinting. Trends Biochem. Sci. **1994**, *19*, 9–15.
8. Shea, K. Molecular imprinting of synthetic network polymers. The De Novo synthesis of macromolecular binding and catalytic sites. Trend Polym. Sci. **1994**, *2*, 166–173.
9. Vulfson, E.; Alexander, C.; Whitocome, M. Assembling the molecule cast. Chem. Britain **1997**: January issue, 23–26.
10. Steinke, J.; Sherrington, D.C.; Dunkin, I.R. Imprinting of synthetic polymers using molecular template. Advan. Polym. Sci. **1995**, *123*, 81–125.
11. Wulff, G. Enzyme like catalysis by moleciularly imprinted polymers. Chem. Rev. **2002**, *102*, 1–28.
12. Klein, J.U.; Whitcombe, M.J.; Mulholland, F.; Vulfson, E.N. Template-mediated synthesis of a polymeric receptor specific to amino acid sequence. Angew. Chem. Int. Ed. **1999**, *38*, 2057–2060.
13. Lubke, M.; Witocombe, M.J.; Vulson, E.N. A novel approach to the molecular imprinting of polychlorinated aromatic compounds. J. Am. Chem. Soc. **1998**, *120*, 13342–13348.
14. Wulff, G.; Schauhoff, S. Racemic resolution of free sugars with macroporous polymers prepared by molecular imprinting. Selectivity dependence on the arrangement of functional group versus spatial requirements. J. Org. Chem. **1991**, *56*, 395–400.
15. Shea, K.J.; Thompson, E.A. Template synthesis of macromolecules. Selective functionalization of an organic polymer. J. Org. Chem. **1978**, *43*, 4253–4255.
16. Whitcombe, M.J.; Rodriguez, M.E.; Villar, P.; Vilson, E.N. A new method for the introduction of recognition site functionality into polymers prepared by molecular imprinting: Synthesis and characterization of polymeric receptors for cholesterol. J. Am. Chem. Soc. **1995**, *117*, 7105–7111.
17. Wulff, G.; Sarhan, A.; Zabrocki, K. Enzyme-analogus built polymers and their use for the resolution of racemates. Tetrahedron Lett. **1973**, *44*, 4329–4332.
18. Wulff, G.; Sarhan, A. The use of polymer with enzyme-analogous structures for the resolution of racemates. Angew. Chem. Int. Ed. **1972**, *11*, 364.
19. Wulff, G.; Heide, B.; Helfmeir, G. Molecular recognition through the exact placement of functional groups on rigid matrices via a template approach. J. Am. Chem. Soc. **1986**, *108*, 1089–1091.
20. Shea, K.J.; Doughery, T.K. Molecular recognition on synthetic amorphous surfaces. The influence of functional group positioning on the effectiveness of molecular recognition. J. Am. Chem. Soc. **1986**, *108*, 1091–1093.

21. Shea, K.J.; Sasaki, D.Y. On the control of microenvironment shape of functionalized network polymers prepared by template polymerization. J. Am. Chem. Soc. **1989**, *111*, 3442–3444.

22. Kempe, M.; Mosbach, K. Molecular imprinting used for chiral separations. J. Chromatogr. A. **1995**, *694*, 3–13.

23. Andersson, L.I.; Ekberg, B.; Mosbach, K. Bioseparation and catalysis in molecularly imprinted polymers. In *Moleculcular Interactions in Bio separations*; Ngo, T., Ed.; Plenum Press, New York, 1993, 383–394.

24. Andersson, L.; Sellergen, B.; Mosbach, K. Imprinting of amino acid derivatives in macroporous polymers. Tetrahedron Lett. **1984**, *25*, 5211–5214.

25. Sellergen, B.; Ekberg, B.; Mosbach, K. Molecular imprinting of amino acid derivatives in macroporous polymers. Demonstration of substrate- and enantio-selectivity by chromatographic resolution of racemic mixtures of amino acid derivatives. J. Chromatogr. **1985**, *347*, 1–10.

26. Kemp, M.; Mosbach, K. Separation of amino acids, peptides and proteins on molecularly imprinted stationary phases. J. Chromatogr. A **1995**, *691*, 317–327.

27. Fischer, L.; Muller, R.; Ekberg, B.; Mosbach, K. Direct enantioseparation of β-adrenergic blockers using a chiral stationary phase prepared by molecular imprinting. J. Am. Chem. Soc. **1991**, *113*, 9358–9360.

28. Vlatakis, G.; Andersson, L.I.; Muller, R.; Mosbach, K. Drug assay using antibody mimics made by molecular imprinting. Nature **1993**, *361*, 645–647.

29. Krotz, J.M.; Shea, K.J. Imprinted polymer membranes for the selective transport of targeted neutral molecules. J. Am. Chem. Soc. **1996**, *118*, 8754–8755.

30. Andersson, L.I. Application of molecualr imprinting to the development of aqueous buffer and organic solvent based radioligand binding assays for (S)-propranolol. Anal. Chem. **1996**, *68*, 111–117.

31. Kempe, M. Antibody-mimicking polymers as chiral stationary phases in HPLC. Anal. Chem. **1996**, *68*, 1948–1953.

32. Jeffrey, G.A.; Saenger, W. In *Hydrogen Bonding in Biological Structure*; Springer-verlag: Berlin, 1991.

33. Guerra, C.F.; Bickelhaupt, F.M.; Snijders, J.G.; Baerends, E.J. Hydrogen bonding in DNA base pairs. Reconciliation of theory and experiment. J. Am. Chem. Soc. **2000**, *122*, 4117–4128.

34. Takeuchi, T.; Matui, J. Recognition of drug and herbicides. Strategy in selection of functional monomers for noncovalent molecular imprinting. In *Recognition with Imprinted Polymers*, ACS Symposium Series, 703; Bartsch, R.A., Maeda, M. Eds.; Oxford University Press: Washington DC, 1998; 119–134.

35. Tanabe, K.; Takeuchi, T.; Matui, J.; Ikebukuri, K.; Yano, K.; Karube, I. Recognition of barbiturates in molecularly imprinted copolymers using multiple hydrogen bonding. J. Chem. Soc. Chem. Commun. **1995**, 2303–2304.

36. Kugimiya, A.; Mukawa, T.; Takeuchi, T. Synthesis of 5-fluorouracil-imprinted polymers with multiple hydrogen bonding interactions. Analyst **2001**, *126*, 772–774.

37. Oikawa, E.; Motomi, K.; Aoki, T. Synthesis and properties of poly(thioether amide)s from 2,6-bis(arylamide9pyridine and dithols. J. Polym. Sci. Part A Polym. Chem. **1993**, *31*, 457–465.

38. Fish, R.H. Metal ion templated polymers, Studies of *N*-(4-vinylbenzyl)-1,4,7-triazacyclononane-metal ion complexes and their polymerization with divinylbenzene: the importance of thermodynamic and imprinting parameters in metal ion selectivity studies of the demetalated, templated polymers. In *Recognition with Imprinted Polymers*, *ACS Symposium Series 703*; Bartsch, R.A., Maeda, M., Eds.; Washington, DC, 1998; 238–250.

39. Vigneau, O.; Pinel, C.; Marc, L. Ionic imprinted resins based on EDTA and DTPA derivatives for lanthanides (III) separation. Anal. Chim. Act. **2001**, *435*, 75–82.

40. Singh, A.; Puranik, D.; Guo, Y.; Chang, E.L. Towards achieving selectivity in metal ion binding by fixing ligand-chelator complex geometry in polymers. React. Func. Polym. **2000**, *44*, 79–89.

41. Asanuma, H.; Kakazu, M.; Shibata, M.; Hishiya, T.; Komiyama, M. Synthesis of molecular imprinted polymer of β-cyclodextrin for the efficient recognition of cholesterol. Super Molecule Sci. **1998**, *5*, 417–421.

42. Sreenivasan, K. Synthesis and evaluation of a beta cyclodextrin-based molecularly imprinted copolymer. J. Appl. Polym. Sci. **1998**, *70*, 15–18.

43. Steed, J.W.; Atwood, J.L. In *Supramolecular Chemistry*. John Wiely & Sons Ltd; Chichester: 2000, 321–324.

44. Hishiya, T.; Shibata, M.; Kakazu, M.; Asanuma, H.; Komiyama, M. Molecularly imprinted cyclodextrins as selective receptors for steroids. Macromolecules **1999**, *32*, 2265–2269.

45. Asanuma, H.; Kajiya, K.; Hishiya, H.; Komiyama, M. Molecular imprinting of cyclodextrin in water for the recognition of peptides. Chem. Lett. **1999**, 665–667.

46. Asanuma, H.; Akiyama, T.; Kajiya, K.; Hishiya, T.; Komiyama, M. Molecular imprinting of cycrodextrin in water for the recognition of nanometer-scaled guests. Anal. Chim. Acta. **2001**, *435*, 25–33.

47. Hafkamp, R.J.H.; Kokke, B.P.A.; Danke, I.M.; Geurts, H.P.M.; Rowan, A.E.; Feiters, C.; Nolte, R.J.M. Organogel formation and molecular imprinting by functionalized gluconamides and their metal complex. Chem. Commun. **1997**, 545–546.

48. See text of Ref. 43, 37–39.

49. Smith, K.M. In *Porphyrins and Metalloporphyrins*. Elsevier Scientific Pub. Com: New York, 1975.

50. Lauceri, R.; Raudino, A.; Scolaro, L.M.; Micali, N.; Purrello, R. From achiral porphyrins to template-imprinted chiral aggregates and further, self-replication of chiral memory from scratch. J. Am. Chem. Soc. **2002**, *124*, 894–895.

51. Karz, A.; Davis, M.E. Molecular imprinting of bulk, microporous silica. Nature **2000**, *403*, 286–287.

52. Ishii, T.; Nakashima, K.; Shinkai, S. Regioselective introduction of two boronic acid groups in to [60]fullerene using sacchrides as imprinting template. Chem. Commun. **1998**, 1047–1048.

53. Rosen, S.L. In *Fundamental Principles of Polymer Materials*, 2nd Ed.; John Wiley & Sons Inc: New York, 28–31.

54. Mulder, M. In *Basic Principles of Membrane Technology*, 3rd Ed.; Kluwer Academic Publishers: Dordrecht, 1996, 29–30.

55. Braco, L.; Dabulis, K.; Klibanov, A.M. Production of abiotic receptors by molecular imprinting of proteins. Proc. Natl. Acad. Sci. USA **1990**, *87*, 274–277.

56. Dabulis, K.; Klibanov, A.M. Molecular imprinting of proteins and other macromolecules resulting in new adsorbents. Biotech. Bioeng. **1992**, *39*, 176–185.

57. Ohya, Y.; Miyaoka, J.; Ouchi, T. Recruitment of enzyme activity in alubumin by molecular imprinting. Macromol. Rapid. Commun. **1996**, *17*, 871–874.

58. Liu, J.; Luo, G.; Gao, S.; Zhang, K.; Chen, X.; Shen, J. Generation of glutathione peroxidase like mimic using bioimprinting and chemical mutation. Chem. Commun. **1999**, 199–200.

59. Israelachvili, J.N. In *Intermoleculer and Surface Forces*. Academic Press: San Diego, 1985, 229–286.

60. Conn, M.M.; Rebek Jr, J. Self-assembling capsules. Chem. Rev. **1997**, *97*, 1647–1668.

61. Richards, E.G. In *An Introduction to the Physical Properties of Large Molecules in Solution*. Cambridge University Press: New York, 1980, 15–35.

62. Geddle, U.L.F.W. In *Polymer Physics*. Chapman & Hall: London, 1995, 48–49.

63. Kobayashi, T.; Wang, H.Y.; Fukaya, T.; Fujii, N. Molecular imprint membranes prepared by phase pnversion of polyacrylonitrile copolymers with carboxylic acid groups. In *Recognition with Imprinted Polymers*, ACS Symposium Series, 703; Bartsch, R.A., Maeda, M., Eds.; Oxford University Press: Washington DC, 1998; 188–208.

64. Kobayashi, T.; Wang, H.Y.; Fujii, N. Molecular imprinting of theophylline in acrylonitrile-acrylic acid copolymer membrane. Chem. Lett **1995**, 927–928.

65. Wang, H.Y.; Kobayashi, T.; Fujii, N. Molecular imprint membranes prepared by phase inversion technique. Langmuir **1996**, *12*, 4850–4865.

66. Wang, H.Y.; Kobayashi, T.; Fujii, N. Molecular imprint membranes prepared by phase inversion technique (II). Influence of coagulation temperature in phase inversion process on the encoding in polymeric membranes. Langmuir **1997**, *13*, 5390–5400.

67. Kobayashi, T.; Wang, H.Y.; Fujii, N. Molecular imprint membranes of polyacrylonirile copolymers with different acrylic acid segments. Anal. Chim. Acta **1998**, *365*, 81–88.

68. Kobayashi, T.; Murawaki, Y.; Reddy, P.S.; Abe, M.; Fujii, N. Molecular imprinting of caffeine and its recognition assay by quartz crystal microbalance. Anal. Chim. Acta **2001**, *435*, 141–149.

69. Trotta, F.; Driori, E.; Baggiani, C.; Lacopo, D. Molecular imprinted polymeric membrane for baringin recognition. J. Membrane Sci. **2002**, *201*, 77–84.

70. Kobayashi, T.; Fukaya, T.; Abe, M.; Fujii, N. Phase inversion molecular imprinting by using template copolymers for high substrate recognition. Langmuir **2002**, *18*, 2866–2872.

71. Reddy, P.S.; Kobayashi, T.; Fujii, N. Molecular imprinting in hydrogen bonding networks of polyamide nylon for recognition of amino acids. Chem. Lett. **1999**, 293–294.

72. Reddy, P.S.; Kobayashi, T.; Abe, M.; Fujii, N. Molecularly imprinted nylon-6 as a recognition materials of amino acids. Eur. Polym. J. **2002**, *38*, 521–526.

73. Reddy, P.S.; Kobayashi, T.; Fujii, N. Recognition characteristics of dibenzofuran by molecularly imprinted polymers made of common polymers. Eur. Polym. J. **2002**, *38*, 779–785.

74. Kobayashi, T.; Reddy, P.S.; Abe, M.; Ohta, M.; Fujii, N. Molecularly imprinted polysulfone membranes with acceptor sites for donor dibenzofuran as novel membrane adsorbents. Charge transfer interaction for recognition origin. Chem. Matter. **2002**, *18*, 2866–2872.

75. Wizeman, W.J.; Kofina, P. Molecularly imprinted polymer hydrogels displaying isomerically resolved glucose binding. Biomaterials. **2001**, *22*, 1485–1491.

76. Huval, C.C.; Bailey, M.J.; Braunlin, W.H.; Holmer-Farley, S.R.; Mandeville, W.H.; Petersen, J.S.; Polomoscanik, S.C.; Sacchiro, R.J.; Cgen, X.; Dhal, P.K. Novel cholesterol lowcring polymeric drugs obtained by molecular imprinting. Macromolecules. **2001**, *34*, 1548–1550.

77. Sreenivasan, K. Synthesis and evaluation of a molecularly imprinted polyurethane-poly (HEMA) semi-interpenetrating polymer networks as membrane. J. Appl. Polym. Sci. **1998**, *70*, 19–22.

78. Matsui, J.; Tamaki, K.; Sugimoto, N. Molecular imprinting in alcohols: utility of a pre-polymer based strategy for synthesizing stereoselective artificial receptor polymers in hydrophilic media. Anal. Chim. Acta **2002**, *466*, 11–15.

11

Imprinting in Inorganic Matrices

Jessica L. Defreese and Alexander Katz   University of California at
Berkeley, Berkeley, California, U.S.A.

I. INTRODUCTION

Inorganic oxides as imprinting matrices have received increasing attention [1] in recent years due in large part to their extreme structural rigidity, which is crucial for retention of imprinted information because it provides the scaffold that mechanically supports the functional group organization. As shown in Fig. 1, the distance between cross-linking points in a silica monomer is much shorter than the distance between cross-linking points in a commonly used polymer monomer, ethylene glycol dimethacrylate (EGDMA). This allows silica to achieve a greater cross-linking density, resulting in a Young's modulus up to 200 times larger than that of cross-linked EGDMA [2,3].

Inorganic oxides are also useful due to other material properties. When prepared by a sol–gel synthesis process, characteristics such as average pore-size, surface area, and matrix polarity can be controlled by varying the processing conditions. Sol–gel chemistry uses acid- and base-catalyzed methodologies to hydrolyze and condense alkoxides to form hybrid organic–inorganic structures [4]. Thin films, monoliths, powders, and fibers can all be prepared with relative ease [5]. A short summary of sol–gel chemistry is provided by Dai in Chapter 13 of this text. The resulting amorphous glasses are thermally stable, optically transparent, and chemically inert to most dry organic solvents and reagents. These characteristics are amenable to a number of applications, and imprinted inorganic oxides have been explored as heterogeneous catalysts, chemical sensors, selective adsorbents, separation media, and scaffolds for solid-state electronic devices.

One of the first ideas on how to synthesize imprinted materials can be attributed to Linus Pauling and his theories on antibody formation. He speculated that antigens functioned as molecular imprints or templates, causing organization of antibody precursors and leading to the construction of specific antibodies [6]. Although this idea is now known to be incorrect, the concept of molecular imprinting was quickly utilized in other systems. The first experimental examples of noncovalent inorganic oxide imprinting were pioneered in silica systems as early as 1933. Polyakov et al. [7] dried silica gels under the vapors of benzene, toluene, xylene, and naphthalene, and discovered that an imprinted gel showed increased adsorptive activity towards the specific vapor in which it was dried. Several years later, Dickey [8] prepared selective

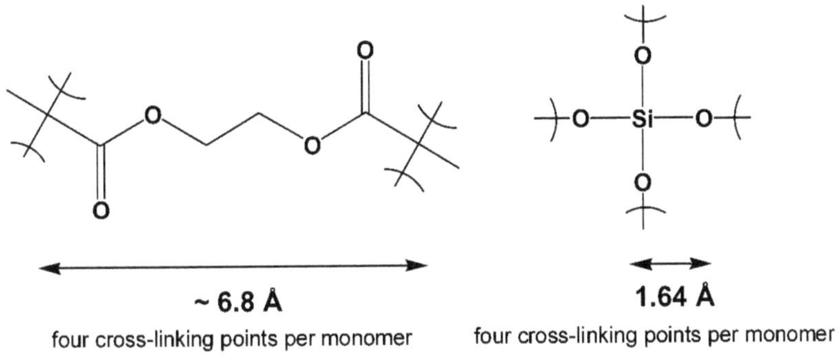

~ 6.8 Å   1.64 Å

four cross-linking points per monomer   four cross-linking points per monomer

Figure 1 The high cross-linking density of silica gives it much greater mechanical stability than cross-linked organic polymers.

adsorbents by introducing homologues of methyl orange dye into sodium silicate solutions prior to gelation of the silica. Removal of the dye homologues by methanol extraction yielded silica gels capable of specific adsorption of the corresponding dyes. Dickey proposed a "footprint" mechanism for the adsorption, where the presence of the dye molecules presumably generated selective micropores by non-covalent interaction with the silanols and silica during gelation [9]. Haldeman and Emmett [10] corroborated the "footprint" theory with nitrogen pore-filling experiments. However, the work of Hodgson and co-workers suggested that the previous adsorption results could be explained by an association mechanism, in which residual dye molecules not removed during washing could serve as nucleation sites, selectively re-adsorbing like dye molecules in a process akin to crystallization [11].

These early mechanistic arguments indicate the controversy associated with lack of site isolation in molecular imprinting with inorganic oxides. In these cases, it was argued that the weak, noncovalent interactions between the imprint and material framework were incapable of producing isolated binding sites because there was no impetus for imprint molecules to favor interaction with framework precursors over those with additional imprint molecules. The energy of the noncovalent interaction between the framework and imprint must be comparable to the heat of vaporization of the imprint in order to avoid phase separation or the need for excessively dilute synthesis conditions. If these energies are not similar, large binding-site hetero-geneity can arise due to sites generated from clusters of imprint molecules, rather than from individual isolated species [12]. As a result, few efficient "footprint" sites may actually exist. Dickey [9] noted that his gels contained a variety of sites with differing affinities for dye readsorption, with the most strongly binding sites being the least numerous type. As shown in Fig. 2, the Dickey data of methyl orange adsorption onto a silica gel templated with methyl orange follows a Freundlich isotherm. The Freundlich isotherm assumes a linear dependence of adsorption enthalpy on surface coverage, consistent with binding sites that are not isolated from each other. A material with truly isolated sites would be expected to follow a Langmuir isotherm.

Following the promising work of Dickey and others [13–20] throughout the 1950s and 1960s, it was not until the mid-1990s that inorganic oxide imprinting was explored again by implementing advances in sol–gel technology for the synthesis of organic–inorganic hybrid materials [21]. Generally employing 1–2% of imprint

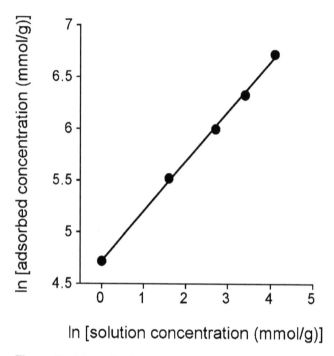

Figure 2 Adsorption isotherm of methyl orange dye on a silica gel templated with methyl orange [9]. The linear least-square regression to the data (•) represents a Freundlich isotherm.

relative to inorganic precursors, materials capable of rebinding a variety of analytes in ranges from 20% to 80% over nonimprinted control materials have been prepared. The noncovalent imprinting technique has since shown success in the production of chemical sensors and selective adsorbents, as described below.

II. NONCOVALENT IMPRINTING

Pinel et al. [22] utilized standard sol–gel techniques to explore the hydrolysis and condensation of tetraethyl orthosilicate (TEOS) to produce amorphous solids imprinted with (−)-menthol and o-cresol. The templates were removed by extraction with methanol, and adsorption experiments were carried out by shaking suspensions of crushed gels with cyclohexane solutions of the appropriate analytes. As measured by GC, they were able to obtain >20% rebinding of menthol to (−)-menthol-imprinted gels. They also achieved significant regiorecognition for o-cresol over p-cresol using an o-cresol-imprinted gel. Although they did not achieve the enantio-recognition they originally anticipated in the (−)-menthol system, their results were encouraging for the ability to generate selective adsorbents from molecularly imprinted inorganic oxides.

Makote and Collinson [23] prepared imprinted silicate films from a composite sol, adding 10 mol% phenyltrimethoxysilane (PTMOS) and 10 mol% methyltri-methoxysilane (MTMOS) to a tetramethoxysilane (TMOS) sol in ethoxyethanol. This composite created more stable films than could be obtained by TMOS alone. After addition of 2 mol% dopamine, the sol was cast onto the surface of a glassy carbon electrode and dried to form a thin uniform film. The dopamine template was leached from the film by soaking the electrode in phosphate buffer. As characterized by cyclic voltammetry, the templated film showed 67% response to the dopamine analyte, compared to zero response from a nontemplated film.

Using a similar TMOS/PTMOS/MTMOS system, Marx and Liron [24] templated thin silica films with propranolol. The films were spin-coated onto glass plates and the imprint removed via Soxhlet extraction with acidic methanol. Compared to a propranolol-imprinted acrylic polymer system, the sol–gel films showed lower capacity but better specificity and reproducibility. Additionally, they used radiolabeling experiments to demonstrate enantioselectivity in which the (S)-propranolol-imprinted film was able to selectively bind 3H-(S)-propranolol over nonradioactive (R)-propranolol, even when the (R) enantiomer was present in approximately 4000-fold excess.

Lam and coworkers [25] have extended noncovalent inorganic oxide imprinting into the chemical sensor field by combining their imprinted silica gels with photo-induced electron transfer (PET) technology. A sol was prepared from TEOS and PTMOS with 0.2 mol% of 9-(chloromethyl)anthracene included as the PET sensor monomer. The sol was subsequently templated with 2,4-dichlorophenoxyacetic acid (2,4-D). Following template removal via Soxhlet extraction, the material showed significant affinity and selectivity for the target analyte, as assessed by spectrofluorometric titrations in aqueous media. Additional work is necessary to improve the fluorescent responses of the sensor, but the study demonstrated the feasibility of incorporating PET technology into imprinted inorganic oxides.

Kunitake and coworkers have explored molecularly imprinted ultrathin titanium dioxide films. Employing a procedure of sequential chemisorption and activation onto quartz crystal microbalance (QCM) resonators, they have been able to produce stable films selective for azobenzene carboxylic acids [26] and protected amino acids [27]. In collaboration with Willner and co-workers, they have also used their synthesis techniques to functionalize ion-sensitive field-effect transistors (ISFET) with imprinted TiO_2 films [28]. The resulting sensors displayed impressive selectivity in distinguishing two chloroaromatic acids. Additionally, Willner and coworkers [29] have used similar ISFET devices to generate enantioselective and enantiospecific sensors for 2-methylferrocene carboxylic acid, 2-phenylbutanoic acid, and 2-propanoic acid.

III. POTENTIAL FOR SELECTIVE CATALYSIS VIA COVALENT IMPRINTING

Given the success of inorganic oxide imprinting for generating selective adsorbents and chemical sensors, it should also be possible to design heterogeneous catalysts via similar procedures. Rather than simply creating a "footprint" for the rebinding of a particular analyte, if one could generate an imprint that corresponds to a transition state analogue, it should be possible to design materials to catalyze particular chemical transformations. Success in the catalytic antibody field demonstrates the viability of using an imprinting type of approach for the synthesis of novel catalysts. Selective catalysis requires homogeneous, isolated sites since only a few nonselective but highly active sites can destroy selectivity due to amplifications in the turnover frequency. In order to prepare single-site materials for catalysis, a covalent imprinting method must be employed. Because the covalent interactions between imprint and framework are generally stronger than the noncovalent interactions between clustered imprint molecules, isolated sites can be generated.

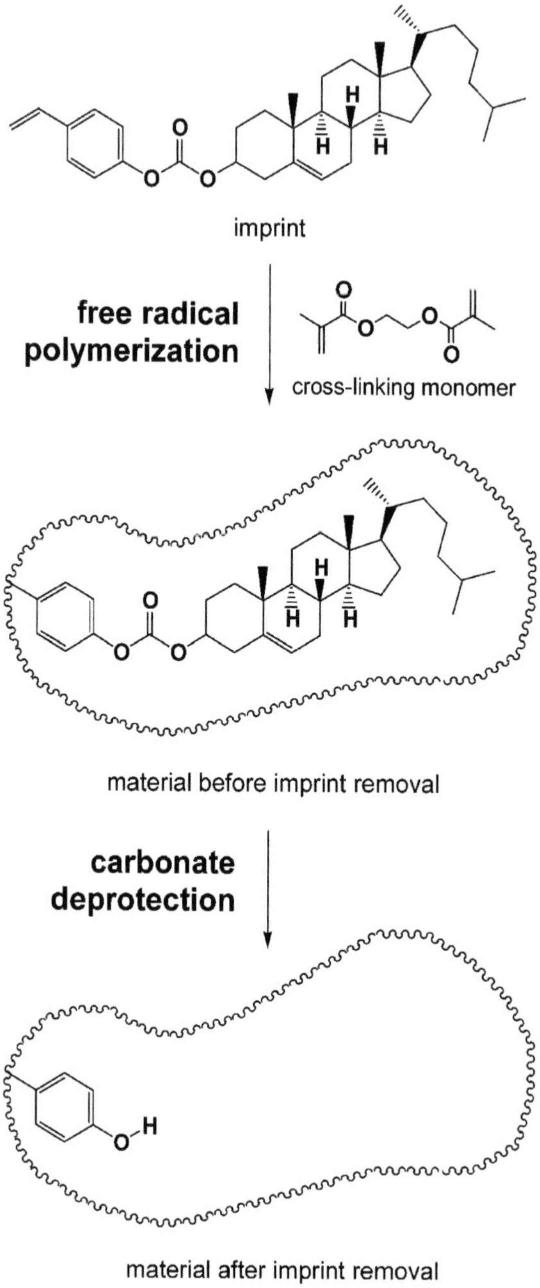

imprint

free radical polymerization

cross-linking monomer

material before imprint removal

carbonate deprotection

material after imprint removal

Figure 3 Covalent imprinting of cholesterol into an organic polymer (adapted from Ref. 34).

Heilmann and Maier attempted a covalent system in inorganic imprinting. They designed a phosphonate ester as a transition state analogue for the *trans*-esterification of ethyl phenylacetate with *n*-hexanol to form hexyl phenylacetate [30]. The phosphonate ester was covalently imprinted into silica gel and then removed by calcination. The remaining silica solid appeared to show significant catalysis and substrate selectivity. However, the system turned out to be more complicated than anticipated, and the results were difficult to rationalize on the basis of imprinting effects alone [31–33]. Shortly thereafter, the first successful creation of isolated sites was achieved in an organic polymer system. As illustrated in Fig. 3, Whitcombe et al. [34] used a covalent imprinting scheme to generate a material containing selective sites for the rebinding of cholesterol from hexane solutions. The presence of isolated sites was confirmed by a Langmuirian adsorption isotherm, as demonstrated by a linear Scatchard plot.

One of the key aspects of covalent imprinting in inorganic matrices is the ability to remove a portion of the imprint molecule without changing the connectivity of the material framework. The mild washing methods used in noncovalent syntheses are insufficient to break covalent bonds holding an imprint within an inorganic matrix. A breakthrough came in the late 1990s, when Corriu and coworkers [35] demonstrated that they could generate porosity in hybrid solids by using selective chemical deprotection to remove the organic moieties from their structures. As illustrated in Fig. 4, use of the Si–C≡C group allows cleavage of the Si–C_{sp} bond under mild conditions by fluoride-catalyzed hydrolysis. However, since the acetylene moieties were used as bulk porogens, no attempt was made to leave organic chemical functionalities organized in the resulting pores.

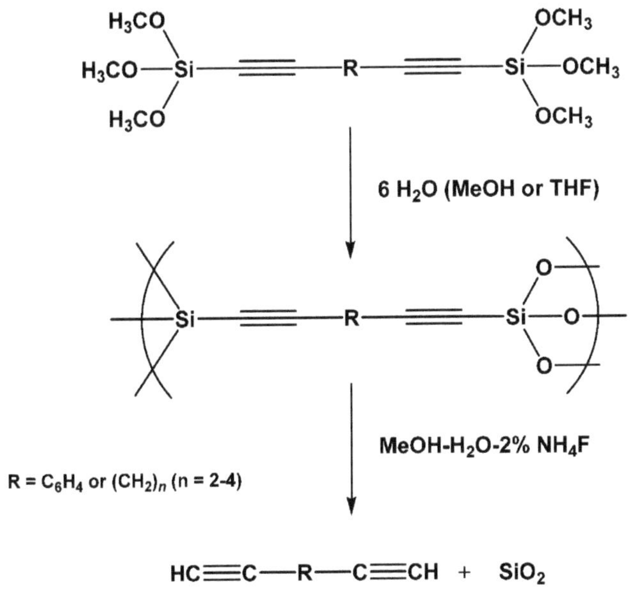

Figure 4 Generation of porosity in silica via chemical removal of organic fragments (adapted from Ref. 35).

If covalent inorganic oxide imprinting is to be used for catalytic applications, systems must be developed by which one can precisely position organic chemical functionalities within designed, isolated pores. Two systems based on covalent imprinting that attempt to accomplish this will be discussed in further detail below.

Recent further progress in the covalent imprinting of bulk silica has introduced a thermolytic method for imprint deprotection [36,37]. This provides a facile procedure for the imprinting of multiple chemical functional groups within a hydrophilic silica framework.

IV. EXAMPLE 1—COVALENT IMPRINTING OF PRIMARY AMINE SITES INTO BULK SILICA

Covalent imprinting of bulk silica was demonstrated by Katz and Davis [38] in 2000. Control of local functional group organization was accomplished by organizing primary amines within pores of controlled size and shape. A novel carbamate protection strategy allowed one, two, or three primary amines to be anchored within a micropore resulting from carbamate deprotection. The porosity generated by imprint core removal was quantified with argon physisorption. The resulting material displayed selective catalysis in the Knoevenagel condensation of isophthalaldehyde and malononitrile.

Figure 5 shows the acid-catalyzed sol–gel hydrolysis and condensation of the imprint with a large excess of the silica source, tetraethylorthosilicate. This creates a hybrid organic–inorganic material with the imprint covalently immobilized within the silica framework. The acidic synthesis conditions were chosen to ensure comparable hydrolysis rates between the organosilane imprint and the TEOS. Additionally, a relatively small ratio of imprint to TEOS was used to further avoid phase separation, which would cause the imprint to cluster and lead to loss of site isolation in material 2. A clear gel with polymer-like microstructure was obtained, in contrast to base-catalyzed sol–gels which are known to be collodial and cloudy [39,40].

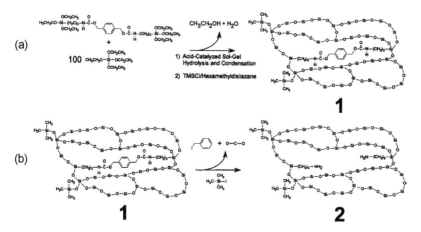

Figure 5 Preparation procedures used to create imprinted silica comprising (a) sol–gel hydrolysis and condensation catalyzed by HCl and (b) carbamate deprotection with trimethylsilyliodide in acetonitrile.

The silica monoliths were ground to a fine powder and subsequently contacted with an equimolar mixture of chlorotrimethylsilane and hexamethyldisilazane in order to convert the framework silanols to trimethylsilyl (TMS) groups. This renders the material hydrophobic and avoids generating large amounts of hydroiodic acid during the deprotection due to trimethylsilyliodide (TMSI) reacting with the silanol groups. The objective of the deprotection process is carbamate bond cleavage to remove the aromatic imprint core, thereby synthesizing a pair of imprinted primary amines separated by a distance slightly larger than 11 Å in **2**. TMSI cleaves benzyl carbamates to amines via a silyl carbamate intermediate, as shown schematically in Fig. 6 [41]. The silyl carbamate protects the latent amine functionality from further side reactions, such as alkylation with benzyl halide, until hydrolysis and spontaneous decomposition of carbamic acid (Fig. 6b–d). Although TMSI is known to readily deprotect carbamates at room temperature in less than 5 min [41], the conditions for carbamate deprotection in imprinted silica are considerably more severe.

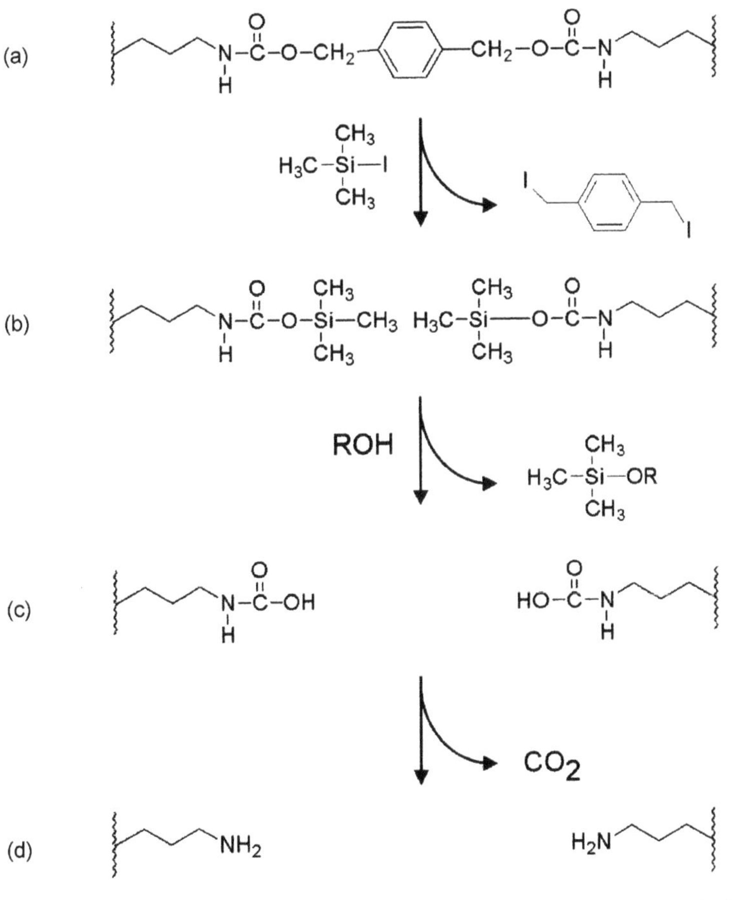

Figure 6   Illustration of carbamate deprotection of **1** with trimethylsilyliodide. The first step is a transesterification reaction to convert the benzylcarbamate in (a) to the trimethylsilyl-carbamate shown in (b). Treatment with a suitable OH such as water yields the free carbamic acid in (c), which spontaneously loses carbon dioxide to provide (d) an imprinted site comprising a pair of free amines in **2**.

One gram of TMS-capped silica was treated with 15 mL of 0.25 M TMSI in dry acetonitrile at 70°C for a period of at least 12 h. Temperatures of less than 40°C (which were sufficient to deprotect systems based on one benzyl carbamate per imprint molecule) did not show appreciable deprotection of material 1. Given the treatment necessary to achieve carbamate deprotection, it is important to emphasize that the TMSI procedure does not change the connectivity of the silica framework, as ascertained by studies on model crystalline materials such as a high-silica zeolite faujasite [42]. Following TMSI treatment, the silica was filtered and washed with acetonitrile, methanol, saturated aqueous sodium bicarbonate, methanol, and acetonitrile. The purpose of the aqueous treatment was hydrolysis of the silyl carbamate intermediate as shown in Fig. 6b–c.

The deprotection process illustrated in Figs. 5 and 6 was followed by solid-state ^{13}C CP/MAS NMR spectroscopy. As illustrated in Fig. 7, the carbonyl (158.2 ppm), aromatic (128.2 and 137.4 ppm), and benzylic (67.4 ppm) resonances decreased due to cleavage of those functional groups by the carbamate deprotection. However, the propyl tether resonances at 10.5, 23.1, and 43.2 ppm remained intact (Fig. 7b). Incomplete hydrolysis left a small amount of residual ethoxy moieties at 54.1 and 17.5 ppm. The degree of deprotection achieved in the material resulted in an amine number density of approximately 0.25 mmol/g, as ascertained by nonaqueous titration of the amines with benzoic acid [42].

The imprinting process (Fig. 5) was also characterized using argon pore-filling experiments at cryogenic temperature. The adsorption isotherms shown in Fig. 8a displayed an increase in pore volume due to the creation of imprinted porosity, as evidenced by the significant increase in argon uptake in going from the protected to the deprotected material. The creation of porosity is expressed by the subtraction isotherm in Fig. 8b (inset). The difference in volume adsorbed between the two materials reached a limiting value of approximately 27 cm^3 STP/g. Using molecular mechanics software, a volume for the imprint fragment removed from the material during deprotection was calculated to be 450 Å, corresponding to a density of slightly

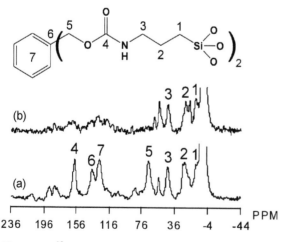

Figure 7 ^{13}C CP/MAS NMR spectra of (a) material 1 and (b) material 2 showing the propyl tethers and lack of carbamate resonances.

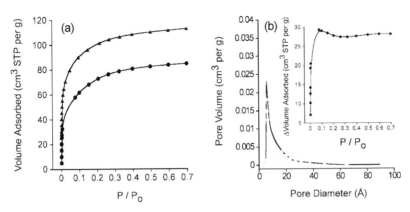

Figure 8 (a) Argon adsorption isotherms for **1** (—●—) and **2** (—▲—). (b) Pore size distribution plot obtained by using the method of Horvath and Kawazoe [43] for **1** (——) and **2** (——). Inset of (b): difference isotherm between **2** and **1** using data from (a) (data from Ref. 38).

larger than 0.7 g/cm³ for imprint in silica. Based on the pore volume synthesized, the increased uptake of argon in the deprotected material was converted to a site density of about 0.25 mmol amines/g (80% deprotection), which was consistent with results from the nonaqueous benzoic acid titration and solid-state NMR spectroscopy study (Fig. 7).

Most of the difference in adsorption capacity occurred below a relative pressure of 0.1, indicating that the synthesized porosity was in the microporous regime, as expected based on the size of the imprint fragment removed. This increase in micro-porosity afforded by the imprinting process was represented in the form of a pore-size distribution by the method of Horvath and Kawazoe [43]. Figure 8b shows pore-size distributions for the material before and after deprotection, and indicates that the two materials are equivalent in their pore-size distributions except at the small length scale corresponding to the imprint fragment removed during deprotection.

As an example of selective molecular recognition with the imprinted silica **2**, a Knoevenagel C–C bond-forming reaction was performed with a bi-functional reactant (Fig. 9). This type of a sequential reaction system is important in numerous industrial applications such as the dehydrogenation of butylene to butadiene or the partial oxidation of naphthalene or *o*-xylene to phthalic anhydride [44]. The ability to suppress the B to C reaction avoids production of undesired products in these cases.

Figure 10a shows the conversion of reactant A, isophthalaldehyde, as a function of time for both the imprinted silica **2** and the surface-functionalized material (comprising a monolayer of aminopropyl groups covalently attached to the surface of a mesoporous silica) [45]. Both the imprinted material and surface-functionalized material catalyzed the reaction of A to B. This was evident in the high turnover frequencies of 367 and 74 turnovers per amine site per hour for the surface-functiona-lized and imprinted silicas, respectively. Although both materials showed catalytic turnover, the turnover frequency observed in the imprinted silica may have been slightly mass-transfer-limited due to the microporosity of the silica framework. However, the imprinted material did not catalyze the production of C to the same

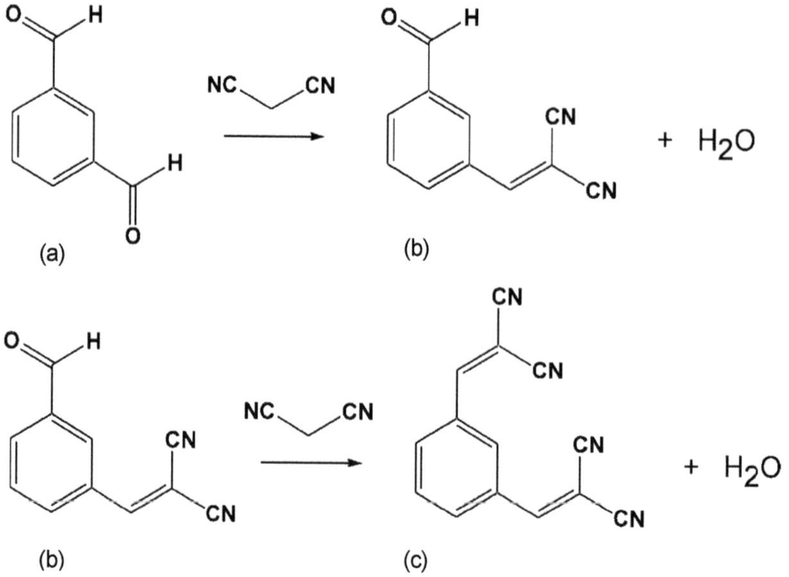

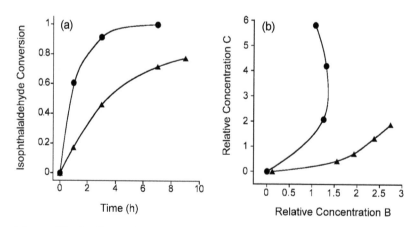

Figure 9 Knoevenagel condensation of isophthalaldehyde (a) and malononitrile to synthesize mono-substituted (b) and di-substituted (c) products via base catalysis.

extent as the surface-functionalized material, which produced a much higher concentration of C. The imprinted material exhibited selectivity for producing product B over C relative to the surface-functionalized material, as shown in Fig. 10b. This selectivity is unlikely to be explained by mass transport effects in the imprinted material and must have been due to selectivity of the imprinted catalytic sites [44]. In other words, this selectivity was a direct result of the imprinted amines, which were not present in the surface-functionalized material.

Figure 10 (a) Conversion of isophthalaldehyde vs. time for the Knoevenagel condensation of isophthalaldehyde and malononitrile using a surface-functionalized silica catalyst (—●—) and material 2 (—▲—). (b) Phase plot representation of concentration of B vs. concentration of C for the Knoevenagel condensation reaction in (a) for a surface-functionalized silica catalyst (—●—) and material 2 (—▲—).

In conclusion, the covalent imprinting of silica with amine functionality can be performed as described above. The materials synthesized by this method have been characterized by solid-state NMR spectroscopy, and the imprinting process has been followed with argon pore-filling experiments. The porosity synthesized via imprinting is on the length scale of the imprint fragment removed, with the final imprinted silica containing pairs of primary amines organized within each site. This silica exhibits selective catalysis in a model system comprising the Knoevenagel condensation of isophthalaldehyde and malononitrile.

A. Example 1—Protocol

1. Methods

(a) Imprint Synthesis. The imprint, [3-(triethoxysilylpropyl)-1,4-phenylenebis (methylene)carbamate, was prepared from the parent diol by standard coupling procedures [46]. A solution of 1,4-benzenedimethanol (1.66 g, 12 mmol) in THF (20 mL) was added drop-wise to a solution of 1,1'-carbonyldiimidazole (3.89 g, 24 mmol) and a catalytic amount of sodium ethoxide in THF (20 mL). The flocculent diimidazolide precipitate was redissolved by gently heating and stirring for an additional hour at room temperature. The reaction was monitored by thin layer chromatography (TLC) using 50 vol% ethyl acetate in hexane. 3-Aminopropyl-triethoxysilane (6.2 mL, 26 mmol) was added to the mixture, which was stirred for an additional 3 h at room temperature. Following solvent removal and dissolution of the reaction mixture in ether, excess imidazole was removed by filtration. The crude product was purified by column chromatography (Silica Gel 60 and 40 vol% ethyl acetate in hexane), and the recovered solid was recrystallized in chloroform/hexane to give white platelets (yield 41%). The shifts for proton, carbon, and silicon NMR spectroscopy are given below, as well as typical elemental analysis characterization. ^{1}H NMR (CDCl$_3$) δ (ppm): 7.32 (s, 4H, aromatic); 5.17 (t, 2H, N*H*); 5.07 (bs, 4H, C*H$_2$*O); 3.81 (q, 4H, OC*H$_2$*CH$_3$); 3.15–3.22 (m, 4H, C*H$_2$*); 1.57–1.67 (m, 4H, C*H*); 1.21 (t, 6H, OCH$_2$C*H$_3$*); 0.624 (t, 4H, C*H$_2$*); and ^{13}C{^{1}H} NMR (CDCl$_3$) δ (ppm): 156.4 (*C*=O); 136.6 (*C*$_q$); 128.2 (*C*H); 66.2 (*C*H$_2$O); 58.5 (O*C*H$_2$CH$_3$); 43.5 (*C*H$_2$); 23.3 (*C*H$_2$); 18.3 (O*C*H$_2$*C*H$_3$); 7.63 (*C*H$_2$); and ^{29}Si NMR (CDCl$_3$) δ (ppm): 45.7 (*Si*(OCH$_2$CH$_3$) ^{3}CH$_2$). Anal. Calc. for C$_{28}$H$_{52}$O$_{10}$N$_2$Si$_2$: C, 53.14; H, 8.28; N, 4.43; Si, 8.87. Found: C, 53.22; H, 8.48; N, 4.37; Si, 8.74.

(b) Material Synthesis and Processing. A typical sol–gel is synthesized by dissolving 0.65 g (1.03 mmol) of the imprint and 23.0 mL (103 mmol) of TEOS in 194 mL absolute ethanol in a 16 ounce jar. Following addition of 64.5 mL of pH 2.0 HCl, the mixture is capped loosely and stirred 24 h at 8°C cooling. It is then covered with weighing paper to facilitate slow evaporation and stirred for 12 h at 15°C followed by 8 days at room temperature. With approximately three-quarters of an inch of liquid head remaining, the clear mixture is transferred to a 40°C oven and loosely capped. The cap is removed after 10 days of gelation, and the glass monoliths are aged in the oven for an additional 8 days. The resulting materials contain approximately 0.28 mmol/g of imprint as determined by a mass balance. Thermogravimetric analysis shows that the gels can be heated in air to over 200°C without detectable decomposition of imprint if further such processing is desired.

The silica monoliths were ground into a fine powder using a ball mill and repetitive wet-sieving in dry ethanol through a 10 μm screen mesh. Residual water and ethanol was removed from the powder by Soxhlet extraction for 24 h in acetonitrile refluxing over calcium hydride, followed by a brief wash with chloroform and pentane. To prepare the material for deprotection with trimethylsilyliodide, the framework silanol groups were capped with trimethylsilyl groups by treating the silica with an excess of an equimolar mixture of 1,1,1,3,3,3-hexamethyldisilazane and chlorotrimethylsilane. The powder was subsequently Soxhlet extracted in acetonitrile to remove residual capping reagents.

One gram of TMS-capped silica was treated with 15 mL of 0.25 M TMSI in dry acetonitrile at 70°C for a period of at least 12 h. Following TMSI treatment, the silica was filtered and washed with 50 mL/g each of acetonitrile, methanol, saturated aqueous sodium bicarbonate, methanol, and acetonitrile. The silica was then Soxhlet extracted with acetonitrile refluxing in calcium hydride and stored dry in a desiccator.

(c) Catalysis Experiments. The Knoevenagel catalysis experiment was performed by adding 1.5 mmol of reactant A (isophthalaldehyde) and 3.0 mmol of malononitrile in 50 mL of acetonitrile containing 8 mg of hexamethylbenzene (internal standard) at 80°C. Following addition of 30 mg of imprinted silica or 6 mg of surface-functionalized silica (both corresponding to 0.006 mmol of catalytic amine sites), the reaction was stirred continuously.

V. EXAMPLE 2—COMPARISON OF NONCOVALENT AND COVALENT IMPRINTING STRATEGIES FOR THE DEVELOPMENT OF CHEMICAL SENSORS

Edmiston et al. [47] utilized inorganic oxide imprinting technology to develop a chemical sensor for the detection of 1,1-bis(4-chlorophenyl)-2,2,2-trichloroethane (DDT), an insecticide with adverse environmental impact. Their first approach was to explore a noncovalent imprinting method similar to the procedures summarized above, with about 4 mol% DDT template relative to silica framework species. Bis (trimethoxysilylethyl)benzene (BTEB) was chosen as the precursor to the inorganic oxide framework, rather than the more common precursors of TEOS or TMOS, in order to produce a highly porous material [4] to facilitate diffusion of DDT into and out of the sensor. By combining the nonpolar BTEB with the relatively polar ethanol solvent, the researchers anticipated that imprinted sites would be generated via van der Waals and π-stacking interactions between the DDT template and the inorganic framework.

The sol–gel solution was prepared under acidic conditions in order to achieve comparable rates of hydrolysis between organosilanes and silanes. This avoids phase separation based on vastly differing condensation rates, since hydrolysis and condensation occur simultaneously during sol–gel processing. Aliquots of the solution were placed into vials and allowed to gel and dry for 18 h under ambient conditions prior to sealing the vials. Following mechanical crushing of the silica gels to pieces about 1 mm in diameter, they were agitated with ethanol, acetone, ethanol, and water in sequential 24 h treatments. These extraction steps removed the DDT template. The binding affinity of DDT and other structural analogues was assessed by shaking

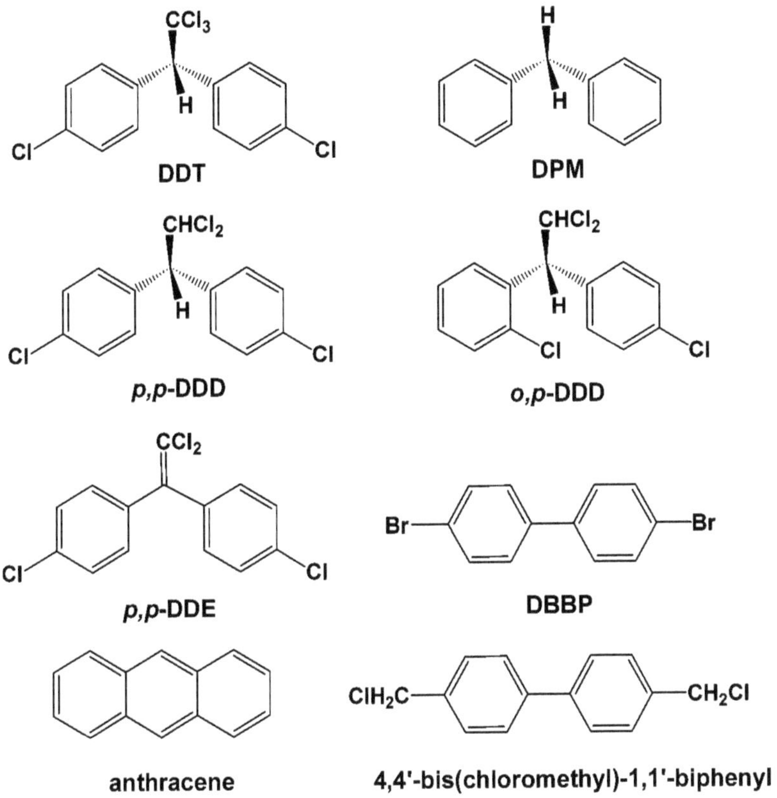

Figure 11 Structure and identities of DDT and the structural analogues used to test the specificity of the imprinted sol–gel materials (adapted from Ref. 47).

the silica gels for 24 h in solutions containing 5 ppm of the appropriate analyte in 60% ethanol/water. Figure 11 shows the structure of DDT and the structural analogues used to test the specificity of the imprinted sol–gel materials. The concentration of each species remaining in solution was analyzed by GC/MS. The partition coefficient, K, for each analyte adsorbing onto the imprinted and nonimprinted gels was calculated as shown in Eq. (1), where $g_{sol-gel}$ is the mass of the material and $g_{solution}$ is the mass of analyte solution used.

$$K = \frac{\text{moles test compound}_{sol-gel}/g_{sol-gel}}{\text{moles test compound}_{solution}/g_{solution}} \tag{1}$$

The selectivity ratio for the imprinted material over the nonimprinted material was calculated as shown in Eq. (2).

$$\text{selectivity ratio} = \frac{K_{MIP}}{K_{NP}} \tag{2}$$

As seen from data in Table 1, binding to a large number of highly selective sites was not achieved. Although each of the analytes partitioned to a greater extent

Table 1 Partition Coefficients for Imprinted and Nonimprinted Sol–Gel Materials Using Noncovalent Molecular Imprinting (Adapted from Ref. 47)

Test compd	K Nonimprinted BTEB sol–gel	Imprinted BTEB sol–gel	Selectivity ratio	Log–P (octanol–water)[a]
DDT	$216 \pm 17 \, (n=5)$	$325 \pm 41 \, (n=5)$	1.50	6.91
DBBP	$364 \pm 40 \, (n=3)$	$513 \pm 30 \, (n=3)$	1.46	5.72
Anthracene	$35.0 \pm 0.4 \, (n=2)$	$41.0 \pm 2.4 \, (n=2)$	1.17	4.45
DPM	$29.9 \pm 3.2 \, (n=2)$	$35.9 \pm 2.0 \, (n=2)$	1.20	4.14

[a]Partition ratio between octanol and water provided as a measure of each compound's polarity [49].

into the imprinted materials, all of them also showed significant affinity for the nonimprinted silica gels. This indicates that the test compounds may be partitioning out of the polar solvent and into the relatively nonpolar silica gel, rather than binding to selective recognition sites in the material. It was determined that DDT lacked sufficiently strong intermolecular interactions with BTEB to create selective-imprinted sites during noncovalent synthesis.

Faced with the data from the noncovalent imprinting method, Edmiston and coworkers developed a covalent imprinting system as an alternate strategy to generate selective binding sites for DDT. Their synthesis scheme is summarized in Fig. 12. An imprint for DDT was synthesized from 4,4'-ethylidenebisphenol (EBP) due to its structural similarity to DDT as well as its ability to undergo modification for covalent attachment to silica. The imprint was incorporated into sol–gel monoliths in a similar manner as for the noncovalent sol–gels, with a 1:150 mole ratio of imprint:BTEB. After polymerization, the monoliths were crushed and then extracted with THF for 72 h. Following the methods of Whitcombe and coworkers [46], removal of the EBP template was accomplished by treatment with $LiAlH_4$. Template removal was verified by IR spectroscopy and elemental analysis. The silica gels were then sequentially rinsed with anhydrous THF, 0.1 M HCl, 0.1 M NH_3, and water prior to shaking with 5 ppm solutions of DDT or structural analogues in 60% ethanol/water. Amounts remaining in solution were quantified by GC/MS.

Silica gel materials without the imprint removed were used as a control. As shown in Table 2, the imprinted material was most selective for EBP, rebinding the template in an amount three times more than the control material. The imprinted adsorbent also bound more DDT, demonstrating that EBP could be used as a structural template for DDT. Anthracene and DPM were bound about equally to both the imprinted and control sol–gels. Overall, the results indicate that imprinting did lead to the binding of target molecules.

However, FT–IR measurements showed a large number of Si–OH groups in the imprinted material. In order to prepare a more nonpolar sol–gel, monoliths were prepared as described above, but treated with neat chlorotrimethylsilane (CTMS) prior to template removal. The binding data of the CTMS-treated materials are shown in Table 3. A significant increase in partition coefficients was observed for DDT and all other nonpolar analytes. The highest selectivity was seen for the rebind-

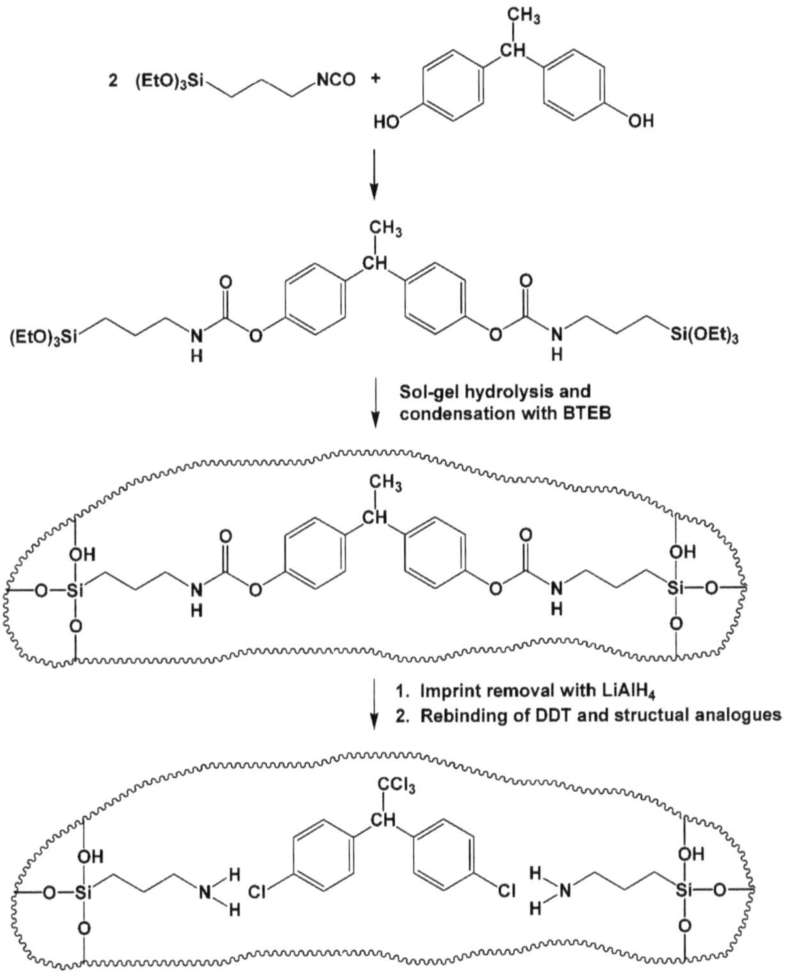

Figure 12 Covalent imprinting strategy for the creation of DDT-selective sites (adapted from Ref. 47).

Table 2 Partition Coefficients for Sol–Gel Materials Prepared Using Covalent Molecular Imprinting (Adapted from Ref. 47)

Test compd	K		Selectivity ratio
	Spacer not removed	Spacer removed	
EBP (template)	$4.6 \pm 1.2\ (n = 3)$	$13.9 \pm 3.1\ (n = 3)$	2.99
DDT	$55 \pm 3.3\ (n = 3)$	$72 \pm 13\ (n = 3)$	1.29
Anthracene	$28 \pm 6.5\ (n = 2)$	$28 \pm 4.2\ (n = 2)$	1.00
DPM	$21.7 \pm 1.8\ (n = 3)$	$16.7 \pm 2.4\ (n = 3)$	0.77

Table 3 Partition Coefficients for Sol–Gel Materials Prepared Using Covalent Molecular Imprinting Combined with CTMS Treatment (Adapted from Ref. 47)

Test compd	K Spacer not removed	K Spacer removed	Selectivity ratio	log–P (octanol-water)[a,b]
EBP	1.4 ± 4.8 (n = 2)	7.4 ± 0.4 (n = 2)	5.22	3.1[b]
DDT	324 ± 64 (n = 2)	438 ± 65 (n = 2)	1.35	6.91
p,p-DDE	158 ± 53 (n = 3)	232 ± 30 (n = 3)	1.46	6.54
p,p-DDD	102 ± 18 (n = 2)	128 ± 6 (n = 2)	1.26	6.02
o,p-DDD	74 ± 30 (n = 2)	82 ± 4 (n = 3)	1.09	5.78
Anthracene	133 ± 16 (n = 3)	122 ± 7 (n = 2)	0.92	4.45
DBBP	935 ± 10 (n = 2)	818 ± 5.6 (n = 2)	0.87	5.72
DPM	58.6 ± 1.4 (n = 2)	58.7 ± 2.5 (n = 2)	1.00	4.14

[a]Partition ratio between octanol and water provided as a measure of each compound's polarity [49].
[b]Estimated from the log–P (octanol–water) values for 4,4′-methylenediphenol (2.91) and 4,4′-isopropylidenediphenol (3.32).

ing of the template, EBP. DDT and its structural analogues 2,2-bis(4-chlorophenyl)-1,1-dichloroethylene (p,p-DDE), and 2,2-bis(4-chlorophenyl)-1,1-dichloroethane (p,p-DDD) were bound with modest selectivity. However, 1-(2-chlorophenyl)-1-(4-chlorophenyl)-2,2-dichloroethane (o,p-DDD) was bound less selectively, presumably due to misalignment of one of the chlorine atoms. Anthracene, DBBP, and DPM were bound nonselectively to both the imprinted and control materials. This indicates that while imprinted sites can be generated to selectively bind a target analyte, imprinting strategies in this instance did not exclude other molecules from partitioning into the hydrophobic silica gel matrix. Additionally, some of the DDT may be bound through nonspecific interactions with the bulk material, rather than in the imprinted sites.

Edmiston and coworkers also explored the use of their DDT-sensing material as a fluorescence sensor. They developed methods for depositing the materials as films and for covalently binding the fluorescence probe 7-nitrobenz-2-oxa-1,3-diazole (NBD) to the silica framework. The resulting sensing films showed a limit-of-detection down to 50 ppt of DDT in water, and were reusable following an acetone rinse. However, observed changes in fluorescence upon binding of the analyte were very small. Improved positioning of the fluorescent probe near the binding pocket may enhance the performance of the sensor.

A. Example 2—Protocol (Excerpted Verbatim from Ref. 47)

1. Reagents

Anthracene, 3-aminopropyltriethoxysilane (APTS), 1,1-bis(4-chlorophenyl)-2,2,2-trichloroethane (DDT), 2,2-bis(4-chlorophenyl)-1,1-dichloroethylene (p,p-DDE), 1-(2-chlorophenyl)-1-(4-chlorophenyl)-2,2-dichloroethane (o,p-DDD), 2,2-bis(4-chlorophenyl)-1,1-dichloroethane (p,p-DDD), diphenylmethane (DPM), 4,4′-ethylidenebisphenol (EBP), 4,4′-ethylenedianiline (EDA), 4,4′-dibromobiphenyl

(DBBP), 4,4'-bis(chloromethyl)-1,1'-biphenyl, chlorotrimethylsilane (CTMS), lithium aluminum hydride (LiAlH4), acriflavin, anhydrous THF, and anhydrous DMSO were obtained from Aldrich. Bis(trimethoxysilylethyl)benzene (BTEB) and 3-isocyanatopropyltriethoxysilane were obtained from Gelest. Chlorobenzene was purchased from J. T. Baker. All solvents were HPLC grade, and chemicals were used as received.

2. Methods

(a)   Preparation and Testing of Noncovalently Imprinted Sol–Gel Monoliths. The sol–gel solution was prepared by combining 900 μL of 0.1% (w/v) DDT in ethanol with 200 μL of BTEB, 200 μL of H_2O, and 30 μL of 12 M HCl. Aliquots of 100 μL were then added to 1.5 mL vials and were allowed to gel and dry for 18 h under ambient conditions before the vials were sealed. The resulting sol–gels were mechanically crushed to pieces approximately 1 mm in diameter. The sol–gels were then rinsed in place by sequential 24 h treatments were ethanol, acetone, ethanol, and water using constant agitation. Test solutions containing 5 ppm of either DDT or structural analogues in 60% ethanol/water were added to the vials and mixed with the sol–gels for 24 h. The amount of each species left in solution was determined using an Agilent 6890/5973 GC/MS.

(b)   EBP Sacrificial Spacer. A 1:2 mole ratio of EBP (0.4283 g) and 3-isocyanato-propyltriethoxysilane (1.000 mL) were added to 5.0 mL of dry THF and allowed to react under dry nitrogen for 20 h at 65°C. The THF was distilled off, leaving a solid with a melting point of 63–66°C. The product was confirmed by infrared spectro-scopy, noting the elimination of the N=C=O peak (2283 cm^{-1}) and the appearance of a C=O peak (1718 cm^{-1}) and a NH peak (3330 cm^{-1}) characteristic of carbamates. Strong absorbance bands near 1100 cm^{-1} indication of silicon ethoxy groups were also confirmed by 400 MHz ^{1}H NMR (DMSO-d_6) δ (ppm): 7.68 (s, 2H, N*H*CO); 7.05 (dd, 8H, aromatic); 4.15 (q, 1H, Ph-C*H*(CH$_3$)-Ph); 3.72 (q, 12H, ethoxy CH$_2$); 3.09 (t, 4H, C*H$_2$*–NH); 1.55 (d, 3H, Ph-CH(C*H$_3$*) -Ph); 1.44 (m, 4H, CH$_2$C*H$_2$*CH$_2$); 1.33 (t, 18H, ethoxy CH$_3$); 0.60 (m, 4H, Si-CH$_2$); and ^{13}C NMR (DMSO-d_6) δ (ppm): 155.4 (OCONH), 142.7 (C1 phenyl), 136.3 (C4 phenyl), 121.5 (C2 phenyl), 114.9 (C3 phenyl), 57.6 (ethoxy CH$_2$), 57.5 (Ph–*C*H–Ph), 55.9 (NHCH$_2$), 42.8 (NH*C*H$_2$CH$_2$), 22.8 (CH$_2$–Si), 18.5 (Ph$_2$CH–*C*H$_3$), 18.1 (ethoxy CH3). Elemental analysis results (calculated (found)): C, 57.6 (57.0); H, 7.96 (7.78); O, 3.95 (3.93).

(c)   Preparation and Testing of Covalently Imprinting Sol–Gel Monoliths. Sol–gel monoliths containing the EBP spacer were prepared in the same manner as the non-covalent sol–gels, except that 1640 μL of ethanol, 400 μL of BTEB, 184 μL of 0.0414 M EBP spacer in ethanol, and 60 μL of 12 M HCl were initially mixed to prepare the sol–gel solution. This resulted in a 1:150 mol ratio of spacer:BTEB. After polymerization, the crushed pieces were then extracted in place with anhydrous THF for 72 h. To remove the EBP template, the sol–gels were treated with 70 mM LiAlH$_4$ for 24 h at 60°C. The sol–gels were then sequentially rinsed with anhydrous THF, 0.1 M HCl, 0.1 M NH$_3$, and water over a period of 48 h. Test solutions of 5 ppm DDT or structural analogues in 60% ethanol/water were then added to the vials containing the sol–gels and allowed to mix for 24 h. The amount of the test com-pound remaining in solution was measured by GC/MS. A second set of sol–gels were prepared and tested in exactly the same manner except that they were incubated in

1.0 mL of neat CTMS for 16 h and rinsed with THF for 48 h prior to the template-removal step. To provide controls when utilizing both procedures, sol–gels were prepared in tandem without performing the $LiAlH_4$ step so that the spacer was not removed.

VI. CONCLUSIONS

The rich history of molecular imprinting, dating back almost 80 years, has its origins in the use of inorganic oxide matrices as a material framework. With the advent of modern sol–gel science and methods for the synthesis of hybrid organic–inorganic materials, the field of imprinting in inorganic oxides has grown substantially. We have critically examined some of the relative advantages of covalent and noncovalent approaches for imprinting. Although synthetically more demanding, the covalent approach offers the capability to produce single-site materials that do not contain heterogeneity even at high site density. The two specific examples analyzed here highlight the importance of a covalent imprinting approach for the formation of imprinted inorganic oxide materials for selective adsorption and catalysis.

ACKNOWLEDGMENTS

We would like to thank the National Science Foundation for a Graduate Research Fellowship to J.L.D. A.K. is grateful to Professor Mark E. Davis, Caltech, for helpful technical discussions on the subject of molecular imprinting.

REFERENCES

1. Sasaki, D.Y. In *Molecular Imprinting Approaches Using Inorganic Matrices. Molecularly Imprinted Polymers: Man-Made Mimics of Antibodies and Their Applications in Analytical Chemistry.* Elsevier Science, Amsterdam, 2001.
2. Lungu, A.; Neckers, D.C. Correlation between the mechanical properties and the homogeneity of some photopolymerized acrylic/methacrylic networks. J. Polym. Sci., Part A: Polym. Chem. **1996**, *34*, 3355–3360.
3. Weast, R.C.; Astle, M.J., Eds.; *CRC Handbook of Chemistry and Physics*, 62nd Ed.; CRC Press, Inc.: Boca Raton, Flo, 1981–1982.
4. Loy, D.A.; Shea, J.K. Bridged polysilsesquioxanes—highly porous hybrid organic–inorganic materials. Chem. Rev. **1995**, *95*, 1431–1442.
5. Schottner, G. Hybrid sol–gel-derived polymers: applications of multifunctional materials. Chem. Mater. **2001**, *13*, 3422–3435.
6. Pauling, L. A theory of the structure and process of formation of antibodies. J. Am. Chem. Soc. **1940**, *62*, 2643–2657.
7. Polyakov, M.V.; Stadnik, P.M.; Paritzkii, M.V.; Malkin, I.M.; Dukhina, F.S. The question of the structure of silica gel. Zh. Fiz. Khim. **1933**, *4*, 454–456.
8. Dickey, F.H. The preparation of specific adsorbents. Proc. Natl. Acad. Sci. USA **1949**, *35*, 227–229.
9. Dickey, F.H. Specific adsorption. J. Phys. Chem. **1955**, *59*, 695–707.
10. Haldeman, R.G.; Emmett, P.H. Specific adsorption of alkyl orange dyes on silica gel. J. Phys. Chem. **1955**, *59*, 1039–1043.
11. Morrison, J.L.; Worsley, M.; Shaw, D.R.; Hodgson, G.W. The nature of the specificity of adsorption of alkyl orange dyes on silica gel. Can. J. Chem. **1959**, *37*, 1986–1995.

12. Katz, A.; Davis, M.E. Investigations into the mechanisms of molecular recognition with imprinted polymers. Macromolecules **1999**, *32*, 4113–4121.

13. Curti, R.; Colombo, U.; Clerici, F. Chromatography with specific adsorbents. Gazz Chim. Ital. **1952**, *82*, 491–502.

14. Klabunovskii, E.I.; Agronomov, A.E.; Volkova, L.M.; Balandin, A.A. Absorption of racemic and (+)-2-butanol on stereospecific silica gels. Izv. Akad. Nauk. SSSR, Otd. Khim. Nauk. **1963**, 228–234.

15. Klabunovskii, E.I.; Balandin, A.A.; Godunova, F.L. Inversion of L-menthone. Izv. Akad. Nauk. SSSR, Otd. Khim. Nauk. **1963**, 886–890.

16. Erlenmeyer, H.; Bartels, V.H. The problem of similarity in chemistry. Thin layer chromatography on silica gel of specific absorptivity. Helv. Chim. Acta. **1964**, *47*, 46–51.

17. Erlenmeyer, H.; Bartcls, V.H. The problem of similarity in chemistry. Specific adsorption on silica gels II. Helv. Chim. Acta. **1964**, *47*, 1285–1288.

18. Bartels, V.H.; Erlenmeyer, H. The problem of similarity in chemistry. Specifically adsorbing silica gels III. Helv. Chim. Acta. **1965**, *48*, 285–290.

19. Bartels, V.H.; Prijs, B.; Erlenmeyer, H. Specifically adsorbing silica gels. Helv. Chim. Acta. **1966**, *49*, 1621–1625.

20. Majors, R.E.; Rogers, L.B. Variables in the preparation of modified silica adsorbents. Anal. Chem. **1969**, *41*, 1052–1057.

21. Brinker, C.J.; Scherer, G.W. In *Sol–Gel Science: the Physics and Chemistry of Sol–Gel Processing*. Academic Press: New York, 1990.

22. Pinel, C.; Loisil, P.; Gallezot, P. Preparation and utilization of molecularly imprinted silicas. Adv. Mater. **1997**, *9*, 582–585.

23. Makote, R.; Collinson, M.M. Template recognition in inorganic–organic hybrid films prepared by the sol–gel process. Chem. Mater. **1998**, *10*, 2440–2445.

24. Marx, S.; Liron, Z. Molecular imprinting in thin films of organic–inorganic hybrid sol–gel and acrylic polymers. Chem. Mater. **2001**, *13*, 3624–3630.

25. Leung, M.K.P.; Chow, C.F.; Lam, M.H.W. A sol–gel derived molecular imprinted luminescent PET sensing material for 2,4-dichlorophenoxyacetic acid. J. Mater. Chem. **2001**, *11*, 2985–2991.

26. Lee, S.W.; Ichinose, I.; Kunitake, T. Molecular imprinting of azobenzene carboxylic acid on a TiO_2 ultrathin film by the surface sol–gel process. Langmuir **1998**, *14*, 2857–2863.

27. Lee, S.W.; Ichinose, I.; Kunitake, T. Molecular imprinting of protected amino acids in ultrathin multilayers of TiO_2 gel. Chem. Lett. **1998**, *12*, 1193–1194.

28. Lahav, M.; Kharitonov, A.B.; Katz, O.; Kunitake, T.; Willner, I. Tailored chemosensors for chloroaromatic acids using molecular imprinted TiO_2 thin films on ion-sensitive field-effect transistors. Anal. Chem. **2001**, *73*, 720–723.

29. Lahav, M.; Kharitonov, A.B.; Willner, I. Imprinting of chiral molecular recognition sites in thin TiO_2 films associated with field-effect transistors: novel functionalized devices for chiroselective and chirospecific analyses. Chem. Eur. J. **2001**, *7*, 3992–3997.

30. Heilmann, J.; Maier, W.F. Selective catalysis on silicon dioxide with substrate-specific cavities. Angew Chem. Int. Ed. Engl. **1994**, *33*, 471–473.

31. Ahmad, W.R.; Davis, M.E. Transesterification on imprinted silica. Catal. Lett. **1996**, *40*, 109–114.

32. Maier, W.F.; BenMustapha, W. Transesterification on imprinted silica—reply. Catal. Lett. **1997**, *46*, 137–140.

33. Hunnius, M.; Rufinska, A.; Maier, W.F. Selective surface adsorption versus imprinting in amorphous microporous silicas. Microporous Mesoporous Mater. **1999**, *29*, 389–403.

34. Whitcombe, M.J.; Rodriguez, M.E.; Villar, P.; Vulfson, E.N. A new method for the introduction of recognition site functionality into polymers prepared by molecular imprinting—synthesis and characterization of polymeric receptors for cholesterol. J. Am. Chem. Soc. **1995**, *117*, 7105–7111.

35. Boury, B.; Corriu, R.J.P.; Le Strat, V.; Delord, P. Generation of porosity in a hybrid organic–inorganic xerogel by chemical treatment. New J. Chem. **1999**, *23*, 531–538.
36. Bass, J.D.; Katz, A. Thermolytic synthesis of imprinted amines in bulk silica. Chem. Mater. **2003**, *15*, 2757–2763.
37. Bass, J.D.; Anderson, S.L.; Katz, A. The effect of outer-sphere acidity on chemical reactivity in a synthetic heterogeneous base catalyst. Angew. Chem. Int. Ed. **2003**, *42*, 5219–5222.
38. Katz, A.; Davis, M.E. Molecular imprinting of bulk, microporous silica. Nature **2000**, *403*, 286–289.
39. Brinker, C.J.; Keefer, K.D.; Schaefer, D.W.; Ashley, C.S. Sol–gel transition in simple silicates. J. Non-Cryst. Solids **1982**, *48*, 47–64.
40. Cao, G.Z.; Tian, H. Synthesis of highly porous organic/inorganic hybrids by ambient pressure sol–gel processing. J. Sol–Gel Sci. Technol. **1998**, *13*, 305–309.
41. Olah, G.A.; Narang, S.C. Iodotrimethylsilane—a versatile synthetic reagent. Tetrahedron **1982**, *38*, 2225–2277.
42. Katz, A. *The Synthesis and Characterization of Molecularly Imprinted Materials.* California Institute of Technology: Pasadena, 1999.
43. Horvath, G.; Kawazoe, K. Method for the calculation of effective pore-size distribution in molecular-sieve carbon. J. Chem. Eng. Jpn. **1983**, *16*, 470–475.
44. Wheeler, A. Reaction rates and selectivity in catalyst pores. Adv. Catal. **1951**, *3*, 249–327.
45. Vansant, E.F.; van der Voort, P.; Vrancken, K.C. In *Characterization and Chemical Modification of the Silica Surface.* Elsevier: New York, 1995.
46. Staab, H.A. New methods of preparative organic chemistry IV. Syntheses using heterocyclic amides (azolides). Angew. Chem. Int. Ed. Engl. **1962**, *1*, 351–367.
47. Graham, A.L.; Carlson, C.A.; Edmiston, P.L. Development and characterization of molecularly imprinted sol–gel materials for the selective detection of DDT. Anal. Chem. **2002**, *74*, 458–467.
48. Lubke, M.; Whitcombe, M.J.; Vulfson, E.N. A novel approach to the molecular imprinting of polychlorinated aromatic compounds. J. Am. Chem. Soc. **1998**, *120*, 13342–13348.
49. Hansch, C.; Leo, A.; Hoekman, D.H. . In *Exploring QSAR*; American Chemical Society: Washington, DC, 1995.

12
Post Modification of Imprinted Polymers

Ken D. Shimizu   University of South Carolina, Columbia,
South Carolina, U.S.A.

I. INTRODUCTION

Molecularly imprinted polymers (MIPs) are highly crosslinked polymers that can be readily tailored with affinity and selectivity for a molecule of interest [1–3]. MIPs compare favorably with other synthetic molecular recognition systems such as engineered antibodies and molecule receptors. For these reasons, MIPs have demonstrated utility in a wide array of applications requiring molecular recognition including separations [4], catalysis [5], and sensing [6,7]. Despite their versatility, MIPs have been limited in their practical utility due to their poor overall binding characteristics. A major contributor to this problem has been the heterogeneity of binding sites in MIPs. Ideally, the imprinting process would generate homogenous polymers in which all the sites were identical with high binding affinities like enzymes and antibodies (Fig. 1). The imprinting process, however, typically proceeds with low fidelity, generating binding sites that vary widely in shape, size, and binding affinity. This binding site heterogeneity is detrimental to the utility of MIPs in almost every application. For example, binding site heterogeneity leads to severe peak asymmetry and tailing that hinders the use of MIPs in chromatographic applications [8]. The heterogeneous distribution is also weighted toward the low affinity low selectivity sites and these low affinity sites dominate the binding properties of MIPs, yielding low overall binding constants [9,10].

A variety of different strategies have been employed to try to improve the binding properties of MIPs. The most common ones have focused on improving the polymerization process. Imprinting variables such as solvent, temperature, stoichiometry, and concentration have all been optimized, leading to considerable improvements in the binding characteristics of MIPs. However, MIPs still, in general, have poor overall binding affinity and binding site heterogeneity. Presented in this chapter will be an alternative methodology for improving the binding properties of MIPs by modification of the preformed MIP matrix. The strategy has a number of attractive attributes. First, the approach is complementary to existing imprinting protocols and has the potential for additive improvements in the existing affinity and selectivity of an MIP. In addition, chemical modification offers a route

Homogeneous   Heterogeneous

Figure 1 Schematic depictions of homogeneous and heterogeneous MIPs that contain binding sites in which the sites are identical or different. The depth of the site corresponds to the binding affinity.

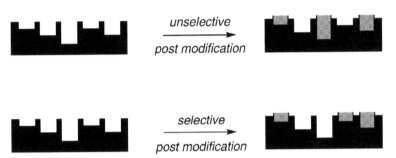

Scheme 1 Schematic depictions of unselective (top) and selective (bottom) postmodification of heterogeneous MIPs.

to introduce new types of affinities and selectivities beyond those generated in the original imprinting process. This provides a method to further tailor and tune the binding properties of MIPs.

Postmodification of MIPs can be grouped into two general classes: thermal (Sec. II) and chemical treatments (Sec. III). Each can be carried out in *unselective* or *selective* manner (Scheme 1). Unselective treatments uniformly modify binding sites irrespective of their binding affinity or selectivity, and therefore, do not change the distribution of sites. The resulting modified MIPs do not demonstrate improvements in average affinity and selectivity. Selective treatments, on the other hand, target only a specific subset of the binding sites. In the example in Scheme 1, the selective post modification targets the low affinity sites, leaving the high affinity sites unmodified. In this manner, selective modification can alter the distribution, potentially leading to improvements in affinity and selectivity.

II. THERMAL POSTMODIFICATION

Thermal treatments of MIPs subsequent to the imprinting process have been fairly well studied [8,11,12]. These experiments were carried out for a variety of reasons including improving binding properties [11], completing the polymerization process [13], and testing the thermal stability of MIPs [14]. To date, all thermal treatments have been carried out in an unselective manner and the resulting polymer typically demonstrates lower affinity and selectivity. Thermal treatments at lower temperatures ($<150°C$), however, have led to some accompanying enhancements in other material properties such as mass transfer kinetics, capacity, and long-term thermal stability.

Sellergren et al. [12,15,16] have most extensively examined the effects of thermal postmodification on the binding properties of MIPs. The polymer that was studied

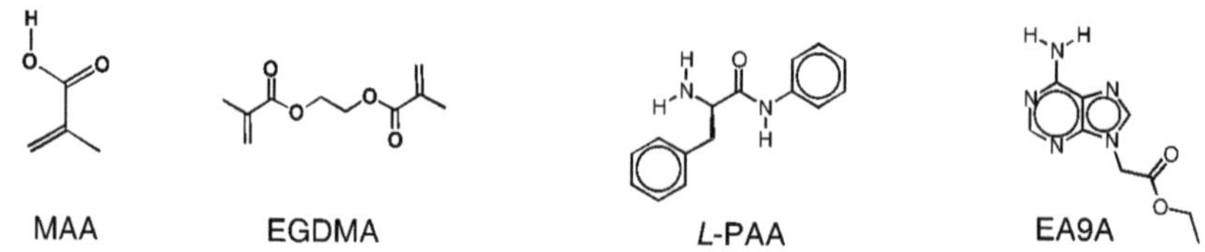

Functional Monomer   Crosslinker   Templates

MAA   EGDMA   *L*-PAA   EA9A

Figure 2 Functional monomers, crosslinkers, and templates for molecularly imprinted polymers.

was an MAA/EGDMA polymer imprinted with L-phenylalanine anilide (L-PAA) (Fig. 2). Enantioselectivity of the MIP arose from hydrogen bonding interactions of the carboxylic acids, lining the binding captivity, with the L-PAA guest molecules. Initial reports showed that MIPs thermally annealed at 120°C after the polymerization process showed reduced affinity and enantioselectivity. Comparison of a series of MIPs thermally treated at varying temperatures (50°C, 120°C, 140°C, 160°C) revealed that the MIPs gradually lost their abilities to differentiate between enantiomers with increasing temperature. Complete loss of enantioselectivity was seen for the MIP annealed at 160°C. The drop in selectivity and affinity of the annealed polymers was attributed to collapse or deformation of the imprinted cavities when the polymer is heated above its glass transition temperature. One piece of evidence that supported the presence of morphological changes upon thermal annealing was the greater susceptibility of high porosity MIPs (>200 m^2/g) to the deleterious effects of thermal treatments in comparison to a lower porosity MIP (3.8 m^2/g) [8]. Presumably, the higher porosity polymer contains more cavities that can collapse or shrink on heating leading to greater changes in binding properties.

Although the affinity and selectivity of the annealed polymers was diminished, the annealed polymers did display improved overall chromatographic performance due to 33–50% higher capacities and a slight improvement in the mass-transfer kinetics [15]. The corresponding increase in capacity, in these experiments, can also be explained by the collapse of cavities in the MIPs, which leads to increased densities and higher numbers of sites per gram of polymer.

Most recently Sellergren et al. [17] have shown that thermal postpolymerization treatment of MIP is a successful method to inhibit the slow leaching of encapsulated template molecules out side the matrix. The slow release of template molecules from MIPs is particularly problematic for sensor applications that are measuring low analyte concentrations ($< \mu$M). Thermal postmodification appears to rigidify the MIP matrix as seen by the lower swelling of thermally treated MIPs. The reduction in the flexibility of the matrix permanently entraps the strongly adsorbed template molecules that would otherwise slowly leach out of the matrix over time.

Nicholls and Svenson [14] have also studied the thermal treatment of MIPs in order to assess the thermal stability of the common MAA/EGDMA matrix of MIPs. The results were consistent with those reported by Sellergren et al. as thermal annealing above 150°C led to a loss in affinity for the template, theophylline. Nicholls' study suggests a further explanation for the loss of selectivity upon thermal treatment. The drop in affinity and selectivity was correlated to the decarboxylation of the key hydrogen bonding carboxylic acids in the matrix which was observed by a decrease in the C=O carbonyl band (1735 cm^{-1}) in the IR spectra. Sellergren and Shea [11] had also tested the thermal stability of the MAA/EGDMA matrix by thermal gravimetric analysis (TGA) and saw no mass loss below 200°C. However, the binding properties of these thermally treated polymers (>200°C) were not reported.

Overall, thermal treatments appear to uniformly alter the MIP matrix and do not target specific classes of sites. The result is that the annealed MIPs show equal or lesser affinity and selectivity. In some cases, the decrease in binding properties was somewhat offset by improvements in material properties such as binding capacity, binding site accessibility, and reduction of the slow leaching of the template molecule from the matrix [13]. To date, no examples have been reported in which thermal treatments of MIPs led to selective alterations of the matrix, and this may be a fruitful area for future studies.

III. CHEMICAL POSTMODIFICATION

In contrast to thermal treatments, chemical treatments make available a wider range of transformations and reaction conditions which would modify the matrix. These chemical modifications can be targeted at the functional monomer (Sec. III.A) or the crosslinker (Sec. III.B).

A. Modifications of the Functional Monomer

The role of the functional monomer in MIPs is to provide binding interactions with the guest molecule via hydrogen bonds, electrostatic interactions, or via kinetically labile covalent bonds. The imprinting process generates binding sites lined with functional monomers. Chemical modification of these functional groups is, therefore, expected to directly affect the binding properties of the MIP. Again these chemical modifications can be grouped into *unselective* and *selective* postmodifications, and these will be addressed individually.

1. Unselective Methods

The first example of the chemical modification of MIPs was reported by Sellergren and Shea ([11]). The carboxylic acids in an MAA/EGDMA matrix were exhaustively esterified with diazomethane (Scheme 2). The goal of the study was to test the hypothesis that the acidic protons of the carboxylic acids provide the key recognition interactions with the guest molecule L-PAA via hydrogen bonding and electrostatic interactions. Consistent with this premise, the esterified MIP showed a dramatic loss

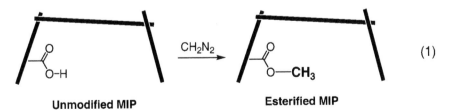

        Unmodified MIP   Esterified MIP

Scheme 2 Chemical modification of MIPs by esterification with diazomethane.

in affinity and selectivity in comparison to the unmodified MIP. Although these reactions were carried out in an unselective manner, the ability to reduce the affinity of modified sites was a key element in later studies that selectively modified MIPs.

Shimizu et al. [18] have further examined the effects of esterification of the carboxylic acids in an MAA/EGDMA matrix imprinted with ethyl adenine-9-acetate (EA9A) (Fig. 2). Consistent with the results of Sellergren and Shea, esterification of the matrix led to a loss of affinity and selectivity. Both diazomethane and phenyldiazomethane were examined, yielding similar results, suggesting that the difference in size between the two esterification reagents was insufficient to change the selectivity of the reaction.

Interestingly, quantitative comparisons of the effects of esterification on high and low affinity sites suggested that esterification with diazomethane was actually slightly more selective for the high affinity sites. This preference was contrary to expectations because it was anticipated that the higher affinity sites would be deeper in the matrix and would react slower due to greater steric hindrance. Two possible explanations can be given for the apparent selectivity of diazomethane for the high affinity sites. The first is that the appearance of selectivity arises from a redistribution of binding sites. For example, esterification of a high affinity site would transform it into a low affinity site. This might generate the appearance of selectivity, as the number of high affinity sites would decrease; whereas the number of low affinity sites would increase. A second explanation is that the better sites contain multiple functional groups [19] and, therefore, are statistically more likely to be esterified than the lower affinity sites that contain only a single carboxylic acid. Regardless of the explanation, the overall effect of esterification of MIPs with diazomethane was a diminishment in capacity and affinity.

2. Selective Methods

Improvements in the binding properties can be achieved by selective chemical modification of the matrix. This strategy takes advantage of the heterogeneity inherent in MIPs and tries to isolate subsets of the population with particularly attractive binding properties. There are two general strategies that have been implemented to selectively modify MIPs, leading to improvements in the existing affinity or selectivity of an MIP (Scheme 3a) or the introduction of new types of selectivity not present in the original MIP (Scheme 3b).

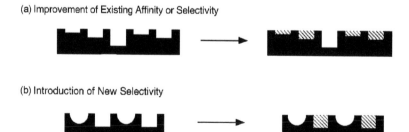

(a) Improvement of Existing Affinity or Selectivity

(b) Introduction of New Selectivity

Scheme 3 Selective chemical modification of an MIP that targets specific classes of sites either (a) low affinity sites (as denoted as the more shallow imprints), or (b) sites with a specific selectivity.

In both cases, the key is to selectively modify a particular subset of binding sites, thereby eliminating its contribution to the overall binding properties. The remaining unmodified sites will then have enhanced overall affinity and selectivity. In the first case (Scheme 3a), the low affinity sites (the more shallow imprints) are targeted for chemical modification, knocking out their contribution to the overall binding properties of the matrix. The binding properties of the modified polymer should then dominated by the unreacted high affinity sites, leading to higher average affinity. In the second strategy (Scheme 3b), selective chemical modification alters the selectivity of the matrix. Initially, the polymer has an equal number of sites selective for circular and rectangular guest molecules and therefore is not able to differentiate between the two. Selective modification of the rectangular sites leaves only the circular sites and now the polymer will preferentially bind the circular guest molecules.

(a)   Enhancements in Existing Affinity and Selectivity.   As mentioned earlier, the binding sites in MIPs are not homogeneous and typically span the entire range of binding affinities from low to high. Selective chemical modification provides one strategy to narrow the distribution of sites in favor of the more desirable high affinity sites. This has been accomplished by one of two methods: (1) a selective reagent, and (2) guest directed selectivity.

Reagent based selectivity.   The first strategy for selectively modifying an MIP is to use a selective reagent. The source of the selectivity is the size exclusion principle as shown in Scheme 3. The unmodified MIP contains considerable binding site heterogeneity as represented by the small, medium, and large binding sites. The system can be made more homogeneous and specific by chemical modification with a large reagent. This reagent is sterically only able to access the large binding sites, leaving the small and medium binding sites unmodified. The resulting MIP should then be more selective for smaller guest molecules (Scheme 4).

This size-selective chemical modification strategy has been successfully applied by Frèchet and Svec [20] to selectively modify a polymeric size exclusion support using a large polymeric reagent. The polymeric reagent preferentially reacted with the larger pores, leaving a more narrow distribution of smaller pores sizes and an overall more selective support. Whitcombe et al. have extended this chemical modification strategy to MIPs (Scheme 5), yielding up to five fold enhancements in selectivity. The unmodified MIPs were covalently imprinted to have affinity and selectivity for various sized N-containing heterocycles: pyridine (small), quinoline (medium), and acridine (large). The unmodified MIPs showed some specificity corresponding to the size of the template molecule used in the imprinting process. However, MIPs were still highly heterogeneous and contained binding sites that could accommodate all the three guest molecules. Selective chemical modification was able to reduce the heterogeneity, leading to higher affinities and selectivities.

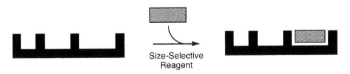

Scheme 4   Chemical modification of an MIP using a size-selective reagent.

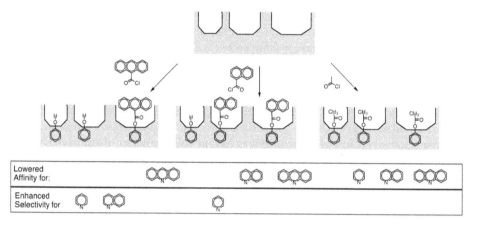

Scheme 5

To improve the specificity of the system, Whitcombe et al. treated the unmodified MIP with various sized acid chlorides to preferentially block the corresponding binding cavities. For example, treatment of the MIP with the largest acid chloride (anthracene-9-carbonyl chloride) reacted with only the largest binding sites. The modified MIP showed improved selectivity for the two smaller guests, pyridine and quinoline. Alternatively, reaction with the intermediate sized 1-naphthoyl chloride blocked the medium and large binding sites. The resulting matrix had greater selectivity for the smallest guest, pyridine. Finally, chemical modification with the smallest reagent, acetyl chloride, esterified all the binding sites equally and no enhancement in selectivity was observed.

Gagné et al. [21] have also utilized this strategy of selectively blocking or "poisoning" sites in an MIP in order to try to improve catalytic activity and selectivity. An MIP was synthesized containing catalytically active Pt(II) centers within a chiral binding pocket. The MIP was shown to be able to catalyze the ene reaction of methylenecyclohexane with ethylglyoxylate (Scheme 6). It was hoped that the chiral pocket of the MIP would catalyze the ene reaction in an enantioselective fashion.

Improvements in catalytic activity and selectivity were attempted using the selective modification strategy as shown in Scheme 7. Initially the polymer was imprinted with an *R*-template ligand, generating a chiral *R*-cavity. The existence of the chiral cavities was demonstrated by "poisoning" experiments. The MIP was treated with chiral ligands that would bind to the catalytic platinum centers and inhibit the reaction at that binding site. The corresponding catalytic activity of the MIP was found to be dependent on the chirality of the "poisoning" ligand. A higher activity was observed for the MIP poisoned with the mismatched *S*-poison in comparison to the MIP poisoned with the matched *R*-poison. This outcome was

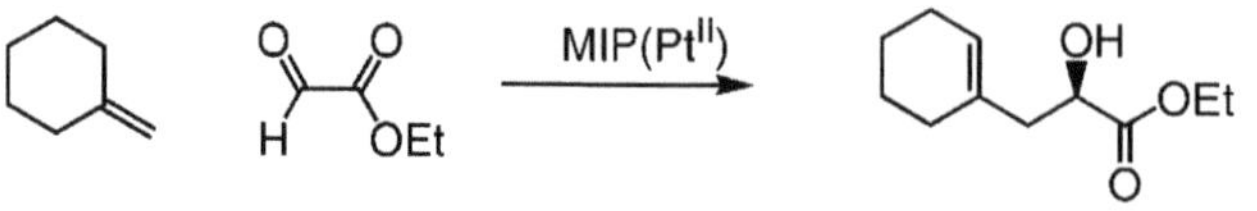

Scheme 6 The MIP catalyzed ene reaction of methylenecyclohexane and ethylglyoxylate.

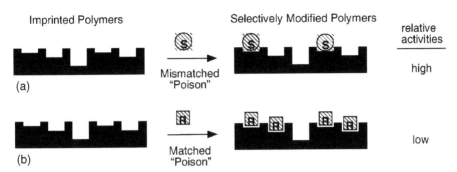

Scheme 7 Representations of "poisoning" experiments of a catalytically active MIP carried out with (a) mismatched, and (b) matched blocking agents.

interpreted as the *R*-poison reoccupying and blocking a higher percentage of binding sites in the *R*-imprinted polymer than the *S*-poison.

Ideally, the sites occupied by the mismatched *S*-poison would also be the sites that also catalyzed the formation of the minor *S*-product. The *S*-poisoned MIP would then show higher selectivity for the major *R*-product. Unfortunately, although the expected changes in activity were observed, the corresponding enhancements in enantioselectivity were not observed.

Guest directed selectivity. A second method for targeting a specific class of binding sites is to protect the desired subset of binding sites and then react the remaining undesired sites. This can be accomplished by using the guest molecule to selectively shield the higher affinity sites. In the example in Scheme 8, the MIP contains binding sites with different binding affinities, represented by imprints of varying depths. The MIP is first equilibrated with a low concentration of guest molecule that preferentially binds to and protects the highest affinity sites. Chemical modification then preferentially eliminates the unoccupied low affinity sites. Finally, removal of the guest molecule unmasks the desired high affinity sites. The entire process can be carried out in a single reaction vessel, since the interactions of the guest molecule with the MIP are noncovalent interactions and are fully reversible.

McNiven et al. [22] first demonstrated this guest directed chemical modification strategy. In their studies, an MAA/EGDMA polymer imprinted with testosterone (Fig. 3) was partially esterified with methyl iodide (MeI) and 1,8-diazabicyclo(5.4.0) undec-7-ene (DBU) both in the presence and absence of the

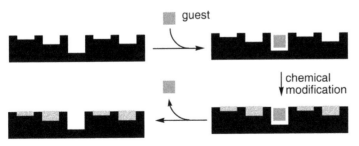

Scheme 8 Representation of the guest directed selective chemical modification strategy. The depth of the imprint in these models corresponds to the binding affinity of that site.

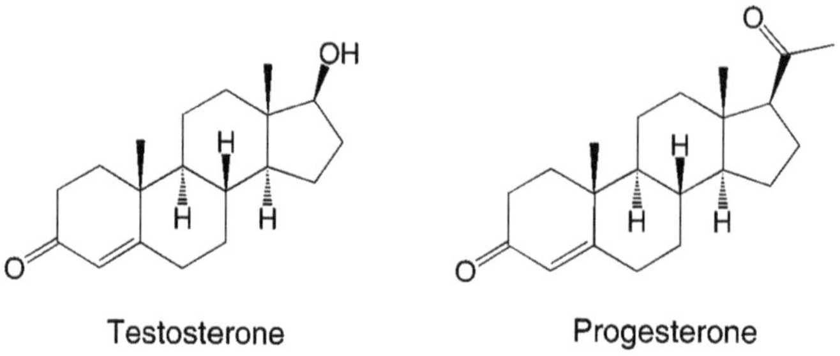

Testosterone   Progesterone

Figure 3 Molecular templates.

template molecule. The affinity and selectivity of these modified polymers was compared by HPLC. The polymers esterified in the presence of the template showed
enhanced separation factors for testosterone vs. progesterone. Under the best conditions, the separation factor improved from 3.87 in the unmodified MIP to 4.38 in the
partially esterified MIP. This enhancement in selectivity was accompanied by
increase in affinity for testosterone ($k'_{unmodified} = 1.58$ to $k'_{modified} = 1.91$). Control
reactions, on the other hand, carried out in the absence of testosterone yielded no
changes in the separation factor and only a slight enhancement in retention times.

The improvements in binding properties were interpreted as the result of
selective protection of the high affinity sites by the testosterone guest molecules
during the esterification process. Sellergren et al. [27] had demonstrated earlier
that esterification of the carboxylic acids in an MAA/EGDMA matrix is an
effective method to eliminate the binding ability of the corresponding binding sites.
Consistent with the above explanation was that complete esterification of the matrix
even in the presence of template led to a drop in affinity with no accompanying
improvement in selectivity. This is presumably because exhaustive esterification of
the matrix ensures that in all the sites both low and high affinity will be modified to
an equal extent.

Shimizu et al. [18] have also utilized the guest directed selective chemical modification strategy to enhance the binding properties of an ethyl adenine-9-acetate
(EA9A) selective MIP. The MAA/EGDMA matrix was esterified with diazomethane or phenyldiazomethane both in the presence and absence of the template
molecule (EA9A). This study was able to more definitively verify the ability of
the template molecule to selectively protect the high affinity sites, using the affinity
distribution analyses (see Chapter 16) which quantitatively measures changes in the
populations of the high and low affinity sites [10].

An affinity distribution is a measure of the number of sites (y-axis) having a
particular association constant (x-axis). For example, the affinity distributions shown
in Fig. 4 are typical of those seen in noncovalent MIPs. They contain binding sites
having association constants that cover the range from high to low affinity. In
addition, the distribution is heavily skewed in favor of the low affinity sites. This
analysis is valuable because it gives a graphical method with which to compare the
number of low and high affinity binding sites before and after chemical modification.
Figure 4 shows an overlay of affinity distributions measured for an unmodified
EA9A imprinted polymer and polymers esterified with diazomethane in the absence
and presence of EA9A. The two chemically modified MIPs showed a reduction in

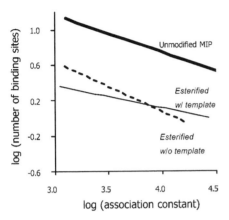

Figure 4 Overlaid affinity distributions of the unmodified MIP, MIP esterified in the presence, and absence of template EA9A.

the numbers of both high and low affinity sites as evidenced by the lower vertical displacement of distributions. However, the different slopes of the affinity distributions of the two esterified MIPs demonstrate that the presence of template affects the selectivity of the reaction. The slope of the distribution in the presence of template had a flatter slope meaning that it retained a higher percentage of high affinity sites in comparison to the polymer esterified in the absence of template. These affinity distributions give direct evidence for the ability of the template to occupy and protect the higher affinity sites and leaving the lower affinity site more susceptible to esterification and inactivation.

IV. GUEST DIRECTED ESTERIFICATION OF AN EA9A MIP WITH PHENYLDIAZOMETHANE

A typical procedure for this guest directed chemical modification strategy using an EA9A selective MIP which is esterified with the safer phenyldiazomethane was adapted from a previous report [18]. The unmodified MIP **1** selective for ethyl adenine-9-acetate (EA9A) was synthesized as previously described from MAA, EGDMA, EA9A and AIBN in acetonitrile at 60°C. To a 250-mL round bottom flask containing MIP **1** (2.0 g) and 73 mg (0.33 mmol) of EA9A in MeCN (110 mL) was added phenyldiazomethane in toluene (3.2 mL of a 2.0 M solution). Wulfman et al. [23,24] have described an excellent synthesis of phenyldiazomethane from phenyl-*p*-toluenesulfonylhydrazone. The reaction mixture was shaken for 2 h on a Burrell wrist action shaker. The acetonitrile solution was decanted. The polymer was washed six times with methanol by shaking the polymer with methanol and decanting the solutions. Subsequent evaporation of the polymer under vacuum afforded a white powder (2.05 g). Control polymers were made in identical fashion, but without EA9A present in the modification reactions. The binding properties of the polymers were tested by batch rebinding studies and curve fitting as described in Chapter 16.

Further evidence for the importance of the template molecule in effecting the selectivity of the reaction was that more harsh reaction conditions that disrupted the hydrogen bonding interactions of the template for the MIP diminished the effect of the template on the esterification reaction. For example the use of the THF instead of acetonitrile as the solvent in the esterification reaction yielded polymers modified both in the presence and absence of template molecule with identical binding properties.

The examples from Karube et al. and Shimizu et al. demonstrate that guest directed site selective chemical modification is a viable strategy for directly improving the binding properties of an MIP. The resulting polymers have a higher percentage of high affinity sites and therefore show improved selectivity and affinity for the template molecule. The methodology is complementary to the existing MIP protocols and yields additive improvements to the existing binding properties of an MIP. Finally, in keeping with the synthetic efficiency of MIPs, the guest directed chemical modification strategy adds only a single postpolymerization step. This efficiency is made possible by the reversible noncovalent interactions of the template molecule.

(a) Introduction of New Types of Selectivities.   More recently, Shimizu et al. have extended the guest directed chemical modification strategy to not only improve the existing binding properties of an MIP matrix but also to introduce new types of selectivity. This was demonstrated by endowing a previously unselective MIP matrix with enantioselectivity as shown in Scheme 9. First a "racemic" MIP containing an equal number of L- and D-selective binding sites was prepared by imprinting a racemic guest. The "racemic" polymer had affinity but no enantioselectivity for the target molecule. The racemic MIP was then equilibrated with a single enantiomer (the L-enantiomer in the example in Scheme 9), which selectively binds to and protects the L-selective binding sites. Chemical modification then preferentially eliminates the D-selective sites. Washing away the noncovalent protecting group then yields an enantioselective polymer, containing an excess of L-selective sites.

The synthesis of an enantioselective polymer surface in this manner represents a new strategy for making chiral polymers. The overall process is similar to the kinetic resolution of racemic molecules. In kinetic resolution, enantiomers are differentiated using a reaction that preferentially reacts with one enantiomer over the other. The outcome is that one enantiomer is chemically modified and the other is unreacted and are now easily differentiated. In contrast to molecular kinetic resolution, the modified and unmodified binding sites in the "resolved" polymer cannot be physically separated. However, they hopefully have been sufficiently differentiated by the enantioselective reaction so that the chemically modified binding sites are no longer able to bind. The modified polymer should then effectively have an enantiomeric excess of binding sites.

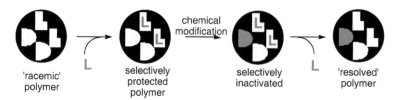

Scheme 9 Selective chemical modification using the template molecule to mask one specific class of sites.

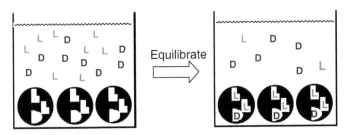

Scheme 10 Testing the enantioselectivity of the selectively chemically modified MIP using batch rebinding studies with the racemic guest molecule.

To extend this strategy to MIPs a well-studied MIP system was used to generate the "racemic" MIP starting material [12,15,16]. Racemic PAA was imprinted in the standard MAA/EGDMA matrix. This "racemic" MIP was equilibrated with enantiomerically pure L-PAA and then treated with diazomethane. Esterification presumably reacts with the unprotected D-selective sites. Verification of the enantioselectivity in the modified polymer was carried out using batch rebinding studies (Scheme 10). The polymer was equilibrated with racemic PAA. The observation of enantiomeric excess in the free solution confirms the enantioselectivity of the modified polymer.

The above assay also allowed quantification of the level of enantioselectivity introduced into the polymer system. A separation factor can be calculated from the ratio of the liquid/solid partition coefficients ($\alpha_B = $([bound L-PAA]/[free L-PAA])/([bound D-PAA]/[free D-PAA])), which is analogous to the separation factor used in chromatography [25]. An α_B value of 1 represents no enantioselectivity, and values greater than 1 represent increasing degrees of enantioselectivity. The enantioselectivity of the chemically modified MIPs ($\alpha_B = 6.2$) was found to be comparable to MIPs synthesized by directly imprinting L-PAA. The only difference was that the directly imprinted polymer had more sites which was consistent with the selective chemical modification methodology eliminating a portion of the original imprinted sites.

Normally the separation factor is a constant. However, in the case of MIPs, separation factors are highly concentration dependent (Fig. 5). The identical trend was observed for the "resolved" chemically modified MIP. At low analyte concentrations, the modified MIP showed a very high selectivity. However, with increasing analyte concentration, the measured enantioselectivity of the MIP rapidly decreased. In both cases, this strong concentration dependence is probably due to the heterogeneity of the binding sites. At low concentration, only the high affinity high selectivity sites are being measured. In contrast, at high concentrations, all the binding sites are being measured and the more numerous low affinity low selectivity sites dominate the properties of the polymer.

The similarity of the directly imprinted polymer and the selectively chemically modified MIP suggests that, in most cases, it will be easier to introduce selectivity by traditional molecular imprinting. However, there are situations in which the selective chemical modification strategy could be used to complement the "normal" imprinting process. For example, some molecules are not easily directly imprinted

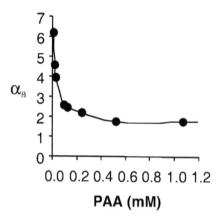

Figure 5 Graph showing the strong concentration dependence of the enantioselectivity of the selectively chemically modified MIP.

such as radical inhibitors, molecules containing reactive double bonds, or macromolecules that become physically entrapped during the polymerization process [26]. A second complementary application would be to use the selective chemical modification procedure to either improve or fine-tune the existing selectivity of an MIP to a degree that could not be accomplished via the imprinting process alone.

A. Chemical Modification of the Crosslinker

In contrast to modifications directed toward the functional monomer, only a few examples have been reported of modification reactions directed toward the crosslinker. The crosslink agents in MIPs are generally considered inert structural units that maintain binding site shape and integrity. Modifications of the crosslinker are expected to indirectly affect the binding properties, but more importantly to also affect materials properties such as the accessibility and mass transfer kinetics of the binding sites.

1. Unselective Methods

Sellergren and Shea [11] have demonstrated that partial hydrolysis of the EGDMA matrix in an MIP with 10% NaOH/MeOH (20/80, v/v) led to a diminishment in affinity and selectivity. This was also accompanied by greater swelling of the hydrolyzed matrix due to the lowered crosslinking percentages. These observations were in line with the generally accepted premise in molecular imprinting that lower levels of crosslinking yield materials of lower affinities and selectivities.

In another example, partial hydrolysis of the EGDMA crosslinker led to an enhancement in the properties of a catalytically active MIP [27]. The MIP had been simultaneously covalently and noncovalently imprinted with ester hydrolysis transition state analogs of an ester hydrolysis reaction. Increases in both rate and selectivity were observed in the catalytic activity of transition state imprinted polymer upon treatment with NaOH/MeOH. These improvements were proposed to arise from greater accessibility to catalytic sites in the MIP. Hydrolysis of the crosslinker apparently uncovers previously inaccessible active sites in the polymer leading to higher site density and catalytic activity.

2. Selective Methods

Only a single example of selective chemical modifications directed at the crosslinker has been reported by Hiratani et al. [28]. The unusual aspect of this system is that the crosslinks can be reversibly broken and reformed. The crosslinker contains disulfide linkages that can be broken under reductive conditions and reformed under oxidative conditions. Selectivity in the reformation of the crosslinks was accomplished by the presence of the guest molecule (metal dications) that are bound to the matrix via carboxylates. The overall procedure was termed "postimprinting" and is outlined in Scheme 11.

   This system has the ability to lose and then regain its imprint. Initially a gel containing a crosslinker with a disulfide linkage is imprinted with a dication (Pb^{2+}). This initial imprinted polymer now shows enhanced affinity for dications (Ca^{2+}). This imprint can be "erased" by reducing the disulfide bonds in the crosslinker, breaking the S–S bonds. Reoxidation and reformation of the disulfide crosslinks in the absence of Ca^{2+} do not regain Ca^{2+} affinity. However, if the thiols are reoxidized in the presence of Ca^{2+}, then the imprint is "rewritten".

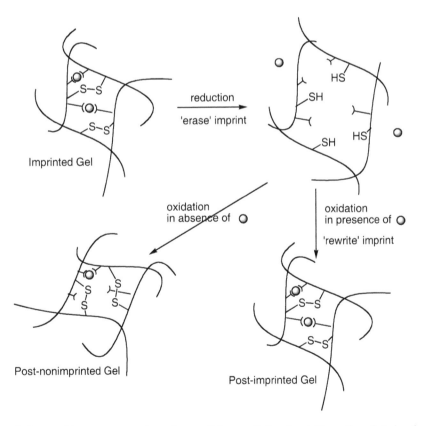

Scheme 11   Representation of reversible crosslinker to yield postimprinted polymers.

V. OUTLOOK FOR POSTMODIFICATIONS OF MIPs

Postmodification of MIPs has been demonstrated as a viable method to improve or fine-tune their binding properties. These strategies are complementary to existing molecular imprinting protocols because the enhancements that they produce are additive to the existing affinities and selectivities of an MIP. In general, post-polymerization treatments that alter or modify the MIP matrix in an unselective manner were found to lead to a loss in affinity and selectivity of the matrix. In contrast, selective chemical modification strategies that target specific subsets of binding sites in MIPs have been demonstrated to be effective in not only enhancing the existing affinity and selectivity of an MIP but also in altering and tailoring in new types of selectivity to an MIP matrix.

One disadvantage of chemical modifications strategies is that they generally yield diminished capacity. This is because the chosen chemical transformations all reduce the binding ability of the modified binding sites. For this reason, these strategies are best applied to applications in which high selectivity and affinity are of primary importance such as in sensors and immuno-type assays. Applications in which capacity is important such as chromatography may not be advantageous. Chemical transformations that enhance the binding affinity of a modified site would have the potential of improving affinity and capacity simultaneously. However, to date, there are no examples of this type of chemical modification in MIPs.

In choosing what type of modification strategy to apply to an MIP is often not so straightforward. There is balance of binding and materials properties in which the improvements in particular properties are offset by the diminishment in others. For example, in the thermal treatments of polymers reported by Sellergren et al. [29], there is an overall reduction in selectivity but a compensating increase in capacity. Alternatively, the guest directed selective chemical modification strategies applied by Karube et al. [22] and Shimizu et al. [18] showed in enhancement in the affinities and selectivities but an overall loss in capacity. Where selective post-polymerization modification may have the most utility is to generate complex types of selectivity that cannot be introduced by imprinting alone. Therefore, the ability to introduce new types of selectivity into MIPs may be the direction with the greatest potential for chemical post-modification strategies.

REFERENCES

1. Sellergren, B. In *Molecularly Imprinted Polymers. Man Made Mimics of Antibodies and their Applications in Analytical Chemistry*; Elsevier: Amsterdam, 2001.
2. Wulff, G. Molecular imprinting in cross-linked materials with the aid of molecular templates—A way towards artificial antibodies. Angew. Chem. Int. Ed. Engl. **1995**, *34*, 1812–1832.
3. Shea, K.J. Molecular imprinting of synthetic network polymers: the de novo synthesis of macromolecular binding and catalytic sites. Trends. Polym. Sci. **1994**, *2*, 166–173.
4. Sellergren, B. Imprinted chiral stationary phases in high-performance liquid chromatography. J. Chromatogr. A **2001**, *906*, 227–252.
5. Wulff, G. Enzyme-like catalysis by molecularly imprinted polymers. Chem. Rev. **2002**, *102*, 1–27.
6. Takeuchi, T.; Haginaka, J. Separation and sensing based on molecular recognition using molecularly imprinted polymers. J. Chromatogr. B **1999**, *728*, 1–20.

7. Haupt, K.; Mosbach, K. Molecularly imprinted polymers and their use in biomimetic sensors. Chem. Rev. **2000**, *100*, 2495–2504.

8. Sellergren, B.; Shea, K.J. Origin of peak asymmetry and the effect of temperature on solute retention in enantiomer separations on imprinted chiral stationary phases. J. Chromatogr. A **1995**, *690*, 29–39.

9. Andersson, L.I.; Muller, R.; Vlatakis, G.; Mosbach, K. Mimics of the binding-sites of opioid receptors obtained by molecular imprinting of enkephalin and morphine. Proc. Natl. Acad. Sci. USA **1995**, *92*, 4788–4792.

10. Umpleby II, R.J.; Bode, M.; Shimizu, K.D. Measurement of the continuous distribution of binding sites in molecularly imprinted polymers. Analyst **2000**, *125*, 1261–1265.

11. Sellergren, B.; Shea, K.J. Influence of polymer morphology on the ability of imprinted network polymers to resolve enantiomers. J. Chromatogr. **1993**, *635*, 31–49.

12. Chen, Y.B.; Kele, M.; Sajonz, P.; Sellergren, B.; Guiochon, G. Influence of thermal annealing on the thermodynamic and mass transfer kinetic properties of D- and L-phenylalanine anilide on imprinted polymeric stationary phases. Anal. Chem. **1999**, *71*, 928–938.

13. Andersson, H.S.; Karlsson, J.G.; Piletsky, S.A.; Koch-Schmidt, A.C.; Mosbach, K.; Nicholls, I.A. Study of the nature of recognition in molecularly imprinted polymers, II [1]—Influence of monomer–template ratio and sample load on retention and selectivity. J. Chromatogr. A **1999**, *848*, 39–49.

14. Svenson, J.; Nicholls, I.A. On the thermal and chemical stability of molecularly imprinted polymers. Anal. Chim. Acta. **2001**, *435*, 19–24.

15. Sellergren, B. Molecular imprinting by noncovalent interactions: tailor-made chiral stationary phases of high selectivity and sample load capacity. Chirality **1989**, *1*, 63–68.

16. Chen, Y.B.; Kele, M.; Quinones, I,; Sellergren, B.; Guiochon, G. Influence of the pH on the behavior of an imprinted polymeric stationary phase-supporting evidence for a binding site model. J. Chromatogr. A **2001**, *927*, 1–17.

17. Ellwanger, A.; Berggren, C.; Bayoudh, S.; Crecenzi, C.; Karlsson, L.; Owens, P.K.; Ensing, K.; Cormack, P.; Sherrington, D.; Sellergren, B. Evaluation of methods aimed at complete removal of template from molecularly imprinted polymers. Analyst **2001**, *126*, 784–792.

18. Umpleby, R.J.; Rushton, G.T.; Shah, R.N.; Rampey, A.M.; Bradshaw, J.C.; Berch, J.K.; Shimizu, K.D. Recognition directed site-selective chemical modification of molecularly imprinted polymers. Macromolecules **2001**, *34*, 8446–8452.

19. Shea, K.J.; Sasaki, D.Y. An analysis of small-molecule binding to functionalized synthetic polymers by 13C CP/MAS NMR and FT-IR spectroscopy. J. Am. Chem. Soc. **1991**, *113*, 4109–4120.

20. Svec, F.; Frechet, J.M.J. Pore-size specific modification as an approach to separation media for single-column, two-dimensional HPLC. Am. Lab. **1996**, *28*, 25–34.

21. Koh, J.H.; Larsen, A.O.; White, P.S.; Gagne, M.R. Disparate roles of chiral ligands and molecularly imprinted cavities in asymmetric catalysis and chiral poisoning. Organometallics **2001**, *21*, 7–9.

22. McNiven, S.; Yokobayashi, Y.; Cheong, S.H.; Karube, I. Enhancing the selectivity of molecularly imprinted polymers. Chem. Lett. **1997**, *12*, 1297–1298.

23. Creary, X. Tosylhydrazone salt pyrolyses: Phenyldiazomethanes. Diazomethane synthesis. Org. Syn. **1986**, *64*, 207–216.

24. Wulfman, D.S.; Yousefian, S.; White, J.M. The synthesis of aryl diazomethanes. Synth. Commun. **1998**, *18*, 2349–2352.

25. Tobler, E.; Lammerhofer, M.; Oberleitner, W.R.; Maier, N.M.; Lindner, W. Enantioselective liquid–solid extraction for screening of structurally related chiral stationary phases. Chromatographia **2000**, *51*, 65–70.

26. Spivak, D.A.; Shea, K.J. Investigation into the scope and limitations of molecular imprinting with DNA molecules. Anal. Chim. Acta **2001**, *435*, 65–74.
27. Sellergren, B.; Karmalkar, R.N.; Shea, K.J. Enantioselective ester hydrolysis catalyzed by imprinted polymers. 2. J. Org. Chem. **2000**, *65*, 4009–4027.
28. Hiratani, H.; Alvarez-Lorenzo, C.; Chuang, J.; Guney, O.; Grosberg, A.Y.; Tanaka, T. Effect of reversible cross-linker, N,N'- bis(acryloyl)cystamine, on calcium ion adsorption by imprinted gels. Langmuir **2001**, *17*, 4431–4436.
29. Sajonz, P.; Kele, M.; Zhong, G.M.; Sellergren, B.; Guiochon, G. Study of the thermo-dynamics and mass transfer kinetics of two enantiomers on a polymeric imprinted stationary phase. J. Chromatogr. A **1998**, *810*, 1–17.

13

Molecular Imprinting Using Hybrid Materials as Host Matrices

Sheng Dai Oak Ridge National Laboratory, Oak Ridge, Tennessee, U.S.A.

I. INTRODUCTION

Imprinting matrices can play crucial roles in enhancing the imprinting effect through control of the environment surrounding the imprinting sites [1–9]. This surrounding environment can be viewed as the *second coordination shell* to the template molecules or ions, which is shown schematically in Fig. 1. Until now, the main matrices employed in molecular imprinting have been either organic or inorganic polymers [1–4, 10–15]. Cross-linked polymethacrylate is an example of an organic-imprinting matrix, while silica sol–gel glass represents a popular inorganic-imprinting matrix (see Chapter 11 for imprinting in inorganic matrices). In this chapter, we discuss a relatively new type of matrix based on hybrid materials [6–8, 16–18].

Hybrid materials are defined as synthetic materials with organic (hydrophobic) and inorganic (hydrophilic) components [19–21]. This combination results in a unique composite material with amphiphilic properties. Hybrid materials can be further divided into two classes: (1) homogeneous systems derived from monomers or miscible organic and inorganic components and (2) heterogeneous systems with separated domains ranging from few nanometers to micrometers in size. This chapter focuses on the use of the first class of the hybrid materials as matrices for conducting imprinting synthesis and specially for regulating the second coordination shell via the imprinting synthesis. The examples used in this chapter are from the published literature from the author's group, although there are many excellent papers that interested readers should also need to consult for related research [4,16,17]. *Sol–gel processing* is widely employed to synthesize this class of hybrid materials [19].

The submitted manuscript has been authored by a contractor of the U.S. Government under contract DE-AC05-00OR22725. Accordingly, the U.S. Government retains a nonexclusive, royalty-free license to publish or reproduce the published form of this contribution, or allow others to do so, for U.S. Government purposes.

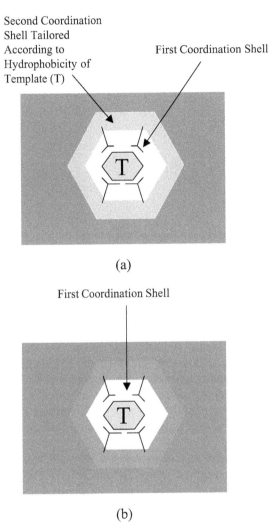

Second Coordination Shell Tailored According to Hydrophobicity of Template (T)

First Coordination Shell

(a)

First Coordination Shell

(b)

Figure 1 Comparison of surrounding environments created by molecular imprinting in (a) hybrid and (b) nonhybrid hosts. In hybrid hosts, there is a regulated distribution of hydrophobic or hydrophilic components around template. Minimum such regulation is expected for the nonhybrid hosts.

II. SOL–GEL PROCESSING

The sol–gel process, which refers to a multitude of reactions that employ a wide variety of alkoxide precursors to prepare many different inorganic oxide-based products, has been applied most often to the production of glasses and ceramic oxides [22]. Typically, the sol–gel process involves a metal alkoxide such as $Si(OR)_4$, water, and a second solvent that acts as a miscibility agent. The particles that develop in the colloidal sol under a basic condition cross-link to form a gel, which is subsequently dried to form a porous glass. The fundamental sequence of the reactions responsible for the ultimate development of a cross-linked three-dimensional solid matrix is as follows: (1) hydrolysis of Si-alkoxide bonds to produce silanol groups and (2) cross-condensation of silanols with (2b) themselves or (2a) unhydrolyzed Si–OR groups to produce Si–O–Si moieties, which form the basis for the growing polymeric matrix. These reactions are illustrated below:

$$Si(OR)_x + H_2O \rightarrow (RO)_{x-1}Si - OH + ROH \tag{1}$$

$$(RO)_{x-1}Si - OH + (RO)_xSi \rightarrow (RO)_{x-1}Si - O - Si(OR)_{x-1} + ROH \tag{2a}$$

$$(RO)_{x-1}Si - OH + HO - Si(OR)_{x-1} \rightarrow (RO)_{x-1}Si - O - Si(OR)_{x-1} + H_2O \tag{2b}$$

In sol–gel solutions, the rate, extent, and even the mechanisms of the reactions are profoundly affected and may be systematically influenced by many factors, such as the size of the alkoxide ligand on silicon, the solution pH, the types and concentrations of solvents, the temperature, and the presence of catalysts. Each factor can affect the course of structure development and subsequent properties of the gel and the materials produced. Therefore, a multitude of variables are available with which to influence and tailor the structure and properties of sol–gel-derived materials. Furthermore, because little or no heating is required in the sol–gel process, the gel can be doped with thermally sensitive molecules that would normally be precluded from incorporation into traditional inorganic matrices (for recent reviews and reports see Ref. 23). The technique of entrapping molecules within the pores of sol–gel-derived glasses is well known and has been applied in a number of applications, including the preparation of chemical sensing materials, wave guides, and solid-state dye lasers [22,23].

A new dimension in sol–gel chemistry has been opened up with sol–gel alkoxide precursors consisting of functional organic groups [R'-Si(OR)₃] or organic spacers directly bound to siloxanes [(RO)₃Si–R'-Si(OR)₃] [19–21]. Figure 2 shows some examples of the precursors with trialkoxyl silanes attached to alkyl and phenyl groups. The hybrid matrices can be readily prepared by the cocondensation of these

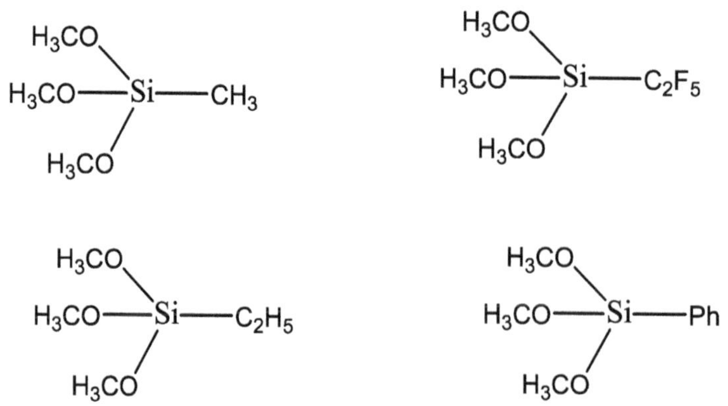

Figure 2 Examples of trialkoxyl silane precursors.

precursors with tetraalkoxide silanes [19]. The matrix properties, such as hydrophobicity and porosity, can be systematically "engineered" according to the degree of interpenetration and mixing ratios of the organic and inorganic components. Figure 3 shows some examples of the second type of hybrid siloxane precursor (i.e., with organic spacers or bridges bonded to trialkoxyl silanes). In this case, hybrid matrices with defined organic and inorganic contents can be directly prepared through the self-condensation reaction without tetraalkoxyl silanes [24]. The above two types of hybrid materials have been extensively described in the literature and are finding increasing applications in many different areas. Materials with compositions and properties that span the entire range from pure organic polymers to pure inorganic oxide matrices are now being prepared and studied. The mild conditions used in the sol–gel process (relative to the preparation of other inorganic oxide-based materials) also lend themselves to the incorporation of a wide variety of hybrid precursors and functionality that may be tailored to the type of reactivity or physical property desired in the final material.

III. SOL–GEL PROCESSING AND MOLECULAR IMPRINTING

The earliest matrix used to conduct molecular imprinting is based on silica sol–gel materials [25]. Recently, there is the resurgence of using the sol–gel materials based on silicon or metal alkoxide precursors as the matrices for conducting molecular imprinting [5, 11, 14–18]. Both direct template synthesis and surface imprinting functionalization have been explored. Specifically, Wulff et al. [15] showed that the rigidity of the silica matrix can be successfully employed to conduct surface imprinting of organic molecules via a covalent approach. More recently, Katz and Davis [5] developed a bulk-imprinting technique on a microporous sol–gel matrix. Markowitz et al. [11] have successfully generated the imprinting sites on novel silica nanoparticles for catalysis applications.

Recent breakthroughs in sol–gel synthesis have resulted in a novel methodology for preparing ordered mesoporous inorganic materials (Fig. 4) with extremely high surface areas ($> 1000 \, m^2 \, g^{-1}$) in which the pore size can be tuned from 1.2 to 35 nm by adjustment of the synthetic parameters [26–28]. The original method was developed by scientists at Mobil Oil Research and Development for synthesis of ordered mesoporous silica [26]. The essence of this new methodology is the use of

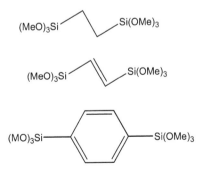

Figure 3 Examples of bridged polysilsesquioxanes.

Figure 4 Honeycomb texture of ordered mesoporous SiO_2.

molecular assemblies of surfactants (i.e., micelles, or related substances such as block copolymers) as structure "directors" during formation of metal oxides such as SiO_2 and $SiO_2–Al_2O_3$ (Fig. 5). The surfactant templates can be removed from the resulting gel by solvent extraction or calcination at 450°C. Although it was initially used to synthesize metal oxides, the surfactant template approach is now used to engineer the porosity of a wide variety of new and novel materials. Hybrid porous materials have been synthesized by template-directed cocondensation of tetraalkoxysilanes and functional organotrialkoxysilanes [29–31]. Notably, Hall et al. [29] have demonstrated that organo-functionalized MCM-41 silica hexagonal mesophases containing binary combinations of covalently linked phenyl, amino, or thiol moieties can be synthesized. The drawback associated with this methodology is that it is very difficult to increase the content of the functional organosilane without disrupting the mesoporous silica networks. Accordingly, both surface areas and pore volumes of these materials decrease considerably with an increase in the functional organosilane content. Recently, Inagaki et al. [32] have developed a novel methodology to synthesize porous hybrid materials based on the sol–gel processing of a single bridged-silsesquioxane precursor. Highly porous and periodic organosilicas with organic groups inside the channel walls were synthesized. The key to the success

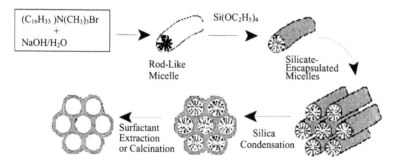

Figure 5 Surfactant self-assembly synthesis of ordered mesoporous SiO_2.

of this new synthetic methodology lies in the unique structural feature of the bridged-silsesquioxane sol–gel precursor [24].

We [7–9] and others [4] have combined the template synthesis of mesoporous materials with the imprinting synthesis, resulting in a methodology for synthesizing functional porous materials with double templates (metal ions or molecules and surfactant micelles). The porous materials derived from such syntheses have hierarchical structures. On the microporous level, removal of the metal ion or molecular template from the corresponding complex leaves ion or molecular sorption cavities that exhibit ionic or molecular recognition. These pores give the sorbent enhanced selectivity for the given ion or molecule. On the mesoporous level, removal of the surfactant micelles results in the formation of relatively large cylindrical pores (diameters of 25–60 Å) that give the gel an overall porosity, including large surface areas and excellent metal-ion transport kinetics. The removal of the surfactant template is usually through solvent extraction. Pore sizes can be easily controlled by the chain length of the surfactant molecules. This combination of high capacity and selectivity, coupled with fast kinetics, makes these materials ideal candidates for many applications. Because the entire process utilizes *template or imprinting synthesis* twice and on different length scales, it can be viewed as a hierarchical double-imprinting process. For example, Shin et al. [4] have successfully demonstrated that the surface imprinting of interesting organic molecules can be achieved on ordered mesoporous matrices.

IV. HYBRID FUNCTIONAL LIGANDS USED TO IMPRINT METAL IONS

Although the hybrid matrices can be used directly to imprint targeted species through the control of the hydrophilic and hydrophobic interactions from silanol and alkyl groups, stronger interactions are usually needed to create high-affinity adsorption sites. Many organic functional ligands can be readily anchored to siloxanes through the Si–C bond. These ligands can be used to construct the *first coordination shell* for the imprinting template. Some of the $(RO_3)Si$-grafted ligands that have already strong affinities toward certain metal ions are shown in Fig. 6. The methodologies for making such functional ligands are well developed and have found applications in the process of anchoring metal ions on silica to produce unique heterogeneous catalysts [21]. The ligand that is selected for construction of the first corrdination shell in the discussion of this chapter is 3-(2-aminoethylamino)propyltrimethoxysilane.

V. TEMPLATES FOR IMPRINTING SYNTHESIS

In order to illustrate the potential of the imprinting synthesis using hybrid materials, Cu^{2+}-selective sorbents were chosen to demonstrate both the basic principles of the concept and the ease with which imprinted sorbents may be prepared. Furthermore, an extensive body of literature already exists concerning imprinted polymers for the separation of copper ions [1,18,33]. A critical comparison of the binding properties of our sorbents with those of imprinted organic polymeric matrices shows the advantages of our imprinted hybrid sorbents (vide infra). To test the selectivity of these sorbents, we conducted competitive ion-binding experiments using aqueous

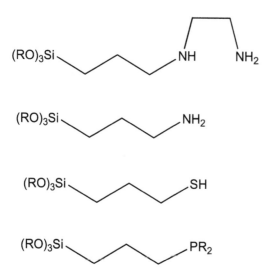

Figure 6 Examples of bifunctional silane ligands.

Cu^{2+}/Zn^{2+} mixtures. This system constitutes one of the most stringent tests for ion-binding selectivity because both ions have identical charges, are of similar sizes, and exhibit high affinities toward amine ligands (see Chapter 6 for dative bonding interaction).

VI. BATCH EQUILIBRIUM TESTS

The selectivity coefficient, k, for the binding of a specific ion in the presence of competitor species can be obtained from equilibrium binding data according to Eq. (4):[9,33]

$$M_1(\text{solution}) + M_2(\text{sorbent}) \rightleftharpoons M_2(\text{solution}) + M_1(\text{sorbent}) \qquad (3)$$

$$k = \{[M_2]_{\text{solution}}[M_1]_{\text{sorbent}}\} / \{[M_1]_{\text{solution}}[M_2]_{\text{sorbent}}\}$$
$$= K_d(\text{Cu})/K_d(\text{Zn}), \qquad (4)$$

where K_d is the distribution coefficient, expressed as

$$K_d = \{(C_i - C_f)/C_f\} \times \{\text{volume solution (mL)}\}/\{\text{mass gel (g)}\}. \qquad (5)$$

Here, C_i is the initial solution concentration and C_f is the final solution concentration. Comparison of the k values for the imprinted and control blank gels can show the effect of imprinting on the metal-ion selectivity for a given material. A measure of the increase in selectivity due to molecular imprinting can be defined by the ratio of the selectivity coefficients of the imprinted (MIP) and nonimprinted (NP) materials:

$$k' = k_{\text{MIP}}/k_{\text{NP}} \qquad (6)$$

VII. IMPRINTING WITH A COCONDENSATION OF RSi(OR)$_3$ AND Si(OR)$_4$ [6]

The hybrid matrices used to imprint the template copper ion were synthesized by the cocondensation of alkyltrimethoxysilane [RSi(OR)$_3$] and tetramethylorthosilicate (TMOS). The complexing ligand for the first shell coordination shell is 3-(2-aminoethylamino)propyltrimethoxysilane (AAPTS). Table 1 summarizes batch adsorption results for the distribution constant (K_d), selectivity coefficient of the sorbent toward Cu^{2+} (k), and the relative selectivity coefficient (k) obtained in these competitive ion-binding experiments between zinc and copper ions for the hybrid sorbents prepared by the cocondensation of RSi(OR)$_3$ and Si(OR)$_4$. As seen in Table 1, K_d(Cu) values of the imprinted and control blank sorbents prepared using only TMOS and AAPTS are very close. However, the K_d(Cu) values of the imprinted and control blank sorbents are very different for the hybrid sorbents prepared using the mixtures of TMOS, AAPTS, and methyltrimethyoxysilane (MTMOS) or phenyltrimethoxysilane (PTMOS). For example, the K_d(Cu) value of sorbent Imp-2 is approximately 28 times greater than that of the corresponding control blank (Nonimp-2). The enhanced imprinting effect for the hybrid sorbents synthesized using the organosilane precursor MTMOS or PTMOS can be attributed to the hydrophobicity imparted by the functional group of CH$_3$–Si or C$_6$H$_5$–Si in the hybrid sorbents. Figure 7 is a schematic illustration of this enhanced imprinting effect for the hybrid sorbents. In the matrices of the nonimprinted control blanks, the complexation capability of AAPTS ligand for the hydrophilic copper ions is greatly reduced by the presence of the hydrophobic CH$_3$–Si or C$_6$H$_5$–Si group. However, the surrounding environment or the second coordination shells of the Cu^{2+} absorption sites in the imprinted sorbents are already optimized by the hydrophilic template (Cu^{2+}). Here, the hydrophobic CH$_3$–Si or C$_6$H$_5$–Si group is located far from the first coordination shell. This matrix preorganization is induced by the unfavorable interaction between the hydrophilic copper ions and the hydrophobic CH$_3$–Si or C$_6$H$_5$–Si group. In the case of the control blank sorbents, the distance between the AAPTS and the CH$_3$–Si or C$_6$H$_5$–Si group should be closer and the adsorption sites are expected to be more hydrophobic. Accordingly, the K_d(Cu) values of the nonimprinted sorbents are significantly lower than those of the imprinted sorbents.

Table 1 Competitive Loading of M$_1$ (Cu^{2+}) and M$_2$ (Zn^{2+}) by Copper-imprinted and Conventional Hybrid Sorbents at pH 5.2 (Acetic Acid/Sodium Acetate Buffer)[a] (Reproduced from Data Reported in Ref. 6, with Permission from Elsevier Science)

Sample[b]	Host precursor composition	Uptake Cu (%)	Uptake Zn (%)	K_d (Cu)	K_d (Zn)	k
Nonimp-1	TMOS	91	8	1000	9	110
Imp-1	TMOS	94	12	1500	14	110
Nonimp-2	1 TMOS:1 MTMOS	38	5	62	6	11
Imp-2	1 TMOS:1 MTMOS	95	5	1800	6	300
Nonimp-3	1 TMOS:1 PTMOS	28	2	39	2	19
Imp-3	1 TMOS:1 PTMOS	91	2	970	2	530

[a]The initial Cu^{2+} and Zn^{2+} concentrations are 1×10^{-4} and 1×10^{-3} mol L^{-1}, respectively.
[b]Nonimp, nonimprinted sorbent; Imp, imprinted sorbent.

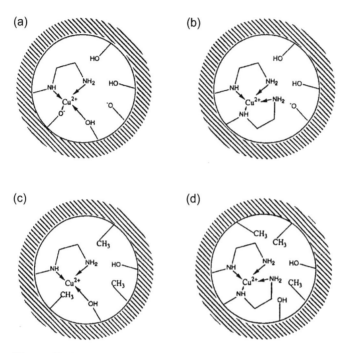

Figure 7 Schematic diagram of adsorption sites for imprinted ("b" and "d") and nonimprinted ("a" and "c") sorbents. Sorbents "a" and "b" are prepared using TMOS and AAPTS. Sorbents "c" and "d" are prepared using TMOS, MTMOS, and AAPTS. (Reproduced from data reported in Ref. 6, with permission from Elsevier Science).

The variation in $K_d(Cu)$ in this case is fundamentally identical to the change in pKa induced by the matrix hydrophobicity [34]. In the control blank sorbents prepared using only TMOS and AAPTS, the presence of abundant hydrophilic Si–OH groups provides adsorption sites for Cu^{2+} in addition to AAPTS. The difference in complexation capability between sites "a" and "b" is less than that between sites "c" and "d" (Fig. 7). Furthermore, the concentration of site "a" is approximately two times greater than that of site "b". Therefore, the $K_d(Cu)$ values of the imprinted and nonimprinted sorbents prepared using only TMOS and AAPTS are very similar.

VIII. HIERARCHICAL IMPRINTING WITH BRIDGED-POLYSILSESQUIOXANE HYBRID HOSTS [7]

The mesoporous organosilica system used in the current discussion was prepared by the combination of the surfactant template synthesis using 1,2-bis(triethoxysilyl)ethane (BTSE) and the Cu^{2+}-imprinting synthesis using AAPTS. Briefly, copolymerization of BTSE with metal-ion complexes of AAPTS around supramolecular assemblies of cetyltrimethylammonium chloride (CTAC) gives a composite polymer. Two important interactions are used in this double-imprinting synthesis. On mesoscale, the interfacial interaction between organosilica and quarternary ammonium surfactant micelles determines the sizes of the mesopores. On a microscale, the template ions determine both the stereochemistry and the location

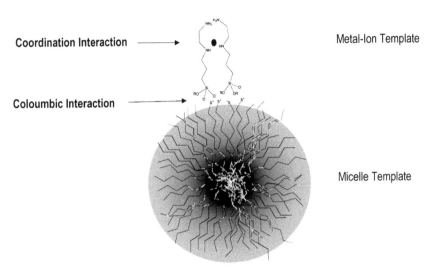

Coordination Interaction ⟶   Metal-Ion Template

Coloumbic Interaction ⟶

                                                                    Micelle Template

Figure 8 Schematic diagram showing two interactions used in hierarchical imprinting synthesis.

of amine ligands (Fig. 8). The hydrophilicity of the template ions also regulates the local environment of the ligands, making it more hydrophilic than those in which the template ions are not used. The difference in the environment of the second coordination shell in the imprinted and nonimprinted materials is provided by the distribution of hydrophobic alkyl spacers and hydrophilic silanol groups.

Plots of the logarithm of the distribution coefficient ves. initial Cu(II) concentration are shown in Fig. 9 for the copper-imprinted and nonimprinted hybrid materials from solutions with concentrations ranging from 6.35 from 63.55 ppm Cu(II) during a standard batch procedure at pH 5.0. A direct comparison shows that at every concentration, the copper-imprinted material removed more of the copper ions than did the nonimprinted hybrid, which provides a clear example of the increased affinity of the imprinted sorbent for the metal-ion template. The results of competitive ion-binding batch tests with Zn(II) as the competitor species are summarized in Table 2. At all concentrations, the copper-imprinted hybrid absorbed more Cu(II) and more Zn(II) than did the nonimprinted material. The enhanced capacity for Cu(II) adsorption due to molecular imprinting is expected, while the reason for the increased uptake of the Zn(II) competitor species is less clear. However, the latter can be explained by the favorable second coordination shell created by the imprinting synthesis. The imprinting synthesis generates hydrophilic environments around the functional amine ligands in the relatively hydrophobic organic–inorganic matrix. An important concept here is that the imprinting process enhances the affinity of the sorbent for the target Cu(II) more than that for the Zn(II) competitor. This is shown by the larger selectivity coefficients (k) as well as relative selectivity coefficients (k') greater than unity for the copper-imprinted sorbent at all solution concentrations with the exception of the lowest $[10^{-4}Cu(II)/10^{-4}Zn(II)]$. At this lowest concentration, the imprinted material should have enough binding sites to remove most of both metal species, decreasing its calculated selectivity. The detailed discussions of this imprinting system and related materials imprinted by other metal templates are available in the literature [7].

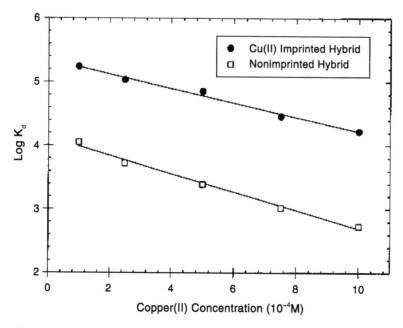

Figure 9 Cu(II) adsorption isotherms for the copper imprinted and nonimprinted sol–gels. Solutions were buffered at pH 5.0 with HAc/NaAc (0.05 M). A 0.1 g amount of sorbent was equilibrated with 10 mL of metal ion solution for 1 h. (Reproduced from data reported in Ref. 7, with permission from the American Chemical Society.)

IX. TYPICAL EXPERIMENTAL PROCEDURE [6,7]

Chemicals. Tetramethylorthosilicate (TMOS), methyltrimethoxysilane (MTMOS), phenyltrimethoxysilane (PTMOS), 1,2-bis(triethoxysilyl)ethane (BTSE), and

Table 2 Competitive Loading of M_1 (Cu^{2+}) and M_2 (Zn^{2+}) by Copper-imprinted and Conventional Hybrid Sorbents at pH 5.0 (Acetic Acid/Sodium Acetate Buffer) (Reproduced from Data Reported in Ref. 7, with Permission from the American Chemical Society)

Sorbent[a]	C_0[b] (Cu /Zn) (M)	Uptake Cu (%)	Uptake Zn (%)	K_d (Cu)	K_d (Zn)	k	k'
Nonimp	10^{-4}	98.99	20.20	9700	25.3	386	
Cu-Imp	10^{-4}	99.93	85.47	153000	588	260	0.67
Nonimp	2.5×10^{-4}	96.60	20.08	2840	25.1	113	
Cu-Imp	2.5×10^{-4}	99.92	74.68	129700	295	440	3.9
Nonimp	5.0×10^{-4}	92.69	14.33	1270	16.7	76	
Cu-Imp	5.0×10^{-4}	99.88	60.38	85400	152.4	561	7.4
Nonimp	7.5×10^{-4}	89.81	6.71	880	7.2	122	
Cu-Imp	7.5×10^{-4}	99.80	43.63	48800	77.4	631	5.2
Nonimp	10^{-3}	80.84	4.30	420	4.5	94	
Cu-Imp	10^{-3}	99.58	31.62	23500	46	507	5.4

[a]Nonimp, nonimprinted sorbent; Cu-Imp, Cu-imprinted sorbent.
[b]C_0, initial concentration for Cu(II) or Zn(II).

3-(2-aminoethylamino)propyltrimethoxysilane (AAPTS) were obtained from Gelest, Inc.; sodium hydroxide (50 wt%) and cetyltrimethylammonium chloride (CTAC, 25 wt%) were obtained from Aldrich. The nitrate salts of copper, nickel, zinc, and cadmium were purchased from Fluka. Toluene is from Aldrich. Nitric acid (69%) and hydrochloric acid (37%) were obtained from J.T. Baker. Absolute ethyl alcohol (EtOH) and methanol (MeOH) were purchased from AAPER Alcohol and Chemical. All chemicals were used as received.

Synthesis of Cu^{2+}-Imprinted Hybrid Hosts Through Cocondensation of RSi(OR)$_3$ and TMOS. In a typical synthesis of imprinted sol–gel silica, CuCl$_2$·2H$_2$O (0.002 mol) was dissolved in 4 mL MeOH. An appropriate amount of AAPTS ligand (0.004 mol) was added to this solution. The copper complex [Cu(AAPTS)$_2$$^{2+}$] was formed immediately. A mixture of MTMOS (x mol) or PTMOS (x mol) and TMOS (0.04– x mol) was then added to the solution, which was followed by addition of 2 mL MeOH and y mL of toluene. The methoxy groups of AAPTS readily undergo hydrolysis and condensation reactions, which anchor it to the polymer. The ethylenediamine moeity contains nitrogen donor atoms that can sequester metal ions from aqueous solutions by forming stable complexes via dative bonding. The value of x was 0–0.02 mol, while the value of y was 0–20 mL. The solution was clear, and no precipitation was observed. Hydrolysis and condensation of silanes were achieved by addition of an aqueous sodium fluoride solution (2% w/v, 4 mL). Blue gels were formed immediately and aged for 4 days at room temperature. Nonimprinted white silica gels were similarly prepared without addition of the copper template. The final silica gels were crushed. Template-free imprinted sorbents were easily obtained by extraction of the Cu^{2+} template with 1 M nitric acid. Control blank sorbents were also treated identically to ensure that both types of sorbents had identical surface treatments. Both imprinted and control blank sorbents were then neutralized to pH 7 using a buffer solution of NaHCO$_3$ and dried using a vacuum oven at 60°C for 24 h before adsorption tests were conducted.

Synthesis of Cu^{2+}-Imprinted Hybrid Hosts Using Bridged Polysilsesquioxane. The BTSE monomer undergoes polymerization by base-catalyzed hydrolysis and condensation reactions. The polymer is functionalized by incorporation of the bifunctional reagent, AAPTS, into the reaction mixture. The hybrid control materials in this study were synthesized according to the following procedures. The appropriate amount of 25% w/w CTAC solution was added to deionized water. Sodium hydroxide was added upon stirring, followed by the addition of BTSE and the bifunctional ligand (AAPTS). The resulting reaction mixture had the following molar composition:

1.0BTSE : 0.12CTAC : 0.15AAPTS : 1.0NaOH : 230H$_2$O

The reaction mixture was stirred in a closed vessel at room temperature for 24 h, and the solid precipitate was recovered via vacuum filtration. The copper-imprinted hybrid sorbents were prepared first by the addition of the appropriate amount of 25 wt% CTAC solution to deionized water. The Cu(NO$_3$)$_2$·3H$_2$O was then dissolved in the micellar solution. AAPTS was added while stirring, followed by the BTSE. Sodium hydroxide was then added to the blue reaction mixture, which had the following molar composition:

1.0BTSE : 0.12CTAC : 0.15AAPTS : 0.075Cu(NO$_3$)2
    × 3H$_2$O : 1.0NaOH : 230H$_2$O

After recovery via vacuum filtration, the solid precipitates were rinsed with deionized water and ethanol and placed under vacuum at 80°C for 4 h. The assynthesized material was placed in an excess of 1.0 M HNO_3 and stirred for 15 min to strip the copper template. This material was recovered by vacuum filtration, rinsed with deionized water, and placed in a large excess ($\sim$150 mL/g) of EtOH/HCl (1.0 M w.r.t. HCl) to extract the surfactant and remaining metal templates. This mixture was stirred well under gentle heating (50°C) for 8 h. The sorbent was again recovered by vacuum filtration, washed with plenty of EtOH, and dried under vacuum at 80°C for 1 h.

Batch Procedures. All metal-ion solutions were buffered to a specific pH with sodium acetate/acetic acid (0.05 M). In a typical run, 0.1 g sorbent and 10 mL metal-ion solution were placed in a capped plastic vials and sonicated for 30 min. The resulting mixture was filtered and both the filtrate and the initial standard solutions were analyzed via inductively coupled plasma (ICP) spectroscopy in order to measure the initial and final metal-ion concentrations. The overall capacity of the sorbent for a given metal ion was then calculated by the change in concentration between the filtrate and the initial metal-ion solution.

X. CONCLUSIONS

In conclusion, hybrid materials provide unique matrix for conducting imprinted material synthesis. This imprinting synthesis has resulted in a new class of porous hybrid sorbents that exhibit favorable arrangement of not only the first coordination shell but also the tailored second coordination shell. We view these new sorbents as heterogeneous analogues to crown ether-type ligands that are tailored for a specific target ion. The simplicity of this technique should lead to a wide variety of new, highly selective sorbents, the properties of which can be optimized for many metal ions with the proviso that they form stable coordination complexes with a suitable bifunctional ligand containing a silane group. Furthermore, this imprinting synthesis methodology should not be limited to the binding of metal ions. If complexes or molecules can be formed between targeted organic molecules and functional groups containing a silane group, application of the above methodology should lead to the synthesis of sorbents that exhibit molecular recognition of organic molecules.

ACKNOWLEDGMENT

The author wants to thank both present and former graduate students and postdoctoral fellows (M. C. Burleigh, Y. H. Ju, C. D. Liang, R. Makote, Y. S. Shin, Z. T. Zhang) for their hard work on this project and conducting all experiments described in this chapter. This work was conducted at the Oak Ridge National Laboratory and supported by the Division of Chemical Sciences, Office of Basic Energy Sciences, US Department of Energy, under contract No. DE-AC05-96OR22464 with UT-Battelle.

REFERENCES

1. (a) Wulff, G. Molecular imprinting in cross-linked materials with the aid of molecular templates. A way towards artificial antibodies. Angew. Chem. Int. Ed. Engl. **1995**, *34*, 1812–1846; (b) Wulff, G. Enzyme-like catalysis by molecularly imprinted polymers. Chem. Rev. **2002**, *102*, 1–27.

2. (a) Haupt, K.; Mosbach, K. Molecularly imprinted polymers and their use in biomimetic sensors. Chem. Rev. **2000**, *100*, 2495–2504; (b) Mosbach, K. Molecular imprinting. Trends Biochem. Sci. **1994**, *19*, 9–14.

3. (a) Shea, K.J. Molecular imprinting of synthetic network polymers: the de novo synthesis of macromolecular binding and catalytic sites. Trends Polym. Sci. **1994**, *2*, 166–184; (b) Hart, B.R.; Shea, K.J. Synthetic peptide receptors: molecularly imprinted polymers for the recognition of peptides using peptide–metal interactions. J. Am. Chem. Soc. **2001**, *123*, 2072–2073.

4. Shin, Y.S.; Liu, J.; Wang, L.Q.; Nie, Z.M.; Samuels, W.D.; Fryxell, G.E.; Exarhos, G.J. Ordered hierarchical porous materials: towards tunable size- and shape-selective microcavities in nanoporous channels. Angew. Chem. Int. Ed. Engl. **2000**, *39*, 2702–2707.

5. Katz, A.; Davis, M.E. Molecular imprinting of bulk, microporous silica. Nature **2000**, *403*, 286–289.

6. Makote, R.D.; Dai, S. Matrix-induced modification of imprinting effect for Cu^{2+} adsorption in hybrid silica matrices. Anal. Chim. Acta. **2001**, *435*, 169–175.

7. Burleigh, M.C.; Dai, S.; Hagaman, E.W.; Lin, J.S. Imprinted polysilsesquioxanes for the enhanced recognition of metal ions. Chem. Mater. **2001**, *13*, 2537–2546.

8. Dai, S.; Burleigh, M.C.; Ju, Y.H.; Gao, H.J.; Lin, J.S.; Pennycook, S.; Barnes, C.E.; Xue, Z.L. Hierarchically imprinted sorbents for the separation of metal ions. J. Am. Chem. Soc. **2000**, *122*, 992–993.

9. Dai, S. Hierarchically imprinted sorbents. Chem-Eur. J. **2001**, *7*, 763–768.

10. (a) Whitcombe, M.J.; Vulfson, E.N. Imprinted polymers. Adv. Mater. **2001**, *13*, 467–482; (b) D'Souza, S. M.; Alexander, C.A.; Carr, S.W.; Waller, A.M.; Whitcombe, M.J.; Vulfson, E.N. Directed nucleation of calcite at a crystal-imprinted polymer surface. Nature **1999**, *398*, 312–316.

11. Markowitz, M.A.; Deng, G.; Gaber, B. Effects of added organosilanes on the formation and adsorption properties of silicates surface-imprinted with an organophosphonate. Langmuir **2000**, *16*, 6148–6155.

12. Umpleby, R.J.; Rushton, G.T.; Shah, R.N.; Rampey, A.M.; Bradshaw, J.C.; Berch, J.K.; Shimizu, K.D. Recognition directed site-selective chemical modification of molecularly imprinted polymers. Macromolecules **2001**, *34*, 8446–8452.

13. Shi, H.Q.; Tsai, W.B.; Garrison, M.D.; Ferrari, S.; Ratner, B.D. Template-imprinted nanostructured surfaces for protein recognition. Nature **1999**, *398*, 593–597.

14. Dai, S.; Shin, Y.; Barnes, C.E.; Toth, L.M. Enhancement of uranyl adsorption capacity and selectivity on silica sol–gel glasses via molecular imprinting. Chem. Mater. **1997**, *9*, 2521–2529.

15. Wulff, G.; Heide, B.; Helfmeier, G. Molecular recognition through the exact placement of functional groups on rigid matrices via a template approach. J. Am. Chem. Soc. **1986**, *108*, 1089–1091.

16. Marx, S.; Liron, Z. Molecular imprinting in thin films of organic–inorganic hybrid sol–gel and acrylic polymers. Chem. Mater. **2001**, *13*, 3624–3630.

17. Makote, R.; Collinson, M.M. Template recognition in inorganic–organic hybrid films prepared by the sol–gel process. Chem. Mater. **1998**, *10*, 2440–2446.

18. (a) Ichinose, I.; Kikuchi, T.; Lee, S.W.; Kunitake, T. Imprinting and selective binding of di- and tri-peptides in ultrathin TiO_2-gel films in aqueous solutions. Chem. Lett. **2002**, 104–105; (b) He, J.H.; Ichinose, I.; Kunitake, T. Imprinting of coordination geometry in ultrathin films via the surface sol-gel process. Chem. Lett. **2001**, 850–851.

19. (a) Loy, D.A. Hybrid organic–inorganic materials. MRS Bull **2001**, *26*, 364–365; (b) Shea, K.J.; Loy, D.A. Bridged polysilsesquioxanes. Molecular-engineered hybrid organic–inorganic materials. Chem. Mater. **2001**, *13*, 3306–3319; (c) Shea, K.J.; Loy, D.A. Bridged polysilsesquioxanes: molecular engineering of hybrid organic–inorganic materials. MRS Bull **2001**, *26*, 368–376; (d) Sanchez, C.; Lebeau, B. Design and

properties of hybrid organic–inorganic nanocomposites for photonics. MRS Bull **2001**, *26*, 377–387; (e) Chujo, Y.; Tamaki, R. New preparation methods for organic–inorganic polymer hybrids. MRS Bull **2001**, *26*, 389–392.

20. Corriu, R.J.P.; Leclercq, D. Recent developments of molecular chemistry for sol–gel processes. Angew. Chem. Int. Ed. Engl. **1996**, *35*, 1420–1436.

21. Schubert, U. Catalysts made of organic–inorganic hybrid materials. New J. Chem. **1994**, *18*, 1049–1058.

22. (a) Brinker, C.J.; Scherer, G.W. *Sol–gel Science*. Academic Press: New York, 1990; (b) Lev, O. Diagnostic applications of organically doped sol–gel porous glasses. Analusis **1992**, *20*, 543–552; (c) Wright, J.D.; Sommerdijk, N.A.J.M. *Sol–gel Materials Chemistry and Applications*. Gordon Breach: Netherlands, 2001.

23. (a) Zink, J.I.; Valentine, J.S.; Dunn, B. Biomolecular materials based on sol–gel encapsulated proteins. New J. Chem. **1994**, *18*, 1109–1112; (b) Dunn, B.; Zink, J.I. Optical properties of sol–gel glasses doped with organic molecules. J. Mater. Chem. **1991**, *1*, 903–912; (c) Avnir, D. Organic chemistry within ceramic matrices: doped sol–gel materials. Acc. Chem. Res. **1995**, *28*, 328–334; (d) Klein, L.C. Sol–gel optical materials. Annu. Rev. Mater. Sci. **1993**, *23*, 437–446; (e) Dunbar, R.A.; Jordan, J.D.; Bright, F.V. Development of chemical sensing platforms based on sol–gel-derived thin films: origin of film age vs performance trade-offs. Anal. Chem. **1996**, *68*, 604–6010.

24. Loy, D.A.; Shea, K.J. Bridged polysilsesquioxanes—highly porous hybrid organic–inorganic materials. Chem. Rev. **1995**, *95*, 1431–1442.

25. (a) Tailor-made compounds predicted by Pauling. Chem. Eng. News **1949**, *27*(March), 913; (b) Dickey, F.H. Specific adsorption. J. Phys. Chem. **1955**, *59*, 695–707.

26. Kresge, C.T.; Leonowicz, M.E.; Roth, W.J.; Vartuli, J.C.; Beck, J.S. Ordered mesoporous molecular-sieves synthesized by a liquid-crystal template mechanism. Nature **1992**, *359*, 710–714.

27. Huo, Q.; Margolese, D.I.; Ciesla, U.; Feng, P.; Gler, T.E.; Sieger, P.; Leon, R.; Petroff, P.M.; Schuth, F.; Stucky, G.D. Generalized synthesis of periodic surfactant/inorganic composite materials. Nature **1994**, *368*, 317–321.

28. Tanev, P.T.; Pinnavaia, T.J. A neutral templating route to mesoporous molecular-sieves. Science **1995**, *267*, 865–869.

29. Hall, S.R.; Fowler, C.E.; Lebeau, B.; Mann, S. Template-directed synthesis of bi-functionalized organo-MCM-41 and phenyl-MCM-48 silica mesophases. Chem. Comm. **1999**, 201–202.

30. Brown, J.; Mercier, L.; Pinnavaia, T.J. Selective adsorption of Hg^{2+} by thiol-functionalized nanoporous silica. Chem. Comm. **1999**, 69–70.

31. Lim, M.L.; Blanford, C.F.; Stein, A. Synthesis and characterization of a reactive vinyl-functionalized MCM-41: probing the internal pore structure by a bromination reaction. J. Am. Chem. Soc. **1997**, *119*, 4090–4091.

32. Inagaki, S.; Guan, S.; Fukushima, Y.; Ohsuna, T.; Terasaki, O. Novel mesoporous materials with a uniform distribution of organic groups and inorganic oxide in their frameworks. J. Am. Chem. Soc. **1999**, *121*, 9611–9614.

33. Kuchen, W.; Schram, J. Metal-ion-selective exchange resins by matrix imprint with methacrylates. Angew. Chem. Int. Ed. **1988**, *27*, 1695–1697.

34. (a) Rottman, C.; Avnir, D. Getting a library of activities front a single compound: tunability and very large shifts in acidity constants induced by sol–gel entrapped micelles. J. Am. Chem. Soc. **2001**, *123*, 5730–5734; (b) Rottman, C.; Grader, G.; Hazan, Y.D.; Melchior, S.; Avnir, D. Surfactant-induced modification of dopants reactivity in sol–gel matrixes. J. Am. Chem. Soc. **1999**, *121*, 8533–8543.

14

Thermodynamic Considerations and the Use of Molecular Modeling as a Tool for Predicting MIP Performance

Ian A. Nicholls   University of Kalmar, Kalmar, Sweden

Sergey A. Piletsky, Biening Chen, Iva Chianella, and Anthony P. F. Turner   Cranfield University at Silsoe, Silsoe, Bedfordshire, United Kingdom

I. INTRODUCTION

The types of molecular recognition events involved in the synthesis of a molecular imprinted polymer (MIP) and in the recognition of a ligand by an MIP are identical to those observed throughout nature [1]. The general principles underlying the synthesis of molecularly imprinted polymers have been outlined in previous chapters (Chapters 3–13).

   The nature of the template, and monomers and the polymerization reaction itself determine the quality and performance of the polymer product. Moreover, the quantity and quality of molecularly imprinted polymer recognition sites is a direct function of the mechanisms and the extent of the monomer–template interactions present in the prepolymerization mixture. Consequently, an understanding of the physical rules governing the formation of monomer–template complexes is fundamental to our ability to rationally design polymerization systems for producing polymers with a given activity. Given that we can have an understanding of the physical parameters governing the formation of intermolecular interactions (i.e., in the polymerization process and in MIP–ligand recognition) we have a basis for the rational design of MIPs.

   The rapid development of computational chemistry over the past three decades has provided the chemist with powerful tools for the prediction of chemical properties [2]. Granted an understanding of the physical rules governing molecular recognition alluded to above, we can potentially utilize computational techniques, in the design of MIPs. With the help of computational techniques, it is now feasible to screen a library of functional monomers to identify that or those that interact most strongly with a given template or undertake the design of novel monomers for a given template. In this chapter, we shall first describe the physical factors underlying molecular recognition and their consequences for the practical design of MIPs, then secondly describe the use of computational techniques as a predictive tool for the analysis of polymerization systems.

II. THERMODYNAMIC CONSIDERATIONS

A. A General Model for Describing Molecular Recognition Phenomena

Many attempts have been made to understand and quantify the fundamental physical basis for molecular recognition, i.e., why two molecules do or do not interact, (Fig. 1). As the imprinting process is a direct consequence of molecular recognition events, i.e., template–monomer interaction, insights into the fundamental physical basis for molecular recognition should prove useful.

Page and Jencks [3] and Jencks [4] developed paradigms that have been of central importance to the approaches used by a number of researchers in efforts to identify these physical factors. In particular, semi-empirical approaches have been independently developed by Andrews et al. [5], a "back of the envelope" approach to the prediction of ligand–receptor binding constants, and Williams et al. [6–8], a detailed factorization of the energetic contributions to binding. The general thermodynamic treatment developed by Williams, Eq. (1), is the more instructive in this regard, and can be utilized as a basis for better understanding the recognition events involved in MIP synthesis and to ligand–MIP binding events [9–11]. We shall begin by describing ΔG_{bind} and its measurement before examining the nature, and consequences for the imprinting process, of the other terms in the following equation:

$$\Delta G_{bind} = \Delta G_{l+r} + \Delta G_r + \Delta G_h + \Delta G_{nb} + \sum \Delta G_p + \Delta G_{conf} + \Delta G_{vdW} \tag{1}$$

where the Gibbs free energy changes are: ΔG_{bind}, complex formation; ΔG_{l+r}, translational and rotational; ΔG_r, restriction of rotors upon complexation; ΔG_h, hydrophobic interactions; ΔG_b, residual soft vibrational modes; $\sum \Delta G_p$, the sum of interacting polar group contributions; ΔG_{conf}, adverse conformational changes; and ΔG_{vdW}, unfavorable van der Waals interactions.

Although molecular imprinting is conceptually elegant in its simplicity, as reflected in the highly schematic representations of the imprinting process depicted in many papers, the molecular level events taking place in the polymerization mixture are very complex. A range of variables play a role in determining the success, or otherwise, of a polymerization. As the basis for the molecular memory of MIPs lies in the formation of template–functional monomer solution adducts in the prepolymerization reaction mixture. The relative strength of these interactions, ranging from weak van der Waals forces to reversible covalent bonds, is critical. The position of the equilibrium for formation of self-assembled solution adducts between

Figure 1   The extent of formation of a noncovalent complex between two molecular entities is a consequence of chemical equilibria.

templates and attendant monomers, ΔG_{bind}, determines the number of resultant sites, and the degree of receptor site heterogeneity. The more stable and regular the template–functional monomer complex, the greater the number, and fidelity, of resultant MIP receptors. The extent of "optimal" template coordination by a functional monomer is dependent upon the nature of each of the chemical components present in the polymerization mixture, and upon the physical environment (temperature and pressure). The relative strengths of monomer(s)–monomer(s), template–template, solvent–template, and solvent–monomer(s) interactions influence the position of equilibrium and thus the extent and quality of functional monomer–template interactions, which in turn govern the quality of the resultant receptor population. In cases utilizing reversible covalent functional monomer–template interactions, complex stability is guaranteed. It is noteworthy, however, that some degree of the shape complementarity exhibited by a polymer arises from cross-linking monomer interactions, which are noncovalent in nature.

B. Determination of ΔG_{bind}

A number of spectroscopic techniques can be employed for experimentally determining the extent of template complexation by functional monomer(s), and the role of other factors such as solvent and temperature upon complex formation. Direct evidence for the formation of noncovalent monomer–template interactions, the extent of which is reflected in the total binding term, ΔG_{bind}, was first presented in an elegant NMR study by Sellergren et al. [12]. A combination of chemical shift and line-broadening arguments were used in this study to establish the formation of functional monomer (methacrylic acid)–template (phenylalanine anilide) complexes. The results also suggested the minor presence of template self-association and higher order complexes.

Complexation has also been studied using UV-spectroscopic titrations in order to calculate the dissociation constants for the solution adducts and the relative concentration of fully complexed template in the polymerization mixture [13]. Although the concentration ranges accessible with UV studies are often below those used in standard imprinting polymerization techniques, the identification of complex formation, or otherwise, is a useful tool. This approach was also used to verify the inert nature of the cross-linking agent ethylene glycol dimethacrylate and for the screening of candidate functional monomers [14].

A similar approach has been adopted by Whitcombe et al. [15], where NMR chemical shift studies allowed the calculation of dissociation constants and a potential means for predicting the binding capacities of MIPs. The NMR characterization of functional monomer–template interactions has also been applied to the study of the interaction of 2,6-bis(acrylamido) pyridine and barbiturates [16], and of 2-aminopyridine and methacrylic acid [17]. Recent NMR work in our laboratory [18] has involved the determination of template–monomer interactions for a nicotine–methacrylic acid system. Significantly, it was shown in this study that template self-association complexes are present in the prepolymerization mixture and that the extent template self-association is dependent both upon solvent and the presence of monomer.

The position of the equilibrium between free template–monomer(s) and their corresponding complexes is a product of both temperature and pressure. Lower temperatures of polymerization can be used advantageously for the preparation of electrostatic interaction-based MIPs, as demonstrated by O'Shannessy et al. [19], though elevated temperatures (120°C) were shown by Sellergren and Shea [20] to yield polymers with improved performance characteristics. In this latter study, the influence of porogen (solvent of polymerization) on template–monomer complexes was also addressed in conjunction with the morphology and properties, temperature stability and swelling capacity, of the resultant polymers. Recently the Sellergren group [21] showed that high pressure (1000 bar) polymerization can be used to enhance the selectivity of the resultant imprinted polymers. However, the influence of pressure and temperature effects on the polymerization process (e.g., reaction rate) and polymer structure (e.g., pore structure and swelling properties) make the drawing of direct conclusions regarding their influence on the mechanisms of complex formation difficult. This is made even harder due to the template-dependent nature of the pressure induced selectivity enhancement.

The use of noncovalent interactions for molecular imprinting is inherently limited by the stability constant(s) for the formation of solution adducts. Moving to higher functional monomer–template ratios, in order to push the equilibrium towards complex formation, leads to increased numbers of randomly distributed functional groups, which in turn lead to increased levels of nonspecific binding. Much lower ratios, a nonimprinted polymer being the extreme case, lead to lower site population densities in the polymer. However, many more detailed optimization studies are warranted in order to better understand the influence of polymerization mixture composition on template complexation, and thus polymer recognition characteristics. That we still have much to learn regarding polymer optimization is reflected in the results from a study by Mayes and Lowe [22]. Here, they demonstrated that the quality and number of high affinity morphine selective sites is of the same order of magnitude in polymers prepared using 2% of the template concentration as in polymers prepared using 4% in a previous protocol [23]. These results have been supported by another recent paper utilizing precipitation polymerization compositions, where the Mosbach group has described the synthesis of imprinted microspheres with binding parameters superior to conventional imprinted particles [24]. The utility of this development for expediting polymer synthesis, and the potential of this new imprinted polymer format in a range of application areas is exciting.

C. The Physical Factors Influencing ΔG_{bind}

1. The Role of ΔG_{t+r}

The ΔG_{t+r} term in Eq. (1) reflects the change in translational and rotational Gibbs free energy associated with combining two or more free entities in a complex, a process that is entropically unfavorable, (Fig. 2). This term carries implications for the order (number of components) of complexes which may be formed. It can be stated that functional monomer systems capable of simultaneous interactions of correct geometry, relative to multiple single point interacting monomers, should yield higher concentrations of complexed template, due to a reduction in the adverse loss of translational and rotational free energy, an entropically unfavorable process. A number of examples of the application of multiple functional monomer–template interactions have recently appeared (see for example Ref. 16).

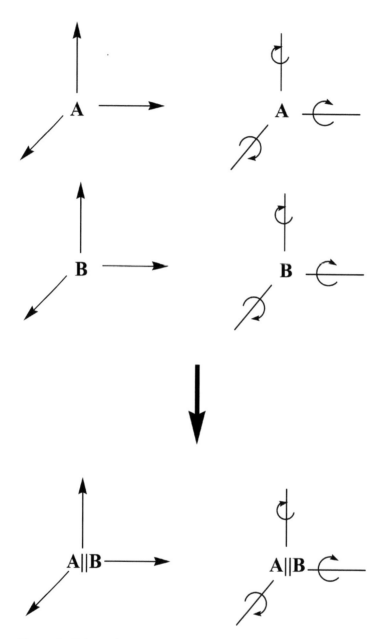

Figure 2 Schematic representation of the consequences of translational (left) and rotational (right) degrees of freedom upon complexation. Prior to interaction, the two interacting structures (A and B) each have three degrees of translational freedom, and three degress of rotational freedom, i.e., 12 degrees of freedom. Upon complexation, this total is reduced to six, an entropically unfavorable process.

2. The Role of ΔG_r

The selectivities observed for MIPs prepared with rigid templates are superior to those of less rigid structures, which is a direct consequence of ΔG_r, the restriction of rotors upon complexation. The origin of this effect lies in the fact that the more rigid the structure, the lower the number of possible solution conformations and thus solution complexes. This in turn renders a narrower site distribution in the resultant polymer. This is supported by the observation that the highest MIP–ligand affinities thusfar observed have been for rigid templates such as the alkaloids morphine [25] and yohimbine [26]. This is further underscored by the affinity of the morphine MIP for its template being superior to that of the opioid receptor binding neuropeptide leucine enkephalin for an enkephalin MIP, even though the peptide has a greater number of points for potential electrostatic interactions.

3.   The Role of $\sum \Delta G_p$

The number and relative strengths of different template–functional monomer interactions, as given by the $\sum \Delta G_p$ term, dictate the degree of selectivity of the resultant polymer [27]. In terms of electrostatic interactions, several reports have been made of enhanced selectivity through the use of more strongly interacting functional monomers that favor template–monomer complex stability. The Takeuchi group [28,29] has presented a series of studies using trifluoromethylacrylic acid as a functional monomer, and demonstrated that it can be used to augment template recognition. The lower pK_a of this monomer enhances ion pairing and ion–dipole interactions.

The use of metal ion coordination of templates has been shown to be advantageous in a number of studies. The possibility of multiple interactions with the template, in conjunction with the greater relative strengths of template–metal ion interactions, lends their use in highly polar solvents, such as methanol, as exemplified by the work from the Arnold group in a study using Cu(II) coordination of imidazole containing templates [30]. In the work by the Murray group [31], the coordination and luminescent properties of Eu(III) have been used to selectively bind hydrolysis products from the nerve agent Soman, and Pb(II) selective matrices have been prepared and evaluated [32]. The capacity of metal ions, e.g., Co(II) [33] and Ti(IV) [34], to engage in multiple simultaneous interactions has been used to coordinate reacting groups in polymers with catalytic activities.

4.   The Role of ΔG_h

The vast majority of MIPs reported to date have described the use of nonpolar organic solvent–soluble templates and polymer systems, and their evaluation in similar solvents. The often impressive ligand selectivities obtained are a reflection of the strengths of electrostatic interactions in nonpolar media. Nonetheless, for many potential applications, in particular those of an environmental or biomedical nature, the templates of interest, e.g., peptides, proteins, oligonucleotides, sugars, are often incompatible with such polymerization media. Another option when using water as the solvent of polymerisation is the use of hydrophobic moiety selective functional monomers, which can contribute to solution adduct formation through the hydrophobic effect, the contribution of which is reflected in the ΔG_h term of Eq. (1). Alternatively, functional monomers capable of sustaining electrostatic interactions at such high polarities, e.g., metal ion chelation can be employed, as discussed above. With this in mind, a number of groups have been working towards the development of water-compatible polymer systems for use in molecular imprinting. To this end, cyclodextrin based functional monomers, (Fig. 4) have been developed in an attempt to provide a general approach to imprinting in water [35–37]. These polymer systems have been effectively applied to the imprinting of steroids [35,36] and organic solvent insoluble species, in particular amino acids [37]. Another, quite unique, related approach has been the use of crown ethers to solubilise highly polar, e.g., *zwitterionic*, species at low polarities [38]. This work, by Andersson and Ramström, provides a promising basis for further studies, perhaps through the use of polymerizable crown ether derivatives.

5. The Role of the Remaining Terms and Summation

In relation to the other terms described in Eq. (1), the extent to which soft vibrational modes, ΔG_{vib}, contribute to solution adduct homogeneity is a direct product of the temperature employed for the polymerization process. The conformational, ΔG_{conf}, and van der Waals, ΔG_{vdW}, energy terms in Eq. (1) reflect the need for compromise of template conformation and effective solvation in order to form solution adducts. Determination of the significance of these terms can be made using molecular modeling.

To summarize, each of the thermodynamic terms contributing to the stability and uniformity of the solution adducts formed during the molecular imprinting prearrangement process play a role in determining ΔG_{bind}. How these terms influence the noncovalent MIP–ligand recognition shall be discussed in the following section. It can be concluded that the exclusive use of noncovalent template–functional monomer interactions is inherently limited by the stabilities of the template–functional monomer complexes. The subsequent heterogeneity of the template-selective recognition sites, their poly-clonal nature, is a direct consequence of the extent of template complexation. A small number of studies have attempted to combine the versatility of noncovalent interaction based molecular imprinting with the benefits afforded by the constraint of functional monomer and template through reversible covalent bonds [39–42]. This approach, although requiring the synthesis of template derivatives, offers much scope for enhancing the fidelity of template-selective recognition sites.

III. APPLICATION OF THERMODYNAMIC PRINCIPLES TO LIGAND–POLYMER BINDING

Although the thermodynamic terms derived from Eq. (1) can also be used to shed light on recognition of an analyte by an MIP; solvent–analyte and analyte–analyte interactions and macroscopic effects (e.g., solvent and analyte diffusion rates and surface areas) also play a critical role in determining the binding characteristics of a polymer and its suitability for use in a given application.

In terms of ligand binding to an MIP recognition site, the conformational, ΔG_{conf}, and van der Waals, ΔG_{vdW}, energy terms in Eq. (1) reflect the need for compromise of template conformation and degree of effective solvation, so long as rebinding takes place in the polymerization solvent. As both the prearrangement phase and ligand rebinding are under thermodynamic control, we may assume that the template population shall not possess, on average, conformational strain, nor adverse van der Waals interactions. This statement relates to the central dogma of molecular imprinting, namely that the solution adducts are reflected in the recognition sites presented in the bulk polymer. Thus for the study of template rebinding, Eq. (1) can be simplified to

$$\Delta G_{bind} = \Delta G_{t+r} + \Delta G_r + \Delta G_h + \Delta G_{vib} + \sum \Delta G_p \tag{2}$$

In terms of the thermodynamics of MIP–ligand binding, the rigidity of templates contributes to the fidelity of the resultant receptor site populations. The binding of a ligand to an MIP recognition site is governed by the same thermodynamic constraints. Thus, all other factors being equal, a more flexible ligand shall incur a larger energetic (entropic) penalty, ΔG_r, in order to bind (5–6 kJmol^{-1}, i.e., one order of magnitude in terms of binding constant).

The role of electrostatic interactions, $\sum \Delta G_p$, on ligand recognition has been examined through the study of the influence of solvent polarity on MIP–ligand selectivity and by using template analogues. By examining enantiomer capacity factors and MIP enantioselectivity as a function of eluent water content, the relative strength of noncovalent electrostatic interactions has been studied in methacrylic acid–ethylene glycol dimethacrylate [43]. The inclusion of increasing amounts of water in the eluent has a more profound effect upon the capacity factor for the imprinted enantiomer than its antipode. This was interpreted in terms of the greater involvement of hydrogen bonding modes in specific template recognition than in nonspecific binding modes. Other studies by the Mosbach group using acrylamide MIPs [44,45] demonstrated similar effects, and in addition a contribution from hydrophobic interactions was observed as the ligand solubilities allowed study at high water concentrations, up to 70% in acetonitrile. In the latter study [45], it was shown that polymer performance improved at higher ionic strength, though the influence of aqueous solvent ionic strength on the swelling characteristics of polymers of this type remains to be examined. Similar rebinding behavior has been observed in other systems, for example in acrylamide–4–vinylpyridine co-polymers imprinted with the sulfamethoxazole [46].

The presence of water, as described above, is adverse with respect to electrostatic interactions, though is conducive to hydrophobic interactions. The group of Andersson [23,47,48] has undertaken a series of thorough studies on the application of MIPs prepared in organic solvents for binding studies in aqueous environments. Through careful optimization of ligand binding studies, it is possible to enhance, in some cases, the polymer selectivities. Most interestingly, converse effects have been reported when using organic solvents on polymers imprinted in water [37]. Thus, when a combination of hydrophobic and electrostatic interacting functional monomers are employed, a balance of entropy motivated hydrophobic effect interactions and enthalpic electrostatic interactions is necessary for optimal binding. As more polar environments lead to a weakening of electrostatic interactions and less polar environments a weakening of the hydrophobic effect, a balance of these leads to entropy–enthalpy compensation effects. This is clearly illustrated in the case of the local anesthetic bupivacaine [49] (Fig. 7). This phenomenon has also been used for the development of synthetic receptor binding assays for herbicides by the Mosbach group [50,51], where they demonstrated that imprinting in vinyl pyridine–ethylene glycol dimethacrylate co-polymers was possible in methanol containing 20% water.

A number of significant quantitative and semi-quantitative chromatographic analyses of MIP–ligand binding phenomena have been presented, which have helped to shed light on the mechanisms of ligand–MIP binding, and augment the information derived from ligand binding data. In one such study, Sellergren and Shea [52] extensively examined the influence of pH, temperature, and polymer heat treatment on chromatographic response. Using the influences of flow rate and sample loading on peak asymmetry they identified that nonlinear binding adsorption isotherms are associated with the binding of template structures. In addition, the influence of polymer heat treatment on the retention characteristics for enantiomers of phenylalanine anilide (PA) on an L-enantiomer MIP showed that the retention of the nonimprinted antipode carries resemblance to that of weak cation exchange resins. Furthermore, they demonstrated using variable temperature chromatographic studies that the nature of the mobile phase and its interaction with the solubilized analyte produce marked differences in chromatographic behaviour. Van't Hoff analyses allowed dissection of the entropic and enthalpic contributions to binding under different elution conditions, from which inference to the nature of the solvation state of the analyte and its influence on binding could be drawn. A similar approach has since been adopted by Lin et al. [53] working with an electrochromatographic format.

In another study Sellergren and Shea [54] modeled the chromatographic behavior of a solute, PA, on imprinted and nonimprinted polymers using a simple cation-exchange model, and were able to describe the behavior as a function of mobile phase pH. Moreover, the model they proposed was sufficient to explain the differences in selectivity between the two polymers.

In a recent study Sajonz et al. [55] have examined in detail the thermodynamics and mass transfer kinetics of the enantiomers of PA on a PA imprinted polymer using staircase frontal analysis. The adsorption data for both enantiomers were fitted to both Freundlich and bi-Langmuir isotherms, the former being indicative of higher template solvation states. They characterized the concentration and enantiomer dependence of the mass flow properties which led to the identification of a low concentration of very high affinity (nM) sites. This case provides a good example of how physical–chemical studies of ligand binding behavior can yield information regarding polymer micro environment. Further studies from the same group [56] have provided further justification for use of the Freundlich isotherm to model ligand–polymer recognition.

To summarize thusfar, the physical factors governing the formation of template–monomer interactions and ligand–polymer binding events have been described. The rational design of MIP systems requires that the physical terms described here be taken into consideration. The significance of a number of these factors is most evident in some of the examples and discussion presented in Chapter 16. Complex (multicomponent) systems such as imprinted polymers, are, however, not particularly amenable to the use of detailed thermodynamic treatments. Nonetheless, the general principles outlined above provide realistic guidelines which can provide a basis for the more empirical approaches often used in new MIP development, and also in conjunction with more accessible theoretical treatments, such as molecular modeling. In the following sections, we shall describe the use of modeling as a predictive tool in the design of molecularly imprinted polymers, and provide some practical illustrations of its use.

IV. MOLECULAR MODELING AND COMPUTER SIMULATION: TOOLS FOR THE DESIGN OF MIPs

In principle, the broad range of functional monomers currently available makes it possible to design an MIP specific for any type of stable chemical compound. Currently the selection of the best monomers for polymer preparation is one of the most crucial issues in molecular imprinting. Thermodynamic calculations and combinatorial screening approaches offer possible solutions, and have already been used successfully for predicting polymer properties and for the optimization of polymer compositions (see Ref. 9,57,58, and Chapter 8), however, in practical terms, application of these methods is not trivial. The problem lies in the technical difficulty of performing detailed thermodynamic calculations on multicomponent systems and the amount of time and resources required for the combinatorial screening of polymers. To check a simple two-component combination of 100 monomers, for example, one has to synthesize and test more than 5000 polymers, a very difficult task. This task will be further complicated by the possibility that these monomers could be used in monomer mixtures in different ratios.

One potential solution to the problem of polymer design lies in molecular modeling and in performing thermodynamic calculations with the aid of a computer. Modern computer technology has made molecular modeling much easier than previously. It is no longer necessary for most modelers to write their own computer programs and maintain their own computer systems. Software can be obtained from commercial software companies and academic laboratories. Molecular modeling and computer simulation can now be performed in any laboratory or classroom. The most widely available commercial software is drug design oriented and involves modeling of macromolecules (QUANTA/CHARMN/MCSS/HOOK, INSIGHT-II/Ludi, CERIUS 2 (MSI, San Diego, USA) and SYBYL (Tripos, San Diego, USA). The software used in all of the following examples was performed using SYBYL.

Molecular modeling has already been widely adopted in chemistry. It helps computational chemists to generate and refine molecular geometry (in terms of bond lengths, bond angles, and torsions), finding low-energies conformations. It facilitates calculation of energy levels such as binding free energy, heat of formation and activation energy, electrostatic properties (moments, charges, ionization potentials, and electron affinity), spectroscopic properties (vibration modes and chemical shifts) and bulk properties (volumes, surface areas, diffusion, viscosity) [59–62]. Here, we present some practical examples of how molecular modeling and computer simulation have been used for the design of MIPs. To understand how it works, we shall first discuss the basics of molecular modeling.

A. Theory—Molecular Mechanics

Among the theoretical methods available for predicting the geometry and behavior of molecules, molecular mechanics is one of the most commonly used. This method was developed out of the need to describe the molecular structures and properties of molecules in as practical a manner as possible.

The theoretical background of the molecular mechanics approach has been discussed in detail elsewhere [63]. Briefly, molecular mechanics uses classical mechanics to represent molecules. Therefore, molecules are viewed as collections of spheres connected by mutually independent, flexible springs representing bonds.

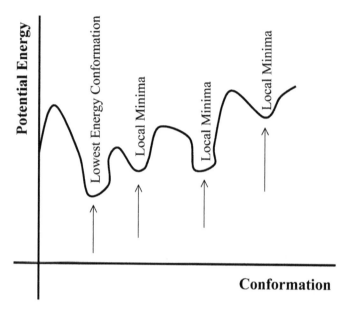

Figure 3 A schematic representation of a much simplified energy hypersurface, obtained by plotting molecular conformation vs. potential energy.

The mathematics of spring deformation, as described by Hookes' law, is used to model the "elastic" forces between these spheres. This treatment requires that only the motions of nuclei are studied and that the electrons are not examined explicitly; electrons are assumed to find an optimal distribution around the nuclei.

The essential objective of molecular mechanics is the prediction of the energy of a molecule or a complex of interacting molecules in a particular conformation. The energy value calculated from molecular mechanics, however, cannot be considered as absolute physical quantities, but only as relative values to compare the energies of different conformations of the same structure/complex or different complexes. The total molecular mechanical energy is a calculated term by comparing the bond parameters of the model with the accepted parameter values that are taken from either experimental data or from ab initio calculations. A simplified molecular mechanics representation for a single molecule is given in Eq. (3)*. The combination of the molecular mechanical equations and a database of empirical parameters of atom and bond types is often referred to as a force field. Examples of force field are TRIPOS [64], MM2 [65] and AMBER [66].

$$E_{\text{Potential Energy}} = E_{\text{stretching}} + E_{\text{bending}} + E_{\text{torsional}} + E_{\text{electrostatic}} + E_{\text{vdW}} \qquad (3)$$

B. Computer Simulation Using Molecular Mechanics

Plotting the potential energy function against the conformation of the molecule, it would be possible to see the surface of peaks and troughs as shown in Fig. 3. Each

---

*Energetic parameters contributing to the prediction of molecule structure by molecular mechanics are given in Eq. (3).

peak corresponds to a high and unfavorable energy conformation, and each trough to a more stable energy conformation [67]. Two types of algorithm are used for the analysis of the potential energy surface and conformations of molecules by molecular mechanics, energy minimization simulations, and molecular dynamics simulations.

Energy minimization considers energy as a function of the atomic coordinates and the energy minimization attempts to generate the coordinates which correspond to a minimum of energy. Several techniques have been used for calculation including Simplex [68], Powell [69], steepest descents, conjugate gradient, and the second derivative Newton–Raphson and the pseudo-second derivative BFGS [70] methods. SYBYL uses the MAXIMIN2 minimization method—a combination of first and nonderivative methods.

The primary optimization methods used in the MAXIMIN2 procedure adjust the atomic coordinates of all the atoms simultaneously based on the first derivative of the energy equation with respect to the degrees of freedom. Aggregates of atoms, within SYBYL, have no degrees of freedom while atoms not in aggregates have three degrees of freedom each (the translational coordinates of the atom). Energy minimization alone does not produce a resulting geometry with the lowest energy conformation or "global conformation". With flexible molecules, it is often observed that several conformations exist with very similar energies [71]. This is due to the possibility of the conformation being in one of several possible local minima with similar energies as shown in Fig. 3. Another powerful technique, molecular dynamics, is used to analyze the time-dependent motions and the conformational flexibility of molecules. Molecular dynamics calculates the "real" dynamics of the system, from which the time average of properties can be calculated. The result is a trajectory that specifies how the positions and velocities of the particles in the system vary with time. The trajectory is obtained by solving the differential equations embodied in Newton's second law[*]:

$$m_i \frac{\delta^2 \overline{x}(t)}{\delta t^2} = m_i \vec{a}_i(t) = \vec{F}_i = -\vec{\nabla}_i E \tag{4}$$

for intermolecular interactions, the force on each particle will change whenever the particle changes its position, or whenever any of the other particles with which it interacts changes position. In a real molecular system, the motions of all particles are coupled together under the influence of a continuous potential resulting in a many-body problem, which cannot be solved analytically by using algorithms developed so far. Various approximations and integration methods were developed to generate molecular dynamics trajectories at the best possible precision. Finite difference methods (adopted by SYBYL software system) and predictor–corrector integration methods are the common ones. In a finite difference method, the integration is broken down into many small stages, each separated in time by a fixed time δt. The total force on each particle in the configuration at a time t is calculated as the

---

[*]In Eq. (4), Newton's second law has been used in the calculation of position of the particle and in the simulation of particle motion. Here m_i is the mass of the atom i; $x_i(t)$ is the position of the atom i; $\vec{F}_i$ is the force acting on atom i; E is the total potential energy function of the system; $\vec{\nabla}_i$ is a gradient with respect to x_i; $\vec{a}_i(t)$ is acceleration.

vector sum of its interactions with other particles. From the force, the accelerations of the particles can be calculated, which are then combined with the positions and velocities at a time t to calculate the positions and velocities at a time $t + \delta t$. After each integration step, the Verlet algorithm is used to calculate the atom velocities using the force and the location of atoms [72]. Using velocity calculations, the new atom locations and the temperature of the ensemble can be calculated. Assembling the "snapshots" together provides the atomic trajectories and allows examination of the "conformational space" of the molecule.

C. Leapfrog

Leapfrog, a component of the SYBYL software package, is a second generation de novo drug discovery program that allows the evaluation of potential ligand structures mainly on the basis of their binding score. This is calculated with electrostatic/ sterical screening by trying repeatedly different ligands (one each time) in different positions of the template and then either keeping or discarding the results. Leapfrog can run in three alternative modes:

- OPTIMIZE: suggests improvements to existing leads.
- DREAM: proposes new molecules expected to have good binding.
- GUIDE: supports interactive design by performing and evaluating user modifications.

Structures of ligands are saved in a SYBYL database and are referenced by a Molecular Spreadsheet containing the Leapfrog binding energies that serve as a vehicle for launching additional structure evaluations.

D. Simulated Annealing

Simulated annealing is a general-purpose global optimization technique based on the constrained molecular dynamics simulation. The simulated annealing process consists of first "melting" the system being optimized at a high effective temperature, then lowering the temperature by slow stages until the system freezes and no further changes occur. At each temperature, the simulation must proceed long enough for the system to reach a steady state. Practically simulated annealing process is a type of molecular dynamics experiment used to obtain several different low energy conformations of a single molecule or to obtain several different low configurations of a system of molecules, for example, a template surrounded by monomers. In simulated annealing, a cost function takes the role of the free energy in physical annealing and a control parameter corresponds to the temperature. To use simulated annealing in conformational analysis, the cost function would be the internal energy. In complex formation, the cost function will be the total energy of the system. At high temperatures, the system is able to occupy high-energy regions of conformational space and to pass over high-energy barriers. As the temperature falls, the lower energy state becomes more probable in accordance with the Boltzmann distribution. At absolute zero, the system should occupy the lowest-energy state (i.e., the global minimum energy conformation). To guarantee that the globally optimal solution is actually reached would require an infinite number of temperature steps, at each of which the system would have to come to thermal equilibrium.

V. COMPUTATIONAL DESIGN OF IMPRINTED POLYMERS

At present the molecular modeling of complex systems such as molecularly imprinted polymers, their structure and possible interactions with template, solvent and other molecules is impossible due to the extremely large computational workload required for such complex systems. We can, however, lower the computational requirements by simplifying the model. The main paradigm of molecular imprinting can be described in statement that "the strength and type of interactions, existing between monomers and template in monomer mixture will determine the recognition properties of the synthesized polymer". The assumption is that the complexes formed in monomer mixture will somehow survive the polymerization stage and their structure will be preserved in synthesized polymer. Thus, instead of modeling the polymer, we can model the monomer mixture and the interactions taking place in solutions between monomers, cross-linker, template, and solvent. The interactions between monomers and template models can be quantified and used for rational selection of monomers for polymer preparation. Although approximately 4000 different polymerizable compounds have been reported, and which can potentially be used as functional monomers, in reality many of them have similar properties and structures. As an assumption, it might be sufficient to test possible interactions between a minimal library of functional monomers (20–30 compounds) and a target template. The design of such a virtual library of functional monomers and its screening against target compounds is relatively easy to perform. Initial selection of monomers can be based upon their strength of interactions with a template, as discussed earlier in this chapter, which involve the terms $\Sigma\Delta G_p$, ΔG_h, ΔG_r and ΔG_{t+r}, as described in Eq. (1) (as described in Sec. II.A). The Leapfrog algorithm was developed for such tasks and it is particularly useful for selection of monomers, which have high affinity to the template.

Despite the valuable information that can be obtained with Leapfrog screening, there are a number of important questions which it cannot address:

What will happen if you have more than one monomer in your system (it is increasingly desirable to have several monomers in monomer mixture, which could interact with different functionalities present in a template structure)? Will monomers interact with template only or will they also bind to each other?

- What is the optimal ratio between your template and monomers?
- What interference might you expect from cross-linker or solvent?
- What is the influence of the temperature on complex formation?

To answer these questions, we have to perform simulated annealing. In order to simulate the prearrangement of functional monomers with template in a monomer mixture, multiple copies of several (typically 2–3 best monomers selected on the basis of their binding score from Leapfrog results) are placed around the template structure in a so-called solvation box using the solvation function in SYBYL. A simulated annealing experiment can then be performed using default functions, which include energy calculations made after each iteration. At the end of the experiment, the number and the position of the functional monomers interacting with template are examined, indicating the optimal type and ratio of the template and the monomers recommended for MIP design. It is possible to fill the solvation box with molecules of cross-linker and solvent and analyze their influence on complex formation.

It is also possible to study the role of the temperature on the complexation process. In the future, it should also be possible to use computer simulation for the prediction of polymer properties (e.g., affinity and selectivity). Our laboratory is actively working in this direction and the following examples serve as illustrations of the results achieved so far and also to provide practical recommendations for scientists interested in using and contributing to the further development of this technology.

VI. CASE STUDY 1: MICROCYSTIN-LR, COMPUTATIONAL DESIGN OF AN IMPRINTED POLYMER FOR MICROCYSTIN-LR

A. Virtual Library Design and Monomer Screening

In the first example, molecular modeling and a Leapfrog searching algorithm were used for the selection of monomers with high affinity to Microcystin-LR (Fig. 4). The practical significance of this work lies in the necessity to develop synthetic receptors, capable of recognizing Cyanobacterial toxin—Microcystin-LR, which could be used in environmental analysis of this toxin as recognition elements in assays and sensors and for solid-phase extraction matrices.

In this and following examples, the workstation used to simulate monomers–template interactions was a Silicon Graphics Octane running the IRIX 6.6 operating system. The workstation was configured with two 195 MHz reduced instruction set processors, 712 MB memory and a 12 GB fixed drive. The system that was used to execute the software packages was SYBYL 6.7 Tripos Inc. (St. Louis, MI, USA).

In the first part of the work, computer modeling was used to develop molecular model of Microcystin-LR. An initial structure in low-energy conformation was found in the Protein Data Bank (file 1fjm.pdb) [73]. The molecule was corrected

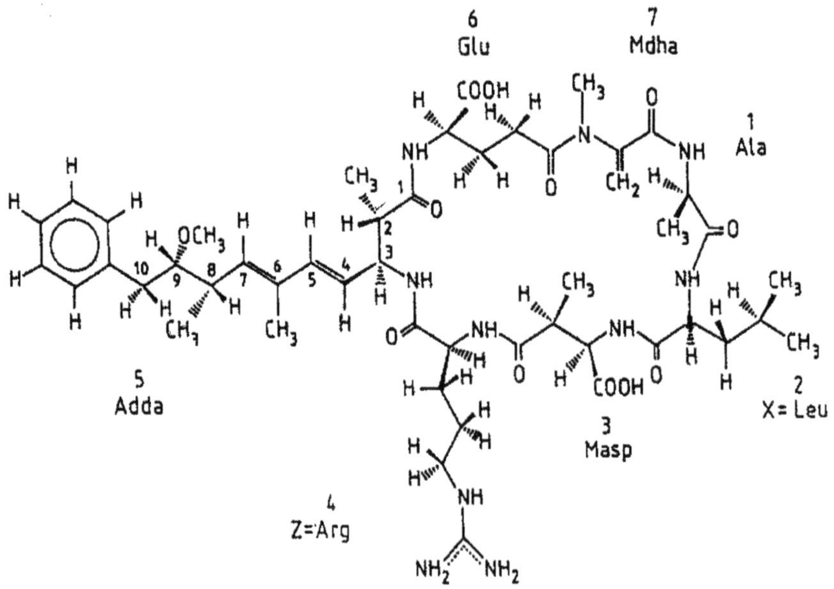

Figure 4 Structure of Microcystin-LR.

(bonds were clarified), charged with the Gasteiger–Hückel method available in SYBYL and a molecular mechanics method was then applied, for a total of 2000 iterations, to perform an energy minimization using the Powell method. All the conditions applied for the energy minimization are shown in Fig. 5a and b.

Two figures, printed directly from SYBYL software, show the windows with parameters used for energy minimization. As a dielectric function and dielectric constant, the default parameters were used (a value of '1' was used for the dielectric constant) as shown in Fig. 5b. The dielectric constant "1" corresponds to vacuum and it adequately represents the binding performed in hydrophobic conditions, favorable for electrostatic interactions. The Gradient termination was used and convergence was reached when the difference in energy between one step and the next was less than 0.001 kcal/mol, achieved within 2000 iterations. The structure of Microcystin-LR obtained (Fig. 6) was utilized for the subsequent computational design.

A virtual library containing commonly used functional monomers was designed for screening for monomer–template interactions (Fig. 7). Firstly, the monomers were drawn, using the sketch molecule command, charged with the Gasteiger–Hückel method and refined by molecular mechanics. The energy minimization of the functional monomers was performed, using the same parameters described previously (see Fig. 5). The functional monomers in the database possess functionalities which are able to interact with template; predominantly through electrostatic (ionic and hydrogen bonds), van der Waals and dipole–dipole interactions.

The Leapfrog program was then used to screen the database of functional monomers. The software identified the possible binding points on the surface of the template molecule, marked with stars in Fig. 6 and probed each of the monomers of the virtual libraries for its possible interactions with Microcystin-LR. The program was applied in Dream Mode for 80,000 iterations, performed in three different runs (20,000, 20,000 and 40,000). The results from each run were examined to evaluate the empirical binding score–the energy of interaction with the template. The final binding scores, reflecting the energy of complexation between the six best monomers and the template, are presented in Table 1. It is noteworthy that the software also permits the evaluation of the relative contribution of electrostatic and other interactions.

The screening clearly indicates that strongest interactions with the template are achieved with acidic monomers, most likely through ionic bonds formed between the carboxylic group of the functional monomer and the positively charged guanidine group of the arginine residue (see Fig. 4). The selection of the best monomers and optimization of polymer composition (determining monomer–template ratio) was performed in the simulated annealing experiment described below.

In order to simulate the prearrangement of functional monomers with the template in the monomer mixture prior to polymerization, multiple copies of the best four monomers (according to Leapfrog results) were placed around Microcystin-LR in a solvation box. The box was then charged with the Gasteiger–Hückel method and its energy was minimized using Powell method (termination gradient, default parameters for dielectric constant and function and convergence at 0.05 kcal/mol in 2000 steps). The energy minimization of the solvation box was performed by applying periodic boundary conditions (Fig. 8a–c) in order to keep the monomers reasonably close to the toxin. A simulated annealing process was applied to optimize the arrangement of functional monomers around the template in the solvation box (Fig. 9). Annealing conditions were fixed as 1000K–300K sweeping in 32,000

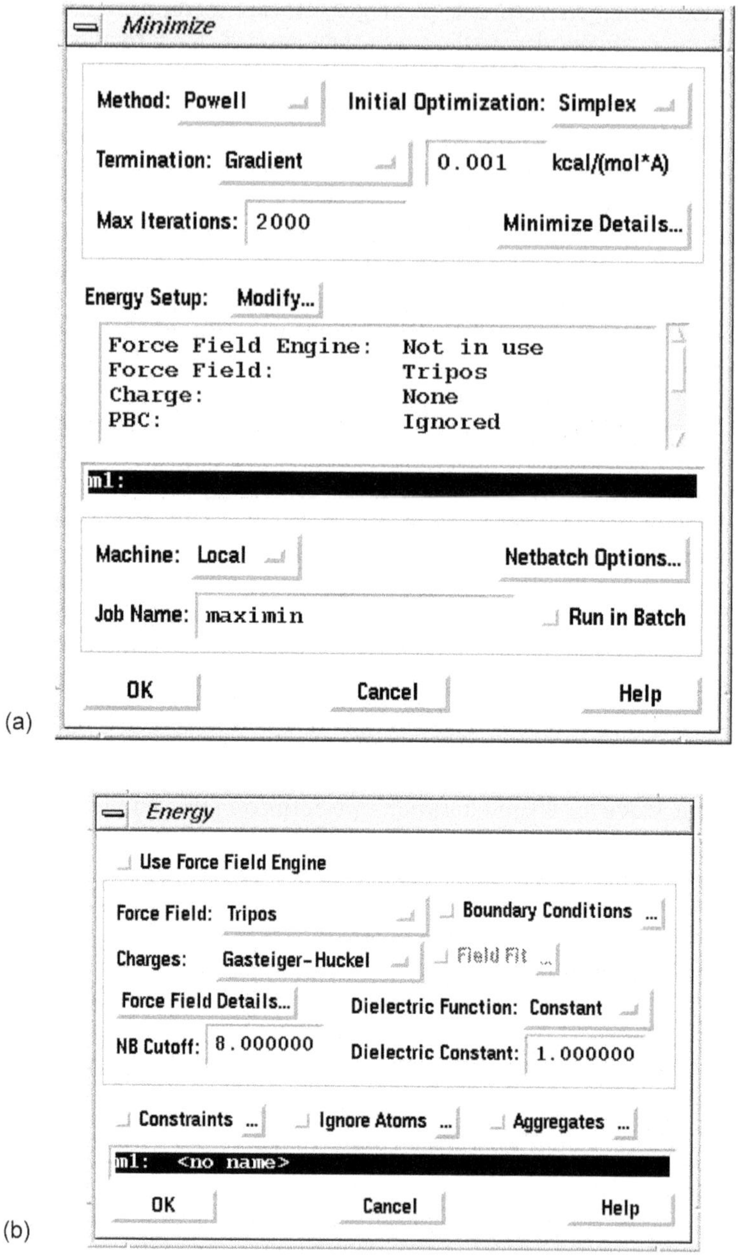

```
┌──────────────────────────────────────────────────────────┐
│ ⊐  Minimize                                                │
├──────────────────────────────────────────────────────────┤
│                                                            │
│   Method: Powell     ⌐    Initial Optimization: Simplex ⌐ │
│                                                            │
│   Termination: Gradient   ⌐   0.001   kcal/(mol*A)        │
│                                                            │
│   Max Iterations: 2000          Minimize Details...       │
│                                                            │
│   Energy Setup:  Modify...                                 │
│   ┌────────────────────────────────────────────────┐     │
│   │ Force Field Engine:  Not in use                 │     │
│   │ Force Field:         Tripos                      │     │
│   │ Charge:              None                        │     │
│   │ PBC:                 Ignored                     │     │
│   └────────────────────────────────────────────────┘     │
│   ██m1:███████████████████████████████████████████        │
│                                                            │
│   Machine: Local  ⌐            Netbatch Options...        │
│                                                            │
│   Job Name: maximin               ⌐ Run in Batch          │
│                                                            │
│      OK            Cancel              Help               │
└──────────────────────────────────────────────────────────┘
```
(a)

```
┌──────────────────────────────────────────────────────────┐
│ ⊐  Energy                                                  │
├──────────────────────────────────────────────────────────┤
│   ⌐ Use Force Field Engine                                 │
│                                                            │
│   Force Field: Tripos        ⌐   ⌐ Boundary Conditions ...│
│                                                            │
│   Charges:   Gasteiger-Huckel ⌐   ⌐ Field Fit ⌐          │
│                                                            │
│   Force Field Details...     Dielectric Function: Constant ⌐│
│                                                            │
│   NB Cutoff: 8.000000    Dielectric Constant: 1.000000    │
│                                                            │
│   ⌐ Constraints ...  ⌐ Ignore Atoms ...  ⌐ Aggregates ...│
│   ██m1:   <no name>███████████████████████████            │
│      OK            Cancel              Help               │
└──────────────────────────────────────────────────────────┘
```
(b)

Figure 5 SYBYL windows for energy minimization. The main window (Minimize) showing the minimization method, termination, iterations, and convergence is reported in (a); the utilized charges and the dielectric function and constant are shown in (b).

consequent steps. Equilibrium length was determined in 2000 fs. The energy minimization, performed after each of the molecular dynamics steps, was carried out using the condition described in Fig. 8. Periodic boundary conditions were applied both during energy minimization and also during molecular dynamics.

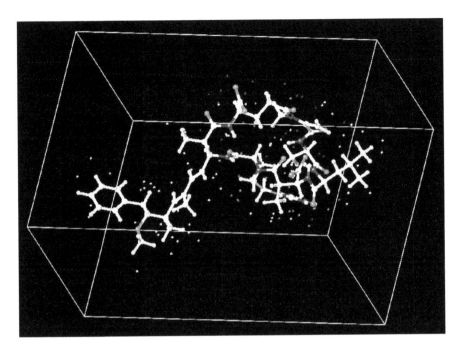

Figure 6 The picture represents the structure of Microcystin-LR and shows all possible sites available for interaction with monomers (marked with stars).

At the end of the program run, the number and the position of the functional monomers were examined. The type and quantity of the monomers participating in the complex with the template (first shell layer), and monomers which interacted with the first shell layer of monomers (second shell layer) thus stabilizing complex, indicated the type and ratio of the template and monomers in an optimized MIP composition.

The structure of the final complex, predicted by computer modelling, is presented in Fig. 10. This complex contains one molecule of Microcystin-LR, one molecule of AMPSA and six molecules of UAEE. The same molar ratio predicted by the modeling program was then used for the synthesis of the polymer. Two of the four monomers selected for the annealing process were unable to establish direct binding with the toxin, probably due to steric factors and internal competition (the monomers favored interaction with one another rather than with Microcystin-LR).

B. Computationally Designed Imprinted Polymer for Microcystin-LR: Synthesis and Testing

The computationally designed MIP was synthesized using the calculated ratio between template and monomers (Microcystin-LR (1 eq.), AMPSA (1 eq.), UAEE (6 eq.)) as described elsewhere [74] and tested for its affinity for the target toxin by using an enzyme-linked competitive assay with a Microcystin–HRP conjugate.

Dissociation constants for polymer–template complexes were obtained from competitive assay experiments, using a double reciprocal plot. These values were compared to those obtained for polyclonal and monoclonal antibodies selective for

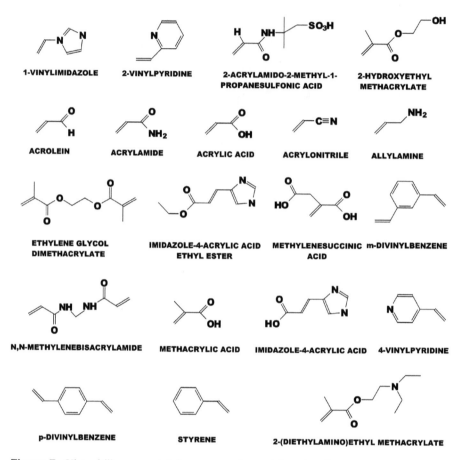

Figure 7 Virtual library containing commonly used functional monomers.

Microcystin-LR (Table 2). The affinity of the computationally designed MIP for Microcystin-LR was found to be slightly higher than that of the polyclonal antibodies, but lower than the affinity of the monoclonal antibodies. The detection range of Microcystin-LR in a competitive assay was broader for MIPs than for antibodies tested. This indicates that the polymers, in contrast to the monoclonal antibodies,

Table 1 Leapfrog Screening Result: Binding Score of the Interactions of Microcystin-LR with the Five Best Monomers and Methacrylic Acid Selected from the Virtual Library

Monomers	Binding score (kcal mol^{-1})
2-Acrylamido-2-methyl-1-propanesulfonic acid (AMPSA)	−50.91
Itaconic acid (IA)	−40.40
Urocanic acid (UA)	−40.02
Urocanic acid ethyl ester (UAEE)	−39.15
N,N'-methylene-bis-acrylamide (Bisacrylamide)	−37.12
Methacrylic acid (MAA)	−26.89

(a)

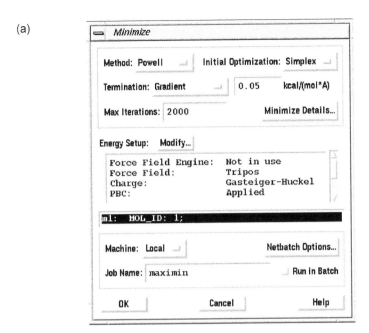

(b)   (c)

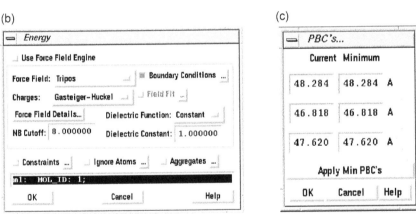

Figure 8 SYBYL windows for energy minimization of the solvated box. The main window is reported in (a); the windows showing charges and active boundary conditions (PBC) are reported in (b) and (c).

have a distribution of high and low affinity binding sites, and a different distrubution than in the polyclonal antibodies tested.

VII. CASE STUDY 2: THE DESIGN, SYNTHESIS, AND EVALUATION OF EPHEDRINE SELECTIVE MIPs

The objective of the present work was screening of virtual library of functional monomers with the aim to identify the monomers capable of interacting with ephedrine (Fig. 11).

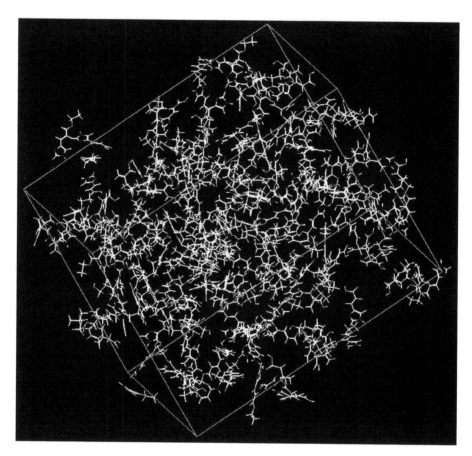

Figure 9   An example of a solvation box in a simulated annealing experiment.

Once again, we hypothesize that the monomers giving the highest binding score with the template should give the polymer (both, MIP and blanks) with higher affinity and specificity. In order to test this concept, a set of MIPs was designed by computational methods, then synthesized and the polymers tested using high-performance liquid chromatography (HPLC) experiments to monitor the separation of ephedrine enantiomers.

In the first step, a molecular model of (−)-ephedrine (template) and a virtual library was created, as described in CASE STUDY 1. In addition, a molecular model of the formally charged (ionized) template was generated and added to the library. All of these structures were then charged with the Gasteiger–Huckel computational method, and refined using the MAXIMIN2 command. In the second step, the LEAP-FROG algorithm was used to screen the library of functional monomers for their possible interaction with the template. The program was run for 30,000 iterations and the results were examined and the empirical binding energy scores evaluated (Table 3).

The best four monomers, those giving the highest binding scores for complexation with the template, were selected for the polymer preparation: itaconic acid, methacrylic acid, hydroxyethyl methacrylate, and acrylamide. For the purpose of

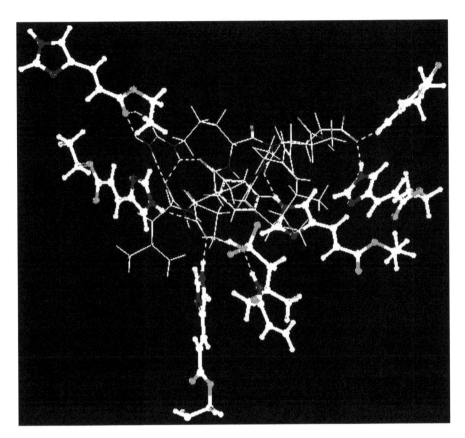

Figure 10 Final result of computer modeling. Interactions between Microcystin-LR and monomers. Microcystin-LR (balls and sticks) interacts with six molecules of UAEE (light gray) and one molecule of AMPSA (dark gray) (from Chianella *et al.*, 2002).

having an additional point for comparison, a polymer was prepared using 2-vinylpyridine. This monomer is found near the bottom of Table 3 having a binding energy of only $-1.82\,\mathrm{kcal/mol}$.

The polymers were synthesized and tested as described elsewhere [75]. By using corresponding nonimprinted (blank) polymers in the HPLC study, we anticipated seeing a correlation between the general affinity of the polymer and the calculated binding energy for template–monomer interactions. To mimic the conditions where a neutral polymer interacts with a neutral and protonated (ionized) template, the

Table 2 Affinity and Sensitivity Range of MIPs and Antibodies for Microcystin-LR, Evaluated by Competitive Assay

Receptor	K_d (nM)	Sensitivity range (μg l^{-1})
Computational MIP	0.30 ± 0.08	0.1–100
Monoclonal antibody	0.03 ± 0.01	0.025–5
Polyclonal antibody	0.50 ± 0.07	0.05–10

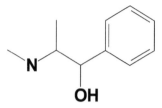

Figure 11 Structure of ephedrine.

free base and HCl-salt of ephedrine were injected separately into the columns containing blank polymers. The experiment conducted with chloroform in the presence of 0.3% acetic acid indicated that a correlation between the calculated binding energy and the capacity factor of the analyte exists (Table 4). The monomers with the highest binding energy scores (itaconic and methacrylic acids) produced the polymers with the strongest affinity for the template, as shown in the HPLC experiments. One exception was found, however, the HEM-based polymer. In this case, the polymer demonstrated an unusually poor ability to bind ephedrine. Moreover, the capacity factor (k') observed, 0.3, did not match the predicted binding energy (−15.72 kcal/mol). Even better correlation was found between the binding energy of monomer–template interaction and imprinting factor (I). Obviously the strength of the monomer–template interaction is important for successful imprinting.

It is interesting to note that computer simulation can also be used to mimic real chromatographic experiments. For example, the interaction of the methacrylic acid based polymer with the HCl salt of ephedrine (binding energy of MA with ionized ephedrine is −22.48 kcal/mol) which is much stronger, k' 7.3, than its interaction with

Table 3 Screening of a Virtual Library of Functional Monomers on Their Interaction with Ephedrine

N	Binding interactions[a]	Binding (kcal/mol)
1	Itaconic acid (neutral)–ephedrine (neutral)	−33.81
2	Itaconic acid (neutral)–ephedrine (charged)	−23.14
3	Methacrylic acid (neutral)–ephedrine (charged)	−22.48
4	Methacrylic acid (neutral)–ephedrine (neutral)	−14.62
5	Hydroxethyl methacrylate (neutral)–ephedrine (neutral)	−15.72
6	Acrylamide (neutral)–ephedrine (neutral)	−13.63
7	Allylamine (neutral)–ephedrine (neutral)	−12.02
8	Urocanic acid ethyl ester (neutral)–ephedrine (neutral)	−6.59
9	4-Vinylpyridine (neutral)–ephedrine (neutral)	−4.99
10	Styrene–ephedrine (neutral)–ephedrine (neutral)	−4.80
11	1-Vinylimidazole (neutral)–ephedrine (neutral)	−4.37
12	Urocanic acid (neutral)–ephedrine (neutral)	−3.60
13	2-Vinylpyridine (neutral)–ephedrine (neutral)	−1.82

[a]The computer ignored the interaction between monomers and template if they are below the cut-off criteria (−0.2 kcal/mol).

Table 4 Chromatographic Evaluation of Polymers Performed in Chloroform. Ten Microliters of the Sample (concentration 1 mg/ml) were Injected for the Analysis. The Eluent was Chloroform with 0.3% Acetic Acid. The Flow Rate was 1 ml/min (from Ref. [75])

Monomer	Template condition	Monomer–template binding energy (kcal/mol)	k' (−)-ephedrine Blank	k' (−)-ephedrine MIP	I (Imprinted factor)
IA	Neutral	−33.81	9.4	∞	∞
	Ionized	−23.14	7.2	−	−
MA	Neutral	−14.62	3.3	9.46	2.87
	Ionized	−22.48	7.3	−	−
HEM	Neutral	−15.72	0.3	0.8	2.67
AA	Neutral	−13.63	1.4	2.59	1.85
2-VP	Neutral	−1.82	0.1	0.1	1

neutral free base ephedrine where the capacity factor was 3.3 (binding energy of MA with neutral ephedrine is −14.62 kcal/mol). A similar correlation also existed in the performance of the itaconic acid-based polymer (Table 4).

Temperature has a complex effect on monomer–template and polymer–template interactions. Lower temperature should be beneficial due to a reduction of the influence of residual vibrational modes (ΔG_{vib}) and an increase in the strength of polar interactions ($\Sigma \Delta G_p$, see Eq. (1). At the same time, the tighter complexes formed at lower temperature could be responsible for an increase in adverse conformational (ΔG_{conf}) and unfavorable van der Waals (ΔG_{vdW}) terms in Eq. (1). Due to this, it is important to have the possibility to model conditions of complex formation where the role of the template can be analyzed in detail.

The objective of this work was to model interactions between ephedrine and hydroxyethyl methacrylate at different temperatures in order to predict the performance of the real polymer in enantioseparation. During the first step, a solvation box was created containing template surrounded by multiple copies of monomer using the XFIT solvation algorithm (an integrated module in the SYBYL 6.7 package) and the energy of the system was minimized. The simulated annealing process was then applied to the box to analyze the arrangement of functional monomers around the template. The high temperature was applied in order to avoid entrapment in local minima. Next the temperature was lowered in order to bring the system into a stable state. The starting annealing temperature was fixed at 983 K (710°C) and lowered to a temperature of 283 K (10°C). Next, the temperature was further lowered to final temperature 253 K (−20°C) in four subsequent 10 K steps (dynamic equilibrium was reached in 2000 fs). After each step, the system was minimized to 0.01 kcal/mol. At the end of the program, the number and the position of functional monomers were examined. The structure of the complex formed between the monomer and template is presented in Fig. 12. It is interesting to note that two molecules of hydroxyethyl methacrylate were found to interact with one molecule of ephedrine, which indicates the presence of three-point binding modes, which are essential for chiral recognition. The distance between hydroxyls in monomer–template complex varied with temperature (Table 5).

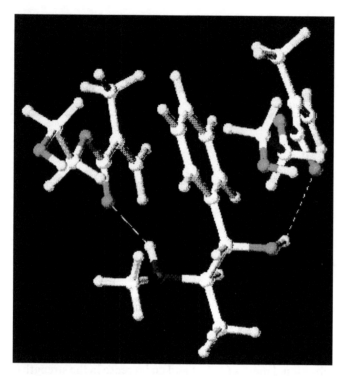

Figure 12 The interactions between two molecules of hydroxyethyl methacrylate and ephedrine in monomer mixture.

The modeling results clearly indicate that the formation of monomer–template complexes is dependent upon the temperature. It was anticipated that polymers synthesized at higher temperatures (> 283 K or 10°C) would be inferior to those prepared at lower temperatures due to the lack of template binding to the second monomer. In order to prove this hypothesis, a series of polymers were synthesized at different temperatures (−30°C, −20°C, −10°C, 0°C, 20°C, and 80°C) and tested in HPLC studies. The details of this work are published elsewhere [76]. The chromatographic evaluations were performed at temperatures varying from −10°C to 55°C. The results of this evaluation, expressed in terms of temperature dependence of separation factors for ephedrine enantiomers, are presented in Fig. 13. As expected, a clear decrease in separation factors (α) from $\alpha = 4.04$ (for polymers made at −30°C

Table 5 Influence of the Temperature on Distance Between Hydroxyls in Monomers-Template Complex

Temperature (K)	Distance
253	5.85 Å
263	6.57 Å
273	6.72 Å
283	∞

to 20°C) to $\alpha = 1.53$ (for polymer made at 80°C) was observed with increase in polymerization temperature. An interesting observation was the ability of the polymer to "memorize" the temperature developed during polymerization. It is possible that during the polymerization stage the microenvironment of the developing imprints adjusts to the solvation by the porogen and also to the temperature conditions. The postulated mechanism of MIP recognition originates, basically, from two factors: shape of the imprints, and the spatial positioning of the functional groups in the polymer which are participating in the complex with the template and are integrated into the polymer network during the polymerization stage. The distance between these groups and their orientation in the polymer can be affected by both solvent induced swelling and also by temperature. Consequently, it would appear as though the ligand recognition capacity of an MIP at a given temperature may be influenced by the temperature employed during the polymer synthesis.

Other simulation experiments performed in our laboratory include design of MIP adsorbents for solid-phase extraction of atrazine [77], DDT, lindane, aflatoxin B1, ochratoxin A, and tylosin (unpublished data) and the development of assay/sensor recognition elements for biotin (unpublished data) and creatinine [78]. In all these cases, molecular modeling proved to be a useful tool for MIP design. It would

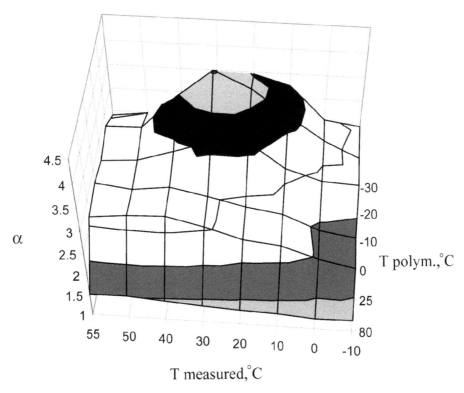

Figure 13 Influence of the polymerization temperature on separation factors (α) for polymers measured at different temperatures. Flow rate, $1\,\mathrm{ml\,min^{-1}}$; mobile phase, 0.05% hexamethylenediamine in chloroform. Injection amounts were $8\,\mu g$ (48.5 nmol) in $40\,\mu L$ injection volume. (From Ref. [76].)

be an overstatement, however, to say that in all our experiments we have had perfect matches between the modeling results and polymer testing. The reason for some "mismatches" undoubtedly lies in the complex nature of polymerization process and in the performance limitations of modeling algorithms. We believe that it would be beneficial, in the future, to combine the molecular modeling of the monomer mixtures with molecular modeling of the polymerization process (e.g., phase separation).

VIII. SUMMARY AND OUTLOOK

We believe that in the near future, the use of computational methods in conjunction with spectroscopic techniques and thermodynamic considerations should allow the in silico simulation to be used broadly for the analysis of the influence of polymerization conditions (solvent, cross-linker, temperature) on the performance of imprinted polymers, for optimization of the monomer composition and for "tailoring" polymer performance for specific applications. The computational approach described here represents a first step towards the truly rational design (tailoring) of MIPs and prediction of polymer properties. Nonetheless, improvements in our capacity to predict polymer performance should be benefited through the combination of molecular modeling, further physical characterization of the imprinting process, and combinatorial strategies (Chapter 8).

REFERENCES

1. Pauling, L.C. In *The Nature of the Chemical Bond and the Structure of Molecules and Crystals*, 3rd Ed.; Cornell University Press: Ithaca, 1960.
2. Leach, A.R. *Molecular Modeling Principles and Applications*, 2nd Ed.; Prentice Hall: Harlow, UK, 2001; 1–25.
3. Page, M.I.; Jencks, W.P. Entropic contributions to rate accelerations in enzymic and intramolecular reactions and the chelate effect. Proc. Natl. Acad. Sci. USA **1971**, *68*, 1678–1683.
4. Jencks, W.P. On the attribution and additivity of binding energies. Proc. Natl. Acad. Sci. USA **1981**, *78*, 4046–4050.
5. Andrews, P.R.; Craik, D.J.; Martin, J.L. Functional group contributions to drug receptor interactions. J. Med. Chem. **1984**, *27*, 1648–1657.
6. Williams, D.H.; Cox, J.P.L.; Doig, A.J.; Gardner, M.; Gerhard, U.; Kaye, P.T.; Lal, A.R.; Nicholls, I.A.; Salter, C.J.; Mitchell, R.C. Towards the semiquantitative estimation of binding constants. Guides for peptide–peptide binding in aqueous solution. J. Am. Chem. Soc. **1991**, *113*, 7020–7030.
7. Searle, M.; Williams, D.H.; Gerhard, U. Partitioning of free-energy contribution in the estimation of binding constants–residual motions and consequences for amide–amide hydrogen-bond strengths. J. Am. Chem. Soc. **1992**, *114*, 10697–10704.
8. Holroyd, S.E.; Groves, P.; Searle, M.S.; Gerhard, U.; Williams, D.H. Rational design and binding of modified cell-wall peptides to vancomycin-group antibiotics—factorizing free-energy contributions to binding. Tetrahedron **1993**, *49*, 9171–9182.
9. Nicholls, I.A. Thermodynamic considerations for the design of and ligand recognition by molecularly imprinted polymers. Chem. Lett. **1995**, 1035–1036.
10. Nicholls, I.A.; Andersson, H.S. Thermodynamic principles underlying molecularly imprinted polymer formulation and ligand recognition. In *Molecularly Imprinted Polymers, Man-Made Mimics of Antibodies and Their Practical Application in Analytical Chemistry*, Sellergren, B., Ed.; Elsevier: Amsterdam, The Netherlands, 2001; 59–70.

11. Nicholls, I.A.; Adbo, K.; Andersson, H.S.; Andersson, P.O.; Ankarloo, J.; Hedin-Dahl-
     ström, J.; Jokela, P.; Karlsson, J.G.; Olofsson, L.; Rosengren, J.P.; Shoravi, S.; Svenson,
     J.; Wikman, S. Can we rationally design molecularly imprinted polymers? Anal. Chim.
     Acta **2001**, *435*, 9–18.
12. Sellergren, B.; Lepistö, M.; Mosbach, K. Highly enantioselective and substrate-selective
     polymers obtained by molecular imprinting utilizing noncovalent interactions. NMR and
     chromatographic studies on the nature of recognition. J. Am. Chem. Soc. **1988**, *110*,
     5853–5860.
13. Andersson, H.S.; Nicholls, I.A. Spectroscopic evaluation of molecular imprinting
     polymerisation systems. Bioorg. Chem. **1997**, *25*, 203–211.
14. Svenson, J.; Andersson, H.S.; Piletsky, S.A.; Nicholls, I.A. Spectroscopic studies of the
     molecular imprinting self assembly process. J. Mol. Recogn. **1998**, *11*, 83–86.
15. Whitcombe, M.J.; Martin, L.; Vulfson, E.N. Predicting the selectivity of imprinted poly-
     mers. Chromatographia **1998**, *47*, 457–464.
16. Tanabe, K.; Takeuchi, T.; Matsui, J.; Ikebukuro, K.; Yano, K.; Karube, I. Recognition
     of barbiturates in molecularly imprinted copolymers using multiple hydrogen bonding.
     J. Chem. Soc. Chem. Comm. **1995**, 2303–2304.
17. Zhou, J.; He, X.W. Study of the nature of recognition in molecularly imprinted polymer
     selective for 2-aminopyridine. Anal. Chim. Acta. **1999**, *381*, 85–91.
18. Svenson, J; Karlsson, J.; Nicholls, I.A. Chromatogr. A **2004**, *1024*, 39–44.
19. O'Shannessy, D.J.; Ekberg, B.; Mosbach, K. Molecular imprinting of amino acid deriva-
     tives at low temperature (0°C) using photolytic homolysis of azobisnitriles. Anal.
     Biochem. **1989**, *177*, 144–149.
20. Sellergren, B.; Shea, K.J. Influence of polymer morphology on the ability of imprinted
     network polymers to resolve enantiomers. J. Chromatogr. **1993**, *635*, 31–49.
21. Sellergren, B.; Dauwe, C.; Schneider, T. Pressure-induced binding sites in molecularly
     imprinted network polymers. Macromolecules **1997**, *30*, 2454–2459.
22. Mayes, A.; Lowe, C.R. Optimization of molecularly imprinted polymers for radio-ligand
     binding assays. Royal Society of Chemistry: Cambridge, UK, 1998; Vol. 25, 28–36.
23. Andersson, L.I.; Müller, R.; Vlatakis, G.; Mosbach, K. Mimics of the binding sites of
     opioid receptors obtained by molecular imprinting of enkephalin and morphine. Proc.
     Natl. Acad. Sci. USA **1995**, *92*, 4788–4792.
24. Ye, L.; Cormack, P.A.G.; Mosbach, K. Molecularly imprinted monodisperse micro-
     spheres for competitive radioassay. Anal. Commun. **1999**, *36*, 35–38.
25. Fischer, L.; Müller, R.; Ekberg, B.; Mosbach, K. Enantioseparation of β-adrenergic
     blockers using a chiral stationary phase prepared by molecular imprinting. J. Am. Chem.
     Soc. **1991**, *113*, 9358–9360.
26. Berglund, J.; Nicholls, I.A.; Lindbladh, C.; Mosbach, K. Recognition in molecu-
     larly imprinted polymer α_2-adrenoceptor mimics. Bioorg. Med. Chem. Lett. **1996**, *6*,
     2237–2242.
27. Andersson, H.S.; Koch-Schmidt, A.-C.; Ohlson, S.; Mosbach, K. Study of the nature of
     recognition in molecularly imprinted polymers. J. Mol. Recogn. **1996**, *9*, 675–682.
28. Matsui, J.; Miyoshi, Y.; Takeuchi, T. Fluoro-functionalized molecularly imprinted
     polymers selective for herbicides. Chem. Lett. **1995**, 1007–1008.
29. Matsui, J.; Takeuchi, T. A molecularly imprinted polymer rod as nicotine selective
     affinity media prepared with 2-(trifluoromethyl)acrylic acid. Anal. Commun. **1997**, *34*,
     199–200.
30. Vidyasankar, S.; Dhal, P.K.; Plunkett, S.D.; Arnold, F.H. Selective ligand-exchange
     adsorbents prepared by template polymerisation. Biotechnol. Bioeng. **1995**, *48*, 431–436.
31. Jenkins, A.L.; Uy, O.M.; Murray, G.M. Polymer-based lanthanide luminescent sensor
     for detection of the hydrolysis product of the nerve agent Soman in water. Anal. Chem.
     1999, *71*, 373–378.

32. Zeng, X.F.; Murray, G.M. Synthesis and characterization of site-selective ion-exchange resins templated for lead(II) ion. Sep. Sci. Technol. 1996, 31, 2403–2418.

33. Matsui, J.; Nicholls, I.A.; Takeuchi, T.; Mosbach, K.; Karube, I. Metal ion mediated recognition in molecularly imprinted polymers. Anal. Chim. Acta 1996, 335, 71–77.

34. Santora, B.P.; Larsen, A.O.; Gagne, M.R. Toward the molecular imprinting of titanium Lewis acids: demonstration of Diels-Alder catalysis. Organometallics 1998, 17, 3138–3140.

35. Sreenivasan, K. Synthesis and evaluation of a beta-cyclodextrin-based molecularly imprinted copolymer. J. Appl. Polym. Sci. 1998, 70, 15–18.

36. Asanuma, H.; Kakazu, M.; Shibata, M.; Hishiya, T.; Komiyama, M. Molecularly imprinted polymer of beta-cyclodextrin for the efficient recognition of cholesterol. Chem. Comm. 1997, 1971–1972.

37. Piletsky, S.A.; Andersson, H.S.; Nicholls, I.A. Combined hydrophobic and electrostatic interaction-based recognition in molecularly imprinted polymers. Macromolecules 1999, 32, 633–636.

38. Andersson, H.S.; Ramström, O. Crown ethers as a tool for the preparation of molecularly imprinted polymers. J. Mol. Recogn. 1998, 11, 103–106.

39. Sellergren, B.; Andersson, L. Molecular recognition in macroporous polymers prepared by a substrate analogue imprinting strategy. J. Org. Chem. 1990, 55, 3381–3383.

40. Byström, S.E.; Börje, A.; Åkermark, B. Selective reduction of steroid 3- and 17-ketones using LiAlH$_4$ activated template polymers. J. Am. Chem. Soc. 1993, 115, 2081–2083.

41. Whitcombe, M.J.; Rodriguez, M.E.; Villar, P.; Vulfson, E.N. A new method for the introduction of recognition site functionality into polymers prepared by molecular imprinting: synthesis and characterization of polymeric receptors for cholesterol. J. Am. Chem. Soc. 1995, 117, 7105–7111.

42. Lübke, M.; Whitcombe, M.J.; Vulfson, E.N. A novel approach to the molecular imprinting of polychlorinated aromatic compounds. J. Am. Chem. Soc. 1998, 120, 13342–13348.

43. Nicholls, I.A.; Ramström, O.; Mosbach, K. Insights into the role of the hydrogen bond and hydrophobic effect on recognition in molecularly imprinted polymer synthetic peptide mimics. J. Chromatogr. A 1995, 691, 349–353.

44. Yu, C.; Mosbach, K. Molecular imprinting utilizing an amide functional group for hydrogen bonding leading to highly efficient polymers. J. Org. Chem. 1997, 62, 4057–4064.

45. Yu, C.; Ramström, O.; Mosbach, K. Enantiomeric recognition by molecularly imprinted polymers using hydrophobic interactions. Anal. Lett. 1997, 30, 2123–2140.

46. Zheng, N.; Li, Y.-Z.; Cheng, W.-B.; Wang, Z.-M.; Li, T.-J. Sulfonamide imprinted polymers using co-functional monomers. Anal. Chim. Acta 2002, 452, 277–283.

47. Andersson, L.I. Application of molecular imprinting to the development of aqueous buffer and organic solvent based radioligand binding assays for S-propranolol. Anal. Chem. 1996, 68, 111–117.

48. Andersson, L.I.; Müller, R.; Mosbach, K. Molecular imprinting of the endogenous neuropeptide Leu-5 enkephalin and some derivatives thereof. Macromol. Rapid Comm. 1996, 17, 65–71.

49. Karlsson, J.G.; Andersson, L.I.; Nicholls, I.A. Probing the molecular basis for ligand-selective recognition in molecularly imprinted polymers selective for the local anaesthetic bupivacaine. Anal. Chim. Acta 2001, 435, 57–64.

50. Haupt, K.; Dzgoev, A.; Mosbach, K. Assay system for the herbicide 2,4-D using a molecularly-imprinted polymer as an artificial recognition element. Anal. Chem. 1998, 70, 628–631.

51. Haupt, K.; Mayes, A.G.; Mosbach, K. Herbicide assay using an imprinted polymer based system analogous to competitive fluoroimmunoassays. Anal. Chem. 1998, 70, 3936–3939.

52. Sellergren, B.; Shea, K.J. Origin of peak asymmetry and the effect of temperature on solute retention in enantiomer separations on imprinted chiral stationary phases. J. Chromatogr. A **1995**, *690*, 29–39.

53. Lin, J.M.; Nakagama, T.; Uchiyama, K.; Hobo, T. Temperature effect on chiral recognition of some amino acids with molecularly imprinted polymer filled capillary electrochromatography. Biomed. Chromatogr. **1997**, *11*, 298–302.

54. Sellergren, B.; Shea, K.J. Chiral ion-exchange chromatography. Correlation between solute retention and a theoretical ion-exchange model using imprinted polymers. J. Chromatogr A **1993**, *654*, 17–28.

55. Sajonz, P.; Kele, M.; Zhong, G.M.; Sellergren, B.; Guiochon, G. Study of the thermodynamics and mass transfer kinetics of two enantiomers on a polymeric imprinted stationary phase. J. Chromatogr A **1998**, *810*, 1–17.

56. Szabelski, P.; Kaczmarski, K.; Cavazzini, A.; Chen, Y.-B.; Sellergren, B.; Guichon, G. Energetic heterogeneity of the surface of a molecularly imprinted polymer studied by high performance liquid chromatography. J. Chromatogr. A **2002**, *964*, 99–111.

57. Lanza, F.; Sellergren, B. Method for synthesis and screening of large groups of molecularly imprinted polymers. Anal. Chem. **1999**, *71*, 2092–2096.

58. Takeuchi, T.D.; Fukuma, D.; Matsui, J. Combinatorial molecular imprinting: an approach to synthetic polymer receptors. Anal. Chem. **1999**, *71*, 285–290.

59. Boyd, D.B.; Snoddy, J.D.; Lin, H.-S. Molecular simulation of DD-peptidase, a model β-lactam-binding protein: synergy between X-ray crystallography and computational chemistry. J. Compu. Chem. **1991**, *12*, 635–644.

60. Chun, P.W.; Jou, W.S. Molecular information of ubiquitinated structures and the implications for regulatory function. J. Mol. Graphics **1992**, *10*, 7–11.

61. Falconi, M.; Rotilio, G.; Desideri, A. Modeling the three-dimensional structure and electrostatic potential field of the two Cu, Zn superoxide dismutase variants from Xenopus laevis. Proteins **1991**, *10*, 149–155.

62. Tallon, M.; Ron, D.; Halle, D.; Amodeo, P.; Saviano, G.; Temussi, P.A.; Selinger, Z.; Naider, F.; Chorev, M. Synthesis, biological activity, and conformational analysis of [pGlu[6],N-MePhe[8],Aib[9]] substance P (6-11): a selective agonist for the NK-3 receptor. Biopolymers **1993**, *33*, 915–926.

63. Holtje, H.D.; Folkers, G.; Molema, R.; Kubinyi, H.; Timmerman, H. Molecular modeling: basic principles and application. In *Methods and Principles in Medicinal Chemistry*; Mannhold, R., Kubiny, H., Timmermann, H., Eds.; VCH Verlagsgesellschaft: Weinheim, 1996; Vol. 5, 194.

64. Clark, M.; Cramer, R.D.; Vanopdenbosh, N. Validation of the general-purpose Tripos 5.2 force-field. J. Comput. Chem. **1989**, *10*, 982–1012.

65. Allinger, N.L. Conformation analysis. MM2 a hydrocarbon force field utilising V1 and V2 tortional terms. J. Am. Chem. Soc. **1977**, *99*, 8127–8134.

66. Weiner, S.J.; Kollman, P.A.; Case, D.A.; Singh, C.S.; Ghio, C.; Alagona, G.; Profeta, S. Jr.; Weiner, P. A new force field for molecular mechanical simulation of nucleic acids and proteins. J. Am. Chem. Soc. **1984**, *106*, 765–784..

67. Fraga, S.; Parker, J.M.R.; Pocock, J.M. *Computer Simulations of Protein Structures and Interactions*, 1st Ed.; Springer-Verlag: New York, 1995, 282.

68. Nelder, J.A.; Mead, R. A simplex method for function minimisation. Computing J. **1965**, *7*, 308–313.

69. Powell, M.J.D. Restart procedures for the conjugate gradient method. Math. Programming **1977**, *12*, 241–254.

70. Press, W.H.; Teukolsky, S.A.; Vetteling, W.T.; Flannery, B.P. *Numerical Recipes in C: The Art of Scientific Computing*, 2nd Ed.; Cambridge University Press, 1993, 994.

71. Klebe, G.; Mietzner, T. A fast and efficient method to generate biologically relevant conformations. J. Comput. Aided Mol. Des. **1994**, *8*, 583–606.

72. Gunsteren, W.F.; Van Berendsen, H.J.C. Algorithms for macromolecular dynamics and constraint dynamics. Mol. Phys. **1977**, *34*, 1311–1327.
73. Goldberg, J.; Huang, H.; Know, Y.; Greengard, P.; Nairn, A.C.; Kuriyan, J. Three-dimensional structure of the catalytic subunit of protein serine/threonine phosphatase-1. Nature **1995**, *376*, 745–753.
74. Chianella, I.; Lotierzo, M.; Piletsky, S.A.; Tothill, I.E.; Chen, B.; Karim, K.; Turner, A.P.F. Rational design of a polymer specific for Microcystin-LR using a computational approach. Anal. Chem. **2002**, *74*, 1288–1293.
75. Piletsky, S.A.; Karim, K.; Piletska, E.V.; Day, C.J.; Freebairn, K.W.; Legge, C.; Turner, A.P.F. Recognition of ephedrine enantiomers by MIPs designed using a computational approach. Analyst **2001**, *126*, 1826–1830.
76. Piletsky, S.A.; Piletska, E.V.; Karim, K.; Freebairn, K.W.; Legge, C.H.; Turner, A.P.F. Polymer cookery: influence of polymerization conditions on the performance of molecularly imprinted polymers. Macromolecules **2002**, *35*, 7499–7504.
77. Piletsky, S.A.; Day, R.M.; Chen, B.; Subrahmanyam, S.; Piletska, E.V.; Turner, A.P.F. Rational design of MIPs using computational approach. PCT/GB01/00324.
78. Subrahmanyam, S.; Piletsky, S.A.; Piletska, E.V.; Chen, B.; Karim, K.; Turner, A.P.F. "Bite-and-Switch" approach using computationally designed molecularly imprinted polymers for sensing of creatinine. Biosensors Bioelectron. **2001**, *16*, 631–637.

15

Selectivity in Molecularly Imprinted Matrices

David A. Spivak Louisiana State University,
Baton Rouge, Louisiana, U.S.A.

I. INTRODUCTION

The experimentalist is often faced with the challenge of selectively separating a desired compound from a mixture and molecularly imprinted polymers (MIPs) can be instrumental in achieving such substrate selective separations. Polymers have been imprinted using organic molecules [1,2], metal ions [3–5], and large biomolecules such as proteins and DNA [6,7]. These different substrate types often require different strategies for optimal imprinting and recognition. However, most studies on molecular imprinting to date have focused on organic molecules, providing most of what is known about selectivity in MIPs. Therefore, treatment in this tutorial will correspondingly focus on imprinting small organic molecules. The category of organic molecule substrates can be further divided into:

1. enantiomeric organic molecules;
2. isomeric organic compounds;
3. different organic compounds.

The difficulty in achieving selectivity for these organic compounds follows the relative order given above. Enantiomeric molecules differ only in the three-dimensional geometry of the constituent atoms in space, while the remaining properties of both molecules are identical. Isomers that are not enantiomeric, whether diastereomeric or geometrical, afford more distinctions for selectivity such as different shapes, different conformations, and different spatial disposition of atoms which correspond to differences in binding energies. This is more so for unrelated organic compounds which may drastically differ in partitioning properties. Before discussing the particulars of preparing selective MIPs, it is important to first understand the analysis methods used to analyze selectivity and how to effectively use and compare them.

II. ANALYSIS OF SELECTIVITY IN IMPRINTED POLYMERS

A. Batch Rebinding

There are primarily two analysis methods for MIP selectivity in order to compare and evaluate MIP host design for selective binding of targets (Scheme 1), chromatography and batch rebinding. Batch rebinding will be discussed first, which involves analysis of a heterogeneous mixture of host MIP in a solution of guest substrate S (Fig. 1). In this case, a known amount of granulated MIP is added to a solution of substrate S, and the amount of substrate remaining in solution after adsorption to the polymer is measured and referred to as C_f (the concentration of free substrate). Subtraction of C_f from the total substrate added (C_t) gives the amount of substrate adsorbed to the MIP (C_b). Following classical adsorption models, the equation for this binding event is written as

$$MIP + S = MIP' \cdot S \tag{1}$$

with the partition coefficient

$$K_p = [MIP' \cdot S]/[S] \tag{2}$$

Selectivity by any host molecule or polymer arises from the differences in the free energy of adsorption of one substrate vs. another. The free energy of binding the guest, ΔG, by the polymer can be determined by

$$\Delta G = -RT \ln K_p \tag{3}$$

The difference in free energies is calculated to be

$$\Delta G_2 - \Delta G_1 = \Delta\Delta G = -RT \ln K_{P2} - (-RT \ln K_{P1}) = -RT \ln (K_{P2}/K_{P1}) \tag{4}$$

Thus, the ratio of the partition coefficients determines the selectivity of a host between two substrates, which is referred to as the *separation factor* (α) or sometimes as the *selectivity factor*

$$\alpha = (K_{P2}/K_{P1}) \tag{5}$$

When applying α to MIPs, the separation factor indicates how many times better substrate 2 binds to the polymers vs. substrate 1. This is of particular interest for comparing enantiomers of the same compound which have the same properties

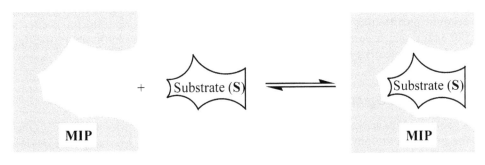

Scheme 1   Binding of a substrate (*S*) to a molecularly imprinted polymer (MIP).

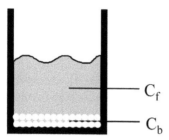

Figure 1 General method of batch rebinding.

except for their three-dimensional topological disposition. However, obtaining α values is not limited to enantiomers, and any two compounds may be compared on the same polymer by computing their α value. The difference in comparing two nonenantiomeric compounds vs. two enantiomers is that the nonenantiomeric compounds have different properties other than three-dimensional geometry, such as structure, size, and hydrophobicity, to name a few. A better way to compare the imprinting effects of nonenantiomeric compounds uses a modified partition coefficient, **I**, which is referred to as the *imprinting factor* [8]. I is obtained from the ratio of the partition coefficient of a substrate on an imprinted polymer, K_{MIP}, and the partition coefficient K_{NP}, using the same substrate on a nonimprinted polymer (NP) with the same monomer formulation.

$$I = (K_{\text{MIP}}/K_{\text{NP}}) \tag{6}$$

The imprinting factor represents how many times better the substrate binds to its imprinted polymer vs. a nonimprinted (generic) polymer. In this manner, binding due to nonspecific interactions is numerically removed, leaving a value for the binding due solely to the imprinting effect, namely the three-dimensional organization of monomers in the MIP. Taking the ratio of imprinting factors for two different substrates, I_1 for substrate one and I_2 for substrate two, we obtain the *specific selectivity factor S* [8]

$$S = (I_1/I_2) \tag{7}$$

It is conventional to put the higher value in the numerator (which is usually the imprinted substrate) and the lower value in the denominator. The specific selectivity factor indicates how many times more the *imprinting effect* is observed for substrate 1 vs. substrate 2 on one polymer. Thus, the specific selectivity factor does not take into account partitioning effects between two molecules due to nonimprinted effects. For enantiomeric compounds, the value for K_{NP} is the same for both substrates, thus the specific selectivity factor simply reduces to α.

B. Chromatography

As solids, MIPs are predisposed for use as chromatographic stationary phases. Once the polymers are obtained in the desired particle size, e.g., by sizing and sieving, columns can be made using traditional chromatographic packing techniques (Scheme 2). Equations for the same analysis using chromatographic parameters

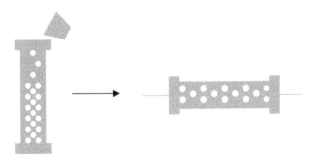

Scheme 2   Once MIPs are sized, they can be packed into HPLC columns.

are proportional to the equations for batch rebinding above [9]. The fundamental equation relating batch rebinding to chromatography is:

$$K = k'/\phi \tag{8}$$

The symbol ϕ is the phase volume ratio (volume of stationary phase/volume of mobile phase), and k' is the capacity factor that is found using the following equation:

$$k' = (R_V - D_V)/D_V \tag{9}$$

where R_V is the retention volume of the sample (retention time × flow rate) and D_V the retention volume of an unretained sample (e.g., acetone).

The relationship between K and k' holds if both batch rebinding and chromatographic methods are under equilibrium conditions, allowing the equations derived for the batch rebinding method to be converted to the following equations for chromatographically derived data:

$$\alpha = (k'_2/k'_1) \tag{10}$$

$$I = (k'_{MIP}/k'_{NP}) \tag{11}$$

$$S = (I_1/I_2) \tag{12}$$

Before using any of the equations presented above, a very important caveat must be put forth. It was realized early on by Wulff et al. [10], and more recently explained by the groups of Shimizu [11–13] and Sellergren [14] that MIPs are formed with a distribution of binding or "active" sites. The distribution is in the quality of the binding sites, i.e., some are good but most are not. The meaning of "good" depends on whether strong binding, good selectivity, or fast catalysis is targeted by the imprinting process. However, in all cases the distribution must be taken into account, and the overall performance of the MIP will be a combined average of the ensemble of sites tested. Therefore, the values found by the equations above are most reproducible and comparable if the same conditions and quantities are used for each analysis.

III. EVALUATION OF TEMPLATES

To evaluate the "imprintability" of different templates, some general rules regarding template structure, composition, and topology will be put forth in a qualitative manner.

A. Types of Interactions

When presented with the task of developing an imprinted polymer, a good place to start is evaluating the binding potential of the template. A first consideration is which types of interactive functional groups on the template are available to form the prepolymer complex during the imprinting process and for molecular recognition during rebinding.

Rule 1. There must be one or more *binding* interactions possible.

In 1906, Paul Ehrlich stated "Corpora non agunt nisi fixata", which translates to "Molecules do not act if they do not bind" [15]. As simple as this may sound, this concept is the basis and first step of any molecular recognition event. Nowhere is this any more true than for molecular imprinting, which utilizes solution binding to create binding sites via a prepolymer complex and for molecular recognition of the target species during rebinding analyses. Examples of the different types of noncovalent interactions are shown in Fig. 2. In addition, binding interactions between the template and functional monomers should be as complementary as possible. For example, hydrogen-bonding interactions can be hydrogen donating (D) or hydrogen accepting (A). A hydrogen-bonding pair, such as template and functional monomer, should align donor groups with acceptor groups and vice versa, as shown in Fig. 2 .

The contribution of Van der Waals interactions between nonpolar segments of the template with the MIP matrix is often overlooked, primarily because it is difficult to measure their contribution to binding. These interactions are rarely effective for formation of the prepolymer complex in creating an MIP binding site. However, in

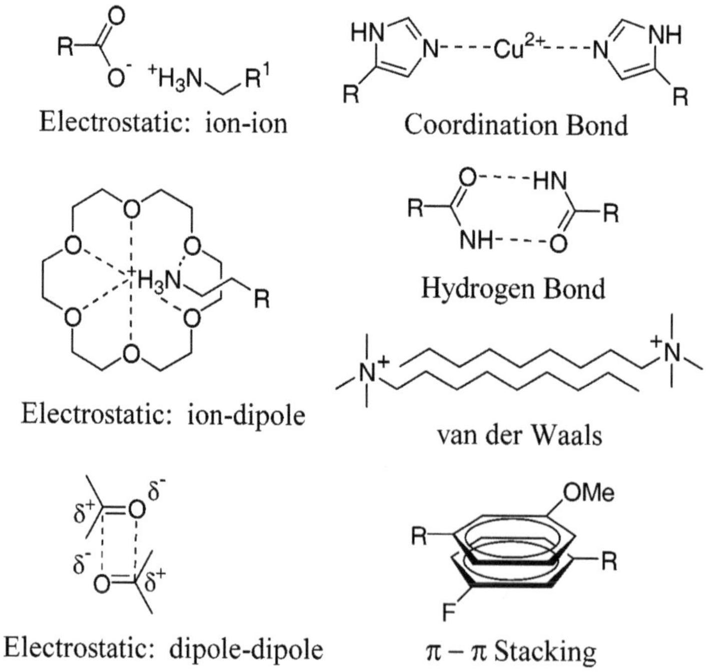

Figure 2 Examples of noncovalent interactions for MIPs.

[Template]   polymerize
                                                                                        Number of
        +   ⇌   [Pre-Polymer Complex]   ⇌   Binding Sites
[Functional Monomer]

Scheme 3 Relationship between solution phase binding and specific binding sites made in the MIP.

many cases MIPs exhibit chiral selectivity for templates that have only one or two identifiable strong binding interactions. The generally accepted Ogsten model for enantioselectivity requires at least three points of contact between substrate and receptor [16]. Therefore, it can be deduced that the other required points of contact come from interactions with the matrix. These interactions can be either Van der Waals interactions or simply steric interactions which will be discussed later in this text.

Rule 2. The *stronger* the binding interaction, the better.

The need for a strong bond is primarily for formation of the prepolymer complex. Studies suggest that the higher the solution binding constant between the template and a functional monomer, the higher the binding affinity of the template for the imprinted polymer. This is because the prepolymerization complex is one factor responsible for the creation of the actual "active sites" in the polymer that bind the substrate. In other words, each complex in the prepolymer solution is postulated to give rise to each binding site. By applying Le Châtelier's principle to the complex formed prior to polymerization, increasing concentration of components or binding affinity of the complex in the prepolymerization mixture will drive the functional monomer and template toward more prepolymerization complex (Scheme 3). This will increase the concentration of prepolymer complex, which then increases the number of final binding sites in the imprinted polymer.

Table 1 gives the ranges of binding energies for several major types of noncovalent interaction. Of course, covalently bound template-functional monomer complexes are the strongest. Covalent prepolymer complexes, however, are usually not found to generate stronger binding sites than noncovalently imprinted polymers. This may be due to stronger affinity sites arising from higher order complexes that

Table 1 Range of Binding Energies for Noncovalent Interactions

Entry	Type of binding interaction	Range of binding energies (kcal/mol)
1	Electrostatic	
	a.ion–ion	20–80
	b.ion–dipole	12–50
	c.dipole–dipole	1–10
2	Coordination bond	20–50
3	Hydrogen bond	1–30
4	π–π stacking	0–12
5	van der Waals interactions	0–1.5

are possible using noncovalent imprinting methods, but not possible for covalent methods. One study has shown that covalently imprinted polymers do, however, provide a narrower distribution of binding sites [13]. In the case of rebinding template to the imprinted polymer, however, stronger interactions are not always desirable. For example, in chromatography, strong interactions can provide slow mass-transfer kinetics for binding that can cause spreading of peak bandwidths. Slow mass-transfer kinetics also hinder real-time measurements for MIPs used in sensor applications. However, a reliable binding interaction must be present for rebinding to take place to an appreciable extent. Thus depending on the application, there needs to be a balance between the binding strength of the interaction between template and functional monomer before and after polymerization.

Rule 3. Interactions with specific directionality provide more selective binding sites.

There are two important contributors to the molecular recognition afforded by molecularly imprinted polymers:

1. binding affinity;
2. selectivity.

Binding affinity refers to the strength of the binding interaction, often given as a binding constant (K, mol^{-1}) or as free energy (ΔG, kcal/mol). Selectivity refers to how much more the MIP binds to one substrate vs. another, often represented as a selectivity factor (α) discussed earlier. Selectivity is often the primary goal of an MIP, which relies on relative binding values and not on achieving any specified absolute binding values. Therefore, optimizing selectivity in MIPs is not only related to Rule 2, but has additional requirements as well. It is often assumed that the imprinting process alone will configure binding groups between the template and polymer that are complementary to the template, preorganizing the specific interactions needed for selectivity of the template over any other analyte. This is not always the case, however, and further fine tuning of complementary interactions is needed to obtain the desired selectivity. This fine tuning can come from directionality of the binding interactions between the template and functional monomers in MIPs. Thus, after the molecular imprinting process provides preorganization of functional groups in the binding cavity, the interactions themselves can increase selectivity in the binding site by limiting the orientation of the interaction. For example, in a survey of MIPs toward selective binding of 2-phenylbutyric acid (1), chiral selectivity was found only with functional monomers such as 2-aminopyridine methacrylamide (2) capable of hydrogen bonding with the template in a single, coplanar direction (Fig. 3) [17]. This is what is meant by an interaction having specific directionality. On the other hand, a primary amine on the MIP, provided by monomers such as *N*-(2-aminoethyl) methacrylamide (3), presents a charge that can be regarded as spherical in nature, which does not provide directionality and does not limit the orientation of interaction with the template, although it does provide a strong binding interaction.

B.　Number of Possible Interactions

Rule 4. The *more interactions* between the template and polymer, the better the binding and selectivity.

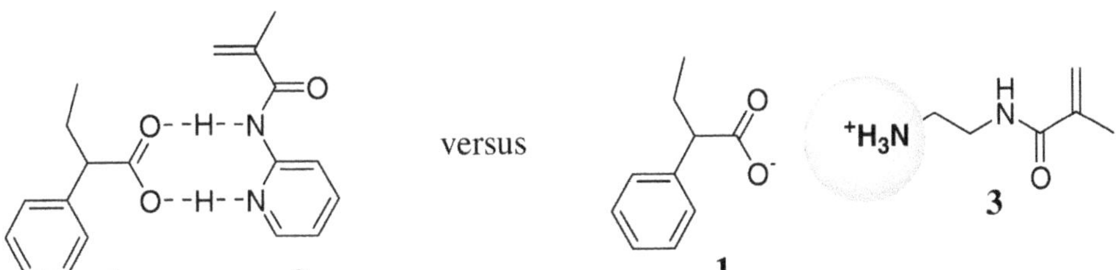

Figure 3 Comparison of directional vs. nondirectional binding interactions.

Increasing the number of interactions between the template and the MIP can increase both the overall binding affinity and the substrate specificity. In the case of binding affinity, each binding interaction between the template and MIP can have more than merely an additive effect on binding energy due to positive cooperativity between binding of two or more groups. This effect is sometimes referred to as the "chelate effect". To optimize the cooperativity between functional groups, the molecular framework between them should be rigid. If there is too much conformational freedom between the multiple functional groups, they will not act in concert, thus not obtaining the desired entropic gain. In the case of selectivity, each unique point of contact between the MIP and the target molecule (vs. any other analyte) increases the specificity of the binding site. A good example by Sellergren and coworkers is shown in Table 2 [18]. As the number of possible interactions between the phenylalanine derivatives 5–7 and methacrylic acid increase, there is an increase in the selectivity exhibited by the MIP. It was previously mentioned that chiral selectivity requires at least three interactions. With enantiomers, the first two interactions with a host molecule are not unique, while the third and higher number interactions include those that distinguish one enantiomer from another. This explains how the roles of binding and specificity are intimately intertwined. If the distinguishing interactions are weak, specificity will be low; conversely, if the distinguishing interactions are strong, specificity will be higher. Multiple weak interactions can add up to a strong interaction. Therefore, the more interactions there are between the template and MIP, the larger the binding affinity and the greater the potential for specificity.

C. Spatial Disposition

Rule 5. Binding and selectivity by multiple interactions are optimized by *increasing the intramolecular distance between functional groups on the template.*

To optimize the potential binding and specificity of an MIP for a target, functional groups should be spaced as far apart as possible to avoid interference between groups. This is especially true for interactive groups or identical groups. Interactive groups within close proximity on a template, for example, a primary amine and a carboxylic acid, may prefer to interact with each other intramolecularly rather than the external functional monomers. If the interactive groups on the template of the same type are too close to each other to be distinguishable, the functional monomer(s) may not distinguish the two sites and interact with both as though they are one. For

Table 2 Increase in Separation Factor (α) for MIPs with Increase in the Number of Interactions Between Template and Functional Monomer

Entry	Name of template derivative	Template and potential interactions with functional monomer (MAA)	Separation factor (α)
1	L-Phenylalanine ethyl ester		1.3
2	L-Phenylalanine anilide		3.7
3	L-(4-Amino-phenylalanine) anilide		5.7

example, binaphthylamine (**8**) is a chiral molecule with two primary amines pointed toward each other. Although binaphthylamine is a premier chiral ligand for organometallic complexes, MIPs developed toward binaphthylamine do not show appreciable chiral recognition. On the other hand, MIPs made against the similar template, 1,2-diphenyl-ethylenediamine [9], exhibit good chiral recognition. This is attributed to a conformation where the amine groups are spaced as far apart as possible and pointing in opposite directions (DA Spivak, H Kim, unpublished results, 2002 (Table 3)).

Rule 6. Selectivity increases as the *proximity of binding interactions* to a "*point of selectivity*" increases.

On a template molecule, a "point of selectivity" is a topological point in the molecule that determines its uniqueness with respect to other molecules being

Table 3 Comparison of Interactions of Methacracrylic acid with Templates That Have Different Spatial Disposition of Two Amine Groups

MIP entry	Pre-polymer complex of bis-amino templates with methacrylic acid	Separation factor (α)
1	**8**	1.05
2	**9**	1.38

compared. For example, in the case of enantiomers, the point of selectivity is the stereogenic center. Therefore, the closer the binding group(s) of an MIP are to the stereogenic center, the better the enantioselectivity. A clear example of this was demonstrated by comparing polymers imprinted with α-methylbenzylamine (**10**) with polymers imprinted using β-methyl phcnylethylamine (**11**) [19] (Table 4).

Table 4 Comparison of Selectivity of MIP Made Using the Template α-Methylbenzylamine (**10**) vs. MIP Made Using β-Methyl Phenethylamine as Template (**11**)

MIP entry	Template	Separation factor (α)
1	**10**	1.33
2	**11**	1.13

Moving the ionic binding event further away from the chiral center resulted in 15% less stereoselectivity of a β-methyl phenylethylamine MIP for its template vs. an α-methylbenzylamine MIP for its template. In this case, all substituents held close to the point of contact would be most susceptible to interactions within the polymer binding site. Thus, distinguishing interactions between the correct substrate and polymer would be maximized leading to optimal recognition, while repulsive interactions between the wrong substrate and polymer would be maximized, thus limiting the binding interaction. This effect would be lessened in both cases the further the initial point of contact is from the spatial differentiation of the molecules (Table 4).

D. Sterics

As previously discussed, the number of interactions necessary for selectivity often requires interactions with the MIP matrix and the template. The matrix can affect selectivity due to steric interactions between the binding cavity features and the template. These interactions are often overlooked because they are hard to measure. In fact, when considering binding interactions and their preorganization, steric interactions are not included since they do not actually provide a binding interaction. Instead, steric interactions are a secondary effect that can only take place after a primary binding event brings the template and MIP together. However, steric interactions should be considered one of the more definitive interactions in determining the selectivity of a receptor for a substrate, since two molecular moieties cannot occupy the same space. An example of chiral selectivity by MIPs that must be due to the influence of steric interactions was demonstrated using α-methylbenzylamine (Fig. 4) [19]. This template has essentially one type of binding group, namely the primary amine. Since the primary amine lacks directionality, enantioselective interactions between the template and the MIP matrix arise from the relative three-dimensional geometric positions of the benzyl, methyl, and hydrogen groups.

In a related example, MIPs toward 1-(1-naphthyl) ethylamine (12) also showed enantioselectivity. In fact, the separation factor was larger than in the case of α-methylbenzylamine (Table 5). The binding affinities of these two primary amine functionalized templates are roughly the same. However, the separation factor for 1-(1-naphthyl) ethylamine is greater than that for α-methylbenzylamine. Therefore, the selectivity enhancement of the 1-(1-naphthyl) ethylamine vs. α-methylbenzylamine must be due to the size difference of the naphthyl group *versus* the phenyl group. From this finding comes rule 7:

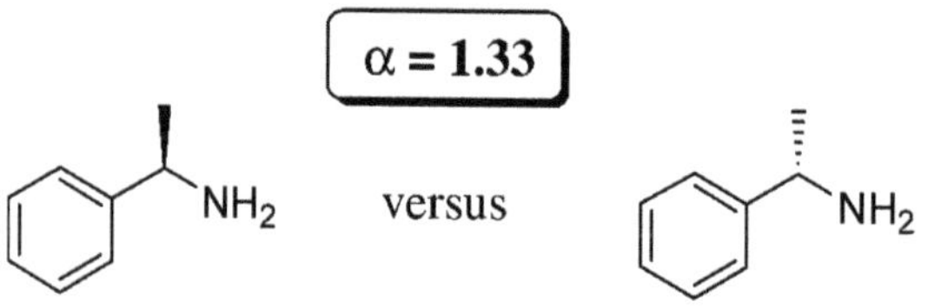

Figure 4 Enantioselectivity for α-methylbenzylamine shown to be due to steric factors.

Table 5 Comparison of Enantioselectivities of α-Methylbenzylamine (**9**) on its Imprinted Polymer vs. 1-(1-Naphthyl) ethylamine (**12**) on its Imprinted Polymer

MIP entry	Template	Separation factor (α) for enantiomers
1	**10**	1.33
2	**12**	1.58

Rule 7. Greater *differences in the size of steric groups* surrounding a "point of selectivity" promote greater binding selectivity.

E. Conformational Effects

The imprinting process gives rise to selective sites based on three-dimensional shape and organization of functional groups. Therefore, the ground-state conformation of a template is an important consideration for MIP selectivity, since this will play a role in shape determination and placement of functional groups. Lepisto and Sellergren [20] have demonstrated conformational selectivity in MIPs by correlating solution NMR studies on L-phenylalanine anilide (L-PheNHPh, **13**) and L-phenylalanine-*N*-methylanilide (L-PheNMePh, **14**) with the performance and cross-reactivity of MIPs elicited toward these templates. Very different solution conformations between L-PheNHPh vs. L-PheNMePh were found by NMR, even though these two compounds differ by only a methyl group (not a very large steric factor). The data in Table 6 show that that there is low selectivity between enantiomers of PheNHPh by the polymer imprinted with L-PheNMePh (i.e., low cross-reactivity). This effect is unlikely to be steric in origin, since the polymer imprinted with L-PheNMePh should be able to easily accommodate L-PheNHPh. The lower cross-reactivity of the polymer imprinted with L-PheNHPh in recognizing enantiomers of PheNMePh is due both to the different spatial organization of interactions built into the MIPs during the imprinting process and steric differences in the templates.

Conformational flexibility between the point of selectivity and the binding groups also plays a role in the loss of recognition. The more flexible the bonds between the point of selectivity and the binding interaction, the greater the entropy, which translates to a decrease in selectivity. The effect of conformational flexibility of a template molecule on binding selectivity is seen comparing α-methylbenzyl amine vs. 1-aminoindane used as a "conformationally locked" model of α-methylbenzylamine (Table 7) [19]. The separation factor for 1-aminoindane on its imprinted polymer was approximately 8% higher than that found for α-methylbenzylamine on its

Table 6 Selectivity and Cross-reactivity of Polymers Imprinted with L-PheNHPh and L-PheNMePh (Adapted From Ref. 20)

MIP entry	Template	Separation factor (α) for L-PheNHPh (**13**)	Separation factor (α) for L-PheNMePh (**14**)
1	**13**	4.18	1.07
2	**14**	1.36	2.03

imprinted polymer, showing increased specificity from a decrease in conformational entropy of binding.

Rule 8. Decreasing conformational energy states of a template increases selectivity.

IV. POLYMER FORMULATION

A. Choice of Functional Monomer–Complementary Interactions

The functional monomer and template should form a complementary interacting pair. Therefore, many of the considerations discussed for evaluating the

Table 7 Influence of Conformational Rigidity on Enantioselectivity by MIPs

MIP entry	Template	Separation factor (α)
1	**10**	1.33
2	**15**	1.44

"imprintability" of the template apply to the choice of functional monomer. Let us look at these in the context of choosing functional monomers.

Corollary 1. Choose a functional monomer that has one or more binding interactions with the template.

Since there are many different types of functional groups on different templates, different functional monomers are needed to obtain prepolymer complexes. Some examples of monomers that have been employed in MIPs are shown in Fig. 5.

Corollary 2.   The stronger the binding interaction, the better.

As previously pointed out, a strong binding interaction will increase the concentration of prepolymer complex, which in turn increases the number of binding sites and leads to more selective MIPs. One example of increasing the selectivity of MIPs using the popular functional monomer methacrylic acid (MAA) is the development of MIPs using 2-(trifluoromethyl) acrylic acid (TFMAA, **19**) [21,22]. Table 8 shows an example of an MIP using TFMAA that exhibits a 44% increase in the selectivity factor of nicotine vs. cotinine (a nicotine metabolite), when compared to a similar MIP made using MAA. A combination of two possible reasons is responsible for the increase in selectivity seen in Table 8. The first is that the lower pK_a of TFMAA increases the concentration of ionic prepolymer complex with basic groups on a template, which would increase the number of specific binding sites in the MIP and thus the selectivity. The second possibility is that stronger interactions are formed for specific interactions, but that the nonspecific interactions remain less affected. Shea and coworkers[23] have shown the second possibility to be operational in a different study comparing benzodiazepine derivatives.

Corollary 3.   Interactions with specific directionality provide more selective sites.

Ionic interactions such as an amine base with a carboxylic acid are good for creating strong noncovalent bonds between the template and functional monomers.

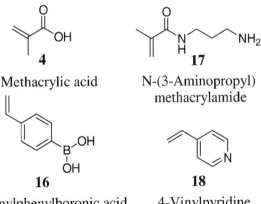

4
Methacrylic acid

17
N-(3-Aminopropyl)
methacrylamide

16
4-Vinylphenylboronic acid

18
4-Vinylpyridine

Figure 5 Commercially available acidic and basic functional monomers that have been commonly used for molecular imprinting.

Table 8 Comparison of Selectivity Factors (*S*) for MIPs Made with TFMAA and MAA (Adapted from Ref. [22])

Polymer entry	Functional monomer	pK_a of functional monomer	Selectivity (*S*) of nicotine vs. cotinine
1		4.8	1.54
2		3.1	2.23

19

However, for the most part these acid–base type interactions are not as directional as hydrogen bonds. Functional monomers with directionality are important for chiral selectivity of templates and analytes; however, this is less important for analytes consisting of different molecules. An example of this was already shown in conjunction with rule 3.

B. Choice of Cross-Linking Monomer

Figure 6 shows the cross-linkers most commonly used for forming MIPs. The first cross-linking monomer to be employed for molecular imprinting in organic polymers was divinylbenzene (usually a mixture of both meta (**20**) and para (**21**) isomers), which is still a useful cross-linker today. Early studies by Wulff et al. [24] have shown that ethyleneglycol dimethacrylate (EGDMA, **22**) forms superior MIPs vs. DVB. A relatively new cross-linker that has shown some improvement over EGDMA is the trifunctional monomer 2,2-bis-(hydroxymethyl)butanol trimethacrylate (TRIM, **23**) [25,26]. At the moment, these are the cross-linkers of choice for molecular

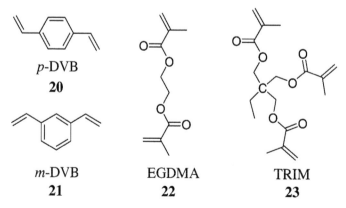

p-DVB	EGDMA	TRIM
20	**22**	**23**

m-DVB
21

Figure 6 Commercially available cross-linking monomers that are commonly used for MIPs.

imprinting due to commercial availability, good performance, and low cost. More specialized cross-linking monomers have been synthesized for MIPs and, in some cases, provide enhanced properties vs. the generally used cross-linkers described above. However the design, synthesis, and application of the specialized cross-linkers are beyond the scope of this tutorial.

C. Ratio of Monomers and Templates

1. Ratio of Cross-Linker to Functional Monomer (X/M)

Early studies by Wulff have shown that for covalent molecular imprinting, maximizing the amount of cross-linker maximizes enantioselectivity of MIPs [24]. For noncovalent molecular imprinting, it is necessary to optimize the ratio of cross-linker to noncross-linking functional monomer, i.e., optimize X/M. For example, Sellergren investigated optimization of X/M for traditional MIPs made using EGDMA as cross-linker, MAA as the noncross-linking functional monomer and employing L-phenylalanine anilide (L-PheNHPh) as the template [27]. Figure 7 shows the separation factors (α) of MIPs made with different ratios of EGD-MA/MAA. Initially, enantioselectivity for L-PheNHPh vs. D-PheNHPh increases as the mol% of MAA increases. This increase starts to diminish above 20 mol% MAA, and enantioselectivity actually decreases above 30 mol% MAA. Loss of selectivity by imprinted polymers having more than 20–30 mol% MAA has been postulated to arise, in part, from the need for a minimum amount of cross-linker (i.e., EGDMA) to form a rigid enough polymer network that will maintain the fidelity of the binding sites. This limits the amount of noncross-linking functional monomer (MAA) that can be used for formation of the MIP binding sites. Although studies such as these should be carried out for optimization of X/M, a good rule of thumb is to use a 4/1 ratio of X/M.

2. Ratio of Functional Monomer to Template (M/T)

According to Le Châtelier's principle, it can be predicted that increasing either component of the complex in the prepolymerization mixture will drive the functional monomer (e.g., MAA) and template toward more prepolymerization complex,

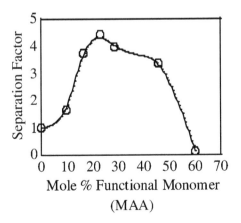

Figure 7 Determination of optimum ratio of X/M (adapted from Ref. [28]).

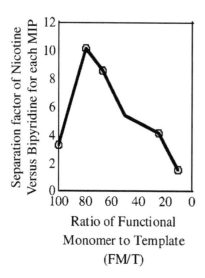

Figure 8 Determination of optimum ratio of M/T (adapted from Ref. 29).

([MAA + template]). The prepolymerization complex is then postulated to be responsible for the creation of the actual "active sites" in the polymer that bind the substrate. As seen in part "a" above, there is a limit to the amount of functional monomer that can be employed; however, the amount of template can be increased until the point where it becomes 100% of the porogen. Nicholls and coworkers have published such a study where a series of MIPs were made with increasing amounts of the template nicotine, while keeping the amount and ratio of MAA/EGDMA constant [28]. In this manner, they were able to change the M/T ratio without changing the M/X composition of the final MIP. Figure 8 shows the selectivity ($\alpha = k'_{nicotine}/k'_{bipyridine}$) of nicotine relative to bipyridine by MIPs made with different M/T (MAA/nicotine) ratios. Initially, low M/T ratios exhibit low selectivity, most likely due to a low number of specific sites from the small concentration of prepolymer complex. As the M/T ratio increases, there is an increase in selectivity up to an optimal M/T ratio which needs to be determined for each system. Increasing the M/T past the optimal ratio results in a decrease in selectivity. The decrease in selectivity is attributed to the greater increase of nonspecific (or less-specific) binding sites vs. the imprinted specific binding sites. Unless very strong interactions are used, it can be empirically stated that for optimum selectivity the minimum ratio of M/T should be greater than 1. The maximum ratio of M/T is system dependent and requires optimization. In the example presented here, the optimum ratio for M/T is 4/1.

Rule of thumb: In the absence of a systematic study to optimize X/M and M/T ratios, a value of **4/1** can be used for both ratios.

D. Choice of Solvent (Porogen)

There are two effects the polymerization solvent, referred to as the porogen, has on the formation of MIPs. The first is in the formation of the prepolymer complex. For weaker noncovalent polar–polar interactions such as ionic or hydrogen bonds, non-polar solvents are preferred in order to drive the template and functional monomer toward complex formation. The use of polar solvents, especially those capable of hydrogen bonding, inhibits the formation of noncovalent prepolymer complexes, decreasing the number of imprinted sites in the polymer. The second effect of the polymerization solvent is on the formation of the porous structure of the polymer, which is why it is referred to as the porogen. The details of how porogens can be used to control structure and morphology of these macroporous polymers can be found in the literature [29]. To help understand the effects of porogen on selectivity by MIPs, a study published by Sellergren and Shea [30] evaluated MIPs templated with L-phenylalanine anilide (L-PheNHPh) using several different types of porogens. The results in Table 9 show that enantioselectivity is better for polymers that are made with relatively nonpolar porogens that have a poor capacity for hydrogen bonding. A decrease in enantioselectivity is seen for MIPs made with polar porogens with moderate capacity for hydrogen bonding and enantioselectivity trends further down with the polar-protic porogens.

Rule 9: For week noncovalent interactions, using nonpolar, nonhydrogen-bonding solvents *improves selectivity* in MIPs.

An interesting observation from Table 9 is that the MIP made with acetonitrile as porogen exhibited higher enantioselectivity than the MIP made with chloroform. A larger concentration of prepolymer complex would be expected using chloroform vs. the more polar acetonitrile as porogen; thus, a higher performance by the chloroform solvated MIP might be expected. However, in the above study, the MIPs were all evaluated in a chromatographic mobile phase made primarily of acetonitrile. Early on in the development of MIPs, Kempe and Mosbach [31] suggested that using the same solvent in the mobile phase that was used as porogen would mimic, in the chromatographic mode, the interactions existing prior to and during the polymerization. A possible explanation for this may lie in a link between conditions during polymerization and those during rebinding analysis on the product network polymer. The origins of specificity in the imprinted polymer are postulated to arise from the positioning of complementary functional groups which are then covalently locked

Table 9   Selectivity of MIPs Against L-Phenylalanine Made with Different Porogens (Adapted from Ref. 30)

Entry	Porogen	Hydrogen-bonding capability[a]	Separation factor (α)
1	Benzene	P	6.8
2	Chloroform	P	4.5
3	Acetonitrile	P	5.8
4	Tetrahydrofuran	M	4.1
5	Dimethylformamide	M	2.0
6	Isopropanol	S	3.5
7	Acetic Acid	S	1.9

[a]P, poor hydrogen bonding solvent; M, moderate hydrogen bonding solvent; S, strong hydrogen bonding solvent.

Table 10 Capacity Factors of Polymers Imprinted with 9-Ethyladenine Employing Chloroform and Acetonitrile as Porogen and Mobile Phase (Adapted from Ref. 17)

Polymer	Porogen	k' for mobile phase 85/15: CH_3CN/CH_3COOH	k' for mobile phase 85/15: $CHCl_3/CH_3COOH$	Swelling in MeCN (Ref. 30) (mL/mL)
1	CH_3CN	7.45	1.96	1.36
2	$CHCl_3$	2.70	16.58	2.11

into place during polymerization. Different swelling properties of different solvents, such as chloroform and acetonitrile, may play a role in determining shape and distance parameters that are locked into the forming polymer. In order to recreate and maintain these shape and distance parameters, it is possible that optimum rebinding conditions require the same, or very similar, swelling conditions used for polymerization. To test this hypothesis, a study was carried using two MIPs templated with 9-ethyladenine, one using chloroform as porogen and the other using acetonitrile. Selectivity was evaluated using two mobile phase systems on each column, 85/15-acetonitrile/acetic acid and 85/15-chloroform/acetic acid to recreate prepolymerization conditions as well as obtain suitable chromatograms. Table 10 shows the capacity factors of 9-EA for all four cases. These results indicate that solvent does affect the microenvironment of the binding sites created in the polymer. It also appears that there is enhanced binding in polymers immersed in the solvent in which they were polymerized. This suggests rule number 10:

Rule 10: Optimum *rebinding conditions* for a given template should *follow* closely the *polymerization conditions*, e.g., mobile phases should include the solvent used as porogen.

Analysis of MIP binding interactions in aqueous media (Nicholls and coworkers [32]) have shown that binding still can take place; however, the rules

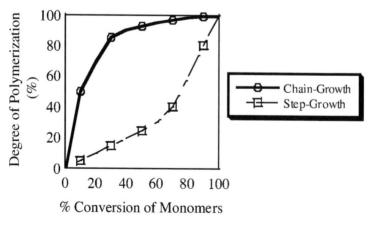

Figure 9 Variation of molecular weight with conversion for chain-growth vs. step-growth polymerization mechanisms.

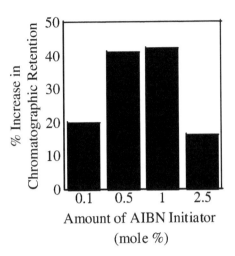

Figure 10 Improvement in binding affinity vs. amount of initiator used in photo-initiated polymerization of MIPs (adapted from Ref. 32).

change to reflect hydrophobic effects instead of the electrostatic and hydrogen-bonding effects usually encountered in organic solvents.

E. Initiator

The process of molecular imprinting most often uses free radical polymerization of monomers that have double bonds. There are two primary reasons for using free radical polymerization, in addition to it being an inexpensive and easy process to carry out. First, free radical polymerization eliminates any interference with a noncovalent prepolymer complex by polar intermediates that would arise from other types of polymerization processes such as anionic, cationic, redox, condensation, or metal-catalyzed polymerization. Second, the kinetics of radical polymerization follows a chain-growth mechanism vs. step-growth kinetics for condensation type polymerizations. The fast, radical chain-growth mechanism produces high molecular weight gel polymer in a short period of time, which could improve the chances for "locking-in" the prepolymer complex (Fig. 9).

The typical radical initiator used is azo-bis(isobutyronitrile) (AIBN), which is useful for organic polymer solutions. There are various derivatives of azo-initiators available, including water-soluble species for aqueous polymerizations. Traditionally, 1 mol% of AIBN is introduced into the prepolymerization mixture. To investigate whether this is an optimum concentration of initiator, several polymers were

Table 11 Comparison of Separation Factors for MIPs Made by Photochemical Initiation vs. Thermal Initiation (Adapted from Ref. 34)

Temperature of polymerization	Separation factor (α)
Thermal, 60°C	1.57
Photochemical, 0°C	2.28

imprinted with 2-aminopyridine employing different concentrations of AIBN initiator [33]. The polymers were evaluated in the chromatographic mode and the results in Fig. 10 show that 1 mol% AIBN provides the maximum increase in binding affinity by the MIP. It is interesting to note that an increase in initiator actually has a positive effect on binding. This result may find its origins in kinetic parameters, i.e., the speed of polymerization, or in the morphology of the polymer.

F. Polymerization Conditions

Radical polymerization by azo-initiators can be thermochemically or photochemically initiated. Several studies have shown that polymerization of MIPs at lower temperatures forms polymers with greater selectivity vs. polymers made at elevated temperatures [17,34]. To polymerize at colder temperatures, it is necessary to use photochemical polymerization. For example, Mosbach and coworkers [35] presented a study on enantioselectivity of L-PheNHPh imprinted polymers, one polymer being thermally polymerized at 60°C, the other photochemically polymerized at 0°C. The results of this study in Table 11 show that better selectivity is obtained at the lower temperature vs. the identical polymers thermally polymerized polymers. The reason for this has again been postulated on the basis of Le Châtelier's principle, which predicts that lower temperatures will drive the prepolymer complex toward complex formation, thus increasing the number and, possibly, the quality of the binding sites formed.

V. CONCLUDING REMARKS

Selectivity in molecularly imprinted polymers is an average effect of a distribution of binding sites. This makes rational design of an imprinted site an inexact science, since there is no single model that can be used to determine quantitative structure-binding interactions. Instead, molecular imprinting is the science of changing the distributions of binding sites to favor more specific sites. The main consequence of this is that all design efforts for MIPs will be less quantitative and more qualitative in their application. The intent of the rules and examples put forth in this review is to give the researcher a qualitative understanding of the origins of selectivity in MIPs in order to effectively design, synthesize, and evaluate imprinted polymers. Furthermore, the approach in this tutorial focuses on using known or commercially available monomers in order to demonstrate molecular imprinting principles using a traditional polymer format. However, there are many novel and possibly more effective monomers and polymer formats yet to be invented that have new and different rules guiding their design and application. Many exciting possibilities lie ahead in this respect, in addition to untapped applications of more traditional MIP technologies.

REFERENCES

1. Sellergren, B. Imprinted chiral stationary phases in high-performance liquid chromatography. J. Chromatogr. A **2001**, *906*, 227–252.
2. Andersson, L.I. Molecular imprinting: developments and applications in the analytical chemistry field. J. Chromatogr. A **2000**, *745*, 3–13.

3. Brunkan, N.M.; Gagné, M.R. Effect of chiral cavities associated with molecularly imprinted platinum centers on the selectivity of ligand-exchange reactions at platinum. J. Am. Chem. Soc. **2000**, *122*, 6217–6225.

4. Bae, S.Y.; Southard, G.L.; Murray, G.M. Molecularly imprinted ion exchange resin for purfication, preconcentration, and determination of UO_2^{2+} by spectrophotometry and plasma spectrometry. Anal. Chim. Acta. **1999**, *397*, 173–181.

5. Dai, S.; Burleigh, M.C.; Shin, Y.; Morrow, C.C.; Barnes, C.E.; Xue, Z. Angew. Chem. Int. Ed. **1999**, *38*, 1235–1239.

6. Ratner, B.D.; Shi, H. Recognition templates for biomaterials with engineered bioreactivity. Curr. Opin. Solid State Mater. Sci. **1999**, *4*, 395–402.

7. Spivak, D.A.; Shea, K.J. Investigation into the scope and limitations of molecular imprinting with DNA molecules. Anal. Chim. Acta. **2001**, *435*, 65–74.

8. Cheong, S.H.; McNiven, S.; Rachkov, A.; Levi, R.; Yano, K.; Karube, I. Testosterone receptor binding mimic constructed using molecular imprinting. Macromolecules *30*, 1317–1322.

9. Ringo, M.; Evans, C. Liquid chromatography as a measurement tool for chiral interactions. Anal. Chem. **1998**, *70*, 315A–321A.

10. Wulff, G.; Grobe-Einsler, R.; Vesper, W.; Sarhan, A. Enzyme-analogue built polymers, 5, On the specificity distribution of chiral cavities prepared in synthetic polymers. Markromol. Chem. **1977**, *178*, 2817–2825.

11. Umpleby II, R.J.; Baxter, S.C.; Chen, Y.; Shah, R.N.; Shimizu, K.D. Characterization of molecularly imprinted polymers with the Langmuir–Freundlich isotherm. Anal. Chem. **2001**, *73*, 4584–4591.

12. Umpleby II, R.J.; Baxter, S.C.; Bode, M.; Berch, J.K.; Shah, R.N.; Shimizu, K.D. Application of the Freundlich adsorption isotherm in the characterization of molecularly imprinted polymers. Anal. Chim. Acta **2001**, *435*, 35–42.

13. Umpleby II, R.J.; Bode, M.; Shimizu, K.D. Measurement of the continuous distribution of binding sites in molecularly imprinted polymers. Analyst **2000**, *125*, 1261–1265.

14. Sajonz, P.; Kele, M.; Zhong, G.; Sellergren, B.; Guiochon, G. Study of the thermodynamics and mass transfer kinetics of two enantiomers on a polymeric imprinted stationary phase. J. Chromatogr. A **1998**, *810*, 1–17.

15. Ehrlich, P. The Harben lectures on experimental researches on specific therapeutics. Lancet **1907**, *172*, 1634–1636.

16. Ogston, A.G. Interpretation of experiments on metabolic processes, using isotopic tracer elements. Nature **1948**, *162*, 963.

17. Spivak, D.A.; Shea, K.J. Evaluation of binding and origins of specificity of 9-ethyladenine imprinted polymers. J. Am. Chem. Soc. **1997**, *119*, 4388–4393.

18. Sellergren, B.; Nilsson, K.G.I. Molecular imprinting by multiple noncovalent host–guest interactions: synthetic polymers with induced specificity. Meth. Molec. Cell. Biol. **1989**, *12*, 59–62.

19. Spivak, D.A.; Campbell, J. Systematic study of steric and spatial contributions to molecular recognition by non-covalent imprinted polymers. Analyst **2001**, *126*, 793–797.

20. Lepisto, M.; Sellergren, B. Discrimination between amino acid amide conformers by imprinted polymers. J. Org. Chem. **1989**, *54*, 6010–6012.

21. Matsui, J.; Nicholls, I.A.; Takeuchi, T. Molecular recognition in cinchona alkaloid molecular imprinted polymer rods. Anal. Chim. Acta. **1998**, *365*, 89–93.

22. Matsui, J.; Doblhoff-Dier, O.; Takeuchi, T. 2-(Trifluoromethyl)acrylic acid: a novel functional monomer in non-covalent molecular imprinting. Anal. Chim. Acta **1997**, *343*, 1–4.

23. Hart, B.R.; Rush, D.J.; Shea, K.J. Discrimination between enantiomers of structurally related molecules: separation of benzodiazepines by molecularly imprinted polymers. Macromolecules **1997**, *30*, 1317–1322.

24. Wulff, G.; Vietmeier, J.; Poll, H.-G. Enzyme-analogue built polymers, 22. Influence of the nature of the crosslinking agent on the performance of imprinted polymers in racemic resolution. Makromol. Chem. **1987**, *188*, 731–740.
25. Kempe, M.; Mosbach, K. Receptor binding mimics: a novel molecularly imprinted polymer. Tetrahedron Lett. **1995**, *36*, 3563–3566.
26. Schweitz, L.; Andersson, L.I.; Nilsson, S. Capillary electrochromatography with predetermined selectivity obtained through molecular imprinting. Anal. Chem. **1997**, *69*, 1179–1183.
27. Sellergren, B. Molecular imprinting by noncovalent interactions. Enantioselectivity and binding capacity of polymers prepared under conditions favoring the formation of template complexes. Makromol. Chem. **1989**, *190*, 2703–2711.
28. Andersson, H.S.; Karlsson, J.G.; Piletsky, S.A.; Koch-Schmidt, A.-C.; Mosbach, K.; Nicholls, I.A. Study of the nature of recognition in molecularly imprinted polymers, II. Influence of monomer–template ratio and sample load on retention and selectivity. J. Chromatogr. A **1999**, *848*, 39–49.
29. Guyot, A. Synthesis and structure in polymer supports. In *Synthesis and Separations using Functional Polymers*; Sherrington, D.C., Hodge, P., Eds.; John Wiley & Sons: New York, 1989, 1–331.
30. Sellergren, B.; Shea, K.J. Influence of polymer morphology on the ability of imprinted network polymers to resolve enantiomers. J. Chromatogr. **1993**, *635*, 31–49.
31. Kempe, M.; Mosbach, K. Binding studies on substrate- and enantio-selective molecularly imprinted polymers. Anal. Lett. **1991**, *24*, 1137–1145.
32. Karlsson, J.G.; Andersson, L.I.; Nicholls, I.A. Probing the molecular basis for ligand-selective recognition in molecularly imprinted polymers selective for the local anaesthetic bupivacaine. Anal. Chim. Acta **2001**, *435*, 57–64.
33. Spivak, D.A. Contributions to the theory and practice of molecular imprinting in synthetic network polymers. Applications to the molecular recognition of biomolecules. PhD dissertation, University of California, Irvine, Irvine, CA, 1995.
34. O'Shannessy, D.J.; Ekberg, B.; Andersson, L.I.; Mosbach, K. Recent advances in the preparation and use of molecularly imprinted polymers for enantiomeric resolution of amino acid derivatives. J. Chromatogr. **1989**, *470*, 391–399.
35. O'Shannessy, D.J.; Ekberg, B.; Mosbach, K. Molecular imprinting of amino acid derivatives at low temperature (0°C) using photolytic homolysis of azobisnitriles. Anal. Biochem. **1989**, *177*, 144–149.

24. Wulff, G.; Vietmeier, J.; Poll, H.-G. Enzyme-analogue built polymers, 22. Influence of the nature of the crosslinking agent on the performance of imprinted polymers in racemic resolution. Makromol. Chem. **1987**, *188*, 731–740.

25. Kempe, M.; Mosbach, K. Receptor binding mimics: a novel molecularly imprinted polymer. Tetrahedron Lett. **1995**, *36*, 3563–3566.

26. Schweitz, L.; Andersson, L.I.; Nilsson, S. Capillary electrochromatography with predetermined selectivity obtained through molecular imprinting. Anal. Chem. **1997**, *69*, 1179–1183.

27. Sellergren, B. Molecular imprinting by noncovalent interactions. Enantioselectivity and binding capacity of polymers prepared under conditions favoring the formation of template complexes. Makromol. Chem. **1989**, *190*, 2703–2711.

28. Andersson, H.S.; Karlsson, J.G.; Piletsky, S.A.; Koch-Schmidt, A.-C.; Mosbach, K.; Nicholls, I.A. Study of the nature of recognition in molecularly imprinted polymers, II. Influence of monomer–template ratio and sample load on retention and selectivity. J. Chromatogr. A **1999**, *848*, 39–49.

29. Guyot, A. Synthesis and structure in polymer supports. In *Synthesis and Separations using Functional Polymers*; Sherrington, D.C., Hodge, P., Eds.; John Wiley & Sons: New York, 1989, 1–331.

30. Sellergren, B.; Shea, K.J. Influence of polymer morphology on the ability of imprinted network polymers to resolve enantiomers. J. Chromatogr. **1993**, *635*, 31–49.

31. Kempe, M.; Mosbach, K. Binding studies on substrate- and enantio-selective molecularly imprinted polymers. Anal. Lett. **1991**, *24*, 1137–1145.

32. Karlsson, J.G.; Andersson, L.I.; Nicholls, I.A. Probing the molecular basis for ligand-selective recognition in molecularly imprinted polymers selective for the local anaesthetic bupivacaine. Anal. Chim. Acta **2001**, *435*, 57–64.

33. Spivak, D.A. Contributions to the theory and practice of molecular imprinting in synthetic network polymers. Applications to the molecular recognition of biomolecules. PhD dissertation, University of California, Irvine, Irvine, CA, 1995.

34. O'Shannessy, D.J.; Ekberg, B.; Andersson, L.I.; Mosbach, K. Recent advances in the preparation and use of molecularly imprinted polymers for enantiomeric resolution of amino acid derivatives. J. Chromatogr. **1989**, *470*, 391–399.

35. O'Shannessy, D.J.; Ekberg, B.; Mosbach, K. Molecular imprinting of amino acid derivatives at low temperature (0°C) using photolytic homolysis of azobisnitriles. Anal. Biochem. **1989**, *177*, 144–149.

3. Brunkan, N.M.; Gagné, M.R. Effect of chiral cavities associated with molecularly imprinted platinum centers on the selectivity of ligand-exchange reactions at platinum. J. Am. Chem. Soc. **2000**, *122*, 6217–6225.

4. Bae, S.Y.; Southard, G.L.; Murray, G.M. Molecularly imprinted ion exchange resin for purfication, preconcentration, and determination of UO_2^{2+} by spectrophotometry and plasma spectrometry. Anal. Chim. Acta. **1999**, *397*, 173–181.

5. Dai, S.; Burleigh, M.C.; Shin, Y.; Morrow, C.C.; Barnes, C.E.; Xue, Z. Angew. Chem. Int. Ed. **1999**, *38*, 1235–1239.

6. Ratner, B.D.; Shi, H. Recognition templates for biomaterials with engineered bioreactivity. Curr. Opin. Solid State Mater. Sci. **1999**, *4*, 395–402.

7. Spivak, D.A.; Shea, K.J. Investigation into the scope and limitations of molecular imprinting with DNA molecules. Anal. Chim. Acta. **2001**, *435*, 65–74.

8. Cheong, S.H.; McNiven, S.; Rachkov, A.; Levi, R.; Yano, K.; Karube, I. Testosterone receptor binding mimic constructed using molecular imprinting. Macromolecules *30*, 1317–1322.

9. Ringo, M.; Evans, C. Liquid chromatography as a measurement tool for chiral interactions. Anal. Chem. **1998**, *70*, 315A–321A.

10. Wulff, G.; Grobe-Einsler, R.; Vesper, W.; Sarhan, A. Enzyme-analogue built polymers, 5, On the specificity distribution of chiral cavities prepared in synthetic polymers. Markromol. Chem. **1977**, *178*, 2817–2825.

11. Umpleby II, R.J.; Baxter, S.C.; Chen, Y.; Shah, R.N.; Shimizu, K.D. Characterization of molecularly imprinted polymers with the Langmuir–Freundlich isotherm. Anal. Chem. **2001**, *73*, 4584–4591.

12. Umpleby II, R.J.; Baxter, S.C.; Bode, M.; Berch, J.K.; Shah, R.N.; Shimizu, K.D. Application of the Freundlich adsorption isotherm in the characterization of molecularly imprinted polymers. Anal. Chim. Acta **2001**, *435*, 35–42.

13. Umpleby II, R.J.; Bode, M.; Shimizu, K.D. Measurement of the continuous distribution of binding sites in molecularly imprinted polymers. Analyst **2000**, *125*, 1261–1265.

14. Sajonz, P.; Kele, M.; Zhong, G.; Sellergren, B.; Guiochon, G. Study of the thermodynamics and mass transfer kinetics of two enantiomers on a polymeric imprinted stationary phase. J. Chromatogr. A **1998**, *810*, 1–17.

15. Ehrlich, P. The Harben lectures on experimental researches on specific therapeutics. Lancet **1907**, *172*, 1634–1636.

16. Ogston, A.G. Interpretation of experiments on metabolic processes, using isotopic tracer elements. Nature **1948**, *162*, 963.

17. Spivak, D.A.; Shea, K.J. Evaluation of binding and origins of specificity of 9-ethyladenine imprinted polymers. J. Am. Chem. Soc. **1997**, *119*, 4388–4393.

18. Sellergren, B.; Nilsson, K.G.I. Molecular imprinting by multiple noncovalent host–guest interactions: synthetic polymers with induced specificity. Meth. Molec. Cell. Biol. **1989**, *12*, 59–62.

19. Spivak, D.A.; Campbell, J. Systematic study of steric and spatial contributions to molecular recognition by non-covalent imprinted polymers. Analyst **2001**, *126*, 793–797.

20. Lepisto, M.; Sellergren, B. Discrimination between amino acid amide conformers by imprinted polymers. J. Org. Chem. **1989**, *54*, 6010–6012.

21. Matsui, J.; Nicholls, I.A.; Takeuchi, T. Molecular recognition in cinchona alkaloid molecular imprinted polymer rods. Anal. Chim. Acta. **1998**, *365*, 89–93.

22. Matsui, J.; Doblhoff-Dier, O.; Takeuchi, T. 2-(Trifluoromethyl)acrylic acid: a novel functional monomer in non-covalent molecular imprinting. Anal. Chim. Acta **1997**, *343*, 1–4.

23. Hart, B.R.; Rush, D.J.; Shea, K.J. Discrimination between enantiomers of structurally related molecules: separation of benzodiazepines by molecularly imprinted polymers. Macromolecules **1997**, *30*, 1317–1322.

16
Binding Isotherms

Ken D. Shimizu University of South Carolina, Columbia,
South Carolina, U.S.A.

I. INTRODUCTION

The molecular recognition abilities of MIPs are clearly their most important material property. The molecular affinity and selectivity form the basis for the utility of MIPs in chromatography [1], biosensors [2], and catalysis applications [3]. Therefore, the ability to accurately quantify and measure these properties is an essential tool for the verification of the imprinting process and for the optimization of MIPs. The binding properties of MIPs can be measured in chromatographic, solid-phase extraction, and membrane transport studies. In each of these analyses, binding properties can be calculated from the corresponding experimental binding isotherms. The extraction of binding properties from isotherms requires the application of specific binding models and assumptions about the distribution of binding sites in MIPs. This chapter will review the use of binding isotherms to characterize MIPs and, in particular, review practical aspects of applying different binding models to MIPs. Other chapters in this volume may also be referred with respect to binding analyses; batch rebinding (Chapters 16, 22), chromatography (Chapter 20).

II. BINDING ISOTHERMS

A binding isotherm is a measure of the concentration-dependent recognition behavior of a system. Binding isotherms are commonly plotted as the concentration of analyte bound to the solid phase (**B**) vs. analyte free in solution (**F**). Figure 1 shows an example of an experimental isotherm for a polymer (MIP1) imprinted with ethyl adenine-9-acetate (EA9A). The concentration of bound EA9A increases with increasing free concentration. When the polymer is saturated and all the sites are filled, the concentration of bound analyte will level off and remain constant. The isotherm in Fig. 1 for MIP1 has not reached saturation, which is typical of most binding isotherms for MIPs.

Binding isotherms are commonly measured in MIPs from batch rebinding [4,5] or frontal chromatography studies [6–8]. In batch rebinding experiments, a constant weight of polymer is equilibrated with solutions of analyte of varying initial concentration (**I**) (Fig. 2). Equilibration of the solution typically takes anywhere from 15 min to 24 h and should be independently measured for each new polymer. The

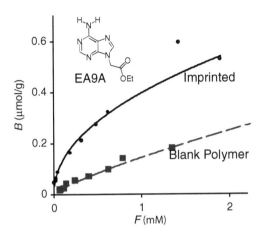

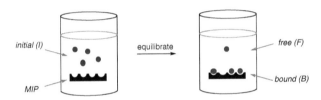

Figure 1 Experimental binding isotherms for an ethyl adenine-9-acetate (EA9A) imprinted polymer (MIP1) and its corresponding nonimprinted polymer.

Figure 2 Representation of a batch rebinding study.

solid polymer is then separated either by centrifugation or filtration, and the concentration of free analyte (**F**) remaining in solution is measured by UV/Vis spectroscopy, fluorescence spectroscopy, HPLC, or scintillation. The corresponding bound concentration (**B**) is calculated as the difference between the initial and final free concentrations (**B** = **I** – **F**). Batch rebinding experiments can also be carried out in a reverse manner in which the concentration of analyte is held constant and the weight of polymer is varied [9].

In frontal chromatography, the MIP to be tested is used as the stationary phase usually in a HPLC column. The MIP column is first equilibrated with pure solvent and then switched over to a solution containing a constant concentration of analyte, which is taken as the free concentration (**F**). The corresponding bound concentration (**B**) is calculated from Eq. (1) in which **V** is the volume of solution it takes for the column to re-establish equilibrium, $\mathbf{V_0}$ is the column volume and **m** is the mass of polymer in the column

$$\mathbf{B} = \frac{\mathbf{F}(\mathbf{V} - \mathbf{V_0})}{\mathbf{m}} \tag{1}$$

Binding isotherms allow for the comparison of binding properties in three ways: (1) graphically; (2) from the calculation of binding parameters such as number of sites (**N**) and association constant (**K**); and (3) finally using affinity distributions. The easiest method is to graphically compare the binding isotherms directly as shown in Fig. 1 that compares the binding isotherms of MIP1 and the corresponding nonimprinted polymer (NP1) [10]. Both polymers were synthesized from the same methacrylic acid/ethylene dimethacrylate (MAA/EDMA) copolymers and differ only in the presence and absence of template molecule (EA9A) during the polymerization process. The overlaid plots of the binding isotherms for MIP1 and NP1 verify the imprinting effect as the imprinted polymer rebinds significantly more EA9A at comparable concentrations than the nonimprinted polymer. More quantitative comparisons such as the calculation of binding parameters or affinity distributions require the application of specific binding models. The application and reasons for selecting different binding models will be discussed in the following section.

There are two additional points to be made regarding the experimental determination of isotherms. First, the concentration range should be as wide as possible. Profound changes in the shape of an isotherm can result from simply extending the range of guest concentrations. Second, the number of data points used to generate the isotherm is important, especially when a new system is being studied. The assumption that an isotherm is described by a certain function, for example, is valid only when an adequate number and spacing of points have been determined. Once such an assumption has been validated, however, a substantial amount of time can be saved by decreasing the number of points.

III. BINDING MODELS

A binding model specifies a particular mathematical relationship between the concentrations of bound (B) and free (F) and therefore can be used to model the binding isotherm. These mathematical relationships have their basis in assumptions about the physical composition of a system, specifically the number of different types of binding sites in the polymer and their relative populations. The ease and accuracy of the calculation are dependent on the selection of the appropriate binding model. The use of binding models is not restricted to experimental methods that generate a binding isotherm. Whether the experimenter is aware of it or not, the comparison of MIPs using common chromatographic separation factors (α) or transport rates all assume a particular binding model, usually the homogeneous Langmuir isotherm.

In general, binding models can be classified into two classes: homogeneous and heterogeneous (Fig. 3). Homogeneous binding models assume that there is only one type of binding site with a single set of binding parameters (N, K). The second class are heterogeneous binding models that assume that a system has two or more different types of binding sites in the polymer, each with a unique binding constant (K_i). Each type of binding model and its application to MIPs will be discussed in the subsequent sections.

Homogeneous   Heterogeneous

Figure 3 Schematic representations comparing homogeneous and heterogeneous MIPs, in which the depth of the imprint corresponds to the affinity (K) of the respective binding sites.

in Fig. 1 that compares the binding isotherms of MIP1 and the corresponding nonimprinted polymer (NP1) [10]. Both polymers were synthesized from the same methacrylic acid/ethylene dimethacrylate (MAA/EDMA) copolymers and differ only in the presence and absence of template molecule (EA9A) during the polymerization process. The overlaid plots of the binding isotherms for MIP1 and NP1 verify the imprinting effect as the imprinted polymer rebinds significantly more EA9A at comparable concentrations than the nonimprinted polymer. More quantitative comparisons such as the calculation of binding parameters or affinity distributions require the application of specific binding models. The application and reasons for selecting different binding models will be discussed in the following section.

There are two additional points to be made regarding the experimental determination of isotherms. First, the concentration range should be as wide as possible. Profound changes in the shape of an isotherm can result from simply extending the range of guest concentrations. Second, the number of data points used to generate the isotherm is important, especially when a new system is being studied. The assumption that an isotherm is described by a certain function, for example, is valid only when an adequate number and spacing of points have been determined. Once such an assumption has been validated, however, a substantial amount of time can be saved by decreasing the number of points.

III. BINDING MODELS

A binding model specifies a particular mathematical relationship between the concentrations of bound (B) and free (F) and therefore can be used to model the binding isotherm. These mathematical relationships have their basis in assumptions about the physical composition of a system, specifically the number of different types of binding sites in the polymer and their relative populations. The ease and accuracy of the calculation are dependent on the selection of the appropriate binding model. The use of binding models is not restricted to experimental methods that generate a binding isotherm. Whether the experimenter is aware of it or not, the comparison of MIPs using common chromatographic separation factors (α) or transport rates all assume a particular binding model, usually the homogeneous Langmuir isotherm.

In general, binding models can be classified into two classes: homogeneous and heterogeneous (Fig. 3). Homogeneous binding models assume that there is only one type of binding site with a single set of binding parameters (N, K). The second class are heterogeneous binding models that assume that a system has two or more different types of binding sites in the polymer, each with a unique binding constant (K_i). Each type of binding model and its application to MIPs will be discussed in the subsequent sections.

Homogeneous   Heterogeneous

Figure 3 Schematic representations comparing homogeneous and heterogeneous MIPs, in which the depth of the imprint corresponds to the affinity (K) of the respective binding sites.

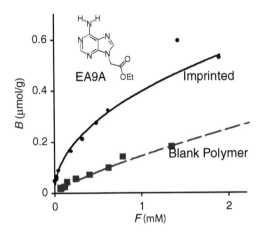

Figure 1 Experimental binding isotherms for an ethyl adenine-9-acetate (EA9A) imprinted polymer (MIP1) and its corresponding nonimprinted polymer.

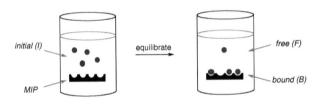

Figure 2 Representation of a batch rebinding study.

solid polymer is then separated either by centrifugation or filtration, and the concentration of free analyte (F) remaining in solution is measured by UV/Vis spectroscopy, fluorescence spectroscopy, HPLC, or scintillation. The corresponding bound concentration (B) is calculated as the difference between the initial and final free concentrations (B = I − F). Batch rebinding experiments can also be carried out in a reverse manner in which the concentration of analyte is held constant and the weight of polymer is varied [9].

In frontal chromatography, the MIP to be tested is used as the stationary phase usually in a HPLC column. The MIP column is first equilibrated with pure solvent and then switched over to a solution containing a constant concentration of analyte, which is taken as the free concentration (F). The corresponding bound concentration (B) is calculated from Eq. (1) in which V is the volume of solution it takes for the column to re-establish equilibrium, V_0 is the column volume and m is the mass of polymer in the column

$$\mathbf{B} = \frac{\mathbf{F}(\mathbf{V} - \mathbf{V_0})}{\mathbf{m}} \tag{1}$$

Binding isotherms allow for the comparison of binding properties in three ways: (1) graphically; (2) from the calculation of binding parameters such as number of sites (N) and association constant (K); and (3) finally using affinity distributions. The easiest method is to graphically compare the binding isotherms directly as shown

A. Homogeneous Binding Models

1. Langmuir Isotherm

The simplest binding models are homogeneous models. These types of binding behavior are typically seen in structurally homogeneous systems such as enzymes, monoclonal antibodies, or synthetic molecular receptors [11]. Experimentally, homogeneous binding models are the easiest to apply and have been extensively used to characterize MIPs, usually in the form of Scatchard plots. The Scatchard equation (2) is a rearranged form of the Langmuir isotherm (Eq. (3)). The variables N and K correspond to the number of binding sites and the association constant of the binding sites, respectively

$$\frac{\mathbf{B}}{\mathbf{F}} = \mathbf{KN} - \mathbf{KB} \tag{2}$$

$$\mathbf{B} = \frac{\mathbf{NKF}}{1 + \mathbf{KF}} \tag{3}$$

The Scatchard analysis is applied by first replotting the binding isotherm in $\mathbf{B}/\mathbf{F}$ vs. $\mathbf{B}$ format (Fig. 4), which is referred to as the Scatchard plot. Systems that are homogeneous will display a linear Scatchard plot [12]. Linear regression analysis yields a slope and x-intercept from which the binding constant (K) and the total number of binding sites (N), using Eqs. (4) and (5) [13] can be calculated. The units of B and F determine the units for the calculated binding parameters K and N. For example, units of µmol/g and mmol/L for B and F, yield K in units of L/mmol and N in µmol/g, respectively

$$\text{slope} = -\mathbf{K} \tag{4}$$

$$x - \text{intercept} = \mathbf{N} \tag{5}$$

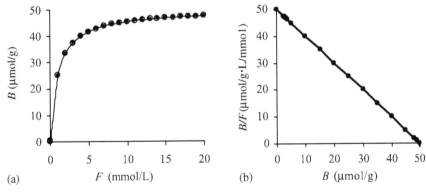

(a) F (mmol/L)   (b) B (µmol/g)

Figure 4 Hypothetical binding isotherm (a) and corresponding Scatchard plot for a homogeneous MIP2 (b).

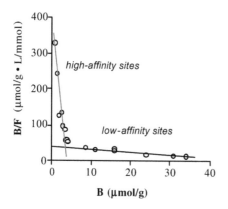

Figure 5 Limiting slopes analysis of Scatchard plot for MIP2 selective for EA9A.

B. Heterogeneous Binding Models

1. Bi-Langmuir Isotherm

Unfortunately, most MIPs cannot be fit to a homogeneous model. The reason is that, unlike enzymes or antibodies, the binding sites in MIPs vary widely in size, shape, and rigidity. The binding properties of the sites likewise vary widely in affinity and selectivity. Evidence for this binding site heterogeneity in MIPs is their highly concentration-dependent selectivities and severe peak asymmetry in chromatographic applications [14]. Heterogeneity also complicates the characterization of binding properties of MIPs. The Scatchard plots of MIPs are commonly curved (Fig. 5) and, therefore, are not well modeled by linear regression analysis [9,15].*

   The most common method to analyze nonlinear Scatchard plots of MIPs has been to use the limiting slopes method [5]. The method is attractive because it is relatively easily applied. Two straight lines are drawn through the curved Scatchard plot to yield two sets of binding parameters (N_1, K_1 and N_2, K_2) corresponding to the low- and high-affinity binding sites (Table 1). The limiting slopes analysis of curved Scatchard plots is a graphical method for applying the bi-Langmuir model (Eq. (6)) in which there are two distinct types of binding sites with separate binding parameters.

$$B = \frac{N_1 K_1 F}{1 + K_1 F} + \frac{N_2 K_2 F}{1 + K_2 F} \qquad (6)$$

   Difficulties, however, arise in applying the limiting slopes method to MIPs. For example, when binding isotherms for a polymer are measured under slightly different concentration ranges, the Scatchard analyses yield different sets of binding parameters [16]. The inability of the bi-Langmuir model to yield consistent binding

---

*Heterogeneity is not the only possible interpretation of curved Scatchard plots. Another possibility is cooperativity in which the first binding event increases or decreases the binding affinity of subsequent binding events. Cooperativity has been observed in a few MIPs (see Refs. 10 and 11), however, heterogeneity is usually cited as the major contributor to the curvature in Scatchard plots of MIPs.

Table 1 Binding Properties of Low- and High-Affinity Sites as Measured by the Limiting Slopes Analysis of Curved Scatchard Plots

	Low affinity sites	High affinity sites
N (µmol/g)	54	3.8
K (M^{-1})	6.2×10^2	1.0×10^5

Table 2 Binding Properties of High-Affinity Sites as Measured by Scatchard Analyses

	MIP2	MIP3
N (µmol/g)	4.0	16.0
K (M^{-1})	1.0×10^5	5×10^4

parameters for a single polymer becomes compounded when attempting to compare two different MIPs (Fig. 6). For example, comparison of the high-affinity sites of MIP2 and MIP3 using the Scatchard method yields ambiguous results. MIP2 has the higher binding constant ($K_a = 1.0 \times 10^5$ vs. 5×10^4 M^{-1}), whereas MIP3 has a greater number of binding sites (N =16.0 vs. 4.0 µmol/g). By this method, it is unclear which is the better polymer, because it is unclear which binding parameter is more important: the number of binding sites (N) or the association constant (K_a).

2. Affinity Distributions

From the above discussion, it is apparent that two-site binding models are too simple to accurately characterize most heterogeneous MIPs. Higher order Langmuir binding models have also been applied to MIPs such as tri- and tetra-Langmuir isotherms [9]. However, all of these models assume that MIPs contain a specific number of different types of sites, which is unlikely given the sources of the structural diversity of

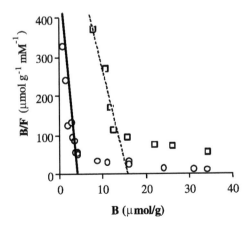

Figure 6 Overlaid Scatchard plots of MIP2 (circles) and MIP3 (squares).

the binding sites in MIPs. More likely, MIPs contain a continuous distribution of binding sites from high to low affinity. The question then arises is: what is the shape of the heterogeneous distribution in MIPs? To answer this question, we analyzed a number of MIPs from the literature and from our own labs using various continuous distribution models. These studies led to the conclusion that the distribution in MIPs can be generally described as a broad unimodal distribution. For example, an affinity distribution calculated for an atrazine imprinted polymer (MIP4) synthesized by Takeuchi and coworkers [17] is shown in Fig. 7.

Figure 7 is an example of an affinity distribution, which is a measure of the number of binding sites (N) having a particular binding affinity (log K). The x-axis is usually plotted in log K format to make this axis proportional to the binding energy (ΔG), and for this reason, affinity distributions are also called site-energy distributions.

MIPs were found to have a continuous heterogeneous distribution containing a population of sites covering the entire range from low to high affinity. This helps to explain how the Scatchard analysis could yield a different set of N's and K's for measurements in different concentration ranges. It also helps to explain the highly concentration-dependent binding behavior of MIPs. At high concentrations, the low-affinity, low-selectivity sites dominate the binding properties and at low concentrations, the high-affinity and high-selectivity sites dominate. The broad unimodal distribution was also physically consistent with the sources of heterogeneity and with previous measures of the heterogeneity in MIPs [18,19].

The use of affinity distributions to characterize MIPs has certain advantages. First, affinity distributions can be generated directly from binding isotherms; therefore, they do not require the development of new experimental methods. In many cases, existing binding isotherms can be re-examined to yield the corresponding affinity distributions. Second, affinity distributions more accurately models the heterogeneous distribution in MIPs. This allows for the more accurate quantitative comparison of the binding properties in MIPs, which is difficult using the Langmuir or bi-Langmuir model.

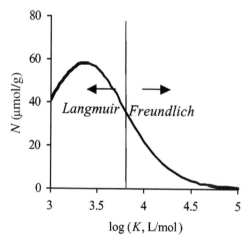

Figure 7 Affinity distribution for a typical imprinted polymer (MIP4).

The major disadvantage of affinity distributions is that they are computationally difficult to generate from the corresponding binding isotherms. In contrast, binding parameters are easily calculated from the homogeneous models, using graphical methods and linear regression analysis. This is perhaps one reason why the homogeneous models have been used to characterize MIPs, despite their difficulties in dealing with the heterogeneous nature of these systems.

3. Freundlich Isotherm

The use of simplified heterogeneous binding models can make the generation of the corresponding affinity distribution easier to apply, comparable even to the homogeneous Scatchard analysis. These models are only accurate within a limited concentration range. However, the binding isotherms of most MIPs have been measured in relatively narrow concentration ranges such that these simplified models are applicable. In MIPs, the binding behavior can be separated into two distinct concentration ranges. The low-affinity sites in the peak of the affinity distribution can often be accurately modeled by the homogeneous Langmuir isotherm. Whereas the high-affinity sites in the tailing region can be modeled by the heterogeneous Freundlich isotherm (Fig. 7).

The distinction between these two regions is more apparent in the examination of the log–log graphs of the corresponding binding isotherms. For example, in the experimental isotherm for MIP4, the low concentration data prior to saturation are well fit by the Freundlich isotherm, as evidenced by their linearity in this region (Fig. 8). However, as the isotherm reaches saturation, the data deviate from the Freundlich isotherm, and this high concentration region is well modeled by the Langmuir isotherm.

Most MIPs have been measured over relatively narrow concentration ranges that cover only one of the two distinct regions. This allows the application of one or the other local isotherm. Polymers that have strong interactions between the monomer and template or have been covalently imprinted to yield a more homogeneous distribution can often be characterized solely with the Langmuir isotherm. These

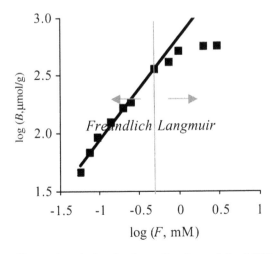

Figure 8 Binding isotherm (log format) for MIP4.

systems are relatively rare. For example, a covalently imprinted polymer synthesized and studied by Whitcombe et al. [20] was well modeled by the Langmuir isotherm as evidenced by its linear Scatchard plot.

Most noncovalently imprinted polymers, on the other hand, have been measured in the tailing region of the distribution affinity distribution and are well fit by the Freundlich isotherm (Eq. (7)) [10,19]. To demonstrate the generality of the Freundlich isotherm, binding isotherms for a series of noncovalently imprinted polymers from the literature were examined and were found to be well fit to the Freundlich isotherm, even though the MIPs varied widely in their template molecules, matrix composition, and binding interactions [21]. Three representative isotherms are shown in Fig. 9 from the research groups of Shea [8], Karube [22] and Mosbach [23]. All three have linear log–log binding isotherms, which are characteristic of the Freundlich isotherm

$$\mathbf{B} = \mathbf{a}\mathbf{F}^m \tag{7}$$

The advantage of using the Freundlich isotherm is that the corresponding affinity distributions can be readily generated using an affinity distribution (8) [24] (Fig. 10). This equation calculates the number of binding sites (N) having binding constant (K), requiring only two fitting parameters a and m

$$\mathbf{N_i} = 2.3\mathbf{am}(1 - \mathbf{m}^2)\mathbf{K_i}^{-m} \tag{8}$$

The overall process of applying the Freundlich binding model to characterize an MIP is as simple as the Scatchard analysis but with the advantage of using a physically more appropriate heterogeneous binding model. The three-step procedure is outlined below. First, the binding isotherm is plotted in log B vs. log F format. Second, the isotherm is fit to a straight line (Eq. (9)), which is the log form of the Freundlich isotherm (Eq. (7)). The slope and the y-intercept of the line yield the Freundlich

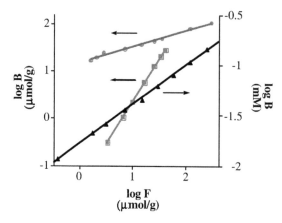

Figure 9 Experimental binding isotherms for noncovalently imprinted polymers synthesized by Shea (circles), Karube (squares), and Mosbach (triangles).

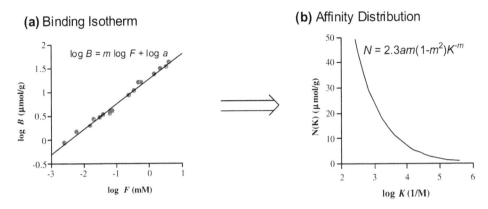

Figure 10   Binding isotherm for MIP3 (a) and the corresponding affinity distribution (b).

fitting parameters **m** and **a**, where **m** = slope and **a** = $10^{(\text{y-intercept})}$. Finally, these fitting parameters can then be used to calculate the affinity distribution using Eq (8).

$$\log \mathbf{B} = \mathbf{m} \log \mathbf{F} + \log \mathbf{a} \tag{9}$$

There are a few practical limitations in using affinity distributions to characterize MIPs. First, the affinity distribution expression allows calculation of the distribution over the entire range of binding constants. However in practice, the concentration ranges of the experimental binding isotherm set the limits for the calculated affinity distribution [25]. Typically the limits ($\mathbf{K}_{min}$ and $\mathbf{K}_{max}$) are set by $\mathbf{F}_{min}$ and $\mathbf{F}_{max}$ by Eq. (10) and (11)

$$\mathbf{K}_{min} = 1/\mathbf{F}_{max} \tag{10}$$

$$\mathbf{K}_{max} = 1/\mathbf{F}_{min} \tag{11}$$

Second, the Freundlich isotherm is only a locally accurate model for MIPs and therefore cannot always be used to model binding behavior over the entire measured concentration range. The Freundlich isotherm is more accurate in the lower concentrations of analyte. This is because the Freundlich isotherm is unable to model saturation behavior; deviations from the Freundlich isotherm are expected at higher analyte concentrations, such as for MIP4 (Fig. 8). It is important, therefore, to check whether the Freundlich isotherm is appropriate for the measured concentration range. The easiest way is to test this graphically by checking if the log–log form of the binding isotherm falls on a straight line. MIPs that have curved log–log binding isotherms may still be able to be fit to the Freundlich isotherm at lower concentrations such as the example in Fig. 8. Alternatively, a different binding model such as the Langmuir or Langmuir–Freundlich may be used.

Third, a general drawback of heterogeneous binding models is that they require significantly more data points in the binding isotherm than the corresponding homogeneous binding models. The Freundlich isotherm, however, is an exception to this generalization because its isotherm can be represented in a linear format. This allows the binding isotherm to be accurately measured with significantly fewer data points than binding models, such as the bi-Langmuir, which yields curved isotherms.

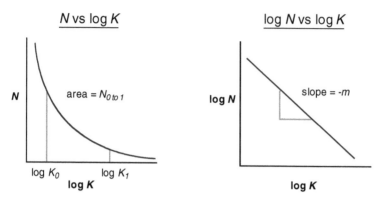

Figure 11 Representations of the two formats for the affinity distributions.

(a) Comparison of Polymers using the Freundlich Isotherm. There are two general graphical formats for affinity distributions: the semi-log (N vs. log K) and the log formats (log N vs. log K) (Fig. 11). In the semi-log format of the affinity distribution, the number of sites can be calculated from the area under the distribution, using Eq. (12). This equation does not give the total number of binding sites (N_T), but rather the number of binding sites between K_0 and K_1.[*] Therefore, in comparing the number of sites by this method, it is important to calculate $N_{(k_0-k1)}$ over the same limits

$$N_{(K_0-K_1)} = a(1 - m^2)(K_0^{-m} - K_1^{-m}) \tag{12}$$

In log format, the exponentially decaying distribution of the affinity distribution becomes a straight line allowing for easy visual comparison of polymers. The vertical displacement of the lines yields a measure of the capacity (number of sites) of the polymer. The slope yields a measure of the ratio of high-affinity sites to low-affinity sites, with flatter slopes corresponding to higher percentages of high-affinity sites.

It is actually not necessary to measure the slope of the affinity distribution because it is equal to the negative of Freundlich fitting parameter (slope $=-m$). In the case of the Freundlich isotherm, m has dual significance. As mentioned earlier, m is a measure of the percentage of high-affinity sites, but it is also a measure of the heterogeneity of the system, and thus, is also called the heterogeneity index. The heterogeneity index varies from one to zero with m = 1 being homogeneous and with m approaching 0.0 becoming increasingly heterogeneous. Unlike N_T, the heterogeneity index can be accurately measured from the Freundlich portion of the affinity distribution. One of the more interesting consequences of the Freundlich fitting parameter

---

[*]In order to accurately measure the total number of sites (N_T), the corresponding isotherm has to be measured in the saturation region. The binding isotherm of most noncovalent MIPs, in contrast, has been measured only in the sub-saturation concentration region. Thus, N_T cannot be accurately measured in systems that are well fit by the Freundlich isotherm, because the Freundlich isotherm is only accurate in the sub-saturation concentration region.

m being both a measure of the ratio of high- to low- affinity sites and the heterogeneity is that polymers that show a more favorable ratio of high- to low-affinity sites ($m \geq 0$) are increasingly more heterogeneous.

A nother useful parameter that can be calculated from the affinity distribution is K_{avg}. K_{avg} is not the average association of all sites but average number of sites within the limits K_0 to K_1. To accurately compare the K_{avg} of two polymers, it is necessary that K_{avg} has been calculated over the same limits

$$K_{avg(K_0-K_1)} = \left(\frac{m}{1-m}\right) \frac{K_0^{1-m} - K_1^{1-m}}{K_0^{-m} - K_1^{-m}} \tag{13}$$

In addition to the various binding parameters that can be calculated from the affinity distribution ($N_{(k_0-k_1)}$, K_{avg}, m), the affinity distributions can be directly compared in graphical format. For the exponentially decaying Freundlich distribution, the differences between affinity distributions are most easily visualized in log format (log N vs. log K). There are several possible outcomes when comparing two polymers by the Freundlich isotherm. The distributions could be: (1) parallel, having different capacities but an equal ratio of high- to low-affinity sites; (2) crossing, having similar capacities but different ratios of high- to low-affinity sites; and (3) divergent or convergent, having different capacities and ratios of low- to high-affinity sites (Fig. 12).

4. Langmuir–Freundlich

Not all imprinted polymers can be modeled using solely Freundlich or Langmuir isotherms. This is because, although both isotherms can be used to characterize MIPs, they are only accurate within limited concentration regions. The Langmuir isotherm best models the saturation behavior of an MIP, usually in the high concentration region; whereas, the Freundlich isotherm is limited to the lower sub-saturation concentration region of the binding isotherm. As MIPs improve, the concentration window that is measure will begin to span both saturation and subsaturation regions. In addition, to more accurately characterize a system, the isotherm would be ideally measured over both concentration regions. These isotherms require hybrid heterogeneous binding models that can span both saturation and sub-saturation regions are required. Along these lines, we have applied the

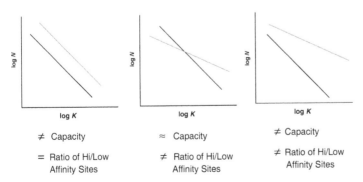

Figure 12 Schematic representations of possible differences in two hypothetical affinity distributions.

Langmuir–Freundlich (LF) isotherm (Eq. (14)) to characterize MIPs [26]. For example, the LF isotherm is superior to the Freundlich isotherm in fitting the experimental binding isotherm of a trimethoprim-selective MIP synthesized and characterized by Sellergren and coworkers (Fig. 13) [27].

The LF isotherm is a heterogeneous binding model based upon a broad unimodal distribution. It was used to generate the distributions shown earlier in this chapter (Fig. 7). The advantage of the LF isotherm is that it can model both saturation and sub-saturation concentration regions of the binding isotherm for an MIP individually or simultaneously. For this reason, the LF can be applied to almost any MIP whether it contains a relatively homogeneous or heterogeneous distribution of sites. In practice, however, relatively few MIPs have been measured in a sufficiently wide concentration range to merit analysis by the LF isotherm. The majority of MIPs can be equally well modeled by the local Langmuir or Freundlich isotherms

$$\mathbf{B} = \frac{\mathbf{N_T}\, \mathbf{a}\mathbf{F^m}}{1 + \mathbf{a}\mathbf{F^m}} \tag{14}$$

Experimentally, the application of the Langmuir–Freundlich isotherm to MIP is similar to that of the Freundlich isotherm. First, the LF isotherm is fit to the experimental binding isotherm. Unlike the Freundlich or Langmuir, however, this cannot be done using simple linear regression analysis. Instead, a curve-fitting program is required. The LF isotherm requires only three fitting parameters $\mathbf{a}$, $\mathbf{m}$, and $\mathbf{N_T}$. The parameters $\mathbf{a}$ and the heterogeneity index $\mathbf{m}$ are identical to those of the Freundlich isotherm. $\mathbf{N_T}$ is the total number of sites, which can be accurately measured since the isotherm usually includes the saturation concentration region. Finally, the mean of the unimodal distribution ($\mathbf{K_0}$) can be determined by the relationship: $\mathbf{K_0} = \mathbf{a^{1/m}}$.

In addition, to these binding parameters the corresponding affinity distribution can likewise be generated from the corresponding affinity distribution equation (Eq. (15)). This expression is more complex than the affinity distribution equation for the Freundlich isotherm (Eq. (8)) [26]. However, it is still a simple algebraic expression that calculates the number of binding site ($\mathbf{N_i}$) having an association

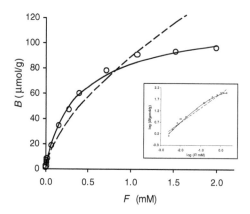

Figure 13 Binding isotherm for a trimethoprim imprinted polymer synthesized by Sellergren et al. fitted to a Langmuir–Freundlich (solid line) and a Freundlich (broken line) isotherms.

constant (K_i), using the LF fitting parameters: $\mathbf{a}$, $\mathbf{m}$, and $\mathbf{N_t}$.

$$N_i = N_{Tam}(1/K_i)^m$$
$$\times \frac{(1 + 2\mathbf{a}(1/K_i)^m + \mathbf{a}^2(1/K_i)^{2m} + 4\mathbf{a}(1/K_i)^m m^2 - \mathbf{a}^2(1/K_i)^{2m} m^2 - m^2)}{(1 + \mathbf{a}(1/K_i)^m)^4}$$

(15)

Like the affinity distribution equations generated using the Freundlich isotherm, the affinity distribution should be calculated over only a specific range of association constants that is set by the concentration range of the experimental binding isotherm using expressions (10) and (11). The primary advantage of the Langmuir–Freundlich isotherm is that it can be generally applied to almost any isotherm as it represents a more universal model for the binding behavior of MIPs based on a physically appropriate unimodal heterogeneous model. The affinity distributions generated from the LF isotherm can yield similar binding parameters as those calculated using the Freundlich isotherm. For example, the area under the affinity distribution when plotted in semi-log format (N vs. log K) yields the number of sites within the measured association constant range. However, the integral equation (not shown) is different for that derived for the Freundlich isotherm.

IV. CONCLUSION

Ultimately, in applying specific binding models, it is important that they are based on physical models that accurately describe the binding sites in the system being measured. Interestingly, this does not always work in reverse. The fact that a particular binding model can be used to fit the experimental isotherm can be given as support, but not proof, for the existence of a particular distribution of sites. This is because, given the experimental error in an isotherm, there is the possibility that other binding models could be equally accurate in fitting the experimental isotherm. However, in most cases, the ability of a particular binding model to fit the binding isotherm is good evidence for the existence of that distribution.

Binding isotherms are an important and very useful method to characterize the binding characteristics of MIPs. In particular, binding isotherms allow for the quantification of binding properties such as the total number of binding sites (N_T) and average binding affinity (K_{avg}). The simplest binding models are homogeneous models based directly on the Langmuir isotherm. These are easily applied and readily yield binding parameters. More recently, the physically more appropriate heterogeneous binding models have been applied to MIPs. This has been facilitated by the recent derivation of affinity distribution equations that allow the calculation of the corresponding affinity distribution from the easily estimated fitting parameters. The affinity distributions allow for more accurate quantitative and graphical comparison of the binding properties of MIPs. In addition, the heterogeneous binding models can yield binding parameters similar to those of the homogeneous models such as total number of sites (N_T) and average binding affinity (K_{avg}), as well as a measure of the heterogeneity of the system from the heterogeneity index ($\mathbf{m}$). For these reasons, the measurement of binding properties of MIPs using binding isotherms has become crucial in the optimization and in the understanding of the mechanism of the imprinting process.

REFERENCES

1. Takeuchi, T.; Haginaka, J. Separation and sensing based on molecular recognition using molecularly imprinted polymers. J. Chromatogr. B **1999**, *728*, 1–20.
2. Haupt, K.; Mosbach, K. Molecularly imprinted polymers and their use in biomimetic sensors. Chem. Rev. **2000**, *100*, 2495–2504.
3. Wulff, G. Enzyme-like catalysis by molecularly imprinted polymers. Chem. Rev. **2002**, *102*, 1–27.
4. Allender, C.J.; Brain, K.R.; Heard, C.M. Binding cross-reactivity d Boc-phenylalanine enantiomers on molecularly imprinted polymers. Chirality **1997**, *9*, 233–237.
5. Shea, K.J.; Spivak, D.A.; Sellergren, B. Polymer complements to nucleotide bases. Selective binding of adenine derivatives to imprinted polymers. J. Am. Chem. Soc. **1993**, *115*, 3368–3369.
6. Sellergren, B. Molecular imprinting by noncovalent interactions—enantioselectivity and binding-capacity of polymers prepared under conditions favoring the formation of template complexes. Makromolek. Chem.—Macromol. Chem. Phys. **1989**, *190*, 2703–2711.
7. Kempe, M.; Mosbach, K. Directed resolution of naproxen on a non-covalently molecularly imprinted chiral stationary phase. J. Chromatogr. **1994**, *664*, 276–279.
8. Spivak, D.; Gilmore, M.A.; Shea, K.J. Evaluation of binding and origins of specificity of 9-ethyladenine imprinted polymers. J. Am. Chem. Soc. **1997**, *119*, 4388–4393.
9. Andersson, L.I.; Muller, R.; Vlatakis, G.; Mosbach, K. Mimics of the binding-sites of opioid receptors obtained by molecular imprinting of enkephalin and morphine. Proc. Natl. Acad. Sci. USA **1995**, *92*, 4788–4792.
10. Umpleby, R.J.; Rushton, G.T.; Shah, R.N.; Rampey, A.M.; Bradshaw, J.C.; Berch, J.K.; Shimizu, K.D. Recognition directed site-selective chemical modification of molecularly imprinted polymers. Macromolecules **2001**, *34*, 8446–8452.
11. Lehn, J.-M. *Supramolecular Chemistry: Concepts and Perspectives*; VCH: New York, 1995.
12. Scatchard, G. The attractions of proteins for small molecules and ions. Ann. N.Y. Acad. Sci. **1949**, *51*, 660–672.
13. Cerofolini, G.F. Localized adsorption of heterogeneous surfaces. Thin Solid Films **1974**, *23*, 129–152.
14. Sellergren, B.; Shea, K.J. Origin of peak asymmetry and the effect of temperature on solute retention in enantiomer separations on imprinted chiral stationary phases. J. Chromatogr. A **1995**, *690*, 29–30.
15. Allender, C.J.; Brain, K.R.; Heard, C.M. Binding cross-reactivity d-Boc-phenylalanine enantiomers on molecularly imprinted polymers. Chirality **1997**, *9*, 233–237.
16. Umpleby, R.J. II; Bode, M.; Shimizu, K.D. Measurement of the continuous distribution of binding sites in molecularly imprinted polymers. Analyst **2000**, *125*, 1261–1265.
17. Matsui, J.; Doblhoffdier, O.; Takeuchi, T. Atrazine-selective polymer prepared by molecular imprinting technique. Chem. Lett. **1995**, *6*, 489.
18. Wulff, G.; Grobe-Einsler, W.; Vesper, W.; Sarhan, A. Enzyme-analogue built polymers, (5) on the specificity distribution of chiral cavities prepared in synthetic polymers. Makromol. Chem. **1977**, *178*, 2817–2825.
19. Chen, Y.B.; Kele, M.; Sajonz, P.; Sellergren, B.; Guiochon, G. Influence of thermal annealing on the thermodynamic and mass transfer kinetic properties of D-and L-phenylalanine anilide on imprinted polymeric stationary phases. Anal. Chem. **1999**, *71*, 928–938.
20. Whitcombe, M.J.; Rodriguez, M.E.; Villar, P.; Vulfson, E.N. A new method for the introduction of recognition site functionality into polymers prepared by molecular imprinting: synthesis and characterization of polymeric receptors for cholesterol J. Am. Chem. Soc. **1995**, *117*, 7105–7111.

21. Umpleby, R.J. II; Baxter, S.C.; Bode, M.; Berch, J.K.; Shah, R.N.; Shimizu, K.D. Application of the Freundlich adsorption isotherm in the characterization of molecularly imprinted polymers. Anal. Chim. Acta **2001**, *435*, 35–42.

22. Cheong, S.-H.; Rachkov, A.E.; Park, J.-K.; Yano, K.; Karube, I. Synthesis and binding properties of a noncovalent molecularly imprinted testosterone-specific polymer. J. Polym. Sci., Part A: Polym. Chem. **1998**, *36*, 1725–1732.

23. Ramström, O.; Ye, L.; Mosbach, K. Artificial antibodies to corticosteroids prepared by molecular imprinting. Chem. Biol. **1996**, *3*, 471–477.

24. Umpleby, R.J. II; Rampey, A.M.; Baxter, S.C.; Rushton, G.T.; Shah, R.N., Bradshaw, J.C., Shimizu, K.D. The Freundlich isotherm/affinity distribution analysis of molecularly imprinted polymers 1: evaluation of the Fl-AD analysis for the characterization the binding properties of molecularly imprinted polymers. *J. Chromatogr. B* **2004**, *804*, 141–149.

25. Thakur, A.K.; Munson, P.J.; Hunston, D.L.; Rodbard, D. Characterization of ligand-binding systems by continuous affinity distributions of arbitrary shape. Anal. Biochem. **1980**, *103*, 240–254.

26. Umpleby R.J. II; Baxter, S.C.; Chen, Y.; Shah, R.N.; Shimizu, K.D. Characterization of molecularly imprinted polymers with the Langmuir–Freundlich isotherm. Anal. Chem. **2001**, *73*, 4584–4591.

27. Quaglia, M.; Chenon, K.; Hall, A.J.; De Lorenzi, E.; Sellergren, B. Target analogue imprinted polymers with affinity for folic acid and related compounds. J. Am. Chem. Soc. **2001**, *123*, 2146–2154.

17

Molecularly Imprinted Polymer Beads

Lei Ye and Ecevit Yilmaz Chemical Center, Lund University, Lund, Sweden

I. INTRODUCTION

The majority of traditional molecularly imprinted polymers (MIPs) are in the format of irregular particulates obtained by grinding cross-linked polymer monoliths. The widely used grinding and sieving procedures are time-consuming, and often lead to a poor yield of useful MIP particles. In chromatographic uses, the nonuniform (wide variation of) particle shape and size make it difficult to achieve an optimum column packing. Also the cross-linking polymerization in bulk is only applicable in small laboratory scale. For mass production of MIPs, scalable preparation methods are needed. The most common methods for preparing polymer beads are emulsion polymerization, dispersion polymerization, and suspension polymerization (Table 1). Emulsion polymerization normally leads to uniform latex particles with diameters smaller than 1 μm. Dispersion polymerization and suspension polymerization give larger beads and a broader distribution in particle size, although the latter can be controlled to certain extent by optimizing the reaction condition. In many cases, an aqueous medium serves as a continuous phase in the polymerization system to both support the resulting polymer beads and to provide easy temperature control during the polymerization. In principle, all the above-mentioned methods can be adopted for making molecularly imprinted polymer beads, though under certain circumstances polar or aqueous media do interfere with weak molecular interactions, as represented by the devastating effect of water on the hydrogen bond interaction between template and functional monomers. In addition to the above-mentioned methods, there have been investigations for new polymerization systems that are suitable for making molecularly imprinted polymer beads, for example by using preformed beads as supporting skeleton to generate hybrid materials. These will be discussed in the following sections.

II. EMULSION POLYMERIZATION

By emulsion polymerization, small MIP beads with a narrow size distribution can be generated. Until now, there have been only limited examples of MIP nanobeads that

Table 1 Basic Techniques Used to Prepare Polymer Beads

Methods	Mechanism for bead formation	Reaction components	Typical particle size
Emulsion polymerization	Emulsion polymerization starts with a dispersion of monomer micelles in a nonsolvent medium stabilized by appropriate surfactant molecules. Polymerization proceeds within each spherical micelle once a radical enters the micelle from the continuous phase, which finally results in a dispersion of latex.	Dispersed phase (monomer), continuous phase (nonsolvent and initiator), surfactant	$< 1\ \mu m$, uniform in size
Dispersion polymerization	Dispersion polymerization starts with a solution of monomer in a reaction medium. When the resulting polymer chain grows to a certain critical point, the polymer precipitates to form spherical particulate that is prevented from coagulation by a polymeric stabilizer.	Monomer, solvent, initiator, stabilizer	$1–10\ \mu m$
Suspension polymerization	Suspension polymerization starts with a dispersion of droplets containing monomer, initiator, and solvent in a continuous phase. The droplet is maintained by a mechanical stirring with the help of appropriate surfactant.	Dispersed phase (monomer, initiator and solvent), continuous phase (nonsolvent), surfactant	$20–200\ \mu m$

were successfully prepared by simple emulsion polymerization. The use of water as the continuous phase in emulsion polymerization requires that relatively strong interactions between template and functional monomers are maintained during the process of bead formation. The relatively stable complex formed between transition metal ions and chelating monomers has enabled preparation of metal-imprinted nanobeads using oil-in-water (O/W) emulsion polymerization. Koide et al. [1] have used a polymerizable surfactant, 10-(p-vinylphenyl)decanoic acid to localize Zn^{2+}, Cu^{2+}, and Ni^{2+} ions on the surface of micelles, and cross-linked the surfactant by a co-polymerization with divinylbenzene (Fig. 1). The carboxylate head group of the surfactant provided metal coordination sites in the final 200–300 nm sized polymer beads. Since the template metal ions were kept at the oil–water interface during the imprinting reaction, this surface imprinting method generated easily accessible binding sites on the resulting nanobeads.

Microemulsion polymerization has also been employed for the preparation of inorganic nanoparticles [2]. By using a functional surfactant carrying at its polar head a α-chymotrypsin transition state analoguc (TSA), Markowitz et al. used a series of silane monomers for the preparation of catalytic silica nanoparticles in a reversed micelle (water-in-oil, W/O) environment. Analogous to the previous metal-imprinting, here the TSA template was also kept at the surface of the reverse micelles, therefore the active sites on the silica nanobeads were also made easily accessible to catalyze the amide hydrolysis reaction (Fig. 2a). Using this method, very uniform silica beads were obtained (Fig. 2b). It should be noted that although the template used was a chymotrypsin TSA, the MIP displayed higher catalytic activity

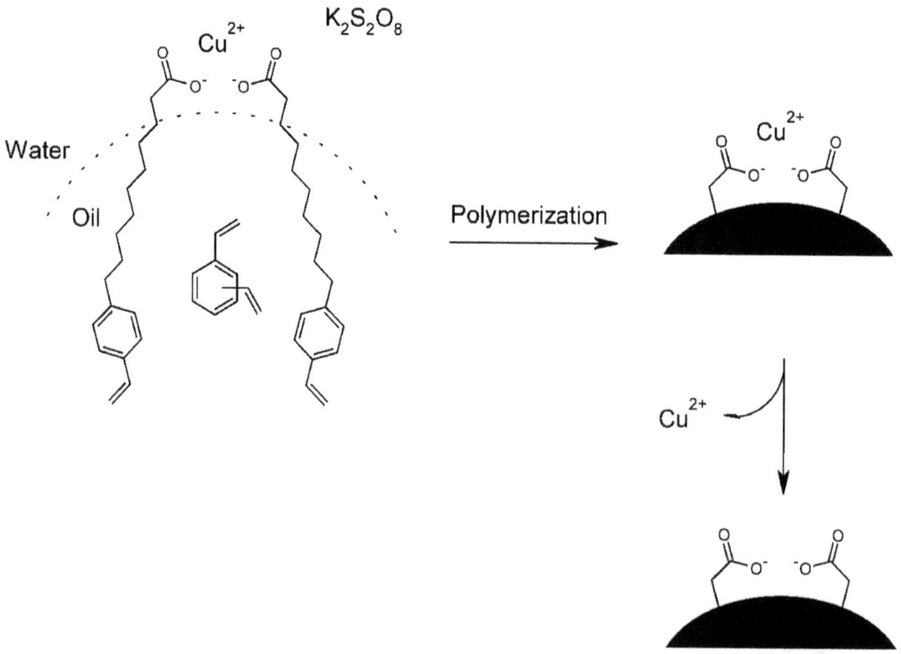

Figure 1   Surface metal imprinting using oil-in-water emulsion polymerization.

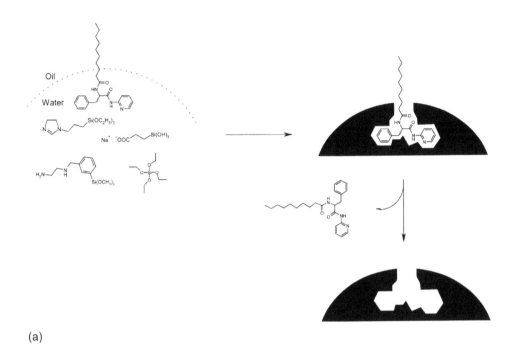

(a)

(b)

Figure 2 Surface imprinting against a transition state analogue (TSA) template. (a) The
template is part of a surfactant and therefore is located at the surface of the reversed micelle
during the polymerization. (b) A transmission electron micrograph of the imprinted
catalytic nanoparticles. Reprinted in part with permission from Ref. 2. Copyright (2000)
American Chemical Society.

against trypsin substrate. Also the chiral selectivity of the MIP was reversed, i.e., the MIP imprinted against the L-form chymotrypsin TSA was more efficient in hydrolyzing the D-form trypsin substrate. An in-depth understanding of the reversed selectivity is still desired.

III. DISPERSION POLYMERIZATION

Dispersion polymerization starts from a homogeneous solution containing monomer, initiator and, also often an appropriate stabilizer. When the polymer chain grows to such an extent that the solvent can no longer provide effective solvation for the resulting polymer, discrete polymer granules are phase-separated from the solution. The added stabilizer, often a polymer modifier, provides an effective steric barrier to prevent aggregation of the polymer granules. Despite the fact that dispersion polymerization has been a common method for preparing commodity polymer beads, and a wide range of monomer and solvent systems have been developed for dispersion polymerization, use of this technique for making MIP beads has been reported in only a few examples.

Molecular imprinting polymerization often involves a large fraction of cross-linking monomer, which is distinct from the traditional dispersion polymerization used for the preparation of noncrosslinked polymer beads. Although the traditional imprinting method giving porous monolith also starts from a homogeneous solution of template, functional monomer, cross-linker, and initiator in an appropriate solvent, we shall reserve the term "dispersion imprinting polymerization" for those that do generate spherical particles. According to this criterion, the traditional grinding and sieving method is standing on the border between bulk polymerization (no solvent) and dispersion polymerization.

In producing conventional imprinted polymer monolith, a total monomer concentration of around 20–50% (v/v) is often used, and the molar fraction of cross-linker in total monomer takes between 50% and 98% (mol/mol). Phase separation occurs during the cross-linking process when the limit of solubility of the growing polymer is exceeded. Imprinted polymers with a wide range of porosities can be obtained by controlling the point of phase separation, which, in turn, depends on the nature and the volume of solvent used and the amount of cross-linker employed. In a more dilute reaction system, the growing polymer chains are unable to occupy the entire volume of the reaction vessel, which results in a dispersion of macrogel particles in the solvent. Further dilution of the system will lead to discrete microgel spheres—microgel powder in the region III of the morphology diagram as previously described by Sherrington [3] (Fig. 3).

By adjusting the amount and type of solvent used in the traditional imprinting recipe, Sellergren [4] has demonstrated that an MIP material containing stable agglomerates of micron-sized particles could be generated, without using a grinding step. However, the obtained agglomerates displayed rather poor shape regularity. Inspired by the precipitation polymerization method for the preparation of monodisperse cross-linked polymer microspheres [5], a cross-linking polymerization method has been developed using a high monomer dilution to give uniform imprinted nanospheres. The method relies on the fact that the growing polymer chains formed during the polymerization reaction aggregate into small nanoshperes that precipitate

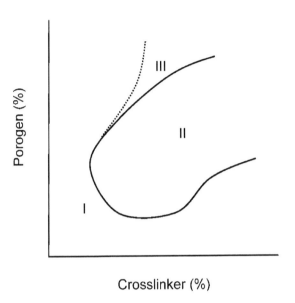

Figure 3 A diagram showing the effect of porogen and cross-linker on the final polymer morphology: (I) gel-type resin; (II) macroporous resin; and (III) microspheres and microgel powder. Adapted from Ref. 3.

out from solution. The high dilution prevents any further aggregation or fusion of the spheres into a bulk polymer. The noncovalent imprinting reaction was carried out in a common imprinting solvent, acetonitrile, using methacrylic acid and trimethylol-propane trimethacrylate as the functional monomer and cross-linker. Uniform nanobeads specific for theophylline and 17β-estradiol were readily obtained by this method [6]. Figure 4a shows a scanning electron micrograph (SEM) of the 17β-estradiol-imprinted nanobeads. The mean diameter of the particles was approximately 300 nm. These nanobeads are ideal to use in drug assays due to their stable suspension in solvent and therefore are very easy to dispense. An advantage of these imprinted nano- or micro-spheres is their easily accessible binding sites, which allowed us to further develop an ELISA type assay for a herbicide, 2,4-dichlorophenoxyacetic acid (2,4-D), where a bulky 2,4-D-horseradish peroxidase conjugate could be used as a tracer. The bound tracer was displaced from the MIP by the 2,4-D present in a sample, and easily quantified by monitoring the enzyme-catalyzed colorimetric and chemilumilescence reactions [7,8]. Using the same preparation method, other groups have generated nanoparticules, and used them as chiral selectors in capillary electrophoretic separation of different chiral drugs [9,10].

By introducing a signaling moiety (a reporter group) into MIP microspheres, we have recently developed scintillation proximity assays (SPA) applicable in both organic solvent and aqueous buffer. In this case imprinted binding sites replaced antibodies or receptors to provide the required specific binding [11,12]. Figure 4b is an SEM for the scintillation MIP microspheres used in the SPA for an adrenergic β-blocker, (S)-propranolol. The MIP was prepared using methacrylic acid and divinylbenzene as the functional monomer and cross-linker, and acetonitrile as the solvent.

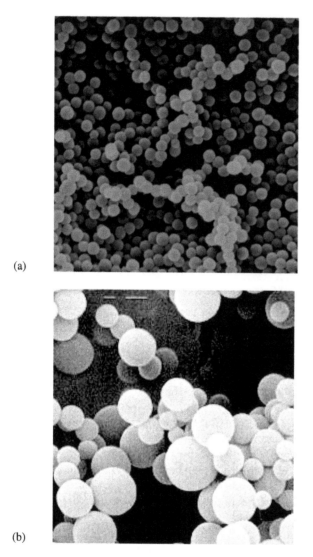

(a)

(b)

Figure 4 Scanning electron micrographs for imprinted polymer microspheres against (a) 17β-estradiol; (b) (S)-propranolol.

Our precipitation polymerization method for imprinting may be defined as a stabilizer-free dispersion polymerization, as no extra polymer stabilizer is needed in the imprinting process. The omission of stabilizer has greatly simplified polymer purification following the imprinting reaction, since no effort is required to remove the otherwise sticky stabilizer on the bead surface. A key point here is to find out the optimal compatibility among template, monomer and solvent, in order that uniform MIP beads are obtained. Although certain physicochemical parameters of the individual components, such as solubility parameters, can be used to guide the choice of correct imprinting composition, there is still a lack of simple, general guidelines ensuring formation of uniform beads in all the circumstances.

IV. EXPERIMENTAL PROTOCOL 1

A. Molecular Imprinting of 17β-Estradiol Using Precipitation Polymerization

1. Materials

17β-Estradiol and 17α-estradiol are purchased from Sigma (St. Louis, MO, USA). [2,4,6,7-^{3}H(N)]Estradiol (specific activity 72.0 Ci/mmol) is from NEN Life Science Products, Inc. (Boston, MA, USA). Scintillation liquid, Ecoscint O, is from National Diagnostics (Manville, NJ, USA). Methacrylic acid (MAA), trimethylolpropane trimethacrylate (TRIM), and azobisisobutyronitrile (AIBN) are from Merk (Darmstadt, Germany). AIBN is re-crystallized from methanol prior to use. Anhydrous acetonitrile used for polymer synthesis is from Lab-Scan (Stillorgan, Ireland).

2. Polymer Preparation

17β-Estradiol (200 mg) and MAA (298 mg) are dissolved in anhydrous acetonitrile (40 mL) in a borosilicate glass tube equipped with a screw cap. TRIM (502 mg) and AIBN (14 mg) are then added, the solution purged with nitrogen for 5 min and the tube sealed under nitrogen. Polymerization is induced by UV irradiation (350 nm) using an RMA-400 Rayonet photochemical reactor from Southern New England Ultraviolet Co. (Bradford, CT, USA) at 20°C and continued for 24 h. The microspheres obtained are collected by centrifugation at 8000 rpm for 10 min using an RC5C superspeed refrigerated centrifuge from BECKMAN (Palo Alto, CA, USA). The template is extracted by washing the polymer repeatedly in methanol containing 10% acetic acid, followed by a final wash in the same volume of acetone. Successive centrifugation and decanting steps are used for solvent exchange. The microspheres are finally dried in vacuo. Non-imprinted microspheres are prepared and treated in exactly the same way, except that no template is used in the polymerization stage.

3. Polymer Characterization

Particle morphology of the imprinted microspheres can be inspected using scanning electron microscope (SEM). Selective binding of the imprinted polymers is characterized using radioligand binding analysis. Both saturation study and competitive binding analysis can be easily carried out using a standard β-counter for quantification of the unbound fraction of the tritium-labeled template.

The binding efficacy of the microspheres is estimated from a saturation study. Varying amounts of microspheres are incubated with [2,4,6,7-^{3}H(N)]estradiol (417 fmol (1110 Bq)) in acetonitrile (1 mL) in polypropylene microcentrifuge tubes at 20°C for 16 h. A rocking table is used to ensure a gentle mixing. The microspheres are then separated by centrifugation at 14,000 rpm for 5 min, supernatant (500 μL) is withdrawn and mixed with 10 mL of scintillation liquid, from which the radioactivity is measured using a model 2119 RACKBETA β-radiation counter from LKB Wallac (Sollentuna, Sweden). A plot of bound fraction vs. polymer concentration reveals that approximately 50 mg of MIP microspheres are required to uptake about half of the radioligand from 1 mL of acetonitrile. At the same polymer concentration, the nonimprinted polymer binds less than 20% of the radioligand by nonspecific adsorption.

The imprinted microspheres are suspended in acetonitrile (150 mg mL^{-1}) and sonicated to form a polymer stock suspension, from which 200 μL is transferred into each microcentrifuge tube. Varying amounts of nonradiolabeled ligand, 17β-estradiol or 17α-estradiol, and the radioligand, [2,4,6,7-^{3}H(N)]estradiol (417 fmol (1110 Bq)) are added, and the final volume adjusted to 1 mL with acetonitrile. The competition binding is allowed to proceed for 16 h by incubation at 20°C, using a rocking table for gentle mixing. The amount of bound radioligand is estimated by measuring the radioactivity from 500 μL supernatant following removal of the microspheres by centrifugation at 14,000 rpm for 5 min. A plot of bound radioligand vs. the amount of nonlabeled 17β- and 17α-estradiol can be established from which the binding cross-reactivity of 17α-estradiol on the MIP can be estimated.

V. SUSPENSION POLYMERIZATION

Suspension polymerization has been investigated to the greatest extent for making molecularly imprinted polymer beads, using both aqueous and nonaqueous continuous phases. This was mainly due to the initiative of using the micron-sized beads for chromatographic separation purposes, which is particularly amenable using suspension polymerization. The dispersed phase comprises a large number of droplets containing standard imprinting components (i.e., template, functional monomer, cross-linker, initiator, and porogenic solvent), which can be considered as individual miniaturized microreactors. With the aid of appropriate surfactant, a mechanical stirring maintains a relatively stable dispersion of the droplets in the continuous phase during the polymerization.

Using a covalent imprinting approach, Byström et al. [13] prepared porous MIP beads using an aqueous continuous phase. The binding sites for steroids were introduced by copolymerization of an acrylated steroid with divinylbenzene, followed by chemically removing the steroid template using reductive cleavage. The obtained polymer beads were then used as a polymer-supported reagent for the regio- and stereo-selective hydride reduction of steroidal ketones. It was demonstrated that the regio- and stereo-selectivity for steroid transformation was dictated by the specific binding sites generated during the imprinting reaction. It is interesting to note that a bulk polymer made in pure organic solvent with the same monomer composition displayed very little imprinting effect. The superior imprinting effect for the suspension MIP beads may be partially contributed by the hydrophobic interaction between the steroid template and the cross-linker, which should be more prominent in an aqueous medium.

Covalent imprinting involving a sacrificial spacer has also been demonstrated in suspension polymerization. The imprinting of cholesterol via the carbonate ester was carried out using both ethyleneglycol dimethacrylate and divinylbenzene as the cross-linker. It was found that uptake of cholesterol from an apolar solvent by the MIP beads was similar to that of the corresponding bulk polymer [14], although the nonspecific binding by the suspension polymer was somewhat higher.

Despite the fact that water interferes with many noncovalent imprinting systems, there have been numerous examples of noncovalent MIP beads prepared by suspension polymerization in water [15–18]. In almost all the cases, either a strong ionic, or hydrophobic interaction, or a combination of both (between template and monomers) were gainfully utilized. To generate larger MIP beads for chromatographic separation of metal ions, Yoshida et al. [19] developed a water-in-oil-in-water (W/O/W) multiple emulsion polymerization method, where a functional host

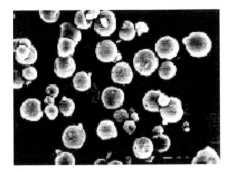

Figure 5 A scanning electron micrograph for MIP beads prepared against Boc-*L*-Phe using suspension polymerization in perfluorocarbon liquid. Reprinted in part with permission from Ref. 21. Copyright (1996) American Chemical Society.

molecule, 1,12-dodecanediol-*O*,*O'*-diphenyl phosphonic acid, was physically entrapped via its aliphatic tail in a cross-linked polyacrylate matrix, while the phosphonic acid head group formed complex with the template metal ion, Zn (II), at the water–oil interface. By this method, porous MIP beads with diameter of 15–40 μm were obtained. In this surface imprinting approach, strong coordination interaction between the water-soluble metal ions and the lipophilic functional hosts at the oil–water interface has been successfully exploited. The imprinting polymerization actually took place in discrete droplets suspended in water, in each of the droplet was contained many stabilized W/O reverse micelles.

The down side of noncovalent imprinting in aqueous suspension system is when hydrophilic template and functional monomer are involved, these species will unavoidably partition into the aqueous continuous phase, therefore seriously reduce the amount of imprinted sites on the resulting polymer beads. For noncovalent imprinting, suspension polymerization can be carried out more feasibly in a nonaqueous medium. The prerequisite of maintaining a nonpartitioning dispersed phase can be achieved using a chemically inert liquid medium, for example using a perfluorocarbon liquid as the continuous phase [20]. The suspension polymerization in perfluorocarbon liquid for noncovalent imprinting was first realized by Mayes and Mosbach [21]. The MIP beads prepared in this manner displayed chromatographic efficiency for chiral separation of amino acid derivative comparable to their "bulk" counterpart, even when a much higher flow rate was used. By adjusting the amount of perfluoronated surfactant, the particle size of the MIP beads could be readily controlled within the range of 10–100 μm, which should fit most of chromatographic applications. Figure 5 shows an SEM for the MIP beads prepared in a perfluorocarbon liquid. In a further study, Ansell and Mosbach [22] incorporated magnetic iron oxide into the MIP beads prepared in perfluorocarbon liquid. These MIP beads could be easily isolated by applying an external magnetic field, therefore opened a new opportunity for imprinted polymers in magnetic separation area.

VI. IMPRINTING ON PREFORMED BEADS

Instead of making molecularly imprinted polymer beads directly, different methods have been developed to graft an MIP layer on preformed spherical particles to give composite beads. The preformed spherical particles are simply used as a "skeleton" to give the final material a desired particle shape. In certain cases, the "skeleton" can even be removed afterwards, to afford MIP materials with controlled pores dictated by the part of the sacrificial supporting material.

Surface grafting of molecularly imprinted polymer on spherical silica and polymer particles has been used to give hybrid materials with controlled shape and size. When silica beads are used, they are often first chemically modified, so that either a polymerizable vinyl group, or an initiator is immobilized on their surface [23–25]. This is to ensure that the part of imprinted layer is covalently grafted on the supporting matrix with a good stability. Typical surface modification chemistry is outlined in Fig. 6, where the immobilized iniferter (initiator-transfer agent-terminator) method [26] was claimed to couple all the imprinting material to the supporting silica beads [27], as the polymer chain propagation was confined to the surface of the support.

Plunkett and Arnold [28] grafted a layer of bis-imidazole-imprinted, metal-coordination polymer on the surface of spherical silica particles, and applied the composite beads for chromatographic separation of a series of model compounds using metal chelating interactions. Porous silica beads (LiChrosphere 1000, 10 μm diameter, 1000 Å pores) were first derivatized with 3-(trimethoxysilyl) propylmethacrylate to introduce covalently bound vinyl group, which was used later in the imprinting step to fix the MIP layer on the support. Coordination between the bis-imidazole template and the functional monomer, diacetic acid was mediated by Cu (II), whereas in the later application, a weaker binding ion, Zn (II) was used to

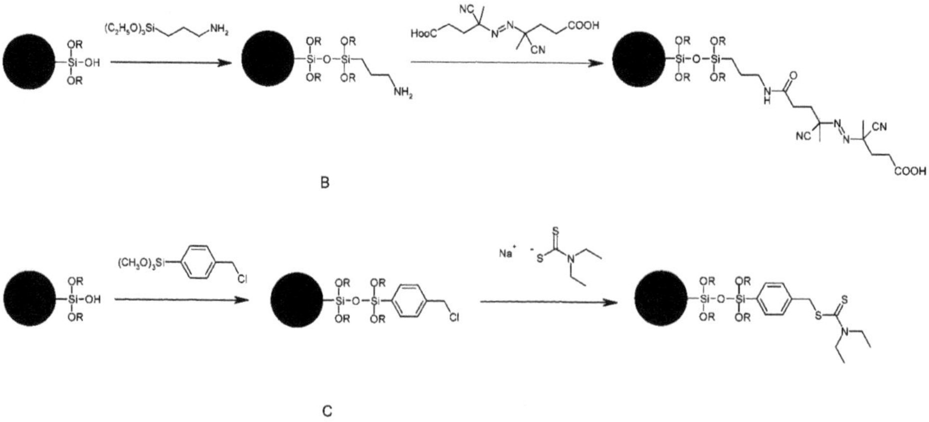

Figure 6 Surface modifications of supporting materials to indroduce: (A) a polymerizable vinyl group; (B) an initiator; and (C) an iniferter.

affect easy chromatographic separation of the template from similar imidazole-containing compounds. Using similar surface modification for introducing vinyl group, silica-supported MIP beads have also been prepared to afford chromatographic separation of proteins [29–31].

The drawback of the vinyl modification method is that it is difficult to control the extent of polymer grafting, i.e., a quite large portion of MIP is not formed on the support surface due to the polymerization in solution. In an alternative way, Sellergren and co-workers [25,32] have carried out surface-initiated imprinting polymerization. This was achieved by first coupling a radical initiator, 4,4'-azo-bis(4-cyanopentanoic acid) to an amino-functionalized silica beads, followed by initiating the radical polymerization on the surface of the support. As the initiator was immobilized on the support, in theory at least 50% of the imprinting components (functional monomer and cross-linker) could be grafted onto the support. To further increase the grafting efficiency on silica beads, the same group used a surface-immobilized iniferter, benzyl-N,N-diethyldithiocarbamate to initiate the imprinting polymerization [27]. It was found that minimal chain propagation occurred in solution, which can be attributed to the quasi-living characteristics of the iniferter-induced polymerization [33,34].

Regarding the use of porous silica as a "shape template," Yilmaz et al.[35] have shown that it is possible to skip the step of surface modification on silica, instead to carry out an imprinting polymerization directly within the interconnected pores in the silica beads. The obtained composite beads can be used directly for chromatographic packing to afford chiral separation of ($\pm$)-isoproterenol. Furthermore, the part of silica skeleton could be chemically removed after the imprinting step, to give MIP beads with enlarged pores and a higher surface area. An SEM for the ($-$)-isoproterenol-imprinted composite beads, as well as the MIP beads following removal of the silica support, is shown in Fig. 7. In other related studies, different templates have been immobilized on porous silica beads prior to initiating the imprinting polymerization. The removal of template and chemical dissolution of silica were carried out by hydrolytic cleavage [36,37].

Molecularly imprinted composite polymers were also prepared using pre-formed poly(trimethylolpropane trimethacrylate) beads as the supporting material [38]. Spherical polymer particles were firstly prepared using standard suspension polymerization of trimethylolpropane trimethacrylate. The residual double bonds in these supporting beads were used to graft up to 64 mol% of molecularly imprinted cross-linked polymethacrylate. The resulting composite polymer beads were used as a chromatographic stationary phase for chiral separation of Boc-D,L-Phe, which gave an enantioselectivity for the template molecule, Boc-L-phe equivalent to the purely imprinted polymer, but with an improved column efficiency.

Noncrosslinked polyacrylate latex beads have been exploited for the preparation of metal ionimprinted nanobeads [39]. The seed latex beads containing mobile carboxyl functional groups were prepare by standard aqueous emulsion polymerization. The seed particles were loaded with divinylbenzene and butyl acrylate, and then exposed in an aqueous medium to the template metal ions, e.g., Cu (II), Ni (II), and Co (II). The metal ions re-organized the carboxyl binding groups at the polymer–water interface. A further cross-linking polymerization fixed the metal–carboxylate complex on the surface of the nanobeads, therefore generated, after removal of the template, surface-exposed binding sites for the metal ion templates.

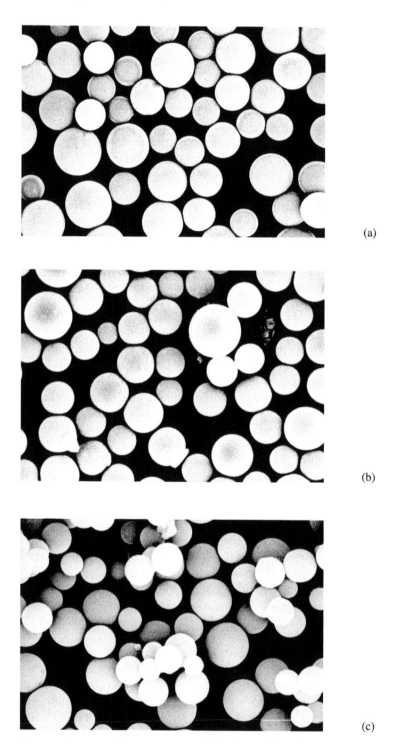

(a)

(b)

(c)

Figure 7 Molecular imprinting using porous silica beads as support. Scanning electron micrographs for (a) porous silica bead support; (b) silica-MIP composite beads; and (c) MIP beads after removal of the support.

Molecularly imprinted core-shell particles can be prepared in water using uniform colloid particles as the support. Although aqueous medium is not typically applicable for non–covalent imprinting due to its very polar and hydrogen bonding characteristics, it does give reasonably good MIP beads when recognition based on hydrophobic and ionic interactions can be gainfully utilized. By utilizing a combination of hydrophobic and electrostatic interactions, Carter and Rimmer [40] used a two-stage emulsion polymerization method to prepare caffeine-imprinted core-shell nanobeads in an aqueous medium. The core latex was prepared by emulsion polymerization of styrene and divinylbenzene, followed by coating an imprinted polymer layer in a second polymerization step.

Similarly, Pérez et al. [41] utilized a covalent imprinting approach to prepare their cholesterol-imprinted core-shell nanoparticles in a two-stage polymerization process. The seed latex particles were synthesized by an oil-in-water emulsion polymerization, which were then used in a second stage imprinting polymerization to introduce the cholesteryl (4-vinyl) phenyl carbonate into the resulting shells. After removal of the sacrificial carbonate moiety and the cholesterol template, specific binding sites for cholesterol were generated, in which a phenol functional group could provide hydrogen bond interaction during the rebinding experiment (Fig. 8). By using a polymerizable surfactant (pyridinium 12-(4-vinylbenzyloxycarbonyl) dodecanesulfate) in combination with a template surfactant (pyridinium 12-(cholesteryloxycarbonyloxy) dodecanesulfate) at the second polymerization stage, the authors obtained hydrophobic cholesterol-binding sites on the surface of the final core-shell nanobeads [42].

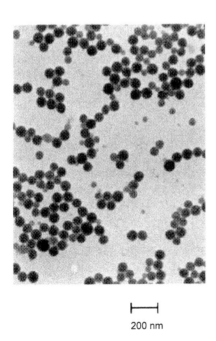

├──┤
200 nm

Figure 8 Cholesterol imprinted core-shell nanoparticles. Reprinted in part with permission from Ref. 41. Copyright (2000) John Wiley & Sons, Inc.

To generate larger MIP beads for chromatographic applications, Haginaka et al. [43,44] developed a multistep swelling and polymerization method. The uniform-sized polystyrene seed particles (diameter of ca. 1 μm) were prepared by an emulsifier-free emulsion polymerization. The seed particles were first activated with a microemulsion of dibutyl phthalate (first swelling), followed by swelling in toluene containing a radical initiator (second swelling). These particles were then exposed to an imprinting mixture (composed of template, functional monomer and cross-linker) dispersed in an aqueous medium (third swelling), and finally polymerized. This multi-swelling and polymerization procedure could give uniform MIP beads with a size of 5–6 μm in diameter, which fits the common size demand for HPLC packing materials.

VII. EXPERIMENTAL PROTOCOL 2

A. Molecular Imprinting Using Silica Beads as Shape Template

In this example, porous silica beads are filled with the liquid imprinting mixture and then polymerized. This process leads to spherical and discrete MIP-composite particles in only one synthetic step.

1. Materials

Commercially available silica particles (Kromasil, 13 μm mean diameter, C4-phase, 0.7 ml/g pore volume, 195 Å pore size) are from Eka Chemicals AB (Bohus, Sweden). ($\pm$)-Isoproterenol·HCl and ($-$)-isoproterenol·HCl are purchased from Sigma (St. Louis, USA). ($-$)-Isoproterenol, trifluoromethacrylic acid (TFMAA), methacrylic acid (MAA) and divinylbenzene (DVB) are obtained from Aldrich (Steinheim, Germany). DVB is of technical grade (80%) and the inhibitor is removed using basic alumina (Merck, Darmstadt, Germany) prior to use. Solvents for polymer preparation are of anhydrous quality and are from Labscan (Dublin, Ireland). Hydrofluoric acid (HF), acetone (p.a.) and acetonitrile (MeCN)(HPLC quality) are obtained from Merck.

2. Methods

A prepolymerization mixture is prepared in the following way: Initiator AIBN (95 mg), template molecule ($-$)-isoproterenol (396 mg, 2.5 mmol) and functional monomer TFMAA (1.05 g, 10 mmol) are weighed in a glass vial, dissolved in MeCN (4.5 ml) and DVB (6.7 ml, 50 mmol) and sonicated. After complete dissolution, the solution is cooled and then sparged with N_2 for 2 min. In order to obtain discrete silica-polymer particles and to prevent sticky particle agglomerates, the amount of prepolymerization mixtures added to the silica is kept slightly lower (0.65 mL added per g silica) than the BET pore volume of the silica (0.7 mL/g silica). Thus 3.25 mL of the pre-polymerization mixture is added to 5 g silica and shaken vigorously for about 5 min. The mixture is then gently stirred with a spatula until all the pre-polymerisation mixture penetrated the silica pores. Finally, the mixture is sonicated to remove any entrapped air bubbles. A free-flowing material is thus obtained. This is flushed with N_2, sealed, and allowed to polymerize in an oven at 65°C for at least 16 h. After polymerization is completed, silica-MIP composite beads are obtained (Fig. 7b).

3. Chromatographic Evaluation

Polymer particles are suspended in water (25% MeCN) and then slurry-packed into stainless steel columns (250 mm × 4.6 mm i.d.) using an air-driven fluid pump (Haskel, Burbank, CA, USA) and water (25% MeCN) as the packing solvent. The packed columns are washed on-line on a Beckman HPLC system (comprising a solvent module 126 and diode array detector 168) using MeCN (20% acetic acid) to remove the print molecule until a stable base line is obtained. The mobile phase is then changed to a citrate buffer (pH 3.0, 25 mM citrate) containing 10% MeCN (v/v) at a flow rate of 1 mL min^{-1}. For a test of chiral resolution, a racemic mixture of (+)- and (−)-isoproterenol (20 μL at 2 mM in the mobile phase) is injected, and the elution monitored at 280 nm. Acetone can be used as a void marker for the calculation of capacitor factor (k') and separation factor (α).

VIII. MISCELLANEOUS METHODS

Selective cell-recognition materials have great potentials for biomedical, environmental, and food analysis of microorganisms. Using a bacteria-mediated lithographic procedure, imprinted cavities specific for whole cells have been created on polymer beads (cf. Chapter 9) [45]. In the published example, a diacid chloride contained in a dispersed organic phase was reacted with a water-soluble polyamine in the presence of a bacterium template located at the organic-water interface. The cells acted as temporary protecting groups and structural templates during the multistep imprinting process. It was shown that the resultant polymeric beads exhibited on their surface functionally anisotropic patches that were defined by the template.

A quite different technique has been applied for the formation of silicone microbeads using cationic aerosol photopolymerization. In the presence of a template (a morphine derivative, thebaine), a very fast cationic polymerization was carried out in aerosol droplets generated from an airbrush. The aerosol droplets contained a bis-epoxy silicone monomer, a diaryliodonium salt photoinitiator, and the template [46]. With this method, no supporting liquid phase (neither organic solvent nor water) is needed, which simplifies greatly the polymerization process and polymer work up. However, it has been so far limited to using specialty monomers and initiators, and cannot be applied to templates that bear many common functional groups (e.g., organic acids and nitrogenous bases).

IX. CONCLUDING REMARKS

It is difficult to enumerate an optimal method for preparing MIP beads. The different approaches leading to spherical MIPs seem to be mutually supplementary, given the fact that they often generate MIP beads in different size levels. In Fig. 9 the common preparation methods that afford the different sizes of the imprinted beads, together with a general comment on their compatibility with molecular imprinting, easiness in controlling size distribution, and their cost-efficiency are summarized. As applications of imprinted polymer beads cover a very broad area, from binding assays to preparative chromatography, different particle sizes are needed to address individual problems. For example, in binding assays nanobeads are easy to handle, even with standard liquid delivery systems. On the other hand, large beads are required for fast

Particle size, μm

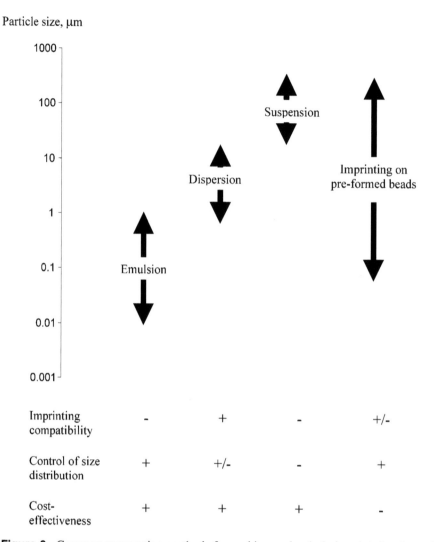

Figure 9 Common preparation methods for making molecularly imprinted polymer beads and the size range of the product beads. A general comment (+: pro; −: con) on the different methods is included.

flow in chromatographic purification and solidphase extractions. In the coming years, we expect novel imprinting chemistries to be developed, which, in combination with the polymerization techniques described in this section, will lead to smarter MIP beads in a more cost-efficient manner.

REFERENCES

1. Koide, Y.; Senba, H.; Shosenji, H.; Maeda, M.; Takagi, M. Selective adsorption of metal ions to surface-template resins prepared by emulsion polymerization using 10-(p-vinyl-phenyl)decanoic acid. Bull. Chem. Soc. Jpn. **1996**, *69*, 125–130.

2. Markowitz, M.A.; Kust, P.R.; Deng, G.; Schoen, P.E.; Dordick, J.S.; Clark, D.S.; Gaber, B.P. Catalytic silica particles via template-directed molecular imprinting. Langmuir **2000**, *16*, 1759–1765.
3. Sherrington, D.C. Preparation, structure and morphology of polymer supports. Chem. Commun. **1988**, 2275–2286.
4. Sellergren, B. Imprinted dispersion polymers: A new class of easily accessible affinity stationary phases. J. Chromatogr. A **1994**, *673*, 133–141.
5. Li, K.; Stöver, D.H. Synthesis of monodisperse poly(divinylbenzene) microspheres. J. Polym. Sci. Part A: Polym. Chem. **1993**, *31*, 3257–3263.
6. Ye, L.; Cormack, P.A.G.; Mosbach, K. Molecularly imprinted monodisperse microspheres for competitive radioassay. Anal. Commun. **1999**, *36*, 35–38.
7. Surugiu, I.; Ye, L.; Yilmaz, E.; Dzgoev, A.; Danielsson, B.; Mosbach, K.; Haupt, K. An enzyme-linked molecularly imprinted sorbent assay. Analyst **2000**, *125*, 13–16.
8. Surugiu, I.; Danielsson, B.; Ye, L.; Mosbach, K.; Haupt, K. Chemiluminescence imaging ELISA using an imprinted polymer as the recognition element instead of an antibody. Anal. Chem. **2001**, *73*, 487–491.
9. Schweitz, L.; Spegel, P.; Nilsson, S. Molecularly imprinted microparticles for capillary electrochromatographic enantiomer separation of propranolol. Analyst **2000**, *125*, 1899–1901.
10. Boer, T.; Mol, R.; Zeeuw, R.A.; Jong, G.J.; Sherrington, D.C.; Cormack, P.A.G.; Ensing, K. Spherical molecularly imprinted polymer particles: a promising tool for molecular recognition in capillary electrokinetic separations. Electrophoresis **2002**, *23*, 1296–1300.
11. Ye, L.; Mosbach, K. Polymers recognizing biomolecules based on a combination of molecular imprinting and proximity scintillation: a new sensor concept. J. Am. Chem. Soc. **2001**, *123*, 2901–2902.
12. Ye, L.; Surugiu, I.; Haupt, K. Scintillation proximity assay using molecularly imprinted microspheres. Anal. Chem. **2002**, *74*, 959–964.
13. Byström, S.E.; Börje, A.; Akermark, B. Selective reduction of steroid 3- and 17-ketones using LiAlH$_4$ activated template polymers. J. Am. Chem. Soc. **1993**, *115*, 2081–2083.
14. Flores, A.; Cunliffe, D.; Whitcombe, M.J.; Vulfson, E. Imprinted polymers prepared by aqueous suspension polymerization. J. Appl. Polym. Sci. **2000**, *77*, 1841–1850.
15. Matsui, J.; Okada, M.; Tsuruoka, M.; Takeuchi, T. Solid-phase extraction of a triazine herbicide using a molecularly imprinted synthetic receptor. Anal. Commun. **1997**, *34*, 85–87.
16. Matsui, J.; Fujiwara, K.; Ugata, S.; Takeuchi, T. Solid-phase extraction with a dibutyl-melamine-imprinted polymer as triazine herbicide-selective sorbent. J. Chromatogr. A. **2000**, *889*, 25–31.
17. Lai, J.P.; Lu, X.Y.; Lu, C.Y.; Ju, H.F.; He, X.W. Preparation and evaluation of molecularly imprinted polymeric microspheres by aqueous suspension polymerization for use as a high-performance liquid chromatography stationary phase. Anal. Chim. Acta. **2001**, *442*, 105–111.
18. Lai, J.P.; Cao, X.F.; Wang, X.L.; He, X.W. Chromatographic characterization of molecularly imprinted microspheres for the separation and determination of trimethoprim in aqueous buffers. Anal. Bioanal. Chem. **2002**, *372*, 391–396.
19. Yoshida, M.; Hatate, Y.; Uezu, K.; Goto, M.; Furusaki, S. Metal-imprinted microsphere prepared by surface template polymerization and its application to chromatography. J. Polym. Sci. Part A: Polym. Chem. **2000**, *38*, 689–696.
20. Zhu, D.W. Perfluorocarbon fluids: universal suspension polymerization media. Macromolecules **1996**, *29*, 2813–2817.
21. Mayes, A.G.; Mosbach, K. Molecularly imprinted polymer beads: suspension polymerization using a liquid perfluorocarbon as the dispersing phase. Anal. Chem. **1996**, *68*, 3769–3774.

22. Ansell, R.J.; Mosbach, K. Magnetic molecularly imprinted polymer beads for drug radio-ligand binding assay. Analyst **1998**, *123*, 1611–1616.

23. Norrlöw, O.; Glad, M.; Mosbach, K. Acrylic polymer preparations containing recognition sites obtained by imprinting with substrates. J. Chromatogr. A. **1984**, *299*, 29–41.

24. Wulff, G.; Oberkobusch, D.; Minarik, M. Enzyme-analogue built polymers, 18. Chiral cavities in polymer layers coated on wide-pore silica. React Polym. **1985**, *3*, 261–275.

25. Sulitzky, C.; Rueckert, B.; Hall, A.J.; Lanza, F.; Unger, K.; Sellergren, B. Grafting of molecularly imprinted polymer films on silica supports containing surface-bound free radical initiators. Macromolecules **2002**, *35*, 79–91.

26. Wang, H.Y.; Kobayashi, T.; Fujii, N. Surface molecular imprinting on photosensitive dithiocarbamoyl polyacrylonitrile membranes using photograft polymerization. J. Chem. Tech. Biotechnol. **1997**, *70*, 355–362.

27. Ruckert, B.; Hall, A.J.; Sellergren, B. Molecularly imprinted composite materials via iniferter-modified supports. J Mater Chem **2002**, *12*, 2275–2280.

28. Plunkett, S.D.; Arnold, F.H. Molecularly imprinted polymers on silica: selective supports for high-performance ligand-exchange chromatography. J. Chromatogr. A. **1995**, *708*, 19–29.

29. Kempe, M.; Glad M.;Mosbach, K. An approach towards surface imprinting using the enzyme Ribonuclease A. J. Mol. Recogn **1995**, *8*, 35–39.

30. Burow, M.; Minoura, N. Molecular imprinting: Synthesis of polymer particles with antibody-like binding characteristics for glucose oxidase. Biochem. Biophys. Res. Commun. **1996**, *227*, 419–422.

31. Hirayama, K.; Sakai, Y.; Kameoka, K. Synthesis of polymer particles with specific lysozyme recognition sites by a molecular imprinting technique. J. Appl. Polym. Sci. **2001**, *81*, 3378–3387.

32. Quaglia, M.; Lorenzi, E.; Sulitzky, C.; Massolini, G.; Sellergren, B. Surface initiated molecularly imprinted polymer films: a new approach in chiral capillary electrochromatography. Analyst **2001**, *126*, 1495–1498.

33. Otsu, T.; Masatoshi,Y. Role of initiator-transfer agent-terminator (iniferter) in radical polymerizations: polymer design by organic disulfides as iniferters. Makromol. Chem. Rapid. Commun. **1982**, *3*, 127–132.

34. Otsu, T. Iniferter concept and living radical polymerization. J. Polym. Sci. A.: Polym. Chem. **2000**, *38*, 2121–2136.

35. Yilmaz, E.; Ramstroem, O.; Moeller, P.; Sanchez, D.; Mosbach, K. A facile method for preparing molecularly imprinted polymer spheres using spherical silica templates. J. Mater. Chem. **2002**, *12*, 1577–1581.

36. Yilmaz, E.; Mosbach, K.; Haupt, K. The use of immobilized templates—A new approach in molecular imprinting. Angew. Chem. Int. Ed. **2000**, *39*, 2115–2118.

37. Titirici, M.M.; Hall, A.J.; Sellergren, B. Hierarchically imprinted stationary phases: mesoporous polymer beads containing surface-confined binding sites for adenine. Chem. Mater. **2002**, *14*, 21–23.

38. Glad, M.; Reinholdsson, P.; Mosbach, K. Molecularly imprinted composite polymers based on trimethylolproprane trimethacrylate (TRIM) particles for efficient enantiomeric separations. React. Polym. **1995**, *25*, 47–54.

39. Tsukagoshi, K.; Yu, K.Y.; Ozaki, Y.; Miyajima, T.; Maeda, M.; Takagi, M. Surface imprinting: preparation of metal ion-imprinted resins by use of complexation at the aqueous-organic interface. In *Molecular and Ionic Recognition with Imprinted Polymers*; Bartsch, R.A., Maeda, M., Eds.; American Chemical Society: Washington, DC, 1998, 251–263.

40. Carter, S.R.; Rimmer, S. Molecular recognition of caffeine by shell molecular imprinted core-shell polymer particles in aqueous media. Adv. Mater. **2002**, *14*, 667–670.

41. Pérez, N.; Whitcombe, M.J.; Vulfson, E.N. Molecularly imprinted nanoparticles prepared by core-shell emulsion polymerization. J. Appl. Polym. Sci. **2000**, *77*, 1851–1859.

42. Pérez, N.; Whitcombe, M.J.; Vulfson, E.N. Surface imprinting of cholesterol on submic-
     rometer core-shell emulsion particles. Macromolecules , **2001**, *34*, 830–836.
43. Haginaka, J.; Takehira, H.; Hosoya, K.; Tanaka, N. Molecularly imprinted uniform-
     sized polymer-based stationary phase for naproxen. Comparison of molecular recogni-
     tion ability of the molecularly imprinted polymers prepared by thermal and redox
     polymerization techniques. J. Chromatogr. A **1998**, *816*, 113–121.
44. Haginaka, J.; Takehira, H.; Hosoya, K.; Tanaka, N. Uniform-sized molecularly
     imprinted polymer for (*S*)-naproxen selectively modified with hydrophilic external layer.
     J Chromatogr. A. **1999**, *849*, 331–339.
45. Aherne, A.; Alexander, C.; Payne, M.J.; Perez, N.; Vulfson, E.N. Bacteria-mediated
     lithography of polymer surfaces. J. Am. Chem. Soc. **1996**, *118*, 8771–8772.
46. Vorderbruggen, M.A.; Wu, K.; Breneman, C.M. Use of cationic aerosol photopolymeri-
     zation to form silicone microbeads in the presence of molecular templates. Chem. Mater.
     1996, *8*, 1106–1111.

18

Molecularly Imprinted Polymer Films and Membranes

Mathias Ulbricht Lehrstuhl für Technische Chemie II,
Universität Duisburg-Essen, Essen, Germany

I. INTRODUCTION

The development of MIPs with a "well-defined size and shape" is very important for extending the fundamental understanding of MIP formation and recognition by MIPs as well as the practical applications of MIPs. The controlled geometry can help to reduce the problems encountered due to the heterogeneity of MIP binding sites including their uneven distribution in "conventional" cross-linked bulk MIPs.

Among the numerous reviews about MIPs, only one had been concerned explicitly with MIP membranes [1]. The intention of this paper is to extend this first attempt including an update about the most recent research, and to discuss comprehensively the work devoted to preparation and characterization of MIP films and membranes with a particular focus onto general implications of the results with those MIPs of well-defined shape.

II. DEFINITIONS AND CONCEPTS

A *film* is a "relatively thin" flat layer or sheet. With respect to the actual dimensions one can distinguish:

- *"ultrathin"* *films* having molecular dimensions (e.g., self-assembled (SA) monolayers or Langmuir–Blodgett (LB) films) up to a few nm;
- *"thin"* *films* with a thickness from a few nm up to a few µm;
- *"thick"* *films* having a thickness of up to a few 100 µm.

For "thick" films, the flat three-dimensional shape requires a macroscopic area thus also enabling their assembly in technical devices such as, e.g., diffusion or filtration cells. "Thin" or "ultrathin" films, however, can also be considered relatively "flat" if they are the wall of a hollow fibre, a microsphere, or even the surface layer of a porous particle (see Scheme 1).

The internal structure of films may be nonporous or porous. Depending on the mechanical stability as a function of internal structure and thickness, either self-supported or supported films can be distinguished.

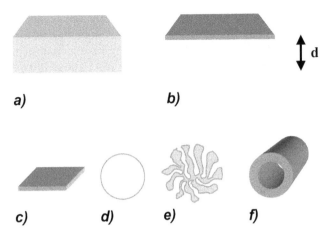

Scheme 1 Shapes of MIP films (characteristic thickness, d=50–500 μm): thick films can have (a) a homogeneous, or (b) a composite structure (cross-section); thin films can be (c) homogeneous and flat, (d) the wall of a capsule, (e) the surface layer of a porous particle, or (f) the wall of a hollow-fiber.

A *membrane* is an interphase between two phases, explicitly excluding simple interfaces. Hence, topologically, all films—"ultrathin", "thin", or "thick"—may also be considered as membranes. More specifically, a membrane is a barrier which separates (organizes) a system into compartments, and which controls or even regulates the exchange between these compartments. The driving force for transport is a difference in the chemical potential across the membrane. Of course, the barrier properties will depend very much on the membrane's structure. The major mechanisms for selectivity in biological membranes are transport via carriers and through channels, both controlled by receptor-mediated signals.

Synthetic separation membranes are either nonporous or porous. For nonporous membranes, permeability and selectivity are based on a solution-diffusion mechanism; examples for technical membrane separations are gas separation, reverse osmosis, or pervaporation. For porous membranes, either diffusive or convective flow can yield a selectivity based on size, for larger pore sizes typically according to a sieving mechanism; examples for technical membrane separations are dialysis, ultrafiltration, or microfiltration. It is important to note that additional interactions between permeand and membrane, e.g., based on ion exchange or affinity, can change the membrane's selectivity completely; membrane adsorbers with a pore structure of a microfiltration membrane are an example.

With synthetic membranes, carrier transport has been realized predominantly in supported liquid membranes. Artificial membranes with synthetic transport channels are still far from any practical relevance.

In order to achieve a high membrane performance, both selectivity and permeability should be high. However, the function of a membrane depends also on its structural integrity and stability, and hence limitations exist for reducing membrane thickness to reduce membrane resistance and increase permeability. This is not a critical problem in microfiltration where the pores enable a size-based selection for species between about 100 and a few 1000 nm, and consequently the technical microfiltration membranes or membrane adsorbers have a thickness of 50–500 μm. However, many other membranes used in membrane technology have a composite structure, for example, a thin nonporous or microporous layer (e.g., about 100 nm thick) on top of a macroporous support membrane (e.g., about 100 μm thick; cf. Scheme 1).

An *MIP film or membrane* is a film or membrane either composed of an MIP or containing an MIP. Hence, specific binding of the template can occur on or in the film or membrane. As a consequence of the *binding selectivity* of the template vs. other species, four categories of MIP film or membrane function can be distinguished (see Scheme 2).

i. The template binding to the MIP site can be monitored directly (by mass, density, or refractive index—quartz crystal microbalance (QCM), surface plasmon resonance (SPR); by spectroscopic properties of the template—fluorescence, UV, or IR spectroscopy, either ex situ or in situ; or by enzyme activity) or indirectly (based on competitive binding between template and a template conjugate with a label, e.g., enzyme activity, fluorescence). All these effects can be used for template detection and quantification in a sensor system; the MIP film is the receptor only.

ii. The template binding can induce local changes of the MIP site or in its immediate vicinity (changed UV or fluorescence spectrum of the MIP, e.g., via changed MIP site conformation or local swelling by complex formation with the template). The MIP film can be used in a sensor system and combines receptor and

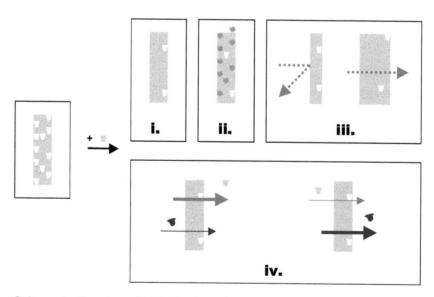

Scheme 2 Functions of MIP films by various responses to specific template recognition in a receptor layer: (i) direct or indirect monitoring of template binding; (ii) monitoring of template binding via local changes of the MIP site; (iii) monitoring of template binding via induced changes of the entire MIP film; and (iv) coupled template binding and membrane permeation selectivity.

transducer functions. The transducer function, however, does not rely on a membrane effect.

iii. The template binding can change the properties of the entire MIP film or membrane, thus "triggering" changes of, e.g., film or membrane permeability (by changed swelling) or conductivity (by changed ion permeability). The effect will be specific with respect to the receptor function, the resulting "general" response can be used as transducer in a sensor system. This is a membrane effect, but without a permeation selectivity.

iv. The template binding to an MIP in a film or membrane can be coupled with a permeation selectivity, thus enabling a membrane separation. The transport "pathways" in the membrane can be either the free volume between polymer chains, the solvent fraction of a swollen polymer gel, or connected pores in a solid polymer. Having MIP binding sites present in the membrane, two major mechanisms for selective template transport can be regarded:

- facilitated permeation driven by preferential sorption of the template due to affinity binding—no or slower transport of other solutes,
- no or retarded permeation due to affinity binding—faster transport of other solutes, until a saturation of MIP sites with template is reached.

In the first case, depending on the membrane structure as well as MIP site concentration and distribution, transport can occur via solution-diffusion or carrier-mediated ("facilitated") transport. In the second case, due to the saturation behavior, separation efficiency will be mainly determined by MIP binding capacity. Because selectivity is imparted by specific adsorption, those MIP membranes can also be considered as membrane adsorbers. In both cases, control of porosity and hence accessibility of imprinted sites is a main precondition for achieving efficient and selective transport through MIP membranes.

III. PREPARATION METHODS

A. General Considerations

Strategies towards MIP films and membranes are adapted either from film/membrane formation technology or from knowledge about preparation of MIPs. Thus, the current research and development activities in this field can be described as attempts to "merge" two more or less established technologies. Consequently, empirical work still plays an important role.

A general problem of the MIP technology—as it had been established during the last decades—is linked to the simultaneous—and widely random—formation of the imprinted "receptor" sites and the polymer matrix. A certain amount of matrix material is required to "host" and stabilize the receptor sites. This polymer can form a nonporous or a porous structure; in the first case receptor sites on the solid surface determine MIP performance, in the latter case pore size and pore size distribution as well as pore connectivity play an additional role. Random distribution and uneven accessibility of imprinted "receptor" sites in the (three-dimensional) volume of an MIP material—typically particles—are characteristic for the state of the art.

Another important issue is related to the three-dimensional nature of the imprinted sites itself. With MIP films—in a two-dimensional projection islands of binding sites in a matrix—the impact of the "third dimension" can be studied. On the other hand, for larger templates, sufficient accessibility—for template removal and rebinding—can only be achieved with adapted three-dimensional MIP film structures, typically on support surfaces [2,3].

Consequently, MIP films have a particular potential if their thickness ("third dimension" of a MIP material) can be well controlled:

- "Ultrathin" MIP films are the boundary ("two-dimensional") case, with maximum accessibility, but low binding capacity per exposed area; immobilized/stabilized monolayers require the lowest amount of "matrix material" per amount imprinted sites.
- "Thin" MIP films may provide relatively unhindered access to binding sites, with relatively low binding capacities; the larger content of matrix material as compared with "ultrathin" films can yield a better stabilization.
- "Thick" MIP films will always be limited by uneven accessibility of imprinted sites; the advantage is the (potentially) much higher binding capacities as compared with thinner films; ideally, "thick" MIP films should have a homogeneous internal structure enabling an evaluation based on integral models for binding.

In order to obtain MIP films or membranes with "well-defined size and shape" (cf. Section I), three strategies can be envisioned:

1. sequential approach—preparation of films or membranes from previously synthesized "conventional" MIPs (i.e., particles);
2. simultaneous formation of MIP structure and film or membrane shape;
3. sequential approach—preparation of MIPs on or in films or membranes with suited shape.

For strategy (1), there are only a few examples yet, such as embedding MIP particles in a film, e.g., with a polymer used as "glue" [4], or creating a "flat three-dimensional" arrangement of MIP particles, e.g., as a filter cake on a membrane or immobilized between two films, membranes, or frits [5]. Until now this route yielded mainly "thick" MIP "films" with low integrity and homogeneity, and thus with poorly defined shape.

Consequently, strategies (2) and (3) will form the core of this review. "Ultrathin" films are typically prepared and characterized on a solid support. "Thin" films may be either supported or free standing. "Thick" films are mainly free standing. However, binding inside a macroscopically "thick" MIP film will take place in an "ultrathin" or "thin" layer on the accessible internal surface. Therefore, on a microscopic level, strategies (2) and (3) will need to fulfill similar criteria with respect to the formation of MIP layers. For strategy (2), it is necessary to achieve macroscopic continuity—the "well-defined shape"—and MIP formation in one synthesis step. The key for success is giving or preserving the desired "flat" shape for the reaction mixture until solidification. For strategy (3), it is the aim to cover the support film or membrane externally or internally with a well-defined "ultrathin" or "thin" MIP film.

B.   Main Synthesis Strategies for MIP Films and Membranes

The repertoire of strategies (see Scheme 3) involves on the one hand the adaptation of the main conventional or "alternative" MIP preparations to thin-film formation (see below), and on the other hand, the adaptation of state-of-the-art thin-film preparation technologies to the synthesis of MIPs (see Section III.C).

1.   In Situ Cross-Linking Polymerization

When established MIP synthesis protocols shall be applied, the control of MIP film thickness as well as the "synchronization" of imprinting and film solidification are of critical importance for the MIP film's structure, shape, and function. It is important to note that during cross-linking polymerization (cf. [6]) several stages of imprint and polymer structure formation—along with establishment of the hierarchical pore structure—are passed, starting with primary aggregates (diameter between 5 and 30 nm, yielding micropore diameters of < 2 nm), followed by aggregation to microspheres (diameter 60–500 nm; yielding mesopore and macropore diameters of 2–200 nm) and eventually concluded by the formation of macroscopic particles [7]. Hence, the intrinsic dimensions of "conventional" MIP materials are in the range of thin MIP films' thickness.

One noteworthy exception with strategy (1) (cf. Section III.A) is the work of Lai et al. [8] who claim the preparation of thin layers of MIPs for theophyllin, caffein, and xanthin by coating a "solution" of the previously prepared MIP—via in situ cross-linking polymerization of the "standard" MAA/EDMA system—on the sensitive

Preparation method	applicable (+) / already applied (✳) to the preparation of:			
	ultrathin MIP films	**thin supported MIP films**	**self-supported MIP films**	**Composite MIP films**
Langmuir-Blodget layer	✳			+
Self-assembled monolayer	✳			+
Self-assembled multilayers	✳	+	(+) *with post-crosslinking*	+
in situ polymerization	(✳) *with surface initiation*	✳ *	✳	✳
Sol/gel process	✳	✳ *	✳	✳
Grafting reactions	✳	✳		✳
Polymer solution phase inversion		✳ *	✳	
Polymer solution casting / leaching		✳	✳	

Scheme 3   Preparation strategies towards MIP films and membranes. *very thin—down to 10–20 nm—and flat films in combination with spin-coating as initial step.

layer of a SPR sensor. Because a specific sensor response was obtained, the thickness of the MIP film must have been less than about 100 nm.

2. Sol–Gel Processes Towards Inorganic and Inorganic/Organic Hybrid Materials

The first demonstration of molecular imprinting had been the synthesis of a silica network by a sol–gel process, which then showed selective binding of the template. However, thereafter it appeared that the precision with respect to the creation of specific molecular recognition sites in bulk inorganic materials seemed to be limited compared to organic polymers. As a consequence, the imprinting attempts with purely inorganic materials have been very much focussed onto creating micropores well-defined in size and shape by using templates in sol–gel technologies [9]. Following that road, there had been also successful preparations of inorganic microporous membranes (with pore sizes below 2 nm) for selective separation and catalysis [9]. Furthermore, large-template (e.g., bacteria) directed sol–gel synthesis of inorganic materials could be used for the preparation of macroporous materials with well-defined macroscopic structure including a hierarchical porosity [10].

Recently, thin inorganic/organic hybrid films have been prepared on supports via the sol–gel technique. For example, porous films imprinted for dopamin [11] or propanolol [12] were obtained as follows: a mixture of alkyl and aryl trimethoxysilanes with tetramethoxysilane in ethoxy ethanol/water/hydrochloric acid was prepared—conditions to achieve a controlled, partial hydrolysis and condensation producing the "sol"—, after template addition spin-coated onto a glass surface, dried—thus producing the "gel"—and finally washed to remove the template. The MIP recognition was based on the microporous cross-linked siloxane network with hydroxyl—from hydrolysis of the reactive silane—and organic functional groups. For DDT, a template that lacks functional groups for strong intermolecular interactions with the matrix, the noncovalent approach was not successful, only a covalent approach using a "sacificial spacer" yielded an MIP film with DDT binding selectivity [13].

3. Polymer Solution Phase Inversion—"Alternative Imprinting"

The formation of membranes from synthetic polymers is typically achieved by the phase inversion (PI) process, which starts with a stable solution of the polymer which is then subjected to controlled demixing [14]. As a result, a porous structure is obtained where the polymer-rich phase forms the matrix of the membrane. The demixing of a previously formed liquid film (either flat or hollow-fiber; cf. Scheme 1) can be achieved by two main processes:

- "wet PI"—at least three components are used: a polymer, a solvent, and a non-solvent, where the solvent and the nonsolvent are miscible with each other; the liquid film of the polymer solution is brought into contact with the precipitation bath containing the nonsolvent (immersion precipitation process).
- "dry PI"—either by evaporation of the solvent or/and changing the temperature.

These phase inversion processes can be applied for molecular imprinting, i.e., the solidification of an existing polymer is used instead of an in situ polymerization. Similar work had been reported before for bulk materials [15,16]. Fundamental investigations towards this "alternative imprinting", with the preparation of MIP membranes as main example, had been done by Yoshikawa and coworkers [17–28] ("dry PI") and Kobayashi and coworkers [29–37] ("wet PI"; cf. Section III.D).

Recently, "alternative imprinting" had also been used for preparing thin films—without a membrane function (cf. Scheme 2)—on supports. For example, porous PAN-co-P4Py films—imprinted for caffein via "wet PI" with water as precipitation bath—showed a specific response for caffein from aqueous solution using an QCM [35]. For PVC-co-PAA as the functional matrix polymer, imprinting for metal ions in porous films was accomplished by using a mixture of volatile solvent and a nonsolvent acting as porogen (THF and water, respectively) for film casting and subsequent solidification due to solvent evaporation [38]. Polyphosphazene films imprinted for the antibiotic rifamycin SV were prepared via the "dry PI" [39].

C. Syntheses of Supported MIP Films

1. Ultrathin Films by Self-Assembly

By self-assembly at interfaces, it is possible to prepare ultrathin and structurally well-defined layers [40]. The mechanism is based on (lateral) supramolecular interactions in one two-dimensional layer providing organization and stabilization; and in addition, molecular interactions or reactions between adjacent layers can be used to create organized and stable three-dimensional structures.

(a) Self-Assembled Monolayers. SA or LB monolayers are not necessarily suitable for molecular imprinting because the adaptable structural modification that is required for the template imprinting process is, in principle, not compatible with the ordered (supra) molecular structure. On the other hand, self-organization of nano structures depends on flexibility of the molecular building blocks, hence it is an intrinsic problem of molecular imprinting to "fix" an adapted structural modification.

Two mechanisms can be used for recognition based on size and shape as well as functionality:

- Well-defined defects or pores ("perforations") in the monolayer.
- Well-defined 2D arrangement of exposed functional groups ("foot-prints").

According to the first approach, imprinted SA films were prepared from alkyl or aryl thiols on gold [41–43]; a stable, "floating" receptor structure based on molecular "spreader bars" had also been proposed [43].

The second approach is only possible with a fixation of the lateral flexibility. Imprinted cross-linked LB layers had been prepared from two polymerizable amphiphils, one of which had a boronic acid group as binding site for recognition of a sugar derivative as template [44]. It should be noted, that a self-assembled layer of an amphiphilic organophosphorous extractand had also been imprinted for enantioselective recognition of an aminoacid at the surface of polymer particles obtained by a cross-linking polymerization of an emulsion [45].

(b) Self-Assembled Multilayers. A step-wise sol–gel reaction—via chemisorption and subsequent activation—towards ultrathin MIP films had been developed by Lee et al. [46]: porous cross-linked titan dioxide gel films with precisely controlled thickness in the nm range were thus prepared and the templates included azobenzene carboxylic acid [46], chloroaromatic acids [47], as well as various chiral carboxylic acids [48].

A step-by-step deposition, such as the noncovalent polyelectrolyte multilayer approach using alternating adsorption of polyelectrolytes with opposite charge according to Decher [49], had also been adapted to MIP layer syntheses [50]: starting with a gold electrode anionically modified by 2-carboxyl ethanthiol, first a polycation (PDADMAC) and then a polyanion—a PAA copolymer containing amide-linked phenyl boronic acid functions as binding group—were sequentially adsorbed in the presence of AMP as the template. Note that this approach is similar to the "alternative imprinting" because presynthesized functional polymers are used (cf. Section III.B); however, a high degree of order in the "third dimension" of the film is achieved by self-assembly.

2. Thin Films by In Situ Polymerization Processes

Besides attempts with sol–gel and phase inversion processes (cf. Section III.B), the majority of work has been devoted to the most established MIP synthesis technology. In situ cross-linking copolymerization in a film of a liquid reaction mixture yields a "flat" MIP material (thin or thick MIP film). It should be noted that this is also similar to the so-called "in situ preparation" of MIPs,* e.g., established by Takeuchi et al. [51] ("thinly coated vials") and Lanza and Sellergren [52] ("mini-MIPs") for screening of MIP libraries. Depending on the monomer mixture, these materials are either flat monoliths or cross-linked and agglomerated particles.

More control of macroscopic shape can be achieved either by controlling the liquid film thickness of the reaction mixture—by enclosing between two supports, or casting or spin coating on a support—and ensuring high conversion or controlling the conversion of a deposition reaction from a solution to the support.

(a) Reactive Coating. In most cases, acrylate-based mixtures had been used at high monomer conversion to prepare MIP films for a great variety of templates [53–57]. Enclosing the reaction mixture between glass plates ("sandwich") along with UV initiation is very versatile because evaporation from the reaction mixture is prevented, and consequently film thickness, homogeneity, and morphology can be controlled. For example, Haupt et al. [54] prepared films with a thickness of about 2 µm, imprinted for S-propanolol, from an MAA/TRIM monomer mixture in toluene with AIBN as initiator. In a comparative study, for amine-imprinted MIP films, both casting (PI) of PAA and in situ polymerization of AA gave similar results, but due to the simple template structure, the MIPs were considered only size selective [56].

Dickert et al. [58–61] had introduced the in situ synthesis of polyurethanes as very versatile system for the preparation of thin MIP films, especially due to the excellent film forming properties of the polymer. These materials had been especially successful for imprinting of templates without a distinct functional group, i.e., recognition was based mainly on shape complementarity. This system could also be useful for imprinting large templates such as cells [61].

For MIP deposition at relatively low monomer conversion, other polymerization mechanisms had also been used. For example, a functional polyaniline layer with about 100 nm thickness, coating the bottom of polystyrene microplates, was obtained from 3-aminophenylboronic acid in aqueous buffer solution, initiated by chemical

---

*See chapters 17–19.

oxidation with persulfate; the method had been successfully applied to the imprinting of proteins [62].

(b)   Grafted Reactive Coating.   An improved adhesion or a covalent fixation to the support as well as a better control of the reaction can be achieved by reactive groups immobilized on the support surface: thus, growing polymer chains can become attached to the surface ("grafting to"; cf. Section III.C.3). Under optimized preparation conditions, such "grafted reactive coating" can yield an MIP film with an appropriate thickness to allow a three-dimensional cross-linked structure to recognize the template and at the same time ensure its accessibility.

There has been work by various groups devoted to this "surface imprinting" [63–72]. Silica [63–65] and polymer [66,71] particles had been modified, so that polymerizable groups were available which could react during a homogeneously initiated polymerization (see Scheme 4a). Particles had been imprinted also for the recognition of proteins [67,68].

When established protocols for bulk MIP syntheses were adapted for coating of flat surfaces, the MIP layers resemble the monolithic or particle aggregate structures obtained in the "in situ preparation" of MIPs (cf. above) [70]. However, homogeneous thin films could also be obtained: using the "sandwich" technology, imprinting of a SPR sensor surface was demonstrated, hence MIP film thickness was less than 100 nm [72].

The main problems with the approach are the required special support preparation before MIP synthesis, and that both imprinting efficiency and final film thickness depend on liquid layer thickness as well as reaction conditions, typically optimized

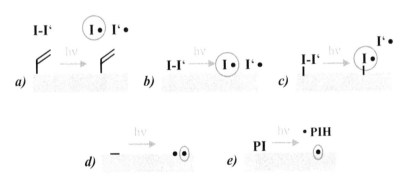

Scheme 4 Surface-selective initiation methods for the synthesis of thin films (photoreactions as example: radical as initiating species is labeled by a circle); (a) support surface is functionalized with a group which can react during MIP synthesis, e.g. via in situ radical polymerization initiated by cleavage of an initiator (1-1′) in the bulk of the reaction mixture; (b) an initiator is coated (adsorbed) to the support, the initiation takes place predominantly close to the support surface, thus facilitating MIP synthesis in a thin film coating the support; (c) an initiator group is covalently bound to the support, the initiation takes place exclusively on the support surface, and "grafting-from" yields a thin MIP film covalently bound to the support; (d) a direct excitation of a suited polymer support can produce initiating species at the surface, and MIP synthesis proceeds via "grafting-from" yielding a thin MIP film covalently bound to the support; (e) a special initiator is adsorbed to the support, and the initiation (hydrogen abstraction of the initiator PhI from the support) takes place exclusively on the support surface yielding a thin MIP film covalently bound to the support.

for a homogeneous bulk MIP preparation. Therefore, control of MIP film structure had been rather poor.

(c)   Surface-Selective Initiation of Reactive Coating.   Surface-selective initiation, i.e., providing a higher initiation/polymerization efficiency close to the surface as compared with the bulk of the reaction mixture, can also increase the performance of a coating procedure, but this is limited to certain monomers. The most straightforward example is an electropolymerization, initiated at electrodes which are then used for a sensor based on the MIP film as recognition element. Such polymers had been prepared from pyrrole [73], phenol [74], or *o*-phenylenediamine [75] and the films were very thin so that they allowed detection with capacity measurements.

Kochkodan et al. [76] had found that thin MIP layers could be prepared via a controlled polymer deposition process at low monomer conversion, using the very fast photoinitiation of a cross-linking polymerization by α-scission photo-initiator such as benzoin ethylether which had been previously coated on the surface (see Scheme 4b). This had been investigated with porous membranes as matrix which had been functionalized very evenly (see Fig. 1). With the macroporous MIP composite membranes, the membrane pore size could be used to "probe" grafted layer thickness, as shown by data from permporometry (see Fig. 2). For the same range of coverage with a cross-linked polyacrylate/acrylamide—estimated

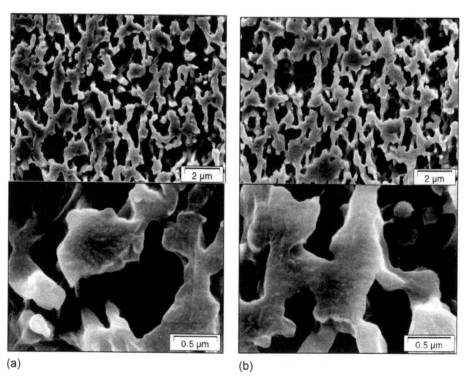

(a)   (b)

Figure 1 SEM micrograph of cross-section details for: (a) unmodified PVDF membranes (0.22 μm nominal pore size); (b) same membrane functionalized with photo-grafted AMPS/MBAA (DG = 400 μg/cm^2) [77a].

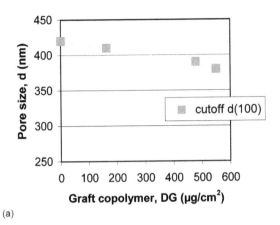

(a)

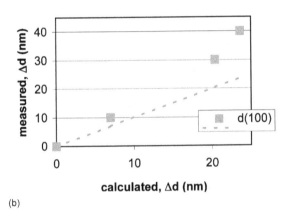

(b)

Figure 2 (a) Trans-membrane pore diameter (maximum value) measured with a Coulter permporometer for PVDF membranes (0.22 μm nominal pore size) functionalized to a different degree with photo-grafted AMPS/MBAA [77a]; (b) measured reduction of pore size (cf. a) as a function of the calculated value assuming an even coverage of the entire specific surface area of the PVDF membrane with grafted copolymer at negligible swelling.

thickness between 5 and 15 nm (cf. Fig. 2)—a template specificity was observed only for a medium degree of coverage and when the photo-initiator had been coated on the surface; a homogenous initiation did not yield a specific material (see Fig. 3). Hence, both *optimum MIP film thickness* and internal structure for molecular recognition by MIP sites can be adjusted by a surface-selective initiation.

3. Thin Films by Surface-Initiated Graft Copolymerization

Macromolecules or polymer layers can be covalently attached ("grafted") to surfaces of various materials by this means altering their interface structure and properties, including the interactions of the materials with their environment such as adhesion, adsorption, or molecular recognition. There are two approaches for preparing grafted surfaces [78]:

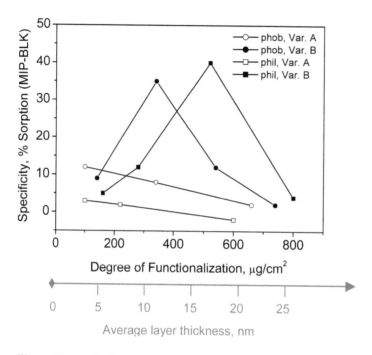

Figure 3 Imprinting effect for desmetryn as function of the degree of surface coverage with photo-grafted AMPS/MBAA for PVDF and precoated PVDF membranes (0.22 μm nominal pore size), prepared with ("Var. A") and without ("Var. B") coating of the photoinitiator onto the membrane surface (cf. Scheme 4b) [76]; estimated average MIP film thickness from permporometry (cf. Fig. 2) is included.

- direct coupling of existing polymers to the surface ("grafting-to");
- graft copolymerization of monomers onto the surface ("grafting-from").

To our knowledge, there is not any explicit example for surface imprinting via the "grafting-to" of presynthesized macromolecules (while "footprinting" with small molecules had been reported, cf. Section III.C.1). This may be due to the conformational flexibility of polymers. However, it may be envisioned that imprinted sites for larger species could be obtained by this route. Note, however, that the "grafted reactive coating" (cf. Section III.C.2) can also be described as "grafting-to" approach.

Advanced "grafting-from" approaches use molecular control of the interface reaction. Pioneering work had been done by Wang et al. [32], who had synthesized a copolymer with photo-initiator side groups, prepared a membrane by "wet PI", and finally performed a surface-initiated imprinting polymerization yielding an MIP membrane with a pronounced selectivity for Tho over Caf.

Covalent immobilization of (photo) initiator groups on the support surface (see Scheme 4c) had been done on silica particles using a silanization with a conjugate of an azo-initiator [79,80]. MIP functionalization with L-phenylalanine anilide as template and MAA/EDMA along with UV-initiation [79] was feasible without significant polymerization in the bulk of the reaction mixture, and the synthesized composite materials were superior compared to conventional MIP particles. Most remarkably, the enantioseparation performance (capacity and separation factor) was a function of the average grafted layer thickness: an *optimum MIP film thickness* was observed at about 7 nm. This was similar to the conclusions for surface-selective initiation by Kochkodan et al. [76] (cf. Section III.C.2). However, depending on the separation conditions, enantioselectivity could still be achieved with materials having an average MIP film thickness of less than 1 nm [79].

Using an intrinsic photo-reactivity of the support polymer (see Scheme 4d) is an alternative for initiation of a surface imprinting polymerization without an added photo-initiator. This "grafting-from" approach offers the highest surface selectivity of initiation, but it is limited to a few polymers. In first experiments with polysulfone, the imprinting efficiency had been rather low [77b].

Piletsky et al. [81] had found that using a coated hydrogen abstraction photo-initiator (see Scheme 4e) very thin MIP films, which were covalently anchored and covered the entire surface of the base material, could be synthesized by a photo-initiated cross-linking graft copolymerization. This approach had been first explored with benzophenone as photo-initiator and a membrane from polypropylene as support. MIP synthesis and recognition were possible in/from water, and significantly less cross-linker than with bulk preparations was necessary to obtain the highest template specificity. Both effects were explained by a contribution of the solid polymer support to the stabilization of the imprinted sites. The approach is very flexible because no premodification is necessary.

This last "grafting-from" method can be adopted to any support material with C–H bonds. Consequently, an analogous MIP functionalization had been used successfully for the grafting-from functionalization of gold electrodes covered with a self-assembled layer of an alkylthiol [82,83].

In conclusion, the "grafting-from" strategy had in the last few years been established as a most versatile approach towards thin-layer MIP materials, which is also very useful for further elucidation of MIP structure and recognition function.

4. Advanced MIP Layer Architecture by Sequential Grafting

"Surface imprinting" of latex particles is an approach to use functional groups present on the surface of a polymer and fix their conformation by a surface modification with another polymer [84,85]. An attempt had been made to reduce the non-specific binding by applying a hydrophilic coating after the MIP preparation [86].

Sergeyeva et al. [87] had discovered that a previously prepared hydrophilic layer on a support can have two functions: (i) forming a matrix for a cross-linking polymerization (via surface-selective initiator [87] (cf. Scheme 4e) or adsorbed "homogeneous" initiator [76] (cf. Scheme 4b) and limiting monomer conversion to "filling" the layer thus forming an interpenetrating network (IPN), (ii) minimizing nonspecific binding (see Scheme 5). A superior MIP material's performance, especially a high template specificity, can be achieved.

D. Preparation of Self-Supported MIP Films and Membranes

1. In Situ Cross-Linking Polymerization

Marx-Tibbon and Willner [88] had prepared an MIP membrane by a cross-linking copolymerization of a mixture of acrylamide and acrylate monomers including a photo-isomerizable functional monomer (merocyanine acrylate). The authors

Layer 1 –
hydrophilic, low-binding

Base polymer
hydrophobic

Scheme 5 Thin MIP composite films with high template affinity at low nonspecific background binding: the matrix for MIP synthesis is a hydrophilic, low binding layer 1 on the support. Only the fraction of the functional monomer (grey dots) in the reaction mixture, which before polymerization had been involved in the pre-organized complex (dark grey dots) with the template (large dots), and which could not fully penetrate into layer 1, was during cross-linking polymerization in layer 1 (IPN formation) fixed in MIP sites on top of the hydrophilic, low binding layer 1.

observed a "poor mechanical stability", obviously due to the swollen structure. Mathew-Krotz and Shea [89] had also prepared free-standing membranes by thermally initiated cross-linking copolymerization of one of the "standard" monomer mixtures (MAA/EDMA) for molecular imprinting. From SEM studies, a regular porous structure built up by 50–100 nm diameter nodules was discussed (cf. Section III.B).

Sergeyeva et al. [90,91] had introduced a very promising improvement by using an oligourethane–acrylate macromonomer in imprinting polymerization mixtures in order to increase the flexibility and mechanical stability of the membranes; self-supported MIP membranes with a thickness between 60 and 120 μm could be prepared. Sreenivasan [92] had also addressed the problem of film stability, and a semi-IPN film was prepared by the polymerization of HEMA in a blend with polyurethane.

Kimaro et al. [93] had prepared free-standing membranes by thermally initiated cross-linking copolymerization of styrene monomers followed by leaching of a polyester present as "pore former" at a concentration of 1.8 wt% in the reaction mixture. SEM pictures suggested the presence of isolated pores with diameters of up to 1 μm at a low density (< 2%). In line with permeation data, it could be speculated that trans-membrane channels had been obtained, induced by the presence of a removable macromolecular pore former in the reaction mixture.

Nevertheless, for achieving a higher separations performance, thin-film MIP composite membranes should be synthesized (see Section III.E).

2. Polymer Solution Phase Inversion—"Alternative Imprinting"

Yoshikawa and coworkers [17–22,24,26–28] were using specifically synthesized polystyrene resins with peptide recognition groups, in a blend with a matrix polymer, for the membrane formation via a "dry PI" process. The resulting membranes seemed to be microporous. The permeability was much higher for the MIP as compared with the blank membranes; hence, the low-molecular weight templates seemed to act also as a pore former.

This "alternative imprinting" had been extended to other chiral peptide side groups [27], and imprinting specificity was indeed influenced by structure and size of the recognition group. However, the matrix polymer was also involved, at least in the formation of the specific MIP micropore structure. Remarkably high molar ratios between template binding capacity and recognition group content in the membrane had been found, indicating a high efficiency of the peptides for chiral template recognition. With these resins, other templates where the recognition did not depend on chirality could also be imprinted [26].

The impact of solvent composition provided further important insights into MIP site structure and function: both increasing and decreasing the water content in an aqueous ethanol mixture decreased binding selectivity [28]. In the first case, the increasing contribution of ionic relative to directed hydrogen bonding interactions seemed to be the reason, a behavior typical for many MIPs. However, in the second case, it had been confirmed experimentally that the MIP membranes lost their "template memory" when exposed to a too organic environment where swelling and chain rearrangement seemed to "erase" the imprinted information.

Kobayashi and coworkers [29–31,33,35] were using functional acrylate copolymers for a "wet" Pi process. Asymmetric porous membranes were obtained. The capacity and selectivity increased with higher content of template (theophyllin) in the casting solution; for the best membranes, however, 4.7 wt% Tho at 10 wt% polymer had been used [30]. Interestingly, a very pronounced effect of fixation temperature onto binding capacity and selectivity (batch binding) had been observed; for PI at $40°C$: $n_{Tho} = 0.26\,\mu mol/g$ and $\alpha_{Tho/Caf} = 11$, for PI at $10°C$: $n_{Tho} = 1.25\,\mu mol/g$ and $\alpha_{Tho/Caf} = 52$ [31]. Remarkably, in single solute experiments, with these Tho-MIP membranes no Caf binding could be detected. On the other hand, imprinting for Caf was not successful because these membranes showed the same binding for Tho and Caf. Hence, the preferential binding of Tho relative to Caf to the membrane polymer via ion exchange could not be overcompensated.

In the last years, the selection of polymers had been extended to most of the commonly used membrane materials: cellulose acetate [25], polyamide [34,36], polyacrylonitrile [37], polysulfone [26,37], and modified polysulfone [23], but also including polystyrene and PVC [37], or PVC-co-PAA [38]. The exceptions are the hydrophobic—and almost nonfunctional—polymers (polyolefines, PVDF, or Teflon). However, because both recognition sites and pore structure are "fixed" at the same time within the same material, a comparison of the efficiency of different MIP membranes, and thus polymer materials, was rather complicated [26].

Nevertheless, in a comparative study, Reddy et al. [37] had found that the affinity of MIP membranes for dibenzofuran made from common synthetic polymers showed the following order: PVC > PSf > PSt > PAN (binding from methanol), while for all polymers higher affinities as compared to blank membranes had been observed.

The formation of porous imprinted membranes from a compatible blend of a matrix polymer—for adjusting pore structure—and a functional polymer—for providing binding groups—is currently being explored using cellulose acetate and weakly sulfonated polysulfone, respectively, and rhodamine dyes as the template [94].

An important conclusion is that the solidification of a presynthesized polymer is indeed an alternative to the commonly used in situ cross-linking polymerization for MIP synthesis (cf. Section III.B). It is most remarkable that all MIPs prepared via "alternative imprinting" had at least acceptable binding performance in aqueous media. For MIPs via "wet PI" with water as precipitation bath, this is related to the preparation conditions, but also the MIPs via "dry PI" from purely organic solvents were at least well adapted to aqueous/organic conditions (cf. above).

However, as with the "conventional" bulk MIP materials, the limited accessibility of MIP binding sites due to a random distribution inside and on the surface of the bulk polymer phase remains one of the major unsolved problems. Thus, the advantage of membrane preparation technology to provide well-defined pore structures is not yet fully exploited for the preparation of self-supported micro and macroporous MIP membranes.

E. Preparation of MIP Composite Membranes

Composite membranes allow to adjust MIP recognition and pore structure by two different materials. All preparation strategies for supported MIP films (cf. Scheme 3) could be applied for achieving MIP-specific separation via successful adaptation to the support membrane structure.

MIP "plugged" porous supports could be one description of the first MIP "membranes" [1,95–97], because established MIP reaction mixtures (e.g., MAA/EDMA) had been polymerized on and in glass filters (e.g., 4 mm thick) [98]. This approach had been extended to other templates as well as to a comparsion of covalent and noncovalent approach [98]. No further characterization of MIP composite structures was reported.

Thin-film MIP composite membranes (cf. Scheme 1b), imprinted for theophyllin and caffein, had been prepared by Hong et al. [99], using photo-copolymerization of a MAA/EDMA mixture on top of an asymmetric 20 nm pore size alumina membrane. Additional gas permeation studies suggested that the membranes were defect ("pinhole") free.

Pore-filling MIP composite membranes had been first prepared by Dzgoev and Haupt [100]. They casted the reaction mixture into the pores of a symmetric microfiltration membrane from polypropylene (cutoff pore size 0.2 μm) and performed a cross-linking copolymerization of a functional polyacrylate for imprinting protected tyrosine. Hattori et al. [101] had used a commercial cellulosic dialysis membrane (Cuprophan) as matrix and applied a two-step grafting procedure by, (i) activation of the cellulose by reaction with 3-methacryloxypropyl trimethoxysilane from toluene in order to introduce polymerizable groups into the outer surface layer, (ii) UV-initiation of an in situ copolymerization of a typical reaction mixture (MAA/EDMA, AIBN) for imprinting theophylline.

Thin-layer MIP porous composite membranes had been specifically developed to achieve high performance as MIP membrane adsorbers [76,81,87] (cf. Sec. V.D). All approaches shown in Scheme 4 could be used to internally coat porous membranes evenly and reproducibly; the specific examples had been already discussed (see Sections III.C.2, III.C.3, III.C.4). Furthermore, the structure of the base membrane can be used as a means to adapt both pore size—permeability—as well as internal surface area—binding capacity—to the desired application.

IV. MIP FILMS AS RECEPTOR LAYERS

A chemo- or biosensor is a system consisting of a receptor coupled with a transducer to a detector thus enabling the conversion of a chemical signal—binding to the receptor—into a physical signal. Regarding the function of the sensor system, the simplest task of an MIP film could be immobilizing the receptor in close contact to the detector [102].

From the four categories of MIP film function, i., ii. and iii. could be used (cf. Scheme 2). While MIP receptor films according to categories i. and ii. provide a direct response to the template binding (see Section IV.A), in films according to category iii. the response is triggered by the structure of the entire film (see Section IV.B). Here, the discussion will be limited to the impact of film shape—supported, self-supported or composite MIP films (cf. Section III)—and thickness, while other aspects of MIP-based sensors or assays will covered in detail elsewhere.[†]

A. Direct Response to Template Binding

In most of the systems described until now, the MIP film had the function to capture specifically the analyte and thus to enable its straightforward detection ("category i."; Scheme 2). For a detection using transmission of the MIP film, its homogeneity and transparency must allow retrieving the information with minimum interference. For a surface-selective detection, the MIP film thickness must be adapted carefully to the sampling depth of the method which is, e.g., about 100 nm for SPR, and around 1 μm for ATR-IR.

If a very sensitive detection is available, even binding to ultrathin, monolayer films could be sufficient [44]. Alternatively, film thickness and its internal structure, especially porosity, are important for enhancing the accessible MIP binding site capacity and, consequently, the sensitivity of the detection. Note, however, that porosity can also reduce the film's transparency and thus reduce the signal-to-noise ratio. Hence, in most cases, it is the aim to optimize film structure and thickness for suited binding capacity in the MIP film.

Using QCM detection, a wide variety of conditions with respect to MIP material, preparation method and film thickness had been applied successfully: few nm thin titan dioxide imprinted for azobenzene derivatives [46] (cf. Section III.C.1), 10 nm thin electropolymer imprinted for glucose [75] (cf. Section III.C.2), 100 nm thin PUR imprinted for polar organic solvents [59] (cf. Section III.C.2), 2 μm thin PMAA-co-TRIM imprinted for S-propanolol [54] (cf. Section III.C.2), 50 μm thick porous PAN-co-AA imprinted for caffein [35] (cf. Section III.D.2), and 30 μm thick macroporous nylon imprinted for L-glutamin [36] (cf. Section III.D.2).

Using SPR detection, a method where sensitivity increases with the size of the bound analyte, about 100 nm thick PTMAEMA-co-HEMA-co-EDMA films (cf. Section III.C.2) had been used to detect the relatively high-molecular weight analyte ganglioside GM_1 while the actual template sialic acid gave no response [72]. Surprisingly, Lai et al. [8] were able to detect the low-molecular weight analyte caffein using their spin-coated crosslinked MIP (cf. Section III.B).

Using fluorescence detection with 3 μm thick PUR films imprinted for polycyclic aromatic hydrocarbons, the obtained high sensitivity had been explained with binding to MIP sites in the bulk of the MIP film [59] (cf. Section III.C.2). The binding

---

[†]See chapters 25–27.

selectivity of 10 μm thick porous PVC-co-PAA films imprinted for copper ions had been quantified using ex situ IR detection [38] (cf. Section III.D.2).

Among the very few examples for chemical signal amplification after analyte binding to the MIP site, a secondary reaction in a competitive ELISA-analogous assay had been established with a "glued" MIP particle film [4]. Furthermore, with an about 100 nm thin protein-imprinted functional polyaniline film, the enzyme activity of bound peroxidase was used to retrieve quantitative data about MIP site capacities [62].

Only few first attempts had been reported about the integration of transducer structures next to the receptor site in MIP films ("category ii."; Scheme 2). MIP films of 300 nm thickness, synthesized by either in situ polymerization of AA or PAA phase inversion brought MIP sites close to a photo-luminescent Cd/Se semiconductor electrode so that amine template binding could be monitored by changed luminescence [56]. The intercalation of a fluorescent dye into 300 nm thick PUR films imprinted for organic vapors enabled the construction of an optrode where quenching could be measured as a function of binding to the MIP sites [58]. A statistical distribution of an environmentally sensitive fluorescence probe in sol/gel films imprinted for DDT had yielded very limited success, further development of synthesis strategies for an addressed incorporation will be necessary [13].

B. Triggered Response to Template Binding

Larger effects can also be triggered by template binding to MIP sites, thus potentially enabling a signal amplification mechanism ("category iii."; Scheme 2). Here, internal structure of the MIP film, including its porosity, is of main importance.

Investigations of ultrathin MIP monolayers (cf. Section III.C.1) with voltammetry suggested that imprinted sites—"perforations" in an insulating matrix—could discriminate the transport of different redox active molecules to the electrode [41]. Furthermore, the binding to imprints in SA monolayers, created by the "spreader-bar" technique, resulted in substance-specific capacity changes [43].

A very efficient transduction after recognition was based on a significant swelling, proportional to the amount of bound template AMP, of an MIP film that had been synthesized via step-wise self-assembly coating [50] (cf. Section III.C.1). Sallacan et al. [57] recorded the response of an MIP film from a polyacrylamide with phenylboronic acid groups (cf. Section III.C.2) to the binding of imprinted nucleotides and sugars by three different methods, QCM, impedance, and potentiometry (ISFET): while microgravimetric and electrochemical detection were based on the swelling of the functional MIP membranes, the ISFET effects were based on the alteration of membrane charge by the formation of the complex with boronate ligand.

In an early approach towards MIP-based sensors using capacitance measurement, thin MIP membranes were prepared by in situ polymerization of MAA/EDMA and then "sandwiched" as a sensing layer in a field effect device: a capacitance decrease was observed due to specific binding of the template L-phenylalanine anilide [103]. Recently, two promising alternative approaches towards ultrathin MIP films for capacitive sensors had been reported: electropolymerization of phenol for imprinting of phenylalanine [74], and photo-initiated graft copolymerization of AMPS/MBAA for imprinting of desmetryn [82] and creatinine [83] (cf. Sections III.C.2, III.C.3).

Even if the detailed mechanism for specific response to analyte binding is not always clear yet, the data for ultrathin and thin MIP films suggest that the permeability of the entire MIP film is involved. Hence, MIP membranes are most versatile in triggering—and potentially amplifying—signals. Beside the supported MIP films described above, self-supported MIP membranes could, in principle, be used in sensors but also for larger-scale separations (see Section V.A).

V. MIP MEMBRANES FOR SUBSTANCE-SPECIFIC SEPARATION

MIP membrane's function in correlation with its structure based on the preparation strategy (cf. Section III) will now be discussed. From the four categories of MIP film function, iii. and iv. are concerned (cf. Scheme 2). For MIP membranes, it is critically important to control both MIP specificity and film porosity (see Scheme 6). With exclusively microporous MIP membranes, template binding to MIP sites can either change the porous network structure significantly thus altering membrane permeability in general ("gate effect", see Section V.A) or the permeation rate is controlled by the interaction with the "micropore walls" (see Section V.B). In membranes with too large trans-membrane pores nonselective transport by diffusion or convection can only be compensated by binding to accessible MIP sites causing a retardation which can be used in membrane adsorbers (see Section V.D).

A. "Gate Effect" of Microporous MIP Membranes

Investigations of conductivity of MIP membranes—separating two cells filled with electrolyte—had led to the first MIP membrane sensor [1,96]: an increase of template concentration in the solution caused a change (mostly an increase) of membrane conductivity. The magnitude of the effect was much higher for substances with the same or very similar structure as compared with the template. Hence, sensor membranes imprinted with L-Phe, atrazine, or cholesterol all showed pronounced selectivity and a sensitivity in the micromolar range. Another important observation, made by UV–Vis spectroscopic analysis in the two cells, was that the addition of template (here: atrazin) increased the permeation rate of another substance (here: p-nitrophenol) through the MIP membrane [1,97].

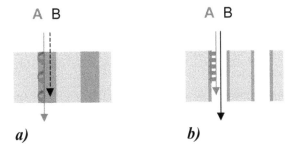

a)   *b)*

Scheme 6 Separation mechanisms for MIP membranes, with MIP sites for a substance A: (a) transport of A is facilitated via binding/desorption to MIP sites ("fixed carrier" membrane); (b) transport of A is retarded either by binding or binding/desorption to MIP sites on the surface of trans-membrane pores (membrane adsorber).

The opposite behavior of noncovalently and covalently imprinted membranes [98] was explained by the effect of template binding onto MIP swelling: a strong shrinking due to binding of template to the covalently imprinted material could be detected macroscopically, while only very little effects were observed for the noncovalently imprinted materials. The response of the membranes showed a saturation behavior as a function of template concentration. Hence, it can be postulated that depending on the specific binding in the membrane, transport pathways through the membrane can be opened or closed ("gate effect"; cf. Section IV.B).

It must be noted that all MIP "membranes" discussed above were prepared by casting reaction mixtures on/in thick glass frits (cf. Section III.E); with those MIP "plugged" porous supports, a detailed discussion of transport mechanisms lacks a sound basis.

However, studies with self-supported and much thinner membranes by Sergeyeva et al. [90,91] (cf. Section III.D.1) provided very clear evidence for the "gate effect": a most remarkable template specificity of conductivity response could be observed (see Table 1). Similar, supporting results had been obtained by another group [104]. One more convincing proof for the "gate effect" was results of Hattori et al. [101] with "pore-filled" MIP composite membranes (cf. Section III.E): The transport rate of another solute (creatinine) increased 1.23-fold in the presence of the template (Tho) while without any additive and with Caf the flux was the same (see Table 1).

B. Facilitated or Retarded Template Transport Through Microporous MIP Membranes

MIP membranes prepared by in situ polymerization in the presence of different templates (cf. Section III.D.1) showed a similar transport behavior because a faster transport of the template as compared with other solutes could be observed (cf. Scheme 6a).

In the study of Marx-Tibbon and Willner [88], membranes from the photoreactive polyacrylate/amide system, in the zwitter-ionic state imprinted with tryptophan, showed a specific tryptophan permeability which could be "erased" by light irradiation of the membrane inducing an isomerization of the functional groups to the uncharged spiropyrane form. According to the solution-diffusion mechanism, this could be taken as evidence that specific sorption in the membrane can yield specific transport through the membrane.

For membranes from a "standard" in situ polymerization MIP mixture, imprinted with 9-ethyl adenine (see Table 1), significant selectivities for the template derivatives adenine and adenosine, both vs. the respective guanine derivatives, were observed. Reducing solvent polarity increased binding strength and at the same time transport selectivity [89].

Membranes prepared by copolymerization and subsequent leaching of the polymeric "pore former" (see Table 1) showed remarkably high selectivities for UO_2^{2+} in competitive experiments with Ni^{2+}, Cd^{2+}, Zn^{2+}, and Cu^{2+} at quite high fluxes. This had been explained by a "selective binding of uranyl ion to imprinted sites along channels that span the membrane" [93].

For thin-film composite MIP membranes, according to gas permeation data defect ("pinhole") free, a higher performance than for the thick free-standing films from similar polymers was observed (see Table 1): fluxes of up to 18 nmol/cm^{2}h at real

Table 1 Microporous MIP Membranes—Preparation and Separation Performance

Matrix	Functionality	Template	Preparation	Membrane thickness (μm)	Separation by	Source conc. (mmol/L)	Solvent	Flux (nmol/cm²·h)	Perm-selectivity[a]	Adsorption selectivity qualitatively	Reference
	EDMA/MAA	9-EA	In situ polymn.	Not given	Diffusion	0.076	CHCl$_3$/MeOH, 94/6	0.2	$\alpha_{As/Gs} \sim 3.4$	n.d.[b]	Mathew-Krotz and Shea, 1996 [89]
	EDMA/MAA	Tho / Caf	In situ polymn.	~0.05 (film on/in support)	Diffusion	0.010 / 0.001 / 0.010	MeOH	17 / 18	$\alpha_{Tho/Caf} \sim 2.6$ $\alpha_{Tho/Caf} \sim 5.0$ $\alpha_{Caf/Tho} \sim 3.0$	n.d.	Hong el al., 1998 [99]
OUA	TEDMA/MAA	Atrazin	In situ polymn.	60 or 120	"Gate effect"	0.005–0.200	Phosphate buffer		"Gate effect": atrazin vs. other triazines $\alpha > 6$	n.d.	Sergeyeva et al., 1999 [90,91]
	TRIM/MAA	CBZ-L-Tyr	In situ polymn.	160 (filled support)	Diffusion	2.0	CHCl$_3$/MeOH, 50/50	117	$\alpha_{L/D} \sim 3.4^c$	n.d.	Dzgoev and Haupt, 1999 [100]
Cellulose	EDMA/MAA	Tho	In situ (graft) polymn.	surface layer (filled support)	"Gate effect"	2.7	H$_2$O	0.8	"Gate effect": for creatinin $\alpha_{Tho/Caf} > 1.23$	n.d.	Hattori et al., 2001 [101]
	PSt-co-DVB-co-vinylbenzoat	UO$_2^{2+}$	In situ polymn. (porogen)	100	Diffusion	0.2	H$_2$O	2.7	$\alpha(UO_2^{2+}/Ni^{2+}) > 100$	n.d.	Kimaro et al., 2001 [93]
PAN-co-St	PSt-DIDE	Boc-L-Trp	Dry PI	~145	Diffusion	1.0	H$_2$O/EtOH 50/50	~5	$\alpha_{D/L} \sim 1.4$	L > D	Yoshikawa et al., 1995 [17]

PAN-co-St	PSt-DIDE	Boc-L-Trp	Dry Pl	~145	Electro-dialysis	1.0	H$_2$O/EtOH 50/50	~3	$\alpha_{L/D} \sim 6.0$	L > D	Yoshikawa et al., 1996 [18]
PAN-co-St	PSt-DIDE	Boc-L-Trp	Dry Pl	~145	Electro-dialysis diffusion	1.0	H$_2$O/EtOH 50/50	~3	$\alpha_{L/D} \sim 6.0$	L > D	Yoshikawa et al., 1999 [24]
PAN-co-St	PSt-FFE	Boc-L-Trp	Dry Pl	~140	Electro-dialysis	1.0 / 1.0	H$_2$O/EtOH 50/50	<1 / ~0.3	$\alpha_{L/D} \sim 0.8$ / $\alpha_{L/D} \sim 4.6$	L > D	Yoshikawa et al., 2001 [26]
PAN-co-St	PSt-DIDE	9-EA	Dry Pl	~145	Diffusion	1.0	H$_2$O/EtOH 50/50	~0.75	$\alpha_{Gs/As} \sim 1.2$	As > Gs	Yoshikawa et al., 2001 [27]
CA		9-EA	Dry Pl	~110	Diffusion	1.0	H$_2$O/EtOH 50/50	~0.8	$\alpha_{Gs/As} \sim 1.2$	As > Gs	
PSf		9-EA	Dry Pl	~105	Diffusion	1.0	H$_2$O/EtOH 50/50	~0.75	$\alpha_{Gs/As} \sim 1.2$	As > Gs	
PSf-COOH		Boc-D-Glu / Boc-L-Glu	Dry Pl	Not given	Electro-dialysis	1.0	H$_2$O/EtOH 50/50	~5	$\alpha_{D/L} \sim 1.2$ / $\alpha_{L/D} \sim 1.2$	D > L / L > D	Yoshikawa et al., 1998 [21,22]
CA		D-Glu / L-Glu	Dry Pl	105	Electro-dialysis	1.0	H$_2$O/EtOH 50/50	10	$\alpha_{D/L} \sim 2.3$ / $\alpha_{L/D} \sim 2.3$	D > L / L > D	Yoshikawa et al., 1999 [25]

a "Real" selectivities; from multisolute (competition) experiments.
b Not determined.
c "Ideal" selectivity from single solute experiments; no selectivity in competition experiments.

theophyllin/caffein selectivities of up to 3.0 were achieved. The inverse dependency of the selectivity factor on the source concentration was in agreement with a "facilitated transport" mechanism [99].

For pore-filling composite membranes (see Table 1), in single solute experiments, enantioselectivity, i.e., an enhanced flux for the template, was observed for the MIP membrane while the blank membrane showed the same flux for D and L enantiomers. However, no selectivity could be observed in experiments with mixtures, i.e., under real enantiomer separation conditions [100]. Presumably, this could be caused by nonselective transport pathways due to uneven/incomplete pore filling, a hypothesis also in agreement with the exceptionally high fluxes.

MIP membranes prepared via "alternative imprinting" (cf. Section III.D.2) showed a more complex transport behavior. Yoshikawa et al. had developed specifically synthesized polystyrene resins with chiral tetrapeptide recognition groups, which had to be used in blends with a matrix polymer for membrane formation via a "dry PI" process. The resulting membranes seemed to be microporous and had a low permeability (see Table 1).

Diffusion studies revealed the role of the template—first chiral amino acid derivatives—as porogen, and the observed transport selectivity—slower transport of the template—was explained by a retardation due to specific template binding to the "pore walls". However, the same membranes showed an opposite selectivity in electrodialysis. Moreover, electrodialysis performance was also very much susceptible to the applied voltage. The MIP membrane behavior—for both dialysis and electrodialysis—was summarized in a phenomenological relationship, where the flux monotonically increased with the difference in chemical potential while the selectivity was around 1 at about 20 kJ/mol (corresponding to a concentration difference of 1 mmol/l), showed a pronounced maximum in the range of 200 kJ/mol and levelled off again to about 1 at very high potential values [24]. The authors also argued [26] that by applying a pressure difference such as in membrane filtration, a similar increase in selectivity as compared with diffusion could be expected. This, however, is hindered by the microporous structure of the thick MIP membranes.

With the chiral peptide resins, achiral templates could also be imprinted (see Table 1). Remarkably, also with other polymers without a pronounced functionality the transport selectivity was opposite to the adsorption selectivity (see Table 1). Hence all membranes prepared by Yoshikawa et al. showed in diffusion experiments the behavior of a membrane adsorber (cf. Section V.D). In fact, the experimental data show for several cases that there was only a *"time lag" for permeation* of the template (or a template derivative) what is in agreement with a saturation of binding sites in the pores and subsequent "breakthrough". However, in other cases, data collected over long periods were in agreement with a *retarded transport rate* for the template.

C. Microporous MIP Membrane's Structure and Transport Mechanism

Porosity data for MIPs, e.g., from BET analyses, provide evidence that imprinted sites constitute a significant part of the micropore fraction which is not present in blank materials. Unfortunately, for microporous MIP membranes, no detailed pore morphology analysis had been performed yet. For MIPs prepared via "dry PI", no porosity data are available at all. Also, information about the through-pore structure, e.g., from permporometry (cf. Fig. 2), could add essential information regarding MIP membrane structure.

A *static model* will be based on binding to the "walls" of *permanent pores* what could either facilitate or retard the transport of the template. One critical parameter is the density of MIP sites: with increasing concentration, the contribution of facilitated transport via fixed binding/"carrier" sites will also increase [105] (cf. Scheme 6a).

Imprinting in microporous materials may also contribute to the connectivity of pores. Data of Piletsky et al. [1,97] and Yoshikawa et al. [17] indicated such an effect: significant permeabilities could only be detected for MIP but not the blank membranes. On the other hand, these results and the confirmation of a "pinhole" free structure even for 50 nm thin layers [99] suggest the absence of large transmembrane pores. Note that the binding of the template but not of other substances substantially decreased the gas permeability [99]. Hence, binding may indeed take place at the "walls" of micropores having diameters less than 2 nm.

Template size- or function-specific transport channels had been proposed as an additional effect contributing to MIP membrane's transport selectivity [1]. The verification of this hypothesis will require further systematic investigations. Note, however, that an analogous template effect for implementing size selectivity is well-known for zeoliths [9].

A *dynamic model*, considering the *adaptation of micropore structure* to environmental conditions due to interactions with solutes, in particular the template, might be much more realistic for understanding liquid separation with MIP membranes. Cross-linked porous polymers have a hierarchical pore structure (cf. Section III.B), and upon solvation their structure is flexible. Significant template-induced polymer shrinking or swelling has been observed [106] (cf. Section IV.B). The "gate effect" for MIP membranes had been confirmed with charged and uncharged spezies: in microporous MIP films, this phenomenon could be used as a trigger for signal amplification in a sensor system (cf. Section IV.B) but also for enhancing separation selectivity and permeability (cf. Section V.A). Hence, the impact of this "gate effect" onto other permeation data (cf. Section V.B) should also be (re)analyzed.

In conclusion, microporous MIP membrane's permselectivity is based on preferential and reversible binding and exchange between template and MIP sites in the membrane, thus providing pathways for selective trans-membrane transport. However, the different behavior of membranes from different materials and preparation methods, imprinted for various templates and studied under various conditions, demonstrates the need for further detailed investigations of membrane structure as well as detailed transport characterization of well-defined membranes from controlled preparations, with a particular focus on dynamic effects onto micropore structure.

D. Macroporous Membrane Adsorbers

With macroporous membranes, convective flow can be used as additional driving force for transport and separation (cf. Section V.B). For separation's performance both binding affinity and capacity will be important, especially when membrane adsorbers (cf. Scheme 6b) shall be compared with other established or competing materials (beads, fibers, capillaries, monoliths). Again, membrane pore structure will have the main impact, because—if present—the micropore fraction will deter-

Table 2 Macroporous MIP Membrane Adsorbers—Preparation and Filtration Separation Performance

Matrix Polymer	Functionality	Template	Preparation	Thickness (μm)	Separation by	Feed conc. (μmol/L)	Solvent	Flux (L/m²·h) (pressure)	Binding selectivity single solute	Binding capacity	Reference
PAN-co-AA (10% AA)		Tho	Wet PI	100	Filtration	3.6	H_2O	5.6 (2.5 kPa)	$\alpha_{Tho/Caf} > 50$	0.52 μmol/g	Wang et al., 1996 [30]
PAN-co-AA (15% AA)		Tho	Wet PI	100	Batch sorption	3.6	H_2O	-	$\alpha_{Tho/caf} = 52$	1.25 μmol/g	Kobayashi et al., 1998 [33]
PAN-co-AA (16.6% AA)		Naringin	Wet PI	500 (casted film)	Filtration	6.7	H_2O	19 (n.d.)		0.13 μmol/g	Trotta et al., 2002 [109]
Nylon		L-gluta-mine	Wet PI	100	Batch sorption	0	H_2O	-	$\alpha_{L/D} = 3.3$	2.8 μmol/g	Reddy et al., 1999 [34]
				30	Filtration	10	H_2O	2.9 (1.0 kPa)	$\alpha_{L/D} = 3.5$	3.5 μmol/g	Reddy et al., 2002 [36]
PAN-co-DTCS	MAA/MBAA	Tho	Photo-grafting	100	Filtration	3.6	H_2O	3.3 (2.0 kPa)	$\alpha_{Tho/caf} = 5.9$	8 μmol/m²	Wang et al., 1997 [32]
PP	AMPS/MBAA	Desme-tryn	Photo-grafting	150	Filtration	10	H_2O	120 (n.d.)	Group specific	6 μmol/m²	Piletsky et al., 2000 [81]
PVDF precoat.	AMPS/MBAA	Terbu-meton	Photo-grafting	125	Filtration	10	H_2O	120 (n.d.)	$\alpha_{Terbumeton/Atrazin} = 15$	3 μmol/m²	Sergeyeva et al., 2001 [87]
PVDF; PVDF precoat.	AMPS/MBAA	Desme-tryn	Photo-polymn.	125	Filtration	10	H_2O	120 (n.d.)	n.d.	3 μmol/m²	Kochkodan et al., 2001 [76]

mine capacity while a connected macropore fraction will be essential for efficient transmembrane transport. The advantages of membrane adsorbers with optimized pore structure are high dynamic capacity—close to the static capacity—at high throughput [107,108].

MIP membranes prepared via "alternative imprinting" (cf. Section III.D.2, see Table 2) from functional poly(acrylates) had an asymmetric pore structure with a microporous skin layer, induced by the immersion step, which largely reduced membrane permeability. Moderate binding capacities at high selectivities in batch experiments had been found in an optimization study [31]; exactly those membranes had not been characterized in filtration experiments. Recently, Trotta et al. [109] had applied the same system to imprinting of the more complex molecule naringin, in an attempt to evaluate the feasibility for removal of a bitter compound from orange juice. Membrane adsorption separation had been performed at a somewhat higher flux, however, only very low binding capacities had been obtained.

MIP membranes from nylon had a rather symmetric pore morphology, but had been applied in binding experiments also at very low flux. Identical moderate binding capacities had been observed for batch equilibrium sorption ("static") and filtration ("dynamic"), but this can be explained by the protocol where Tho had been accumulated over a very long filtration time, i.e., under quasi-"static" conditions [36].

MIP membranes based on the photo-reactive copolymer PAN-co-DTCS and finally imprinted by photo-grafting, had also the typical asymmetric, not yet optimized pore structure, so that only very low permeabilities could be used. Due to the low flux and the long filtration time, the binding capacities were quite high [32].

Thin-layer MIP composite membranes (cf. Section III.E, see Table 2) had been prepared by surface functionalization of various commercial porous membranes, which had already been optimized towards high performance in microfiltration. These membranes have also a moderate specific surface area (e.g., 0.22 μm PVDF: $4\,m^2/g$, 0.2 μm PP: 20 m^2/g). By photo-initiated graft copolymerization [81] or cross-linking polymerization [76], the entire internal pore structure could be coated evenly with thin MIP layers without formation of agglomerates. Most important, at suited degrees of functionalization, no pore blocking occurred as indicated by the preserved high membrane permeability and specific surface area [81].

MIP membranes imprinted for herbicides, had been characterized at very high flow rate which would also enable an efficient processing of large volumes containing a dilute valuable or toxic compound. Compared with static capacities for the PI membranes (cf. above) the obtained dynamic binding capacities, normalized to the amount of grafted copolymer ($\sim$1 μmol/g), were in the same order of magnitude. Note that binding capacities increased with decreasing flux [76]. In addition, for the MIP membranes based on PP, maximum binding capacities of up to 128 nmol/cm^2 had been measured at 1 mM terbumeton. This corresponded to an about 40% use of the functional monomer AMSP in the formation of MIP sites [81], what could be explained by the thin-layer MIP structure. For IPN-MIP membranes (cf. Scheme 5), a very high selectivity at reasonable capacities had been achieved (note that the specific surface area of the PVDF was smaller than that of PP) [87].

Furthermore, quantitative template recovery by elution from the MIP membranes was possible at lower pH; and it been shown that the MIP membranes were reusable in several subsequent bind–wash–elute cycles [81]. Hence, tailored MIP materials for membrane SPE could already be envisioned.

VI. APPLICATIONS

Many applications can be foreseen when the unique recognition properties of MIPs can be successfully combined with adapted material formats and respective manufacturing technologies [110]. First, the special demands of sensor and separation technology will largely stimulate the development of MIPs with well-defined size and shape. Second, such MIPs will soon occupy also niches in broader fields, such as the LifeSciences or the chemical, pharmaceutical, biotechnological, and environmental industries.

Regarding the adaptation of MIPs to applications, the particular advantages of MIP films and membranes are:

a. MIP films of well-defined thickness and internal structure yield, compared with conventional MIP materials, superior performance in affinity technologies;
b. thin MIP films can, in principle, be immobilized on any base material thus tailoring its affinity;
c. MIP films or membranes can be easily and efficiently integrated into various already existing or novel technical processes.

A. Sensors and Assays

MIP films have already been adapted to various sensor and assay formats, fulfilling the minimum requirement—immobilization of the receptor—but also fitting to the need of various detection formats (cf. Section IV). MIP films could also be integrated into micro-systems or -arrays; here, potential manufacturing technologies based on the synthesis of self-assembled or grafted MIP layers (cf. Sections III.C.1, III.C.3) will have particular benefits. The integration of transducer functions into MIP films will be another "hot" issue; ordered two- and three-dimensional film topologies and architectures will offer great opportunities for implementing improved detection specificity and signal amplification. Also, the "gate effect", an example for a transducer function of the entire MIP film (cf. Section V.A), will find use in advanced MIP-based chemo-sensors.

B. Separations

The development of MIP composite materials based on thin MIP films for improving separation's performance has attracted increasing attention. Examples include tailored particles for SPE or chromatography and coated tubes or capillaries for electrophoretic and other separations. The latter formats will also gain importance in lab-on-a-chip systems [111].

C. Membrane Separation

MIP membranes are on their way to combine the potential of MIPs with the unique advantages of membrane technology [1], e.g., continuos separation processes without a phase change and at low temperature and energy consumption [14]. Also, membrane separations can provide high throughput in combination with an efficient up scaling (kg-scale with technical membrane modules) or down scaling (e.g., ng-scale with 384-well membrane filter plates) of separation capacity.

The main problem is to optimize MIP recognition and membrane transport properties at the same time (cf. Section V). The most promising routes are innovative preparation strategies based on novel materials, e.g., polymer blends, block copolymers, or inorganic/organic composites (cf. Section III.D), and the preparation of composite membranes (cf. Section III.E). Surface functionalization, by self-assembly or controlled grafting (cf. Scheme 3), can be used for either coating the pore surface or controlled filling of pores. The latter route, applied to asymmetric ultrafiltration membranes, could ultimately enable the application of the MIP "gate effect" also for efficient separations.

MIP membrane adsorbers for the specific sample enrichment from large volumes by membrane SPE, and for the specific decontamination of large process streams will be among first examples for applications (cf. Section V.D). Other promising continous separations are the resolution of enantiomers or the product removal from bioreactors, both feasible by electrodialysis or dialysis (cf. Section V.B).

In the future, porous catalytic MIP membranes could also serve as a key element for advanced integrated "bio-mimetic" processes in reaction engineering [112,113].

D. LifeSciences and Other Applications

Biocompatibility of materials—"biomaterials"—in contact with cells or tissue, relies on specific molecular recognition processes, especially at the interfaces. Imprinted surfaces are expected to play a key role in this field in the future [114]. Ultrathin or thin MIP layers for recognition of proteins, but also for cell-specific recognition based on surface-marker structures or cell shape could be envisioned.

Controlled release or delivery from or through MIP films or membranes—including isotropic or hollow fibers or capsules (cf. Scheme 1)—will be a field of attractive potential applications. Targets could be drugs but also other technically or environmentally interesting substances. Release from MIP-based depots could occur passive, with the film as barrier dictating the transport kinetics, but also triggered by a stimulus from the environment, e.g., via recognition of a specific signal molecule at a MIP site (cf. [101]).

VII. CONCLUSIONS

Molecular imprinting is an interdisciplinary research subject, inspired by biological principles such a self-organization and recognition, with synthetic and analytical chemistry as two main fields. Research devoted to the synthesis and characterization of MIP films and membranes includes significant additional efforts in material's science and engineering because such "well-defined" MIP formats can largely facilitate both innovative manufacturing and application technologies for MIPs. However, much more information about detailed MIP structure will be required, in particular for MIPs prepared via novel, e.g., "alternative" approaches.

Ultrathin and thin MIP films have already proved to be important model systems for elucidation of MIP structure and function. Important information—e.g., regarding critical dimensions to achieve molecular recognition—and superior performance—e.g., high selectivity via improved accessibility—can be achieved by a well-defined two- and three-dimensional organization of MIP films. This is especially true when those MIP films are prepared by self-assembly or controlled surface grafting. In many cases, an additional stabilization of MIP sites by a solid support is also involved in specific recognition.

The unique feature of MIP membranes is the interplay of selective binding and permeation. Receptor and transport properties of microporous MIP membranes can be based on template-specific binding sites in trans-membrane pores, which serve as fixed carriers for "facilitated" transport. Furthermore, template binding in microporous membranes can also lead to a "gate effect", which either increases or decreases membrane permeability. Alternatively, MIP membranes can also function as adsorbers. Currently, mainly empirical data are available. Again, a more detailed structure characterization will be necessary in order to be able to rationally design perm-selective MIP membranes. Those MIP membranes could serve as model systems for cellular transmembrane transport and natural receptors.

Ultimately, investigations of MIP films and membranes will open the way to the design of complex, but nevertheless well-defined supramolecular devices that can perform highly selective recognition, transformation, transfer, and regulation functions.

NOMENCLATURE

9-EA	9-ethyl adenine
AA	Acrylic acid
AIBN	Azo-bis-isobutyronitrile
AMP	Adenosine monophosphate
AMPS	2-acrylamido-2-methyl-1-propane sulphonic acid
As	Adenosine
Boc	Tert.-butyloxycarbonyl
CA	Cellulose acetate
Caf	Caffein
CBZ	Carboxybenzoyl
DDT	1, 1-bis-(4-chlorphenyl)-2,2,2-trichlorethylene
Des	Desmetryn
DIDE	*Tetrapeptide:*
	H-Asp(OcHx)-lie-Asp(OcHx)-Glu(Obz)-CH_2
DTCS	*N,N*-Diethylaminodithiocarbamoylmethylstyrene
DVB	Divinylbenzene
EDMA	Ethylenglycol dimethacrylate
FFE	*Tripeptide:*
	H-Phe-Phe-Glu(Obz)-CH_2
Glu	Glutamic acid
Gs	Guanosine
HEMA	2-hydroxyethyl methacrylate
MAA	Methacrylic acid
MBAA	*N,N'*-methylene bisacrylamide
OUA	Oligourethane acrylate
P4Py	Poly-4-vinylpyridine
PAN	Polyacrylonitrile
PDADMAC	Poly (diallyl-dimethylammonium chloride)
Phe	Phenylalanine
PP	Polypropylene
PSf	Polysulfone
PSt	Polystyrene
PUR	Polyurethane
PVC	Polyvinylchloride

PVDF	Polyvinylidene fluoride
St	Styrene
TEDMA	Triethylenglycol dimethacrylate
Tho	Theophylline
TMAEMA	N,N,N-trimethylaminoethyl methacrylate
TRIM	1,1,1-(trishydroxymethyl)propane trimethacrylate
Trp	Tryptophan
Tyr	Tyrosine

REFERENCES

1. Piletsky, S.A.; Panasyuk, T.L.; Piletskaya, E.V.; Nicholls, I.A.; Ulbricht, M. Receptor and transport properties of molecularly imprinted polymer membranes—A review. J. Membr. Sci. **1999**, *157*, 263–278.

2. Shi, H.; Tsai, W.B.; Garrison, M.D.; Ferrari, S.; Ratner, B.D. Template-imprinted nanostructured surfaces for protein recognition. Nature **1999**, *398*, 593.

3. Aherne, A.; Alexander, C.; Payne, M.J.; Perez, N.; Vulfson, E.N. Bacteria-mediated lithography of polymer surfaces. J. Amer. Chem. Soc. **1996**, *118*, 8771–8772.

4. Surugiu, I.; Danielsson, B.; Ye, L.; Mosbach, K.; Haupt, K. Chemiluminescence imaging ELISA using an imprinted polymer as the recognition element instead of an antibody. Anal. Chem. **2001**, *73*, 487–491.

5. Lehmann, M.; Brunner, H.; Tovar, G. Enantioselective separations: a new approach using molecularly imprinted nanoparticle composite membranes. Desalination **2002**, *149*, 315–321.

6. Okay, O. Macroporous copolymer networks. Prog. Polym. Sci. **2000**, *25*, 711–779.

7. Sellergren, B.; Hall, A.J. Fundamental aspects on the synthesis and characterization of imprinted network polymers. In *Molecularly Imprinted Polymers—Man-made Mimics of Antibodies and Their Application in Analytical Chemistry*; Sellergren, B., Ed.; Elsevier, **2001**; 21–57.

8. Lai, E.P.C.; Fafara, A.; VanderNoot, V.A.; Kono, M.; Polsky, B. Surface plasmon resonance sensors using molecularly imprinted polymers for sorbent assay of theophylline, caffein, and xanthine. Can. J. Chem. **1998**, *76*, 265–273.

9. Raman, N.K.; Anderson, M.T.; Brinker, C.J. Template-based approaches to the preparation of amorphous, nanoporous silicas. Chem. Mater. **1996**, *8*, 1682–1701.

10. Mann, S.; Burkett, S.L.; Davis, S.A.; Fowler, C.E.; Mendelson, N.H.; Sims, S.D.; Walsh, D.; Wilton, N.T. Sol–gel synthesis of organized matter. Chem. Mater. **1997**, *9*, 2300–2310.

11. Makote, R.; Collinson, M.M. Template recognition in inorganic–organic hybrid films prepared by the sol–gel process. Chem. Mater. **1998**, *10*, 2440–2445.

12. Marx, S.; Liron, Z. Molecular imprinting in thin films of organic–inorganic hybrid sol–gel and acrylic polymers. Chem. Mater. **2001**, *13*, 3624–3630.

13. Graham, A.L.; Carlson, C.A.; Edmiston, P.L. Development and characterization of molecularly imprinted sol–gel materials for the selective detection of DDT. Anal. Chem. **2002**, *74*, 458–467.

14. Mulder, M. In *Basic Principles of Membrane Technology*; Kluwer Academic Publishers: Dordrecht, 1991; 71–108.

15. Stahl, M.; Manson, M.O.; Mosbach, K. The synthesis of a D-amino acid ester in an organic medium with a-chymotrypsin modified by a bio-imprinting procedure. Biotechnol. Lett. **1990**, *12*, 161–166.

16. Dabulis, K.; Klibanov, A.M. Molecular imprinting of proteins and other macromolecules resulting in new adsorbents. Biotechnol. Bioeng. **1992**, *39*, 176–185.

17. Yoshikawa, M.; Izumi, J.; Kitao, T.; Koya, S.; Sakamoto, S. Molecularly imprinted polymeric membranes for optical resolution. J. Membr. Sci. **1995**, *108*, 171–175.

18. Yoshikawa, M.; Izumi, J.; Kitao, T. Enantioselective electrodialysis of N-a-acetyltrypto-phans through molecularly imprinted polymeric membranes. Chem. Lett. **1996**, *26*, 611–612.

19. Yoshikawa, M.; Izumi, J.; Kitao, T.; Sakamoto, S. Molecularly imprinted polymeric membranes containing DIDE derivatives for optical resolution of amino acids. Macromolecules **1996**, *29*, 8197–8203.

20. Yoshikawa, M.; Izumi, J.; Kitao, T. Enantioselective electrodialysis of amino acids with charged polar side chains through molecularly imprinted polymeric membranes containing DIDE derivatives. Polym. J. **1997**, *29*, 205–210.

21. Yoshikawa, M.; Izumi, J.; Kitao, T.; Sakamoto, S. Alternative molecularly imprinted polymeric membranes from tetrapeptide residue consisting of D- or L-amino acids. Macromol. Rapid Commun. **1997**, *18*, 761–767.

22. Yoshikawa, M.; Fujisawa, T.; Izumi, J.; Kitao, T.; Sakamoto, S. Molecularly imprinted polymeric membranes involving tetrapeptide EQKL derivatives as chiral-recognition sites toward amino acids. Anal. Chim. Acta **1998**, *365*, 59–67.

23. Yoshikawa, M.; Izumi, J.; Ooi, T.; Kitao, T.; Guiver, M.D.; Robertson, G.P. Carboxylated polysulfone membranes having a chiral recognition site induced by an alternative molecular imprinting technique. Polymer Bull. **1998**, *40*, 517–524.

24. Yoshikawa, M.; Izumi, J.; Kitao, T. Alternative molecular imprinting, a facile way to introduce chiral recognition sites. React. Funct. Polym. **1999**, *42*, 93–102.

25. Yoshikawa, M.; Ooi, T.; Izumi, J. Alternative molecularly imprinted membranes from a derivative of a natural polymer, cellulose acetate. J. Appl. Polym. Sci. **1999**, *72*, 493–499.

26. Yoshikawa, M.; Izumi, J.; Guiver, M.D.; Robertson, G.P. Recognition and selective transport of nucleic acid components through molecularly imprinted polymeric membranes. Macromol. Mater. Eng. **2001**, *286*, 52–59.

27. Yoshikawa, M.; Shimada, A.; Izumi, J. Novel polymeric membranes having chiral recognition sites converted from tripeptide derivatives. Analyst **2001**, *126*, 775–780.

28. Kondo, Y.; Yoshikawa, M. Effect of solvent composition on chiral recognition ability of molecularly imprinted DIDE derivatives. Analyst **2001**, *126*, 781–783.

29. Kobayashi, T.; Wang, H.Y.; Fujii, N. Molecular imprinting of theophylline in acrylonitrile-acrylic acid copolymer membrane. Chem. Lett. **1995**, *24*, 927–928.

30. Wang, H.Y.; Kobayashi, T.; Fuji, N. Molecular imprint membranes prepared by the phase inversion technique. Langmuir **1996**, *12*, 4850–4856.

31. Wang, H.Y.; Kobayashi, T.; Fukaya, T.; Fuji, N. Molecular imprint membranes prepared by the phase inversion technique. 2. Influence of coagulation temperature in the phase inversion process on the encoding in polymeric membranes. Langmuir **1997**, *13*, 5396–5400.

32. Wang, H.Y.; Kobayashi, T.; Fuji, N. Surface molecular imprinting on photosensitive dithio-carbamoyl polyacrylnitrile membrane using photo graft polymerization. J. Chem. Technol. Biotechnol. **1997**, *70*, 355–362.

33. Kobayashi, T.; Wang, H.Y.; Fujii, N. Molecular imprint membranes of polyacrylonitrile copolymers with different acrylic acid segments. Anal. Chim. Acta. **1998**, *365*, 81–88.

34. Reddy, P.S.; Kobayashi, T.; Fujii, N. Molecular imprinting in hydrogen bonding networks of polyamide nylon for recognition of amino acids. Chem. Lett. **1999**, *28*, 293–294.

35. Kobayashi, T.; Murawaki, Y.; Reddy, P.S.; Abe, M.; Fujii, N. Molecular imprinting of caffein and its recognition assay by quartz-crystal microbalance. Anal. Chim. Acta **2001**, *435*, 141–149.

36. Reddy, P.S.; Kobayashi, T.; Abe, M.; Fujii, N. Molecularly imprinted nylon-6 as recognition material of amino acids. Eur. Polym. J. **2002**, *38*, 521–529.

37. Reddy, P.S.; Kobayashi, T.; Fujii, N. Recognition characteristics of dibenzofuran by molecularly imprinted polymers made from common polymers. Eur. Polym. J. **2002**, *38*, 779–785.

38. Kanekiyo, Y.; Sano, M.; Ono, Y.; Inoue, Y.; Shinkai, S. Facile construction of a novel metal-imprinted surface without a polymerisation process. J. Chem. Soc., Perkin Trans. **1998**, *2*, 2005–2008.

39. Gutierres-Fernandez, S.; Lobo-Castanon, M.J.; Miranda-Ordieres, A.J.; Tunon-Blanco, P.; Carriedo, G.A.; Garcia-Alonso, F.J.; Fidalgo, J.I. Molecularly imprinted polyphosphazene films as recognition elements in a voltammetric rifamycin SV sensor. Electroanalysis **2001**, *13*, 1399–1404.

40. Ulman, A. In *An Introduction to Ultrathin Organic Films: from Langmuir–Blodget to Self-Assembly*; Academic: Boston, 1991.

41. Chailapakul, O.; Crooks, R.M. Synthesis and characterization of simple self-assembling, nanoporous monolayer assemblies: a new strategy for molecular recognition. Langmuir **1993**, *9*, 884–888.

42. Lahav, M.; Katz, E.; Doron, A.; Patolski, F.; Willner, I. Photochemical imprint of molecular recognition in monolayers assembled on Au electrodes. J. Amer. Chem. Soc. **1999**, *121*, 862–863.

43. Mirsky, M.V.; Hirsch, T.; Piletsky, S.A.; Wolfbeis, O. A Spreader-bar approach to molecular architecture: formation of stable artificial chemoreceptors. Angew. Chem. Int. Ed. **1999**, *38*, 1108–1110.

44. Miyahara, T.; Kurihara, K. Two-dimensional molecular imprinting: binding of sugars to boronic acid functionalized, polymerized Langmuir–Blodgett films. Chem. Lett. **2000**, *29*, 1356–1357.

45. Yoshida, M.; Uezu, K.; Goto, M.; Furusaki, S.; Takagi, M. An enantioselective polymer prepared by surface molecular-imprinting technique. Chem. Lett. **1998**, *27*, 925–926.

46. Lee, S.W.; Ichinose, I.; Kunitake, T. Molecular imprinting of azobenzene carboxylic acid on a TiO_2 ultrathin film by the surface sol–gel process. Langmuir **1998**, *14*, 2857–2863.

47. Lahav, M.; Kharitonov, A.B.; Katz, O.; Kunitake, T.; Wilner, I. Tailoring chemosensors for chloroaromatic acids using molecular imprinted TiO_2 thin films on ion-sensitive field-effect transistors. Anal. Chem. **2001**, *73*, 720–723.

48. Lahav, M.; Kharitonov, A.B.; Katz, O.; Kunitake, T.; Wilner, I. Imprinting of chiral recognition sites in thin TiO_2 films associated with field-effect transistors: novel functionalized devices for chiroselective and chirospecific analyses. Chem. Eur. J. **2001**, *7*, 3992–3997.

49. Decher, G. Fuzzy nanoassemblies: toward layered polymeric multicomposites. Science **1997**, *277*, 1232–1236.

50. Kanekiyo, Y.; Inoue, K.; Ono, Y.; Sano, M.; Shinkai, S.; Reinhoudt, D.N. "Molecular-imprinting" of AMP utilizing the polyion complex formation process as detected by a QCM system. J. Chem. Soc., Perkin Trans. **1999**, *2*, 2719–2722.

51. Takeuchi, T.; Fukuma, D.; Matsui, J. Combinatorial molecular imprinting: an approach to synthetic polymer receptors. Anal. Chem. **1999**, *71*, 285–290.

52. Lanza, F.; Sellergren, B. Method for synthesis and screening of large groups of molecularly imprinted polymers. Anal. Chem. **1999**, *71*, 2092–2096.

53. Kriz, D.; Ramström, O.; Svensson, A.; Mosbach, K. Introducing biomimetic sensors based on molecularly imprinted polymers as recognition elements. Anal. Chem. **1995**, *67*, 2142–2144.

54. Haupt, K.; Noworyta, K.; Kutner, W. Imprinted polymer-based enantioselective acoustic sensor using a quartz microbalance. Anal. Commun. **1999**, *36*, 391–393.

55. Jakoby, B.; Ismail, G.M.; Byfield, M.P.; Vellekoop, M.J. A novel molecularly imprinted thin film applied to a Love wave gas sensor. Sens. Actuat. **1999**, *76*, 93–97.

56. Nickel, A.M.L.; Seker, F.; Ziemer, B.P.; Ellis, A.B. Imprinted poly (acrylic acid) films on cadmium selenide. A composite sensor structure that couples selective amine binding with semiconductor substrate photoluminescence. Chem. Mater. **2001**, *13*, 1391–1397.

57. Sallacan, N.; Zayats, M.; Bourenko, T.; Kharitonov, A.B.; Willner, I. Imprinting of nucleotide and monosaccharide recognition sites in acrylamidephenylboronic acid–acrylamide copolymer membranes associated with electronic transducers. Anal. Chem. **2002**, *74*, 702–712.

58. Dickert, L.F.; Thierer, S. Molecularly imprinted polymers for optochemical sensors. Adv. Mater. **1996**, *8*, 987–990.

59. Dickert, F.L.; Besenböck, H.; Tortschanoff, M. Molecular imprinting through van der Waals interactions: fluorescence detection of PAHs in water. Adv. Mater. **1998**, *10*, 149–151.

60. Dickert, F.L.; Forth, P.; Lieberzeit, P.; Tortschanoff, M. Molecular imprinting in chemical sensing—Detection of aromatic and halogenated hydrocarbons. Fresenius J. Anal. Chem. **1998**, *360*, 759–762.

61. Dickert, F.L.; Hayden, O.; Halikias, K.P. Synthetic receptors as sensor coatings for molecules and living cells. Analyst **2001**, *126*, 766–771.

62. Bossi, A.; Piletsky, S.A.; Piletska, E.V.; Righetti, P.G.; Turner, A.P.F. Surface-grafted molecularly imprinted polymers for protein recognition. Anal. Chem. **2001**, *73*, 5281–5286.

63. Wulff, G.; Oberkusch, D.; Minarik, M. Enzyme-analogue built polymers, 18: Chiral cavities in polymer layers coated on wide-pore silica. React. Polym. **1985**, *3*, 261–275.

64. Norrlöw, O.; Glad, M.; Mosbach, K. Acrylic polymer preparations containing recognition sites obtained by imprinting with substrates. J. Chromatogr. **1984**, *299*, 29–41.

65. Plunkett, S.D.; Arnold, F.H. Molecularly imprinted polymers on silica: selective supports for high-performance ligand-exchange chromatography. J. Chromatogr. A **1995**, *708*, 19–29.

66. Glad, M.; Reinholdsson, P.; Mosbach, K. Molecularly imprinted composite polymers based on trimethylolpropane trimethacrylate (TRIM) particles for efficient enantiomeric separation. React. Polym. **1995**, *25*, 47–54.

67. Kempe, M.; Glad, M.; Mosbach, K. An approach towards surface imprinting using the enzyme ribonuclease. A. J. Molec. Recogn. **1995**, *8*, 35–39.

68. Hirayama, K.; Borow, M.; Morikawa, Y.; Minoura, N. Synthesis of polymer-coated particles with specific recognition sites for glucose oxidase by the molecular imprinting technique. Chem. Lett. **1998**, 731–732.

69. Schweitz, L.; Andersson, L.I.; Nilsson, S. Capillary electrochromatography with predetermined selectivity obtained through molecular imprinting. Anal. Chem. **1997**, *69*, 1179–1183.

70. Brüggemann, O.; Freitag, R.; Whitcombe, M.J.; Vulfson, E.N. Comparison of polymer coatings of capillaries for capillary electrophoresis with respect to their applicability to molecular imprinting and electrochromatography. J. Chromatogr. A **1997**, *781*, 43–53.

71. Joshi, V.P.; Karode, S.K.; Kulkarni, M.G.; Mashelkar, R.A. Novel separation strategies based on molecularly imprinted adsorbents. Chem. Eng. Sci. **1998**, *53*, 2271–2284.

72. Kugimiya, A.; Takeuchi, T. Surface plasmon resonance sensor using molecularly imprinted polymer for detection of sialic acid. Biosens. Bioelectr. **2001**, *16*, 1059–1062.

73. Spurlock, L.D.; Jaramillo, A.; Praserthdam, A.; Lewis, J.; Brajter-Toth, A. Selectivity and sensitivity of ultrathin purine-templated overoxidized polypyrrole film electrodes. Anal. Chim. Acta. **1996**, *336*, 37–46.

74. Panasyuk-Delaney, T.; Mirsky, V.M.; Piletsky, S.A.; Wolfbeis, O.S. Electropolymerized molecularly imprinted polymers as receptor layers in capacitive chemical sensors. Anal. Chem. **1999**, *71*, 4609–4613.

75. Malitesta, C.; Losito, I.; Zambonin, P.G. Molecularly imprinted electrosynthesized polymers: new materials for biomimetic sensors. Anal. Chem. **1999**, *71*, 1366–1370.

76. Kochkodan, V.; Weigel, W.; Ulbricht, M. Thin layer molecularly imprinted microfiltration membranes by photofunctionalization using a coated α-cleavage photoinitiator. Analyst **2001**, *126*, 803–809.

77. (a) Koch, M. Diplomarbeit, Technische Fachhochschule Berlin, 2001; (b) Kochkodan, V.; Ulbricht, M. ELIPSA GmbH, Berlin, 2000, unpublished.

78. Uyama, Y.; Kato, K.; Ikada, Y. Surface modification of polymers by grafting. Adv. Polym. Sci. **1998**, *137*, 1–39.

79. Sulitzky, C.; Rückert, B.; Hall, A.J.; Lanza, F.; Unger, K.; Sellergren, B. Grafting of imprinted polymer films on silica supports containing surface-bound free radical initiators. Macromolecules **2002**, *35*, 79–91.

80. Schweitz, L. Molecularly imprinted polymer coatings for open-tube capillary electrochromatography prepared by surface initiation. Anal. Chem. **2002**, *74*, 1192–1196.

81. Piletsky, S.A.; Matuschewski, H.; Schedler, U.; Wilpert, A.; Piletskaya, E.V.; Thiele, T.A.; Ulbricht, M. Surface functionalization of porous polypropylene membranes with molecularly imprinted polymers by photografting polymerization in water. Macromolecules **2000**, *33*, 3092–3098.

82. Panasyuk-Delaney, T.; Mirsky, V.M.; Ulbricht, M.; Wolfbeis, O.S. Impedometric herbicide chemosensors based on molecularly imprinted polymers. Anal. Chim. Acta **2001**, *435*, 157–162.

83. Panasyuk-Delaney, T.; Mirsky, V.M.; Wolfbeis, O.S. Capacitive creatinine sensor based on a photografted molecularly imprinted polymer. Electroanalysis **2002**, *14*, 221–224.

84. Tsukagoshi, K.; Yu, K.Y.; Maeda, M.; Takagi, M. Metal ion-selective adsorbent prepared by surface-imprinting polymerization. Bull. Chem. Soc. Jpn. **1993**, *66*, 114–120.

85. Tsukagoshi, K.; Yu, K.Y.; Maeda, M.; Takagi, M.; Miyajima, T. Surface imprinting. Characterization of a latex resin and the origin of the imprinting effect. Bull. Chem. Soc. Jpn. **1995**, *68*, 3095–3103.

86. Haginaka, J.; Takehira, H.; Hosoya, K.; Tanaka, N. Uniform-sized imprinted polymer for (*S*) -naproxen selectively modified with hydrophilic external layer. J. Chromatogr. A **1999**, *849*, 331–339.

87. Sergeyeva, T.A.; Matuschewski, H.; Piletsky, S.A.; Bendig, J.; Schedler, U.; Ulbricht, M. Molecularly imprinted polymer membranes for substance-selective solid-phase extraction from water by surface photo-grafting polymerisation. J. Chromatogr. A **2001**, *907*, 89–99.

88. Marx-Tibbon, S.; Willner, I. Photostimulated polymers: a light-regulated medium for transport of amino acids. J. Chem. Soc., Chem. Commun. **1994**, 1261–1262.

89. Mathew-Krotz, J.; Shea, K.J. Imprinted polymer membranes for the selective transport of targeted neutral molecules. J. Am. Chem. Soc. **1996**, *118*, 8154–8155.

90. Sergeyeva, T.A; Piletsky, S.A.; Brovko, A.A.; Slinchenko, L.A.; Sergeeva, L.M.; Panasyuk, T.L.; El'skaya, A.V. Conductometric sensor for atrazine detection based on molecularly imprinted polymer membrane. Analyst. **1999**, *124*, 331–334.

91. Sergeyeva, T.A.; Piletsky, S.A.; Brovko, A.A.; Slinchenko, L.A.; Sergeeva, L.M.; El'skaya, A.V. Selective recognition of atrazine detection by molecularly imprinted polymer membranes. Development of conductometric sensor for herbicides detection. Anal. Chim. Acta **1999**, *392*, 105–111.

92. Sreenivasan, K. Synthesis and evaluation of a molecularly imprinted polyurethane-poly (HEMA) semi-interpenetrating network as membrane. J. Appl. Polym. Sci. **1998**, *70*, 19–22.

93. Kimaro, A.; Kelly, L.A.; Murray, G.M. Molecularly imprinted ionically permeable membrane for uranyl ion. Chem. Commun. **2001**, 1282–1283.

94. Ramamoorthy, M.; Ulbricht, M. Molecular imprinting of cellulose acetate—sulfonated polysulfone blend membranes for Rhodamine B by phase inversion technique. J. Membr. Sci. **2003**, *217*, 207–214.

95. Piletsky, S.A.; Dubey, I.Y.; Fedoryak, D.M.; Kukhar, P.V. Substrate-selective polymeric membranes Selective transfer of nucleic acid components. Biopolim. Kletka **1990**, *6*, 55–58.

96. Piletsky, S.A.; Butovich, I.A.; Kukhar, V.P. Design of molecular sensors on the basis of substrate-selective polymer membranes. Zh. Anal. Khim. **1992**, *47*, 1681–1684.

97. Piletsky, S.A.; Piletskaya, E.V.; Elgersma, A.V.; Yano, K.; Karube, I.; Parhometz, Y.P.; El'skaya, A.V. Atrazine sensing by molecularly imprinted membranes. Biosens. Bioelectron. **1995**, *10*, 959–964.

98. Piletsky, S.A.; Panasyuk, T.L.; Piletskaya, V.E; El'skaya, A.V.; Levi, R.; Karube, I.; Wulff, G. Imprinted membranes for sensor technology: opposite behavior of covalently and noncovalently imprinted membranes. Macromolecules **1998**, *31*, 2137–2140.

99. Hong, J.M.; Anderson, P.E.; Qian, J.; Martin, C.R. Selectively-permeable ultrathin film composite membranes based on molecularly imprinted polymers. Chem. Mater. **1998**, *10*, 1029–1033.

100. Dzgoev, A.; Haupt, K. Enantioselective molecularly imprinted polymer membranes. Chirality **1999**, *11*, 465–469.

101. Hattori, K.; Yoshimi, Y.; Sakai, K. Gate effect of cellulosic dialysis membrane grafted with molecularly imprinted polymer. J. Chem. Eng. Jpn. **2001**, *34*, 1466–1469.

102. Haupt, K.; Mosbach, K. Molecularly imprinted polymers and their use in biomimetic sensors. Chem. Rev. **2000**, *100*, 2495–2504.

103. Hedborg, E.; Winquist, F.; Lundstrom, I.; Andersson, L.; Mosbach, K. Some studies of molecularly imprinted polymer membranes in combination with field-effect devices. Sens. Actuat. A **1993**, *37–38*, 796–799.

104. Yoshimi, Y.; Ohdaira, R.; Iiyama, C.; Sakai, K. "Gate effect" of thin layer of molecularly imprinted (poly methacrylic acid-co-ethyleneglycol dimethacrylate). Sens. Actuat. B **2001**, *73*, 49–53.

105. Noble, R.D. Generalized microscopic mechanism of facilitated transport in fixed site carrier membranes. J. Membr. Sci. **1992**, *75*, 121–129.

106. Wulff, G. Molecular imprinting in cross-linked materials with the aid of molecular templates—a way towards artificial antibodies. Angew. Chem. Int. Ed. Engl. **1995**, *34*, 1812–1832.

107. Roper, D.K.; Lightfoot, E.N. Separation of biomolecules using adsorptive membranes. J. Chromatogr A **1995**, *702*, 3–26.

108. Svec, F.; Frechet, J.M.J. New designs of macroporous polymers and supports: from separation to biocatalysis. Science **1996**, *273*, 205–210.

109. Trotta, F.; Drioli, E.; Baggiani, C.; Lacopo, D. Molecularly imprinted polymeric membrane for naringin recognition. J. Membr. Sci. **2002**, *201*, 77–84.

110. Piletsky, S.A.; Alcock, S.; Turner, A.P.F. Molecular imprinting: at the edge of the third millenium. Trends Biotechnology **2001**, *19*, 9–12.

111. He, B.; Tait, N.; Regnier, F. Fabrication of nanocolumns for liquid chromatography. Anal. Chem. **1998**, *70*, 3790–3797.

112. Brüggemann, O. Catalytically active polymers obtained by molecular imprinting and their application in chemical reaction engineering. Biomol. Eng. **2001**, *18*, 1–7.

113. Ulbricht, M.; Belter, M.; Langenhangen, U.; Schneider, F.; Weigel, W. Novel molecularly imprinted polymer (MIP) composite membranes via controlled surface and pore functionalizations. Desalination **2002**, *149*, 293–296.

114. Shi, H.; Ratner, B.D. Template recognition of protein-imprinted polymer surfaces. J. Biomed. Mater. Res. **2000**, *49*, 1–11.

19

Micromonoliths and Microfabricated Molecularly Imprinted Polymers

Jennifer J. Brazier and Mingdi Yan   Portland State University,
Portland, Oregon, U.S.A.

I. WHY MOLECULARLY IMPRINTED POLYMER MICROMONOLITHS?

Traditional MIP preparation involves polymerization in bulk followed by a grinding and sieving process to obtain particles smaller than 30 μm dimension. While this technique has shown much success, it also has several recognized limitations. The tedious grinding and sieving process commonly yields nonuniform particles in shape and size, which causes numerous problems such as limitations in chromatographic efficiency. Furthermore, the act of sieving may result in material loss as particles too small are discarded.

In response to these limitations, many research groups have focused on developing novel synthetic techniques to produce molecularly imprinted polymers with well-defined sizes and shapes such as beads, membranes, films, and monoliths (see Chapters 17 and 18 for details regarding beads, membranes and films). Unlike other polymer forms, which use the polymerization conditions to dictate polymer shape, micromonoliths are "container"-defined. These monoliths are fabricated in situ through either a thermal- or photo-initiated polymerization process within a reaction vessel. This greatly simplifies polymer preparation. Through this method, the tedious grinding and sieving procedure as well as the problems of costly particle loss and particle inhomogeneity are avoided.

The impetus for MIP monolith investigation began with the success of Fréchet and Svec in 1992 when they devised a procedure for the in situ preparation of macro-porous polymer monoliths for use in flow-through applications. Although their studies did not directly include imprinted polymers, their findings are absolutely relevant to understanding the morphological issues involved with polymer monolith fabrication. Due to their influence on MIP monolith research, this chapter will begin with a discussion of their findings. Following this discussion, focus will be placed on the unique issues and successes that arose when the procedures designed by Fréchet and Svec were adapted into protocols for the synthesis of MIP monoliths used in capillary electrochromatography and HPLC. Finally, this chapter will present MIP micromonoliths prepared using microfabrication techniques. This development may lead to emerging possibilities for on-chip separation and sensor applications.

II. THE IN SITU PREPARATION OF NONIMPRINTED MONOLITHS

The ability of a monolith to permit liquid to flow through the polymer matrix is critical to its function in applications such as ion exchange, catalysis, adsorption, and chromatographic separations. In contrast to common cross-linked polymers that must be swollen in a solvent to achieve porosity, macroporous polymers possess a fixed pore structure even when dry. Invented in the late 1950s, these polymers were first prepared as macroporous beads through a suspension polymerization technique [1–3]. However, in 1992, Fréchet and Svec, [4] detailed a new procedure to make macroporous polymer monoliths for applications such as capillary electrochromatography and HPLC. Additional progress for continuous bed applications has been advanced in extensive research by Hjertén and others [5–9] as well as detailed work by Tanaka and coworkers [10–15] on silica-based monoliths. The development of macroporous polymers in the monolith format has provided its own unique challenges. Procedures, which had been previously used to obtain macroporous polymer beads, could not be directly translated to making macroporous polymer monoliths. The pore size distributions created for monoliths were shown to be different than those for macroporous beads prepared from identical polymerization mixtures [16,17]. Differences between the two processes, such as the absence of interfacial tension between the aqueous and organic phases and the absence of stirring, created completely different morphologies. Therefore, new synthetic protocols were required in order to produce macroporous monoliths with desirable pore size distributions.

The pore size distribution within a monolith has a direct effect on the performance of the material. Large macropores are required for the mobile phase to flow through the monolith at low pressures, while mesopores (2–50 nm) and micropores (< 2 nm) afford a high surface area for increased capacity. Optimizing this distribution is required for different applications. Pore sizes ranging from a few to tens of microns have shown to be appropriate for micromonoliths used in microfluidic applications [18].

Through various studies, Fréchet and Svec and coworkers [4,16–20] determined the effects of different factors on the porosity of a monolith. In order to discuss these factors, however, it is first necessary to understand the actual mechanism for pore formation during the polymerization process. The classical mechanism for pore formation depends on the porogen used [1,20,21]. In this mechanism, an organic phase is present with monomers, cross-linkers, an initiator, and a porogenic solvent. The polymerization process begins as the initiator decomposes through either thermal or photolytic degradation. The polymer chains that are formed precipitate out of solution and form insoluble gel-like nuclei either due to their extensive cross-linking or because the porogen present becomes a poor solvent for the growing polymer chain. Once these nuclei are formed, the monomers from the solution flood into the nuclei because they are thermodynamically better solvating agents for the polymer than the porogen. Therefore, the polymer nuclei begin to swell. Polymerization still occurs in the solution, however, it is kinetically preferred inside the nuclei since there is a greater local concentration of monomers. These nuclei continue to grow and become cross–linked to other nuclei via branched or cross-linking polymer chains. In this manner, clusters of nuclei (globules) are formed. The clusters eventually contact one another to make a matrix filled with cross-linked globules and voids (pores).

At the end of polymerization, the voids between cross-linked polymer globules are filled with the porogen. However, from this mechanism little else can be predicted about the sizes of these pores. Through many studies, Fréchet and Svec and coworkers [4,16–20] identified various factors that appeared to control the average pore size and pore size distributions. These include temperature of polymerization, composition of pore forming solvent mixture, and content of cross-linker.

By varying the polymerization temperature (or irradiation power), average pore sizes within a range of two orders of magnitude have been created from a single composition of polymerization mixture [20]. In general, increasing the temperature/ irradiation power results in an increase in the volume fraction of smaller pores present in the monolith. This effect can be explained in terms of nucleation rates and the number of polymer nuclei that form during polymerization [20].

Polymerization begins with the decomposition of the initiator. An increase in polymerization temperature causes an increased number of free radicals and therefore a larger number of growing nuclei and globules. The increase in number of nuclei is at the cost of their size. Since the globules obtained are smaller and more numerous, smaller voids are obtained resulting in a shift in the pore size distribution.

However, the effect of temperature can be reversed if a very poor solvent is used, such as dodecanol for the polymerization of styrene and divinylbenzene. In such a case, the temperature affects the solvency, which in turn controls the phase separation of the polymers from solution. As the temperature increases, the solubility of the polymer in the porogen increases. The precipitation may occur when the nuclei reach a higher molecular weight. Therefore, the nuclei and the pores between them are larger. Fréchet and Svec et al. found that this effect was not apparent, however, when mixtures of a very good solvent and a poor solvent were used. In these cases, the pore size is again controlled by nucleation rates [20].

The role that temperature plays can become especially apparent when attempting to make larger monoliths. Inside a large, unstirred mold the dissipation of the heat of polymerization can be ineffective. This causes temperature gradients within the mold, which in turn lead to heterogeneous pore structures throughout the monolith. For example, in a study by Fréchet and Svec et al., a methacrylate solution polymerized at 55°C within a 26-mm diameter mold showed the absence of a significant reaction exotherm and displayed a homogeneous pore distribution [22]. However, the same reaction within a 50-mm diameter mold exhibited a 25°C temperature differential across the column radius resulting in significant difference in the pore distributions between the center core and the outer shell of the monolith. In order to better control the temperatures within larger molds, one could therefore run the reactions at rates slow enough to allow dissipation of the heat of polymerization. This may be accomplished by either running reactions at lower polymerization temperatures or through the gradual addition of the polymerization mixture into the reaction vessel once polymerization has started.

The composition of the porogenic solvent also plays an important role in the pore size distribution. Adjustment of this composition is the most commonly used approach for controlling monolith porosity. In general, larger pores are obtained in a poorer solvent because of an earlier onset of phase separation. This control is explained through solvation of the polymer chains in the reaction medium during the early stages of polymerization. Remember that during polymerization, the growing polymer chains precipitate out of solution either because of extensive cross-linking or because their molecular weights have exceeded the solubility limit of the porogen. If a poor solvent is used, an earlier phase separation occurs and the nuclei that form swell with monomers. The globules that are formed are larger and thus the voids between them are larger. On the other hand, if the solvent is good, it will compete with the monomers to solvate the nuclei. Therefore, the local monomer concentration within the nuclei is smaller and the nuclei tend not to swell as much. This leads to overall smaller globules and thus smaller pores.

A study by Santora et al. [23] provides a good illustration of this point. For the nonpolar system divinylbenzene/styrene, the nonpolar porogen n-hexane generated a smaller average pore size, smaller globules, and higher surface areas. The polar porogen methanol, on the other hand, gave a larger average pore size, larger globules, and thus lower surface areas. In the more polar ethylene glycol dimethacrylate/methacrylic acid (EDMA/MMA) system, this role was reversed. Nonpolar n-hexane now yielded low surface areas, while polar methanol produced high surface area materials [23]. Figure 1 demonstrates the difference in porosity that can be attained simply through the choice of porogen.

Finally, the cross-linking composition plays a large role in the pore size distribution of the monoliths. A higher cross-linking content decreases the average pore size within the monolith. This can be explained by the earlier formation of highly

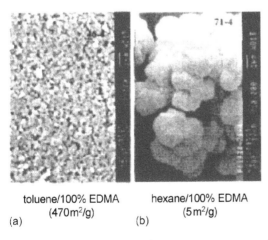

toluene/100% EDMA (470 m²/g) (a)

hexane/100% EDMA (5 m²/g) (b)

Figure 1 SEM images of two macroporous polymers formed from polymerization of EDMA. The two polymers exhibit completely different pore size distributions depending on the porogen used. Image (a) illustrates the large surface area (small pore) obtained from a polymer using toluene as the porogen, while (b) displays the smaller surface area (large pore) that results from the use of hexane as the porogen. The scale bar represents 1 μm. (Reproduced with permission from Ref. 20.)

cross-linked globules with a reduced tendency to coalesce and swell [20]. The extensive cross-linking within the nuclei prevents them from swelling with the monomers, so they remain small. Furthermore, this extensive cross-linking inhibits the coalescence between polymer nuclei.

All these factors must be taken into account when attempting to fabricate monoliths with not only good flow-through characteristics but also optimized pore distributions for a specific application. The novel work by Fréchet and Svec in 1992, which optimized the porosity of a methacrylate-based polymer for chromatography application, was a jumping point for many subsequent studies on macroporous monoliths. As discussed in the following sections, their protocol was absolutely influential on the first attempts at combining monoliths with molecularly imprinted polymers.

III. IN SITU PREPARATION OF MIP MONOLITHS FOR CAPILLARY ELECTROCHROMATOGRAPHY

Original attempts at integrating capillary electrophoresis with molecularly imprinted polymers utilized two distinct formats (see Chapter 20, Sec. II). The first configuration, open-tubular chromatography, used thin coatings of MIP on the inside walls of a capillary tube [24–27]. As the mobile phase advanced through the column, different strength adsorptive interactions between the molecules present in the mobile phase and the thin coating cause chromatographic separations between the molecules. Separations are further facilitated by differences in electrophoretic mobilities of the molecules under an applied electric field. Despite the ease of fabrication, this method has a serious limitation. The low surface-to-volume ratio provided by such open capillaries leads to a low loading capacity. Therefore, only small sample volumes may be used.

The second configuration sought to increase the available surface area present inside the capillaries by packing them with conventionally ground and sieved MIP particles approximately 10 µm in diameter. These particles are generally slurry packed into fused-silica capillary columns of approximately 75-µm inner diameter [28]. To prevent MIP particles from being eluted out of the column during electrochromatography, small retaining frits were fabricated at the ends of the column either by burning silica polymer particles or through the use of a polyacrylamide plug (see Chapter 20, Sec. II for details).

Unfortunately, this method also has several major limitations. First, the process of slurry packing the MIP particles into the capillary column can be very tedious and limits the lengths of capillary columns that may be used. Second, the retaining frits necessary for keeping the MIP particles within the column are not only difficult to fabricate, but are also a major contributor to bubble formation within the column, which stops electro-osmotic flow and effectively prevents the column from working. Furthermore, the use of frits creates the potential for unwanted adsorptive interactions between the analyte and the frit material. Finally, the irregular shape of ground and sieved MIP particles limits chromatographic resolution and efficiency.

In response to these limitations, there have been many efforts to improve these columns. Through use of techniques such as dispersion polymerization [29], researchers have synthesized MIP beads to improve the geometrical uniformity of

MIP particles in CEC. Other groups have attempted to remove the need for problematic retaining frits by first packing the column with ground MIP particles and subsequently polymerizing polyacrylamides [30] or silicates [31] around these particles so they will not be eluted from the column.

The macroporous monolith approach, introduced by Fréchet and Svec, seemed to address many of the problems associated with open-tubular and particle packed columns. First, the adsorptive capacity of capillary monoliths has been found to be 3–5 orders of magnitude larger than that of both open channel and bead-packed columns [32]. Next, since the polymerization takes place within the fused-silica capillary, the tedious process of packing the capillary columns may be avoided. Furthermore, the limitations in chromatographic efficiency caused by irregularities in particle packing and by the nonuniformity of particle sizes are eliminated.

MIP monoliths also solve the difficulties associated with retaining frits. Typically, the capillaries used for monolith fabrication have been derivatized prior to filling so that covalent attachments may form between the capillary walls and the polymer monolith during the polymerization. For example, capillaries that are to be filled with a methacrylate-based polymer may be treated with [(methacryloxy)-propyl]trimethoxysilane according to the procedure developed by Hjertén [33]. This treatment allows the silane-coated walls to copolymerize with the monomers and cross-linkers during MIP formation. The increased mechanical strength provided by these attachments allows a relatively large volume of solution to be pumped through the column within a short period of time, thus increasing the flow rate of the device. Without these covalent attachments, the polymer may be eluted from the capillary column during hydrodynamic pumping and electrochromatography. With the surface treatment, no frits are necessary.

To successfully prepare MIP monoliths in fused-silica capillaries, several conditions must be met. First, the polymerization must be performed under conditions that are favorable to imprint formation (see Chapters 7 and 15). Second, the electrochromatography conditions must favor rapid association and dissociation of the analyte to the imprinted cavities within the column. The electrolytic composition, pH, and temperature play a large role in the selectivity and separation efficiency of these columns [34]. Finally and most importantly, the monolith must possess good flow-through characteristics so that the solvent used in polymerization may be easily exchanged for the electrophoresis electrolyte via hydrodynamic pumping. If the polymer is too dense as shown in Fig. 2a, the electrolyte can no longer be pumped through it. On the other hand, too porous a polymer may result in a low capacity or only a thin film of MIP as shown in Fig. 2b, which compromises the separation ability of the column [35].

The early attempts at fabricating molecularly imprinted capillary monoliths adapted the procedure set forth by Fréchet and Svec [4] for the in situ preparation of non-MIP macroporous polymer rods for LC separation. In this procedure, porogenic solvents cyclohexanol and dodecanol (80:20 v/v) were used with a methacrylate-based polymer system to produce porous monoliths. When this system was applied to the fabrication of molecularly imprinted monoliths for CEC, the polymers obtained were sufficiently porous but resulted in poor enantiomeric separations [36]. It is thought that the polar-protic nature of the porogens used may have inhibited the formation of well-defined imprints. Polar-protic solvents such as these are often poor porogens for the noncovalent imprinting approach because they interfere

(a)

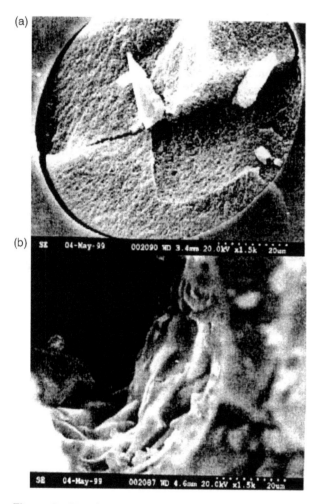

(b)

Figure 2 SEM images of MIP monoliths fabricated inside capillary columns. Image a represents a polymer which is too dense and cannot undergo hydrodynamic pumping. Image b illustrates the other side of the spectrum where only a thin film of polymer remains, thereby compromising the capacity of the polymer. (Reproduced with permission from Ref. 35.)

with ionic and hydrogen-bonding interactions often responsible for the strong template–monomer complexation (see Chapter 3). A less polar, aprotic solvent such as chloroform is a better choice for the porogenic agent and has been successfully used in a methacrylate-based MIP monolith fabricated for the separation of phenylalanine anilide enantiomers [37].

Unfortunately, the commonly used imprinting solvents such as chloroform and toluene may not work for all systems. For example, in preparing an CEC monolith composed of functional monomers MAA, cross-linking monomer trimethylolpropane trimethacrylate (TRIM), and radical initiator 2,2'-azobis(isobutyronitrile) (AIBN), Schweitz et al. found that the use of toluene as the only porogenic solvent failed to produce sufficiently porous monoliths [34]. The polymers produced using toluene were very dense and hydrodynamic pumping was not possible. Under these circumstances, two methods may be used to achieve acceptable porosity. The first method involves interrupting the polymerization midway through the process and flushing the capillary to remove all unreacted species, thereby leaving behind a porous framework [38,39]. The second method involves adjusting the composition of the porogenic agent to control porosity [34,40–42]. Both methods have been used with the system mentioned above to obtain porous MIP monoliths capable of separating enantiomers of propranolol and metoprolol.

In order to use the first method, the polymerization time and temperature must be carefully optimized through trial and error in order to achieve good flow-through properties. Too much polymerization results in a monolith that cannot be flushed via hydrodynamic pumping; while lower degree of polymerization yields a small amount of polymer that the capacity is compromised. The tedious work necessary to optimize the polymer porosity is the main drawback of this method. The polymerization is often difficult to control. However, once conditions for polymerization are determined, the success rate for making the monoliths appropriately porous has been shown to be 100% [39].

In the second method, the porogen is adjusted to achieve better flow-through properties. Changing the composition and percentage of the porogenic agent allows for the adjustment of the pore size and overall morphology of the monolith. For example, while toluene alone yielded monoliths too dense for hydrodynamic pumping, the addition of isooctane in toluene (1–25% v/v) produced monoliths of suitable porosity [34,40–42]. Of course, the composition/percentage of porogen must be carefully optimized not only for the porosity but also for the selectivity of the MIP system. However, once an appropriate porogen is chosen and its volume optimized, the reaction may be allowed to go to completion. Therefore, careful timing of the reaction is not an issue. For this reason, porogens have been used more commonly to achieve good flow-through properties than relying on interrupted polymerization [34,40–42].

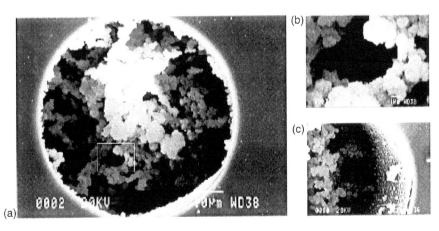

Figure 3 SEM images of a polymer-filled capillary column. (a) Micrometer-sized globular units of macroporous molecularly imprinted polymer surrounded by 1–20 μm wide interconnected superpores. (b) A superpore, about 7 μm wide, magnified from the square present in (a). (c) The covalent attachments of the polymer to the capillary wall can be seen. (Reproduced with permission from Ref. 38.)

In studies by Schweitz et al., capillaries were judged successful if uniform, rigid monoliths were observed upon visual inspection with a microscope and if the capillary was capable of being flushed at low pressures using a syringe. When viewed with scanning electron microscopy, successful monoliths showed aggregates of micrometer-sized globular particles (0.5–1 μm) throughout the capillary as shown in Fig. 3.

The small size of these aggregates and the macroporous structure (200 Å) helped with analyte diffusion and access to imprinted sites [38]. Wide interconnecting superpores (1–20 μm) surrounded the aggregates and permitted bulk flow and easy electrolyte exchange by hydrodynamic pumping. The protocol below details the procedures set forth by Schweitz to successfully prepare these columns [34,40–42].

IV. IN SITU PREPARATION OF MONOLITHIC MIP CHIRAL STATIONARY PHASES FOR CAPILLARY ELECTROCHROMATOGRAPHY

A. Reagents and Equipment

UV transparent fused-silica capillaries (50, 75, or 100-μm i.d., 375-μm o.d.) (TSU075; Polymicro Technologies, Pheonix, AZ) were used. Trimethylolpropane trimethacrylate (TRIM) was purchased from Aldrich. AIBN and (S)-propranolol hydrochloride were obtained from Sigma. The (S)-propranolol hydrochloride was changed into its free-base form (1) through extraction in a solution of ethyl acetate and saturated NaHCO$_3$. Following this treatment, the free-base was washed once with water, dried and stored at −20°C until use.

The toluene was dried and stored over 4 Å molecular sieves. All other chemicals were purchased from Merck and were used without further purification.

B. Method

The first step was to clean the capillary by flushing it with 1 M NaOH for 30 min followed by water for 30 min. The capillary was then derivatized by filling it with a mixture of 4-μL of methacryloxy propyltrimethoxysilane in 1 mL of 6 mM acetic acid and the solution was kept in the capillary for at least 1 h. Then the capillary was rinsed with water and dried with nitrogen.

An UV detection window was prepared on the capillary by removing 0.5 cm of the protecting polymer layer from one end of the capillary by burning. The detection window was covered with aluminum foil so that photopolymerization would not occur in this area. Plastic tubing was then placed on both ends of the capillary and it was filled by syringe with a polymerization mixture consisting of monomer (0.72 mol/L MAA), cross-linker (0.72 mol/L TRIM), radical initiator (0.022 mol/L

(S)-Propranolol (1)

AIBN), and template molecule (0.030 mol/L (*S*)-propranolol free base) in toluene/ isooctane (95/5 v/v). The capillary ends were finally sealed using clips on the plastic tubing.

An alternative method for preparing the UV detection window has been suggested by Lin et al. [37]. In this method the column consisted of two capillary columns that were joined. One capillary column contained the MIP monolith. The other capillary column housed the UV detection window and was filled with electrophoretic buffer. The two sections possessed the same inner and outer diameters. Therefore, they could be connected by a Teflon tube as shown in Fig. 4.

Polymerization was generally carried out in a freezer at −20°C under a 350 nm UV source. Following polymerization, capillaries were rinsed with a solvent such as acetonitrile and electrolyte such as acetonitrile in a low pH buffer solution of 2M acetic acid/triethanolamine (pH 3.0 (80% acetonitrile to 20% buffer, v/v)) in order to wash out any remaining initiator, unreated monomers, and template molecules. This entire process is shown in Fig. 5.

Using this protocol, capillaries were ready to use within 3 h of the start of capillary preparation. They were used continuously during several weeks with repeatable results. When not in use, capillaries were stored at room temperature for at least 12 weeks without any observable changes in selectivity or chromatographic performance. Partially or completely dried capillaries were regenerated through hydrodynamic pumping of fresh electrolyte through the capillary.

Once sufficiently porous monoliths are created, it is still necessary to optimize the MIP system to achieve good CEC separations. Choice of functional monomer, cross-linking monomer, and molar ratio of imprint molecule to monomers will all affect the separation ability of the column. Furthermore, the electrophoretic conditions must be optimized including electrolyte composition and pH (see Chapter 20.II).

Under such optimizations, MIP micromonoliths have shown much promise in CEC (see Chapter 20, Sec. II). The ease of fabrication combined with the minimal consumption of materials (approximately 10–100 nmol of template for a typical 1.5-μL capillary tube) makes this method especially convenient [38]. Since polymerization is in situ, long columns may be prepared without the problems associated with packing.

However, it is still unclear how the efficiencies and resolutions of MIP monoliths compare to other methods of capillary fabrication using molecularly imprinted polymers. Future studies that directly compare the chromatographic results obtained using the same imprinting system under different preparation techniques (i.e., packed columns, coated thin films, immobilized particles, and monoliths) have

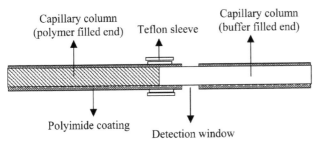

Figure 4 Alternative preparation scheme for capillary columns filled with polymer monoliths in CEC. (Reproduced with permission from Ref. 37.)

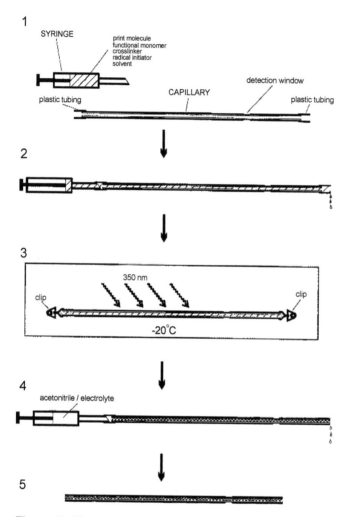

Figure 5 Illustration of the preparation procedure for capillary columns with molecularly imprinted polymer monolith. (1) Polymerization mixture is prepared and plastic tubing is placed on both ends of a capillary derivatized with methacryloxy propyltrimethoxysilane. (2) The capillary is filled with the mixture using a syringe. (3) The capillary is sealed by placing clips on the plastic tubings. The polymerization is performed under an UV source (350 nm) at −20°C for 80 min. (4) The polymerization reaction is interrupted by flushing remaining monomer, radical initiator, and imprint molecule out of the capillary. (5) The capillary column is then ready for CEC. (Reproduced with permission from Ref. 38.)

yet to be performed. Furthermore, tests regarding the column-to-column reproducibility of monolith synthesis should be investigated.

V. PREPARATION OF MIP MONOLITHIC RODS FOR HPLC

Until the early 1990s, column liquid chromatography was performed almost exclusively with small diameter spherical beads. The procedure presented in 1992 by

Fréchet and Svec for fabricating continuous, macroporous polymer rods for HPLC [4] generated much interest in the imprinting field. The simplicity of fabrication is the main attraction of this structure. These monoliths also hold the potential for increasing the efficiencies obtained from columns traditionally packed with irregularly shaped ground and sieved MIP particles. Furthermore, since no polymer is thrown away without being packed into the column, expensive reagent loss may be avoided.

The main requirement for these monoliths is again attaining a porous network with good flow-through characteristics. Similar to early attempts at CEC imprinted monoliths, fabrication of HPLC monoliths directly adapted the procedures of Fréchet and Svec, including the use of the polar-protic solvents cyclohexanol and dodecanol, which have been shown to give optimal porosity [1]. Although enantiomeric separations were achieved [42], it was still thought that these polar-protic solvents were compromising the imprinting process. However, unlike capillary electrochromatography, the limited research in HPLC imprinted monoliths has focused mainly on improving the functional monomers rather than finding different porogens or attempting to interrupt the polymerization [43–45].

Functional monomers that interact strongly with the template molecules even in the presence of these polar-protic solvents are desirable as well as monomers capable of simultaneous multiple interactions. For example, Matsui has found 2-(trifluoromethyl) acrylic acid (TFMAA) to be a better alternative to MAA in imprinting monoliths for antimalarial cinchona alkaloids (45). TFMAA is a stronger acid than MAA due to its strongly electron withdrawing trifluoromethyl substituent, therefore stronger complexation was achieved when dealing with basic template molecules.

Once the imprinting system has been devised to yield favorable monomer–template complexation and the necessary porosity, the preparation of monolithic polymer rods for HPLC is relatively simple. The general protocol detailed below uses an in situ polymerization method developed by Fréchet and Svec [4]. This technique was used by Matsui in the preparation of MIP monolith rods for HPLC separation of antimalarial cinchona alkaloids, (−) cinchonidine and (+) cinchonine, as well as the structural analogues quinidine and quinine [45].

VI. IN SITU PREPARATION OF MOLECULARLY IMPRINTED POLYMER RODS FOR HPLC

A. Reagents and Equipment

A stainless steel tube (50 mm × 4.6 mm i.d. or 100 mm × 4.6 mm i.d.) may be used as the HPLC column. Chemicals were purchased from Katayama (Osaka, Japan). Monomers were purified before use [46]. All other chemicals were used without further purification.

B. Method

An in situ thermal polymerization technique was used to make MIP rods for HPLC. The template molecule (−) cinchonidine (2) or (+) cinchonine (3) (0.2 mmol), cross-linking monomer ethylene glycol dimethacrylate (850 mg), functional monomer 2-(trifluoromethyl)acrylic acid (1.2 mmol), and initiator 2,2′-azobis

(-) Cinchonidine (2)   (+) Cinchonine (3)

(dimethylvaleronitrile) (85 mg) were dissolved in a mixture of cyclohexanol (1.2 g) and 1-dodecanol (0.3 g). The mixture was then degassed by sonication and poured into a stainless steel tube (50-mm × 4.6-mm i.d.). Polymerization was initiated by incubating the column tube at 45°C for 6 h. The column was then fitted with LC connections and subsequently washed with methanol at a flow rate of 0.01–0.1 mL/min for 12 h to remove the porogenic solvents, unreacted monomers, and the template compound. Finally, washing with methanol–acetic acid (8:2, v/v) ensured removal of the template, as determined by UV baseline stability. Nonimprinted blank polymer rods were prepared similarly without the template molecule. Figure 6 summarizes this technique.

Using these conditions, stereoselective recognition of templates and structural analogues quinidine and quinine was observed. However, the separation factors were lower than those obtained using columns packed with ground MIP particles. The separation ability of these columns may be improved through the use of less polar porogens. Furthermore, photolytic polymerization at low temperatures rather than the thermal polymerization used may result in more stable complexes. This would involve a change in column design since the stainless steel columns used are not transparent to UV.

Until column designs allow photopolymerization, one could continue to focus on the following factors: (a) increasing monomer–template interactions; (b) using less polar porogens to enhance these interactions while still achieving the necessary porosity; and (c) lowering the polymerization temperature. For example, McNiven et al. [47] examined all three of these factors when developing an LC-based sensor for the antibiotic chloramphenicol (CAP, 4). This sensor utilizes the displacement of a chloramphenicol dye conjugate (chloramphenicol-methyl red) from the imprinted sites to detect trace amounts of CAP via optical detection at 456 nm.

In this study, three different functional monomers were tested for their imprinting effectiveness. Diethylaminoethyl methacrylate (DAM) had been proven successful before. Allylamine (AA) was thought to be promising due to its basic amino functionality. Methacrylic acid (MAA) is a widely used functional monomer. In these tests it was found that DAM was the most effective monomer while MAA was not effective at all. DAM is a stronger Lewis base than allylamine, therefore, it is better able to form a complex with the hydroxyl groups on the 1,3-diol group of CAP.

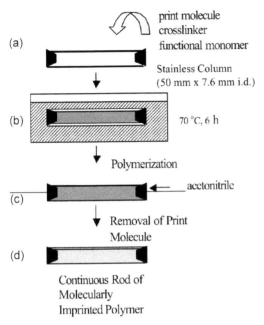

Figure 6 (a) Stainless steel column is filled with polymerization mixture and capped at both ends. (b) Column is incubated in a water bath to allow polymerization to occur. (c) Column is flushed with solvent to remove print molecule. (d) Column is ready to use. (Reproduced with permission from Ref. 43.)

To achieve high fidelity imprints, it was desirable to have as apolar a solvent as possible, but CAP was not sufficiently soluble in dodecanol alone. So enough octanol was added until CAP dissolved. It was also found that lower polymerization temperatures resulted in stronger complexes, so the mixtures were polymerized at 50°C for 24 h.

Micromonoliths used as sensor recognition elements provide a quick and easy fabrication scheme. One limitation to these monolith sensors was their capacity. The capacity of rod polymers was found to be very low and lower than LC columns packed with MIP particles. However, their ability to separate templates such as chloramphenicol from structurally similar analogues has been shown to be slightly better [47]. This may be an indication that while the number of viable imprinted sites is less, they are of similar selectivity.

Chloramphenicol (**4**)

VII. MICROFABRICATED MIP MONOLITHS

Recently, there has been interest in shrinking imprinted polymers into micromono-liths that may be later integrated into miniaturized systems capable of performing on-chip chromatographical separations and sensing. Advantages in speed, portabil-ity, sample/reagent consumption, and efficiency may be gained through the applica-tion of such miniature systems. Currently, the vast majority of microfluidic chips possess open channel formats [48,49]. These systems are best suited when interac-tions with a solid phase is not required such as in electrophoresis. For applications such as solid-phase extraction, separation, or catalysis, a solid phase inside the chan-nels is desirable. One solution is to coat the inside of the channel with a thin coating of a solid phase. The drawback to this design is the low surface-to-volume ratios provided by the thin coatings, which results in low loading capacities. Other config-urations such as beads are rarely used in these devices due to the difficulty of packing the particles into the channels of a microchip, especially if the channels possess com-plex architectures [50]. Monoliths provide an alternative design with benefits that include ease of fabrication and higher loading capacities.

VIII. MIP MICROMONOLITHS FABRICATED THROUGH MICROMOLDING IN CAPILLARIES

Recently, the soft lithography [51,52] technique of micromolding in capillaries (MIMIC) [53,54] has been used to fabricate MIP micromonoliths on silicon wafers [55,56]. In MIMIC, an elastomeric stamp that possesses recessed features is placed in intimate contact with a solid substrate. The recessed microchannels on the stamp form a network of empty capillaries. When a low-viscosity fluid precursor is placed at one end, it spontaneously fills the channels by capillary action. Curing of the fluid leaves patterned microstructures on the substrate surface as shown in Fig. 7.

The technique of MIMIC has a number of advantages. First, it allows the use of a liquid precursor, the imprinting solution, which is otherwise difficult with conven-tional photolithography techniques. Second, the shape and size of MIPs can be read-ily controlled and altered by those on the stamp. Third, stamps are generally made from poly(dimethyl siloxane) (PDMS), which is transparent to UV light down to 300 nm and it is therefore compatible with the photochemical polymerization proce-dures employed in MIP synthesis. Fourth, once the master mold is made, the process can be carried out conveniently in a chemical laboratory. No special facility or equipment is needed.

PDMS stamps are fabricated by casting a mixture of an PDMS prepolymer and the corresponding curing agent on a master mold, which has been treated with a fluorinated silane to ensure facile release of the PDMS stamp. The master mold con-tains the inverse features of the stamp, and is made by conventional photolithogra-phy. Although this procedure is typically performed in labs specifically designed for photolithography, Whitesides and coworkers [57] have developed a procedure for creating master molds with feature sizes larger than 20 μm using equipment available to most chemistry labs such as an office printer, UV lamp, and photoresists. Depend-ing on the desired feature sizes, certain photoresists may be required. For example, to create thick features (20–100 μm) a specialized photoresist, SU-8 negative resist

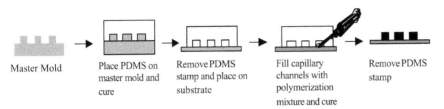

Master Mold   Place PDMS on   Remove PDMS   Fill capillary   Remove PDMS
               master mold and stamp and place on channels with   stamp
               cure   substrate   polymerization
                                                  mixture and cure

Figure 7 Schematic illustration for the fabrication of polymer microstructures using the technique of micromolding in capillaries.

(MicroChem Corp.), has been used [55]. Master molds may also be commercially purchased, but are rather expensive.

A major limitation of this technique, however, is that the cross-linked PDMS stamps tend to swell in many organic solvents that include chloroform, methylene chloride, toluene, and tetrahydrofuran. These solvents have been widely used as porogens in the synthesis of molecularly imprinted polymers. The PDMS stamps swelled dramatically in these solvents and subsequently lifted off from the substrate [55]. Once a stamp has lost the conformal contact with the substrate, the assembly can no longer be used to make polymer structures.

To date, MIP microstructures have been successfully fabricated with dimethyl-formamide (DMF) as a solvent in both an acrylate-based [55] and a polyurethane-based [59] imprinting system. While this solvent still swells PDMS, its effect is small. Other imprinting systems that utilize water and alcohols as the solvent may be possible since these solvents do not swell the PDMS stamps. Alcohols are the best as they have a low surface tension and wet the surface of the PDMS well. Water, however, has a high surface tension and does not wet the surface of the hydrophobic PDMS stamps. Possible solutions may be to render the stamp more hydrophilic by O_2 plasma or UV irradiation to add a small percentage of alcohol ($\sim$5%) to the aqueous solution to wet the surface, or to use a small vacuum-assisted pump to fill the channels.

The protocol below outlines the steps used to create polyurethane-based MIP micromonoliths covalently attached to a silicon wafer support. The imprinting system was adapted from Dickert and Thierer [60], Dickert and Tortschanoff [61] and has not been optimized for the porosity required for flow-through monolith applications.

IX. MICROMOLDING IN CAPILLARIES FOR THE GENERATION OF MOLECULARLY IMPRINTED POLYMER MICROMONOLITHS

A. Reagents and Equipment

PDMS (Slygard 184 elastomer, Dow Corning) and its corresponding curing agent were purchased from KR Anderson (Kent, WA). SU-8 photoresist was purchased through Microchem Corporation (Newton, MA). Silicon wafers were obtained from University Wafers (Boston, MA). Anthracene was purified through recrystallization prior to use. Dimethylformamide was dried and stored over 4 Å molecular sieves. A mixture of p,p'-diisocyanatodiphenylmethane (**5**) and p,o,p'-triisocyanatodiphenyl-

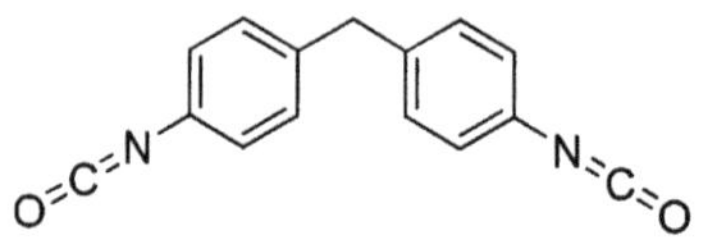

p,p'-diisocyanatodiphenylmethane **(5)**

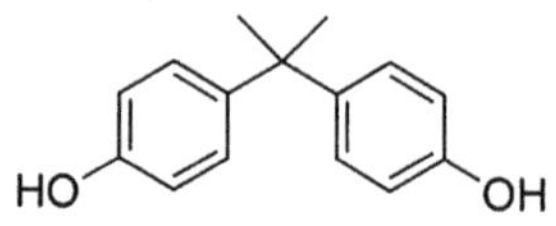

bisphenol A **(7)**

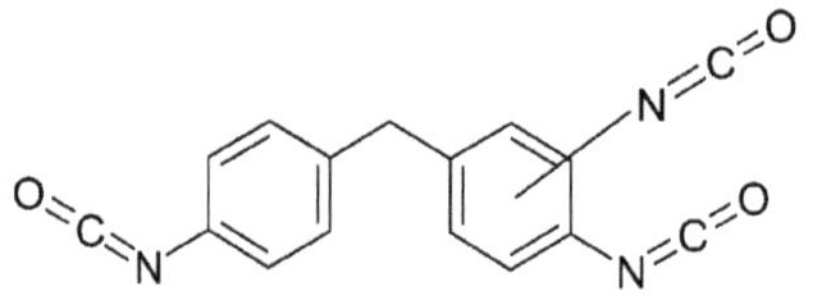

p,o,p'-triisocyanatodiphenylmethane **(6)**

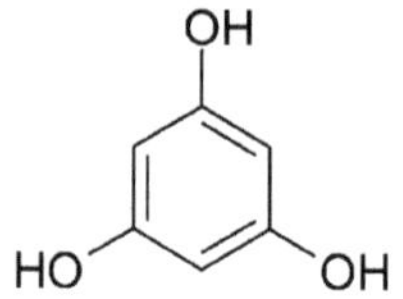

trihydroxybenzene **(8)**

methane **(6)** was obtained from Merck (Darmstadt, Germany) and was used as received. Additional functional monomer, bisphenol A **(7)** and cross-linking monomer, trihydroxybenzene **(8)** were purchased from Aldrich and were used without further purification.

B. Method

Master molds were prepared through conventional photolithography using SU-8 [25] negative photoresist. In this process, a 100 μm thick layer of SU-8 was spin-coated at 525 rpm onto a clean silicon wafer. This wafer was then soft baked at 95°C for 45 min to remove the solvent from the resist layer. It was necessary to slowly ramp up the temperature (2°C/min) from room temperature in order to avoid bubble formation during all baking procedures. Following the soft bake, a patterned quartz/iron oxide mask possessing transparent 100 μm lines was placed in conformal contact with the photoresist-coated wafer. Subsequent UV illumination through the transparent features on the mask initiated cross-linking of the photoresist in the exposed regions. Further baking at 95°C for 15 min completed the cross-linking of the polymer. All the unexposed photoresist was then washed away using SU-8 developer (Microchem Corporation), thereby leaving behind microstructures of 100×100 μm dimension. This wafer was finally treated with a solution of 74 μL tridecafluoro-1,1,2,2-tetrahydrooctyltrichlorosilane in 100 mL toluene for 5 min to prevent the PDMS from sticking to the wafer later.

Elastomeric stamps were produced by combining PDMS and its curing agent in a 1:0.07 ratio (wt/wt) and pouring this mixture over a master pattern produced above. The elastomer was cured at 70°C for 4 h before being peeled off of the master. Both ends of the stamp were cut to open up the microchannels, sonicated twice in ethanol for cleaning, and dried under nitrogen before being placed on the wafer substrate.

The substrates, silicon wafers, were cleaned in piranha solution (conc $H_2SO_4/30\%H_2O_2$; 4:1 v/v), washed thoroughly with boiling water, and dried under nitrogen. In order to ensure covalent attachment of the MIP microstructures to the wafer support, the silicon wafers were previously treated with 3-aminopropyltrimethoxysilane.

Capillaries were then filled with the imprinting mixture containing monomers (0.375 mmol bisphenol A and 0.455 mmol p,p'-diisocyanatodiphenylmethane), cross-linkers (0.250 mmol trihydroxybenzene and 0.195 mmol p,o,p'-triisocyanato-diphenylmethane, template molecule (0.0485 mmol anthracene), and porogen (DMF). Overnight polymerization under ambient conditions was used to cure the polymer. A quartz plate was placed on top of the PDMS to avoid lift-off of the stamp during polymerization. Once cured, the elastomeric stamps were peeled from the wafer support leaving behind the polymer microstructures.

The technique of MIMIC relies on the movement of fluid inside the capillary channels. The channels need to be interconnected in order for the fluid to flow through. For noninterconnected patterns, access holes may be drilled through the PDMS stamp [62]. However, if the device contains a large amount of such noninter-connecting patterns, the filling process can be extremely slow and inefficient.

Freestanding micromonoliths can also be obtained following a similar procedure. In this case, instead of silanizing the wafer, a thin layer of PDMS/curing agent was spin cast on the wafer surface and allowed to cure. Therefore, the polymers were sandwiched between two PDMS, and were released after the PDMS was removed by either dissolving PDMS with tetrabutylammonium fluoride [55], or by swelling PDMS with a solvent such as toluene to physically deform it [58]. The polymers were centrifuged, extracted with solvents, and dried for subsequent analysis. As shown in Fig. 8, isolated 20 μm polyurethane filaments imprinted against anthracene have been prepared using this fabrication technique.

This freestanding configuration may be useful for examining the porosity, morphology, and binding abilities of the micromonoliths before their integration onto microchips. One consideration when using this method, however, is that compared to MIPs prepared by bulk polymerization in large test tubes, this technique produces a small quantity of materials. With a cross-sectional dimension of 20 μm × 20 μm and a spacing of 20 μm between channels, an PDMS stamp of 1 cm × 1 cm produces approximately 1.5 mg of MIP, assuming the density of the MIP to be ~1 g/cm³. To prepare large amounts of materials for full binding assays is time consuming. Solutions may include using larger stamps and to use several PDMS stamps simultaneously on a substrate.

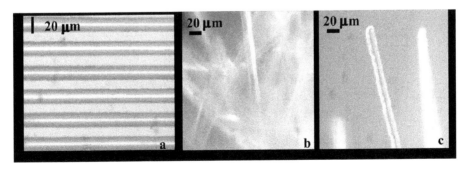

Figure 8 Optical micrographs of 20 μm MIP micromonoliths (a) on an PDMS coated wafer and after isolation to form freestanding micromonoliths (b, c). (Reproduced with permission from Ref. 59.)

X. MIP MICROMONOLITHS GENERATED BY MICROSTEREOLITHOGRAPHY

MIP microstructures have also been fabricated using an emerging microfabrication technique, microstereolithography [63]. Microstereolithography (μSL) is a method for generating complex three-dimensional structures by means of localized photopolymerization using a sharply focused laser beam. In this process an UV beam is used to cure a layer of liquid monomer solution. By lowering the elevator, additional polymer may be built layer by layer to create complex three-dimensional microstructures. A schematic illustration of the μSL process is shown in Fig. 9. A typical microstereolithographic system integrates a UV laser, a beam delivery system, computerized CAD program, and precision x–y–z mobile sample stage to produce high aspect ratio microstructures.

Shea et al. fabricated MIP microstructures imprinted with 9-ethyl adenine (9-EA) by directing a 1–2 μm focused laser beam onto the surface of a glass microslide which was slightly submerged in a liquid monomer solution such that only a thin layer of the solution covered the glass surface. Localized photopolymerization was then controlled by movement of the precision x–y stage to create two-dimensional microstructures. The three-dimensional microstructures were subsequently produced in a layer-by-layer fashion when the stage was lowered further into the liquid monomer solution stepwise as shown in Fig. 10.

An UV absorber is added to the system in order to control curing depth as well as resolution [64]. The spatial resolution may be compromised by light scattering, which can occur in the polymerization mixture due to the high intensity of the laser as well as the phase separation that occurs during polymerization. The addition of an UV absorber with a high molar absorptivity tends to decrease the amount of scattering thus increases the resolution of the features. However, it is important that the UV absorber chosen does not interfere with the imprinting process.

Achieving desirable resolution, however, is a complicated process that extends beyond the simple addition of an UV absorber. Resolution may be affected by what is essentially a coupled optical–chemical–thermal process that occurs during polymerization in the μSL environment. Initial photolysis of the photo-initiators

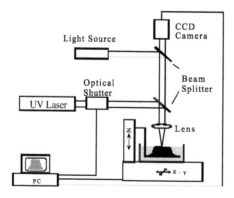

Figure 9 The μSL system includes: an UV solid laser, a beam delivery system, computer controlled precision x–y–z stage and a CAD design tool, and in situ process monitoring system.

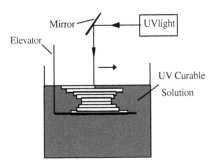

Figure 10 The principle of μSL. Complex 3-D microparts are fabricated layer by layer by curing the thin layer suspension of 1–10 μm thick with an UV focused beam of 1–2 μm.

generates radicals that initiate polymerization. Two subsequent physical phenomena follow the polymerization: heat generation and transport, as well as radical diffusion. The heat generation causes the local temperature to rise, which in turn accelerates the polymerization process. Therefore, thermal influences as well as radical diffusion may enlarge the polymerized line width. The radical diffusion is not an issue in the macroscale stereolithography since the diffusion length is negligible to macroparts dimension. In μSL, however, the radical diffusion can be significant if the diffusion length is comparable to the line width designed.

Another consideration during the microstereolithographic process is that the experimental conditions must be compatible with the imprinting process. The laser used must not disrupt the monomer–template interactions necessary for successful imprinting. In the microstructures fabricated by Shea and coworkers [63] this was shown not to be an issue. Shea used a system comprised of methacrylic acid as the functional monomer, 9-ethyl adenine as the template, Tinuvin as the UV absorber, benzoin ethyl ether as the initiator, and trimethylolpropane trimethacrylate as the cross-linker. Using these components, two-dimensional grid-like microstructures as shown in Fig. 11 were created to test the binding ability of such μSL generated polymers.

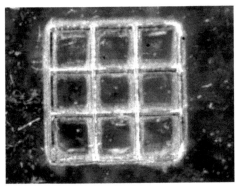

Figure 11 A waffle-like two-dimensional microstructure (600 μm × 600 μm) fabricated through microstereolithography. (Reproduced with permission from Ref. 63.)

Shea and coworkers developed a direct imaging technique to evaluate the binding capacity of these MIP microstructures. The 2-D grids were treated with a fluorescent analogue of 9-EA, 9-dansyl adenine (9-DA), and the binding was measured through increased fluoresence of the imprinted structures. These studies showed a 5:1 preference for 9-DA using the imprinted structure over the control polymer. Such results were compatible with previous binding studies using 9-DA and MIPs prepared under bulk conditions [63].

Shea et al. was then able to use this technique to create the three-dimensional waffle-like microstructures imprinted with 9-EA as shown in Fig. 12. Rebinding studies were conducted using 9-DA and the fluorescent intensity was measured for both the MIP and control 3-D structures. The results showed ∼4.5- fold increase in fluorescent intensity as compared to the control, indicating that the 3-D MIP structures selectively bind 9-DA (Fig. 13).

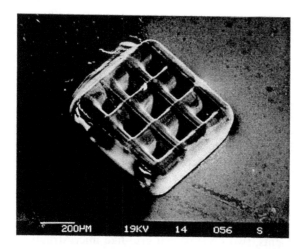

Figure 12   SEM image of 3-D microstructure ($600\,\mu m \times 600\,\mu m \times 100\,\mu m$). The image on the bottom shows the wall thickness to be approximately $20\,\mu m$. (Reproduced with permission from Ref. 63.)

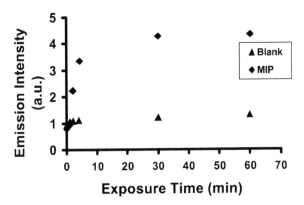

Figure 13 Fluorescence intensity as a function of exposure time for 9-EA-imprinted (MIP) and control (Blank) 3-D microstructures as shown in Fig 12. Rebinding experiments were carried out using 9-DA in chloroform (1.08×10^{-4} M).

XI. MICROSTEREOLITHOGRAPHY FOR THE GENERATION OF MOLECULARLY IMPRINTED POLYMER MICROSTRUCTURES

A. Reagents and Equipment

9-Ethyl adenine (**9**) and 9-dansyl adenine (**10**) were synthesized according to well-established protocols [65]. In this process, adenine was derivatized at the 9-position using dansyl chloride in order to produce 9-dansyl adenine.

B. Method

All glass slides were silanized with 3-(trimethoxysilyl)propyl methacrylate prior to use in order to ensure covalent attachment of the microstructures to the glass substrate. Typical imprinting solutions were obtained by mixing porogen chloroform (6.271 g), cross-linker TRIM (6.274 g), functional monomer MAA (0.327 g), UV absorber Tinuvin (**11**) (0.013 g), photoinitiator benzoin ethyl ether (**12**) (0.131 g), and imprint molecule 9-ethyl adenine (0.050 g). Blank solutions were prepared similarly with the exclusion of 9-ethyl adenine. Laser parameters such as power, scanning speed, and attenuation were varied in order to yield curing depths of ~20 μm. The z-axis was thus lowered 20 μm after each scan until a grid of 100 μm height was created.

The microstructures were extracted with chloroform for 12 h in order to remove the imprint molecule as well as any unreacted reagents from the material. The structures were then exposed to a 1.08×10^{-4} M solution of 9-DA in chloroform for rebinding. The rebinding curves showed that the 3-D microstructures was saturated after only ~15 min.

The 9-DA absorbed on the samples was excited with the 488 nm line from an Ar$^+$ laser. This wavelength was chosen such that it did not overlap with the absorption from any remaining Tinuvin in the sample. The emitted fluorescence at 600 nm from 9-dansyl adenine was imaged on a CCD camera using a 530 nm long pass filter. The rebinding ability of imprinted and control polymers was judged through the intensity of the fluorescence.

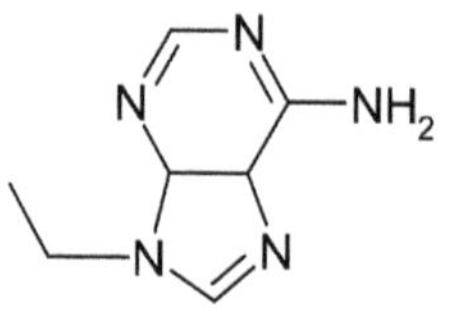

9-Ethyl adenine (9)

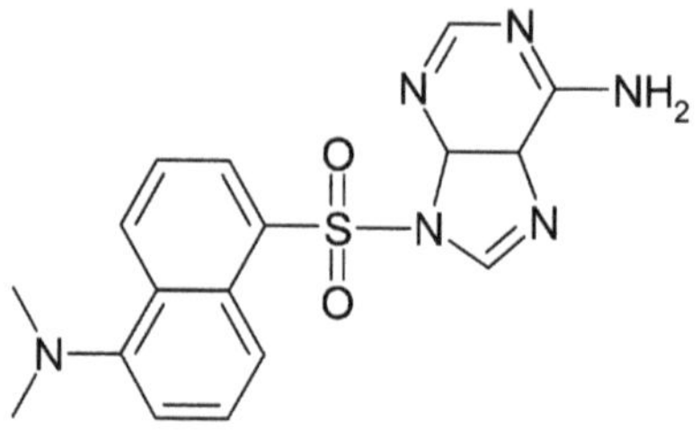

9-Dansyl adenine (10)

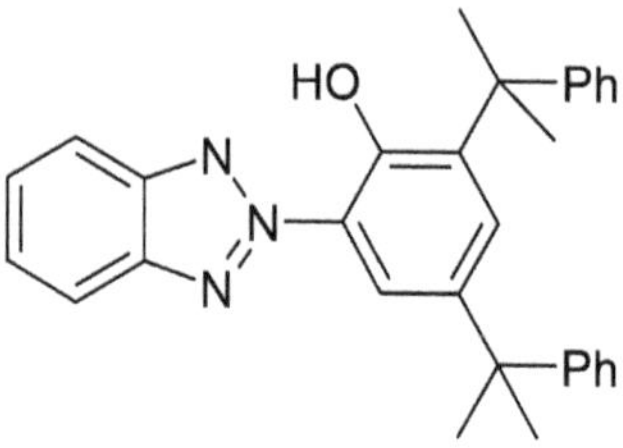

Tinuvin (11)

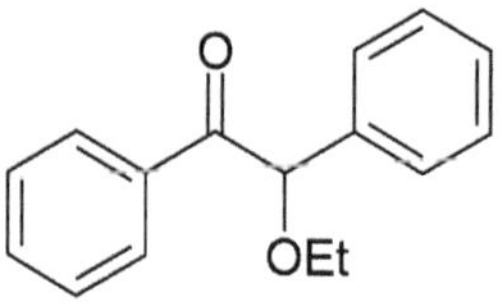

Benzoin ethyl ether (12)

XII. CONCLUSIONS

The emerging design of polymer micromonoliths has already shown much promise. Although more studies are needed to completely evaluate the performance of MIP micromonoliths in comparison to conventionally synthesized MIP particles and their biologically based counterparts, MIP monoliths boast benefits such as ease of fabrication, control of polymer geometry, and possible integration into microdevices. MIP monoliths have already been successfully integrated into separation techniques such as capillary electrochromatography and HPLC. While both systems require the careful optimization of porogens and reaction conditions to introduce the porous network necessary for flow-through applications, the benefits of in situ fabrication and particle uniformity make this mode of synthesis an avenue worthy of study.

New studies are also emerging to produce MIP micromonoliths for their eventual integration into microdevices to be used for diagnostic and clinical testing. Techniques such as micromolding in capillaries and microstereolithography have been employed to create micromonoliths and waffle-like microstructures capable of preferentially rebinding imprint molecules. Considering the demonstrated applications of molecularly imprinted materials in separation, catalysis, assays, and sensing, the integration of microfabrication techniques with molecular imprinting provides opportunities to create spatially resolved functional materials for molecular recognition in miniaturized devices.

REFERENCES

1. Seidl, J.; Malinsky, J.; Dusek, K.; Heitz, W. Macroporous styrene–divinylbenzene copolymers and their use in chromatography and for the preparation of ion exchangers. Adv. Polym. Sci. 1967, 5, 11.

2. Kun, K.A.; Kunin, R. Macroreticular resins. III. Formation of macroreticular styrene–divinylbenzene copolymers. J. Polym. Sci. Polym. Chem. Ed. **1968**, *6*, 2689–2701.

3. Sederel, W.L.; DeJong, G.J. Styrene–divinylbenzene copolymers. Construction of porosity in styrene–divinylbenzene matrixes. J. Appl. Polym. Sci. **1973**, *17*, 2835–2846.

4. Svec, F.; Fréchet, J.M.J. Continuous rods of macroporous polymer as high-performance liquid chromatography separation media. Anal. Chem. **1992**, *64*, 820–822.

5. Mihelic, I.; Krajnc, M.; Koloini, T. Kinetic model of a methacrylate-based monolith preparation. Ind. Eng. Chem. Res. **2001**, *40*, 3495–3501.

6. Dulay, M.T.; Quirino, J.P.; Bennett, B.D.; Kato, M.; Zare, R.N. Photopolymerized sol-gel monoliths for capillary electrochromatography. Anal. Chem. **2001**, *73*, 3921–3926.

7. Hjertén, S. Standard and capillary chromatography, including electrochromatography, on continuous polymer beds (monoliths), based on water-soluble monomers. Ind. Eng. Chem. Res. **1999**, *38*, 1205–1214.

8. Hjertén, S. Continuous beds: high resolving, cost effective, chromatographic matrices. Nature **1992**, *356*, 810–812.

9. Liao, J.L.; Chen, N.; Ericson, C.; Hjertén, S. Preparation of continuous beds derivatized with one-step alkyl and sulfonate groups for capillary electrochromatography. Anal. Chem. **1996**, *68*, 3468–3472.

10. Tanaka, N.; Nagayama, H.; Kobayashi, H.; Ikegami, T.; Hosoya, K.; Ishizuka, N.; Minakochi, H.; Nakanishi, K.; Cabrera, K.; Lubda, D. Monolithic silica columns for HPLC, micro-HPLC and CEC. J. High Resolut. Chromatogr. **2000**, *23*, 111–116.

11. Kobayashi, H.; Tanaka, N.; Ishizuka, N.; Minakuchi, H.; Nakanishi, K. The characterization of porous silica monolithic columns in HPLC and CEC. Chromatography **2000**, *21*, 404–405.

12. Tanaka, N.; Kobayashi, H.; Nakanishi, K.; Minakuchi, H.; Ishizuka, N. Monolithic LC columns. Anal. Chem. **2001**, *73*, 420A–429A.

13. Ishizuka, N.; Kobayashi, H.; Minakuchi, H.; Nakanishi, K.; Hirao, K.; Hosoya, K.; Ikegami, T.; Tanaka, N. Monolithic silica columns for high-efficiency separations by high-performance liquid chromatography. J. Chromatogr. A **2002**, *960*, 85–96.

14. Tanaka, N. An approach to high-performance packing materials for HPLC. Chromatography **1999**, *20*, 190–193.

15. Ishizuka, N.; Minakuchi, H.; Nakanishi, K.; Soga, N.; Nagayama, H.; Hosoya, K.; Tanaka, N. Performance of a monolithic silica column in capillary under pressure-driven and electrodriven conditions. Anal. Chem. **2000**, *72*, 1275–1280.

16. Svec, F.; Fréchet, J.M.J. Kinetic control of pore formation in macroporous polymers. The formation of "molded" porous materials with high flow characteristics for separations or catalysis. Chem. Mater. **1995**, *7*, 707–715.

17. Svec, F.; Fréchet, J.M.J. Temperature, a simple and efficient tool for the control of pore size distribution in macroporous polymers. Macromolecules **1995**, *28*, 7580–7582.

18. Yu, C.; Xu, M.; Svec, F.; Fréchet, J.M.J. Preparation of monolithic polymers with controlled porous properties for microfluidic chip applications using photoinitiated free-radical polymerization. J. Polym. Sci. Part A: Polymer Chem. **2002**, *40*, 755–769.

19. Svec, F.; Fréchet, J.M.J. Molded rigid monolithic porous polymers: an inexpensive, efficient, and versatile alternative to beads for the design of materials for numerous applications. Ind. Eng. Chem. Res. **1999**, *38*, 34–48.

20. Viklund, C.; Svec, F.; Fréchet, J.M.J.; Irgum, K. Monolithic, "molded", porous materials with high flow characteristics for separations, catalysis, or solid-phase chemistry: control of porous properties during polymerization. Chem. Mater. **1996**, *8*, 744–750.

21. Guyot, A.; Bartholin, M. Design and properties of polymers as materials for fine chemistry. Prog. Polym. Sci. **1982**, *8*, 277–331.

22. Peters, E.C.; Svec, F.; Fréchet, J.M.J. Preparation of large diameter "molded" porous polymer monoliths and the control of pore structure homogeneity. Chem. Mater. **1997**, *9*, 1898–1902.

23. Santora, B.P.; Gagne, M.R.; Moloy, K.G.; Radu, N.S. Porogen and cross-linking effects on the surface area, pore volume distribution, and morphology of macroporous polymers obtained by bulk polymerization. Macromolecules **2001**, *34*, 658–661.

24. Tan, Z.J.; Remcho, V.T. Molecular imprint polymers as highly selective stationary phases for open tubular liquid chromatography and capillary electrochromatography. Electrophoresis **1998**, *19*, 2055–2060.

25. Bruggemann, O.; Freitag, R.; Whitcombe, M.J.; Vulfson, E.N. Comparison of polymer coatings of capillaries for capillary electrophoresis with respect to their applicability to molecular imprinting and electrochromatography. J Chromatogr. A **1997**, *781*, 43–53.

26. Schweitz, L. Molecularly imprinted polymer coatings for open-tubular capillary electrochromatography prepared by surface initiation. Anal. Chem. **2002**, *74*, 1192–1196.

27. Quaglia, M.; De-Lorenzi, E.; Sulitzky, C.; Massolini, G.; Sellergren, B. Surface initiated molecularly imprinted polymer films: a new approach in chiral capillary electrochromatography. Analyst **2001**, *126*, 1495–1498.

28. Lin, J.M.; Nakagama, T.; Uchiyama, K.; Hobo, T. Enantioseparation of D,L-phenylalanine by molecularly imprinted polymer particles filled capillary electrochromatography. J. Liq. Chrom. Rel. Technol. **1997**, *20*, 1489–1506.

29. Hosoya, K.; Yoshizako, K.; Shirasu, Y.; Kimata, K.; Araki, T.; Tanaka, N.; Haginaka, J. Molecularly imprinted uniform-size polymer-based stationary phase for high-performance liquid chromatography. Structural contribution of cross-linked polymer network on specific molecular recognition. J. Chromatogr. A **1996**, *728*, 139–148.

30. Lin, J.M.; Nakagama, T.; Uchiyama, K.; Hobo, T. Molecularly imprinted polymer as chiral selector for enantioseparation of amino acids by capillary gel electrophoresis. Chromatographia **1996**, *43*, 585–591.

31. Chirica, G.; Remcho, V.T. Silicate entrapped columns—new columns designed for capillary electrochromatography. Electrophoresis **1999**, *20*, 50–56.

32. Yu, C.; Davey, M.H.; Svec, F.; Fréchet, J.M.J. Monolithic porous polymer for on-chip solid-phase extraction and preconcentration prepared by photoinitiated in situ polymerization within a microfluidic device. Anal. Chem. **2001**, *73*, 5088–5096.

33. Hjertén, S. High-performance electrophoresis: elimination of electroendoosmosis and solute adsorption. J. Chromatogr. **1985**, *347*, 191–195.

34. Schweitz, L.; Andersson, L.I.; Nilsson, S. Capillary electrochromatography with molecular imprint-based selectivity for enantiomer separation of local anaesthetics. J. Chromatogr. A **1997**, *792*, 401–409.

35. de Boer, T.; de Zeeuw, R.A.; de Jong, G.J.; Ensing, K. Selectivity in capillary electrokinetic separations. Electrophoresis **1999**, *20*, 2989–3010.

36. Nilsson, K.; Lindell, J.; Norrlow, O.; Sellergren, B. Imprinted polymers as antibody mimetics and new affinity gels for selective separations in capillary electrophoresis. J. Chromatogr. A **1994**, *680*, 57–61.

37. Lin, J.M.; Nakagama, T.; Uchiyama, K.; Hobo, T. Capillary electrochromatographic separation of amino acid enantiomers using on-column prepared molecularly imprinted polymer. J. Pharm. Biomed. Analysis **1997**, *15*, 1351–1358.

38. Schweitz, L.; Andersson, L.I.; Nilsson, S. Capillary electrochromatography with predetermined selectivity obtained through molecular imprinting. Anal. Chem. **1997**, *69*, 1179–1183.

39. Schweitz, L.; Petersson, M.; Johansson, T.; Nilsson, S. Alternative methods providing enhanced sensitivity and selectivity in capillary electroseparation experiments. J. Chromatogr. A **2000**, *892*, 203–217.

40. Nilsson, S.; Schweitz, L.; Petersson, M. Three approaches to enantiomer separation of β-adrenergic antagonists by capillary electrochromatography. Electrophoresis **1997**, *18*, 884–890.

41. Schweitz, L.; Andersson, L.I.; Nilsson, S. Rapid electrochromatographic enantiomer separations on short molecularly imprinted polymer monoliths. Anal. Chim. Acta **2001**, *435*, 43–47.

42. Schweitz, L.; Andersson, L.I.; Nilsson, S. Molecularly imprinted CEC sorbents: investigations into polymer preparation and electrolyte composition. Analyst **2002**, *127*, 22–28.

43. Matsui, J.; Kato, T.; Takeuchi, T.; Suzuki, M.; Yokoyama, K.; Tamiya, E.; Karube, I. Molecular recognition in continuous polymer rods prepared by molecular imprinting technique. Anal. Chem. **1993**, *65*, 2223–2224.

44. Matsui, J.; Takeuchi, T. A molecularly imprinted polymer rod as nicotine selective affinity media prepared with 2-(trifluoromethyl)acrylic acid. Anal. Comm. **1997**, *34*, 199–200.

45. Matsui, J.; Nicholls, I.A.; Takeuchi, T. Molecular recognition in chinchona alkaloid molecular imprinted polymer rods. Anal. Chim. Acta **1998**, *365*, 89–93.

46. Perrin, D.D.; Armarego, W.L.F. In *Purification of Laboratory Chemicals*, 3rd. Ed.; Oxford: Pergamon Press, 1988.

47. McNiven, S.; Kato, M.; Levi, R.; Yano, K.; Karube, I. Chloramphenicaol sensor based on an in situ imprinted polymer. Anal. Chim. Acta **1998**, *365*, 69–74.

48. Hadd, A.G.; Raymond, D.E.; Halliwell, J.W.; Jacobson, S.G.; Ramsey, J.M. Microchip device for performing enzyme assays. Anal. Chem. **1997**, *69*, 3407–3412.

49. McDonald, J.C.; Metallo, S.J.; Whitesides, G.M. Fabrication of a configurable, single-use microfluidic device. Anal. Chem. **2001**, *73*, 5645–5650.

50. Hadd, A.G.; Jacobson, S.C.; Ramsey, J.M. Microfluidic assays of acetylcholinesterase inhibitors. Anal. Chem. **1999**, *71*, 5206–5212.

51. Xia, Y.; Whitesides, G.M. Soft lithography. Angew. Chem. Int. Ed. **1998**, *37*, 550–575.

52. Zhao, X.M.; Xia, Y.; Whitesides, G.M. Soft-lithographic methods for nano-fabrication. J. Mater. Chem. **1997**, *7*, 1069–1074.

53. Kim, E.; Xia, Y.; Whitesides, G.M. Micromolding in capillaries: applications in materials science. J. Am. Chem. Soc. **1996**, *118*, 5722–5731.

54. Kim, E.; Xia, Y.; Whitesides, G.M. Polymer microstructures formed by moulding in capillaries. Nature **1995**, *376*, 581–584.

55. Yan, M.; Kapua, A. Fabrication of molecularly imprinted polymer microstructures. Anal. Chim. Acta **2001**, *435*, 163–167.

56. Yan, M.; Kapua, A. Microfabricated molecularly imprinted polymers. Polym. Preprint **2000**, *41*, 264–265.

57. Qin, D.; Xia, Y.; Whitesides, G.M. Rapid prototyping of complex structures with feature sizes larger than 20 µm. Adv. Mater. **1996**, *8*, 917–919.

58. Deng, T.; Wu, H.; Brittain, S.T.; Whitesides, G.M. Prototyping of masks, masters, and stamps/molds for soft lithography using an office printer and photographic reduction. Anal. Chem. **2000**, *72*, 3176–3180.

59. Brazier, J.; Yan, M. Molecularly imprinted polymers used as optical waveguides for the detection of fluorescent analytes. Mater. Res. Soc. Symp. Proc. **2002**, *723*, 115–120.

60. Dickert, F.L.; Thierer, S. Molecularly imprinted polymers for optochemical sensors. Adv. Mater. **1996**, *8*, 987–989.

61. Dickert, F.L.; Tortschanoff, M. Molecularly imprinted sensor layers for the detection of polycyclic aromatic hydrocarbons in water. Anal. Chem. **1999**, *71*, 4559–4563.

62. Bao, Z.; Rogers, J.A.; Katz, H.E. Printable organic and polymeric semiconducting materials and devices. J. Mater. Chem. **1999**, *9*, 1895–1904.

63. Conrad II, P.G.; Nishimura, P.T.; Aherne, D.; Schwartz, B.J.; Wu, D.; Fang, N.; Zhang, X.; Roberts, M.J.; Shea, K. Functional molecularly imprinted polymer microstructures fabricated using microstereolithography. Adv. Mater. **2003**, *15*, 1541–1544.

64. Jacobs, P.F. Rapid Prototyping and Manufacturing: Fundamentals of Stereolithography; Society of Manufacturing Engineers: Dearborn, 1992.

65. Tipson, R.S. Synthetic Procedures in Nucleic Acid Chemistry; Interscience Publishers: New York, 1986; Vol. 1 and 2.

20
Chromatographic Techniques

Claudio Baggiani   Università di Torino, Torino, Italy

I. INTRODUCTION

Liquid chromatography and molecular imprinting have been tightly interlaced from the very beginning. Just few years after Dickey's paper on imprinting of silica gels with alkyl-orange dyes [1], Curti and Colombo [2] published another paper about the separation of racemic mixtures on silica gel columns imprinted with a single enantiomer. The rebirth of the noncovalent molecular imprinting technique in the early 1980s by Mosbach and coworkers [3] (noncovalent approach, see this book, Chapter 3) and Wulff and Sarhan (covalent approach, see this book, Chapter 4) was marked by papers mainly concerning the use of high pressure liquid chromatography to characterize the presence of an imprinting effect [4,5].

Today, even if the molecular imprinting paradigm has spread widely into many fields of chemistry far away from separation science such as catalysis or structured nanomaterials, each year chromatography on imprinted phases involves about one-third of all the published papers on this issue [6]. The experimental application of modern liquid chromatographic techniques is of fundamental significance to characterize important properties of imprinted materials such as selectivity, affinity towards the template molecule, amount of binding sites or binding kinetics. Not only is chromatography an essential tool for studying these properties, but it is at the basis of some of the most significant applications, such as racemic mixture resolution or solid phase extraction.

II. BASICS: THE CHROMATOGRAPHIC PROCESS

Chromatographic separation can be defined as the process that involves the distribution of different substances between two phases, a stationary phase and a mobile phase. Solutes distributed preferentially in the mobile phase will move more rapidly through the column than those distributed preferentially in the stationary phase. As a consequence of this, the solutes will elute in order of their increasing partition coefficients towards the stationary phase. During the separation process, it can be assumed that in each point of the column a chemical equilibrium involving the solute molecule will be established between the stationary and the mobile phase. This equilibrium is called "partition equilibrium", and it is governed by noncovalent forces mutually acting between solute molecules, mobile phase and the surface of the stationary phase. The stronger the forces between the solute molecules and the stationary phase, the more the solute will be retained. Conversely, the stronger the interactions between the solute molecules and the mobile phase, the more rapidly the solute will elute from the column (for a comprehensive text on fundamentals of chromatography see Ref. 7 and for definitions and abbreviations in chromatography see Ref. 8).

As a solute is eluted, the distribution of its molecules along the longitudinal axes of the column changes, generating a Gaussian-like profile; a band broadening effect whose maximum is known as "time of retention" (t_r) as depicted in Fig. 1. This parameter, corrected for the time of retention of a solute, that is, not retained by the column (dead time, t_d) is related to the solute "capacity factor" (k) by the relation:

$$k = (t_r/t_d) - 1 \tag{1}$$

The capacity factor is strictly correlated to the extent of the partition equilibrium, thus to the strength of the interaction between the solute molecules and the stationary phase, by the relation:

$$k = K(V_S/V_m) \tag{2}$$

where K is the partition thermodynamic constant, and V_s/V_m the phase ratio, i.e., the ratio between the volume of the stationary phase and the volume of the mobile phase contained in the column.

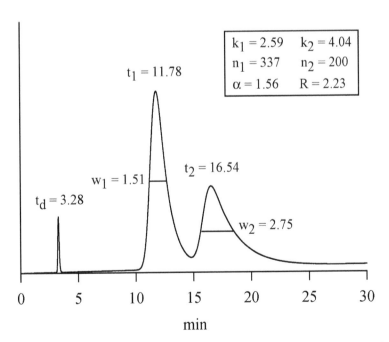

Figure 1   A chromatographic separation of two analytes. Note that despite calculated resolution will be acceptable (α and $R > 1$), the separation is not complete because of the tailed peaks.

The direct comparison of the capacity factor values for two solutes is called "separation factor", α:

$$\alpha = k_2/k_1 \tag{3}$$

and it could be considered a rough estimate of the chromatographic resolution between two peaks. A better way to calculate the resolution includes the width of each peak, and is represented by the equation:

$$R = 2\,|t_{r2} - t_{r1}|/(w_1 + w_1) \tag{4}$$

where w_1 and w_1 are the peak widths at half-height as depicted in Fig. 1.

The extent of a solute zone broadening determines the chromatographic efficiency of the separation, which is expressed in terms of the number of theoretical plates (n), or the height equivalent to a theoretical plate (H) given by the ratio of the column length (L) to the number of theoretical plates ($H = L/n$). The column efficiency can be easily calculated from the profile of a chromatographic peak as:

$$,n = 5.54(t_r/w)^2 \tag{5}$$

The concept of theoretical plate has its origin in the plate model of the chromatographic separation, for which it is assumed that the column can be ideally divided into a number of equally distributed sections, called plates. At each plate the solute partition is sufficiently fast to let equilibrium be established before the solute moves into the next plate. Thus, the theoretical plate is the smallest section of the column in which equilibrium partition is possible.

For a couple of peaks, the plate number is directly related to capacity factor (k), selectivity factor (α), and column resolution (R) by the subsequent equation:

$$n = 16R^2(\alpha/\alpha - 1)^2(1 + 1/k_2)^2 \tag{6}$$

where k_2 is the capacity factor of the later eluted peak (Fig. 1).

Considering a chromatographic process controlled by a partition equilibrium and neglecting extracolumn effects (i.e., band broadening caused by factors outside the column, e.g., tubings, detector etc.), several factors can contribute to the overall solute band broadening: eddy diffusion, longitudinal diffusion, and resistance to mass transfer in mobile and stationary phase.

Eddy diffusion (multipath effect). (Fig. 2a) As the solute molecules are transported by the flow through a packed column some molecules will take a longer path through the column than the average, while others will take a shorter path. The corresponding contribution to the height equivalent of a theoretical plate, H_e, depends only on the homogeneity of the column packing and the average particle diameter d_p. It is defined by

$$H_e = 2\lambda d_p = A \tag{7}$$

where λ, geometric factor, is a dimensionless number between 0.5 and 1 that accounts for the inhomogeneity of packing. In a stationary phase constituted by a molecularly imprinted polymer, this term assumes significant importance due to the difficulty of obtaining regular and monodisperse-imprinted particles by using the widespread technique of bulk polymerization. This produces a large band broadening for all the solutes, including those not recognized by the polymer.

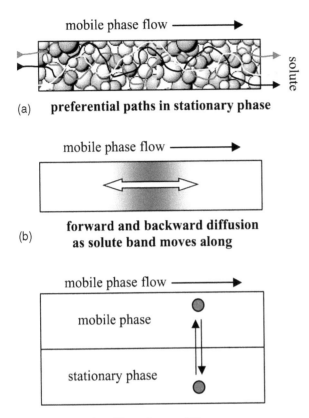

mobile phase flow ⟶

(a)   **preferential paths in stationary phase**

mobile phase flow ⟶

(b)   **forward and backward diffusion
as solute band moves along**

mobile phase flow ⟶

mobile phase

stationary phase

(c)   **kinetic effects in partition process**

Figure 2   Factors affecting band spreading in liquid chromatography.

Longitudinal diffusion. (Fig. 2b) The solute molecules have the tendency to migrate along the longitudinal axes of the column, diffusing from high concentration regions to low concentration regions. The corresponding contribution to the height equivalent of a theoretical plate, H_l, is proportional to the time spent by the sample in the column and to the diffusion coefficient of the solute molecules in the mobile phase (D_m). It is defined by

$$H_l = (2\gamma D_m)/u = B/u \tag{8}$$

where γ is a dimensionless constant (about 0.6) that consists in an obstruction factor to diffusion in the extraparticle space, and u is the mobile phase linear velocity. For all the liquid chromatography techniques, including those based on molecular imprinting, this term can be considered of secondary importance.

Mass transfer resistance. (Fig. 2c) The migration of the solute molecules from the mobile phase to the stationary phase and vice versa is not instantaneous, and it prevents the existence of true partition equilibrium. The corresponding contribution to the height equivalent to a theoretical plate, H_m for the mass transfer resistance in mobile phase, and H_s for the mass transfer resistance in stationary phase, depends on the homogeneity of the column packing, the average particle diameter of the stationary phase, the diffusion coefficients in the mobile and the stationary phase (D_m and D_s) and the thickness of the stationary phase actively involved in the partition process (d). They are defined by

$$H_m = (\Omega d_{p^2} u)/D_m = Cu \qquad (9)$$

$$H_S = (Q U d^2 u)/D_S = Du \qquad (10)$$

where Ω, column factor, Q, homogeneity factor, U, velocity factor, are dimensionless constants. As for the eddy diffusion, in a stationary phase constituted by a molecularly imprinted polymer, this term is increased by the presence of a less-than-ideal column packing, due to the irregular and polydispersed stationary phase.

The sum of the distinct contributions to the solute band broadening is described by the Van Deemter equation (Fig. 3):

$$H = H_e + H_1 + H_m + H_S = A + B/u + (C + D)u \qquad (11)$$

Besides these contributions, it is necessary to take into account that solute band broadening and peak asymmetry in a polymer-imprinted column are also significantly influenced by other factors, normally not present or negligible in conventional liquid chromatography, such as slow desorption kinetics between solute and imprinted binding sites, nonlinear binding isotherms and a significant heterogeneity of the binding sites.

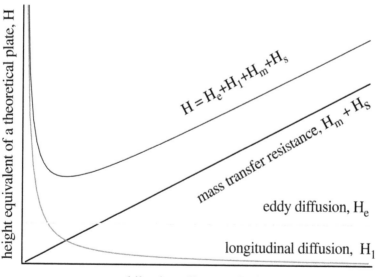

Figure 3   The Van Deemter equation. H and u are expressed in arbitrary units.

A. Chromatographic Mode of MIP

When we speak about liquid chromatography on molecularly imprinted polymers, which mode of chromatography do we mean? Are we referring to a specific interaction mechanism, such as ion-exchange or reverse phase chromatography, or are we thinking about an innovative technique, with nothing in common with other types of chromatography modes? The nature of interactions between the imprinted binding sites and the template could be of several kinds: ionic, hydrogen bond, charge transfer, dipole–dipole. Thus it is quite difficult to characterize MIP-based chromatography on the basis of the noncovalent interaction responsible for the separation process. On the other hand, it is also difficult to consider this kind of technique as something different from the more traditional chromatographic techniques. In fact, we can perform isocratic or gradient elutions, we can make analytical or preparative separations, we can use the same type of detectors and HPLC instruments, and we can use solvents, buffers, and additives exactly like reverse phase or ion-exchange chromatography. Moreover, neglecting the presence of imprinted binding sites, the stationary phase support is quite ordinary: made of cross-linked methacrylate or styrene–divinylbenzene particles having an irregular or spherical form, or possibly silica-supported thin films of polymers.

MIP-based chromatography is not so different from common liquid chromatography. If we consider the presence of sites characterized by binding reversibility and high levels of selectivity, the most similar technique is affinity chromatography, or better still, immunoaffinity chromatography. Applications of immunoaffinity chromatography are widely used in several disciplines: clinical, forensic, and food and environmental chemistry. Problems of analyte isolation from "difficult" samples have been successfully solved using this kind of chromatography. The same has been done (or it is an hypothetical possibility) with molecular-imprinted polymers. Immunoaffinity chromatography uses immobilized natural receptors, produced by cell culture or immunization in animals (immunoglobulins), to isolate specific target molecules (antigens), exactly like MIP-based chromatography uses artificial receptor obtained by radical copolymerization of vinyl-based monomers (imprinted binding sites) to capture template molecules.

In Table 1, the techniques are compared. It is possible to observe that MIP-based chromatography despite some drawbacks has several advantages compared to immunoaffinity chromatography. Very low costs of production and utilization, high resistance to chemical and denaturing conditions, batch-to-batch reproducible binding properties, and a general simplicity of preparation makes this technique appealing and potentially very competitive towards immunoaffinity chromatography.

III. THE MOLECULARLY IMPRINTED STATIONARY PHASE

A MIP-based stationary phase should have several properties that make it a "true" HPLC material. First of all, this hypothetical column is composed of monodispersed and very regular beads, with an average diameter from 2 (analytical purposes) to 10 μm (preparative purposes), having a controlled and defined nano/meso/micro-porosity. These pressure-resistant beads are ordered in highly regular, perhaps isotropic packing, without preferential paths and side-wall effects. Alternatively, the column consists of a highly homogeneous monolithic polymer, characterized by

Table 1 A Comparison Between Immunoaffinity Chromatography and MIP-Based Chromatography

Characteristic features	Immunoaffinity chromatography (including antiserum production)	MIP-based chromatography (including polymer preparation)
Binding of target molecules with low mass (<5000 Da)	Yes, but necessity of a spacer arm to raise antibodies could change the specificity	Yes
Binding of target molecules with high mass (>5000 Da)	Yes	Very difficult
Binding mechanism	Well known	Known, but some aspects under debate
Binding affinity spectrum	Narrow for monoclonal, broad for polyclonal	Narrow for covalent imprinting, broad for non-covalent imprinting
Mean affinity constant	Frequently above $10^9 \, M^{-1}$	Rarely exceeds $10^7 \, M^{-1}$
Binding site density	Very low for polyclonal	High
Binding kinetics	Slow dissociation	Slow dissociation
Specificity	High, fine tuning for monoclonals feasible	High, fine tuning very difficult
Reproducibility	Limited	High
Nonspecific binding	Negligible	Depending of elution conditions
Resistance to extreme environments	Limited	Yes
Reuse	Difficult, some stationary phases are monouse	Yes
Cost for single run	Medium to high	Very low
Commercial availability	Medium	Very rare
Scale-up	Very expensive	Economical if compared to immunoaffinity
In-house feasibility	No, a stabularium, well trained people, a dedicated laboratory are necessary	Yes, simple to make
Health risks (excluded the use of common chemicals for chromatography)	Not significative	Grinding produces sub-micrometric particles, dangerous if inhaled. Some monomers (acrylamide, styrene) are toxic
Literature	Very large	Large and rapidly growing
State of the art	Mature	In continuous evolution

a marked superporosity to support a flow. These conditions are batch-to-batch reproducible and valid for all column formats: capillary (length > 100 mm, ø ≤ 1 mm), micro (length < 50 mm, ø between 1 and 3 mm), classical (length between 50 and 250 mm, ø between 2 and 4 mm) and preparative (variable length, ø > 5 mm). The imprinting effect is based on specific molecular recognition interactions between the analyte and a single class of binding sites, with a well-defined adsorption isotherm, fast associative–dissociative kinetics and negligible nonspecific binding. Thus, the chemical basis of the partition equilibrium are well defined, well known, and can be controlled by acting on the mobile phase composition. Last but not least, the column performance in terms of selectivity and efficiency is well described as validated methods in the scientific literature and are certified when the column is purchased from a commercial source. It is clear that, nowadays not all these requirements are fully satisfied by MIP-based chromatograpic columns described in literature.

Imprinted polymers are frequently prepared by radical copolymerization of two of the most used cross-linkers: ethylene dimethacrylate (EDMA) and trimethylolpropane trimethacrylate (TRIM) with one of the many functional monomers reported in literature (mainly methacrylic acid, acrylamide, 4-vinylpyridine, or trifluoromethacrylic acid, but also 2-hydroxyethylmethacrylate and N,N-diethylaminoethylmethacrylate (see this book, Chapter 7) The polymers obtained as bulk monoliths or dispersed microparticles have properties that are very suitable for liquid chromatographic applications.

Firstly, these polymers are mechanically very stable and are able to withstand high pressures without cracking or collapsing. The majority of the prepolymerization mixture is constituted by the cross-linker and the resulting polymer has a very high degree of cross-linking—often higher than commercially available polymers for chromatographic applications.

Similarly, the resistance to extreme chemical environments is remarkable. High salt concentrations in the mobile phase do not cause noticeable swelling or shrinking of the polymeric matrix, and exposure to very acidic (pH < 2) or basic (pH > 10) eluents does not degrade the polymer. The latter being a result of the high resistance to hydrolysis of the pivalic ester that constitutes the core of the cross-linking molecular structure.

Imprinted polymers can be preserved as a dry powder, without preservatives, and can be quickly dispersed in a suitable solvent and packed into a chromatographic column without any particular preparations. After the utilization, the column can be dismantled, the polymer recovered and stored for future applications without loss of efficiency.

Imprinted polymers for chromatographic purposes are often prepared as bulk materials, by both thermal and photopolymerization. The imprinted polymers are taken from the polymerization vessels, ground by mechanical mortars and sieved into particles of 25–90 μm, suitable for packing into chromatographic columns. This method, by far the most popular, presents many attractive properties, especially to newcomers. It is fast and simple in its practical execution, it does not require particular skills of the operator, it is widely reported in literature for many different templates and it does not require sophisticated or expensive instrumentation (the method requires a water bath, some sieves and a mortar, not necessarily mechanical).

However, it should be considered that the bulk polymerization method presents many significant drawbacks. First of all, the particles obtained after the last sieving step are highly irregular and of variable dimensions. In fact, the sieving process lets us obtain fractions of materials with a broad granulometry only, without an effective possibility to isolate less heterogeneous fractions. Acting on the H_e and H_m terms of the Van Deemter equation, this dramatically reduces the efficiency of an HPLC column packed with such a material. It should be taken into account that, compared to a traditional column packed with particles characterized by the same irregular and polydisperse granulometry, the reduction of the column efficiency for an imprinted column is also increased by increased peak tailing typical of solutes recognized by the imprinted stationary phase.

The procedure of grinding and sieving is cumbersome, and it causes a substantial loss of useful polymer, that can be estimated between 50% and 75% of the initial amount of bulk material. Most of the lost polymer is very fine sub-micrometric powder, which could adhere to the bigger particles and cause excessively high back pressures in the column during and after the packing procedure if multiple washings and careful sedimentation of the polymer are not performed.

The bulk polymerization cannot be scaled-up. The process consists of an exothermal radical reaction that can be controlled only when the amount of polymerization mixture does not exceed a few hundredths of milliliters. After that, heat dissipation becomes difficult and dangerous overheating of the sample can occur, with significant risks for operators.

As stated above, columns packed with irregular materials are less than ideal in terms of chromatographic performance. Thus, in recent years much effort has been dedicated to develop alternative methods to prepare imprinted stationary phases that are superior in terms of efficiency, mass transfer characteristics, and sample load capacity. Micrometer-sized spherical-imprinted polymers with narrow size distribution have been prepared through several techniques reported in Table 2. It should be considered that all these procedures show serious limitations. These are: high sensitivity to small changes in polymerization conditions, a polymerization medium that is not compatible with weak noncovalent interactions between functional monomers and their template, high costs or procedure complexity (which can hinder a wide application of these techniques as valid substitutes to bulk polymerization method).

An alternative to these procedures is the surface coating of supporting beads. Imprinted thin layers and some related applications in chromatography have been used as coatings on chromatography-grade porous silica or styrene–divinylbenzene beads [10–12]. This technique, however, is intrinsically more difficult and it involves many steps (surface activation or functionalization with polymerization promoting agents) that could result in low column reproducibility if the experimental procedures are not fully optimized. Related to these surface-coated materials are imprinted columns suitable for open-tubular liquid chromatography [13]. They were prepared as thin films by in situ thermal radical polymerization inside 25 μm bare silica capillaries. The low back pressure in these columns enabled chiral separations at very low pressure with high efficiency.

In addition, the possibility of obtaining superporous monolithic materials by direct thermal polymerization in the column should be mentioned. This technique, introduced by Svec and Frechet [14] for the preparation of styrene–divinylbenzene columns useful for fast protein separation, has been shown affordable also for molecular imprinting of small molecules [15–17]. The synthesis in the column of a single giant monolithic "bead" should produce packing characterized by a greatly reduced mass transfer resistence and low back pressures even with high velocities. Therefore, this is particularly suitable for applications requiring a micro-liquid or capillary electrochromatographic apparatus. Limitations come from the necessity of the

Table 2 Alternative Methods to Bulk Polymerization (See Also Chapter 17, and Ref. 9)

Method	Product obtained	Method complexity	Advantages	Disadvantages
Stabilized dispersion in organic solvent	Aggregates, sometimes monodispersed spherical beads	Low	Mature technology, scalable to large batches	Very sensitive to experimental conditions, aggregates poorly suitable for chromatography
Suspension in water	Polydispersed spherical particles	Medium	Mature technology, very reproducible, scalable to large batches	Water poorly compatible with noncovalent interactions between functional monomers and templates
Suspension in perfluorocarbons	Polydispersed spherical beads	Medium	No water interference	Very expensive
Two-stage swelling	Monodisperse beads	High	Highly monodisperse beads with good chromatographic properties	Surface-active agents used can interfere with imprinting and are difficult to eliminate

particular combinations of porogenic solvents, usually cyclohexanol–dodecanol that can be detrimental to the formation and stability of noncovalent interactions between the functional monomers and the template molecules.

It should also be mentioned that a variant of the dispersion polymerization technique, in which a porous stationary phase imprinted with the drug pentamidine was prepared directly in a column [18,19]. The polymerization procedure used did not produce a porous monolith but a macro-aggregate of micron-sized beads. A chromatographic evaluation of this column when eluted with a polar mobile phase showed good efficiency and molecular recognition properties comparable with more traditional imprinted columns.

Concluding, present-day stationary phases based on molecularly imprinted materials are affordable and can be considered "true" HPLC stationary phases, even if some drawbacks make them quite far from "ideal" HPLC columns.

IV. RETENTION MECHANISMS ON MOLECULARLY IMPRINTED STATIONARY PHASES

The molecular recognition effect at the basis of the specific interaction between the imprinted stationary phase and the template is based on several kinds of noncovalent bond—ionic, hydrogen bond, charge transfer, dipole-dipole—and its strength depends both on the nature of the imprinted binding sites and on the composition of the mobile phase.

For the stationary and the mobile phase, the presence of ionizable or polar groups and the relative hydrophobicity are determining factors to modulate the specific interactions with the template. Thus, to obtain well optimized separations, it is necessary to take into the account not only the mobile phase composition, but also the structure and the physico-chemical properties of the template and of the stationary phase.

Even if it is quite risky to draw up comprehensive rules to classify retention mechanisms acting on the binding sites of molecularly imprinted stationary phases, some general examples can be seen considering organic and aqueous mobile phases. Some examples of different noncovalent interactions between template, stationary phase, and mobile phase are reported in Fig. 4.

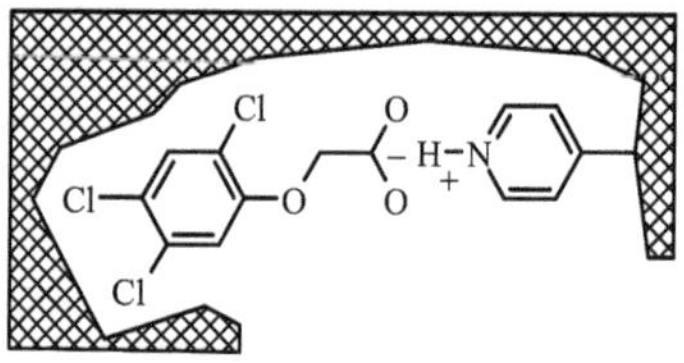

ion-exchange / hydrophobic mixed mode (30)
2,4,5-trichlorophenoxyacetic acid / 4-vinylpyridine
MeCN / water / AcOH 70+29+1 v/v

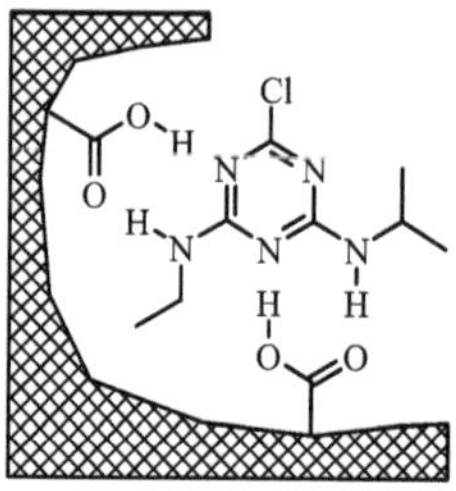

hydrogen bond / hydrophobic mixed mode (26)
atrazine / trifluoromethacrylic acid
phosphate buffer / MeCN 7+3 v/v

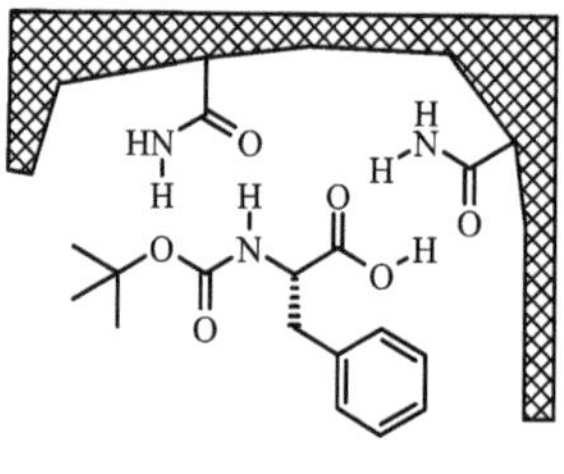

hydrogen bond mode (20)
Boc-L-phenylalanine / acrylamide
MeCN / AcOH 99.7+ 0.3 v/v

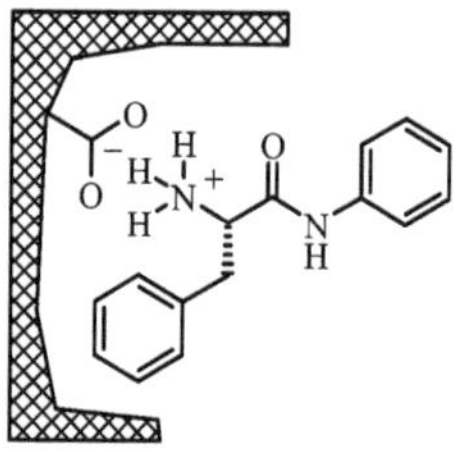

ion-exchange mode (29)
L-phenylalanine anilide / methacrylic acid
MeCN / phosphate buffer 7+3 v/v

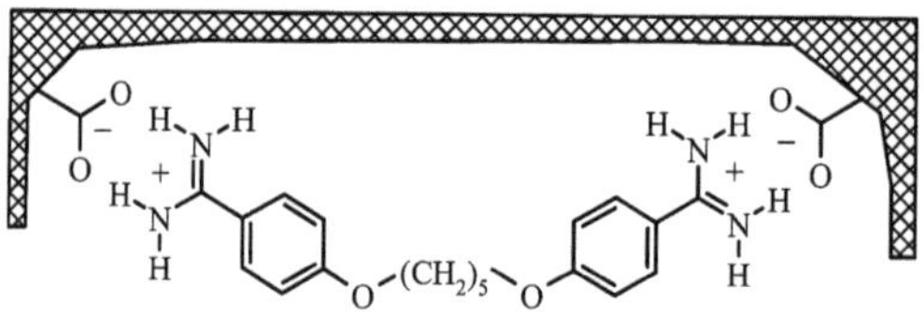

ion-exchange mode (16)
pentamidine / methacrylic acid
phosphate buffer pH 3

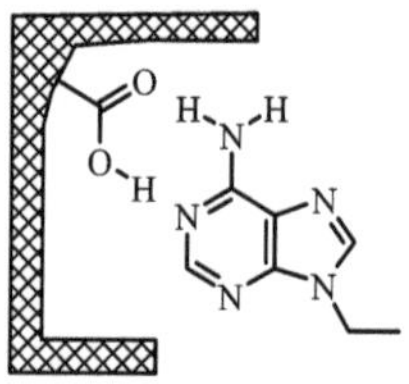

hydrogen bond mode (28)
9-ethyladenine / methacrylic acid
MeCN / AcOH 99+1 v/v

Figure 4 Examples of retention mode, stationary phase, and typical mobile phase used in MIP-based liquid chromatographic separations (MeCN: acetonitrile; AcOH: acetic acid).

A. Hydrogen Bonding Interactions

When templates are characterized by low or moderate polarity, good molecular recognition effects can be seen when organic mobile phases are used. In this case, the interaction of the template with the mobile and stationary phase can be visualized as a partition equilibrium involving hydrogen bonding between the solvated template and the solvated binding site. Optimal retention and selectivity can be obtained by using organic solvents and by modulating their polarity. It is no coincident that many imprinted polymers are initially evaluated by eluting the template with an organic solvent, generally the same that is used as porogen during the polymerization process.

Several chromatographic system models for weak polar templates have been studied using N-protected amino acids as a template [20–26] and substituted s-triazines [27–29], weak hydrogen bonding porogen such as chloroform, or acetonitrile, polar functional monomer such as methacrylic acid, 4-vinylpyridine, or acrylamide and the same porogen as mobile phase, eventually together with a polar modifier.

A marked influence of the amount of the modifier on the retention and selectivity can be seen by using methanol, acetic acid, or pyridine as mobile phase modifier. Selectivity is increased with increasing amount of modifier, whereas retention is decreased. These effects can be attributed to the perturbation of the noncovalent interactions between the template and the stationary phase, due to the competitive interaction between the modifier and binding site. For example, for a Boc-L-phenylalanine-imprinted column eluted with dichloromethane, it was observed that the retention of the template inversely correlated with the hydrogen bond donor parameter of the polar modifier [24]. Acetic acid, methanol, isopropanol, and tetrahydrofuran showed increased column capacity factors, indicating that noncovalent interactions involving the hydrogen bond acceptor sites of the template or the polymer functional groups negatively affect the recognition.

The molecular recognition strength for the template and, in a final analysis, its partition between the stationary and the mobile phase, is conditioned not only by hydrogen bond acceptor/donor properties, but also by relative Brönsted basicity. Considering s-triazines-imprinted polymers [29], it was seen that using acetonitrile as the mobile phase, the strength of the interaction between the template and the binding site was strongly conditioned by the template and functional monomer basicity. In fact, ametryn (pK = 4.1) was best retained by a polymer prepared with the strongly acidic trifluoromethacrylic acid, whereas it was sufficient to use the less acidic methacrylic acid as functional monomer to recognize the more acidic atrazine (pK = 1.7).

It should also be mentioned that a "memory effect" is present when the mobile phase is composed by the solvent used as porogen. In that case, the partition equilibrium is shifted towards the stationary phase, increasing the molecular recognition effect. A column packed with 9-ethyladenine-imprinted polymers showed higher capacity factors when the mobile phase and the porogen were the same, irrespective of the solvent polarity and hydrogen-bond properties [30]. The same has been reported when simple aromatic molecules such as benzene, toluene, and xylenes were used as porogens, where a preferential retention between solvents was seen for the molecule used as porogen [31]. This "memory effect" could be rationally explained by postulating that the binding sites will be complementary towards the (perhaps partially) solvated template.

B. Ion-Exchange Interactions

When templates and functional monomers are characterized by the presence of ionizable functional groups, the column packed with the imprinted polymer shows good efficiency and selectivity in mixed aqueous–organic mobile phases, and the partition equilibrium between mobile phase and binding sites is controlled by ion-exchange interactions. For example, for a L-phenylalanine anilide-imprinted column [32], selectivity was observed to be high and constant when the mobile phase had an apparent pH lower than the pK of the template, falling rapidly when the apparent pH became greater than the pK. In addition, the column capacity factor for the template resulted strictly dependent on the apparent acidity of the mobile phase, reaching a maximum value when pH = pK. These data suggest a retention mechanism involving an ion-exchange interaction between the amine of the template and the carboxylic group present in the imprinted binding site.

However, it is necessary to take into account that interactions between carboxylic and amino groups are not necessarily based completely on ion-exchange mechanism. Mixed-mode partition equilibrium involving both ion-exchange and hydrogen bonding cannot be excluded when columns are eluted using solvent-rich mobile phases obtained by mixing aqueous buffers and polar nonionizable organic solvents, such as acetonitrile, in which the overall dielectric constant is greatly reduced and hydrogen bonding partially favored.

C. Hydrophobic Interactions

When templates are polar, the increase of the aqueous content in the mobile phase causes a marked decreasing of the column capacity factor, whereas templates of moderate or low polarity are increasingly retained and the column behaves as a "true" reverse phase. Such increase in retention can be attributed to a shift of the partition equilibrium towards the stationary phase (bulk of the polymer + binding sites) due to hydrophobic interactions with the template.

For example, when considering a polymer imprinted with the hydrophobic 2,4,5-trichlorophenoxyacetic acid [33], it was seen that by using increasing amounts of aqueous buffer in mobile phase the column capacity factor increased proportionally, both for the imprinted and the nonimprinted polymers. It is worthwhile noting that, in this case, the nonimprinted polymer did not show significant retention for the template using mobile phase mixtures with a water content lower than 30% v/v to elute the column. However, the imprinted polymer showed enhanced retention also with mobile phase mixtures with a water content greater than 30% v/v. Thus, the partition of the template appears to be strongly enhanced not only by a nonspecific hydrophobic effect, but also by the presence of highly hydrophobic binding sites generated by the imprinting process and complementary to the most hydrophobic part of the imprinted molecule.

It should be stressed again that in many cases the increase of hydrophobic interactions could involve not only the binding sites, but also the polymeric backbone, resulting in pronounced nonspecific binding, often in the form of total retention of all hydrophobic compounds with a complete loss of selectivity. Such nonspecific adsorption can be reduced by the addition of organic modifiers or surface-active agents. This effect can be used when imprinted columns are prepared for on-line solid phase extraction, where a preliminary nonselective step of preconcentration is necessary, before washing unwanted substances and recovering the selectively retained analyte.

V. MOBILE PHASE OPTIMIZATION FOR IMPRINTED COLUMNS

In view of the complex partition mechanisms that control the template retention in liquid chromatography when imprinted stationary phases are used, chromatographic process optimization must be carried out to set up robust and efficient separations. Assuming that the imprinted polymer has been prepared correctly by using suitable functional monomer and porogenic solvent, the optimization should be carried out by working on the composition of the mobile phase. Template elutions should be done by involving a pair of columns, one packed with the imprinted polymer (MIP column), and one packed with a blank polymer prepared without the template (NIP column). In fact, beside the classic chromatographic parameters such as capacity factor, number of theoretical plates, and efficiency, the imprinting factor (IF $= k_{\text{MIP column}}/k_{\text{NIP column}}$) should also be considered.

Robust and fully optimized elution conditions could be rapidly obtained by utilizing chemiometric techniques, in analogy with what can be done to optimize the imprinting procedures (see Chapter 8.V). Sub-optimal elution conditions that are to be fully optimized through the use of experimental design techniques can be obtained by starting from an elution with the porogenic solvent and thoroughly considering the nature of the template as reported in Fig. 5: presence of acid or basic groups, possibility of hydrogen bond-based polar interactions, and relative hydrophobicity.

Multifactor experimental designs involving many variables such as percentage composition of solvent mixtures, pH, and molar concentrations of additives, can be easily implemented for chromatographic separations, and semi-empirical models can be obtained by studying the effect of these variables on the chromatographic response (capacity factor, number of theoretical plates, efficiency, and imprinting factor). However, it should be considered that chromatographic techniques (and liquid chromatography on molecularly imprinted polymers) can be characterized by multiple optimal conditions, essentially due to the possibility that under different experimental conditions a group of peaks elutes with high resolution but different elution order. As a consequence, techniques such as sequential simplex optimization, based on the implicit existence of a single optimal condition, should be avoided or used with great prudence.

VI. APPLICATIONS OF MOLECULARLY IMPRINTED STATIONARY PHASES

A. Enantioseparation

The preparation of the conventional chiral stationary phases can be realized both by surface immobilization of natural chiral selectors (polysaccharides, proteins) on chromatographic supports, and by direct synthesis of stationary phases composed of polyacrylates with pendant chiral groups, amides, or helical polymers. Even if they are quite expensive and poorly resistant to chemical and biological attack, these materials are largely used to separate racemic mixtures for preparative and

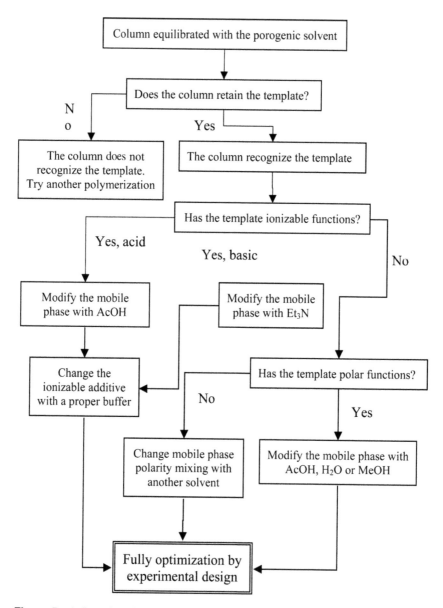

Figure 5 A flow chart for the preliminary optimization of a liquid chromatographic method using a molecular imprinted polymer as stationary phase.

analytical purposes. These techniques suffer from a very limited degree of predictability with regard to elution order and separability of the enantiomers. The effective resolution of a pair of enantiomers can be obtained only with a long trial-and-error process, making screening of stationary phase libraries a necessity. On the other hand, molecular imprinting technique is simple, especially when the noncovalent imprinting approach is used. The selectivity of the imprinted columns is predictable and can be customised for a given separation.

The most extensive studies on molecular-imprinted polymers as stationary phases for enantioseparations were conducted using amino acids and their derivatives as the template, but many other classes of compounds have been studied, including peptides, β-blockers, and nonsteroidal anti-inflammatory drugs (see Table 3). Examples of separations are reported in Figs. 6–8.

In many cases separation factors are comparable or higher than those observed for many of the commercial chiral stationary phases. High selectivity is generally observed in these separations, and relies on the chiral molecular recognition on multiple interaction points; the more functionalized the template is, the more selective recognition is possible. Thus, selectivity is caused by a marked shape matching between the binding site and the template, and conformational or minor structural differences between template and related derivatives can suppress the enantioseparation completely. For example, for different polymers imprinted with L-phenylanilide and L-phenylalanine-N-methylanilide the raceme corresponding to the template was well resolved on the corresponding imprinted polymer, whereas the other raceme was less retained and poorly resolved [34,35]. Similar results were obtained when comparing polymers imprinted with templates different in one methylene group in the alkyl chain, such as N-protected aspartic and glutamic acid [20,41], or ephedrine, and pseudoephedrine [46].

Despite this, there are examples of molecularly imprinted chiral stationary phases that are capable of resolving more than the raceme corresponding to the template. In these cases, minor structural differences are possible without compromising the separation. For example, a polymer imprinted with L-phenylalanine anilide efficiently separated the protected amino acids with different side chains or amide substituents [36]. Anilides of all aromatic amino acids were resolved as well as β-naphthylamides and p-nitroanilides of leucine and alanine [40].

However, among the drawbacks, it is necessary to take into the account that the intrinsically "polyclonal" nature of the noncovalently imprinted polymers causes the presence of a small number of highly enantioselective binding sites and a large number of less selective binding sites, and it is associated to nonlinear adsorption isotherms. As a consequence, enantioseparations are frequently affected by asymmetric peaks and slow association–dissociation kinetics that result in low peak resolution and a marked sensitivity to the amount of sample introduced in the column.

B. On-Line Solid Phase Extraction

A molecularly imprinted column for liquid chromatography can be used not only to separate analytes, but also to selectively extract analytes from complex samples. This technique is called "on-line Molecularly Imprinted Solid Phase Extraction" (on-line MISPE), and it combines the high extraction efficiency of reverse phase SPE for aqueous samples with the high selectivity of the molecular–imprinted polymers. Examples of successful selective extraction and clean-up are reported in Figs. 9 and 10.

In this format, a small column (typically no longer than 1–2 cm, and 2–5 mm of internal diameter), packed with the imprinted material is placed in the loop of the injector or immediately before the analytical column (typically a C18 reverse phase). The imprinted column is loaded with the sample and the interfering substances are washed out by maintaining the analytical column off-line. Then, the analyte is eluted by the mobile phase out of the MISPE column and separated

Table 3 Examples of Molecular Imprinted Polymers Used as Stationary Phases in Chiral Chromatography

Template	Polymer	Separation factor	Ref.
Amino acids			
L-Phe-OH	Cu(II)-VBIDA/EDMA	1.5	10
L-Phe-OEt	MAA/EDMA	1.3	4
L-Phe-NHPh	MAA/EDMA	3.4–3.5–4.9	34–36
L-Phe-NHEt	MAA/EDMA	2.0	4
L-Phe-NMePh	MAA/EDMA	2.0	37
Fmoc-L-Phe-OH	MAA/EDMA	1.4	38
Z-L-Phe-OH	MAA/TRIM	2.3	22
Boc-L-Phe-OH	MAA + 2VP/EDMA	2.0	20
Dansyl-L-Phe-OH	MAA + 2VP/EDMA	3.2	20
L-Arg-OEt	MAA/EDMA	1.5	39
L-Pro-NHPh	MAA/EDMA	4.5	40
Ac-D-Trp-OMe	MAA/EDMA	3.9	21
Ac-L-Trp-OH	AM/EDMA	3.2	23
Boc-L-Trp-OH	MAA + 2VP/EDMA	4.4	20
L-Trp-OEt	MAA/EDMA	1.8	36
Z-L-Asp-OH	MAA + 2VP/EDMA	1.9	20
Z-L-Asp-OH	4VP/EDMA	2.8	41
Z-L-Glu-OH	MAA + 2VP/EDMA	2.5	20
Z-L-Tyr-OH	MAA/PETRA	2.86	22
Peptides			
L-Phe-Gly-NHPh	MAA/EDMA	5.1	40
Boc-L-Phe-Gly-OEt	MAA/TRIM	3.04	41
Z-L-Ala-L-Ala-OMe	MAA/TRIM	3.19	22
Ac-L-Phe-L-Trp-OMe	MAA/EDMA	17.8	42
Z-L-Ala-Gly-L-Phe-OMe	MAA/TRIM	3.6	22
Pharmaceuticals			
(S)-Propranolol	MAA/EDMA	2.8	43
(S)-Timolol	MAA/EDMA	2.9	43
(S)-Naproxen	4VP/EDMA	1.65–1.74	44,45
(S,R)-Ephedrine	MAA/TRIM	3.42	46
(S,S)-Pseudoephedrine	MAA/TRIM	3.19	46
(S)-Benzylbenzodiazepine	MAA/EDMA	3.03	47
Miscellaneous			
(R)-Mandelic acid	4VP/EDMA	1.5	21
(R)-Phenylsuccinic acid	4VP/EDMA	3.6	21
(R)-α-methylbenzylamine	MAA/EDMA	1.6	48
p-NH$_2$-Ph-β-galactoside	MAA/EDMA	1.27	49
(−)-Cinchonidine	MAA/EDMA	5.3 – 31	17,50

AM: acrylamide, EDMA: ethylene dimethacrylate, MAA: methacrylic acid, PETRA: pentaerythritol triacrylate, TRIM: trimethylolpropane trimethacrylate, VBIDA: 4-vinylbenzyliminodiacetic acid, 2VP: 2-vinylpyridine, 4VP: 4-vinylpyridine.

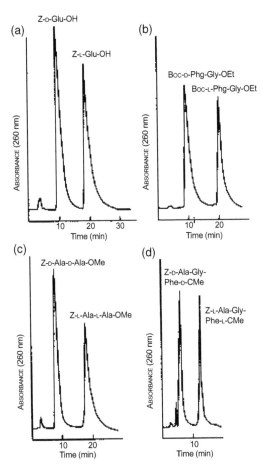

Figure 6 Chiral HPLC separation of amino acid derivatives on imprinted stationary phases packed with Z-L-Glu-OH (a), Boc-L-phe-Gly-OEt (b), Z-L-Ala-L-Ala-Ome (c) Z-L-Ala-Gly-L-Phe-OMe (d). Mobile phase: chloroform – acetic acid. Column 250 × 4.6 mm. Flow rate, 1 mL/min. Detection, UV 260 nm. Reproduced from Ref. 22, with permission.

on-line in the analytical column. Alternatively, an C18 reverse phase precolumn can be placed before an MISPE column, to preconcentrate the analyte and the interfering substances of comparable hydrophobicity. Then, these substances can be coeluted and separated on an MISPE column. This technique has been used for the detection of triazine herbicides in aqueous samples, urine, and apple extracts [51], and coupled with a restricted access material column in the analysis of the analgesic tramadol in serum samples [52].

In a different format, called "pulsed MISPE" (Fig. 11), the sample is preconcentrated on an MISPE column directly connected to the detector and the interfering substances are washed away. Then a small amount of solvent (20 μL), able to disrupt the interaction between the template and the stationary phase is used to quantitatively recover the retained analyte. This format has been applied successfully to the detection and quantification of theophylline [53] and 4-aminopyridine [54] in extracted serum, and nicotine in chewing gum [55] and tobacco [56].

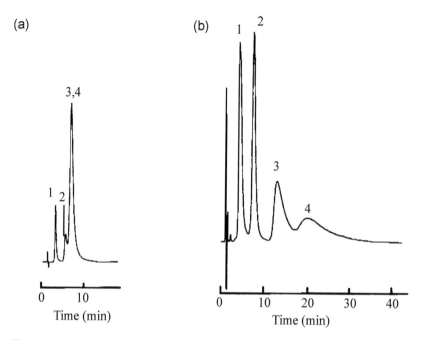

Figure 7 Chiral HPLC separation of 2-arylpropionic acid derivatives on nonimprinted (a) and (*S*)-naproxen-imprinted stationary phase (b). (1) Racemic ketoprofen, (2) racemic ibuprofen, (3) (*R*)-naproxen, (4) (*S*)-naproxen. Mobile phase, 20 mM phosphate buffer pH 3.2 – acetonitrile 1 + 1 v/v. Columns 100 × 4.6 mm. Flow rate, 1 mL/min. Detection, UV 254 nm. Reproduced from Ref. 45, with permission.

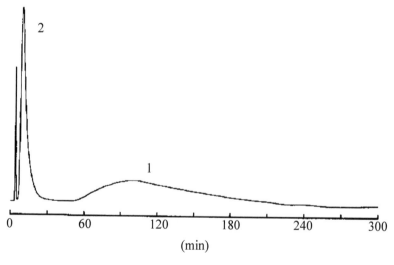

Figure 8 Chiral HPLC separation of a equimolar mixture of (−)-cinchonidine (1) and (+)-cinchonidine (2) on a (−)-cinchonidine-imprinted stationary phase. Mobile phase, methanol – acetic acid 7 + 3 v/v. Columns 150 × 4.6 mm. Flow rate, 0.5 mL/min. Detection, UV 280 nm. Reproduced from Ref. 48, with permission.

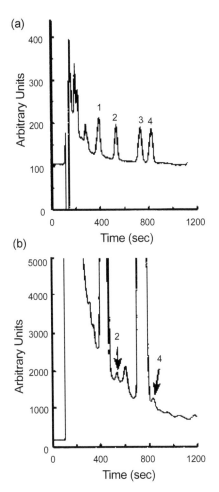

Figure 9 Selective extraction of simazine (1), atrazine (2), propazine (3), and terbutylazine (4) from 200 mL of aqueous sample containing 20 ppm of humic acid and spiked with 0.5 ng/mL of each triazine. Chromatogram (a) and (b): extractions performed with and without a MIP column. Reproduced from Ref. 48, with permission.

C. Affinity Screening of Combinatorial Libraries

Drug discovery methods based on combinatorial chemistry need high-throughput screening techniques to examine rapidly large libraries of synthetic compounds for biological activity. The primary purpose of the screening analysis is to discover if a compound present in the library is better to interact with a receptor relative to other compounds. Liquid chromatography on molecularly imprinted polymers has been proposed as an alternative to in vitro bioassays and traditional affinity chromatography on biological receptors bound to stationary phases. This approach has the advantage that imprinted polymers are economical, simple to prepare, and stable in extreme chemical environments. This is important when the natural receptor is difficult to isolate or is unstable. Utilizing imprinted polymers for an initial screening could serve to identify compounds potentially able to bind the natural

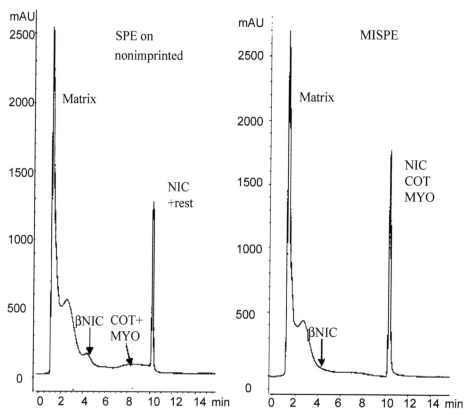

Figure 10 Solid phase extraction of 100 µL of a chewing gum extract spiked with nicotine (0.5 mg/mL), β-nicotyrine, cotinine, and myosmine (0.05 mg/mL). Reproduced from Ref. 55, with permission.

receptor, thereby enabling a limited amount of precious biological material to be conserved and used to examine a more restricted group of preselected substances.

An artificial mimic of the receptor is prepared by molecular imprinting with a ligand of known affinity as a template. Then, substances constituting the library are eluted singularly or in a mixture, and a selectivity index is calculated for each of these substances by direct comparison between the capacity factor of the compounds and the capacity factor of the ligand. Assuming that the imprinted binding site will be a faithful copy of the receptor binding site, the compounds that better fit the natural receptor could be identified as those that are better retained by the imprinted column.

Examples of this technique are described for artificial receptors for the alkaloid yohimbine binding peptides obtained from a phage display library [57], for the steroid libraries related to 11α-hydroxyprogesterone [58], corticosterone [58] (reported in Fig. 12), and cortisol [59]. A molecularly imprinted polymer working as a synthetic receptor for a series of chiral benzodiazepines [47], artificial receptors for the tricyclic antidepressant drug nortriptyline—obtained by covalent and noncovalent molecular imprinting and studied by capillary liquid chromatography with a simulated combinatorial library [60,61]—were also examined.

Step 1

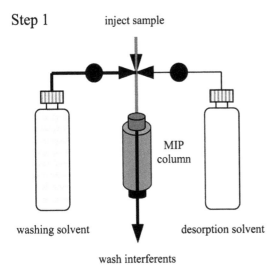

Step 2

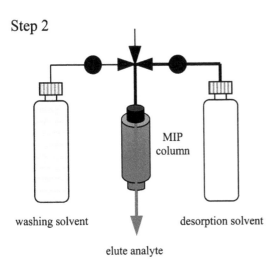

Figure 11 On-line MISPE procedure. Step 1: sample loading and column washing. Step 2: analyte elution with a plug of desorbing eluent.

D. Sensor Based on Liquid Chromatography

Optical sensing systems based on molecularly imprinted columns for liquid chromatography have been described for the quantitative determination of the antibiotic chloramphenicol [62] of L-phenylalanine amide [63] and for the endogenous steroid β-estradiol [64].

The technique is based on the in-column competition for the specific binding sites of the imprinted polymer between the analyte and a less retained marker molecule, resulting in a displacement of this analogue proportional to the amount of analyte introduced in the column. In the case of chloramphenicol, the colored derivative chloramphenicol–methyl red was used as a marker (λ_{abs} 460 nm), for

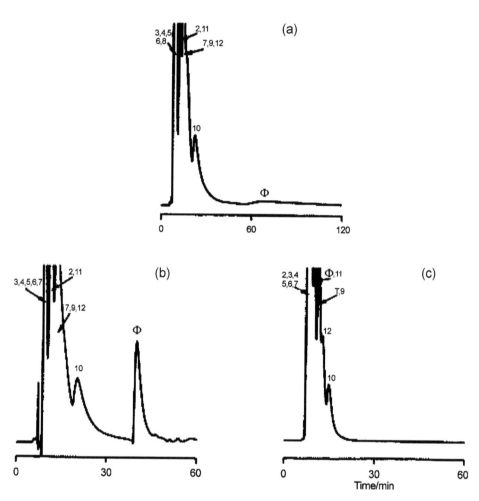

Figure 12 Screening of a steroid library. (a) An imprinted polymer prepared for 11α-hydro-xyprogesterone. Mobile phase: dichloromethane (DCM) – 0.1% acetic acid v/v, Flux 0.5 mL/min. (b) An imprinted polymer prepared for 11α-hydroxyprogesterone. Gradient elution, 0–25 min, DCM 0.1% acetic acid v/v; 25–30 min, DCM 0.1–5% acetic acid v/v; 30–40 min, DCM 5% acetic acid v/v; 40–45 min, DCM 5–0.1% acetic acid v/v. Flux 0.5 mL/min. (c) A control polymer prepared in the absence of template molecule. Isocratic elution, DCM 0.1% acetic acid v/v. Flux 0.5 mL/min. Sample component. (1) 11α-hydroxyprogesterone, (2) 11α-hydroxyprogesterone, (3) 17α-hydroxyprogesterone, (4) progesterone, (5) 4-androsten-3,17-dione, (6) 1,4-androstadiene-3,17-dione, (7) corticosterone, (8) cortexone, (9) 11-deoxy-cortisol, (10) cortisone, (11) cortisone-21-acetate, (12) cortisol-21-acetate. Reproduced from Ref. 58, with permission.

L-phenylalanine amide the dye Rhodamine B was used (λ_{abs} 551 nm) and for β-estra-diol the fluorescent derivative was β-estradiol-3-dansylate (λ_{ex} 283 nm – λ_{em} 307 nm).

This optical detection system showed good selectivity for the analytes, and the analysis time was about 15–20 min. Linear responses over a range of 5–160 μg/mL for chloramphenicol and 0.03–5 μg/L for β-estradiol were obtained, while a sensitivity of up to 500 μM was observed for L-phenylalanine amide.

VII. FRONTAL CHROMATOGRAPHY WITH MOLECULARLY IMPRINTED STATIONARY PHASES

Among the various methods available to determine single-component adsorption isotherms of molecularly imprinted polymers, frontal chromatography is one of the most efficient. It consists of running mobile phases—which contains increasing amounts of the template within a large concentration interval—through an imprinted column, and record the breakthrough curves with a detector. Mass conservation of the solute between the time when the mobile phase containing the template solution enters the column and the final time for which the plateau concentration is reached allow the calculation of the bound amount of template, B, in the stationary phase at equilibrium with a given concentration, C, in the mobile phase (or the free template $F = C - B$)

The bound template B could be measured by integration of the frontal chromatogram (see Fig. 13). The area on the left of the breakthrough point (well visible as a maximum in the chromatogram's first derivative) is the mass of solute constantly present in the column, i.e., the sum of the solute in the mobile phase and the solute partitioned in the stationary phase. The amount of bound template is given by the equation:

$$B = C \frac{V_{eq} - V_d}{V_g}$$

where V_{eq} and V_d are the elution volumes of the equivalent area and the column dead volume, and V_g is the volume of the column.

The experimental plot of the equilibrium concentrations of bound vs. free template defines the adsorption isotherm, from which quantitative information concerning the binding affinity constant (K), binding site concentration (q), and site heterogeneity (ν) of the imprinted stationary phase can be obtained.

Adsorption isotherms can be fitted by using models in which different assumptions about the template–binding site interaction are made. The most simple model is the Langmuir-type adsorption isotherm, in which it is assumed that the stationary phase contains only one class of binding sites (MIP "monoclonality", which are typical of polymers obtained by covalent imprinting) with a well-defined adsorption energy and where template–template interactions do not occur. Other models, such as bi-Langmuir, Freundlich, or Freundlich–Langmuir (see Table 4), assume that the stationary phase has a more or less heterogeneous surface, composed of a multiplicity of different binding site classes (MIP "polyclonality", typical of polymers obtained by noncovalent imprinting).

The frontal chromatographic analysis of a molecular-imprinted column allows not only the determination of adsorption energies and saturation capacities [65,66], but also determination of kinetic data for the association/dissociation process involved in the interaction between template and imprinted binding site, and mass transfer data for the chromatography process itself [67–69].

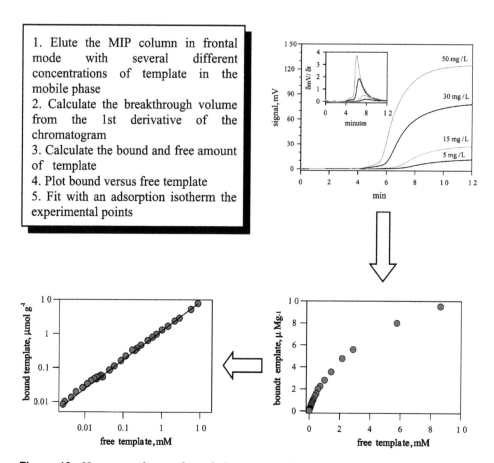

Figure 13 How to perform a frontal chromatography on an MIP-based chromatograpic column.

Table 4 Isotherm Models Used to Fit Binding Data from Molecular Imprinted Polymers

	Isotherm model	Ref.
Freundlich	$B = aF^n$	66–69,71,73
Langmuir	$B = \frac{qKF}{1+KF}$	67,70,73
Bi-Langmuir	$B = \frac{q_1 K_1 F}{1+K_1 F} + \frac{q_2 K_2 F}{1+K_2 F}$	65,67–69,73
Freundlich–Langmuir	$B = \frac{qKF^v}{1+KF^v}$	72
Jovanovich	$B = q(1 - \exp^{-KF})$	73
Bi-Jovanovich	$B = q_1(1 - \exp^{-K_1 F}) + q_2(1 - \exp^{-k_2 F})$	73
Freundlich–Jovanovich	$B = q(1 - \exp^{-KF^v})$	73
Bi-Freundlich–Jovanovic	$B = q_1(1 - \exp^{-K_1 F^v}) + q_2(1 - \exp^{-k_2 F^v})$	65

VIII. CONCLUSIONS

As seen in the previous sections, high pressure liquid chromatography and molecular imprinting technology are tightly interlaced, and imprinted stationary phases find interesting applications, not only to characterize the binding properties of these materials, but also in practical applications ranging from solid phase extraction to chiral chromatography. In spite of the exponential growth in scientific literature, it seems that MIP-based chromatography is at an early stage of development today, with a lot of innovative applications but without common acceptance from people involved in separation science. Even if MIP-based materials are potentially competitive with protein-based chiral stationary phases and immunoaffinity-based SPE materials, the acceptance of this technology in the scientific community, other than the people that work on molecular imprinting, seems to be low.

Many reasons could be invoked to explain this fact. Generally speaking, the acceptance of new technology meets resistance from users of old technology (if my method works well, why change?). It should also be considered that modern HPLC stationary phases spring from a long history of continuous progress towards separation efficiency: efforts sustained by big economic interests and great attention by the scientific community also aid this progress. Against this, MIP technology in separation science is affected by many technical drawbacks. Slow column kinetics, binding site heterogeneity, difficulty in obtaining monodispersed stationary phases with isotropic packings, substantial difficulties in dealing with aqueous mobile phases and the absence of validated separation methods pose severe restriction to widespread chromatographic applications. MIP-based chromatography is however a very promising and productive field in fundamental and applicative research. We are convinced that in a near future many of the underlined problems will be solved.

IX. PROTOCOL 1: PACKING A COLUMN WITH A THEOPHYLLINE-IMPRINTED POLYMER

A. Reagents and Materials

Theophylline (template), caffeine (template analogue), methacrylic acid (MAA, functional monomer), ethylene dimethacrylate (EDMA, cross-linker), 2,2′-azobis-2-methylpropionitrile (AIBN, radical initiator). Analytical-grade chloroform, ethanol (96% v/v), methanol, acetic acid.

B. Method

In a 25 mL borosilicate glass flask dissolve 250 mg of theophylline (1.39 mmol) in 10 mL of chloroform (note 1). Add 0.47 mL (5.56 mmol, molar ratio template – functional monomer 1:4) of methacrylic acid, 8.4 mL of ethylene dimethacrylate (44.5 mmol, molar ratio functional monomer – cross-linker 1:8) and 63 mg of AIBN (note 2). Sonicate the mixture until complete dissolution of theophylline. Purge the flask with analytical-grade nitrogen for almost 5 min, seal with an airtight plug and leave it to polymerize overnight at 60°C in a waterbath. A blank polymer can be prepared in the same manner, omitting the theophylline.

After complete polymerization of the mixture, break the flask and grind the colorless white monolith obtained in a mechanical mortar (note 3). Separate the grinded powder through a double sieve (90 μm upper screen – 25 μm lower screen) to isolate a fraction suitable for column packing (note 4). Wet the fraction retained by the 25 μm-screen with ethanol and wash carefully with abundant deionized water to eliminate the sub-micrometric dust that adheres to the polymer particles (note 5).

Wash the obtained particle fraction with methanol by overnight extraction in a Soxhlet apparatus to eliminate the residual theophylline (note 6). Dry the polymeric particles at 50°C.

Mix 2 g of polymer with 2 mL of methanol–water (1 + 1 v/v) in a test tube and sonicate until complete homogeneity of the mixture is reached. Transfer the suspension rapidly into an empty 100 × 3.9 mm HPLC column mounted on-line with a second empty column to serve as a column packaging device (Fig. 14). Close the column and run the HPLC in pressure-constant mode with methanol–water (1 + 1 v/v) as a mobile phase, gradually increasing the pressure from 1 to 25 MPa by properly increasing the flow rate. Flush the system for about 20 column volumes. Disassemble the packing device carefully by cutting the packed stationary phase at the outlet of the double female joint with a spatula or a knife (note 7).

Close the column, connect it to the detector and run the HPLC in flow-constant mode, elute the column at 1 mL/min with methanol–acetic acid (9 + 1 v/v). Follow

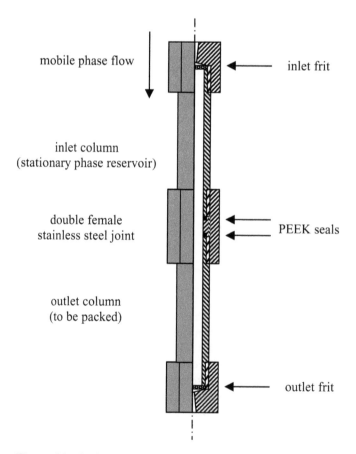

Figure 14 An home-made HPLC column packing device.

the absorbance at 210 nm until it falls below 0.001 UAFS and is stable for more than 10 min (note 8).

Equilibrate the column with chloroform–acetic acid $99 + 1\,v/v$, monitoring the absorbance at 270 nm. Measure the dead time by repeated injections of 5 μL of a 50 μg/mL solution of acetone in chloroform. The performance parameters of the packed column (number of theoretical plates, efficiency, selectivity) can be measured by repeated injections of 5 μL of a 50 μg/mL solution of theophylline and caffeine in chloroform. Calculate the imprinting factor, IF, as the ratio between the capacity factors for the theophylline eluted on the imprinted and nonimprinted columns (note 9)

C. Comments

Note 1. Commercial chloroform is stabilized with ethanol to suppress light-induced decomposition to toxic phosgene. Such stabilizer is detrimental to imprinting because of its good hydrogen bonding properties. Ethanol-free chloroform can be obtained by fractional distillation from the commercial product. For the same reason, before use the chloroform must be dried on 4.5 Å molecular traps to eliminate any trace of water.

Note 2. Commercial methacrylic acid and ethylene dimethacrylate are stabilized with alkylphenols to suppress polymerization. They must be purified by distillation under reduced pressure immediately before use. Liquid–liquid extraction with diluted sodium hydroxide solutions is not practicable with the water-soluble methacrylic acid and less efficient with ethylene dimethacrylate.

Note 3. A porcelain mortar can also be used for manually grinding the polymer. However, this option is suitable only with a limited amount of material and take a longer time!

Note 4. The grinding/sieving step produces a lot of very fine, sub-micrometric particles. This dust is potentially dangerous for the lungs. Undertake the operation under a hood or wear a suitable anti-dusk mask.

Note 5. Very fine, sub-micrometric powder could cause excessive high back pressures in the column during and after the packing procedure. Multiple washings are fundamental to obtain dust-free polymeric particles.

Note 6. The Soxhlet extraction gives a material which is sufficiently free from residual template and suitable for chromatographic applications, considering that subsequent column washing will further reduce the amount of residual template. Do not use this material for solid-phase extraction applications, because this technique requires more exhaustive template extraction methods.

Note 7. A well-packed column is essential to achieve high-efficiency elutions. There are two different methods to pack a column: the "dry" and the "wet" method. In the first case the column is filled with the dry stationary phase, closed and the mobile phase is then pumped in. This method is suitable for silica-based beads, but not for polymeric beads, because this kind of material is prone to swelling when wet. Thus, high column back pressures, preferential channels and flow limitations due to a nonhomogeneous wetting of the stationary mobile phase could deteriorate the chromatographic performances of the column. When the "wet" packing method is used, the stationary phase is swollen and suspended in a solvent of similar density, and introduced into the column as a slurry. In this case the main drawback is the necessity to provide a reservoir for the column to keep the beads in suspension.

It should be in volumetric excess compared to the column's geometric volume. Many suppliers sell (expensive) column packing devices that are suitable for preparative and semi-preparative columns. A simple, home-made, column packing device can be assembled by joining two HPLC columns with a double female joint—the first column acts as a reservoir, the second is the column to be packed.

Note 8. Washing and equilibration of the packed column are very important. To obtain the release of the template, a mobile phase able to interfere with the noncovalent interaction between the template, and the stationary phase should be used. In this tutorial methanol and acetic acid are very efficient, disrupting the hydrogen bond between theophylline and methacrylic acid residues. The column washing could be considered complete when the UV signal falls below 0.001 units of absorbance.

Note 9. It is very important to calculate the imprinting factor, because it is a measure of how much the imprinted binding sites of the polymer are able to recognize the template molecule, compared to non-specific interactions with the polymeric backbone. In fact, whereas the possible presence of nonspecific interactions between the polymer and the template could not be excluded a priori for both the imprinted and the nonimprinted columns, a molecular recognition effect due to imprinted binding sites could only be possible for the imprinted polymer.

X. PROTOCOL 2: PREPARATION OF A THEOPHYLLINE-IMPRINTED MONOLITHIC COLUMN

A. Reagents and Materials

Theophylline (template), caffeine (template analogue), methacrylic acid (MAA, functional monomer), ethylene dimethacrylate (EDMA, cross-linker), 2,2′-azobis-2-methylpropionitrile (AIBN, radical initiator). Analytical-grade cyclohexanol, dodecanol, chloroform, methanol, acetic acid.

B. Method

In a 5 mL test tube dissolve 25 mg of theophylline (0.139 mmol) in 1.6 mL of a cyclohexanol – dodecanol mixture (25 + 7 v/v) (note 1). Then, add 47 μL (0.556 mmol, molar ratio template – functional monomer 1:4) of methacrylic acid, 0.84 mL of ethylene dimethacrylate (4.45 mmol, molar ratio functional monomer – cross-linker 1:8) and about 8 mg of AIBN (protocol 1, note 2). Sonicate the mixture until complete dissolution of the theophylline. Purge the flask with analytical-grade nitrogen for 5 min, then transfer 0.600 mL of prepolymerization mixture into a 50 × 3.9 mm stainless-steel HPLC column, close it with airtight plugs (note 2) and leave it to polymerize overnight at 60°C in a waterbath (note 3). A blank polymer can be prepared in the same manner, omitting the theophylline.

Substitute the airtight plugs with frits and stainless-steel end-fittings, close the column, connect it to the detector and run the HPLC in flow-constant mode, by eluting the column with methanol–acetic acid (9 + 1 v/v) and increasing the flow gradually from 0.1 to 0.25 mL/min (note 4). Follow the absorbance at 210 nm till it falls below 0.001 UAFS and is stable for more than 10 column volumes (protocol 1, note 8).

Equilibrate the column with chloroform–acetic acid (99 + 1 v/v), monitoring the absorbance to 270 nm. Measure the dead time by repeated injections of 5 μL of a

50 μg/mL solution of acetone in chloroform. The performance parameters of the packed column (number of theoretical plates, efficiency, selectivity) can be measured by repeated injections of 5 μL of a 50 μg/mL solution of theophylline and caffeine in chloroform. An example of theophylline – caffeine separation is reported in Fig. 15.

C. Comments

Note 1. A variable amount of isooctane (2–20% v/v) can be added to the porogen mixture to help the formation of the superporous structures and to reduce the flow-resistance of the resulting polymer. However, the optimum amount of isooctane is highly influenced by the composition of the prepolymerization mixture, and trial-and-error experiments are necessary.

  Note 2. Self-made air-tight plugs can be obtained by substituting stainless-steel or carbon frits with silicone rubber disks cut to the same diameter.

  Note 3. It is important that polymerization will be effected with the column positioned vertically. Possible voids in the monolithic polymer will form in the upper end of the column, and can be easily filled using small glass beads without significantly deteriorating the column efficiency.

  Note 4. The porogenic mixture is viscous, and initially column back pressure is high even when eluted slowly. A rapid increase of the elution velocity could produce an excessive pressure build-up, deleterious for the polymer structure.

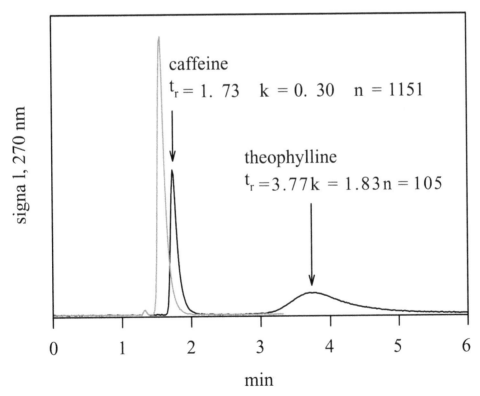

Figure 15 Separation of theophylline and caffeine on a monolithic stationary phase. Black line: column imprinted with theophylline. Gray line: not imprinted column.

XI. PROTOCOL 3: ON-LINE MISPE OF THEOPHYLLINE IN COFFEE EXTRACTS

A. Reagents and Materials

Theophylline. Analytical-grade ethanol-free chloroform (see protocol 1, note 1) methanol, acetic acid. A theophylline-imprinted column.

B. Method

Pack an empty 50 × 3.9 mm HPLC column with the theophylline-imprinted polymer (see protocol 1) or use a monolithic-imprinted column (see protocol 2). To eliminate the residual tamplate wash the column with methanol–acetic acid (9 +1 v/v). Follow the absorbance at 210 nm until it falls below 0.001 UAFS and is stable for more than 10 column volumes. Equilibrate the column with chloroform untill a stable baseline at 270 nm is obtained.

Extract 20 mL of coffee with 20 mL of chloroform (note 1). Separate and centrifuge the organic layer for 10 min at 8000g. Filter the extract through 0.22 μm teflon membranes.

Dilute 100 μL aliquots of chloroform extract (1 +1 v/v) with chloroform solutions of theophylline, using analyte concentrations ranging from 0.2 to 10 μg/mL. Prepare also a sample dilution (1 +1) without adding theophylline.

Perform three injections of 5 μL for each sample, eluting the column with chloroform and follow the absorbance at 270 nm untill it falls below 0.001 UAFS. Switch the eluent to methanol – acetic acid (9 +1 v/v) to elute the retained theophylline (note 2) (Fig. 16). After each elution, re-equilibrate the column with chloroform.

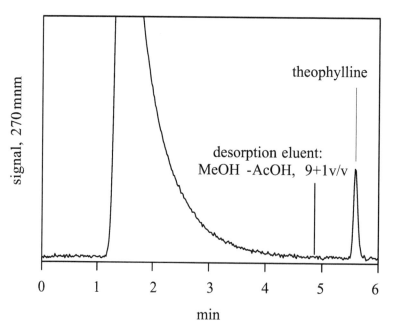

Figure 16 Extraction of theophylline from coffee extract using an on-column MISPE system.

Measure the theophylline peak area for each sample elution, averaging the single values. Build a calibration line plotting measured areas against amount of theophylline added to the samples. Calculate the amount of theophylline present in the coffee, extrapolating the value from the area of the peak corresponding to the sample diluted with chloroform only.

C. Comments

Note 1. A direct extraction of theophylline from coffee is not practical. Direct solvent switch between the aqueous sample containing the analyte and the organic mobile phase is not possible because of the nonmiscibility of the liquids. Acetonitrile could substitute chloroform, but the stationary phase performances in terms of analyte recovery will be badly compromised.

Note 2. Theophylline could also be recovered by eluting the column with chloroform, but the peak will be affected by large tailing. The use of methanol/acetic acetic acid mixtures as eluents assures a fast desorption of the analyte, with narrow peak and a very limited tailing effect.

REFERENCES

1. Dickey, F.H. The preparation of specific adsorbents. Proc. Natl. Acad. Sci. USA **1949**, *35*, 227–229.
2. Curti, R.; Colombo, U. Chromatography of stereoisomers with "tailor made" compounds. J. Am. Chem. Soc. **1952**, *74*, 3961.
3. Andersson, L.I.; Sellergren, B.; Mosbach, K. Imprinting of amino acid derivatives in macroporous polymers. Tetrahedron Lett. **1984**, *25*, 5211–5214.
4. Sellergren, B.; Ekberg, B.; Mosbach, K. Molecular imprinting of amino acids derivatives in macroporous polymers. Demonstration of substrate- and enantio-selectivity by chromatographic resolution of racemic mixtures of amino acid derivatives. J. Chromatogr. **1985**, *347*, 1–10.
5. Glad, M.; Norrlöw, O.; Sellergren, B.; Siegbahn, N.; Mosbach, K. Use of silane monomers for molecular imprinting and enzyme entrapment in polysiloxane-coated porous silica. J. Chromatogr. **1985**, *347*, 11–23.
6. ISI's Web of Science, http://www.isinet.com.
7. (a) Poole, C.F.; Poole, S.K. *Chromatography Today*, 5th Ed.; Amsterdam: Elsevier, **1991**; (b) Snyder, L.R. Theory of chromatography. In *Chromatography*; Heftmann, E. Ed.; Elsevier: Amsterdam, **1992**.
8. IUPAC, Nomenclature in chromatography. Pure Appl. Chem. **1993**, *65*, 819–872.
9. Mayes, A.G. Polymerization techniques for the formation of imprinted beads. In *Molecularly Imprinted Polymers. Man-made Mimics of Antibodies and Their Applications in Analytical Chemistry*; Sellergren, B., Ed.; Elsevier: Amsterdam, **2001**, 305–324.
10. Vidyasankar, S.; Ru, M.; Arnold, F.H. Molecularly imprinted ligand-exchange adsorbents for the chiral separation of underivatized amino acids. J. Chromatogr. A **1997**, *775*, 51–63.
11. Sulitzky, C.; Ruckert, B.; Hall, A.J.; Lanza, F.; Unger, K.; Sellergren, B. Grafting of molecularly imprinted polymer films on silica supports containing surface-bound free radical initiators. Macromolecules **2002**, *35*, 79–91.
12. Ruckert, B.; Hall, A.J.; Sellergren, B. Molecularly imprinted composite materials via iniferter-modified supports. J. Mater. Chem. **2002**, *12*, 2275–2280.

13. Tan, X.J.; Remcho, V.T. Molecular imprint polymers as highly selective stationary phases for open tubular liquid chromatography and capillary electrochromatography. Electrophoresis **1998**, *19*, 2055–2060.

14. Svec, F.; Frechet, J.M.J. Continuous rods of macroporous polymer as high-performance liquid-chromatography separation media. Anal. Chem. **1992**, *64*, 820–822.

15. Matsui, J.; Kato, T.; Takeuchi, T.; Suzuki, M.; Yokoyama, K.; Tamiya, E.; Karube, I. Molecular recognition in continuous polymer rods prepared by a molecular imprinting technique. Anal. Chem. **1993**, *65*, 2223–2224.

16. Matsui, J.; Takeuchi, T. A molecularly imprinted polymer rod as nicotine selective affinity media prepared with 2-(trifluoromethyl)acrylic acid. Anal. Comm. **1997**, *34*, 199–200.

17. Matsui, J.; Nicholls, I.A.; Takeuchi, T. Molecular recognition in cinchona alkaloid molecular imprinted polymer rods. Anal. Chem. Acta **1998**, *365*, 89–93.

18. Sellergren, B. Direct drug determination by selective sample enrichment on an imprinted polymer. Anal. Chem. **1994**, *66*, 1578–1582.

19. Sellergren, B. Imprinted dispersion polymers: A new class of easily accessible affinity stationary phases. J. Chromatogr. A **1994**, *673*, 133–141.

20. Andersson, L.I.; Mosbach, K. Enantiomeric resolution on molecularly imprinted polymers prepared with only non-covalent and non-ionic interactions. J. Chromatogr. **1990**, *516*, 313–322.

21. Ramström, O.; Andersson, L.I.; Mosbach, K. Recognition sites incorporating both pyridinyl and carboxy functionalities prepared by molecular imprinting. J. Org. Chem. **1993**, *58*, 7562–7564.

22. Kempe, M. Antibody-mimicking polymers as chiral stationary phases in HPLC. Anal. Chem. **1996**, *68*, 1948–1953.

23. Yu, C.; Mosbach, K. Molecular imprinting utilizing an amide functional group for hydrogen bonding leading to highly efficient polymers. J. Org. Chem. **1997**, *62*, 4057–4064.

24. Allender, C.J.; Brain, K.R.; Heard, C.M. Binding cross-reactivity of boc-phenylalanine enantiomers on molecularly imprinted polymers. Chirality **1997**, *9*, 233–237.

25. O'Brien, T.P.; Snow, N.H.; Grinberg, N.; Crocker, L. Mechanistic aspects of chiral discrimination on a molecular imprinted polymer phase. J. Liq. Chromatogr. Relat. Technol. **1999**, *22*, 183–204.

26. Meng, Z.; Zhou, L.; Wang, J.; Wang, Q.; Zhu, D. Molecule imprinting chiral stationary phase. Biomed. Chromatogr. **1999**, *3*, 389–393.

27. Matsui, J.; Doblhoff-Dier, O.; Takeuchi, T. Atrazine-selective polymer prepared by molecular imprinting technique. Chem. Lett. **1995**, *6*, 489.

28. Dauwe, C.; Sellergren, B. Influence of template basicity and hydrophobicity on the molecular recognition properties of molecularly imprinted polymers. J. Chromatogr. A **1996**, *753*, 191–200.

29. Takeuchi, T.; Fukuma, D.; Matsui, J. Combinatorial molecular imprinting: an approach to synthetic polymer receptors. Anal. Chem. **1999**, *71*, 285–290.

30. Spivak, D.; Gilmore, M.A.; Shea, K.J. Evaluation of binding and origins of specificity of 9-ethyladenine imprinted polymers. J. Am. Chem. Soc. **1997**, *119*, 4388–4393.

31. Yoshizako, K.; Hosoya, K.; Iwakoshi, Y.; Kimata, K.; Tanaka, N. Porogen imprinting effects. Anal. Chem. **1998**, *70*, 386–389.

32. Sellergren, B.; Shea, K.J. Influence of polymer morphology on the ability of imprinted network polymers to resolve enantiomers. J. Chromatogr. **1993**, *635*, 31–49.

33. Baggiani, C.; Giraudi, G.; Giovannoli, C.; Trotta, F.; Vanni, A. Chromatographic characterization of a molecularly imprinted polymer binding the herbicide 2,4,5-trichlorophenoxyacetic acid. J. Chromatogr. A **2000**, *883*, 119–126.

34. Sellergren, B.; Lepistö, M.; Mosbach, K. Highly enantioselective and substrate-selective polymers obtained by molecular imprinting utilizing noncovalent interactions. NMR and chromatographic studies on the nature of recognition. J. Am. Chem. Soc. **1988**, *110*, 5853–5860.

35. Sellergren, B. Molecular imprinting by noncovalent interactions—enantioselectivity and binding capacity of polymers prepared under conditions favoring the formation of template complexes. Makromol. Chem. **1989**, *190*, 2703–2711.

36. Sellergren, B. Molecular imprinting by noncovalent interactions: tailor-made chiral stationary phases of high selectivity and sample load capacity. Chirality **1989**, *1*, 63–68.

37. Lepistö, M.; Sellergren, B. Discrimination between amino acid amide conformers by imprinted polymers. J. Org. Chem. **1989**, *54*, 6010–6012.

38. Kempe, M.; Mosbach, K. Chiral recognition of N^α-protected amino acids and derivatives in non-covalently molecularly imprinted polymers. Int. J. Peptide Protein Res. **1994**, *44*, 603–606.

39. Sellergren, B.; Nilsson, K.G.I. Molecular imprinting by multiple noncovalent host-guest interactions: synthetic polymers with induced specificity. Methods. Mol. Cell Biol. **1989**, *3/4*, 59–62.

40. Andersson, L.I.; O'Shannessy, D.J.; Mosbach, K. Molecular recognition in synthetic polymers: preparation of chiral stationary phases by molecular imprinting of amino acid amides. J. Chromatogr. **1990**, *513*, 167–179.

41. Kempe, M.; Fischert, L.; Mosbach, K. Chiral separation using molecularly imprinted heteroaromatic polymers. J. Mol. Recogn. **1993**, *6*, 25–29.

42. Ramström, O.; Nicholls, I.A.; Mosbach, K. Synthetic peptide receptor mimics. Highly stereoselective recognition in non-covalent molecularly imprinted polymers. Tetrahedr. Asymm. **1994**, *5*, 649–656.

43. Fischer, L.; Muller, R.; Ekberg, B.; Mosbach, K. Direct enantioseparation of β-adrenergic blockers using a chiral stationary phase prepared by molecular imprinting. J. Am. Chem. Soc. **1991**, *113*, 9358–9360.

44. Kempe, M.; Mosbach, K. Direct resolution of naproxen on a non-covalently molecularly imprinted chiral stationary phase. J. Chromatogr. A. **1994**, *664*, 276–279.

45. Haginaka, J.; Takehira, H.; Hosoya, K.; Tanaka, N. Molecularly imprinted uniform-sized polymer-based stationary phase for naproxen. Comparison of molecular recognition ability of the molecularly imprinted polymers prepared by thermal and redox polymerization. J. Chromatogr. A **1998**, *816*, 113–121.

46. Ramström, O.; Yu, C.; Mosbach, K. Chiral recognition in adrenergic receptor binding mimics prepared by molecular imprinting. J. Mol. Recognit. **1996**, *9*, 691–696.

47. Hart, B.R.; Rush, D.J.; Shea, K.J. Discrimination between enantiomers of structurally related molecules: separation of benzodiazepines by molecularly imprinted polymers. J. Am. Chem. Soc. **2000**, *122*, 460–465.

48. Matsui, J.; Nicholls, I.A.; Takeuchi, T. Highly stereoselective molecularly imprinted polymer synthetic receptors for cinchona alkaloids. Tetrahedron—Asymm. **1996**, *7*, 1357–1361.

49. Nilsson, K.G.I.; Sakaguchi, K.; Gemeiner, P.; Mosbach, K. Molecular imprinting of acetylated carbohydrate derivatives into methacrylic polymers. J. Chromatogr. A **1995**, *707*, 199–203.

50. Meng, Z.H.; Zhou, L.M.; Wang, Q.H.; Zhu, D.Q. Molecular imprinting of (*R*)-(+)-a-methylbenzylamine for chiral stationary phase. Chin. Chem. Lett. **1997**, *8*, 345–346.

51. Bjarnason, B.; Chimuka, L.; Ramström, O. On-line solid-phase extraction of triazine herbicides using a molecularly imprinted polymer for selective sample enrichment. Anal. Chem. **1999**, *71*, 2152–2156.

52. Boos, K.S.; Fleischer, C.T. Multidimensional on-line solid-phase extraction (SPE) using restricted access materials (RAM) in combination with molecular imprinted polymers (MIP). Fresenius J. Anal. Chem. **2001**, *371*, 16–20.
53. Mullett, W.M.; Lai, E.P.C. Determination of theophylline in serum by molecularly imprinted solid-phase extraction with pulsed elution. Anal. Chem. **1998**, *70*, 3636–3641.
54. Mullett, W.M.; Dirie, M.F.; Lai, E.P.C.; Guo, H.S.; He, X.W. A 2-aminopyridine molecularly imprinted polymer surrogate micro-column for selective solid phase extraction and determination of 4-aminopyridine. Anal. Chim. Acta. **2000**, *414*, 123–131.
55. Zander, A.; Findlay, P.; Renner, T.; Sellergren, B.; Swietlow, A. Analysis of nicotine and its oxidation products in nicotine chewing gum by a molecularly imprinted solid phase extraction. Anal. Chem. **1998**, *70*, 3304–3314.
56. Mullett, W.M.; Lai, E.P.C.; Sellergren, B. Determination of nicotine in tobacco by molecularly imprinted solid phase extraction with differential pulsed elution. Anal. Comm. **1999**, *36*, 217–220.
57. Berglund, J.; Lindbladh, C.; Nicholls, I.A.; Mosbach, K. Selection of phage display combinatorial library peptides with affinity for a yohimbine imprinted methacrylate polymer. Anal. Comm. **1998**, *35*, 3–7.
58. Ramström, O.; Krook, L.Y, M.; Mosbach, K. Screening of a combinatorial steroid library using molecularly imprinted polymers. Anal. Comm. **1998**, *35*, 9–11.
59. Baggiani, C.; Giraudi, G.; Trotta, F.; Giovannoli, C.; Vanni, A. Chromatographic characterization of a molecular imprinted polymer binding cortisol. Talanta **2000**, *51*, 71–75.
60. Vallano, P.T.; Remcho, V.T. Affinity screening by packed capillary high-performance liquid chromatography using molecular imprinted sorbents I. Demonstration of feasibility. J. Chromatogr. A **2000**, *888*, 23–34.
61. Khasawneh, M.A.; Vallano, P.T.; Remcho, V.T. Affinity screening by packed capillary high performance liquid chromatography using molecular imprinted sorbents II. Covalent imprinted polymers. J. Chromatogr. A **2001**, *922*, 87–97.
62. Levi, R.; McNiven, S.; Piletsky, S.A.; Cheong, S.H.; Yano, K.; Karube, I. Optical detection of chloramphenicol using molecularly imprinted polymers. Anal. Chem. **1997,** *69*, 2017–2021.
63. Piletsky, S.A.; Terpetschnig, E.; Andersson, H.S.; Nicholls, I.A.; Wolfbeis, O.S. Application of non-specific fluorescent dyes for monitoring enantio-selective ligand binding to molecularly imprinted polymers. Fresenius J. Anal. Chem. **1999**, *364*, 512–516.
64. Rachkov, A.; McNiven, S.; Elskaya, A.; Yano, K.; Karube, I. Fluorescence detection of β-estradiol using a molecularly imprinted polymer. Anal. Chim. Acta. **2000**, *405*, 23–29.
65. Chen, Y.B.; Kele, M.; Quinones, I.; Sellergren, B.; Guiochon, G. Influence of the pH on the behavior of an imprinted polymeric stationary phase—supporting evidence for a binding site model. J. Chromatogr. A **2001**, *927*, 1–17.
66. Szabelski, P.; Kaczmarski, K.; Cavazzini, A.; Chen, Y.B.; Sellergren, B.; Guiochon, G. Energetic heterogeneity of the surface of a molecularly imprinted polymer studied by high-performance liquid chromatography. J. Chromatogr. A **2002**, *964*, 99–111.
67. Sajonz, P.; Kele, M.; Zhong, G.M.; Sellergren, B.; Guiochon, G. Study of the thermodynamics and mass transfer kinetics of two enantiomers on a polymeric imprinted stationary phase. J. Chromatogr. A **1998**, *810*, 1–17.
68. Chen, Y.B.; Kele, M.; Sajonz, P.; Sellergren, B.; Guiochon, G. Influence of thermal annealing on the thermodynamic and mass transfer kinetic properties of D- and L-phenylalanine anilide on imprinted polymeric stationary phases. Anal. Chem. **1999**, *71*, 928–938.
69. Miyabe, K.; Guiochon, G. Kinetic study of the concentration dependence of the mass transfer rate coefficient in enantiomeric separation on a polymeric imprinted stationary phase. Anal. Sci. **2000**, *16*, 719–730.

70. Baggiani, C.; Trotta, F.; Giraudi, G.; Moraglio, G.; Vanni, A. Chromatographic characterization of a molecularly imprinted polymer binding theophylline in aqueous buffers. J. Chromatogr. A. **1997**, *786*, 23–29.

71. Umpleby, R.J.; Baxter, S.C.; Bode, M.; Berch, J.K.; Shah, R.N.; Shimizu, K.D. Application of the Freundlich adsorption isotherm in the characterization of molecularly imprinted polymers. Anal. Chem. Acta. **2001**, *435*, 35–42.

72. Umpleby, R.J.; Baxter, S.C.; Chen, Y.; Shah, R.N.; Shimizu, K.D. Characterization of molecularly imprinted polymers with the Langmuir-Freundlich isotherm. Anal. Chem. **2001**, *73*, 4584–4591.

73. Baggiani, C.; Giovannoli, C.; Tozzi, C.; Anfossi, L. Absorption isotherms of a MIP prepared in presence of a polymerisable template, indirect evidence of formation of template clusters in the binding site. Anal. Chem. Acta. **2004**, *504*, 43–52.

21

Capillary Electrophoresis

Oliver Brüggemann   Technische Universität Berlin, Berlin, Germany

I. INTRODUCTION

Electrophoresis as a technique to separate proteins based on their different mobilities was mentioned for the first time by Tiselius in 1930 [1]. The development of capillary electrophoresis (CE) started decades later in the 1960s with first experiments on displacement electrophoresis in glass tubes with hydroxyethylcellulose [2]. Hjerten [3] investigated the electroosmotic flow in tubes and found a way of eliminating this phenomenon by simply coating the wall of the tubes. Virtanen [4] used for electrophoresis in tubes for the first time glass capillaries with inner diameters (ID) of 200–500 µm. And when applying 75 µm open tubular glass capillaries in electrophoretic separations, Jorgenson and Lukacs [5] created the term of high performance capillary electrophoresis—HPCE.

HPCE nowadays is a versatile technique allowing not only the separation of ionic analytes, but also of neutral compounds. A typical HPCE system is made up of a fused silica capillary with an inner diameter of 25–100 µm immersing with its two ends into buffer (electrolyte) vials which are also connected to two electrodes (Fig. 1). The separated analytes generally are detected at a detection window near the outlet of the capillary, i.e., after covering the distance of the so-called effective length of the capillary. Two major effects are responsible for the separation and detectability of all the analytes. Electrophoresis is of course one principle leading to the actual separation, the other is called electroosmosis resulting in the migration of all analytes towards the detector [6–8]. The electroosmotic flow (EOF) is based on the fact that the inner silica walls of the capillary are negatively charged above pH values of about 2.5 due to the dissociation of the silanol end-groups (Fig. 2). This promotes first of all dipolar water molecules to orient with their positive poles towards the wall creating an immobile inner Helmholtz layer. Beside that, a number of hydronium cations migrate to the wall from the solution to counterbalance the negative charges, however, sojourning in the mobile outer Helmholtz layer. An applied electric field leads to the migration of these hydronium cations towards the cathode forcing other molecules such as water or even anions as well to move into the same direction. This phenomenon allows cations, anions, and neutral analytes passing by the detection window. Nevertheless, the actual separation is based on different electrophoretic behaviors of these compounds [6–9].

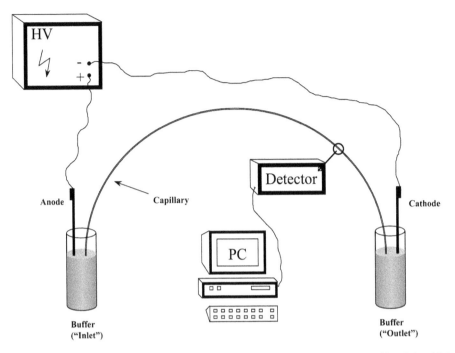

Figure 1   A typical basic HPCE system. The capillary is tipped into two buffer vials which are connected with two electrodes for applying the electric field. After filling the capillary with buffer, injecting the sample either hydrodynamically or electrokinetically, the applied voltage leads to a separation of ions which all migrate towards the detector due to electroosmosis. Detection is performed at a detection window where the capillary is freed from its protecting outer polymer coating. Usually, a personal computer is used to receive and store analytical data.

The simplest HPCE mode is called capillary zone electrophoresis (Fig. 3) where a small volume of the sample is injected into the inlet end of the buffer-filled capillary, and the separation is based on different mobilities of the solutes, i.e., their different mass-to-charge ratios. Although the anions migrate electrophoretically towards the anode, the electroosmotic flow ensures their transport towards the cathode. Only anions with a very high electrophoretic velocity are not detectable. However, this procedure does not permit the separation of neutral compounds, although passing the detector as a group. For such a purpose, surfactants have to be used resulting in the generation of micelles which are able to separate uncharged analytes due to their different distribution factors between the mobile phase and the micelles called the pseudo-stationary phase [6–8]. The analytes have to have different hydrophobicities which lead to different degrees of affinity for the hydrophobic interior of the micelle. But differences in other characteristics of the analytes can be used as well for separation, such as different sizes of molecules in gel electrophoresis with immobilized polyacryl amide gels, or different isoelectric points (pI) in isoelectric focussing [6,7]. Beside these specific techniques, a new branch of HPCE has been developed in the past years called capillary electrochromatography (CEC) [9]. In this procedure, the capillary is filled with a solid stationary phase similar to columns in HPLC. Usually, the capillary has to be packed carefully with relatively small particles (1–3 μm

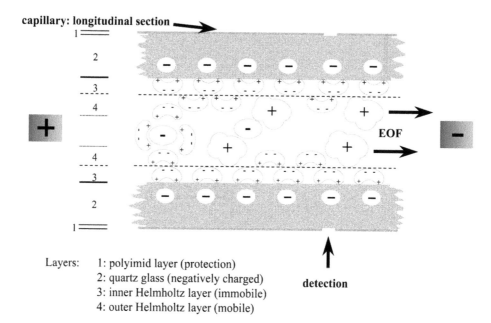

capillary: longitudinal section

Layers: 1: polyimid layer (protection)
         2: quartz glass (negatively charged)
         3: inner Helmholtz layer (immobile)
         4: outer Helmholtz layer (mobile)

detection

Figure 2 Electroosmotic flow (EOF) within a capillary. Due to negatively charged groups on the capillary inner wall (SiO^--functionalities) and the presence of Helmholtz layers, the first mobile film is a positively charged one migrating towards the cathode, drawing all the rest of the capillary filling into the same direction, i.e., cations, anions, and neutral compounds.

diameter) (Fig. 4) and frits have to be installed at both ends of the capillary to retain the particle package. As in the basic HPCE procedure, an applied electric field enables an electroosmotic flow and the analytes to migrate also electrophoretically. Beyond it, the interactions of the solutes with the particles lead to a separation effect as well. This combination of chromatography and electrophoresis results in high numbers of plates, however requires a little more time. Therefore, a slight pressure can be applied on the buffer accelerating the process [7–10].

CEC offers naturally the applicability of all kinds of stationary phases used in HPLC, such as molecularly imprinted polymers (MIP) [11–13]. Thus, also the implementation of MIP in CEC was just a matter of time. In the past few years, different attempts were made to establish MIP as affinity phases in capillaries. Generally, the imprinted materials should selectively recognize the analytes used as template molecules and separate them from other sample components due to prolonged retardation. This system is derived from affinity chromatography which is based on immobilized antibodies allowing the specific recognition of their antigens. However, the drawback of using bio-receptors is their sensitivity towards extreme pH or elevated temperatures. This lack of durability can be overcome by the use of MIP in different formats [14–18].

II. CAPILLARY ELECTROCHROMATOGRAPHY

Beside the actual separation mechanisms, in liquid chromatography, in general a pressure driven mode allows transportation of analytes through the column. In

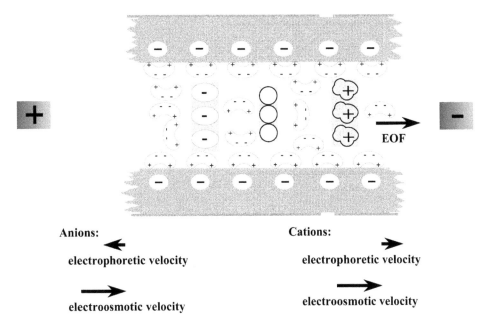

Figure 3 Principle of capillary zone electrophoresis. The EOF leads to the electroosmotic movement of all compounds towards the cathode. However, electrophoretically the anions migrate towards the anode, the higher charged the faster. If the electrophoretic velocity is superior to the electroosmosis, these anions would not be detected. Neutral compounds migrate solely due to EOF.

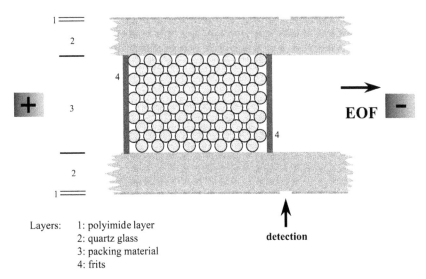

Figure 4 Capillary packed with beads as stationary phase. Additionally, a mobile phase has to be rinsed through the packed bed. During separation, a slight pressure has to be applied to avoid too long retention times of the analytes.

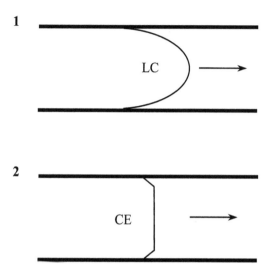

Figure 5 Transport of analytes in LC and CE: pressure driven flow profile in liquid chromatography (1); electroosmotic flow driven plug flow in capillary electrophoresis (2).

contrast, in capillary electrophoresis, the basic driving force is not a mechanical pump, instead it is the phenomenon of the electroosmotic flow within the capillary. Consequently, the analyte zones passing LC-columns or CE-capillaries show different patterns, i.e., they are broadened in LC due to diffusion effects as described by the van Deemter relationship [19] or appear more compact due to the plug flow behavior in CE (Fig. 5).

Thus, capillary electrochromatography represents a combination of these two principles. The bed structure of packed particles within a capillary can be considered as network of interconnected capillaries. In case nonderivatized silica beads are used as stationary phase, Helmholtz layers not only occur at the inner wall of the capillary, but as well around the particles. The total EOF can be described by the following equation:

$$u_{EOF} = \gamma \varepsilon_r \varepsilon_0 E \zeta \frac{1}{\eta} \tag{1}$$

with u_{EOF} as the EOF based eluent velocity, γ as a factor covering tortuosity and porosity of the packed material, ε_r as relative permitivity and ε_0 as permitivity of free space, ζ as zeta potential at the particle/electrolyte interface, E as the applied field, and η as viscosity of the mobile phase. Beyond, the effective thickness δ of the Helmholtz layer can be calculated for a typical electrolyte as followed:

$$\delta = \sqrt{\left(\frac{\varepsilon_r \varepsilon_0 RT}{2cF^2}\right)} \tag{2}$$

with R as gas constant, T the absolute temperature, F as Faraday constant, and c as the electrolyte concentration. Equation (2) helps to understand the reciprocal correlation between particle size and electrolyte concentration. When choosing for instance an electrolyte concentration of 10 mmol L^{-1}, the respective value for δ is

3 nm, i.e., a minimum particle size of 0.12 μm results, based on the demand that the particle diameter should not be smaller than 40δ. In case, 1 μm particles are in use, the maximum concentration of the electrolyte should not exceed 0.1 mmol L^{-1}. In other words, the concentration and composition of the electrolyte should be balanced with the particle size. That has to be done in order to avoid on the one hand too high concentrations leading to self-heating, and consequently to undesired bubble formation within the buffer, and to ensure on the other hand high enough concentrations for a required minimal conductivity of the buffer system, and concomitantly, to allow small enough values of δ, i.e., the applicability of small particles [19].

III. MIP IN CAPILLARIES

So far, molecularly imprinted polymers have been generated in a variety of different formats and with different classes of template compounds (Table 1). In this section, the focus is directed on the different approaches to introduce MIP in diverse configurations.

A. In Situ Generation of MIP

1. MIP as Agglomerated Globular Particles

The first publication in the area of CEC + MIP described the in situ preparation of MIP in capillaries by firstly activating a fused silica capillary with trimethoxysilyl-propyl methacrylate, followed by rinsing a mixture of ethyleneglycol dimethacrylate (EGDMA) as cross-linker, methacrylic acid (MAA) as functional monomer, azobis-(isobutyronitrile) (AIBN) as initiator, 2-propanol or cyclohexanol/dodecanol (4:1, v/v) as porogen, and L-phenylalanine anilide, pentamidine or benzamidine as template, through the capillary [29]. After polymerization at 60°C, the polymer within the capillary was washed with ethanol and finally dried. Samples of the analytes used

Table 1 Compounds used as templates for generating MIP for CEC

Class of template	Template	Format of MIP	Reference
Herbicides, pesticides	2-Phenylpropionic acid	Coating	[20]
Food components	Dansyl-phenylalanine	Coating	[21]
(and derivatives)		Packed particles	[22]
	L-Phenylalanine anilide	Monolith	[23,24]
		Gel additive	[25]
		Packed particles	[26–28]
Pharmaceuticals	Benzamidine, pentamidine	Agglomerated globular particles	[29]
	(R)-propranolol	Monolith	[30–32]
	(S)-metoprolol	Monolith	[30,33]
	(S)-atenolol	Monolith	[33]
	(S)-ropivacaine	Monolith	[34]
	(S)-propranolol	Buffer additive	[35–37]
		Monolith	[33,38]
		Coating	[39]
	nortryptiline	Packed particles	[16]

as templates were injected electrokinetically (e.g., 3 s at 5 kV), separation carried out with an applied voltage of 5 kV and detection performed at 254 and 280 nm. Although not spectacular, the authors were able to separate benzamidine and pentamidine on a capillary equipped with an MIP imprinted with pentamidine. The latter occurred as the second and most retarded peak, as expected for this highly selective affinity separation technique [29].

2. MIP as Coating

One way to implement MIP in capillaries is the in situ generation of polymer coatings on capillary inner walls. Figure 6 demonstrates the principle where a capillary equipped with an MIP coating on the inner wall allows both selective interaction with an analyte and an open-tubular flow mode which enables fast analyses and a quick re-equilibration of the capillary. Brüggemann et al. [20,40] generated these thin molecularly imprinted polymer coatings by simply injecting a mixture of template/monomer/cross-linker in a porogen with an initiator into a capillary activated with covalently linked γ-methacryloxypropyl trimethoxysilane. The sealed capillary was placed into an oven and after two days of polymerization, the coating could be freed of the template by rinsing the tube with an acidic methanol solution. Although prediction of successful imprinting recipes and appropriate polymerization conditions were not possible, the desired coatings could be established by screening different monomer–cross-linker combinations (MAA, vinylpyridine (VPy), EGDMA, divinylbenzene (DVB)). Figure 7 shows one example of such a polymer formed at the inner wall of a 100 μm ID capillary. A 10% (v/v) solution of DVB/4-VPy (85:15 mole ratio) in DMSO was used containing the template S-(+)-phenylpropionic acid. Such a capillary was applied for the separation of the two enantiomers of phenylpropionic acid. Figure 8a displays an electropherogram of

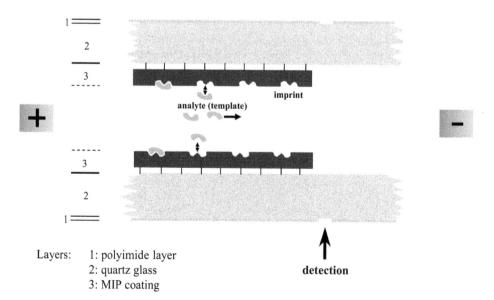

Layers: 1: polyimide layer
2: quartz glass
3: MIP coating

Figure 6 Capillary with molecularly imprinted polymer coating covalently attached to the inner wall.

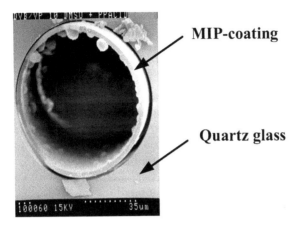

Figure 7 SEM of an MIP coated CEC capillary. Imprinting mixture: 10 % (v/v) polymerizable groups in DMSO; molar ratio of the cross-linker to functional monomer, 85:15; template: 5 mol% S-(+)-2-phenylpropionic acid. Thickness of the coat: 4 μm. Capillary inner diameter: 100 μm. (From Ref. 20.)

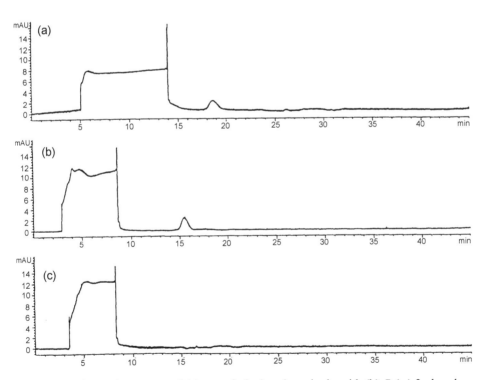

Figure 8 Electropherograms of (a) racemic 2-phenylpropionic acid, (b) R-(−)-2-phenylpropionic acid and (c) S-(+)-2-phenylpropionic acid (using in all three experiments the MIP coated capillary prepared with S-(+)-2-phenylpropionic acid as template (Fig. 7)). Mobile phase: 50 mmol L^{-1} NaH_2PO_4 (pH 4.65). CEC: capillary: 43.5 cm × 100 μm ID, effective length: 35 cm. Temperature: 25°C. Pressure injection: 3 s, 50 mbar. Separation: 10 kV. Detection: 200 nm. (From Ref. 20.)

the racemate showing at first a major baseline drift determined as a systematic signal, followed by a clear peak at 19 min caused by the nontemplate molecule R-$(-)$-phenylpropionic acid and a flat and very broad signal starting at 25 min representing the template S-$(+)$-phenylpropionic acid. Below (Fig. 8b), the result of analyzing pure R-$(-)$-phenylpropionic acid is given showing a similar peak already seen in Fig. 8a. Finally, in Fig. 8c, the electropherogram of pure S-$(+)$-phenylpropionic acid is presented, which shows again a very broad signal nearly vanishing in the baseline noise. A broad run of the template peak was expected similar to the results obtained in affinity HPLC based on MIP stationary phases, however not that plain. This flat template signal indicates on the one hand a relatively high affinity of the MIP coating towards the template S-$(+)$-phenylpropionic acid. On the other hand, it is a proof of the heterogeneity of imprint qualities usually leading to peak broadening [20].

This open tubular approach was improved by Tan and Remcho [21] who generated molecularly imprinted polymer films attached to walls of 25 μm ID capillaries. They injected as well a mixture of porogen, template, functional monomers, cross-linker, and initiator into a capillary activated with a polymerizable silane and performed the polymerization at elevated temperatures. A capillary with a total length of 100 cm and a separation length of 85 cm, respectively, coated with a polymer imprinted with dansyl-L-phenylalanine was applied for a separation of the racemate of dansyl-phenylalanine. Using acetonitrile/10 mmol L^{-1} (pH 7.0) phosphate buffer (10:1, v/v), a baseline separation of the two enantiomers could be achieved (Fig. 9).

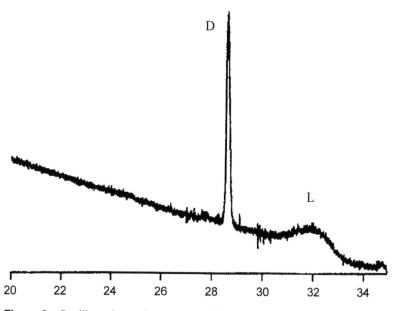

Figure 9 Capillary electrochromatographic separation of a mixture of D- and L-dansyl phenylalanine. Capillary coated with dansyl-L-phenylalanine imprinted polymer. Mobile phase: acetonitrile/10 mmol L^{-1} pH 7.0 phosphate buffer (10:1, v/v). CEC: capillary: 100 cm × 25 μm ID, effective length: 85 cm. Pressure injection: 3 s, 40 mbar. Separation: 30 kV. Detection: 280 nm. (From Ref. 21.)

Again, the nontemplate analyte causes a sharp peak eluting firstly from the capillary, whereas the template occurs as a broad signal, almost indistinguishable from the baseline [21].

3. MIP as Monoliths

A different approach was presented by Schweitz et al. [30–33,38] who developed a technique to generate capillaries filled with molecularly imprinted polymer monoliths, rather than coatings. The imprinting mixture with methacrylic acid as functional monomer, trimethylolpropane trimethacrylate (TRIM) as cross-linker, AIBN as initiator, toluene as porogen and (R)-propranolol as template, was injected into the capillary and polymerization started under a UV-source (350 nm) at −20°C. TRIM had been shown to be an efficient cross-linker leading to polymers with improved load capacity and selectivity. The low polymerization temperature was chosen, because the authors expected more stable complexes of the functional monomer with the template during the imprinting procedure. After only 80 min of polymerization time, the capillary was flushed with acetonitrile and several times with the electrolyte acetonitrile/acetate (80:20, v/v, pH 3.0) to remove remaining monomers and the template, and to allow the establishment of a nondense, but superporous phase in the capillary [30]. In order to attach the polymer covalently to the capillary in this approach, the capillary inner walls had been activated as well with γ-methacryloxypropyl trimethoxysilane, after etching the wall with 1 mol L^{-1} NaOH solution, following Hjertens approach [40]. Figure 10 demonstrates the network of globular units of the macroporous molecularly imprinted polymer and the interconnected superpores allowing a relatively simple flushing of the capillary. This porous structure could be observed within the capillary when investigating a sample via scanning electron microscopy (Fig. 11) [30]. The actual chromatography

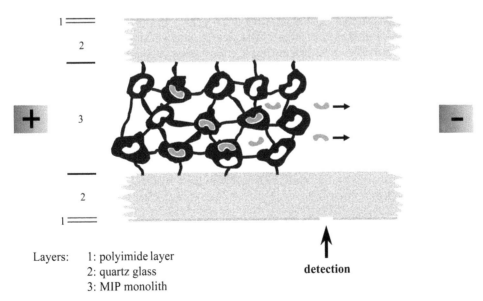

Layers: 1: polyimide layer
          2: quartz glass
          3: MIP monolith

detection

Figure 10 Capillary with molecularly imprinted polymer monolith covalently attached to the inner wall.

Figure 11 Scanning electron micrograph of a polymer-filled capillary column. Micrometer-sized globular units of macroporous molecularly polymer surrounded by 1–20 μm wide interconnected superpores. (From Ref. 30.)

was performed by rinsing the capillary with the electrolyte (acetonitrile/acetate, 80:20, v/v, pH 3.0), injecting the sample electrokinetically at 5 kV for 3 s, and establishing an electric field of 30 kV (857 V cm^{-1}) for separation. With this kind of polymer, baseline separation of the two enantiomers could be achieved when analyzing the racemic propranolol (Fig. 12) [32]. Similar to other enantioseparations based on MIP, the analysis of the racemate leads to two peaks. The first sharp peak could be identified as coming from the nontemplate enantiomer, the second signal appearing much broader corresponded to the template. With this method, the racemate of propranolol could be separated with an R_s value of 1.26 [30]. Analog results were shown in analyses based on MIP imprinted with the other enantiomer (S)-propranolol [33]. Similar effects were found with, e.g., $R_s = 1.17$ for racemic metoprolol separated with an (S)-metoprolol–MIP [33]. Beyond it, the authors demonstrated that this technique could be applied on other templates such as local anesthetics. They used the free base form of (S)-ropivacaine as template, MAA and/or 2-VPy as functional monomers, as cross-linkers TRIM, PETRA, PETEA or EGDMA, as initiator AIBN, and toluene as porogen with contents of isooctane. Best results with respect to chiral separation performances were achieved with an MAA/TRIM copolymer and 10% isooctane in toluene (v/v) [34].

Lin et al. [23,24] realized a similar idea when firstly derivatizing the inner quartz glass wall of the capillary via a Grignard reaction, and thus, covalently binding a vinyl group on the surface of the capillary. This preconditioning step was followed by flushing the polymerization mixture with L-phenylalanine anilide as template, MAA, and VPy as functional monomers, EGDMA as cross-linker and the initiator AIBN into the capillary, and executing polymerization in a water bath at 60°C for 24 h. The resulting capillary filled with a covalently anchored polymer was connected

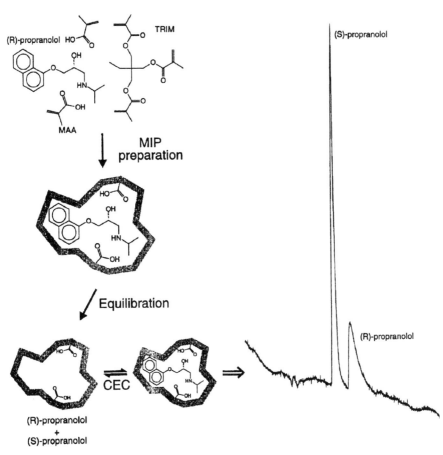

Figure 12 Molecular imprinting of (*R*)-propranolol using methacrylic acid (MAA) as functional monomer and trimethylolpropane trimethacrylate (TRIM) as cross-linker. Since the imprinted enantiomer possesses the higher affinity for the polymer, in the electrochromatogram the (*S*)-propranolol occurs first, and (*R*)-propranolol appears as the longer retarded analyte, i.e., as the second and broader peak. (From Ref. 32.)

to a polymer-free capillary where the detection had to take place, then rinsed with a buffer and loaded with the sample via electrokinetic injection (10–12 kV, 5–8 s). After establishing an electric field (voltage up to 20 kV) which was realizable due to a continuous conducting polymer matrix within the capillary, the authors achieved chiral separation of a mixture of D- and L-phenylalanine with a resolution of 1.25, however, the two optical antipodes did occur only as small signals (Fig. 13) [23].

Parameters for in situ generated MIP applied in CEC are listed in Table 2.

B. External Preparation of MIP

1. MIP as Buffer Additive

Beside generating MIP in the in situ mode, these selective polymers can be produced outside and introduced into the capillaries in the form of small particles

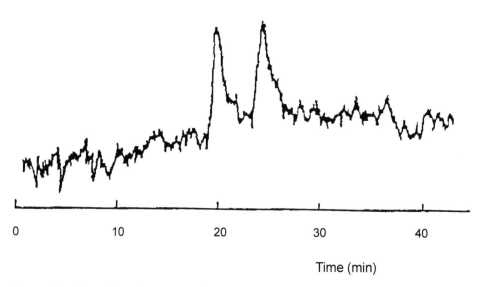

Figure 13 Separation of D,L-phenylalanine with an capillary equipped with an MIP monolith imprinted against L-phenylalanine anilide. Mobile phase: 70:20:10 AcN/HAc/H$_2$O (v/v).
CEC: capillary: 50 cm × 75 μm ID, effective length: 25 cm. Electroinjection: 5 s, 10 kV. Separation: 400 V cm^{-1}. Detection: 254 nm. (From Ref. 23.)

suspended in the mobile phase. Walshe et al. [35] used molecularly imprinted
particles as such an additive within the buffer. First of all, they generated the
polymer by mixing the template (S)-propranolol with N-acryloyl-alanine as functional monomer, EGDMA as cross-linker, AIBN as initiator and chloroform as
porogen, and polymerizing at 60°C for 12 h. After grinding, sieving, sedimentation
of the resulting 20–30 μm particles and extracting the template from the matrix,
the beads were suspended in a 5 mmol L^{-1} phosphate buffer, pH 7.0, up to a concentration of 0.1% w/v. This solution was flushed into the capillary (ID 100 μm)
and the samples were injected by applying a pressure for 3 s. By this means, at
25°C and an applied voltage of 15 kV with 0.05% w/v MIP, a baseline resolution
was observed when analyzing a mixture of (S)- and (R)-propranolol (Fig. 14).
Again, the analyte used as template ((S)-propranolol) occurred as the last peak
broader than the peak caused by the (R)-enantiomer. However, it has to be stated
that the chiral selectivity of the MIP was based not only on the imprinting effect
but also on the presence of a chiral center provided by the N-acryloyl-alanine
anchor [35].

Schweitz et al. [36] used a partial filling method for introducing molecularly
imprinted microparticles suspended in an electrolyte. The MIP was first of all
produced by mixing the template (S) -propranolol with the functional monomer
MAA, the cross-linker TRIM, the initiator AIBN and the solvent acetonitrile. Polymerization was performed using a UV-light-source (350 nm) at −26°C, resulting
directly in polymer particles with sizes of 0.2–0.5 μm. Interestingly, the control polymer particles were found to be smaller in size with diameter of 0.1 μm. After washing
and extracting the microparticles with methanol:acetic acid (9:1, v/v), the MIP beads
were suspended at a concentration of 5 mg ml^{-1} in the electrolyte acetonitrile

Table 2 Parameters for in situ generated MIP applied in CEC

		MIP as agglomerated globular particles [29]
MIP	Template	L-Phenylalanine anilide, pentamidine or benzamidine
	Functional monomer	MAA
	Cross-linker	EGDMA
	Porogen	2-Propanol or cyclohexanol/ dodecanol (4:1, v/v)
	Initiator	AIBN
	Polymerization conditions	60°C
CEC	Capillary	25 cm × 100 μm ID, activated with trimethoxysilylpropyl methacrylate
	Mobile phase	Acetonitrile/50 mmol L^{-1} potassium phosphate buffer (7:3, v/v), pH 2–4
	Injection	Electroinjection: 3 s, 5 kV
	CEC	Separation: 5 kV. Detection: 254 + 280 nm

		MIP as coating [20]
MIP	Template	S-(+)-phenylpropionic acid
	Functional monomer	4-VPy
	Cross-linker	DVB
	Porogen	DMSO
	Initiator	AIBN
	Polymerization conditions	65°C for 48 h
CEC	Capillary	43.5 cm × 100 μm ID, effective length: 35 cm; activated with γ-methacryloxypropyl trimethoxysilane
	Mobile phase	50 mmol L^{-1} NaH_2PO_4 (pH 4.65)
	Injection	Pressure injection: 3 s, 50 mbar
	CEC	Temperature: 25°C. Separation: 10 kV. Detection: 200 nm

		MIP as coating [21]
MIP	Template	Dansyl-L-phenylalanine
	Functional monomer	MAA/2-VPy mixture
	Cross-linker	EGDMA
	Porogen	Toluene/acetonitrile mixture
	Initiator	AIBN
	Polymerization conditions	75°C for 10 min
CEC	Capillary	100 cm × 25 μm ID, effective length: 85 cm, activated with trimethoxysilylpropyl methacrylate

(Continued)

Table 2 (*Continued*)

MIP as coating [21]		
	Mobile phase	Acetonitrile/10 mmol L^{-1} pH 7.0 phosphate buffer (10:1, v/v)
	Injection	Pressure injection: 3 s, 40 mbar
	CEC	Separation: 30 kV. Detection: 280 nm

MIP as monoliths [30,32]		
MIP	Template	(R)-propranolol
	Functional monomer	MAA
	Cross-linker	TRIM
	Porogen	Toluene
	Initiator	AIBN
	Polymerization conditions	350 nm at $-20°$C
CEC	Capillary	35 cm × 75 μm ID, effective length: 26.5 cm, activated with γ-methacryloxypropyl trimethoxysilane
	Mobile phase	Acetonitrile/acetate, 80:20, v/v, pH 3.0
	Injection	Electroinjection: 3 s, 5 kV
	CEC	Separation: 30 kV

MIP as monoliths [23]		
MIP	Template	L-Phenylalanine anilide
	Functional monomer	MAA/4-VPy
	Cross-linker	EGDMA
	Porogen	Chloroform
	Initiator	AIBN
	Polymerization conditions	60°C for 24 h
CEC	Capillary	50 cm × 75 μm ID, effective length: 25 cm, activated by Grignard reaction (SOCl$_2$, MgBr$_2$)
	Mobile phase	70:20:10 AcN/HAc/H$_2$O (v/v)
	Injection	Electroinjection: 5 s, 10 kV
	CEC	Separation: 400 V cm^{-1}. Detection: 254 nm

(25 mmol L^{-1} phosphoric acid/triethanolamine buffer, pH 3.5) (9:1, v/v). This suspension was partially filled into the capillary (100 μm ID, total length: 35 cm, effective length: 26.5 cm) at 50 mbar for 4 s, which corresponds to 11.8 cm of the capillary length. Based on these MIP, a baseline separation of the two enantiomers of propranolol was achieved within 75 s. The electrochromatogram shows firstly the peak of the (R)-enantiomer, secondly the (S)-propranolol used as template, and finally the MIP filling itself which migrated due to the presence of the electroosmotic flow (Fig. 15). In this manner, the analysis was performed only with an amount of maximal 5.8 μg MIP per run [36].

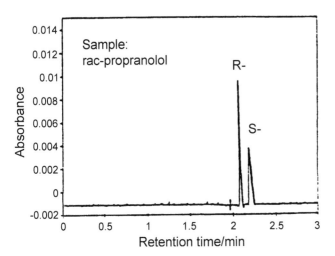

Figure 14 Separation of (*R*)- and (*S*)-propranolol using MIP particles as a chiral additive in the background electrolyte, MIP imprinted with (*S*)-propranolol. Electrolyte: 5 mmol L^{-1} Phosphate, pH 7.0, with 0.05% w/v MIP particles. CEC: capillary: 47 cm × 100 μm ID, effective length: 40 cm. Pressure injection: 3 s. Separation: 15 kV. Temperature: 25°C, Detection: 210 nm. (From Ref. 35.)

2. Capillaries Packed with MIP Particles

Whereas packing LC-columns is a common technique, packing particles into CE-capillaries requires more efforts. First of all, the down-scaled slurry method is sensitive itself, although average particle sizes are small compared to the LC stationary

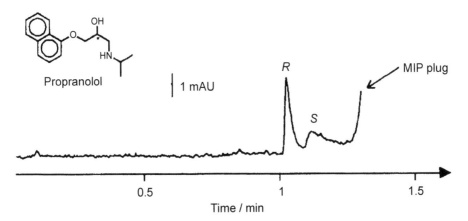

Figure 15 Separation of *R,S*-propranolol. The mobile phase contains particles imprinted against *S*-propranolol. Mobile phase: 90:10 AcN/(25 mmol L^{-1} H$_3$PO$_4$/triethanolamine pH 3.5) (v/v) with 5 mg mL^{-1} MIP particles (0.2–0.5 μm) filled partially into capillary at 50 mbar for 4 s, which corresponds to 11.8 cm of the capillary length. CEC: capillary: 35 cm × 100 μm ID, effective length: 26.5 cm. Electroinjection: 4 s, 5 kV, 25 μmol L^{-1} sample. Separation: 15 kV, i.e., 429 V cm^{-1}, at 5 bar. Detection: 254 nm. (From Ref. 36.)

phases. Due to the narrow inner diameter of the capillary the beads must not exceed a size of 10 μm. Secondly, the particles have to be hindered being rinsed out of the capillary by establishing retaining frits at both ends of the package. Many researchers try to generate these frits for instance by burning packed silica particles at a defined area within the capillary. This of course is quite problematic with polymer particles based on pure polymeric materials, since they would just decompose. Beside that, frits tend to cause bubble formation leading to disturbances in the applied electric field.

Nevertheless, a simple approach was presented by Lin et al. [26,27] who packed MIP particles imprinted with, e.g., L-phenylalanine or the respective anilide derivative into capillaries using polyacrylamide gel plugs as frits. The authors first of all polymerized a mixture of MAA, EGDMA, AIBN, and L-phenylalanine anilide dissolved in chloroform, ground the bulk into particles of 2–10 μm diameter and slurry packed these into a capillary using a liquid chromatograhic pump. The capillary itself was combined of two different capillaries: a nonpacked, buffer filled capillary end where the detection took place, and the actual separation capillary filled with the polymer particles using acrylamid gel plugs at the ends of the polymer package. When searching for the best polymer recipes, for MAA and L-phenylalanine anilide in the imprinting mixture, a functional monomer to template ratio in a range of 5:1 to 4:1 was found to be most efficient. It could be demonstrated that compared to an HPLC-procedure with the same MIP (particle sizes: 20–25 μm) (Fig. 16(1)), CEC

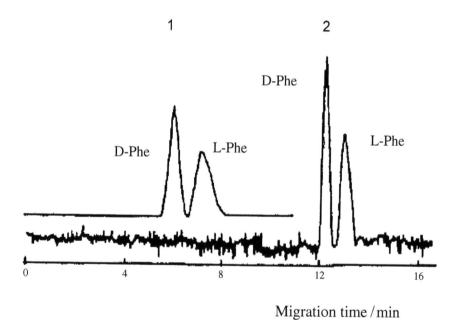

Migration time / min

Figure 16 Comparison of L-phenylalanine anilide imprinted polymer particles in HPLC (1) and CEC (2). Particles sizes for HPLC: 20–25 μm, for CEC: 2–10 μm. Mobile phase: 90:5:5 AcN/HAc/H$_2$O (v/v/v). HPLC: column: 15 cm × 2.1 mm ID, flow rate: 0.2 mL min^{-1}. Detection: 254 nm. CEC: Capillary: 40 cm × 75 μm ID, effective length: 20 cm. Electroinjection: 5 s, 300 V cm^{-1}. Separation: 350 V cm^{-1}. Detection: 254 nm. (From Ref. 26.)

led to better peak shapes and better resolution of D,L-phenylalanine (Fig. 16(2)). Choosing a 75 μm 40 cm capillary (effective length: 20 cm) and an electrolyte solution of acetonitrile/acetate/water (90:5:5, v/v/v), injecting the racemic mixture electro-kinetically at 300 V cm^{-1} for 5 s, and applying voltage of 350 V cm^{-1} for separation, much sharper peaks of both enantiomers were obtained [26]. Beyond, the authors observed that the resolution of D,L-phenylalanine was higher when using the L-phe-nylalanine anilide imprinted polymer compared to the L-phenylalanine MIP [26], similar to the findings in another approach applying MIP as composite gel additives [25]. This new approach focused on the incorporation of an MAA–EGDMA copoly-mer imprinted with L-phenylalanine anilide in a polyacrylamid gel which was filled into a capillary. Primarily, polymer particles were generated at 4°C by polymerizing MAA and EGDMA using AIBN as initiator (UV initiation at 366 nm) and chloro-form as porogen in the presence of the template. The polymer then was ground, wet sieved and the fraction of particles sized smaller than 5 μm mixed at a concentration of 10 mg ml^{-1} with a 6% acrylamide and 5% bisacrylamide gel in acetonitrile. After pumping this mixture into the capillary and heating the capillary at 40°C for 4 h they were used in CEC [25]. The samples were injected electrokinetically for 5 s applying 200 Vcm^{-1}. It could be shown that chiral separation could be achieved with a

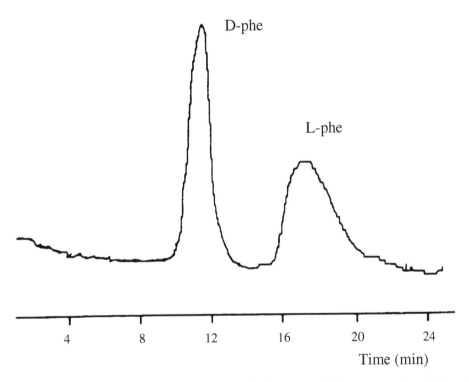

Figure 17 Electropherogram of D,L-phenylalanine using MIP particles (imprinted with L- phenylalanine anilide) as a chiral additive in a polyacrylamide gel. Gel: 5% bisacryl-amide, 6% acrylamide, in 50 mmol L^{-1} (Tris–15% AcN) pH 2.5 (adjusted with citric acid), with 10 mg mL^{-1} MIP particles. CEC: capillary: 40 cm × 75 μm ID, effective length: 25 cm. Electro-injection: 5 s, 200 V cm^{-1}. Separation: 340 V cm^{-1}. Detection: 254 nm. (From Ref. 25.)

resolution of 1.45 when analyzing a mixture of D- and L-phenylalanine (Fig. 17). The chromatogram shows baseline separation with the typical smaller and broadened peak for the compound having the highest affinity towards the stationary phase, in this case L-phenylalanine. Again, the original template had been a derivative of this analyte, i.e., L-phenylalanine anilide. By comparison, the authors found that particles imprinted with the actual L-phenylalanine did lead to a poorer separation of D- and L-phenylalanine [25].

In contrast, Quaglia et al. [28] firstly prepared silica particles surface coated with molecularly imprinted polymers. Silica particles of $10 \mu m$ diameter with a porosity of 1000 Å were linked with the azoinitiator 4,4'-azo-bis(4-cyano pentanoic acid) via triethoxyaminopropylsilane to the glass surface. This was followed by a photoinitiated grafting in the presence of the template L-phenylalanine anilide, the functional monomer MAA, and the cross-linker EGDMA. After polymerization, the enantioselective beads subsequently were treated in a Soxhlet apparatus to remove the template from the coated particles. In a second step, the authors pumped the particles pneumatically into the capillary. Capillaries filled with such beads showed clear chiral separations when analyzing electrochromatographically racemates of phenylalanine anilide. Selectivity (or separation) factors of up to 3.3 were observed [28]. Chirica and Remcho [22] also packed capillaries with MIP particles, however, they were the first not using frits in capillaries. The detailed outcome is discussed later on in Section IV.D.

Parameters for externally generated MIP applied in CEC are listed in Table 3.

IV. USING MIP IN CAPILLARIES—POTENTIAL PROBLEMS

A. Detection of Analytes

Capillaries in all forms of HPCE are mostly made of fused silica, i.e., quartz glass which of course is a brittle material. Thus, these glass tubes usually are coated with a layer of polyimide which allows bending the capillaries and handling them without any special precaution. However, in order to be able to detect the separated migrating analytes, the polyimide coating has to be removed at a specific detection position, i.e., a detection window has to be established [7–9]. At this point, the pure glass is unprotected and has to be handled with extreme caution. Commonly, this section of the capillary is put into a detection cell which protects the window. By this means, detection problems are solved for simple capillary zone electrophoresis techniques and the used diluted buffers normally do not cause any difficulties. However, when polymers are present inside the capillary, e.g., packings of MIP-particles or MIP-coatings at the inner surface, the detection problem in most cases reappears due to the high UV-absorbance of the filling. This can be solved by two methods: either a particle package exists only in a separation part—the nondetection area—of the capillary (by partial filling with MIP) and is locked in by frits [9]. Alternatively, the complete capillary has to be assembled from two individual tubes: a longer packed (or coated) part and a nontreated short part with the normal detection window. This requires a reliable connection, for instance a simple teflon sleeve [26] or a microtight connector (with microfingertight fittings and tubing sleeves) which is provided for instance by the company Upchurch. This latter approach of course is a quite sensitive one, because scaling problems may occur.

Table 3 Parameters for externally generated MIP applied in CEC

MIP as buffer additive [35]		
MIP	Template	(*S*)-propranolol
	Functional monomer	*N*-acryloyl-alanine
	Cross-linker	EGDMA
	Porogen	Chloroform
	Initiator	AIBN
	Polymerization conditions	60°C for 12 h
CEC	Capillary	47 cm × 100 µm ID, effective length: 40 cm
	Mobile phase	5 mmol L^{-1} phosphate buffer, pH 7.0, with suspended MIP particles (0.05% w/v)
	Injection	Pressure injection: 3 s
	CEC	Temperature: 25°C. Separation: 15 kV. Detection: 210 nm
MIP as buffer additive [36]		
MIP	Template	(*S*)-propranolol
	Functional monomer	MAA
	Cross-linker	TRIM
	Porogen	Acetonitrile
	Initiator	AIBN
	Polymerization conditions	350 nm at −26°C
CEC	Capillary	35 cm × 100 µm ID, effective length: 26.5 cm
	Mobile phase	90:10 AcN/(25 mmol L^{-1} H$_3$PO$_4$/ triethanolamine pH 3.5) (v/v) with 5 mg mL^{-1} MIP particles (0.2–0.5 µm) filled partially into capillary at 50 mbar for 4 s, which corresponds to 11.8 cm of the capillary length
	Injection	Electroinjection: 4 s, 5 kV
	CEC	Separation: 15 kV, i.e., 429 V cm^{-1}, at 5 bar. Detection: 254 nm
Capillaries packed with MIP particles [26]		
MIP	Template	L-phenylalanine anilide
	Functional monomer	MAA
	Cross-linker	EGDMA
	Porogen	Chloroform
	Initiator	AIBN
	Polymerization conditions	60°C for 24 h
CEC	Capillary	40 cm × 75 µm ID, effective length: 20 cm
	Mobile phase	90:5:5 AcN/HAc/H$_2$O (v/v/v)
	Injection	Electroinjection: 5 s, 300 V cm^{-1}
	CEC	Separation: 350 V cm^{-1}. Detection: 254 nm

(*Continued*)

Table 3 (*Continued*)

	MIP in composite gels [25]	
MIP	Template	L- phenylalanine anilide
	Functional monomer	MAA
	Cross-linker	EGDMA
	Porogen	Chloroform
	Initiator	AIBN
	Polymerization conditions	366 nm at 4°C
CEC	Capillary	40 cm × 75 µm ID, effective length: 25 cm
	Gel	5% Bisacrylamide, 6% acrylamide, in 50 mmol L^{-1} (Tris–15% AcN) pH 2.5 (adjusted with citric acid), with 10 mg mL^{-1} MIP particles; heating the capillary at 40°C for 4 h
	Injection	Electroinjection: 5 s, 200 Vcm^{-1}
	CEC	Separation: 340 Vcm^{-1}. Detection: 254 nm

B. Reduction of EOF

The electroosmotic flow is commonly used as the driving force for the transportation of separated analytes towards the detector [6–10]. Especially those ions migrating electrophoretically fast to their respective electrode away from the detector are forced by the EOF to migrate anyhow towards the other end of the capillary and, thus, are detectable. The EOF is based on the presence of ionic groups at the capillary inner surface, typically anionic silica groups. If such ionic groups are absent, caused, e.g., by the production of nonionic polymer fillings, the EOF is reduced nearly down to 0 [7,8]. This of course means that the analysis itself takes longer time and some of the analytes may not be detectable at all. In such a case, beside the applied field, an additional pressure has to be established during the separation which means a step towards LC resulting in undesired peak broadening and thus, to a certain extent, the renouncement of the advantage of HPCE.

C. Reproducibility of Characteristics of MIP in Capillaries

So far, little information has been published about the reproducibility of coatings or monoliths in capillaries. It would be interesting to know whether chosen parameters of these polymers, such as coating thickness or monolith density can be generated reliably, and of course whether the performances of different batches of capillaries are similar. Furthermore, it is also difficult to receive similar packing qualities for different capillaries when packing externally generated MIP particles into the capillary.

D. Frits in Capillaries

When packing MIP particles into capillaries, usually frits have to be generated at both ends of the package to prevent the particles to migrate out of the capillary [9]. This is often combined with a partial filling of the capillary ensuring that light absorbing polymer particles are retained from the detection window. However, frit

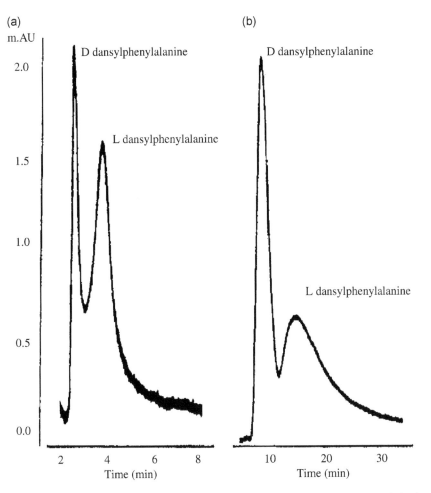

Figure 18 Comparison of dansyl-L-phenylalanine imprinted polymer particles in CEC (a) and HPLC (b). CEC: capillary: 25 cm × 75 μm ID, effective length: 17 cm. Electroinjection: 10 s, 10 kV. Separation: 30 kV. Detection: 280 nm. Mobile phase: AcN/(100 mmol L^{-1} HAc) (80:20, v/v) (pH 3.0). HPLC: Column: 15 cm × 4.6 mm ID, flow rate: 1.0 mL min^{-1}. Detection: 280 nm. Mobile phase: AcN/HAc (98:2, v/v). Sample volume: 30 μL. (From Ref. 22.)

generation is a procedure relatively difficult to handle. In case of silica or composite silica/polymer particles, the particles have to be burned [28] which calls for a sensitive treatment, because simultaneously, the glass protecting outer polyimide coating is removed. Thus, at the frit segments, the capillary is quite fragile, comparable to the detection window zone. If hydrocarbon polymer particles (the typical MIP matrix) are used as stationary phase, the generation of frits is even more complicated, because burning would simply result in the appearance of polymer embers—unsuitable to be used as a frit. What happens, when no frits are used and thus, a package is not hindered to migrate is demonstrated in Fig. 15 where the MIP particles leave the capillary as a plug [36]. An alternative method is presented by Lin et al.

[26] who applied acrylamide to entrap the MIP package. Beside, the generation of polymer monoliths or coatings allows working in a frit-less CEC mode.

Another disadvantage of frits is the resulting high back pressure when applying the capillary. Thus, Chirica and Remcho [22] developed an entrapment technique leading to MIP-filled fritless capillaries. They used MIP particles generated outside the capillary. Nevertheless, the authors first of all had to generate an inlet frit by sintering silica particles, followed by packing molecularly imprinted polymer particles into the capillary, flushing this package with a potassium silicate solution (the "entrapment solution"), heating the capillary at 160°C for a few days to generate the immobile silica matrix containing MIP particles, and finally removing the inlet frit which no longer was required. By this means, they achieved faster and more efficient chiral separations of D-/L-dansylphenyl alanine on L-dansylphenyl alanine imprinted capillaries, when comparing with respective LC-results on the same MIP stationary phase (Fig. 18) [22].

V. DISCUSSION

Different approaches to combine molecular imprinting and capillary electrochromatography have been presented. They can be divided into two major sections: on the one hand, the in situ generation of MIP in capillaries and on the other hand the preparation of MIP outside the capillary followed by insertion into it. The in situ manufacture seems to be the smarter idea, because after a simple preactivation of the capillary inner wall, this technique allows the quick and easy installation of polymers covalently attached to the capillary. However, the major drawback is the lack of predictability of the exact configuration and quality of the resulting polymer, either in form of agglomerates, coatings, or monoliths. A slight change of the MIP recipe can result in drastic differences in the properties of the resulting MIP-equipped capillary [20]. Furthermore, coatings made inside the capillary although of an even structure and permitting easy flushing of the capillary seem to be not as efficient as monoliths which provide a higher load capacity and a larger surface resulting in better separation efficiencies. With in situ generated monolith MIP, baseline separations of racemats could be achieved [32].

When producing the MIP outside the capillary, the polymers first of all can be characterized and selected with respect to highest efficiencies prior to their introduction into the capillary. This allows the use of clearly defined stationary phases. However, the actual transfer of MIP particles into capillaries is a crucial and sometimes difficult procedure. A simple method of course is the use of MIP particles suspended in the background electrolyte [35,36] or mixed with a gel [25] which is to be pumped into the capillary. However, such a capillary does not provide a high load capacity due to the relatively low amount of polymer. In general, in capillary electrochromatography, stationary phases are inserted simply by packing micrometer beads into the capillary which can be done with MIP particles as well. Nevertheless, the packing is more difficult than packing an HPLC column due to the narrow inner diameter of capillaries and due to the fact that capillaries are not made of steel, but of glass which does not allow extreme pressures. Beside, the polymer packing has to be hindered from being forced out of the capillary during the application of an electric field. This can be accomplished by frits whose generation means another difficult approach.

In the future, probably the in situ techniques will attract more focus due to the much easier fabrication mode—especially after finding optimized conditions for the manufacture of more predictable and reproducible MIP.

VI. NEW APPROACHES AND FUTURE OUTLOOK

Presently, when MIP are applied in capillary electrophoresis—or more precisely: capillary electrochromatography—they are used in different configurations, such as coatings, monoliths, packings or buffer additives. Nearly all the publications are focused on chiral separation, last but not least in order to demonstrate obvious imprint-based separation effects. This of course appears to be only the beginning of MIP applications in the broad field of HPCE. One recent example demonstrates the combination of capillaries with MIP as affinity phase applied in an competitive assay. In this flow injection approach, the capillary is coated with a film molecularly imprinted with 2,4-dichlorophenoxyacetic acid (2,4-D). In this specific kind of an enzyme-linked immunosorbent assay (ELISA), the MIP act as the antibody, the template 2,4-D as antigen, and 2,4-D conjugated with tobacco peroxidase (2,4-D-TOP) as competitor. Mixed solutions of 2,4-D and 2,4-D-TOP were drawn into the capillary, incubated for 1 h, followed by washing and the addition of a chemiluminescent substrate. The amount of converted substrate, i.e., the quantity of the released chemiluminescent signal, was detected via a CCD camera [41]. Thus, a direct correlation between the signal and the amount of bound 2,4-D could be achieved. In other words, the more 2,4-D was in the sample the more it could compete with 2,4-D-TOP for the imprints, resulting in less bound 2,4-D-TOP, in less conversion of the substrate and thus, a lower chemiluminescent signal, similar to immunoassays based on antibodies [42] or even MIP in other configurations [43]. Although in this example the capillary is not exposed to an electric field which normally is part of CEC, it shows new ways of utilizing MIP in this area.

It may be expected that the different kind of MIP already in use in various areas of chemical analysis will find their application as well on the field of capillary electrophoresis, e.g., MIP in CEC for screening combinatorial libraries, as firstly published by Vallano and Remcho [16] who used packed capillaries for screening tricyclic antidepressants.

REFERENCES

1. Tiselius, A. The moving boundary method of studying the electrophoresis of proteins. Ph.D. Thesis, Nova Acta Regiae. Soc. Sci. Ups. Ser. IV **1930**, *17* (4), 1–107.
2. Martin, A.J.P.; Everaerts, F.M. Displacement electrophoresis. Anal. Chim. Acta. **1967**, *38*, 233–237.
3. Hjerten, S. Free zone electrophoresis. Chromatogr. Rev. **1967**, *9*, 122–219.
4. Virtanen, R. Zone electrophoresis in a narrow-bore tube employing potentiometric detection. Acta Polytech. Scand. **1974**, *123*, 1–67.
5. Jorgenson, J.W.; Lukacs, K.D. Zone electrophoresis in open-tubular glass capillaries. Anal. Chem. **1981**, *53*, 1298–1302.
6. Li, S.F.Y. Capillary Electrophoresis. J. Chromatogr. Library **1996**, *52*, 2nd reprint.

7. Engelhardt, H.; Beck, W.; Schmitt, T. In *Kapillarelektrophorese*; 1st Ed.; Vieweg: Braunschweig, 1994.

8. Jandik, P.; Bonn, G. In *Capillary Electrophoresis of Small Molecules and Ions*; 1st Ed.; VCH: New York, 1993.

9. Krull, I.S.; Stevenson, R.L.; Mistry, K.; Swartz, M.E. In *Capillary Electrochromatography and Pressurized Flow Capillary Electrochromatography*; 1st Ed.; HNB Publishing: New York, 2000.

10. Landers, J.P. In *Handbook of Capillary Electrophoresis*; 1st Ed.; CRC Press: Boca Raton, 1994.

11. Kempe, M.; Mosbach, K. Separation of amino acids, peptides and proteins on molecularly imprinted stationary phases. J. Chromatogr. **1995**, *691*, 317–323.

12. Skudar, K.; Brüggemann, O.; Wittelsberger, A.; Ramström, O. Selective recognition and separation of β-lactam antibiotics using molecularly imprinted polymers. Anal. Commun. **1999**, *36*, 327–331.

13. Brüggemann, O.; Haupt, K.; Ye, L.; Yilmaz, E.; Mosbach, K. New configurations and applications of molecularly imprinted polymers. J. Chromatogr. **2000**, *889*, 15–24.

14. Fujimoto, C. Packing materials and separation efficiencies in capillary electrochromatography. Trends Anal. Chem. **1999**, *18*, 291–301.

15. Wistuba, D.; Schurig, V. Enantiomer separation of chiral pharmaceuticals by capillary electrochromatography. J. Chromatogr. **2000**, *875*, 255–276.

16. Vallano, P.T.; Remcho, V.T. Highly selective separations by capillary electrochromatography: molecular imprint polymer sorbents. J. Chromatogr. **2000**, *887*, 125–135.

17. Gübitz, G.; Schmid, M.G. Chiral separation by capillary electrochromatography. Enantiomer **2000**, *5*, 5–11.

18. Schweitz, L.; Petersson, M.; Johansson, T.; Nilsson, S. Alternative methods providing enhanced sensitivity and selectivity in capillary electroseparation experiments. J. Chromatogr. **2000**, *892*, 203–217.

19. Grant, I.H. Capillary electrochromatography. In Capillary Electrophoresis Guidebook. Methods in Molecular Biology; Altria, K.D.; Ed.; Humana Press Inc.: Totowa, NJ, 1996; Vol. 52, Chapter 15, 197–209.

20. Brüggemann, O.; Freitag, R.; Whitcombe, M.J.; Vulfson, E.N. Comparison of polymer coatings of capillaries for capillary electrophoresis with respect to their applicability to molecular imprinting and electrochromatography. J. Chromatogr. **1997**, *781*, 43–53.

21. Tan, Z.J.; Remcho, V.T. Molecular imprint polymers as highly selective stationary phases for open tubular liquid chromatography and capillary electrochromatography. Electrophoresis **1998**, *19*, 2055–2060.

22. Chirica, G.; Remcho, V.T. Silicate entrapped columns—new columns designed for capillary electrochromatography. Electrophoresis **1999**, *20* (1), 50–56.

23. Lin, J.M.; Nakagama, T.; Wu, X.-Z.; Uchiyama, K.; Hobo, T. Capillary electrochromatographic separation of amino acid enantiomers with molecularly imprinted polymers as chiral recognition agents. Fresenius' J. Anal. Chem. **1997**, *357*, 130–132.

24. Lin, J.M.; Nakagama, T.; Uchiyama, K.; Hobo, T. Capillary electrochromatographic separation of amino acid enantiomers using on-column prepared molecularly imprinted polymer. J. Pharm. Biomed. Anal. **1997**, *15*, 1351–1358.

25. Lin, J.M.; Nakagama, T.; Uchiyama, K.; Hobo, T. Molecularly imprinted polymer as chiral selector for enantioseparation of amino acids by capillary gel electrophoresis. Chromatographia **1996**, *43*, 585–591.

26. Lin, J.M.; Nakagama, T.; Uchiyama, K.; Hobo, T. Enantioseparation of D,L-phenylalanine by molecularly imprinted polymer particles filled capillary electrochromatography. J. Liq. Chrom. Rel. Technol. **1997**, *20*, 1489–1506.

27. Lin, J.M.; Nakagama, T.; Uchiyama, K.; Hobo, T. Temperature effect on chiral recognition of some amino acids with molecularly imprinted polymer filled capillary electrochromatography. Biomed. Chrom. **1997**, *11*, 298–302.

28. Quaglia, M.; De Lorenzi, E.; Sulitzky, C.; Massolini, G.; Sellergren, B. Surface initiated molecularly imprinted polymer films: A new approach in chiral capillary electrochromatography. Analyst **2001**, *126*, 1495–1498.

29. Nilsson, K.; Lindell, J.; Norrlöw, O.; Sellergren, B. Imprinted polymers as antibody mimetics and new affinity gels for selective separations in capillary electrophoresis. J. Chromatogr. **1994**, *680*, 57–61.

30. Schweitz, L.; Andersson, L.I.; Nilsson, S. Capillary electrochromatography with predetermined selectivity obtained through molecular imprinting. Anal. Chem. **1997**, *69*, 1179–1183.

31. Nilsson, S.; Schweitz, L.; Petersson, M. Three approaches to enantiomer separation of β-adrenergic antagonists by capillary electrochromatography. Electrophoresis **1997**, *18*, 884–890.

32. Schweitz, L.; Andersson, L.I.; Nilsson, S. Molecular imprint-based stationary phases for capillary electrochromatography. J. Chromatogr. **1998**, *817*, 5–13.

33. Schweitz, L.; Andersson, L.I.; Nilsson, S. Molecularly imprinted CEC sorbents: investigation into polymer preparation and electrolyte composition. Analyst **2002**, *127*, 22–28.

34. Schweitz, L.; Andersson, L.I.; Nilsson, S. Capillary electrochromatography with molecular imprint-based selectivity for enantiomer separation of local anaesthetics. J. Chromatogr. **1997**, *792*, 401–409.

35. Walshe, M.; Garcia, E.; Howarth, J.; Smyth, M.R.; Kelly, M.T. Separation of the enantiomers of propranolol by incorporation of molecularly imprinted polymer particles as chiral selectors in capillary electrophoresis. Anal. Commun. **1997**, *34*, 119–122.

36. Schweitz, L.; Spégel, P.; Nilsson, S. Molecularly imprinted microparticles for capillary electrochromatographic enantiomer separation of propranolol. Analyst **2000**, *125*, 1899–1901.

37. Spégel, P.; Schweitz, L.; Nilsson, S. Molecularly imprinted microparticles for capillary electrochromatography: studies on microparticle synthesis and electrolyte composition. Electrophoresis **2001**, *22*, 3833–3841.

38. Schweitz, L.; Andersson, L.I.; Nilsson, S. Rapid electrochromatographic enantiomer separations on short molecularly imprinted polymer monoliths. Anal. Chim. Acta. **2001**, *435*, 43–47.

39. Schweitz, L. Molecularly imprinted polymer coatings for open-tubular capillary electrochromatography prepared by surface initiation. Anal. Chem. **2002**, *74*, 1192–1196.

40. Hjerten, S. High-performance electrophoresis: elimination of electroendosmosis and solute adsorption. J. Chromatogr. **1985**, *347*, 191–198.

41. Surugiu, I.; Svitel, J.; Ye, L.; Haupt, K.; Danielsson, B. Development of a flow injection capillary chemiluminescent ELISA using an imprinted polymer instead of the antibody. Anal. Chem. **2001**, *73*, 4388–4392.

42. Hock, B.; Dankwardt, A.; Kramer, K.; Marx, A. Immunochemical techniques: antibody production for pesticide analysis. A review. Anal. Chim. Acta **1995**, *311*, 393–405.

43. Haupt, K.; Dzgoev, A.; Mosbach, K. Assay system for the herbicide 2,4-dichlorophenoxyacetic acid using a molecularly imprinted polymer as an artificial recognition element. Anal. Chem. **1998**, *70*, 628–631.

22

Metal Ion Selective Molecularly Imprinted Materials

George M. Murray and Glen E. Southard Johns Hopkins University, Laurel, Maryland, U.S.A.

I. INTRODUCTION

The production of selective metal ion sequestering and separations materials is a growing field with broad application and critical importance. Industry requires vast quantities of metals and generates tons of metal wastes. Nuclear energy production and past weapons production facilities have created unique challenges in the area of metal ion separations. Ultimately, technology must reach a point where all metal containing waste streams are treated as recoverable metal resources. Metal ion selective molecularly imprinted materials may be the means to realize this goal. An additional application of metal ion imprinted polymers is as sensors. The ability to detect a specific metal ion in a complex matrix is keenly appreciated.

The purpose of this discussion is to describe the subject of metal ion imprinting and to provide a guide to the current methods for making molecularly imprinted metal ion complexing polymers. The related topic of metal ion mediated molecular imprinting is covered in Chapter 6. The discussion will attempt to follow the progress in metal ion imprinting and relay some experimental detail about the preparations whenever the author detects an innovation or other significant difference in approach. There are several general approaches to preparing metal ion imprinted resins and each will be discussed. Ion complexing polymers differ from the bulk of molecularly imprinted polymers since neither normal covalent nor noncovalent strategies are employed. Rather these sites are composed of ionic groups, coordinating groups, or the combination of the two. The history of metal complexation chemistry has provided the motivation to pursue metal ion imprinting as a means to produce metal ion coordinators by design.

A. Prelude to Metal Ion Imprinting

The traditional techniques of metal ion separation and sequestration are precipitation, solvent extraction, and ion exchange. The introduction of multidentate ligand molecules with ligating atoms arranged to define a cation coordination site with a specific size intruded a new selectivity factor to the traditional techniques. The differing sizes of metal cations were accommodated by ligands with predefined cavities. Two examples are the coronand 18-crown-6 [a cyclic polyether $(-O-CH_2-CH-)_6$] and the cryptand 2.2.2 [a macrobicyclic polyether with N bridgeheads $N(-CH_2-CH_2-O-CH_2-CH_2-O-CH_2-CH_2-)_3N$]. The selectivity that 18-crown-6 shows toward the K^+ cation as compared to other alkali metal cations is illustrated by the following values of the association constants ($\log K$) for the 1:1 cation-to-coronand complexes in water: $Na^+(0.5)$, $K^+(2.1)$, $Rb^+(1.6)$, $Cs^+(1.0)$. Similar data for the 1:1 cation-to-cryptand 2.2.2 complexes are $Li^+(1.8)$, $Na^+(7.3)$, $K^+(9.7)$, $Rb^+(8.4)$, $Cs^+(3.5)$. In both series, correlation of the cavity diameter and the ion sizes clearly shows that the enhanced stability in the case of K^+ is due the size of the cavity. Many other examples of such cavity-ion fits have been reported as a wide variety of coronands and cryptands has been synthesized and investigated. The data show that remarkable enhancements in selectivity can be obtained by size selective coordination sites. The multidentate coronands and cryptands have therefore been used as a complexing agent in nonaqueous solubilization, as an extractant in solvent extraction, and as a complexing agent or extractant group affixed to a polymer in ion exchange [1].

Many variations in coronands and cryptands have been made by altering the number and types of coordinating atoms, the number of atoms between coordinating groups, the total number of atoms in the rings, the molecular groupings containing the coordinating atoms, and the molecular groupings between the coordinating atoms. The effects of these changes on the stabilities, selectivities, and complexation rates have been investigated in numerous instances and considerable understanding of the factors involved has resulted [2]. The major factors producing high stability are a proper-sized cavity containing the optimum number, type, and arrangement of coordinating atoms, a maximum surrounding of the cation by ligand rings, the proper cation-to-coordinating-atom interactions (hard–hard or soft–soft, strong electron donating atoms), and a rigid ligand framework. These factors generally favor slower complexation rates (from rapid diffusion-controlled values of around 10^9/s down to 10^2, in some cases) and much slower decomplexation rates [3]. Such rates can be expected of any complexant that has good size discrimination.

Coronands and cryptands have been incorporated in polymeric structures both as regular and irregular repeated units so as to produce solid extractants. These ligands have been placed both in the polymer backbone and as pendant groups. This tends to modify the cation binding behavior of the ligands, especially if the polymer structure promotes cooperative action among the ligands (that is, formation of n to 1 ligand to cation complexes with $n > 1$). Such cooperative action usually results in enhanced extraction and in some cases enhanced selectivity, particularly if the polymer structure permits only certain ligand geometries (in addition to the preset geometry within the ligand).

In the investigation of the coronands and cryptands, a compound is prepared and then experiments are conducted to ascertain the size of the resulting coordination site and its selectivity characteristics toward cations. The coronand or the cryptand would only accidentally have the precise coordination site size to fit a given cation, since the coronand or cryptand brings a more or less predefined cavity size to the situation. In contrast to these coronands and cryptands, an ideal multidentate compound with coordination sites highly selective for a given cation should be approached differently. Separate coordinating groups (not yet linked into rings or chains or a lattice) would be permitted to arrange themselves around a selected cation to approach the cation with the precise distances, proper charge neutralization, and exact coordination number and geometry that is required. Then, the separate coordinating groups would be linked into a rigid framework to fix the coordinating site. Subsequently, the cation could be removed, leaving sites having a precise fit for the selected cation. In other words, rather than fitting cations to preformed sites, the sites are fit to the desired cations. A succinct and detailed explanation of the improvements in coordination afforded by preorganization is given by Martell and Hancock [2].

B. Overview of the Synthesis of Metal Ion Selective Molecularly Imprinted Polymers

The process of producing a metal ion selective polymer typically includes the following steps: (1) the selection and preparation of ligand monomers; (2) the synthesis of metal ion complexes of the monomers or the preparation of linear copolymers of the complexing monomer and matrix monomer; (3) the preparation of cross-linked copolymers from the monomeric complexes or linear copolymer complexes; (4) the removal of imprint ions from the copolymer; and (5) the testing of the polymers for ion rebinding selectivity. In most cases, the process is iterative and includes the optimization of polymer selectivity. The optimization may include varying the ligand monomers, varying the amount of cross-linking, changing the order of preparation steps, changing the post processing or any other pertinent parameter. Each of these steps is the basis of a current research effort.

1. Monomer Selection and Preparation

The selection of appropriate monomers requires knowledge of ion–ligand association. High selectivity and large capacity are the primary goals for metal sequestering agents. In the case of metal ion sensors, capacity is of lesser importance and high selectivity is vital. Still, the complexation constant need not be overly large as long as a substantial differentiation exists between the target and most other ions. A very useful compilation of association data is found in Sillen and Martell's, "*Stability Constants of Metal-Ion Complexes*" [1]. This compilation is sufficiently comprehensive to show data for a variety of ligands that likely includes either the intended ligand or a very similar compound. There is also sufficient information to investigate selectivity for a diverse collection of metals, providing clues to selectivity. Other compilations of data are available, such as the CRC Handbook and Yatsimirskii and Vasiliev's, "*Instability Constants of Complex Compounds*" [4]. An especially useful source of data on complex compounds is Gmelin and meyer [5]. The attraction of Gmelin for this sort of search is that it often provides enough information that it may be unnecessary to seek the primary literature. The availability of Gmelin as a searchable on-line database is another favorable factor. Data from the above sources were used to create Table 1. The groups of Shea and Wulff have prepared a large number of vinyl-substituted metal ion complexing monomers as detailed in Chapter 7.

The preparation of ligand monomers requires the inclusion of one or more polymerizable substituents. Most imprinted metal ion complexants are made using free radical polymerization, thus requiring that the ligands have a vinyl or allyl

Table 1 Literature Values for log K_1 for a Variety of Divalent Metal Ions with Carboxylic Acids

Metal ion	Acetic acid	Benzoic acid	Formic acid	Chloro-acetic acid
Pb^{2+}	1.39–2.70	2.0	0.74–0.78	1.52
Cd^{2+}	1.30–2.00	1.4	1.04–1.73	1.2
Cu^{2+}	1.61–2.40	1.6–1.92	1.53–2.02	0.9–1.61
Zn^{2+}	0.76–1.59	0.9	0.70–1.20	0.4–0.56
Ni^{2+}	0.41–1.81	0.9	0.46–0.67	0.23
Co^{2+}	0.32–1.36	0.55	0.68	0.23
Ca^{2+}	0.53–0.77	0.2	0.48–0.80	0.14
Mg^{2+}	0.51–0.82	0.1	0.34	0.23

substituent. Some vinyl-substituted organic acids and neutral coordinators such as vinylbenzoic acid and vinyl-18-crown 6 are commercially available. However, depending on the storage history, much of the material may have already polymerized, so it is necessary to purify the ligand monomer prior to use.

2. Preparation of Complexes

If the process of ligand selection has been carefully pursued, the synthesis of metal ion complexes with polymerizable ligands is relatively straightforward. This is usually accomplished by mixing stoichiometric amounts of a metal salt and the complexing ligand in an aqueous or alcohol solution and evaporating to near dryness. The detailed descriptions shown below indicate that water or alcohol/water mixtures of the metal and ligand in stoichiometric ratios, evaporated to dryness, result in near quantitative yield of many complex compounds. Linear copolymer complexes of metal ions can be prepared in a similar fashion and are more easily precipitated from aqueous solutions.

3. Preparation of Polymers

Metal ion binding polymers and metal ion functionality anion binding polymers are made in a similar fashion. There are two approaches to the creation of an ion selective imprinted polymer, based on metal ion coordination. One approach relies on the imprinting process to provide the geometric arrangement needed for complexation of a specific ion. In this approach, the ligand monomers are little more than vinyl-substituted ligating atoms or functional groups attached to a supra-molecular polymeric support. The second approach is to use a chelator or macrocyclic ligand that already has geometric constraints predisposed to a certain metal ion or coordination geometry [2]. The imprinting process is then used to make small changes in the geometry to enhance the selectivity. The primary resource for information about any molecularly imprinted polymer is the database of the Society for Molecular Imprinting, maintained by Hakan Andersson. The efforts of Lars I. Andersson, Ian A. Nicholls, Olof Ramstrom (Sweden), Andrew G. Mayes, Joachim Steinke, Richard J. Ansell, Michael J. Whitcombe (England), and Sergey A. Piletsky (Ukraine) have contributed to the usefulness of this extensive database that can be accessed at http://www.smi.tu-berlin.de/SMIbase.htm.

There are two variations of the first technique, as seen in Fig. 1. The two approaches differ in the sequence of steps to make a polymer. In the first version a linear copolymer is produced, allowed to complex metal ions and the resultant crosslinked to obtain rigidity and site preservation. Examples of this approach are found in some of the works of the groups of Kabanov et al. [6] and Nishide and Tsuchida [7]. The Kabanov group prepared a linear polymer of equal amounts of diethyl vinyl-phosphonate and acrylic acid. This polymer was then equilibrated with Cu^{2+} ions and the resulting compound separated. The copper complex was cross-linked with N,N'-methylene diacrylamide. The Cu^{2+} ions was subsequently desorbed from the cross-linked polymer with 1 N acid. The Nishide group prepared poly (4-vinylpyridine) and equilibrated the polymer with Cu^{2+} ions to produce a complex. The resulting complex was cross-linked using 1,4-dibromobutane, and the Cu^{2+} was removed by treating the polymer with acid. Polymers of this type were also prepared using Fe^{3+}, Co^{2+}, Cd^{2+}, and Zn^{2+} ions, and using no cation [8]. The metal ion used as a template was taken up more effectively and preferentially by a given polymer than the other ions. Visible spectral and equilibrium data indicated that the association constant of the Cu^{2+} complex with the Cu^{2+}-conditioned polymer was higher than with the other cation-conditioned and the nonconditioned polymers. The investigators concluded that the ligand positions in the conditioned polymers

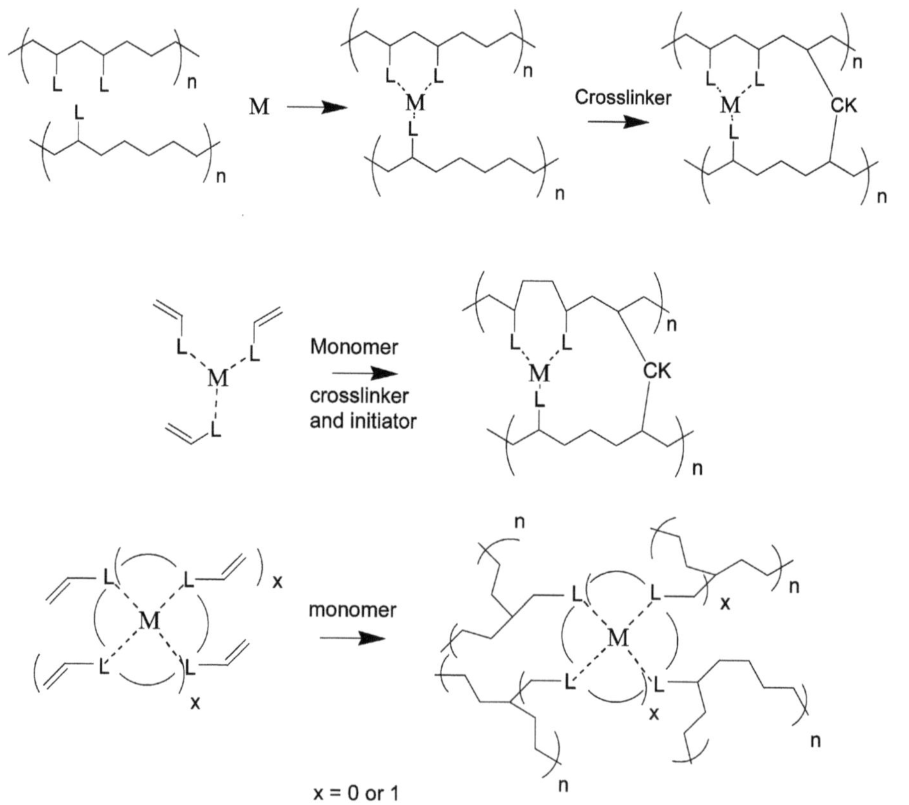

Figure 1 The three significant methods of metal ion imprinting.

were maintained at the optimum conformation for the template cation [8]. Similar experiments were conducted with poly(1-vinylimidazole), using Ni^{2+}, Co^{2+}, Zn^{2+}, and 1-vinyl-2-pyrrolidone with comparable results [9].

II. POLYMERIC METAL COMPLEXATION FOLLOWED BY CROSS-LINKING

A. Reagents and Equipment

Poly(4-vinylpyridine) (PVP)
1,4-dibromobutane
Cupric chloride
Cobaltous chloride
Zinc chloride
Cadmium chloride
Deionized water

B. Method

PVP (0.1 equiv.) was dissolved in methanol, and cupric chloride (5 mmol) in methanol was added with vigorous stirring. To the solution was added the cross-linking agent, 1,4-dibromobutane (40 mmol). The mixture was vigorously agitated for 5 days at 65°C. Subsequently, the resin was washed with methanol and hydrochloric acid repeatedly, followed by repeated washings with dilute alkaline solution and distilled water. The resin was dried and ground to a mesh size of ca. 100. The cobalt^{2+}, zinc^{2+}, and cadmium^{2+} resins were prepared by identical means.

The second variation on the technique is to form metal ion complexes with polymerizable ligands and copolymerize the complexes in the presence of a cross-linker and a noncomplexing matrix monomer. Examples of this approach are the works of Braun and Kuchen [10], Kuchen and Schram [11], Harkins and Schweitzer [12] and our own efforts [13]. This approach may be preferable for sensor applications since all sites are imprinted and there is less chance of making sites that would allow nonspecific binding. The risk in pursuing this path is that metal complexation may adversely affect the reactivity of the monomer toward copolymerization. This problem has been reported with simple monomer ligands [12] and with macrocyclic ligands [14].

III. METAL COMPLEX FORMATION FOLLOWED BY POLYMERIZATION AND CROSS-LINKING

A. Reagents and Equipment

4-Vinylbenzoic acid (VBA)
Styrene
1,4-Divinylbenzene (DVB)
AIBN
Pyridine
Uranyl nitrate
Deionized water
Ultrasonic bath

B. Method

4-Vinylbenzoic acid (1.35 mmol) was added to a solution of D. I. Water (2 mL) and ethanol (1 mL). A small amount of 1.0 M NaOH was also added to aid in dissolution and to make the solution slightly basic. A solution of 0.147 M uranyl nitrate in 1% nitric acid (4.57 mL, 0.67 mmol) was then added. The addition of the uranyl solution immediately caused the precipitation of the $UO_2(VBA)_2$ complex. The mixture's pH was adjusted to neutral by addiotion of 1.0 M NaOH and heated to 60°C for 30 min. The yellow precipitate was isolated by filtration, washed with hot water, and dried in a vacuum desiccator. The yield based on uranyl was 97%.

The imprinted polymers were produced using styrene as the matrix monomer and DVB as the cross-linking agent. The uranyl complex, $UO_2(VBA)_2$, was polymerized by addition of $UO_2(VBA)_2$ (0.48 mmol) to a solution of styrene (5 mL), DVB (0.22 mL), AIBN (50 mg), and pyridine (5 mL) used as a solvent and porogen. The solution was agitated for 2 min before being degassed by nitrogen gas. The vial containing the solution was placed in an ultrasonic bath and was sonicated for 4 h followed by thermal treatment at 60°C overnight.

In many cases, the synthetic steps involved in the production of the polymerizable complex are given in the literature for the analogous compound without vinyl substituted ligands. There is a sizable literature on metal containing polymers. Metal ion inclusion can affect reactivity ratios and care must be taken to select complexes that will incorporate well into the polymeric matrix [16]. Optical sensing strategies tend to limit the relative loading of the chromophoric complex in the copolymer to avoid spectral broadening and self-quenching in the case of luminescence. This requires inclusion of a noncoordinating matrix monomer. Selection of the matrix monomer is based on finding a monomer that will not result in nonspecific binding sites. The avoidance of nonspecific binding sites is the key to making a selective sensor. This is especially important in the area of metal ion sensing with an ion selective electrode because most polar functional groups exhibit metal ion affinity.

The common requirement for the production of a metal ion imprinted polymer is adequate cross-linking. The conventional approach has been to use a large amount of cross-linker, making the cross-linking monomer the bulk of the copolymer. When used with a large amount of porogen, highly cross-linked polymers have been reported showing high capacity but limited selectivity [13]. Another approach uses lower levels of cross-linking with a noncomplexing matrix monomer [12,17]. In this case, the polymers produced are more surface active (lower capacity) and should exhibit faster exchange kinetics. While speed is an issue in sensing, the main reason for this approach is to reduce mechanical stress and strain on the coordination site. In this way, when the metal ion is desorbed, the recoil exhibited by the functional groups making up the rebinding site should be minimized. Again, the purpose is to avoid the production of nonspecific binding sites that could adversely affect polymer selectivity. A common side benefit is that the polymers are less crystalline and exhibit better optical properties.

1. Testing for Selectivity

The classical means for determining the selectivity of an ion exchange resin is to measure the amount on binding as a function of pH. Nishide and coworkers [7,18] showed that in the case of imprinted resins metal ion binding, selectivity can be evaluated by the relative capacities for binding various metals. This may be accomplished by employing batch extraction for measuring the uptake of various cations by a polymer prepared with a given cation. The exchange equilibria are approximated by measuring the capacity of the resin for the various cations involved in the following general reaction:

$$(1/n)M^{n+} + HR \leftrightarrow (1/n)MR_n + nH^+ \tag{1}$$

where HR represents the protonated resin and M^{n+} represents the metal ion. The steps involved in acquiring the capacity data are given in Fig. 2. The data acquired in Fig. 2 are used in the equations that follow to calculate the metal ion capacity and selectivity parameters

$$\text{capacity} = \frac{[M^{n+}]_1 V_2 + [M^{n+}]_2 V_3}{W_1} \tag{2}$$

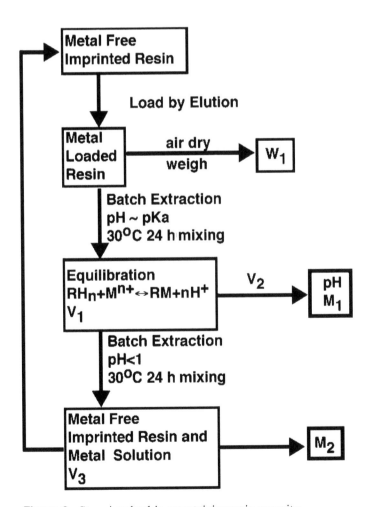

Figure 2 Steps involved in ascertaining resin capacity.

$$K = \frac{[H^+]^n \left(\text{capacity} - [M^{n+}]_1 V_1 \big/ W_1 \right)}{[M^{n+}]_1 \left(n[M^{n+}]_1 V_1 \big/ W_1 \right)^n} \tag{3}$$

$$\alpha_{M_1,M_2} \text{ (selectivity)} = \frac{\text{capacity} M_1}{\text{capacity} M_2} \tag{4}$$

$$\alpha' = \frac{\alpha_{M_1,M_2} \text{ (imprinted)}}{\alpha_{M_1,M_2} \text{ (unimprinted)}} \tag{5}$$

Unimprinted control polymers will exhibit selectivity toward metal binding due to the thermodynamic affinity of the ligand monomer. Equation (5) was suggested by Kuchen and Schram [11] as a means to account for the natural affinities of the ligating monomers so as to better illustrate the selectivity afforded by the imprinting process.

In addition to the cation used to prepare the polymer, other cations with differing charges, sizes, coordination numbers, and/or coordination geometries are used in these selectivity quotient measurements to verify specificity. Measurements are also made using polymers prepared with no metal cation (H^+ or NH_4^+) as experimental controls. The measurements required for these studies are made using a pH meter for $[H^+]_a$ and elemental analysis (ICP-AES or ICP-MS) for $[M^{n+}]_a$.

One useful aspect of the ion selective electrode approach is the ease in which selectivity testing can be performed. Once the polymer is employed as the active ingredient in a polymer membrane electrode, the binding can be examined by measuring the potential of a cell as outlined below. The selectivity obtained by batch extraction procedures gives the same affinity series as that measured by using the polymer in an electrode [19].

IV. SPECIFIC METAL ION IMPRINTING METHODS

The systematic production of metal ion imprinted ion echange resins began in earnest in the seventies. The work was performed primarily by two groups: Kabanov and co-workers and Nishide and coworkers. The Kabanov group's work was initially driven by an interest in the mechanism of ion uptake by complex-forming sorbents [8,9]. They felt that the use of chelating resins, such as those based on iminodiacetic acid, while useful for chemical analysis, had no broad commercial application due to slow exchange kinetics. They reasoned that the rate of exchange was hampered by the time it took for the coordinating groups to find the correct orientation for complexation. They further reasoned that if the coordinating groups could be prearranged to the proper position and held in place by crosslinking, the rate of ion exchange would be increased. The Nishide group was more directed toward improvements in selectivity. They were the first to recognize that the selectivity of metal ion imprinted polymers would be manifest by an increase in imprint ion loading capacity [7].

A. The Beginnings

The Kabanov group produced a series of papers throughout the seventies, describing their approach first evidenced by a patent in 1974 [6]. In one example, they prepared a linear polymer of 7 g of diethyl vinyl-phosphonate [H_2C=CH-$P(OEt)_2O$] and 3 g of acrylic acid [H_2C=CH-$COOH$] [20]. The polymerization was carried out by photochemical initiation in an evacuated vial (10^{-4} Torr) with the addition of 0.1 g cumene hydroperoxide as initiator. Ultra-violet irradiation was provided by a 300 W high pressure mercury lamp. The reaction was complete in 2 h and the resultant copolymer was dissolved in ethanol and precipitated dropwise in excess diethyl ether. After two precipitations, the copolymer was dried in a vacuum desiccator at a slightly elevated temperature. The copolymer was analyzed by several techniques: viscometry, infrared spectrometry, nuclear magnetic resonance spectroscopy, and osmometry.

A solution of 0.8 g/100 mL of the copolymer in ethanol was mixed with the twice the amount of 0.075 M $CuSO_4$ solution whose pH was adjusted to 1.1 by sulfuric acid. The resulting mixture was then slowly titrated with a copper ammonia solution (0.05 M $CuSO_4$, 5 M NH_4OH) until a pH of 3.95 was established. The rate of the titration was such that the pH did not increase at a rate greater than 0.2 units per hour. The precipitated metal–polymer complex was filtered, repeatedly washed with distilled water until the washings tested negative for copper and dried in vacuo at 35–40°C. The copper-complexed resin was cross-linked with N,N'-methylenedi-acrylamide by mixing the dry materials and grinding them together in a ball mill. The resultant powders were pressed into pellets (0.8 cm in diameter and 0.025 cm thick), placed in a vial evacuated to 10^{-4} Torr, and heated to 150°C for 5 h. The Cu^{2+} ion was subsequently desorbed from the cross-linked polymer with 1 M HCl. Similar polymers were prepared using Co^{2+} and Ni^{2+} instead of Cu^{2+}. The three "adjusted" polymers and a polymer prepared similarly with no cation were compared with regard to their cation sorption properties. The Cu^{2+} adjusted polymer showed kinetics, uptake, and selectivity in the order Cu > Co > Ni, the Ni adjusted polymer Ni > Co–Cu, and the Co^{2+} adjusted polymer Co > Ni > Cu [21,22]. As a result of the large amount of coordinating monomers, the polymers had very high capacity of 1.5–4.0 meq/g.

The Nishide group prepared resins based on poly(4-vinylpyridine) [7]. Poly (4-vinylpyridine) was prepared by radical polymerization and purified twice before using. The polymers were then treated with Cu^{2+} ion to produce a complex. The resulting complexes were cross-linked using varying amounts of 1,4-dibromobutane by shaking methanol solutions at 65°C for 5 days. The Cu^{2+} was removed by repeated washing with methanol and 1 M HCl. Polymers of this type were also prepared using Fe^{3+}, Co^{2+}, Cd^{2+}, and Zn^{2+} ions, and using no cation as a control. The metal ion used as a template was taken up more effectively and preferentially by a given polymer than the other ions. Comparison between the Cu^{2+} uptakes of a polymer prepared with Cu^{2+} and a polymer prepared with no cation shows strong enhancement of Cu^{2+} uptake by the former, and little change or a decrease in uptake of the other cations. Similar results were seen with the polymers prepared with Fe^{3+}, Co^{2+}, Cd^{2+}, and Zn^{2+}. Visible spectral and equilibrium data indicated that the association constant of the Cu^{2+} complex with the Cu^{2+}-conditioned polymer is higher than with the other cation-conditioned and the nonconditioned polymers. The investigators concluded that the ligand positions in the conditioned polymers were maintained at the optimum conformation for the template cation.

Kato et al. [23] prepared polymers using vinylimidazole as coordinator and 1-vinyl-2-pyrrolidone and divinylbenzene (DVB) as cross-linkers. Unlike Nishide's earlier work, these polymers were formed by making dissolved metal–monomer complexes and then copolymerizing the complexes in the presence of the cross-linker. For example, the Ni^{2+} complex was prepared by adding 1.0 mmol of $Ni(CH_3COO)_2 \cdot 4H_2O$ to 1.0 mL of 0.1 N acetate buffer containing 4.0 mmol of vinylimidazole. To the complex monomer solution were added 4.4 mL of 1-vinyl-2-pyrrolidone and 0.3 mL of DVB. The mixture was deaerated with nitrogen and irradiated with ^{60}Co gamma rays at a dose rate of 1 Mrad/h for 10 h at ambient temperature. The resin was ground to about 50 mesh and the metal ion was removed using 0.1 N HCl. The removal of all the template was verified by analysis of the wash solution. The resin was then ground to 100–200 mesh and dried in vacuo prior to use. Resins were prepared with Ni^{2+}, Co^{2+}, and Zn^{2+}. Attempts to prepare resins with Fe^{3+} and Cu^{2+} were unsuccessful. The imprinted resins gave higher capacities and larger complexation constants for the imprint ion but not much enhancement in binding selectivity.

B. Expanding the Scope

In the eighties, a few new investigators were attracted to the methodology, primarily, the groups of Neckers and Kuchen. Gupta and Neckers introduced an innovation to the production of metal ion imprinted polymers by using a chelate forming ligand [24]. Like Kato et al. above, the metal ion complexes were not isolated or characterized prior to polymerization. They prepared Cu^{2+}, Ni^{2+}, and Co^{2+} chelate complexes of monomeric of 4-vinyl-4'-methylbipyridine [$H_3C-C_5NH_3-C_5NH_3-CH=CH_2$] by dissolving the metal nitrate in methanol. The appropriate amount of ligand monomer and varying amounts of DVB were added to the methanol solutions. Azobisisobutyronitrile (AIBN) was added to the solutions to give a concentration of 5×10^{-3} M. The mixtures were vacuum-sealed in vials and polymerized at 60°C for 18 h. The polymers were precipitated from the solvent, filtered, washed with methanol, and dried. Metal ions were removed from the polymers by washing with 6–8 N HCl until all the metal was removed and a white polymer was obtained. Metal absorption studies used dried weighed fractions (0.05–0.08 g) of metal free polymers placed in culture tubes with 10 mL solutions of known metal concentration in 0.1 M acetate buffer, pH 4.6. The tubes were capped and shaken for 24 h. The tubes were then centrifuged and the solutions removed by filtration. The concentrations of metal ions in the solutions were measured by titrimetry, spectrophotometry, or atomic absorption spectroscopy. It was stated that the concentrations attributed to the resins were obtained by the difference in concentration of the used solution and the same solution evaluated prior to use. It was found that polymers made with a given template cation take up that cation more than the other two cations. Some studies were also made of the metal absorption as a function of the degree of cross-linking, and visible spectral data led to the conclusion that the metal (M) 4-vinyl-4'-methyl-bipyridine (L) complex on the polymer was ML.

Braun and Kuchen [10] reported a preparation for imprinted resins using dithiophosphonate ligands. The ligand monomers all possessed two vinyl groups: divinyl, bis ortho-styryl, or bis para-styryl. Complexes were prepared using Ni^{2+}, Co^{2+}, and Cr^{3+}. Copolymers were prepared using a solution of 3.0 g of the metal complex, 30.0 g of styrene (Sty), 67.0 g of ethylene glycol dimethacrylate (EGDMA) and 0.4 g of AIBN initiator in 100 g of toluene. This solution was freeze–pump–thaw degassed using liquid nitrogen, then warmed to 65°C prior to being added to a suspension solution consisting of 1.5 g of polyvinyl alcohol in water. The suspension was heated for 24 h at 80°C while being stirred at 600 rpm. The particles were removed by filtration, rinsed with ethanol and dried at 50°C under vacuum. The process produces particles with an average diameter of about 40 μm. The resins had a capacity of about 20 μmol/g. Selectivity was determined by batch equilibration. The imprint ion was seen to load at about twice the amount as a competing ion.

Later, Kuchen and Schram [11] prepared a variety of resins using methacrylic acid alone and in combination with pyridine or 4-vinyl pyridine for Cu^{2+}. Unimprinted controls were prepared as the ammonium salts. The metal complexes were of a bridged type, tetrakis(μ-methacrylato-O,O' bis[aquacopper(II)], tetrakis (μ-methacrylato-O,O' bis[(pyridine) copper(II)] and tetrakis(μ-methacrylato-O,O' bis[(4-vinylpyridine) copper(II)]. Resins were prepared using 3 mmol of complex, 40 mL of solvent (benzene or 50:50 benzene–methanol), 40.0 g of EGDMA and 0.200 g of AIBN. The mixtures were polymerized in an ampoule and the resultant bulk polymers were dried in vacuum, ground and washed free of unreacted monomers using methanol. The template metals were removed using a solution of NH_3/NH_4Cl (1 M in each).

Characterization was performed on a fraction of the powder with size ranging from 0.2 to 0.8 mm. The capacity of the resins was markedly lower than the resins of Kabanov et al. [20] and Nishide et al. [18] due to the smaller amount of complexing monomer relative to the cross-linking monomer. These resins were characterized in greater detail than resins presented previously. The recovery of metal was analyzed and found to be in the range of 60–70%. The resins were further characterized by explicitly determining the effective loading capacity of the template ion, Cu^{2+}, as well as some likely interferents, Zn^{2+}, Cd^{2+}, and Pb^{2+}. The intrinsic selectivity of the unimprinted polymers was evaluated in order to better ascertain the effectiveness of imprinting. The resins capacity was found to remain stable with repeated cycles of loading and unloading. The most significant aspect of the characterization process was the demonstration of an application of the resin to a separation. Using a column containing 10 g of resin, 70% of the copper from 25 l of a solution containing 2×10^{-5} M each of Cu^{2+}, Pb^{2+}, Cd^{2+}, and Zn^{2+} at pH 4.8 (rate of elution 15 mL/min) was recovered while almost none of the other metals was removed from solution.

In the nineties, there were more groups interested in metal ion imprints. There was extensive work by several Japanese groups that will be described separately to reflect their interests in polymer morphology. Harkins and Schweitzer [12] introduced several innovations. The two most prominent changes were the use of metal ion complexes that were synthesized and isolated prior to copolymerization and the investigation of much lower amounts of covalent cross-linking. Harkins and Schweitzer also performed more extensive characterization of their polymers. Free radical copolymerizations were carried out in bulk. The matrix monomers were styrene (Sty) or vinylbenzylchloride (VBC), and the cross-linking agent was DVB. The amount of imprinting complex and the amount of DVB were varied from 1 to 5 mol%. Imprints were made for Ni^{2+}, Cu^{2+}, and Ag^+. The Cu^{2+} polymers gave poor yields. Polymerization was initiated by addition of AIBN followed by heating to 60°C. Solvents were required for solution of the metal ion compounds in the styrenic matrix monomers. The solvents employed included chloroform, toluene, dimethyl formamide, methyl isobutyl ketone, dimethyl sulfoxide, and pyridine. All polymerizations were performed with reaction mixtures sealed in screw cap glass vials under inert atmosphere. Resins were ground as finely as possible and washed with solvents, such as acetone, to remove soluble components. They were gradually exposed to water by gradient elution of increasing water concentration in acetone. When fully hydrated, the resins were exposed to acid solutions until metal ions were no longer removed. The hydrogen form of the resins was then water washed to remove excess acid, and were ready for metal ion loading and further experimentation. Washed resins were loaded with Cu^{2+}, Ni^{2+} both simultaneously, or with each after the other, using batch mode equilibrations. They were water washed until eluted water was found to be metal free. After drying in air for 24 h, samples were weighed into plastic bottles and combined with water. The pH of each mixture was adjusted by addition of dilute HNO_3 and an overnight equilibration was allowed with the mixture gently stirring in a controlled-temperature chamber maintained at 30°C. A portion of the solution was removed from the mixture, and the pH of what remained was further adjusted by addition of concentrated HNO_3. After a second overnight equilibration period, another portion of the solution was removed. Resins were re-used after water washing. Repeated extraction without loading indicated that all available metal ions were removed by this procedure. Metal ion content of removed solutions was determined by atomic absorption spectroscopy (AAS). The resins exhibited selectivity for the imprint ion; however, the degree of selectivity is uncertain due to batch to batch variability.

The first imprinted polymer ion selective electrodes (ISE) were prepared for calcium and magnesium ions by Mosbach's group [25]. The monomer used in the fabrication of the electrode was a neutral ionophore, N,N'-dimethyl-N,N'-bis (4-vinylphenyl)-3-oxapentadiamide. This was chosen as a Ca^{2+} and Mg^{2+} selective ionophore to produce a polymer cavity that would resemble a polyether site with imprinting the means to acquire size selection. The polymer was formed using 80 mol% DVB, 5 mol% imprint ion, 15 mol% monomer, and 1 mol% AIBN per vinyl group dissolved in chloroform. The solution was polymerized at 60°C for 36 h under a nitrogen atmosphere. The membranes were prepared using a mixture of 32.1 wt% PVC, 65.3 wt% bis(2-ethylhexyl) sebacate, 1.85 wt% imprinted polymer, and 0.7 wt% potassium tetrakis (p-chlorophenyl) borate. The imprinting process enhanced the selectivity by factors of 6 and 1.7 for Ca^{2+} binding using Ca^{2+} and Mg^{2+} imprinting, over an unimprinted polymer blank, as measured by batch extraction. A calibration curve (slope 27.15 mV/decade) and selectivity coefficients against alkali metals and alkaline earths for the Ca^{2+} electrode were provided. The electrode was evaluated with metal chloride solutions in a 1:1 methanol–water mixture (v/v) with an unspecified pH. The cell schematic was:

$$Hg|Hg_2Cl_2, KCl(sat.)|test, sol', n|PVC, membrane|10\,mM\;CaCl_2|AgCl|Ag$$

The first paper from our laboratory described the preparation of resins for Pb^{2+} ion due to its toxicity and environmental significance [13]. A predetermined amount of template complex was weighed into a screw cap vial. For the untemplated polymers (blank), an equivalent amount of vinylbenzoic acid was used. The matrix monomer Sty and the cross-linker DVB were then added to the vial. The DVB concentration varied from 1 to 4 mol%. As little pyridine as possible was added to dissolve the complex compound mixture. Finally, 1 mol% of the initiator, AIBN was added and the solution was thoroughly mixed by a Vortex-Genie mixer for about 1 min. Nitrogen was bubbled through the mixture to remove oxygen from the solution and the vial was filled with nitrogen to serve as an inert atmosphere. The vial was then sealed by closing it with a screw cap, placed in an ultrasonic bath, and sonicated for ca. 4 h. The vial was moved to an oil bath at 60°C for another 20 h to complete the polymerization. The polymerized product was ground as finely as possible in a stainless steel mill vial with a Wig-L-Bug Model 6 ball mill after first freezing with liquid nitrogen. Ground polymer was washed with acetone, then slowly exposed to water by subjecting it to a gradient elution of acetone with a steadily increasing water content. Hydrated polymer was washed with 1 N HNO_3 until no Pb^{2+} was detected in the wash solution. The protonated form of the polymer was washed with water to remove excess acid until the pH of the wash solution became greater than 4.

The resins were characterized taking care to avoid detected unbound metal ions. Metal free, resins were loaded with metal ion by the following procedure. The polymers were weighed into 30 mL glass fritted Buchner funnels and a solution of 3.0 mL of a 0.1 M metal nitrate solution diluted to 15.0 mL was allowed to drip through. This process was repeated and the polymers were left standing overnight in contact with the adsorbed metal ion solution. The resulting resins were rinsed with water until the eluted water was found to be metal free. After drying in air, the samples were weighed into Nalgene bottles and combined with water. The pH of each mixture was adjusted to 2.6–2.8 by addition of dilute HNO_3. The mixture was equilibrated at 30°C for 24 h. A portion of the solution was removed for the determination of acid and metal ion content. The pH of the remaining solution was further adjusted to less than 1.0 by addition of concentrated nitric acid. After a second overnight equilibration, the concentration of the metal ion in solution was determined by AAS. After washing with de-ionized water, the resins can be reused. The metal ion recovery of the resins was determined by loading the resin with a predetermined, accurate amount of metal ion, which should be lower than the capacity of the resin. After drying in air, the sample was weighed into Nalgene bottles and combined with water. The pH of each solution was adjusted to less than 1 by addition of concentrated nitric acid. After an overnight equilibration at 30°C, the concentration of metal ion in solution was determined by AAS.

The Pb^{2+} imprinted resins were then applied to chemical analysis [26]. Experiments were carried out to determine and compare percent recoveries of Pb^{2+} from the seawater samples by using different ion exchange resins, such as Chelex-100, Duolite GT-73, a proprietary NASA resin, and the Pb^{2+} imprinted ion exchange resin. The percent recoveries from the Pb^{2+} imprinted resin were greater than 95% over a broad range of pH. The Pb^{2+} imprinted ion exchange resin did not suffer from interferences from other metal ions in seawater matrix. The Pb^{2+} imprinted resin gave superior performance when used for separation and preconcentration prior to analysis by either AAS or spectrophotometry. The utility of the Pb^{2+} imprinted resin was demonstrated by analysis of a standard reference material, Coastal reference seawater (CASS-3). The resin extract was of sufficient purity to be analyzed by spectrophotometry with the nonspecific indicator dithizone.

Following the success of Rosatzin et al. [25], the Pb^{2+}-imprinted ion exchange resin was also used as an ISE [27,28]. The selectivity, longevity, and stability of the imprinted polymer gave this electrode advantages over traditional Pb ISEs. The imprinted polymer displayed greater selectivity for the target ion than either the monomeric ionophore or the polymer prepared without Pb^{2+}. An extensively used electrode did not display loss of function resulting from solubility of the active ingredient. Unlike electrodes based on insoluble salts, this electrode is immune to poisoning from chemically similar ions. The best prototype electrode exhibited a linear potential response to Pb^{2+} ion within the concentration range of 1.0×10^{-5} to $0.1 M Pb^{2+}$ with a near Nernstian slope of 28.6 mV per decade ($R^2 = 0.9998$) and a strong preference for the Pb^{2+} ion over other cations. More pronounced selectivity was obtained when the methodology was applied to the uranyl ion. ISEs produced using 4-vinylbenzoic acid showed minimal interference and when a more selective-ligating monomer, 5-vinylsalicyladoxime, was used there was no response to any of the interferent ions tested. The resins used to make these electrodes were applied to chemical analysis in a manner analogous to the lead resins described above [29].

As mentioned earlier, the groups of Arnold [15] and Fish [14] prepared polymers using a macrocyclic coordinator, vinyl-substituted forms of the ligand TACN. Since macrocyclic ligands already have constraints on cavity size and geometry, it is of interest to see if imprinting can affect the ligands associations. The ligand was functionalized using VBC to produce both mono- and in Fish's case, tri-substituted versions. Fish's group prepared Zn^{2+} imprinted polymers with one and two equivalents of coordinating monomer and attempted to prepare Cu^{2+} imprinted polymers. In subsequent work, Hg^{2+} imprinted polymers were prepared [30]. In this work, a great deal of attention was placed on the complete characterization of the ligand monomer complexes to the extent of obtaining single crystal X-ray structures. The polymers were prepared with a fairly high degree of covalent cross-linking, 44 mol% of the resultant copolymer. A typical polymerization involved preparing a solution of 85 mmol of complex, 66 mmol of DVB, and 6 mmol of AIBN in 4 mL of methanol. The solution was placed in a 12 mL vial and purged with argon for 7 min. The vial was then capped, heated to 78°C over a period of an hour and kept at that temperature for 24 h. The resultant powder was filtered, washed with hot methanol, dried for analysis, and then demetalated using 6 N HCl.

Several unusual results occurred. When attempting to make resins with equal amounts of metal ion and ligating monomer, some of the metal ion was excluded from the polymer. It was determined that a sandwich complex with two equivalents of ligand was being formed. This was verified by making polymers with two equivalents of ligating monomer. The Zn^{2+} imprinted polymer was most selective to Cu^{2+} due to an overwhelming thermodynamic affinity. As expected, the Hg^{2+} imprinted polymer was most selective for Hg^{2+}, even when compared against Cu^{2+} and Fe^{3+}. A polymer without an imprinting ion did not form, but comparison was made to a resin made with a pendant TACN ligand. The imprinted resins had a much greater exchange capacity than did the pendant TACN resin.

Another approach to metal ion imprinting was performed by the group of Lemaire. This work involved first using acrylic acid and later, amino-carboxylates as imprinting monomers for lanthanide ions. The initial work involved developing a medium for separating Gd^{3+} from La^{3+} [31]. This first effort found a more substantial imprinting effect on the selectivity of a Gd^{3+} imprinted resin formed with a ligand monomer derived from diethylene triamine pentaacetic acid (DTPA) than resins made with acrylic acid. Later, in more involved study [17], both an ethylene diamine tetraacetic acid (EDTA) monomer derivative and the DPTA. The general method for the preparation of the imprinted polymers was by mixing 500 mg of ligand monomer, varying amounts of DVB, a molar equivalent of gadolinium nitrate hexahydrate in 5 mL of methanol. The DPTA derivatives included varying amounts of Sty matrix monomer. The solutions were placed in tubes containing AIBN (5%), flushed with argon, capped, sonicated for 15 min then heated to 65°C and polymerized for 72 h. The amount of cross-linking was varied by using DVB of differing qualities with respect to ethylstyrene. The polymers were tested in both the acid form and a Na^+ form. The acrylic acid and EDTA based polymers did not exhibit an imprinting effect. The DPTA polymers gave significant imprinting enhancements. The first study showed an increase in selectivity with a concomitant increase in cross-linking. The second study included Sty as a noncomplexing matrix monomer. In this study, the selectivity was higher for resins with larger amounts of Sty relative to the amount of DVB but capacities were lower. These results agree qualitatively with the above studies for Pb^{2+} and UO_2^{2+} where Sty was used as a matrix monomer.

C. Metal Ion Imprinted Sol Gels

Sol gel matrices has a long association with molecular imprinting as described in Chapter 11. Metal ion imprinting of sol gel materials has generally been avoided since the nonspecific adsorption of metal ions on glass surfaces has long been a pro-

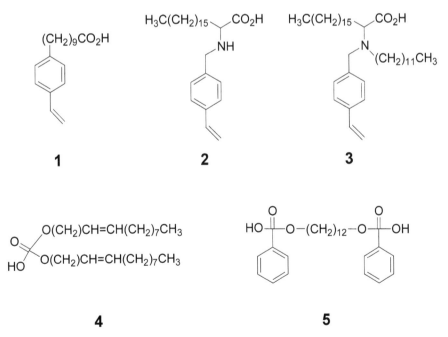

Figure 3 Ligating monomers used for surface imprinting: (**1**) 10-(*p*-vinylphenyl)decanoic acid, (**2**) 2-(*p*-vinylbenzylamino)octadeacanoic acid, (**3**) *N*-dodecyl-(*p*-vinylbenzylamino) dodecanoic acid, (**4**) dioleyl phosphoric acid, and (**5**) 1,12-dodecanediol-*O*,*O*′-diphenyl phosphonic acid.

blem for analytical chemists. Recently, a series of papers has appeared describing metal ion imprinting in sol gels. This area is much less developed than the use of organic polymers and has started to follow the same evolutionary path, proceeding from a simple mixing of metal and matrix materials [32] to the incorporation of specific metal complexing ligands, Fig. 3 [33]. Mixing uranyl nitrate with tetramethyl orthosilicate in methanol creates a sol gel with increased uranyl ion capacity [32]. In competition with Ca^{2+} or Cs^+ there is no concomitant loss of uranyl capacity, but no other "selectivity" data were presented. In the subsequent work, N-[3-(trimethoxysilyl) propyl]ethylenediamine ((TMS) en) was used as a ligating monomer. Sol gels were prepared for Cu^{2+}, Ni^{2+}, and Zn^{2+}. The imprinted materials had higher capacities than unimprinted controls. As expected, the Cu^{2+} imprinted material showed the highest selectivity when used with a mixture of Cu^{2+}, Ni^{2+}, Zn^{2+}, and Cd^{2+}. There was not an equivalent test for a blank material, so the effect of imprinting as opposed to the thermodynamic affinity of the ligand could not be estimated. More recently, surfactants have been added to make pores in the sol gel matrix (see Chapter 13) [34].

V. METAL ION IMPRINTED POLYMERS WITH SPECIFIC MORPHOLOGIES

A. Surface Imprinting

Investigators in Japan have developed a molecular imprinting technique called surfaceimprinting or surface-template polymerization [35]. They believe that the technique allows them to overcome fundamental disadvantages of conventional polymer imprinting methods. These disadvantages include slow rebinding kinetics and difficulty in handling hydrophilic metal ion imprints. Surface imprinted polymers are typically prepared by emulsion polymerization techniques using a ligating monomer, an emulsion stabilizer, a matrix monomer, and an imprint molecule. Many of the ligating monomers are molecules like surfactants with a hydrophilic head and a hydrophobic tail, Fig. 3. The result is that the imprint-ligating monomer complexes will form at the oil and water interface. After the matrix is polymerized, the coordination structure is "imprinted" on the polymer surface, not in the polymer matrix. The evolution of the surface imprinting techniques has been driven by seeking to balance the need to maintain the proximity and orientation of the ligating atoms after removal of the template to the vinyl tether with the need to for ligating monomers to orient properly in the emulsion. The simplest means of holding the ligating atoms in place at the surface is to locate the ligand tether (vinyl group) closer to the ligating atoms. This was done by Tsukagoshi et al. [36] by using butyl acrylate and methacrylic acid. Koide et al. [37] explored several surfactant like ligands. A large variety of ligating monomers and polymerization conditions have been explored using gamma irradiation [36], chemical initiation [37] and chemical initiation followed by post irradiation for greater resistance to swelling [38]. The surface imprinting method usually produces uniform spherical micro-particles with sub-micron diameters. Yoshida et al. [39] have produced slightly larger particles (25 μm) and demonstrated an application.

The polymers of Yoshida et al. [40,41] were produced using water in oil (W/O) and water-in-oil-water (W/O/W) emulsions. A polymer prepared by the W/O/W was synthesized by the following procedure. 1,12-Dodecanediol-O-O'-diphenyl phosphonic acid (60 mM) and L-glutamic acid dioleylester ribitol (20 mM) were dissolved in 40 cm^3 of trimethoylpropane trimethacrylate (TRIM). This was mixed with 20 cm^3 of toluene that was 5 volume % 2-ethylhexanol. An aqueous solution (30 cm^3) containing 10 mM Zn(NO$_3$)$_2$ buffered with 100 mM acetic acid–sodium acetate buffer at pH 3.5 was added. The mixture was sonicated for 4 min to give a W/O emulsion. The emulsion was placed in 500 cm^3 of an aqueous buffer solution (100 mM acetic acid–sodium acetate buffer at pH 3.5) containing 15 mM sodium dodecylsufate and 10 mM Mg(NO$_3$)$_2$ (an ionic strength adjuster) stirring at 500 rpm, forming the W/O/W emulsion. An initiator, 2,2″-azobis(2,4″-dimethyl-valeronitrile), 0.01 weight % relative to TRIM, was added and the solution was polymerized at 55°C for 4 h, under a flow of nitrogen. The polymer was applied to the separation of Cu^{2+} and Zn^{2+} ions using column chromatography. The imprinted resin exhibited better selectivity than a commercial unimprinted resin but was unable to provide a chromatographic separation of Zn^{2+} and Cu^{2+} [39]. More information is available in Chapter 9.

B. Metal Ion Permeable Membranes

Metal ion imprinting was applied to the preparation of a selective ion permeable membrane by Kimaro et al. [42]. Membranes were prepared using uranyl vinyl-benzoate, UO$_2$(VBA)$_2$ as the ion imprinting complex. Sty was used as the matrix monomer and DVB was used as the cross-linking monomer. Membrane synthesis was carried out in a screw top vial by dissolving the uranyl vinylbenzoate complex (20–150 mg) in 400 µL of 2-methoxyethanol. Varying amounts of Sty (1.9 mL) and DVB (50 µL) were employed. Nitrophenyl octyl ether (NPOE, 100 µL) as a plasticizer and a polyester (22 mg), prepared from diglycolic acid and 1,6-hexanediol, were added to the polymerization mixture. After deaeration with dry nitrogen, 20 mg of AIBN was added. The vial was sealed and placed in a sonicator at 60°C. The solution was sonicated until viscous, and the viscous solution was poured into a Teflon mold. The resultant mold was kept in a sealed container and placed in an oven at 60°C for 18 h to complete the polymerization. The thickness of the resulting membranes was approximately 100 µm. A reference membrane imprinted with Ni^{2+} was prepared in the same manner as the imprinted membrane. The metal templates and the polyester were removed using a 0.1 M acetic acid solution followed by a 5% nitric acid solution. Metal ions were removed using acid until the entire template was recovered. The membrane was then washed with deionized water until the acid was removed.

Transport studies were carried out in a U-shaped tube consisting of two detachable parts. The membrane, with an exposed cross-sectional area of 0.613 cm^2, was placed between the two halves of the tube. The halves were held together with a screw-actuated clamp that compresses an o-ring seal to tightly secure the connection. Experiments were performed under quiescent conditions and also by stirring the solutions. The concentration of uranium in the receiving phase was determined by ICP-MS. An experiment performed while stirring the solutions (both the source and the receiving solution) showed that higher fluxes could be obtained by convection. After 24 h of stirring, 25% of UO$_2$$^{2+}$ in the source solution containing 42 µM UO$_2$$^{2+}$ was transported through the membrane, compared to 6.5% when the solution was unstirred.

The selectivity of the membrane was evaluated by carrying out competitive transport experiments. A solution containing 0.2 mM of UO$_2$$^{2+}$, Ni^{2+}, Cd^{2+}, Cu^{2+}

and Zn^{2+} was used as a source solution. After 22 h, a portion of the receiving solution was analyzed by ICP-MS to determine the amount of each ion that was transported across the membrane. Fluxes of the competing ions were found to be very small. UO_2^{2+} was transported at higher rate with selectivity factor (α) ranging from 114 to 152. This selectivity is defined as the ratio of the molar concentration of uranyl ion to molar concentration of the competing metal ions measured in the receiving solution. The origin of selective transport can be attributed to the selective binding of uranyl ion to imprinted sites along channels that span the membrane. The reference membrane prepared by imprinting with Ni^{2+} showed little permeation of uranyl ions, but higher permeation of some of the competing metal ions. The transport fluxes of UO_2^{2+}, Cu^{2+}, and Co^{2+} through the reference membrane were 0.015 ± 0.002, 0.142 ± 0.003, and $0.045 \pm 0.001 \, nmol/cm^2h$, respectively. No Ni^{2+} or Zn^{2+} was detected in the effluent, suggesting that the conditions for membrane preparation may need to be established in a case by case basis. The results do show that the Ni^{2+}-imprinted membrane does not have sites selective for uranyl ions. The selective transport observed in the uranyl ion-imprinted membrane arises from a process that involves preferential and reversible complexation for uranyl ion. Metal ion transport across the membrane requires a counter flow of cations in the reverse direction to maintain electro-neutrality. A surplus of protons was maintained in the receiving solution by the addition of acid. A scanning electron micrograph of the membrane shows that the surface of the membrane has pores in the micron and sub-micron range. Micrographs of the inside of the membranes show an open porous structure. Energy dispersive X-ray emission spectra of the pore area show larger amounts of uranium in

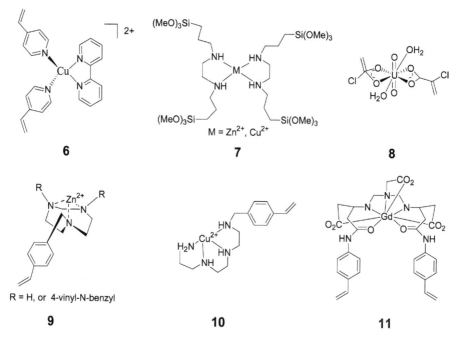

Figure 4 Metal complex monomers, (**6**) Zheng et al. [44], (**7**) Kunitake and coworkers [45], (**8**) Port and coworkers [46], (**9**) Chen et al. [14], (**10**) Singh et al. [47], (**11**) Lemaire and coworkers [31].

the pores relative to the surrounding area, indicating that the pores are involved in metal ion transport.

VI. CONCLUSIONS

The evolution of metal ion imprinting has been toward increasing selectivity at the expense of some exchange capacity. This trend is directed by the predominantly analytical applications of metal ion-imprinted polymers. The application of metal ion imprinted resins to problems in chemical analysis as chemical sensors is still just beginning. There have been some dramatic successes, but the field is still small and not very well recognized. Additional metal complexes available for imprinting are seen in Fig. 4 and 5.

Metal ion imprinted polymers have yet to be characterized in sufficient detail. This lack of characterization is starting to be remedied. The associated field of ionomers [43] gives clues toward the direction some of these new efforts should take. This is especially relevant since characterization of metal ion-imprinted polymers has revealed that some of the metal ions are trapped in inaccessible locations

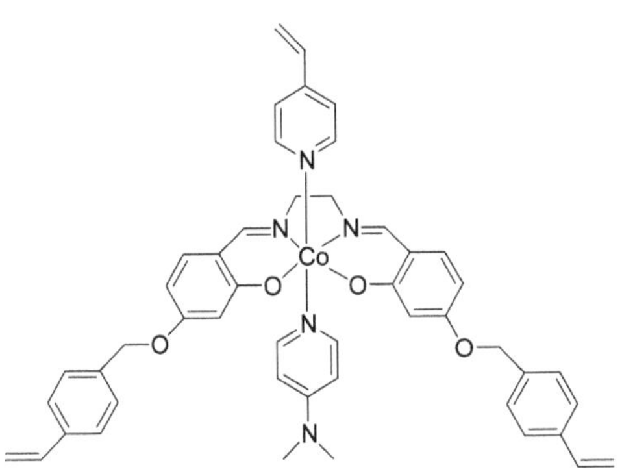

12

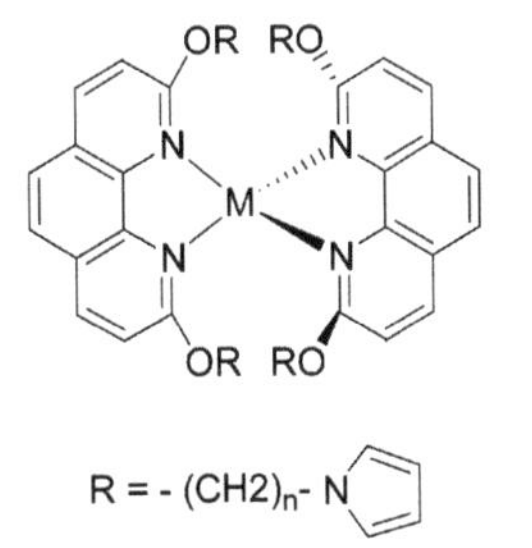

n = 3,6,or 11

M = Cu^{2+}, Co^{2+}, Zn^{2+}, and Ag^{+}

14

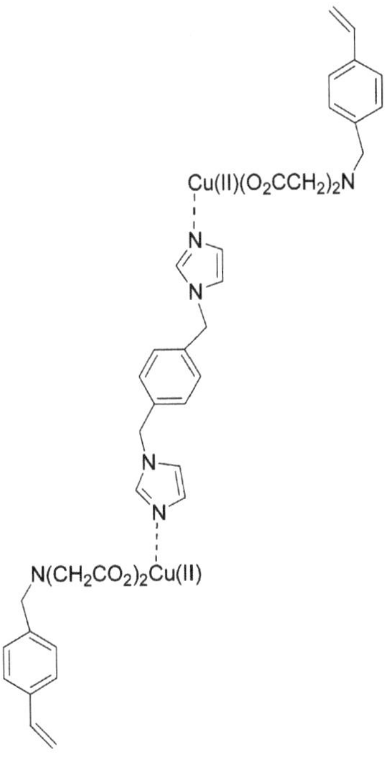

13

Figure 5 More metal complex monomers, (**12**) Borovik and coworkers [48], (**13**) Dhal and Arnold Dhal and [49], (**14**) Sauvage and coworkers [50].

or are bound too strongly to be released. This is especially true of resins formed with a noncomplexing matrix monomer. Such resins possess residual metal ion cross-links that may have a dramatic effect on the polymer properties. The effects of metal ion cross-links on molecular imprinting are not well understood. The effects may help to explain why selectivity of some metal ion-imprinted polymers is increased when the amount of covalent cross-linking is decreased.

LIST OF ABBREVIATIONS

Atomic absorption spectroscopy	AAS
2,2'-Azobisisobutyronitrile	AIBN
Chloroform	Chl
Diethylene triamine pentaacetic acid	DPTA
Dimethyl formamide	DMF
Dimethyl sulfoxide	DMSO
Divinylbenzene	DVB
Ethylene diamine tetraacetic acid	EDTA
Ethylene glycol dimethacrylate	EGDMA
Ion selective electrode	ISE
Inductively coupled plasma atomic emission spectroscopy	ICP-AES
Inductively coupled plasma mass spectroscopy	ICP-MS
Methyl isobutylketone	MIBK
National Aeronautics and Space Administration	NASA
Nitrophenyl octyl ether	NPOE
Polyvinylchloride	PVC
Pyridine	Py
Styrene	Sty
1,4,7-triazacylononane	TACN
N-[3-(trimethoxysilyl)propyl]-ethylenediamine	(TMS)en
Trimethoylpropane trimethacrylate	TRIM
Toluene	Tol
Vinyl benzoic acid	VBA
Vinyl benzylchloride	VBC

REFERENCES

1. Sillen, L.G.; Martell, A.E. In *Stability Constants of Metal-Ion Complexes*, The Chemical Society: London, 1964.
2. Martell, A.E.; Hancock, R.D. The chelate, macrocyclic, and cryptate effects. In *Coordination Chemistry: A Century of Progress*; Kaufman, G.B., Ed.; American Chemical Society: Washington, 1994, 241–254.
3. Martell, A.E.; Calvin, M. In *Chemistry of the Metal Chelate Compounds*; Prentice Hall, Inc: New York, 1952.
4. Yatsimirskii, K.B.; Vasiliev, V.P. In *Instability Constants of Complex Compounds*; Plenum: New York, 1960
5. Gmelin, L., Meyer, R., Eds.; *Gmelins Handbuch der Anorganischen Chemie*, Deutsche Chemische Gesellschaft Verlag Chemie G.M.B.H.: Leipzig-Berlin, 1995.
6. Kabanov, V.A.; Efendiev, A.A.; Orujev, D.D. Auth. Cert., USSR, 1974.
7. Nishide, H.; Tsuchida, E. Selective adsorption of metal ions on poly(4-vinylpyridine) resins in which the ligand chain is immobilized by crosslinking. Makromol. Chem. **1976**, *177*, 2295–2301.

8. Kabanov, V.A.; Efendiev, A.A.; Orujev, D.D.; Samedova, N.M. Synthesis and study of a complexing sorbent with a macromolecule arrangement "adjusted" for ion sorption. Dokl. Akad. Nauk SSSR **1978**, *23*, 356–358.

9. Kabanov, V.A.; Efendiev, A.A.; Orujev, D.D. Complex-forming polymeric sorbents adjusted to an ion being adsorbed. Vysokomol. Soedin **1979**, *A21*, 589.

10. Braun, V.U.; Kuchen, W. Ionenselektive Austauscherharze durch Vernetzende Copolymeisation Vinylsubstituierter Metallkomplexe. Chem.-Zeit. **1984**, *108*, 255–257.

11. Kuchen, W.; Schram, J. Metal ion selective exchange resins by matrix imprint with methacrylates. Angew. Chem. Int. Ed. Engl. **1988**, *27*, 1695–1697.

12. Harkins, D.A.; Schweitzer, G.K. Preparation of site-selective ion-exchange resins. Sep. Sci. Technol. **1991**, *26*, 345–354.

13. Zeng, X.; Murray, G.M. Synthesis and characterization of site-selective ion-exchange resins templated for lead(II) ion. Sep. Sci. Technol. **1996**, *31*, 2403–2418.

14. Chen, H.; Olmstead, M.M.; Albright, R.L.; Devenyi, J.; Fish, R.H. Metal-ion templated polymers: synthesis and structure of *N*-(4-vinylbenzyl)-1,4,7-triazcyclonanezinc(II) complexes, their copolymerization with divinylbenzene, and metal-ion selectivity studies of the demetalated resins—evidence for a sandwich complex in the polymer matrix. Angew. Chem. Int. Ed. Engl. **1997**, *36*, 642–645.

15. Chen, G.; Guan, Z.; Chen, C.-T.; Fu, L.; Sundaresan, V. A glucose-sensing polymer. Nat. Biol. **1997**, *15*, 354–357.

16. Sterling, L.H. In *Introduction to Physical Polymer Science*; Wiley-Interscience: New York, 2001.

17. Vigneau, O.; Pinel, C.; Lemaire, M. Ionic imprinted resins based on EDTA and DPTA derivatives for lanthanides(III) separation. Anal. Chim. Acta **2001**, *435*, 75–82.

18. Nishide, H.; Deguchi, J.; Tsuchida, E. Adsorption of metal ions on crosslinked poly (4-vinylpyridine) resins prepared with a metal ion as template. J. Polym. Sci. **1977**, *15*, 3023–3028.

19. Zeng, X.; Bzhelyansky, A.; Uy, O.M.; Murray, G.M. A Metal ion templated polymeric sensor for lead. Proceedings of the Scientific Conference on Chemical and Biological Research, US Army Edgewood Research, Development and Engineering Center; 1996; 19–22.

20. Kabanov, V.A.; Efendiev, A.A.; OruJev, D.D. Complex-forming polymeric sorbents with macromolecular arrangement favorable for ion sorption. J. Appl. Poly. Sci. **1979**, *24*, 259–267.

21. Efendiev, A.A.; Kabanov, V.A. Selective polymer complexons prearranged for metal ions sorption. Pure Appl. Chem. **1982**, *54*, 2077–2092.

22. Efendiev, A.A.; OruJev, D.D.; Kabanov, V.A. Study of sorption properties of complexing polymer sorbents adjusted to different ions. Dokl. Akad. Nauk SSSR **1980**, *255*, 1393.

23. Kato, M.; Nishide, H.; Tsuchida, E.; Sasaki, T. Complexation of metal ion with poly (1-vinylimidazole) resin prepared by radiation-induced polymerization with template metal ion. J. Polym. Sci. Polym. Chem. Ed. **1981**, *19*, 1803–1809.

24. Gupta, S.N.; Neckers, D.C. Template effects in chelating polymers. J. Polym. Sci. Polym. Chem. Edn. **1982**, *20*, 1609–1622.

25. Rosatzin, T.; Andersson, L.; Simon, W.; Mosbach, K. Preparation of Ca^{2+} selective sorbents by molecular imprinting using polymerisable ionophores. J. Chem. Soc. Perkin. Trans. **1991**, *8*, 1261–1265.

26. Bae, S.Y.; Zeng, X.; Murray; G.M. A photometric method for the determination of Pb^{2+} following separation and preconcentration using a templated ion exchange resin. J. Anal. At. Spec. **1998**, *10*, 1177–1181.

27. Zeng, X.; Bzhelyansky, A.; Bae, S.Y.; Jenkins, A.L.; Murray, G.M. Templated polymers for the selective sequestering and sensing of metal ions. In *Molecular and Ionic*

Recognition with Imprinted Polymers; Bartsch, R.A., Maeda, M., Eds.; American Chemical Society: Washington, 1998, 218–237.

28. Murray, G.M.; Jenkins, A.L.; Bzhelyansky, A.; Uy, O.M. Molecularly imprinted polymers for the selective sequestering and sensing of ions. JHUAPL Tech. Digest **1997**, *18*, 432–441.

29. Bae, S.Y.; Southard, G.E.; Murray, G.M. Molecularly imprinted ion exchange resin for purification, preconcentration and determination of UO_2^{2+} by spectrophotometry and plasma spectrometry. Anal. Chim. Acta **1999**, *397*, 173–181.

30. Fish, R.H. Studies of *N*-(4-vinylbenzyl)-1.4.7-triazocyclonone–metal ion complexes and their polymerization with divinylbenznene. In *Molecular and Ionic Recognition with Imprinted Polymers*; Bartsch, R.A., Maeda, M, Eds.; American Chemical Society Washington, 1998, 238–250.

31. Garcia, R.; Pinel, C.; Madic, C.; Lemaire, M. Ionic imprinting effect in gadolinium/lanthanum separation. Tetrahedron Lett. **1998**, *39*, 8561–8654.

32. Dai, S.; Shin, Y.S.; Barnes, C.E.; Toth, L.M. Enhancement of uranyl absorption capacity and selectivity on silica sol–gel glasses via molecular imprinting. Chem. Mater. **1997**, *9*, 2521–2525.

33. Burleigh, M.C.; Dai, S.; Hagaman, E.W.; Lin, J.S. Imprinted polysilsesquioxanes for the enhanced recognition of metal ions. Chem. Mater. **2001**, *13*, 2537–2546.

34. Dai, S.; Burleigh, M.C.; Shin, Y.; Marow, C.C.; Barnes, C.E.; Xue, Z. Imprint coating: a novel synthesis of selective functionalized ordered mesoporous sorbents. Angew. Chem. Int. Ed. Engl. **1999**, *38*, 1235–1239.

35. Yu, K.Y.; Tsukagoshi, K.; Maeda, M.; Takagi, M.; Maki, H.; Miyajima, T. Metal ion-imprinted microspheres prepared by the reorganization of the coordinating groups on the surface. Anal. Sci. **1992**, *8*, 701–703.

36. Tsukagoshi, K.; Yu, K.Y.; Ozaki, Y.; Miyajima, T.; Maeda, M.; Takagi, M. Surface imprinting: preparation of metal ion-imprinted resins by use of complexation at the aqueous–organic interface. In *Molecular and Ionic Recognition with Imprinted Polymers*; Bartsch, R.A., Maeda, M., Eds.; American Chemical Society: Washington, 1998, 251–263.

37. Koide, Y.; Sohsenji, H.; Maeda, M.; Takagi, M. Selective absorption of metal ions to surface-templated resins prepared by emulsion polymerization using a functional surfactant. In *Molecular and Ionic Recognition with Imprinted Polymers*; Bartsch, R.A., Maeda, M., Eds.; American Chemical Society: Washington, 1998, 264–277.

38. Ueza, K.; Nakamura, H.; Kanno, J.-I.; Sugo, T.; Goto, M.; Nakashio, F. Metal ion-imprinted polymer prepared by the combination of surface template polymerization with postirradiation by gamma rays. Macromolecules **1997**, *30*, 3888–3891.

39. Yoshida, M.; Hatate, Y.; Ueza, K.; Goto, M.; Furusaki., S. Metal-imprinted microsphere prepared by surface template polymerization and its application to chromatography. J. Polym. Sci. A: Polym. Chem. **1999**, *38*, 689–696.

40. Yoshida, M.; Ueza, K.; Goto, M.; Furusaki, S. Required properties for functional monomers to produce a metal template effect by a surface molecular imprinting technique. Macromolecules **1999**, *32*, 1237–1243.

41. Yoshida, M.; Ueza, K.; Nakashio, F.; Goto, M. Metal ion imprinted microsphere prepared by surface molecular imprinting using water-in-oil-in-water emulsions. J. Appl. Polym. Sci. **1998**, *73*, 1223–1230.

42. Kimaro, A.; Kelly, L.A.; Murray, G.M. Molecularly imprinted ionically permeable membrane for uranyl ion. Chem. Comm. **2001**, 1282–1283.

43. Eisenberg, A.; Kim, J. In *Introduction to Ionomers*. John Wiley & Sons: New York, 1998.

44. Zheng, N.; Li, Y.Z.; Wang, Z.M.; Chang, W.B.; Li, T.J. Molecular recognition of Cu complex imprinted polymer characteristics. Acta Chim. In. **2001**, *59*, 1572–1576.

45. He, J.; Ichinose, I.; Kunitake, T. Imprinting of coordination geometry in ultrathin films via the surface sol–gel process. Chem. Lett. **2001**, *9*, 850–851.

46. Sauders, G.D.; Foxon, S.P.; Walton, P.H.; Joyce, M.J.; Port, S.N. A selective uranium extraction agent prepared by polymer imprinting. Chem. Comm. **2000**, *4*, 273–274.

47. Singh, A.; Paranik, D.; Guo, Y.; Chang, E.I. Towards achieving selectivity in metal binding by fixing ligand–chelator complex geometry. React. Funct. Polym. **2000**, *44*, 79–79.

48. Sharma, A.C.; Joshi, V.; Borovik, A.S. Surface grafting of cobalt complexes on polymeric supports: evidence for site isolation and applications to reversible dioxygen binding. J. Polym. Sci. A: Polym. Chem. **2001**, *39*, 888–897.

49. Dhal, P.K.; Arnold, F.H. Template mediated synthesis of metal-complexing polymers for molecular recognition. J. Am. Chem. Soc. **1991**, *113*, 7417–7418.

50. Bidan, G.; Divisia-Blohorn, B.; Lapkowski, M.; Kern, J.-M.; Sauvage, J.-P. Electroactive films of polypyrroles containing complexing cavities prefomed by entwining ligands on metallic centers. J. Am. Chem. Soc. **1992**, *114*, 5986–5994.

23

Solid Phase Extraction and By-Product Removal

Lei Ye Chemical Center, Lund University, Lund, Sweden

I. INTRODUCTION

In the field of analytical chemistry, solid phase extraction (SPE) is becoming increasingly adopted for sample preparation. This is partly due to that highly specific adsorbents are continuously being developed, which makes it possible to directly isolate tiny amount of target compounds from complex sample matrices. Another driving force is the demand of decreasing use of organic solvents for environmental concern. SPE is nowadays routinely utilized in the step of sample preparation for target enrichment and clarification to assist analytical quantification. In addition, SPE based on specific solid adsorbents is also used to remove small amount of impurities from chemical and biochemical synthetic products, where no efficient purification can be achieved using the conventional methods, e.g., liquid–liquid extraction [1].

Solid phase extraction exploits the differential partitioning of compounds between a solid and a solvent phase. For analytical purpose, a typical SPE operation includes the steps of: (i) Sorbent conditioning, (ii) sample application, (iii) washing, and (iv) elution (Fig. 1). Both off-line and on-line SPE are routinely used in analytical laboratories. In the on-line setup the SPE unit is directly coupled to an analytical unit such as high-performance liquid chromatography (HPLC) or gas chromatography (GC) to greatly simplify the whole analytical process [2]. However, the off-line method provides more adjustable operation parameters that can be utilized to gain optimal extraction of targeted analytes, and can be combined with a variety of analytical methods.

In SPE, choice of an appropriate solid phase is based on their possible molecular interactions with the targeted compounds. For historical reasons, reversed-phase type adsorbents have been largely utilized to extract compounds from aqueous samples, given the fact that many of them have been developed already for use in reversed-phase chromatography applications and are commercially available. Ion exchange adsorbents have been used to retain counter ions based on electrostatic interactions. Solid phase adsorbents combining hydrophobic and ionic interactions are also present, for example graphitized carbon black (GCB) obtained by

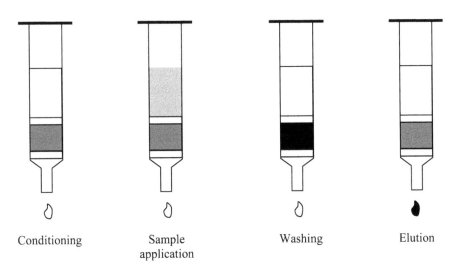

Conditioning   Sample   Washing   Elution
                   application

Figure 1 Typical solid phase extraction steps for sample preparation prior to analyte quantification.

high-temperature treatment of carbon blacks in an inert atmosphere [3]. In general, these solid phases are non-specific and frequently extract a group of compounds having similar physicochemical characteristics, which may not be resolved by later liquid or gas chromatographic separations. To this end, it is favorable to use highly specific solid phases, for example affinity adsorbents obtained by antibody immobilization [4] and molecularly imprinted polymers (MIPs).

Due to their physical and chemical robustness, SPE adsorbents made by molecular imprinting have long shelf life and can be combined with various mobile phases including very harsh elution conditions (organic solvents, pH extremes, etc.). Compared to affinity adsorbents based on antibodies, MIP adsorbents are much easier to produce. Molecular imprinting for SPE application has attracted great research interests in the past few years [5], several start-up companies are focusing on providing custom-made SPE units using molecular imprinting technique.

II. CHARACTERIZATION OF MIP ADSORBENTS FOR SPE APPLICATION

For solid phase extraction, an ideal MIP adsorbent should have the following characteristics: High binding affinity, specificity and capacity; fast association and disassociation kinetics; broad solvent compatibility; and long-term stability.

Compared to most commercial SPE adsorbents, preparation of MIP for SPE normally starts from having a targeted compound in mind. The "rational design" of an imprinting recipe is based on consideration of the functional groups existing in the target compound and its chemical structure. Recent "combinatorial methods" may also be used for fine-tuning the preparation condition (cf. Chapter 8) [6–8]. The obtained MIP has to be checked for its capability of selectively retaining the target compound. In most of the publications related to MIP for SPE, positive imprinting effect has been demonstrated by comparing the retention factors of an imprinted polymer with that of a nonimprinted control polymer, both being used as chromatographic stationary phases. For better evaluation, the breakthrough volume for the MIP cartridge or precolumn should be experimentally determined. In a simple evaluation, a dilute analyte solution can be percolated through an MIP cartridge until the concentration in the effluent reaches the original value (c_0). The plot of concentration (c) in each fraction of effluent vs. sample volume (V) gives the breakthrough curve, where V_b and V_m are defined as the sample volume by which the analyte concentration in effluent reaches 1% and 99% of c_0, respectively (Fig. 2).

Suppose the MIP has equally accessible and homogeneous binding sites (under ideal conditions), the breakthrough curve should become bilogarithmic, and the retention volume V_r correspond to the inflection point of the breakthrough curve. When the same MIP cartridge is used in elution chromatography, injection of a small volume of concentrated analyte in the same solvent will generate a peak at the same volume V_r. The breakthrough volume V_b can be predicated by the following equations:

$$V_b = V_r - 2.3\sigma_v \tag{1}$$

$$\sigma_v = (V_0/N^{1/2})(1 + k) \tag{2}$$

where V_0 is the void volume of the cartridge, N the number of theoretical plates, and k is the retention factor of the analyte as determined in the elution chromatography [9].

In practice, however, the typical binding site heterogeneity of noncovalently imprinted polymers [10], combined with most frequently their irregular particle shapes, leads to asymmetric peak shape in elution chromatography. This has largely prevented any reliable prediction of breakthrough volume for MIP cartridges using elution chromatography data. Instead, the percolation experiment is still the method of choice for determining breakthrough volume.

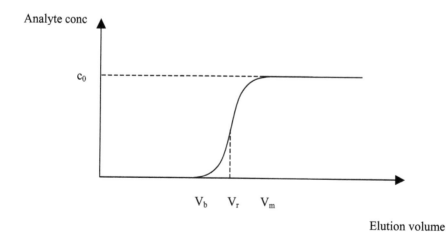

Figure 2 A theoretical breakthrough curve obtained by percolating an idealized MIP cartridge with a dilute analyte solution.

Table 1 Representative MIP Adsorbents for Solid Phase Extraction

Target analyte	Template	Functional monomer	Physical configuration	Notes and references
4-Aminopyridine	2-Aminopyridine	Methacrylic acid	Irregular particles	Potassium channel blocker [11]
Atrazine	Atrazine	Methacrylic acid	Beads	Herbicide [12]
Bupivacaine	Pentycaine	Methacrylic acid	Irregular particles	Local anesthetic drug [13]
Clenbuterol	Bromoclenbuterol	Methacrylic acid	Irregular particles	β-Agonist and growth promoter [14]
Darifenacin	Darifenacin	Methacrylic acid	Irregular particles	Muscarinic receptor antagonist [15]

N-Glutamyl-rubropuctamine/ N-Glutamyl-monascorubramine N-Glutamyl-rubropuctamine/ N-Glutamyl-monascorubramine	4-Aminostyrene	Irregular particles	Secondary metabolites obtained by fermentation [16]
7-Hydroxycoumarin 7-Hydroxycoumarin	Methacrylic acid	Irregular particles	Drug metabolite [17]
Indole-3-acetic acid Indole-3-acetic acid	Methacrylic acid	Irregular particles	Plant hormone [18]
Nicotine Nicotine	Methacrylic acid	Irregular particles	Pharmaceutical formulation [19]
4-Nitrophenol 4-Nitrophenol	4-Vinylpyridine	Irregular particles	Environmental pollutant [20]

(Continued)

Table 1 (*Continued*)

Target analyte	Template	Functional monomer	Physical configuration	Notes and references
Pentamidine		Methacrylic acid	Micro-agglomerates	Drug used for the treatment of AIDS-related pneumonia [21]
Phenytoin		Methacrylamide	Irregular particles	Anticonvulsant drug [22]
Pinacolyl methylphosphonate		Methacrylic acid	Irregular particles	Degradation product of nerve agent [23]
Propranolol		Methacrylic acid	Irregular particles	β-Adrenergic antagonist [24]
Sameridine	Sameridine analogue	Methacrylic acid	Irregular particles	Local anesthetic and analgesic drug [25]

Compound	Structure	Functional monomer	Format	Application
Succinyl-L-tyrosine		Methacrylic acid Vinylbenzyl trimethylammonium chloride $(CH_3)_3N^+$ Cl^-	Irregular particles	Fermentation byproduct [1]
Tamoxifen		Methacrylic acid (Tamoxifen citrate)	Irregular particles	Antiestrogenic drug [26]
Terbumeton		2-Acrylamido-2-methyl-1-propane sulfonic acid SO_3H	Supported membrane	Herbicide [27]
Theophylline		Methacrylic acid	Irregular particles	Bronchodilator [28]
2,4,5-Trichlorophenoxyacetic acid		4-Vinylpyridine	Irregular particles	Herbicide [29]

III. TEMPLATE, FUNCTIONAL MONOMERS, AND CROSS-LINKING MONOMERS

Due to their easy preparation and fast binding kinetics, noncovalently imprinted polymers have been the major adsorbents used in the MIP-based SPE. Both acidic and basic functional monomers have been used to prepare MIP adsorbents targeting, respectively, basic and acidic analytes. Table 1 summarizes some of the reported examples emphasizing the targeted analytes and the functional monomers giving effective extractions. It should be noted that use of the traditional functional monomer, methacrylic acid has in many cases given satisfactory SPE adsorbents. Commonly used cross-linking monomers include ethyleneglycol dimethacrylate (EDMA), trimethylolpropane trimethacrylate (TRIM), and divinylbenzene (DVB). An MIP adsorbent may be genuinely designed to first bind a target analyte through a fast, however, nonspecific mechanism, for example by a hydrophobic interaction provided by the large excess of the cross-linking monomer. This nonspecific binding may result in coadsorption of other compounds from the applied sample. In the washing step, a solvent differentiating specific and nonspecific binding can be applied to remove the nonspecifically bound compound, which results in efficient analyte enrichment and purification [12].

It has been observed in a few cases that SPE adsorbents made by molecular imprinting could release residual template at the elution step of operation. This becomes especially prominent and detrimental when analyzing highly diluted samples (e.g., analyte concentration at the ppt level), as the amount of target compound (which was used as the template for MIP preparation) released may be so high as to make the resulting quantification unreliable. This "template leakage" (also called bleeding and leaching) problem may be eliminated at the MIP preparation stage, using a close structural analogue of the target as the template [25]. Given an appropriate cross-reactivity, the analogue-imprinted polymer can still efficiently extract the target analyte from a sample solution. In a later GC or HPLC analysis, the target analyte can be readily separated from any released template, and be accurately quantified. The use of target analogue as template for preparing MIP-based SPE adsorbents has been successfully demonstrated in a few examples [11,13,14,25]. The capability of the MIPs to bind selectively the target analyte is mainly determined by the molecular similarity (space and functional group distribution) of the analogue with respect to the target analyte.

IV. SORBENT CONFIGURATIONS AND ELUTION CONDITION

Various configurations of MIPs can be used for SPE. These include irregular particles, porous polymer beads, membranes, and monoliths (Table 1). While in batch mode SPE, different configurations often give comparable results, SPE using an MIP-packed cartridge in a column mode is largely affected by physical configurations of the MIPs. Given the most efficient packing and fast flow rate, uniform imprinted MIP beads (50–100 μm) [12], membranes [27], and porous monoliths are expected to find more applications.

To obtain high enrichment factor, minimal volume of the eluting solvent should be applied to take out the MIP-retained analyte with reasonable recovery. This is often achieved by applying a solvent that efficiently diminishes the analyte–MIP interactions. The solvent should also be chosen to allow good swelling of the MIP, therefore provide a fast dissociation kinetic for analyte release. To this end, the SPE of theophylline with pulsed elution is advantageous, as the low solvent volume allows on-line preconcentration and direct determination of the target analyte [28].

The most commonly used MIPs are hydrophobic in general. In SPE for environmental monitoring and for clinical analysis, aqueous samples are largely encountered. In some cases the samples need to be preextracted into a nonpolar organic solvent, and then loaded to the MIP cartridge, in order that the high binding specificity of the MIP for the targeted analyte can be most efficiently exploited. For aqueous samples, the current trend is also directed to the development of more hydrophilic MIP adsorbents, where functional monomers able to produce higher binding strengths can be utilized for imprinting and re-binding the target molecules [27].

V. EXPERIMENTAL PROTOCOL 1

A. Solid Phase Extraction of 7-Diethylamino-4-Methylcoumarin: Characterization of SPE Adsorbents

1. Materials

7-Diethylamino-4-methylcoumarin is available from ICN Biomedical Research Products (Costa Mesa, CA, USA). 2-(Trifluoromethyl) acrylic acid is purchased from Aldrich (Steinheim, Germany). Ethylene glycol dimethacrylate is from Merck (Darmstadt, Germany). 2,2'-Azobis(2,4-dimethylvaleronitrile) is from Wako Pure Chemicals Industries (Osaka, Japan). Anhydrous toluene for polymer preparation is obtained from Lab-Scan (Dublin, Ireland).

2. Polymer Preparation

An MIP adsorbent is prepared for the extraction of 7-diethylamino-4-methylcoumarin (Fig. 3). The print molecule, 7-diethylamino-4-methylcoumarin (4 mmol, 0.925 g), a functional monomer, 2-(trifluoromethyl) acrylic acid (12 mmol, 1.681 g), a cross-linking monomer, ethylene glycol dimethacrylate (60 mmol, 11.893 g) and a polymerization initiator, 2,2'-azobis(2,4-dimethylvaleronitrile) (0.140 g) are dissolved in anhydrous toluene (18 mL) in a 50-mL borosilicate PYREX tube. The solution is briefly purged with dry nitrogen for 5 min and sealed with a screw cap. The PYREX tube is transferred to a water bath preset at 45°C and maintained for 16 h. After polymerization, the polymer monolith is taken from the PYREX tube and fractured. This is further ground with a mechanical mortar (Retsch, Haan, FRG) and wet-sieved with 5% ethanol (v/v), and subjected to repetitive sedimentation in

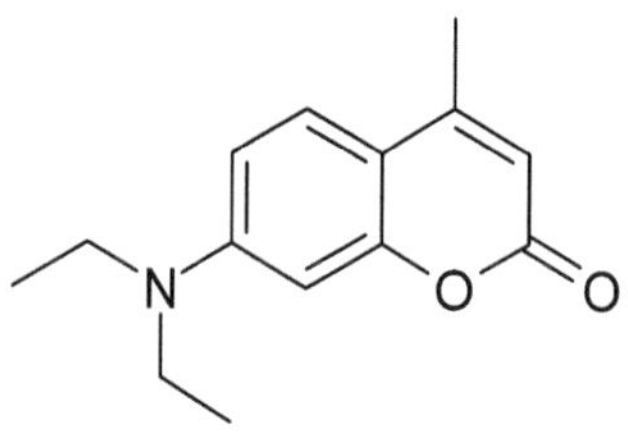

Figure 3   Chemical structure of 7-diethylamino-4-methylcoumarin.

acetone to give particles between 10–25 µm in diameter. The print molecule is removed by repetitive washing in methanol:acetic acid (90:10, v/v), using centrifugation to collect polymer particles for changing new washing solvent. This is repeated until no print molecule can be detected in solvent supernatant after the final centrifugation, using a UV spectrometer at 330 nm for monitoring. The MIP particles are rinsed with acetone and air-dried.

As a control, a nonimprinted polymer is prepared in the same way except that no print molecule is used in the polymerization step.

3. Characterization of SPE Adsorbents

An SPE column is assembled by packing polymer particles (100 mg) between two Teflon frits in a 6-mL glass cartridge. The column is mounted on a BAKER spe™-12G Column Processor (containing a 12-port vacuum manifold and a vacuum gauge controller) from J.T.Baker (Deventer, Holland). Prior to sample application, the column is equilibrated by passing 5 mL of toluene at approximately 2 mL min^{-1} flow rate, by applying an appropriate vacuum to the manifold.

A sample solution of 7-diethylamino-4-methylcoumarin in toluene at 1 µg mL^{-1} is prepared. The sample solution (10 mL) is passed through the SPE column, and each fraction (1 mL) is collected separately. After sample application, 5 mL of methanol:acetic acid (90:10, v/v) is applied to elute 7-diethylamino-4-methylcoumarin from the SPE column, during which each fraction (1 mL) is collected. Solvent in all the fractions collected are evaporated, and the residue is re-dissolved in 1 mL of toluene. The concentration of 7-diethylamino-4-methylcoumarin in the obtained solutions is measured using UV spectrometry at 330 nm, at which a calibration curve has been preestablished.

Figure 4 represents typical breakthrough curves obtained using the MIP column and the control column for the SPE of 7-diethylamino-4-methylcoumarin in the organic solvent. A simple calculation based on these curves gives concentration

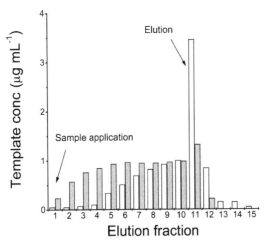

Figure 4 Experimental breakthrough curves for the SPE of 7-diethylamino-4-methylcoumarin using an imprinted polymer (empty bar) and a nonimprinted polymer (shaded bar) as the adsorbents.

factors for the first 1 mL elution (fraction 11) of 3.5 for the MIP column, and 1.3 for the control column, respectively. Cumulative recovery for the elution (fractions 11–15) is 86% for the MIP column, and 84% for the control column.

To simplify analyte detection, the present example extracts 7-diethylamino-4-methylcoumarin from toluene solution at the ppm level, without involving a washing step. At this concentration, early breakthrough occurs for both the MIP and the control cartridges, although the latter is much more pronounced.

VI. BY-PRODUCT REMOVAL

Although most SPE using MIP adsorbents is to simplify sample preparation for analytical purposes, the same technique may eventually be used in a preparative scale to purify synthetic products from a chemical or biochemical transformation process. Given the fact that most MIPs have a relatively high binding affinity but low capacity, MIP based SPE is more applicable to the removal of a small portion of persistent byproduct formed during synthetic reactions [1,30]. An ideal MIP for this purpose should have a very high binding affinity and specificity towards the targeted byproduct, so that it can efficiently remove the impurity within even a single batch operation. At the same time, nonspecific binding of the main product should be minimized, so that the MIP based by-product removal will not cause any obvious product loss. The solvent involved in the SPE should be chosen to maximally satisfy the above criteria, but also compatible with the existing production procedure, so that no extra solvent exchange need to be carried out. In real applications, a compromise is often made between the binding performance of MIPs and the complexity of operation conditions.

VII. EXPERIMENTAL PROTOCOL 2

A. Solid Phase Extraction for Product Purification

In this example, the high specific binding characteristic of MIPs is utilized to remove small amount of impurities from synthetic products. The MIP acts as an efficient byproduct scavenger in the chemical synthesis for the artificial sweetener, aspartame (Fig. 5).

1. Materials
Z-L-Asp(OH)-L-Phe-OMe is available from Indofine Chemical Co. (Somerville, NJ, USA). Z-L-Aspartic acid, Z-D-aspartic acid, L-phenylalanine methyl ester hydrochloride and D-phenylalanine methyl ester hydrochloride are from Sigma (St. Louis, MO, USA). Acetic anhydride, methacrylic acid, 4-vinylpyridine, and ethylene glycol dimethacrylate are from Merck (Darmstadt, Germany). 2,2'-Azobis(2,4-dimethylvaleronitrile) is from Wako Pure Chemicals Industries (Osaka, Japan). Anhydrous acetonitrile for polymer preparation is obtained from Lab-Scan (Dublin, Ireland).

2. Z-L-Aspartic Anhydride
To a solution of acetic anhydride (22 mmol) in ethyl acetate (6 mL) is added Z-L-aspartic acid (15 mmol). The mixture is stirred at 50–60°C for 2 h. The solvent is removed under reduced pressure in a rotary evaporator, which leaves oily product.

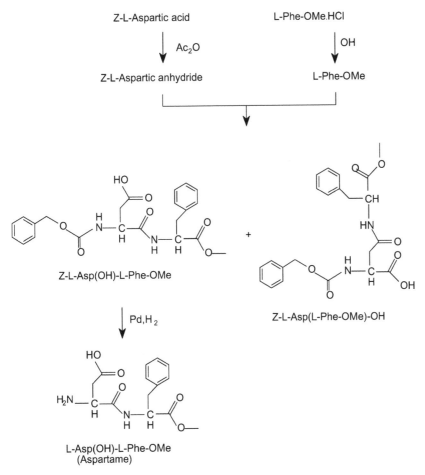

Figure 5 A synthetic route to the artificial sweetener, aspartame. The by-product, Z-L-Asp(L-Phe-OMe)-OH, is formed during the first coupling reaction.

Addition of diethyl ether (10 mL) gives a white precipitate. The precipitate is washed for another two times with diethyl ether (2–10 mL), and re-crystallized from ethyl acetate.

3. Z-D-Aspartic Anhydride

Z-D-aspartic anhydride is synthesized from Z-D-aspartic acid in the same way as that for Z-L-aspartic anhydride.

4. L-Phenylalanine Methyl Ester

To a solution of L-phenylalanine methyl ester hydrochloride (10 mmol, 2.157 g) in water (10 mL) is added sodium carbonate decahydrate (10 mmol, 2.861 g). The solution is extracted with chloroform (2×20 mL). The organic phase is separated and dried over anhydrous sodium sulfate. Removal of solvent under reduced pressure gives an oily product, which is stored at −20°C until further use.

5. D-Phenylalanine Methyl Ester

D-Phenylalanine methyl ester is prepared from D-phenylalanine methyl ester hydrochloride in the same way as that for L-Phenylalanine methyl ester.

6. Z-L-Asp(L-Phe-OMe)-OH

To a solution of L-phenylalanine methyl ester (8 mmol, 1.434 g) in dimethyl sulfoxide (60 mL) is added Z-L-aspartic anhydride (8 mmol, 1.994 g) under stirring. The reaction continues at 20°C for 5 h. The reaction mixture is then poured into distilled water (1.2 L). Acidification with 1 M hydrochloride acid (7–8 drops) gives a crude product as white precipitate. This is separated by filtration, dried in a vacuum and re-crystallized from ethyl acetate. Optical activity ($[\alpha]_D$) in methanol: +10.0.

7. Z-D-Asp(D-Phe-OMe)-OH

Z-D-Asp(D-Phe-OMe) -OH is synthesized from Z-D-aspartic anhydride and D-phenylalanine methyl ester in the same way as for Z-L-Asp(L-Phe-OMe)-OH. Optical activity ($[\alpha]_D$) in methanol: −10.0.

8. Preparation of Molecularly Imprinted Polymer Specific for Z-L-Asp(L-Phe-OMe)-OH (MIP-LL)

Z-L-Asp(L-Phe-OMe)-OH (1 mmol) is dissolved in anhydrous acetonitrile (12 mL) in a 50-mL borosilicate PYREX tube. To the solution is then added freshly vacuum-distilled 4-vinylpyridine (8 mmol), methacrylic acid (16 mmol), ethylene glycol dimethacrylate (40 mmol) and 2,2′-azobis(2,4-dimethylvaleronitrile) (100 mg). The solution is briefly purged with dry nitrogen for 5 min and sealed with a screw cap. The PYREX tube is transferred to a water bath preset at 45°C and maintained for 16 h. After polymerization, the polymer monolith is taken from the PYREX tube and fractured. This is further ground with a mechanical mortar (Retsch, Haan, FRG) and wet-sieved with 5% ethanol (v/v), and subjected to repetitive sedimentation in acetone to give particles between 10–25 μm in diameter. The print molecule is removed by repetitive washing in methanol:acetic acid (90:10, v/v), using centrifugation to collect polymer particles for changing new washing solvent. This is repeated until no print molecule can be detected in solvent supernatant after the final centrifugation, using a UV spectrometer at 260 nm for monitoring. The MIP particles are rinsed with acetone and air-dried.

9. Preparation of Molecularly Imprinted Polymer Specific for Z-D-Asp (D-Phe-OMe)-OH (MIP-DD)

This polymer is prepared in the same way as that for MIP-LL, except that Z-D-Asp (D-Phe-OMe)-OH is used as the template instead of Z-L-Asp(L-Phe-OMe) -OH.

10. Purification of Chemically Synthesized Z-L-Asp(OH)-L-Phe-OMe using MIP as a byproduct Scavenger

Z-L-aspartic anhydride (2 mmol) is dissolved in acetonitrile (20 mL). To the solution is added L-phenylalanine methyl ester (2 mmol) under stirring. The reaction continues at 20°C for 2 h. After the reaction, the solution is diluted 10 times with acetonitrile. The composition of the mixture is analyzed by RP HPLC using an ODS column and UV detection at 210 nm. Mobile phase: 50% (v/v) acetonitril in 10 mM phosphate buffer, pH 4.15; flow rate: 0.5 mL min^{-1}.

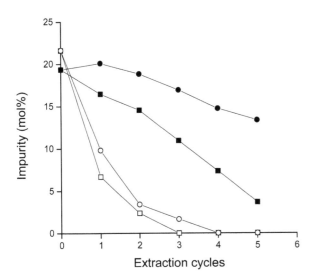

Figure 6 Removal of Z-L-Asp(L-Phe-OMe)-OH (solid symbol) and Z-L-aspartic acid (open symbol) from the coarse product using MIP-LL (square) and MIP-DD (circle) as the solid phase adsorbents.

Solid phase extraction is carried out by gently mixing the polymer particles (50 mg mL^{-1}) in the resulting solution for 1 h. After centrifugation, the supernatant is collected and treated with another batch of polymer, and this batch-mode extraction is repeated for several times. At each step the composition of the supernatant is analyzed by RP HPLC. A plot of byproduct content after each extraction vs. the number of extraction cycles is established.

Figure 6 shows representative polishing curves when the MIPs are applied at a dosage of 150 mg mL^{-1}. In this particular example the crude product contains 59% of Z-L-Asp(OH)-L-Phe-OMe, 19% of Z-L-Asp(L-Phe-OMe)-OH and 22% of Z-L-aspartic acid (mol/mol). Compared to the control polymer that is specific to Z-D-Asp(D-Phe-OMe)-OH, the by-product imprinted polymer (MIP-LL) not only removed the byproduct, but also extracted out more efficiently the related impurity, Z-L-aspartic acid.

In this example we use a polymer imprinted against the antipode of the target as a control. In this way, the selective binding, as evaluated by the adsorption difference between the two polymers, is solely determined by the stereo fitting of the targeted molecules in the imprinted sites.

VIII. CONCLUSION AND OUTLOOK

Molecular imprinting for solid phase extraction has attracted increasing interest in the past few years. MIP-based SPE is now considered to be the first commercialization effort to market the imprinting technology. As in other related applications, success in this aspect largely depends on the performance of MIP adsorbents under real conditions. In addition to strive for very high binding affinity and specificity, perhaps more important issues, but not so much foreseen by many of the research scientists, are the reproducibility of the MIP-based SPE, especially for applications of trace analysis. Using the same operation procedure, can we achieve the same or comparable recovery of analyte in solid phase extraction that is carried out at different times, or by different laboratories? How feasible is it to utilize internal standard? Is there a general preparation method that can mass-produce MIP adsorbents with minimal batch-to-batch variation regarding their binding characteristics? It is the answers to these questions that determine if the MIP-based SPE will achieve a wide acceptance in the real world.

REFERENCES

1. Yu, Y.; Ye, L.; Biasi, V.; Mosbach, K. Removal of the fermentation by-product succinyl L-tyrosine from the beta-lactamase inhibitor clavulanic acid using a molecularly imprinted polymer. Biotechnol. Bioeng. **2002**, *79*, 23–28.
2. Bjarnason, B.; Chimuka, L.; Ramström, O. On-line solid-phase extraction of triazine herbicides using a molecularly imprinted polymer for selective sample enrichment. Anal. Chem. **1999**, *71*, 2152–2156.
3. Di Corcia, A.; Marchese, S.; Samperi, R. Evaluation of graphitized carbon black as a selective adsorbent for extracting acidic organic compounds from water. J. Chromatogr. **1993**, *642*, 163–174.
4. Pichon, V.; Bouzige, M.; Miege, C.; Hennion, M.-C. Immunosorbents: natural molecular recognition materials for sample preparation of complex environmental matrixes. Trends Anal. Chem. **1999**, *18*, 219–235.
5. Sellergren, B.; Lanza, F. Molecularly imprinted polymers in solid phase extractions. Tech. Instrum. Anal. Chem. **2001**, *23*, 355–375.
6. Lanza, F.; Sellergren, B. Method for synthesis and screening of large groups of molecularly imprinted polymers. Anal. Chem. **1999**, *71*, 2092–2096.
7. Takeuchi, T.; Fukuma, D.; Matsui, J. Combinatorial molecular imprinting: an approach to synthetic polymer receptors. Anal. Chem. **1999**, *71*, 285–290.
8. Piletsky, S.A.; Karim, K.; Piletska, E.V.; Day, C.J.; Freebairn, K.W.; Legge, C.; Turner, A.F.P. Recognition of ephedrine enantiomers by molecularly imprinted polymers designed using a computational approach. Analyst **2001**, *126*, 1826–1830.
9. Hennion, M.C. Solid-phase extraction: method development, sorbents, and coupling with liquid chromatography. J. Chromatogr. A **1999**, *856*, 3–54.
10. Umpleby II, R.J.; Rushton, G.T.; Shah, R.N.; Rampey, A.M.; Bradshaw, J.C.; Berch, J.K.; Shimizu, K.D.; Recognition directed site-selective chemical modification of molecularly imprinted polymers. Macromolecules **2001**, *34*, 8446–8452.
11. Mullett, W.M.; Dirie, M.F.; Lai, E.P.C.; Guo, H.; He, X. A 2-aminopyridine molecularly imprinted polymer surrogate micro-column for selective solid phase extraction and determination of 4-aminopyridine. Anal. Chim. Acta. **2000**, *414*, 123–131.
12. Matsui, J.; Okada, M.; Tsuruoka, M.; Takeuchi, T. Solid-phase extraction of a triazine herbicide using a molecularly imprinted synthetic receptor. Anal. Commun. **1997**, *34*, 85–87.
13. Andersson, L.I. Efficient sample pre-concentration of bupivacaine from human plasma by solid-phase extraction on molecularly imprinted polymers. Analyst **2000**, *125*, 1515–1517.
14. Crescenzi, C.; Bayoudh, S.; Cormack, P.A.G.; Klein, T.; Ensing, K. Determination of clenbuterol in bovine liver by combining matrix solid-phase dispersion and molecularly imprinted solid-phase extraction followed by liquid chromatography/electrospray ion trap multiple-stage mass spectrometry. Anal. Chem. **2001**, *73*, 2171–2177.

15. Venn, R.F.; Goody, R.J. Synthesis and properties of molecular imprints of darifenacin: the potential of molecular imprinting for bioanalysis. Chromatographia **1999**, *50*, 407–414.

16. Ju, J.Y.; Shin, C.S.; Whitcombe, M.J.; Vulfson, E.N. Imprinted polymers as tools for the recovery of secondary metabolites produced by fermentation. Biotechnol. Bioeng. **1999**, *64*, 232–239.

17. Walshe, M.; Howarth, J.; Kelly, M.T.; O'Kennedy, R.; Smyth, M.R. The preparation of a molecular imprinted polymer to 7-hydroxycoumarin and its use as a solid-phase extraction material. J. Pharm. Biomed. Anal. **1997**, *16*, 319–325.

18. Kugimiya, A.; Takeuchi, T. Application of indoleacetic acid-imprinted polymer to solid phase extraction. Anal. Chim. Acta **1999**, *395*, 251–255.

19. Zander, A.; Findlay, P.; Renner, T.; Sellergren, B.; Swletlow, A. Analysis of nicotine and its oxidation products in nicotine chewing gum by a molecularly imprinted solid-phase extraction. Anal. Chem. **1998**, *70*, 3304–3314.

20. Masque, N.; Marce, R.M.; Borrull, F.; Cormack, P.A.G.; Sherrington, D.C. Synthesis and evaluation of a molecularly imprinted polymer for selective on-line solid-phase extraction of 4-nitrophenol from environmental water. Anal. Chem. **2000**, *72*, 4122–4126.

21. Sellergren, B. Direct drug determination by selective sample enrichment on an imprinted Polymer. Anal. Chem. **1994**, *66*, 1578–1582.

22. Bereczki, A.; Tolokan, A.; Horvai, G.; Horvath, V.; Lanza, F.; Hall, A.J.; Sellergren, B. Determination of phenytoin in plasma by molecularly imprinted solid-phase extraction. J. Chromatogr. A **2001**, *930*, 31–38.

23. M, Z.-H.; Qin, L. Determination of degradation products of nerve agents in human serum by solid phase extraction using molecularly imprinted polymer. Anal. Chim. Acta **2001**, *435*, 121–127.

24. Martin, P.; Wilson, I.D.; Morgan, D.E.; Jones, G.R.; Jones, K. Evaluation of a molecular-imprinted polymer for use in the solid phase extraction of propranolol from biological fluids. Anal. Commun. **1997**, *34*, 45–47.

25. Andersson, L.I.; Paprica, A.; Arvidsson, T. A highly selective solid-phase extraction sorbent for preconcentration of sameridine made by molecular imprinting. Chromatographia **1997**, *46*, 57–62.

26. Rashid, B.A.; Briggs, R.J.; Hay, J.N.; Stevenson, D. Preliminary evaluation of a molecular imprinted polymer for solid-phase extraction of tamoxifen. Anal. Commun. **1997**, *34*, 303–305.

27. Sergeyeva, T.A.; Matuschewski, H.; Piletsky, S.A.; Bendig, J.; Schedler, U.; Ulbricht, M. Molecularly imprinted polymer membranes for substance-selective solid-phase extraction from water by surface photo-grafting polymerization. J. Chromatogr. A **2001**, *907*, 89–99.

28. Mullett, W.M.; Lai, E.P.C. Determination of theophylline in serum by molecularly imprinted solid-phase extraction with pulsed elution. Anal. Chem. **1998**, *70*, 3636–3641.

29. Baggiani, C.; Giovannoli, C.; Anfossi, L.; Tozzi, C. Molecularly imprinted solid-phase extraction sorbent for the clean-up of chlorinated phenoxyacids from aqueous samples. J. Chromatogr. A **2001**, *938*, 35–44.

30. Ye, L.; Ramström, O.; Mosbach, K. Molecularly imprinted polymeric adsorbents for byproduct removal. Anal. Chem. **1998**, *70*, 2789–2795.

24

Applications of Molecularly Imprinted Materials as Enzyme Mimics

Kay Severin   Swiss Federal Institute of Technology Lausanne (EPFL), Lausanne, Switzerland

I. INTRODUCTION

As described in previous chapters, MIPs prepared by modern protocols are robust materials capable of highly specific molecular recognition. Since specific molecular recognition is the hallmark of enzyme catalysis, it is apparent that MIPs are potentially suited for catalytic applications. And indeed, investigations have shown that catalytic MIPs can be produced and successively more active and selective MIP catalysts were presented over the last years. In the following, the basic ideas and principles of catalytic MIPs are described using selected examples from the literature. For supplementary reading on this topic, review articles by Wulff [1], Whitcombe et al. [2] and Davis et al. [3] are recommended.

Before specific examples are discussed, we want to briefly recall the fundamental principles of enzyme catalysis. The astonishing efficiency of protein-based catalysts is the result of precisely positioned functional groups, which constitute a dynamic binding pocket for the substrate(s) and the transition state of the reaction. For a unimolecular reaction (S→P), the simplified energy profile is depicted in Fig. 1.

The activation barrier of the uncatalyzed reaction is represented by the energy difference of the substrate S and the transition state $S^{\#}$. In the presence of an enzyme (E), a complex ES is formed initially. During the course of the reaction, the enzyme is bound to the transition state ($ES^{\#}$) and to the product (EP). For catalysis to work, the difference in energy between ES and $ES^{\#}$ has to be smaller than the activation energy of the uncatalyzed reaction, meaning that the enzyme *preferentially stabilizes the transition state* of the reaction. To achieve catalytic turnover, it is furthermore important that the complex between the enzyme and the substrate (ES) is thermodynamically more stable than the complex with the product (EP). Any catalysts operating that way will be distinguished by a unique kinetic behavior: the rate of the reaction levels off at high substrate concentrations because the catalyst is saturated ("saturation kinetics"). Mathematically such a behavior can be described by the Michaelis–Menten equation [4].

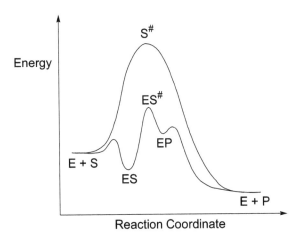

Figure 1 Energy profile for a hypothetical unimolecular enzyme catalyzed reaction (S = substrate, P = product, E = enzyme).

It is understandable that the construction of synthetic enzyme mimics has attracted an enormous amount of interest in the chemical community. The molecular scaffolds used to build artificial enzymes are very diverse and range from cyclic sugars (cyclodextrins) to porphyrin assemblies [5–7]. Not only the building blocks are very diverse but also the design strategies employed. First attempts were mainly based on a more or less accurate imitation of the key features thought to be responsible for enzyme catalysis. But alternative strategies are increasingly being followed, a promising one being the MIP-based approach described in this chapter.

II. STEREO- AND REGIO-SELECTIVE REACTIONS WITH MIPs

Before catalytic MIPs are discussed we want to focus on a closely related topic: the utilization of MIPs as microreactors and as protecting groups. The basic idea is outlined in Fig. 2. The reaction between two chemicals A and B can give rise to two different products P1 and P2, with P1 and P2 being regio- or stereoisomers. Using the product P1 or a structurally related template, a molecularly imprinted polymer is generated (Fig. 2b). Subsequent binding of A and B to this polymeric microreactor will result in the selective formation of P1 due to steric restrictions. Upon release of the product the MIP can be recycled which would allow catalytic turnover.

For selective transformations inside microreactors, some basic requirements have to be fulfilled. Importantly, solution phase reactions outside the MIP must be avoided. This can be achieved by adding the MIP in at least stoichiometric amounts and by providing very high affinity binding sites. For the reaction depicted in Fig. 2, for example, it would be ideal if all of the reactant A is bound to the MIP prior to the reaction with B. In reality, this tight binding is often achieved by covalent attachment of one or both reactants to the polymer.

Conceptually related is the utilization of an MIP as a selective protecting group (Fig. 2c). Here, a close analogue of the reactant B is used as the template in such a fashion that the resulting MIP is able to protect one of the two reactive sites. Again, an imprinting procedure using reversible covalent chemistry is advantageous.

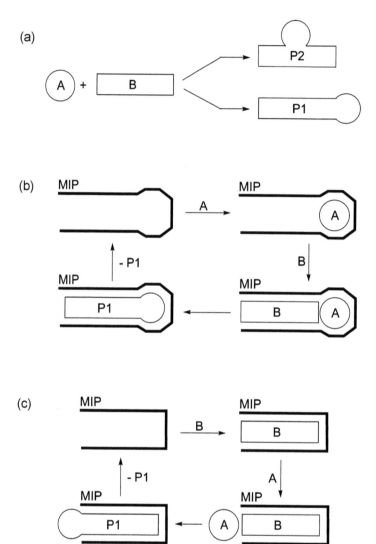

Figure 2 (a) Reaction of A and B leading to the isomeric products P1 and P2; (b) selective reaction with the help of an MIP microreactor; (c) selective reaction with the help of an MIP protecting group.

Following the strategy shown in Fig. 2b, the enantioselective synthesis of amino acids was accomplished [8,9]. As the template monomer, an L-DOPA derivative was employed. The synthetic procedure is outlined in Fig. 3. After hydrolysis, the binding site of the MIP contains a reactive aldehyde and specific cavities for the amino acid side chain and the C-terminus. Addition of glycine gives a Schiff base which can be deprotonated with the help of a base. The alkylation agent RX will preferentially attack from the side, which was previously occupied by the DOPA side chain. Using this methodology, amino acids with an enantiomeric excess of up to 36% ee were obtained. It should be noted that this excess is solely the result of the chiral cavities obtained by imprinting.

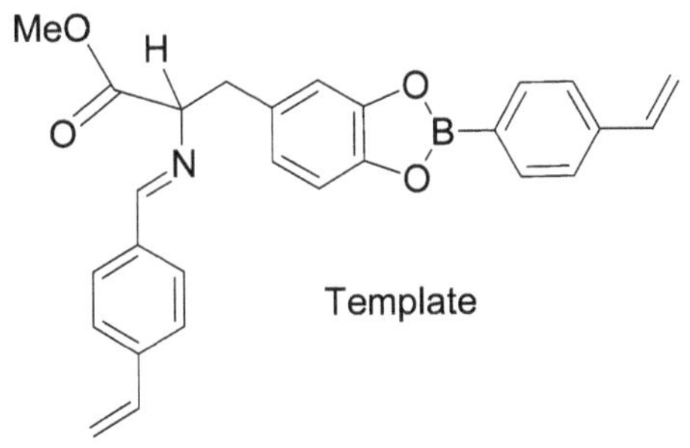

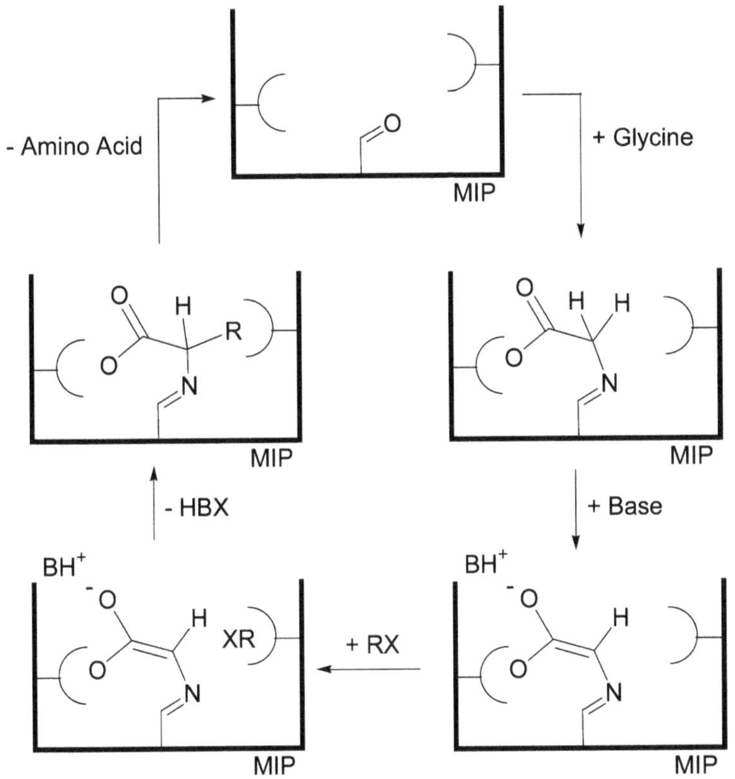

Template

Figure 3 Enantioselective synthesis of amino acids with the help of an MIP, prepared by covalent imprinting.

In extension of this concept, MIPs have been used to synthesize new inhibitors for the proteinase kallikrein [10]. A known inhibitor was employed as the template to generate an MIP following a noncovalent imprinting strategy (Fig. 4). The resulting MIP was shown to act as a microreactor, facilitating the syntheses of the original as well as of new inhibitors. This approach, in which an MIP imitates a biological receptor, represents a new method to synthesize bioactive molecules. The method is especially interesting if the structure of the natural receptor is not known.

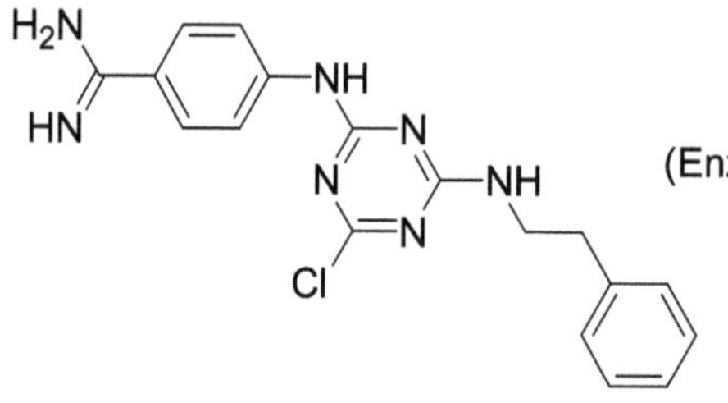

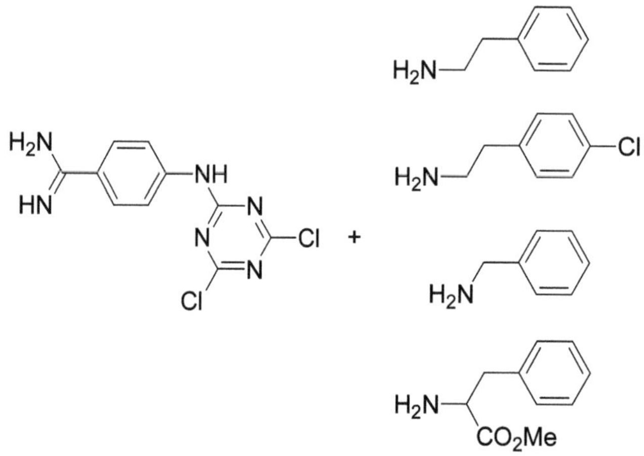

Figure 4 Imprinting with an enzyme inhibitor to generate an MIP which is able to facilitate the production of new inhibitors.

The utilization of MIPs as protecting groups for regioselective reactions (Fig. 2c) was realized using polyfunctional steroids as the substrate [11]. An example is given in Fig. 5. A steroid, functionalized with two polymerizable side chains via boronic ester linkages, is used as the functional monomer for a covalent imprinting process. After polymerization and cleavage of the ester groups an MIP is obtained with binding sites for steroids having hydroxy groups in 3α and 12α position. If the MIP is incubated with the substrate chendeoxycholan-24-ol, it acts as a protecting group for the two secondary alcohols leaving the third, primary alcohol accessible for functionalization. Consequently, reactions with acetic acid anhydride in the presence of pyridine give almost exclusively the monoacylation product with the acetate group at the aliphatic side chain. In control reactions with a nonimprinted polymer, on the other hand, the 3α-hydroxy group is preferentially acylated.

It should be noted that in all examples described above, the MIP was used in stoichiometric and not in catalytic amounts. But after removal of the product, the MIP can be recycled allowing repetitive reactions.

III. IMPRINTING WITH A TRANSITION STATE ANALOGUE

As described above [1], the selective stabilization of the transition state of the reaction is crucial for enzyme catalysis. This was first proposed by Pauling and later

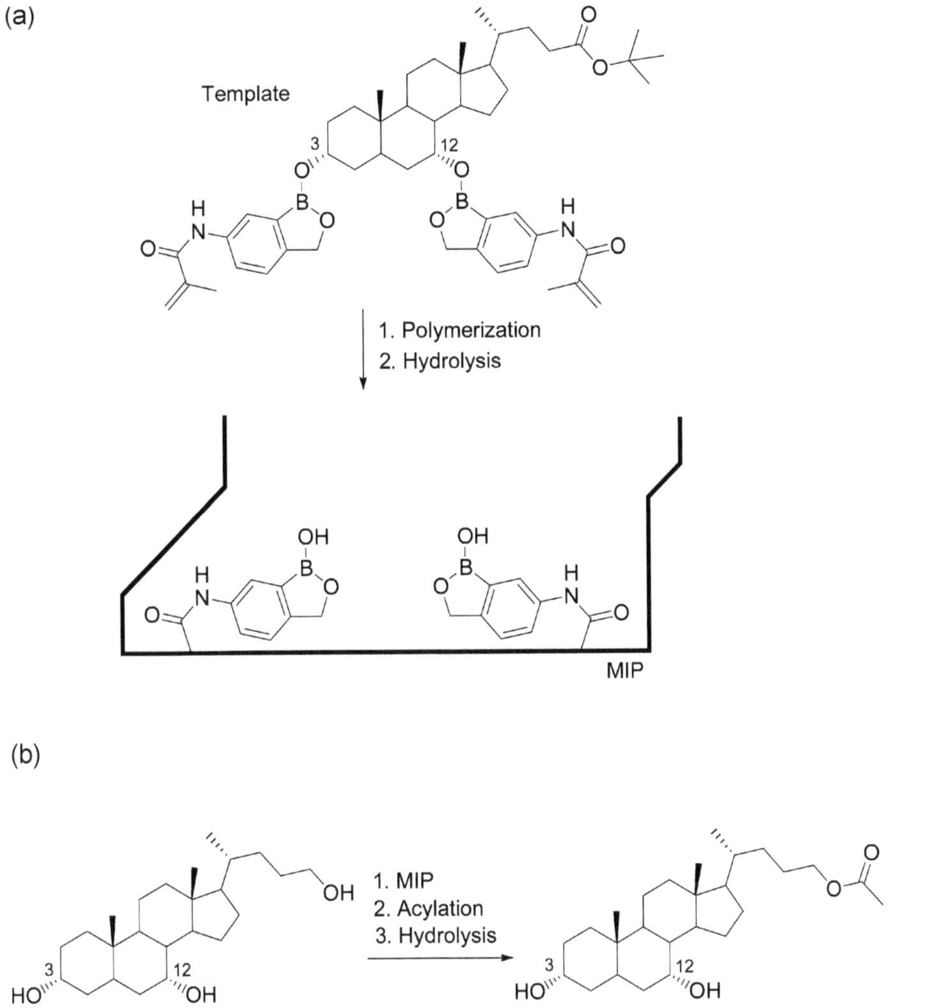

(a)

(b)

Figure 5 (a) Generation of an MIP with specific binding sites for 3α,12α-dihydroxy steroids; (b) the resulting polymer can be used as a protecting group in the selective acylation of a trifunctional steroid.

expanded by Jencks [12]. To some extent, the problem of making enzyme mimics can therefore be reduced to the question of how to make a good and specific receptor for the transition state of the reaction. In this context, the so-called "selection approach" is very appealing. It consists of the following steps: first, a molecule is synthesized that imitates the transition state of the reaction. Then, a high affinity receptor for this transition state analogue (TSA) is determined, preferentially by fast screening of a large receptor library. The reasoning behind this is that if the receptor has a high affinity for the TSA, it should also stabilize the real transition state.

So far, the most successful implementation of this strategy is based on antibodies [13]. A pool of structurally different antibodies is produced by the immune system as a response to a foreign molecule. If the TSA is part of the foreign molecule, then antibodies are generated which bind to the TSA. Using modern methods of molecular biology, antibodies with high TSA affinity can be isolated and characterized. After more than 15 years of research in this area, catalytic antibodies have been found for a variety of transformations such as hydrolysis reactions, Diels–Alder reactions, cyclopropanations and cyclizations [13]. The rate enhancements observed are among the best obtained for artificial enzymes (typically between 10^3 and 10^5 over the background) but they are still low compared to what is found for enzymes.

A very similar strategy has been applied to generate catalytic MIPs: TSAs have been used as a template to generate specific binding sites in polymers using covalent and noncovalent imprinting techniques. As for catalytic antibodies, hydrolysis reactions of esters have been the focus of interest. One of the most successful systems described so far will be our first example [14].

To imitate the transition state of the alkaline hydrolysis reaction depicted in Fig. 6 a phosphonic monoester was employed. This kind of TSA had previously been used to generate catalytic antibodies. It has a geometry similar to the tetrahedral intermediate of the hydrolysis reaction. An amidinium group containing styrene derivative served as the functional monomer. The basic advantage of this monomer is that the interaction with carboxylate and phosphonate groups is very strong. This ensures that the template interacts with the functional monomer during the imprinting

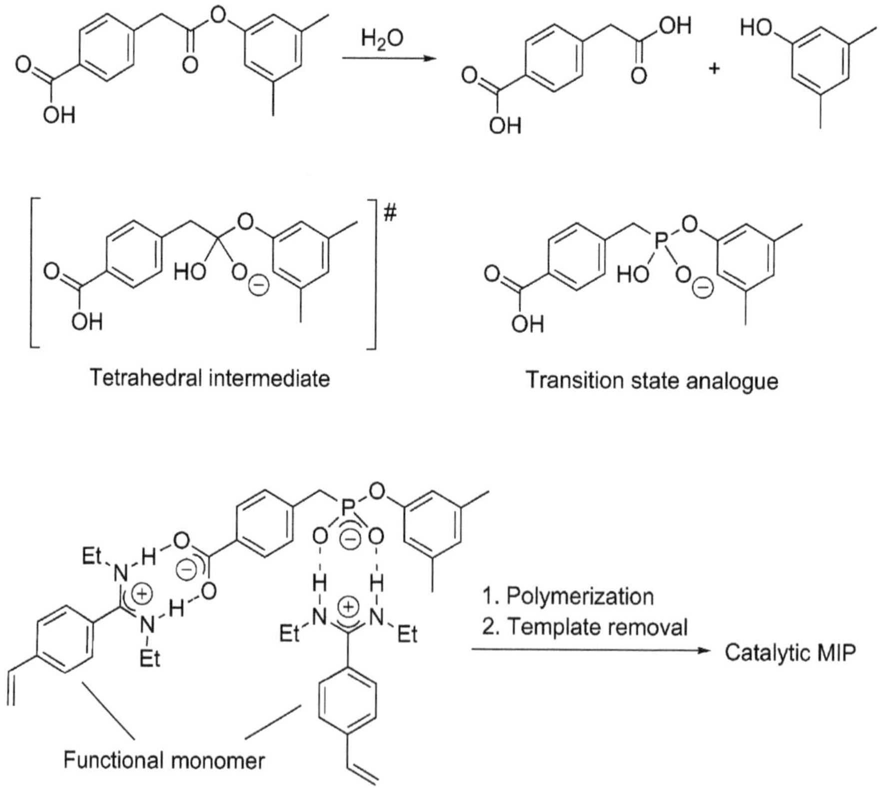

Tetrahedral intermediate   Transition state analogue

1. Polymerization
2. Template removal
→ Catalytic MIP

Functional monomer

Figure 6 Generation of a catalytic MIP using a phosphonic acid monoesters as a template which imitates the transition state of the reactions.

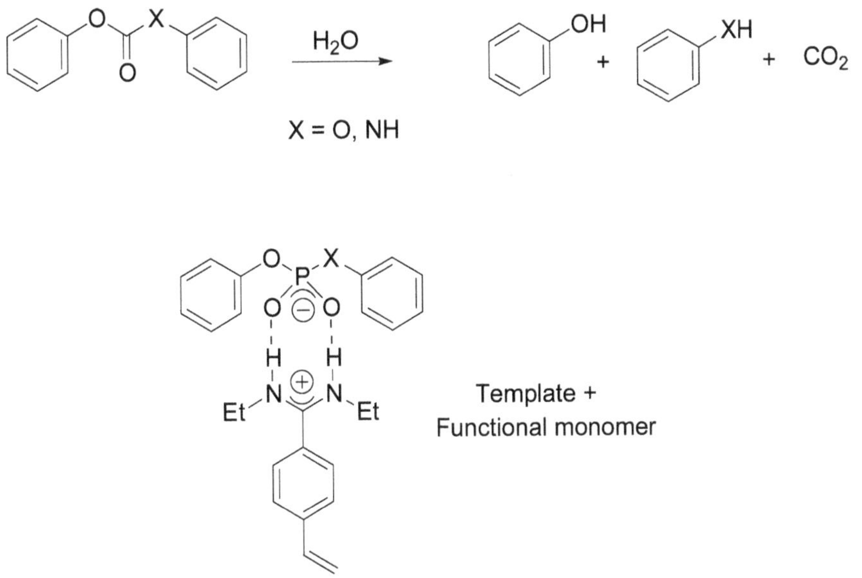

Figure 7 Imprinting with amidinium–phosphonate complexes to generate catalytic MIPs for the hydrolysis of carbonates and carbamates.

process. As a result, the amidinium groups are correctly placed in the binding site imitating the arginine side chains of hydrolytic enzymes or catalytic antibodies. After polymerization using EGDMA as the cross-linker and removal of the template, a catalytically active MIP was obtained. The rate of the MIP catalyzed reaction was approximately 100 times higher than the rate of a control reaction in homogeneous solution. Compared to reactions carried out in the presence of a control polymer imprinted with benzoate, a rate enhancement of factor 5 was observed. Furthermore, the catalytic MIP was shown to display Michaelis–Menten kinetics, substrate selectivity and it can be inhibited by the TSA. Unfortunately, the system shows low turnover due to substrate inhibition of the resulting carboxylic acid.

The problem of product inhibition was addressed in subsequent work by the same group [15]. Instead of esters, the hydrolysis of diphenyl carbonates and carbamates was investigated (Fig. 7). Here, the resulting reaction products show a low affinity for amidinium groups. Using the catalytic MIP obtained by imprinting with a diphenyl posphate template, the pseudo first-order rate constants were enhanced by a factor of 588 (carbonate) and 1435 (carbamate), respectively, when compared to reactions in solution at the same pH. The latter value approaches the regime found for catalytic antibodies. Compared to reactions carried out in the presence of a control polymer prepared without template, rate enhancements of up to a factor of 10 were observed for carbamates. Interestingly, this value is increased to 24 when a catalytic MIP prepared by suspension—instead of bulk polymerization was employed. This highlights the fact that the polymer morphology is of central importance for the activity and selectivity of catalytic MIPs.

Elimination reactions have also been studied extensively. A representative example is depicted in Fig. 8. Using a dicarboxylic acid as a template in combination

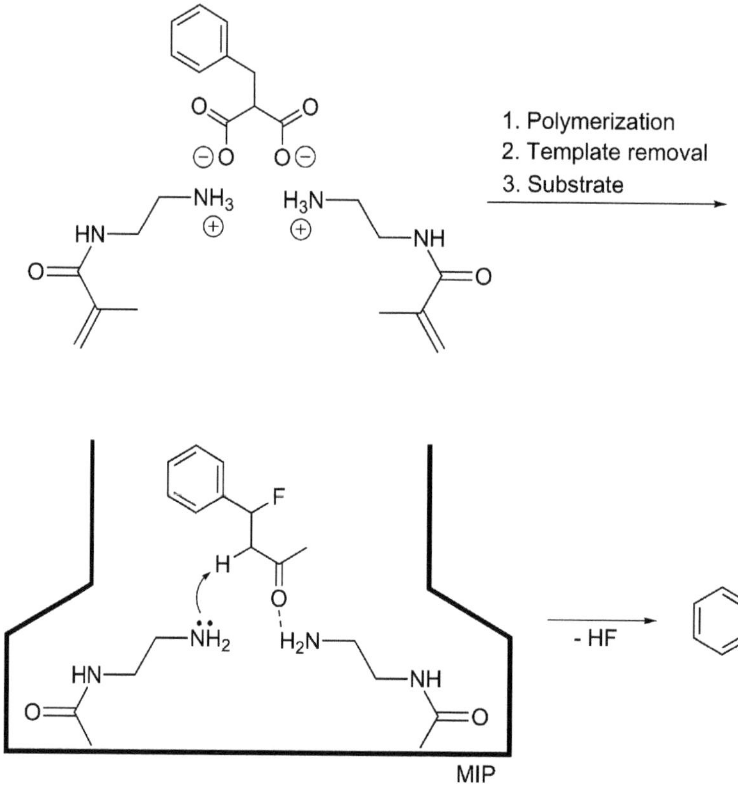

Figure 8 Generation of a catalytic MIP for a dehydrofluorination reaction.

with a primary amine as the functional monomer, a catalytic MIP for the dehydro-fluorination of 4-fluoro-4-(*p*-nitrophenyl) butan-2-one was generated [16]. The amine groups of the resulting polymer are able to act as a substrate binding site due to hydrogen bonding to the keto group and as a basic site for the reaction. After optimization of the polymerization conditions (initiator/porogen level), the MIP was able to enhance the rate of the dehydrofluorination by a factor of 7.5 when compared to a control polymer prepared without template. The kinetics of the reaction follow the Michaelis–Menten equation.

The construction of catalysts for bimolecular reactions represents a special challenge. Due to entropic reasons, the product–catalyst complex is likely to be more stable than the ternary substrate–catalyst-complex. Consequently, turnover is often low or not even observed. For Diels–Alder reactions, the difficulty to obtain turnover is further increased by the fact that the transition-state and the final product are similar in shape. Nevertheless, a catalytic MIP for a Diels–Alder reaction has successfully been prepared [17]. The "trick" employed to overcome the problem of similarity between TSA and product is the utilization of a reaction in which the product spontaneously decomposes (Fig. 9). The same reaction had been previously studied with catalytic antibodies. For the catalytic MIP, significant rate enhancements and Michaelis–Menten kinetics were observed. Addition of the template reduces the rate of the reaction to 41% of the original value whereas the control

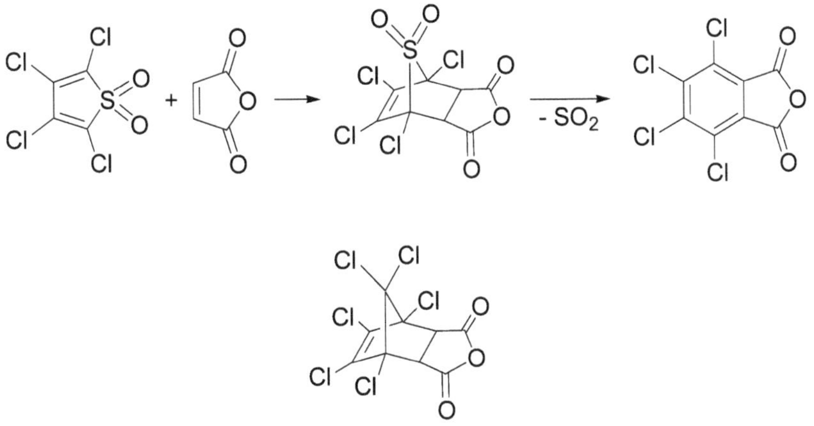

Intermediate analogue

Figure 9 Chlorenic anhydride as the intermediate analogue of the Diels–Alder reaction between maleic anhydride and tetrachlorothiophene dioxide.

polymer is not affected. This indicates that a good part of the catalytic transformation takes place inside the cavities generated by imprinting.

The question, whether imprinted polymers can be used as catalysts for enantio-selective reactions, is of special interest. One of the most comprehensive studies published so far has focused on the selective hydrolysis of D- and L-phenylalanine esters [18,19]. To generate catalytic MIPs, a number of chiral templates were investigated. The template that gave the best results was a chiral phosphonate analogue of phenylalanine. A vinyl monomer is connected to this TSA via a labile ester bond. After co-polymerization with EGDMA and MAA, the template is removed by hydrolysis generating a phenolic-, an imidazole- and an acidic group in the binding site (Fig. 10). This arrangement was chosen because similar functional groups are found in the active site of chymotrypsin, an enzyme that is able to selectively cleave peptide bonds [4]. Since an optical active template was employed, the binding site should be able to differentiate between enantiomers. When tested for ability to catalyze the hydrolysis of an N-protected phenylalanine-p-nitrophenyl ester, a rate enhancement of a factor of 10 was found when compared to the corresponding reaction in solution at the same pH and under pseudo first-order conditions. Importantly, the hydrolysis of the D-phenylalanine derivative was 1.85 times faster than the hydrolysis of the L-phenylalanine derivative. Slightly lower values were observed for the nonactivated substrate D/L-phenylalanine ethyl ester. In addition, control experiments were performed which showed that polymers of this kind are able to differentiate between the ground state and the transition state. This was done by comparing the affinity of MIPs that were imprinted with a simple ethyl ester (ground state) or with a phosphonate template (tetrahedral transition state), respectively.

IV. IMPRINTING WITH A METAL CATALYST

MIP catalysts, prepared with organic templates imitating the transition state of a reaction, show significant rate enhancements when compared to suited control

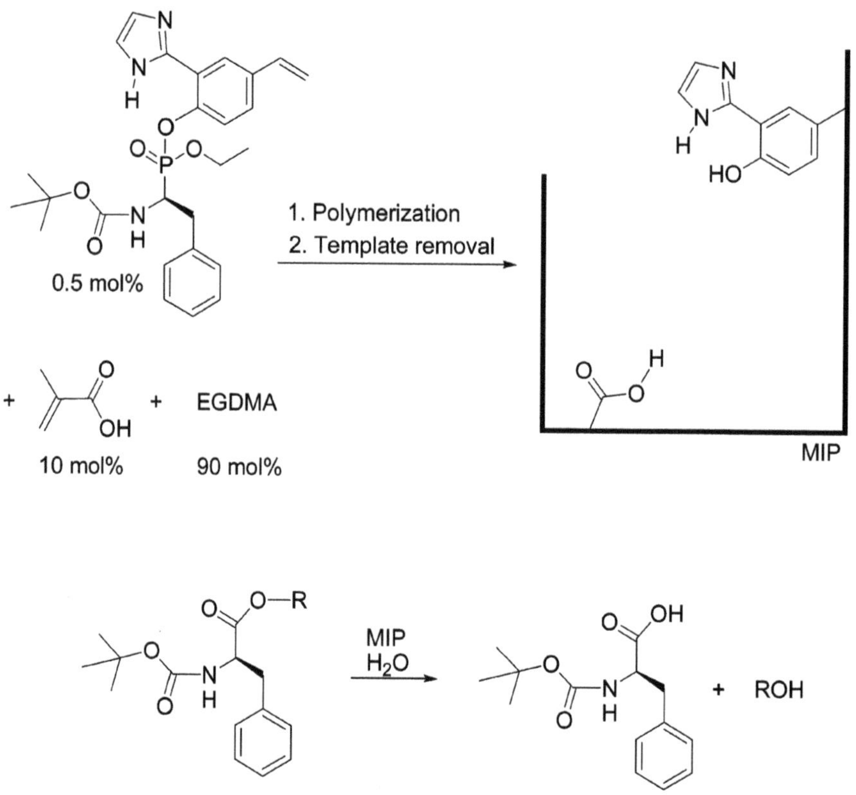

Figure 10 Enantioselective hydrolysis of phenylalanine esters in the presence of an MIP catalyst.

reactions. The general activity of these catalysts, however, is in most cases too low to be of interest for organic synthesis. A promising approach to improve the overall efficiency of MIP catalysts is the utilization of highly active transition metal catalysts [20]. In this context, it should be noted that in terms of activity, modern transition metal complexes start to approach the regime found for enzymes. The selectivity of these catalysts, however, especially the substrate- and regio-selectivity, is often very low. The basic idea is, therefore, to use the power of transition metal catalysts and impose additional selectivity by embedding them in imprinted polymers. The necessary steps are schematically shown in Fig. 11.

First, polymerizable side chains are attached to the ligand periphery of a transition metal catalysts. Then, a pseudo-substrate is coordinated to the active site of the catalyst. This pseudo-substrate should resemble the real substrate(s) finally used in the catalytic transformation. Ideally, the geometry of the catalyst-pseudo-substrate conjugate imitates the transition state of the reaction. It is important that the pseudo-substrate and the catalyst form a thermodynamically stable complex in order to ensure a sufficiently high concentration during imprinting. In this case, it should also be possible to isolate and characterize the conjugate, an important step to rationalize the imprinting results. After polymerization, the pseudo-substrate is selectively cleaved off, thereby generating a specific cavity in proximity to the active

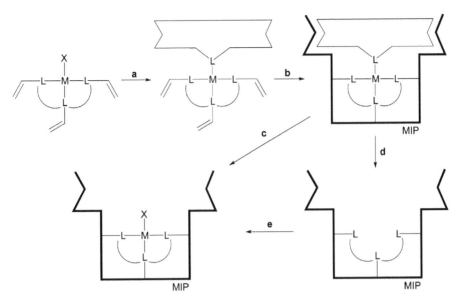

Figure 11 Generation of immobilized metal catalysts with a substrate pocket by imprinting with a pseudo-substrate: (a) attachment of the pseudo-substrate; (b) polymerization; (c) selective cleavage of the pseudo-substrate; (d) removal of the pseudo-substrate and the metal ion; (e) addition of the catalytically active metal ion.

center of the catalyst. If the selective cleavage is not possible, the metal ions are also removed from the polymer under enforced conditions and the catalyst is assembled in a final step on the solid support. The advantage of this additional step is that the metal ions can be switched. A kinetically stable Ru(II) complex, for example, could be used for imprinting and a labile Cu(I) complex for catalysis.

It should be noted that the highly cross-linked polymers generated in an imprinting procedure are well suited for immobilized catalysts. Due to the high surface area, the accessibility of the metal centers inside the polymer is very good. And in contrast to classical supports with a low degree of cross-linking, the choice of the solvent is less crucial since good swelling of the polymer is not a prerequisite for catalysis. Rhodium catalysts which were immobilized in an EGDMA polymer, for example, were shown to function even in protic solvents such as methanol [21].

The strategy described in Fig. 11 was used to generate substrate- and regio-selective MIP catalysts for the transfer hydrogenation of ketones [22,23]. Ruthenium halfsandwich complexes with amino group containing chelate ligands are known to be highly active catalysts for the reduction of ketones using formic acid or isopropanol as the reducing agent. In the proposed transition state of the reaction, the substrate is bound and activated by a hydrogen bond to the amine part of the ligand (Fig. 12a). Hydrogen is then transferred more or less simultaneously. To imitate this transition state with a coordinated pseudo-substrate, a phosphinato-complex was synthesized (Fig. 12b).

With the help of this organometallic TSA, an imprinted polymer was generated by co-polymerization with EGDMA. When tested for the catalytic reduction of benzophenone, the MIP was shown to be seven times more active than a control

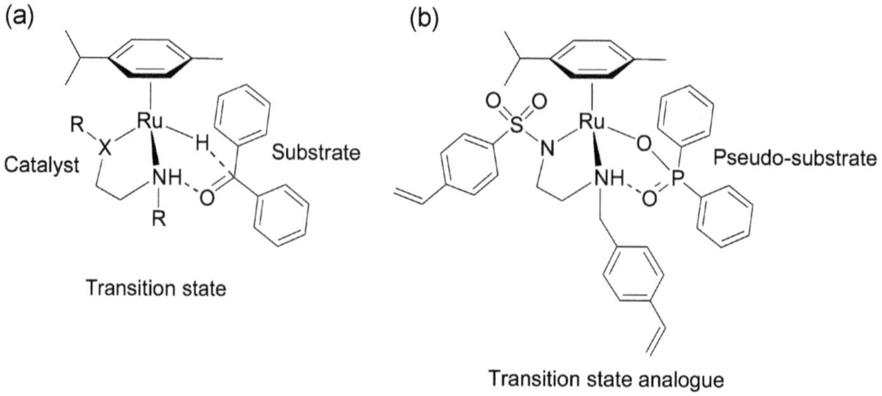

(a)   (b)

Figure 12   (a) Transition state of the ruthenium catalyzed reduction of benzophenone; (b) a ruthenium–phosphinato complex as a TSA.

polymer prepared without pseudo-substrate but having the same amount of ruthenium complex. The initial turnover frequency was determined to be 134 h^{-1}, a value that is sufficient for applications in organic synthesis. Furthermore, the MIP catalyst showed a pronounced substrate-selectivity: out of a mixture of seven different aliphatic and aromatic ketones, benzophenone—the substrate that was imitated by the pseudo-substrate—was clearly the fastest reduced (Fig. 13).

In extension of this work, a chiral rhodium(III) complex having a methyl-phenyl-phosphinato ligand was synthesized [24]. This pseudo-substrate mimics the

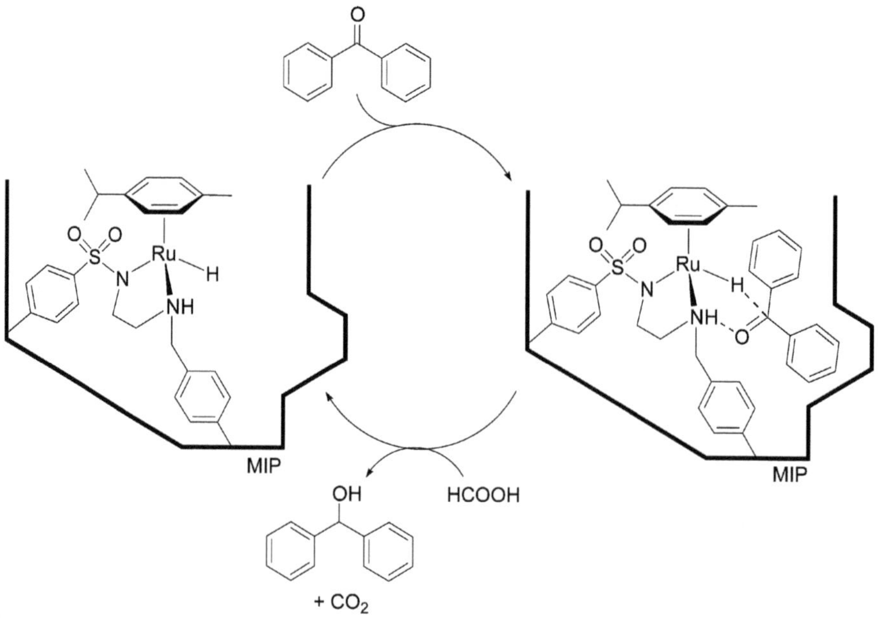

Figure 13   Transfer hydrogenation of benzophenone catalyzed by a ruthenium containing MIP.

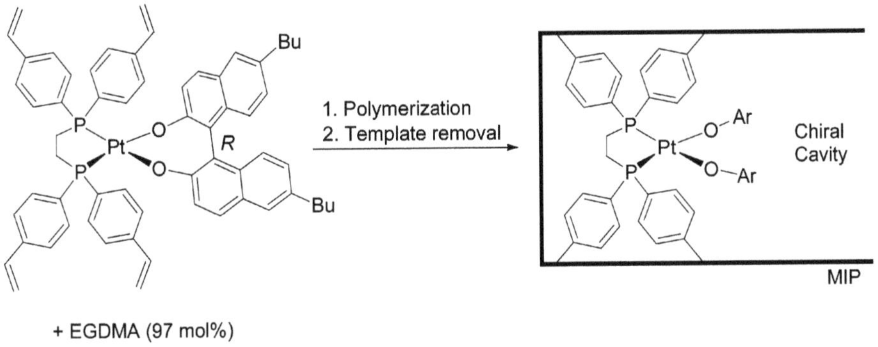

+ EGDMA (97 mol%)

Figure 14 A platinum containing MIP with chiral cavities next to the metal center.

transition state of the asymmetric reduction of acetophenone. The corresponding MIP was shown to be not only more active than a control polymer but also more selective: the enantiomeric excess increased by 2–9% ee (enantiomeric excess) due to imprinting with a maximum selectivity of 95% ee for acetophenone. The same asymmetric reaction was investigated using a catalytic MIP containing polyurea supported rhodium(I) complexes [25,26]. Here, the optically pure reaction product—1-(S)-phenylethanol—was used as the template. Although the overall enantioselectivity of this system was lower (up to 70% ee), a relative increase in selectivity due to imprinting was also observed.

For the application of metal catalyst containing MIPs in asymmetric catalysis, the question to which extent chiral cavities can be prepared by imprinting is of central importance. This issue was investigated in detail using platinum phosphine complexes with vinyl side chains in combination with optically active BINOL derivatives as imprinting ligands [27,28]. In a first experiment, a platinum metallomonomer was co-polymerized with EGDMA in the presence of chlorobenzene as the porogen (Fig. 14). The chiral imprinting ligand was then removed by addition of an aromatic alcohol. The effect of the chiral cavity was investigated in rebinding experiments with rac-BINOL. As expected, sites with different reactivity and selectivity were observed but interestingly there was a strong correlation: the least reactive sites were found to be the most selective (89–94% ee in rebinding experiments).

In extension of this work, a platinum complex with a chiral phosphine ligand and a BINOL template was employed as a metallomonomer [28]. After removal of the template and activation of the complex with silver salts (generation of Lewis acidic platinum centers) the MIP was used as a catalyst for an asymmetric ene reaction. Enantioselectivities of 72% ee were observed (Fig. 15). This compares well to

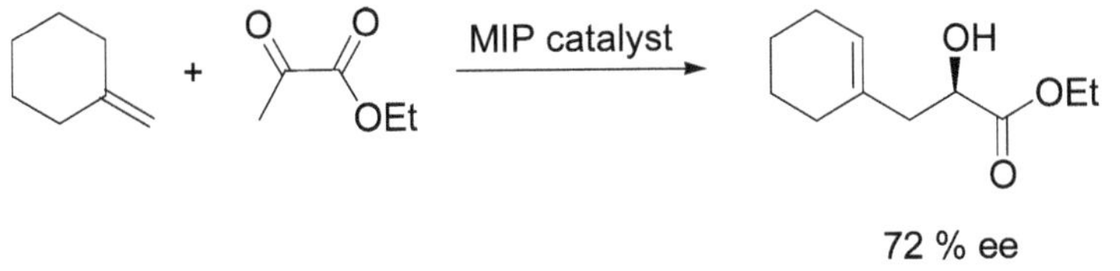

72 % ee

Figure 15 Asymmetric ene reaction with a platinum(II) containing MIP catalyst.

the same reaction carried out in homogeneous solution (75% ee). Polymeric catalysts prepared without template or with a template having the opposite configuration gave lower yields and selectivities (67% ee and 25% ee, respectively).

In view of the fact that many metalloenzymes have porphyrin type ligands (cytochrome P450, catalases, peroxidases, etc.) the construction of MIP catalysts with metalloporphyrin sites is of special interest. First experiments towards this goal have recently been presented [29]. A ruthenium–tetraarylporphyrin complex with four vinyl side chains was synthesized. After co-polymerization with EGDMA (99 mol%) the resulting polymer was tested for its ability to catalyze oxygen transfer reactions. Good to excellent results were obtained in epoxidation reactions and in the oxidation of secondary alcohols. Even the unreactive ethylbenzene could be oxidized to give acetophenone. In several cases the activity of the polymeric catalyst was significantly higher ($>20\times$) than that of the corresponding catalyst in solution. Apparently, the polymeric matrix is able to isolate the reactive metal complexes and thus prevent destructive autoxidation. For many metalloenzymes, the protein matrix provides a similar protective mechanism.

Molecular imprinting cannot only be used to create specific substrate cavities around a metal catalyst but also to enforce a certain ligand geometry. Again, this is similar to metalloproteins for which unusual geometries and coordination numbers have been observed repetitively. Square planar palladium(II) phosphine complexes of the general formula $[PdX_2(PR_3)_2]$ are preferentially found in the thermodynamically more stable *trans*-form. For certain palladium catalyzed reactions, however, a *cis*-configuration is thought to be advantageous. Therefore, an imprinting procedure has been employed to generate a semi-rigid polymeric Pd-catalyst in which the *cis*-form is "fixed" by the polymer matrix [30]. This was achieved by using a catecholate co-ligand which forces the vinyl-containing triphenylphosphine ligands in *cis*-position (Fig. 16). For some cross-coupling reactions such as the Suzuki reaction between *p*-bromoanisole and phenyl boronic acid, the MIP catalyst was shown to be superior to the corresponding homogeneous catalyst.

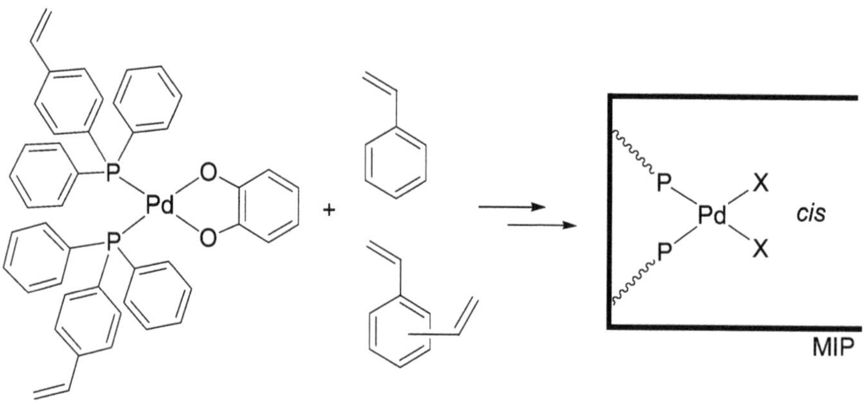

Figure 16 A palladium(II) MIP catalyst for which the a *cis*-configuration was enforced by molecular imprinting.

V. INORGANIC POLYMERS AS CATALYTIC MIPs

The possibility to imprint the surface of commercial silica gel order to generate a catalytically active material was first investigated in 1988 [31,32]. In a procedure termed "footprint catalysis" a small amount of alumina is formed on silica gel in the presence of stable TSAs, which are subsequently removed by washing. The alumina acts as the matrix containing specific cavities of the imprint molecules and provides Lewis acidic aluminum centers for catalysis. Initially, benzenesulphonamide templates were used as TSAs to generate catalytic sites for the reaction of benzoic acid anhydride with alcohols. Improved results were later obtained with a phosphonic acid diamide as the template (Fig. 17) [33]. In acyl transfer reactions with 2,4-dinitrophenolate, the imprinted silica gel was shown to be significantly more active than a control catalyst prepared without template. A direct comparison with other catalytic MIPs is difficult because the amount of active sites is not known. The methodology was subsequently applied to asymmetric reactions using optically active templates [34,35]. Enantioselective catalysis was accomplished with differences in catalytic efficiency for the R- and the S-substrate of a factor of 2.5.

(a)

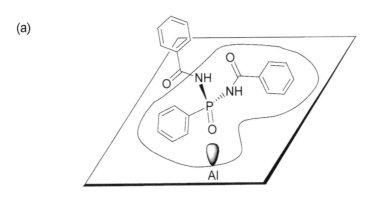

(b)

Figure 17 (a)"Footprint catalyst" containing a Lewis acidic aluminum site with a tetrahedral phosphonic acid diamide template; (b) acyl transfer reaction between benzoic acid anhydride and 2,4-dinitrophenole, catalyzed by imprinted silica/alumina gel.

An alternative approach to surface imprinting of silica was recently introduced [36,37]. The process is based on a sol–gel reaction inside reverse micelles (Fig. 18a). The template molecule acts as the headgroup of a surfactant. After mixing with a nonionic surfactant, cyclohexane, water and ammoniated ethanol, a water-in-oil microemulsion is formed. The co-condensation is then started by addition of $Si(OEt)_4$ and amine-, dihydroimidazole- and carboxylate-terminated trialkoxysilanes to give silica particles with a diameter between 400 and 600 nm. Since the stemplate molecule was located at the interface of the micelle during the condensation reaction, the resulting imprints are formed at the surface of the particle.

N-α-decyl-L-phenylalanine-(2-aminopyridine)amide can be regarded as a surfactant with a headgroup imitating the transition state of the α-chymotrypsin catalyzed hydrolysis of peptides. Using this TSA as a template, catalytically active silica particles were obtained. To test the reactivity, the hydrolysis of an arginine derivative was investigated (Fig. 18b). Compared to the control material prepared without template, a rate enhancement of a factor of 5 was observed. Furthermore, the catalytic silicas were shown to be enantioselective. Remarkably, the preferential

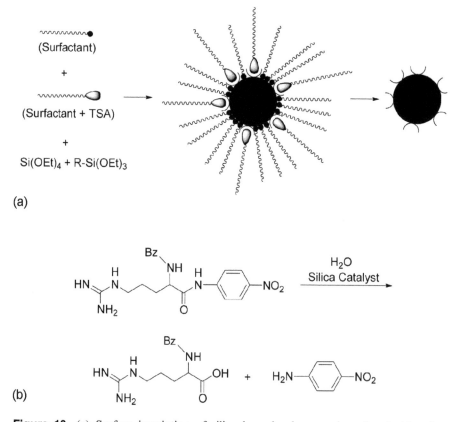

Figure 18 (a) Surface imprinting of silica by sol–gel co-condensation inside of reverse micelles. The TSA represents the headgroup of a surfactant; (b) the resulting silica particles are able to catalyze the hydrolysis of activated amides.

hydrolysis of the D-enantiomer was observed although an imprint molecule with L-configuration was employed.

Bulk imprinting of metal oxide materials can be achieved by sol–gel co-condensation of alkoxides. If the template (or TSA) is attached to a polymerizable side chain (e.g., a triethoxysilane group) the co-condensation with tetraalkoxysilanes leads to a porous amorphous solid with covalently attached templates. The imprint molecule can finally be removed by chemical and/or physical means.

Following this methodology, microporous silica with binding sites containing spatially organized amino groups was generated [38]. Using a benzene derivative, functionalized in 1,3,5-position with triethoxysilane groups via a labile carbamate linkage, imprinted silica was obtained in a sol–gel process with tetraethoxysilane. The binding site was generated by cleavage of the carbamate linker (Fig. 19). Preliminary results have shown that materials of this kind are able to catalyze the Knoevenagel condensation of malononitrile with isophthalaldehyde.

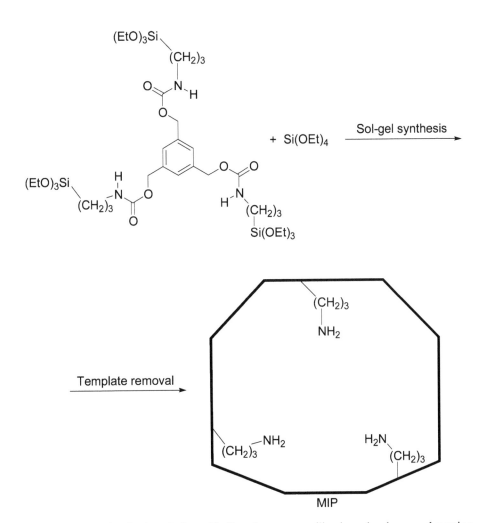

Figure 19 Molecular imprinting of bulk, microporous silica by sol–gel co-condensation.

For the synthesis of enzyme mimics, control experiments are of central importance, especially if the catalytic effects are relatively small and if the precise nature of the catalyst is difficult to establish. This proved to be of special relevance for the imprinting of silica. Among the problems encountered were the lack of reproducibility, side reactions due to impurities, diffusion limited reaction rates and difficulties to attach and to remove the template molecules [39–41].

VI. SUMMARY AND OUTLOOK

In the preceding sections, the main strategies to generate catalytically active MIPs have been summarized. Most of the concepts are inspired by the way nature performs catalytic transformations. That is why these polymers have been termed "biomimetic catalysts" or "enzyme mimics". Compared to their natural counterparts, catalytic MIPs offer a variety of advantages: (1) The polymers are thermally stable and can be used in a variety of solvents ranging from water to unpolar organic solvents. (2) The preparation is fast, easy, and cheap (given that the template synthesis is not too complicated). (3) The catalyst–product separation requires a simple filtration step after which the catalyst can be reused. (4) MIP catalysts can be generated for reactions, which are not catalyzed by enzymes.

Although these are important characteristics for potential applications, one should be aware that so far, the observed rates as well as the achieved turnover numbers of purely organic MIP catalysts are very low compared to what is found for enzymes. In most cases reported, the activity of the imprinted polymer is between 2 and 20 times higher than that of a control polymer prepared without a template. Compared to the corresponding reaction in solution, significantly higher rate enhancements have been obtained ($>10^3$; e.g., Ref. 15). But here, the beneficial situation of a reaction center inside a polymer (site-isolation, hydrophobic effects, etc.) contributes to the increase in activity.

An attractive alternative to improve the general activity of catalytic MIPs is the utilization of transition metal catalysts. In this case, polymeric catalyst with turnover frequencies of $>100 \ h^{-1}$ and turnover numbers of >100 can be prepared. For this type of MIP catalysts, the imprinting procedure should be regarded as a method to

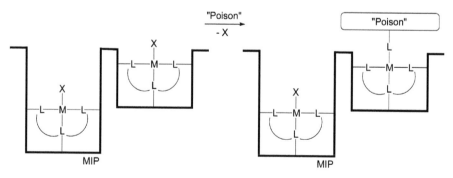

Figure 20   A metal containing MIP catalyst with a selective (left) and an unselective reaction site (right). Upon addition of a catalyst "poison", the less selective but more reactive site is occupied.

influence the substrate-, regio- and enantioselectivity in a controlled fashion. This is very similar to what is found for some metalloenzymes, in which the metal ion and its first coordination sphere is responsible for the activity and the specific pocket provided by the protein for the selectivity (second coordination sphere).

So far, the activity of MIP catalysts has been the center of interest but the issue of selectivity is increasingly being addressed. If the selectivity is expressed in terms of relative rates of product formation, the MIP catalysts reported so far are up to four times more selective (with some exceptions, e.g., Ref. 37). Depending on the reaction studied, this increase in selectivity might already be of interest. For potential applications in asymmetric synthesis, however, significantly higher values are clearly needed.

If the selectivity of the MIP catalyst is the main objective, the partial poisoning of active centers might be a way to improve the performance of the system. The imprinting procedure generates a statistical distribution of selective and less selective reactions centers. Studies indicate that the least selective sites are the most reactive [27]. The reaction of an MIP catalyst with sub-stoichiometric amounts of a catalyst "poison" under kinetic control should, therefore, result in a less active but more selective MIP catalyst. As a poisoning reaction, the covalent modification of functional groups or the irreversible complexation of a metal center could be employed (Fig. 20).

The optimization of an imprinting procedure is a difficult task, because several interdependent parameters have to be adjusted. It is to be expected that combinatorial methods will increasingly be used to speed up such an optimization process. For MIP catalysts, parallel synthesis and parallel catalyst evaluation will soon give rise to polymers with a significantly better catalytic performance. First steps in this direction have already been reported [1]. But besides better imprinting protocols, completely new methods are likely to be developed. Whether MIP based catalysts will ever be competitive to existing catalyst (or even to enzymes) remains to be seen, but without doubt, exciting new results in this area can be expected in the near future.

REFERENCES

1. Wulff, G. Enzyme-like catalysis by molecularly imprinted polymers. Chem. Rev. **2001**, *102*, 1–28.
2. Whitcombe, M.J.; Alexander, C.; Vulfson, E.N. Imprinted polymers: versatile new tools in synthesis. Synlett **2000**, *6*, 911–923.
3. Davis, M.E.; Katz, A.; Ahmad, W.R. Rational catalyst design via imprinted nanostructured materials. Chem. Mat. **1996**, *8*, 1820–1839.
4. Fersht, A. In *Structure and Mechanism in Protein Science: A Guide to Enzyme Catalysis and Protein Folding*; WH Freeman and Company: New York, 1999.
5. Motherwell, W.B.; Bingham, M.J.; Six, Y. Recent progress in the design and synthesis of artificial enzymes. Tetrahedron **2001**, *57*, 4663–4686.
6. Rowan, S.R.; Sanders, J.K.M. Enzyme models: design and selection. Curr. Opin. Chem. Biol. **1997**, *1*, 483–490.
7. Kirby, A.J. Enzyme mechanisms, models and mimics. Angew. Chem. Int. Ed. **1996**, *35*, 707–724.
8. Wulff, G.; Vietmeier, J. Enzyme-analogue built polymers. 25. Synthesis of macroporous copolymers from α-amino acid-based vinyl compounds. Makromol. Chem. **1989**, *190*, 1717–1726.

9. Wulff, G.; Vietmeier, J. Enzyme-analog built polymers. 26. Enantioselective synthesis of amino acids using polymers possessing chiral cavities obtained by an imprinting procedure with template molecules. Makromol. Chem. **1989**, *190*, 1727–1735.

10. Mosbach, K.; Yu, Y.; Andersch, J.; Ye, L. Generation of new enzyme inhibitors using imprinted binding sites: the anti-idiotypic approach, a step toward the next generation of molecular imprinting. J. Am. Chem. Soc. **2001**, *123*, 12420–12421.

11. Alexander, C.; Smith, C.R.; Whitcombe, M.J.; Vulfson, E.N. Imprinted polymers as protecting groups for regioselective modification of polyfunctional substrates. J. Am. Chem. Soc. 1999, *121*, 6640–6651.

12. Jencks, W.P. In *Catalysis in Chemistry and Enzymology*; Dover Publications: New York, 1987.

13. Hilvert, D. Critical analysis of antibody catalysis. Annu. Rev. Biochem. **2000**, *69*, 751–793.

14. Wulff, G.; Groß, T.; Schönfeld, R. Enzyme models based on molecularly imprinted polymers with strong esterase activity. Angew. Chem. Int. Ed. Engl. **1997**, *36*, 1962–1964.

15. Strikovski, A.G.; Kasper, D.; Grün, M.; Green, B.S.; Hradil, J.; Wulff, G. Catalytic molecularly imprinted polymers using conventional bulk polymerization or suspension polymerization: selective hydrolysis of diphenyl carbonate and diphenyl carbamate. J. Am. Chem. Soc. **2000**, *122*, 6295–6296.

16. Beach, J.V.; Shea, K.J. Designed catalysts. A synthetic network polymer that catalyzes the dehydrofluorination of 4-fluoro-4-(*p*-nitrophenyl)butan-2-one. J. Am. Chem. Soc. **1994**, *116*, 379–380.

17. Liu, X.-C.; Mosbach, K. Studies towards a tailor-made catalyst for Diels–Alder reactions using the technique of molecular imprinting. Macromol. Rapid Commun. **1997**, *18*, 609–615.

18. Sellergren, B.; Shea, K.J. Enantioslective ester hydrolysis catalyzed by imprinted polymers. Tetrahedron: Asymmetry **1994**, *5*, 1403–1406.

19. Sellergren, B.; Karmalkar, R.N.; Shea, K.J. Enantioslective ester hydrolysis catalyzed by imprinted polymers 2. J. Org. Chem. **2000**, *65*, 4009–4027.

20. Severin, K. Imprinted polymers with transition metal catalysts. Curr. Opin. Chem. Biol. **2000**, *4*, 710–714.

21. Taylor, R.A.; Santora, B.P.; Gagné, M.R. A polymer-supported rhodium catalyst that functions in polyer protic solvents. Org. Lett. **2000**, *2*, 1781–1783.

22. Polborn, K.; Severin, K. Molecular imprinting with an organometallic transition state analogue. Chem. Commun. **1999**, 2481–2482.

23. Polborn, K.; Severin, K. Biomimetic catalysis with immobilised organometallic ruthenium complexes: substrate- and regioselective transfer hydrogenation of ketones. Chem. Eur. J. **2000**, *6*, 4604–4611.

24. Polborn, K.; Severin, K. Biomimetic catalysis with an immobilised chiral rhodium (III) complex. Eur. J. Inorg. Chem. **2000**, 1687–1692.

25. Locatelli, F.; Gamez, P.; Lemaire, M. Molecular imprinting of polymerized catalytic complexes in asymmetric catalysis. J. Mol. Catal. **1998**, *135*, 89–98.

26. Gamez, P.; Dunjic, B.; Pinel, C.; Lemaire, M. Molecular imprinting effect in the synthesis of immobilized rhodium complex *catalyst* (IRC *cat*). Tetrahedron Lett. **1995**, *36*, 8779–8782.

27. Brunkan, N.M.; Gagné, M.R. Effect of chiral cavities associated with molecularly imprinted platinum centers on the selectivity of ligand-exchange reactions at platinum. J. Am. Chem. Soc. **2000**, *122*, 6217–6225.

28. Koh, J.H.; Larsen, A.S.; White, P.S.; Gangné, M.R. Disparate roles of chiral ligands and molecularly imprinted cavities in asymmetric catalysis and chiral poisoning. Organometallics **2002**, *21*, 7–9.

29. Nestler, O.; Severin, K. A ruthenium porphyrin catalyst immobilized in a highly cross-linked polymer. Org. Lett. **2001**, *3*, 3907–3909.

30. Cammidge, A.N.; Baines, N.J.; Bellingham, R.K. Synthesis of heterogeneous palladium catalyst assemblies by molecular imprinting. Chem. Commun. **2001**, 2588–2589.

31. Morihara, K.; Kurihara, S.; Suzuki, J. Footprint catalysis. I. A new method for designing tailor-made catalysts with substrate specificity: silica (alumina) catalysts for butanolysis of benzoic anhydride. Bull. Chem. Soc. Jpn. **1988**, *61*, 3991–3998.

32. Morihara, K.; Nishihata, E.; Kojima, M.; Miyake, S. Footprint catalysis. II. molecular recognition of footprint catalytic sites. Bull. Chem. Soc. Jpn. **1988**, *61*, 3999–4003.

33. Shimada, T.; Makanishi, K.; Morihara, K. Footprint catalysis. IV. structural effects of templates on catalytic behavior of imprinted footprint cavities. Bull. Chem. Soc. Jpn. **1992**, *65*, 954–958.

34. Morihara, K.; Kurokawa, M.; Kamata, Y.; Shimada, T. Enzyme-like enantioselective catalysis over chiral molecular footprint cavities on a silica (alumina) gel surface. J. Chem. Soc. Chem. Commun. **1992**, 358–360.

35. Matsuishi, T.; Shimada, T.; Morihara, K. Definitive evidence for enantioselective catalysis over molecular footprint catalytic cavities chirally imprinted on a silica (alumina) gel surface. Chem. Lett. **1992**, 1921–1924.

36. Markowitz, M.A.; Kust, P.R.; Deng, G.; Schoen, P.E.; Dordick, J.S.; Clark, D.S.; Gaber, B.P. Catalytic silica particles via template-directed molecular imprinting. Langmuir **2000**, *16*, 1759–1765.

37. Markowitz, M.A.; Kust, P.R.; Klaehn, J.; Deng, G.; Gaber, B.P. Surface-imprinted silica particles: the effect of added organosilanes on catalytic activity. Anal. Chim. Acta. **2001**, *435*, 177–185.

38. Katz, A.; Davis, M.E. Molecular imprinting of bulk, microporous silica. Nature **2000**, *403*, 286–289.

39. Heilmann, J.; Maier, W.F. Problems in selective catalysis with molecular imprints in silica—selective lactones formation from hydroxyesters in micropores. Z Naturforsch **1995**, *50b*, 460–468.

40. Ahmad, W.R.; Davis, M.E. Transesterification on imprinted silica. Catal. Lett. **1996**, *40*, 109–114.

41. Maier, W.F.; Mustapha, W.B. Reply to Transesterification on imprinted silica. Catal. Lett. **1997**, *46*, 137–140.

25
Application of MIPs as Antibody Mimics in Immunoassays

Richard J. Ansell   University of Leeds, Leeds, United Kingdom

I. INTRODUCTION

Immunoassays are a class of analytical methods based on the selective affinity of an antibody for an analyte [1–3]. Biological antibodies are highly useful analytical reagents, which may be produced by innoculation of laboratory animals with the analyte or with a derivative. The archetypal immunoassay, the radioimmunoassay or RIA, involves competition for a limited number of antibody binding sites between the analyte, present at unknown concentration in the sample, and a radiolabelled probe. Immunoassays are widely employed in clinical analysis in particular, but also in other fields including environmental analysis and research.

MIPs have attracted increasing interest as substitutes for biological antibodies in such procedures, which are then commonly referred to as Molecular Imprint Sorbent Assays or MIAs [4–8]. Like antibodies, MIPs can exhibit selective binding for a chosen analyte. Unlike antibodies they are not individual soluble macromolecules but, most often, insoluble particles or films. However, this is not a problem since in most immunoassay formats the biological antibodies are in any case coupled first to a solid phase.

The sensitivities and selectivities of MIAs are comparable to immunoassays employing biological antibodies. In a number of cases complete analytical procedures starting from raw samples (blood, plasma, or urine) have been demonstrated. Some perceived weaknesses of MIPs do not in fact hinder their application in MIAs: Many have been applied in aqueous sytems and a heterogenous distribution of binding sites is not problematic, as long as the recognition sites that bind the probe most strongly are selective.

II. IMMUNOASSAYS BASED ON BIOLOGICAL ANTIBODIES

A. Biological Antibodies

The immune systems of mammals produce antibodies in response to any detected foreign body, or *antigen*, in the bloodstream. This has been known for many years,

641

although the mechanism by which it occurs is a newer discovery. The story of how the earlier *instructional* theory, developed in the 1930s by Breinl, Haurowitz and Pauling, was recognized to be false and superseded in the 1950s by the (now accepted) *clonal selection* theory, provides an interesting case study in the history of science [9]. According to Pauling antibodies were produced by an "apo" antibody or antibody precursor, comprising an unfolded protein which folds around the antigen, producing a binding site [10]. This proposal is often considered as the intellectual dawn of Molecular Imprinting — although as an explanation of the immune response it was subsequently shown to be false. The theory of clonal selection is attributed to Burnet [11], but there were other protagonists involved including Jerne, Talmadge, and Macfarlane [12]. In the now accepted theory, B-cells or lymphocytes, each displaying an individual line of antibodies, lie dormant in the body until called to action. A few cells, producing the antibodies which bind the antigen most strongly, are amplified and mutated to produce antibodies binding more strongly, until a range of new B-cells are working, each producing plasma cells that in turn produce large excesses of their own antibody in the bloodstream [13].

This process can be exploited by the scientist, by injecting an analyte of interest into a laboratory animal, such that the analyte acts as an antigen and complimentary antibodies are produced. The antibodies may be collected from the blood of the animal (Fig. 1). Depending on the strength of the immune response, the serum may contain a sufficiently high antibody concentration to be used as an analytical reagent as is, or the antibodies can be purified from the serum by separation methods.

Antibodies collected in this way will comprise all the different types of antibody that have been produced, with varying strengths of binding and selectivities. Hence the heterogenous antibody mixture is known as a *polyclonal antibody*.

An alternative approach, yielding antibodies that are all identical, and hence called a *monoclonal antibody*, is shown schematically in Fig. 1. In order to obtain a monoclonal antibody, the scientist would like to obtain the cell producing a particular antibody and propagate this cell in vitro on a large scale via cell culture, so that (s) he would have a cell line, all producing the same antibody. Unfortunately B-cells themselves cannot readily be propagated in this way. The method requires that the B-cells be fused in vitro with cancerous cells to produce *Hybridoma* cells, which combine the antibody-generating properties of B-cells with the ability of carcinoma cells to propagate and thrive in cell culture [14]. The hybridomas are each assayed for the strength with which their antibody binds the antigen, and the strongest binders cloned and cultured. This method for the production of monoclonals earned Koehler and Milstein the Nobel Prize for Medicine in 1984.

For small molecules, the immune response is often small or nonexistent [the immune system has evolved to deal with larger invaders such as bacteria]. However, the scientist can still use the immune system of a laboratory animal to produce antibodies to small molecule analytes: (S) he need only first chemically couple the analyte to a larger molecule, a carrier protein. The analyte is now known as a *hapten* (Fig. 2). Only a minority of the antibodies produced may be selective for the hapten, rather than the carrier protein, but the number still usually exceeds that which can be obtained by injecting the analyte alone [13].

Mammals produce a number of different classes of antibodies, of which the most abundant, and most commonly employed in analytical methods, is *immunoglobulin G* or *IgG*. IgG is a protein of molecular weight 150,000 comprising four subunits

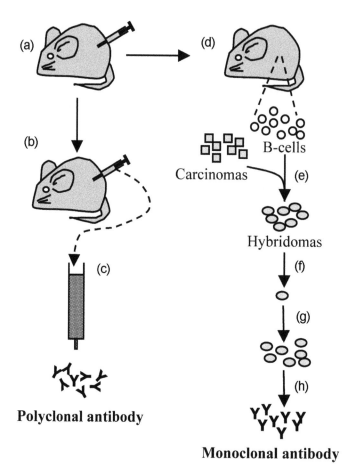

Figure 1 Schematic of production of biological antibodies. (a)–(c) illustrate production of polyclonal antibody, (a) and (d)–(h) illustrate production of monoclonal antibody. (a) inoculation (b) collect blood (c) extract antibodies (d) extract spleen cells (e) cells fused to form hybridomas (f) best antibody-producing hybridoma selected (g) hybridoma cloned (h) antibody produced and collected in cell culture.

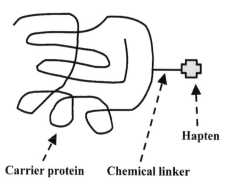

Figure 2 Schematic of a carrier protein–hapten conjugate.

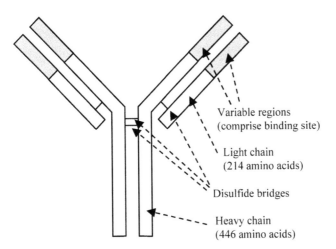

Figure 3 Schematic of structure of an IgG molecule.

connected by disulfide bridges (Fig. 3). The B-cells of a mammal can produce different IgG molecules because the DNA coding for the variable regions shown in the figure is readily mutated as the B-cells multiply. Each cell thus carries slightly different DNA coding for antibodies with slightly different variable regions. These regions form the binding sites of the IgG. Hence, antibodies are often drawn schematically as Y's, with the binding sites at the top of the branches of the Y. IgG binding sites can exhibit remarkably strong binding of their analyte, K_ds typically range from 20 μM to 10 pM under physiological conditions [13].

B. Immunoassay Formats

The polyclonal or monoclonal antibody may be exploited in many different analytical methods to compare unknown concentrations of analyte with known standards: immunoassay is not a single method but a class thereof [1–3]. Table 1 attempts to demonstrate this, different methods being classified according to the physical process behind the measurement of the analyte, and whether the antibody is employed in a homogenous format (individual molecules freely diffusing in solution) or a heterogenous format (antibody molecules physically or chemically anchored to a solid phase).

The first row includes assorted methods where the analyte binds to the antibody, is physically separated from the sample matrix, and then measured. These include immunoaffinity solid-phase extraction, where after separation of the bound antigen from the sample matrix, the antigen is subsequently released and measured, and also several methods that depend on two antibodies binding independently to the analyte, such as the immunoaffinity chromatographic methods behind several home-test diagnostic kits such as pregnancy tests.

The second row comprises methods where the interaction of analyte and antibody is detected directly via a physical measurement, such as in mass-based sensors.

The methods commonly known as binding assays are shown in the last two rows, divided into noncompetitive and competitive assays. The principles behind

Table 1 The plethora of formats of immunoassay. Ab = antibody, Ag = antigen. Noncompetitive and competitive heterogenous binding assays where the probe is labelled with an enzyme are known as enzyme-linked immunosorbent assay (ELISA)

Principle	Homogenous formats (Ab in solution)	Heterogenous formats (Ab on solid-phase)
Physical separation of bound and unbound	Agglutination assays, immunodiffusion	Immunoaffinity chromatography, immunoaffinity solid-phase extraction
Interaction generates signal		Immunosensors
Noncompetitive binding assays (labelled 2nd antibody)	Homogenous immunoradiometric assay (IRMA) Homogenous immunoenzymometric assay (IEMA)	Heterogenous IRMA Heterogenous IEMA (sandwich ELISA)
Competitive binding assays (labelled antigen)	Homogenous radioimmunoassay (RIA) e.g., Scintillation proximity assay (SPA) Homogenous enzyme immunoassay (EIA) e.g., Enzyme monitored immunotest (EMIT)	Heterogenous RIA Heterogenous EIA (competitive ELISA)

these are demonstrated in Fig. 4. In the *non-competitive* format (Fig. 4a), the sample analyte binds to one antibody, and simultaneously or in a subsequent step to another, second, antibody, this one carrying some form of label, such as a radioisotope, fluorophore, or enzyme. The antibodies in this case should be present in excess such that the sites are not saturated. In the most usual version of this approach, the first antibody is coupled to a solid phase as shown. The bound label can then be physically separated from the unbound label (by discarding the solution phase) and the bound label measured. In the widely used ELISA (enzyme-linked immunosorbent assay, Fig. 5) the label is an enzyme and the solid phase is usually a microtiter plate well. After discarding unbound second antibody–enzyme by emptying the plate wells, a substrate of the enzyme is added, which in the presence of the enzyme produces a colored product. The signal (color) produced is then proportional to the analyte concentration originally present. Heterogenous noncompetitive binding assays like ELISA are thus *separation assays*. Homogenous noncompetitive binding assays are also possible, where physical separation of bound and unbound second antibody is not needed (or indeed possible)—thus *nonseparation assays*. Homogenous IEMA (immunoenzymometric assay) formats depend simply on bringing the first and second antibodies into physical proximity as a means of generating the signal. However, all noncompetitive assays depend on an excess of antibody and produce a signal proportional to the analyte concentration.

In the *competitive* format (Fig. 4b), the sample analyte and a probe, most often a labelled form of the analyte itself, compete for a limited number of antibody binding sites: in this case there must be a deficit of antibody present. Heterogenous competitive binding assays are also usually separation assays: the bound probe is separated

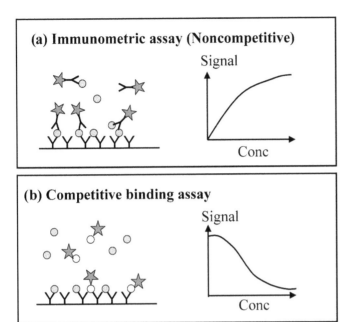

Figure 4 Schematic representation of (a) heterogenous noncompetitive and (b) heterogenous competitive immunoassays. The noncompetitive format involves the formation of antibody–antigen–antibody "sandwich". Signal is proportional to the sites occupied by analyte. The competitive format involves the analyte competing with probe (here, labelled antigen) for the available sites. Signal is proportional to sites not occupied by analyte.

from nonbound and quantified. In the radioimmunoassay (RIA), the probe consists of radiolabelled analyte and can be quantified via radiometric counting of either the bound or nonbound fraction. The bound activity is inversely related to the concentration of analyte present in the sample: the more unlabelled analyte present in the sample, the less sites are available for the labelled form. Homogenous, nonseparation forms of competitive binding assay are also possible, where the physical proximity of antibody and probe is sufficient to generate signal, such as the scintillation proximity assay (SPA). However, all competitive assays depend on a shortfall of antibody and produce a signal inversely related to the analyte concentration.

C. The First RIA

The first heterogenous RIA was reported for human growth hormone (HGH) by Catt et al. in 1967 [15] (Fig. 6). Anti-HGH antibodies were immobilized on polystyrene. Plasma containing standard or unknown HGH concentrations together with a probe of radiolabelled HGH was added to the immobilized antibodies, and the "hot" and "cold" HGH competed for the available binding sites. Finally, the radioactivity present on the polymer was measured and correlated with the cold HGH concentration: the concentration of cold HGH in unknown samples was determined by comparing the bound radioactivity with a standard curve determined for standards of known cold HGH concentrations, similar to the curve shown in Fig. 4b.

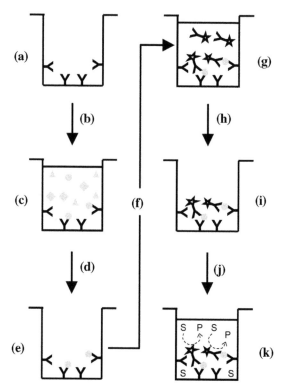

Figure 5 Schematic representation of the most widely used noncompetitive biological immunoassay, the enzyme-linked immunosorbent assay (ELISA), (a) Antibody, selective for analyte, is immobilized on microtiter plate well surface (b) sample added (c) analyte in sample binds to antibody while other compounds in matrix remain in solution (d) sample solution containing nonbound molecules discarded and wells washed (e) analyte remains bound to antibody (f) second antibody–enzyme conjugate added (g) conjugate binds to bound analyte (h) solution containing nonbound conjugate discarded and wells washed (i) conjugate remains bound (j) enzyme substrate added (k) substrate S converted to colored or fluorescent product P at a rate which is proportional to the amount of bound enzyme and hence to the concentration of analyte in the sample.

D. The First MIA

The first MIAs were reported by Vlatakis et al. in 1993 [16] (Fig. 7). Two assays were described, one for the sedative diazepam based on a diazepam-imprinted MIP, and one for the bronchodilator theophylline based on a theophylline-imprinted MIP. Unlike the biological RIA described above, sample plasma containing the drugs was not added directly to the MIP (/antibody), but rather the drug was first extracted from the plasma into organic solvent, this solvent was evaporated, the residue reconstituted in acetonitrile/acetic acid (99:1 v/v) for theophylline or toluene/heptane (3:1 v/v) for diazepam, and this organic solution was added together with radiolabelled drug to the plasma. The extra solvent extraction and solvent exchange steps were necessary because of the superior performance of the MIPs in the organic solvents. Again hot and cold drug competed for the available binding sites, and in

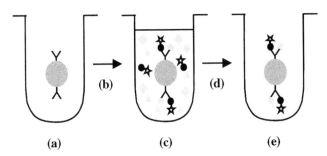

Figure 6 Schematic of the first RIA, for insulin [15]. (a) Polystyrene spheres with anti-insulin antibody immobilized on surface. (b) Sample matrix containing analyte and interferents, together with radiolabelled insulin probe, are added. (c) Analyte insulin and probe radio-labelled insulin compete for antibody binding sites. (d) Matrix containing interferents, unbound analyte and unbound probe, is discarded. (e) Radioactivity of polystyrene spheres is recorded. The more analyte present in the sample, the less radioactivity is recorded.

this case the MIP was subsequently removed by centrifugation and the radioactivity present in the supernatant was determined by liquid scintillation counting. The bound radioactivity could be calculated simply by subtracting the radioactivity remaining in the supernatant (so, unbound) from the radioactivity added. Again,

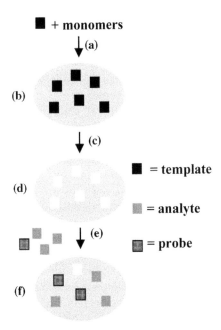

Figure 7 Schematic of molecular imprint sorbent assay (MIA). (a) molecular imprinting process, (b) imprinted polymer containing trapped template–monomer complexes (c) extraction of template, (d) analyte and probe are added to the MIP, (e) analyte and probe compete for the available binding sites. In the conventional radiolabel MIA, the analyte is identical to the template, and the probe is the radiolabelled form of the analyte. In later, alternative MIA designs, template, analyte, and probe are not necessarily identical.

comparing the bound radioactivity for unknown samples with the values for samples of known cold drug concentration enabled calculation of the concentrations of cold drug in the unknowns. The results were impressive: for theophylline, the cold drug was measured in blood in the range 14–224 µM, the assay results correlated excellently with a commercial EMIT (enzyme-multiplied immunotest) based on biological antibodies, and cross-reactivity of related molecules was similar to that of the biological antibodies.

Since this first work many more MIAs have been developed and the approach has been reviewed elsewhere [4–8].

E. Potential Advantages of MIPs

The potential advantages of MIPs as replacements for biological antibodies in immunoassays include

  i. Whereas it is difficult to produce and select antibodies for small molecules (they must be coupled to a carrier protein before inoculation, as described in Section I.A above), MIPs are most readily obtained against small molecules.
 ii. Similarly, it is hard to raise natural antibodies against highly toxic compounds or immunosupressants. But there are no additional problems in making MIPs for such analytes [17].
iii. In some cases, it may be desirable to perform binding assays in nonaqueous media. Protein antibodies are unlikely to function well in such conditions but MIPs may function better in organic solvents than in aqueous conditions.
 iv. MIPs are stable and rugged in comparison with biological antibodies.
  v. Production of MIPs, unlike biological antibodies, does not require the sacrifice of animals.The potential disadvantages include
 vi. Reliable methods for imprinting large molecules are still not available: for protein analytes then, antibodies remain the receptor of choice.
vii. Because the binding sites of MIPs (at least of conventional macroporous acrylate/vinyl-based MIPs) tend to be within the macroporous structure they are expected to be inaccessible to large molecules. This might be expected to prohibit the use of an enzyme-labelled molecule, for instance, as a probe, since the probe would be blocked from binding: However, recent work has shown that microsphere or film formats of MIP, where the binding sites are more readily accessible, can be used in binding assays with enzyme-labelled probes (Section IV.B) [18,19].
viii. Similarly, MIPs are probably not compatible with noncompetitive binding assays such as ELISA in which a first antibody (/MIP) – analyte – second antibody (/MIP) sandwich must be formed since such a sandwich for a MIP will be sterically very hindered. However, even with biological antibodies, methods like ELISA are limited to larger analytes such as proteins since the analyte must have two independent binding sites. Since MIA is better suited to small molecules, the competitive assays are likely to be preferred anyway.
 ix. A common misconception is that MIPs only work in organic solvents. In the first MIA the drug to be measured was extracted from plasma into one solvent, which was evaporated, then dissolved in another for the assay [16]. The laborious extraction and reconstitution steps required in Vlatakis et al.'s assay might be prohibitive. However, in more recent work binding assays have been

performed either directly after the first extraction step in the same solvent (which may in some applications be advantageous in that it serves to concentrate the analyte) [17], or directly in aqueous systems including diluted plasma [20] (Section III.B).

x. The inefficiency of molecular imprinting (i.e., the distribution of binding sites in an MIP, ranging from very strong and selective for the analyte to very weak and nonspecific) has been perceived as a problem, but although it may be so for other applications such as sensors, it is not in fact a problem at all for MIA: as long as the sites that bind the probe most strongly are selective (Section III.C).

xi. Another perceived weakness has been the quantity of template required to prepare a useful amount of MIP for the conduct of large numbers of assays. Dramatic advances have been achieved though in reducing the quantity of template used [21,22] and furthermore MIPs unlike antibodies can be cleaned (even by autoclave) and reused many times.

Thus, it may be seen from (i), (vi), and (viii) in particular that biological antibodies and MIPs are in many ways complementary, biological antibodies having the upper hand at present for immunoassays for large molecules such as proteins, but MIPs having several advantages for assays for small molecules. (ix), (x), and (xi) are only perceived disadvantages and in fact need not be so.

III. SETTING UP AN MIA

It is important to distinguish initially the *template*, which is used to produce the MIP, the *analyte*, and the *probe* which competes with the analyte for binding to the MIP and of which the bound fraction can be quantified (Fig. 7). In this section a typical procedure is outlined for establishing a conventional MIA, such as those developed by Vlatakis et al. [16], where the analyte is used as template, and the probe is a radio-labelled form of the analyte. Essentially, similar approaches have been used to set up most of the MIAs listed in Table 2. Other approaches will be discussed later.

First, the binding of probe alone to the MIP must be optimized. The ideal conditions for an MIA (at least in the case where probe and analyte are chemically identical) will be such that there is a large excess of binding sites compared to the amount of probe, but the solvent strength is such that only about 50–80% of the probe (in the absence of any competitor) binds to the MIP—and only to the strongest binding sites as shown in Fig. 8. If too little MIP is used, sufficient binding of the probe can only be achieved with a solvent in which the nonspecific sites are also employed and hence the MIA will be less specific. If a solvent is used in which the binding interactions are extremely weak, an impractically large amount of MIP will be required for each assay. Thus, the solvent and amount of MIP have to be optimized in parallel. Further, the amount of MIP required such that the 50–80 % bound probe only occupies good binding sites will inherently depend on the quality of the imprinting — if the imprinting is poor and only a few good recognition sites are present, it will be necessary to use more polymer than with an optimized MIP with many high-fidelity sites.

Having stated in Section I.B that competitive immunoassays rely on a shortfall of binding sites, it may appear contradictory now to suggest that an excess of MIP binding sites is required. The reason of course is that in the situation described as

Table 2 Reported MIAs employing radiolabelled probes

Template	Probe	Assay solvent	Competitors	Reference
Diazepam	³H-diazepam	MeCN/AcOH (99:1)	Diazepam, related diazepine derivatives	[16]
Theophylline	³H-theophylline	(a) MeCN/AcOH (99:1), (b) Toluene/THF (9:1), (c) AcN, (d) Toluene	Theophylline, theobromine, caffeine, related xanthines	[16,22,30–33]
Octyl-α-d-glucoside	¹⁴C-Methyl-α-D-glucoside	MeCN/AcOH (199:1)	Methyl-α-D-glucoside, other sugar derivatives	[34]
Atrazine	¹⁴C-atrazine	(a) MeCN, (b) toluene, (c) phosphate pH 7/Tween 20 (9985:15)	Atrazine, related triazines	[35,36]
Morphine	³H-morphine	(a) Toluene, (b) citrate pH 6/EtOH (9:1)	Morphine, related opiates	[21,37]
Leu-enkephalin anilide	³H-Leu-enkephalin	(a) MeCN, (b) citrate pH 4.5/EtOH (9:1)	Leu-enkephalin, related peptides	[37]
S-propranolol	³H-S-propranolol	(a) Toluene, (b) various aqueous buffers, (c) 60% plasma	S-propranolol, R-propranolol, racemates of related drugs	[25,38–41]
Cortisol	³H-cortisol	THF	Cortisol, related steroids and sterols	[42]
Cortisone	³H-cortisone	THF	Cortisone, related steroids and sterols	[42]
Yohimbine	³H-yohimbine	(a) MeCN/AcOH (199:1), (b) phosphate pH 5.0	Yohimbine, corynanthine	[43]
Cyclosporin A	³H-cyclosporin A	Diisopropyl ether	Cyclosporin A, metabolites, unrelated drugs and proteins	[17]
2,4-D	¹⁴C-2,4-D	Phosphate pH 7/Triton X-100 (999:1)	2,4-D, related acids and esters	[18,23]

(Continued)

Table 2 (*Continued*)

Template	Probe	Assay solvent	Competitors	Reference
17β-estradiol	³H-17β-estradiol	MeCN	17β-estradiol and related diols	[30,31,33]
Caffeine	¹⁴C-caffeine	(a) Heptane/THF (3:1), (b) MeCN	Caffeine, theophylline	[26,31,44]
NacGIIIB	¹⁴C-NacGIIIB	H₂O? Unspecified	NacGIIIB, related peptides	[45]
17α-ethynylestradiol	³H-17α-ethynylestradiol	Toluene	17α-ethynylestradiol and related steroids	[46]
Bupivacaine	³H-bupivacaine	(a) Toluene/AcOH (995:5), (b) MeCN/H₂O (2:8), (c) MeCN/H₂O (9:1), (d) Citrate buffer pH 5/ethanol/tween 20 (9495:500:5)	Bupivacaine, related local anaesthetics	[47]
4-Nitrophenol	¹⁴C-4-nitrophenol	MeCN	4-Nitrophenol	[48]

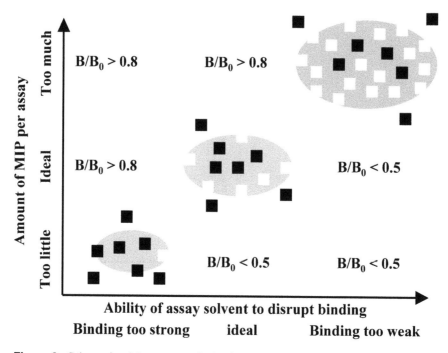

Figure 8 Schematic of the process of selecting the ideal conditions for probe binding to MIP for a MIA. The figure represents a series of conditions investigated with in each case the same amount of total probe, the aim being to optimize the amount of polymer and assay solvent. MIP particles are represented by gray ovals. An increase in amount of MIP is represented as an increase in the size of MIP particles for simplicity. Along the diagonal, conditions are such that $B/B_0 = 4/7$, which is in the range desired. At the lower left-hand corner, too little MIP is used and although the solvent is such that binding interactions are strong and $B/B_0 = 4/7$, some of the binding sites occupied by the probe are poorly imprinted, weak, and nonselective. These conditions will give an MIA with poor selectivity and sensitivity. At the top right-hand corner, too much MIP is used and although the solvent is such that binding interactions are weak and $B/B_0 = 4/7$, these conditions would give a good MIA but are wasteful of MIP. The conditions in the center are ideal: $B/B_0 = 4/7$, but the probe binds only to the best imprinted, strongest, and most selective sites.

ideal in Fig. 8 there is a shortfall of *high-affinity binding sites*. These are the only sites of interest in the MIA—if analyte or an interferent is present and displace the probe from these sites, it will not bind to the weaker recognition sites, but will be truly displaced. We can consider just the high-affinity sites to be examined by the probe and the weaker ones ignored.

A. Preparation of the MIP

The MIP can be produced in whatever format is desired, though most MIAs reported are still based on conventional acrylate/vinyl based particulate polymers. A reference polymer, which may be a nonimprinted polymer or NIP made similarly to the MIP but without template or a control polymer or CIP imprinted with a very different template but otherwise similar to the MIP, is also required for assay optimization. It is important to wash the MIP (and CIP) thoroughly in order to remove as much of the template as possible before use: the strongest binding sites are likely to be the most selective, and are preferred for the MIA, but if they are occupied by unremoved template the analyte and probe will compete only for the weaker and less-selective sites. Washing can be achieved by Soxhlet extraction or repeated flushing (on a filter, or via suspension, centrifugatuion and decanting) with a solvent system capable of displacing the template—for methacrylic acid (MAA)-based polymers, acetic acid/ammonium acetate in ethanol may be used. Finally, to remove traces of the extracting solvent (which will lessen the affinity of probe and analyte for the MIP), extensive washing is repeated with a solvent such as methanol, and the MIP is dried.

Since only small quantities of MIP (mg) may be required for each assay, if particles are to be used it is preferable that they are as small as possible, in order that they can be transferred accurately by making a suspension of known weight per volume and pipetting small amounts thereof [23].

B. Initial Choice of Solvent

A solvent must be chosen for the initial experiments (even if the solvent is later to be optimized as in Section II.E): if the polymer has been produced by noncovalent imprinting, it is usual to begin by studying binding in a nonaqueous solvent. The aim is to maximize the proportion of the probe binding to highly selective, well-imprinted sites (specific sites) and minimize the proportion binding to nonspecific sites. If the template/analyte has a number of groups which are expected to interact strongly with the polymer then a moderately polar solvent such as acetonitrile, even with a modifier to reduce the strength of the interactions, may be chosen in order to minimize nonspecific binding. If on the other hand, the template/analyte includes few groups capable of interacting, it may be better to choose a very nonpolar solvent such as toluene to maximize the binding interactions.

If the assay is to be optimized for performance in an aqueous solution, it is usual to begin with a buffer solution such as phosphate pH 7. Since the MIP itself is often relatively hydrophobic, its surface may be poorly wetted if aqueous buffer alone is used as the solvent (this is certainly true of conventional macroporous MIPs crosslinked with large fractions of ethyleneglycoldimethacrylate (EDMA) or divinylbenzene (DVB)). Thus, a small amount of a nonionic surfactant such as Triton X-100 (c 0.5% w/v) or a miscible organic solvent such as ethanol (c 5% v/v) is usually added to reduce the surface tension. For MIPs based on EDMA or DVB, the additive also serves to reduce hydrophobic interactions, which may be less specific than hydrogen bonds or ion pairs.

C. Calculating the Amount of Probe per Experiment

Conventional MIAs are most usually performed in c 1 mL volumes in plastic microtubes. Eventually, after incubating the MIP with the analyte and probe, the tubes will be centrifuged and a fixed amount (e.g., 0.8 mL) of the supernatant removed for liquid scintillation counting. To minimize expense and exposure to radioactivity and to maximize sensitivity, it is desirable to add as little probe as possible to each assay. However, the practical minimum amount can be determined: in the final assay, very low concentrations of analyte should correspond to 50–80% of the probe remaining bound (i.e., 20–50% in the supernatant, or 16–40% in the 0.8 mL). High concentrations of analyte in the final assay will correspond to less than 50% bound (so more than 16–40% in the 0.8 mL). Thus, the amount of probe added per assay should be such that 16% thereof can be determined with relative uncertainty/standard deviation of less than 5%.

D. Establishing the Binding of the Probe

To establish the binding of the probe to the MIP, a series of experiments are set up, containing the calculated amount of probe and varying amounts of MIP in the 1 mL of chosen solvent, for instance nine different amounts in the interval 3 μg to 30 mg. This is most often achieved by weighing out quantities in excess of 5 mg, while smaller quantities are measured out by pipetting appropriate volumes of for instance a well-stirred 5 mg mL^{-1} suspension of polymer particles in the assay solvent. A solution of the probe in the assay solvent is made up and a volume corresponding to the amount calculated above is added to each experiment. The volume in each microtube is made up to 1 mL with assay solvent. Three experiments at each polymer concentration for both the MIP and the control NIP are set up, and three controls with probe but no polymer. The 57 microtubes are then incubated usually overnight on a rocking platform at room temperature.

After incubation, the tubes are centrifuged at a minimum of 13000 rpm for 15 min (more centrifuging may be necessary for smaller particles). 0.8 mL of supernatant is withdrawn, added to scintillation fluid and counted in a scintillation counter. For each experiment, the fraction of probe bound is calculated as

$$B/T = (\text{control (no polymer) cpm} - \text{sample cpm})/\text{control (no polymer) cpm}$$
$$(1)$$

where B is the amount bound and T the total amount of probe. The average B/T at each concentration for both MIP and CIP or NIP is plotted as in Fig. 9. The figure

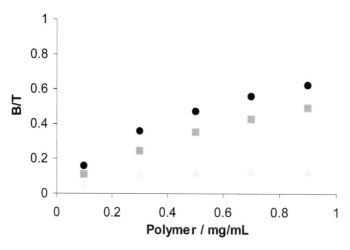

Figure 9 Graph of B/T vs. amount of polymer for ^{14}C-atrazine binding to an atrazine MIP in 25 mM citrate buffer pH 6.0/EtOH (9:1 v/v) [24]. Circles, MIP; squares, control polymer (caffeine-imprinted, CIP). Triangles: Specific binding (B/T(MIP) - B/T(CIP)).

shows binding of ^{14}C-atrazine to an atrazine-imprinted MIP and equivalent caffeine-imprinted CIP in toluene [24].

For the final assay procedure, an B/T of 0.5–0.8 for the MIP, in the absence of added analyte, is ideal.

E. Optimizing the Assay Solvent

Once the appropriate amount of MIP to give B/T c 0.5 is known, the solvent can be further optimized. It is desirable to maximize the difference in B/T between the MIP and CIP or NIP, binding to the latter being entirely nonspecific. The difference is considered to represent the fraction of specific binding:

$$\text{specific binding fraction} = B/T(\text{MIP}) - B/T(\text{CIP or NIP}) \qquad (2)$$

Hence if B/T for the MIP is 0.5 but B/T for the NIP at the same amount of polymer and in the same solvent is 0.49, this indicates the binding of the probe to the MIP is almost entirely nonspecific, and there is little point proceeding further without changing the solvent. While if B/T for the NIP under the same conditions is 0.1, most of the binding to the MIP is probably occurring at selective-imprinted sites.

Thus, a series of assays are set up with the MIP and probe at the concentrations determined in Sections II.C and II.D, but in slightly differing solvents. Assays are again set up in triplicate, together with identical assays with NIP replacing MIP and with controls containing no polymer for each of the assay solvents, incubated overnight, centrifuged and the supernatant measured for activity. B/T for MIP and NIP in each solvent are calculated using Eq. (1) and compared. If the MIA is being optimized to work in aqueous media, the variables that are examined may include pH, buffer salt, ionic strength, and the amount of miscible organic solvent/detergent which is added to enable wetting of the MIP. Fig. 10 shows B/T for a S-propranolol MIP and an NIP incubated with ^{3}H-S-propranolol in aqueous buffer at different

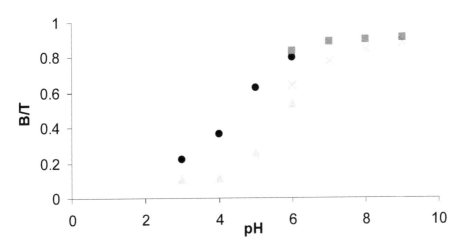

Figure 10 Graph of B/T vs. buffer pH for an aqueous MIA for S-propranolol in 25 mM citrate or phosphate buffer/EtOH (9:1 v/v) [25]. Circles, MIP in citrate buffer; triangles, NIP in citrate buffer; squares, NIP in phosphate buffer; crosses, NIP in phosphate buffer.

pH. The greatest difference in binding to the MIP and the NIP is at pH 5.0 [25], hence this pH was chosen for further optimization.

After optimizing the solvent, B/T (MIP) in the optimized solvent may be below 0.5 or above 0.8, in which case the amount of polymer employed in each experiment should again be varied until B/T (MIP) is in the desired range.

F. The Competition Assay with Analyte and Possible Interferents

The competition assay is usually performed only with the MIP: the quantities of MIP and probe determined above are added to microtubes, followed by varying amounts of analyte or interferent, and the volume made up to 1 mL with assay solvent. The concentrations of analyte or interferent in the assays should vary over as wide an interval as possible, for instance, 15 different concentrations in the interval 1.0 nM to 10 mM (i.e. six decades of concentration). Each assay should be set up in triplicate, and in addition controls with probe only (no MIP, analyte, or interferent) and with MIP and probe only (no analyte or interferent). Ideally, a range of possible interferents should be examined, compounds structurally related to the analyte and also compounds that would be present at significant concentrations in a real analytical application of the assay: for instance, if the MIA is being developed to measure a drug in plasma, common plasma proteins and nutrients should be considered. However, it is often considered sufficient to demonstrate the potential of an MIA by considering only one or two close structural analogues: e.g., if the analyte is chiral, it may be sufficient to study the analyte and its enantiomer. If the assay distinguishes the two, this does not necessarily demonstrate the MIA is a feasible analytical method, but it does show the MIP contains selective binding sites.

The 45 microtubes for the analyte and for each interferent plus the six controls are then incubated overnight, centrifuged and the supernatant measured for activity. The amount of bound probe, B, is calculated first for the controls with probe and MIP but no analyte or interferent, the average of these controls is called B_0:

$$B_0 = \text{counts for control with probe only} \\ - \text{counts for control with probe and MIP} \tag{3}$$

Then for each concentration of analyte or interferent the ratio of bound probe to B_0 is calculated:

$$B/B_0 = \frac{\substack{\text{counts for control with probe only} \\ - \text{counts for assay with analyte/interferent}}}{\substack{\text{counts for control with probe only} \\ - \text{counts for control with probe and MIP}}} \tag{4}$$

B/B_0 is plotted against the concentration of analyte or interferent as shown in Fig. 11. The figure indicates that as the concentration of analyte, caffeine in this case, is increased, it competes more effectively with the probe and more is displaced from the MIP so B falls [26]. Furthermore, the interferent, theophylline in this case, also serves to compete with the probe and depress B, but only at higher concentrations than the analyte: theophylline is less effective at displacing the probe than is caffeine.

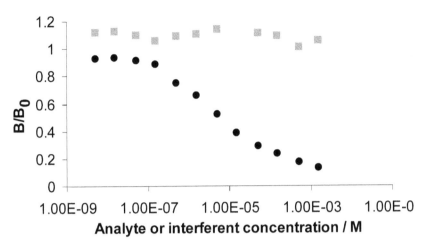

Figure 11   Results of a MIA for caffeine performed in organic solvent (heptane/THF 3:1 (v/v)) [26]. Curves of B/B_0 vs. competitor concentration. Competitor = unlabelled caffeine (squares) or theophylline (circles). Assays performed in 1 mL volume using 8 mg of anti-caffeine MIP. It can be observed that the IC_{50} value for caffeine is about 3 µM while theophylline even at 1 mM concentration does not displace 50% of the bound probe.

If the decrease in B observed as the concentration of analyte is increased is very small, i.e., B/B_0 does not fall below 0.5 even at very high analyte concentration, the MIA is poor and unlikely to be a useful method for measuring real samples. The MIP and/or assay solvent should be redesigned. Similarly, if interferents depress B more easily than does the analyte, the MIA is nonspecific and again the MIP and/or assay solvent should be redesigned.

To obtain quantitative measures of the sensitivity and selectivity of the assay, the IC_{50} values of the analyte and interferents may be calculated.

$$IC_{50} = \text{concentration of analyte or interferent at which } B/B_0 = 0.5 \qquad (5)$$

These values may be estimated from a graph such as Fig. 11. For a greater degree of accuracy, the graph can be replotted in log/logit form, with log(concentration) on the x-axis and $\text{logit}(B/B_0)(= \log((B/B_0)/(1 - B/B_0)))$ on the y-axis. This transformation should produce a straight line as in Fig. 12, from which the IC_{50} value may be calculated by fitting the points to a straight line and taking the antilog of the x-intercept

$$IC_{50} = \text{antilog of } x - \text{intercept of } \text{logit}(B/B_0)\text{vs. log(concentration) plot}$$

$$(6)$$

From Fig. 12, the IC_{50} for the analyte caffeine in this MIA is antilog(−5.070) = 8.5 µM. The cross-reactivity of interferents may be expressed as the ratio of IC_{50} values as shown in Eq. (7)

$$\text{Cross-reactivity} = IC_{50}(\text{analyte})/IC_{50}(\text{interferent}) \qquad (7)$$

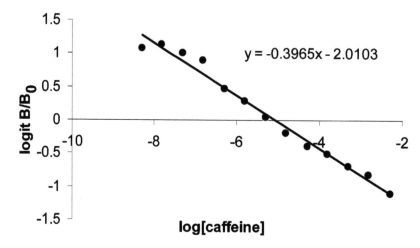

Figure 12 Log–logit plot of the data for caffeine from the experiment described in Fig. 11.

In the MIA of Fig. 11, the IC_{50} value for the interferent theophylline cannot be reliably determined but certainly lies above 1 mM Thus the cross-reactivity of theophylline in this MIA is $< \%0.6\%$.

G. Assessing the Heterogeneity of Binding Sites

When the probe is the radiolabelled form of the analyte, and so the two are chemically identical, further information about the heterogeneity of the MIP binding sites can be obtained from the results of the assays with MIP, probe and analyte in Sec. III.F above. This is based on the assumption that

$$\text{bound probe/total probe} = B/T = \text{bound analyte} + \text{probe/total analyte}$$
$$+ \text{probe} = \text{bound analyte/total analyte} \qquad (8)$$

This is not valid if the probe is in any way chemically different from the analyte. In any case, an analysis such as the following is not essential to the development of a MIA, but is often performed in order to gain an estimate of binding site numbers and affinities.

Using Eq. (8), the average amount of bound analyte and concentration of free analyte can be calculated for each total analyte concentration. The plot of amount of bound analyte, n_B, vs. concentration of free analyte [free analyte] is the binding isotherm (Chapter 16) (Fig. 13). Although since probe and analyte are chemically identical, the isotherm should strictly show bound analyte + probe vs. free analyte + probe, the amount of probe is usually so small in relation to even the lowest concentration of analyte that it can effectively be ignored.

The values of n_B vs. [free analyte] can be fitted to a number of models. The simplest model of adsorption, the Langmuir isotherm, which assumes all binding sites are identical, is

$$n_B = \frac{d \times m[\text{free analyte}]}{1/K_a + [\text{free analyte}]} \qquad (9)$$

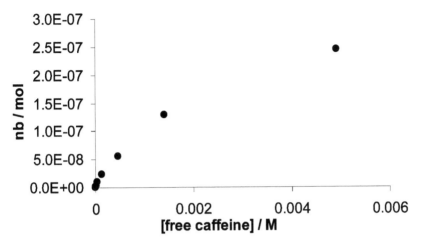

Figure 13 Binding isotherm for caffeine binding to the MIP in the experiment shown in Fig. 11, derived from the B/T ratios and the added "cold" caffeine concentration.

where n_B is the amount of bound analyte in mol, d the density of binding sites in mol g^{-1}, m the mass of MIP in g and K_a is the association constant. The Langmuir isotherm can also be expressed as

$$\frac{n_B}{[\text{free analyte}]} = K_a \times d \times m - K_a \times n_B \qquad (10)$$

Thus, a plot of n_B / [free analyte] (y-axis) vs. n_B (x-axis) should yield a straight line with gradient $-K_a$ and y-intercept $K_a \times d \times m$. This is called a Scatchard plot. Fig. 14 shows a Scatchard plot for the caffeine MIP data derived from Fig. 13. The data clearly do not fit a straight line, and similar results are usually determined for binding isotherms measured with noncovalently imprinted MIPs. The curved data indicate the binding sites are not identical, so the simple Langmuir isotherm is inappropriate.

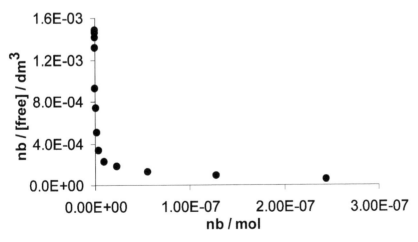

Figure 14 Scatchard plot for caffeine binding to the MIP in the experiment shown in Fig. 11, derived from the B/T ratios and the added "cold" caffeine concentration.

In reality, a noncovalent MIP will contain a distribution of binding sites of continuously varying K_a values, from very small (weak sites) to very high (strong sites). This has been recognized since the earliest work with noncovalent MIPs, and is responsible for effects such as the tailing often observed when MIPs are used as stationary phases in chromatography. It has led to the analogy between MIPs and polyclonal antibodies.

A variety of alternative models more realistic than the Langmuir isotherm may be used to fit the isotherm data: the most accurate but most demanding (in terms of amount and accuracy of data required) is the affinity spectrum approach, which yields a continuous spectrum of number of binding sites vs. K_a [27]. The simplest approach, however, is to model the isotherm as resulting from just two classes of binding site:

$$n_B = \frac{d_1 \times m[\text{free analyte}]}{1/K_{a1} + [\text{free analyte}]} + \frac{d_2 \times m \times [\text{free analyte}]}{1/K_{a2} + [\text{free analyte}]} \tag{11}$$

this is the bi-Langmuir isotherm and d_1 and K_{a1} are the site density and association constant for one class of sites, d_2 and K_{a2} for the second class. Data such as in Fig. 13 may be fit directly to Eq. (11) using a fitting routine. In this way we obtain for the caffeine MIA data in Fig. 12, $d_1 = 1.09 \pm 0.15\,\mu\text{mol g}^{-1}$ $K_{a1} = 3.3 \pm 1.2 \times 10^5\,\text{M}^{-1}$, $d_2 = 48.8 \pm 0.4\,\mu\text{mol g}^{-1}$ and $K_{a2} = 311 \pm 8\,\text{M}^{-1}$. The site populations calculated are usually significantly lower than the theoretical site density on the basis of the amount of template originally included in the MIP mixture (in the above case this was $151\,\mu\text{mol g}^{-1}$). This reflects the choice of conditions such that the probe binds only to the few strongest sites as indicated in Fig. 8.

Some authors have derived an estimate of the parameters d_1, K_{a1}, d_2 and K_{a2} by fitting two tangents to the data in the Scatchard plot (Fig. 14). Each tangent is treated as though it were the Scatchard fit for a Langmuir isotherm for a single class of sites: data points on the left-hand side of the graph (low n_B) correspond to data for assays with low concentrations of analyte and are considered to be largely due to a small population of strong binding sites (d_1 and K_{a1}), those on the right-hand side correspond to data for assays with high concentrations of analyte and are considered to be largely due to a large population of weak binding sites (d_2 and K_{a2}). However, this approach does not result in correct estimates of the parameters (see Ref. [28] and references cited therein), although the estimates obtained are often adequate initial estimates to input into modelling software.

H. Validating the Method with Real Samples

In practice this step has often been omitted where proof-of-principle of an MIA for a new analyte, or of a new MIA format, has been the main goal. Only in a few cases has a full assay procedure for real samples been developed, all involving drug measurements in serum, plasma, or urine and most the work of Andersson [16,17,20,29].

The first full procedures developed relied on extraction of the drug from plasma into an organic solvent. This solvent was either evaporated so the extracted residue could be redissolved in the assay solvent [16] or was chosen to be the assay solvent itself [17]. In both these cases, the assay solvent and MIP quantity per assay could be optimized separately, as in Sections II.D and II.E, then the extraction step optimized. The latter could be done via taking a plasma sample spiked with a known quantity of drug, employing different extraction conditions (amounts of plasma and solvent, mixing time, temperature), performing the MIA on the extracted samples and with controls as optimized in Sections II.D and II.E and measuring B/B_0. The extraction conditions corresponding to the lowest B/B_0 (i.e., the most efficient extraction) could then be used for the full MIA. In the full MIA, drug-free plasma spiked with different known quantities of drug would be used as standards, and plasma samples with unknown drug concentrations compared. For the standards, a calibration curve could be drawn in the form of B/B_0 against drug concentration in the plasma, or, better, logit(B/B_0) against log(drug concentration) which can be fitted to a straight line. The unknowns are determined from the calibration graph and the values obtained compared against those obtained via another method.

Only in one case [20] has the full MIA procedure been developed for measurement directly in plasma/urine. The plasma/urine is in fact diluted by the addition of MIP suspension and probe, so the amount of MIP and composition of the plasma/diluent mixture (the relative volumes of plasma, MIP suspension and probe solution, the pH, buffer salt, and ionic strength of the diluents and amount of surfactant/miscible cosolvent) can be optimized as in Sections II.D and II.E. The optimization of probe binding is performed using drug-free plasma, to ensure $B/T = 0.5$–0.8 in the absence of added analyte/interferent and the binding of the probe is specific as indicated by Eq. (2): thus, Bengtsson et al. [20] found that with 50 µg of polymer, when 600 µL of plasma were added to 200 µL of MIP suspension in 400 mM phosphate buffer pH 7.4 and 200 µL ^{3}H-S-propranolol in ethanol-water (1:1 v/v), B/T for MIP and NIP were approximately 0.58 and 0.14, respectively, suggesting an adequate amount of specific binding to the MIP. For the urine samples, slightly different conditions were required. The full MIA was then developed with spiked plasma standards and real samples as described above.

The work of Levi et al. involved measurement of chloramphenicol in bovine serum but was based on a nonstandard MIA format (Section IV.C) [29].

IV. CONVENTIONAL RADIOLIGAND MIAs

Since the initial work by Vlatakis et al. [16], great advances have been made in applying MIA to diverse analytes (Table 2) [16–18], [20–23,25,26,30–48] and in improving the basic design, to enable assays in aqueous solvents and directly in real samples, to reduce the quantity of MIP required, and improve the selectivity and sensitivity. This work is reviewed in the following section. The number of groups working on the development of MIAs has expanded rapidly. Progress has been made on the development of new MIP compositions and morphologies, both "traditional" MIPs and alternative-imprinted materials, on alternatives to radiolabelled probes which may be more readily accepted by the analytical community, and on assays in different formats. The latter work is reviewed in Section IV.

A. In Organic Solvents

"Traditional" noncovalent MIPs employ functional monomers such as methacrylic acid (MAA) or 4-vinylpyridine (4-Vpy) and cross-linking monomers such as divinyl benzene (DVB), ethylene glycol dimethacrylate (EDMA), or trimethylolpropane trimethacrylate (TRIM) (chapters 3 and 7). Molecular recognition relies on a combination of weak noncovalent interactions to create the binding sites during MIP fabrication and subsequently to rebind the template. Since these interactions are strongest in nonpolar organic solvents, solvents such as toluene and acetonitrile are commonly used for MIP synthesis and are expected to be optimal for MIP application also. Thus, most early MIAs were performed in such solvents. Where MIAs were applied to "real" samples such as plasma, these were first extracted into an organic solvent [16].

The use of an organic solvent as the assay medium should not automatically be considered as a disadvantage. Many medical and environmental analytes may be present at such low concentrations that a preconcentration and clean-up step is required for any assay including biological immunoassay. For nonpolar molecules, liquid–liquid extraction may be the most convenient method.

Since the initial demonstration of the principle using theophylline and diazepam [16], organic-phase MIAs based on 'traditional' MIPs have been developed for a number of other drugs and medical targets including morphine [21,37], Leu-enkephalin [37], cortisol and cortisone [42], yohimbine [43], caffeine [27,31,44], and bupivacaine [48]. In each case selectivity over related compounds has been demonstrated, and limits of detection are typically in the 100 nM to 1 µM range. Most of these studies were not optimized, but a thorough study was made of an MIA for S-propranolol and the detection limit lowered to 5.5 nM [38].

If a liquid–liquid extraction clean-up step is employed, it would clearly be most advantageous to be able to perform the MIA directly in the solvent into which the drug has been extracted. This was demonstrated in an MIA for cyclosporin where plasma was extracted into diisopropylether, then MIP and radiolabelled cyclosporin were added directly [17]. This MIA was also noteworthy because cyclosporin is an immunosuppressant drug, so that preparation of biological antibodies is not trivial. The assay had an impressive limit of detection of 4 nM, but although nonrelated structures exhibited minimal cross-reactivity, four first generation metabolites of the parent drug cross-reacted to 100%. This reflects the complexity of the imprint species, which is a cyclic peptide of molecular mass 1101 and much larger than the small molecules typically considered good imprinting targets. Thus, when applied to patient blood samples and compared with a commercial EMIT method, the MIA consistently overestimated the cyclosporin concentration. However, because only the metabolites cross-reacted to any great extent, the MIA could be considered as complimentary to the EMIT method, in that the difference between the two measurements provided information regarding the concentration of drug metabolites which could also be clinically relevant.

Organic phase MIA has also been applied to the herbicide atrazine, the groups of Stanker [35] and Mosbach [36] publishing similar assays simultaneously. In both cases, selectivity was demonstrated over related substituted triazines. The former assay performed in acetonitrile had a detection limit of 4.6 µM, the latter performed in toluene 250 nM. Other analytes of environmental interest for which conventional radiolabel MIAs have been developed include 2,4-dichlorophenoxyacetic acid [23] and 4-nitrophenol [48], although in the latter case the influence of interferents was not studied.

The stability of MIPs, in comparison to biological antibodies, was demonstrated in a recent study on a theophylline MIP, of identical composition to that used by Vlatakis et al. in the original MIA [44]. The affinity of ^{3}H-theophylline for the MIP in acetonitrile/acetic acid (99:1 v/v) was shown to be essentially unaffected by exposure of the MIP to temperatures up to 150°C, to 5 M HCl or 15% ammonia solution: however, the selectivity of the MIP after treatment was not assessed.

B. In Aqueous Solvents

For many applications, an extraction step into organic solvent is unwarranted and assays would be simpler, and more widely accepted, if they could be performed direct in the sample matrix, which whether for medical or environmental applications is usually aqueous. Thus, from the start there has been a driving force to adapt MIAs to aqueous solvents or cosolvent mixtures. In applications such as chromatography and sensors, traditional MIPs have been harder to adapt to aqueous conditions: This is in part because specific (polar) interactions between good-imprinted sites and analyte are weakened, and in part because nonspecific (hydrophobic) interactions between other small molecules and the polymer matrix are strengthened. However, for MIA nonspecific interactions at weak binding sites are actually relatively unimportant provided the strongest recognition sites retain their selectivity (Sections II.G and III.C). Thus, in most cases MIAs have been adapted to aqueous conditions successfully.

Early demonstrations of aqueous MIAs included those for morphine, Leuenkephalin, atrazine, and S-propranolol [36,37] (Table 2). In each case, the MIA was first developed in organic solvent, then investigations were conducted to find aqueous pH and additive conditions such that the difference between specific binding and nonspecific binding (as measured by B/T (MIP) − B/T (NIP)) was maximized. This was generally found to require a pH where the analyte and polymer carried complementary charges (often pH 5–6 for MAA-based polymers and basic analytes) and the addition of ethanol as cosolvent (1–10%) or a surfactant (0.1–1%) to wet the polymer surface. Sensitivities and selectivities comparable to the organic assay were usually achieved. The most thorough investigation of an aqueous MIA was made for S-propranolol [38]. Both specific and nonspecific binding was reduced by increasing ethanol concentration and by increasing ionic strength. Optimum conditions for the aqueous assay were found at 25 mM sodium citrate pH 6.0 containing 2% (v/v) ethanol. The IC$_{50}$ and limit of detection values were similar as in toluene/acetic acid, but interestingly a different pattern of selectivity was observed (Table 3). Compared to the organic MIA, lower enantioselectivity but increased species selectivity is

Table 3 Cross-reactivities observed in organic and aqueous MIAs for S-propranolol [38]

Assay and competitor	Cross-reactivity (organic MIA)	Cross-reactivity (aqueous MIA)
S-propranolol MIA	Toluene/AcOH (199:1 v/v)	25 mM sodium citrate pH 6/EtOH (49:1 v/v)
S-propranolol	100	100
R-propranolol	1.5	17.3
R,S-atenolol	18.1	< 0.1
R,S-metoprolol	6.3	0.7
R,S-timolol	0.8	0.3

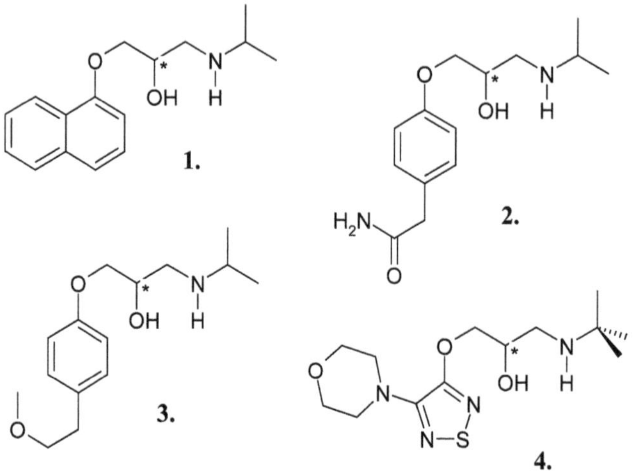

Figure 15 Propranolol and related β-blockers studied in MIAs. 1, propranolol; 2, atenolol; 3, metoprolol; 4, timolol.

observed (i.e., *R*-propranolol cross-reacts more but the related atenolol, metoprolol, and timolol cross-react less) (Fig. 15). This can be understood because the recognition sites that bind the probe strongest are likely to be different in the different solvents, and so exhibit different patterns of selectivity. In organic solvents, the polar interactions are most responsible for recognition and alterations around the amine and hydroxyl groups of the molecule are less likely to be tolerated. In aqueous solvents, hydrophobic contacts between the aromatic group and the polymer at the recognition sites are also likely to be important, and the emphasis on the position of the hydroxy and amine groups is reduced while molecules not having the same aromatic moiety will be less tolerated. Similar changed patterns of selectivity have been observed for a morphine MIA [37] and a bupivacaine MIA [47]. This illustrates that the selectivity of an MIA may be tailored by judicious choice of the solvent.

The *S*-propranolol MIA was also applied to real blood plasma and urine samples. Samples were diluted with ethanol and phosphate buffer, and MIP and ^{3}H-*S*-propranolol added directly. The assays measured plasma and urine concentrations in the range 20–1000 nM with accuracies of 89–107% and 91–125% and precisions of 3–13% and 1–7%, respectively.

Bupivacaine possesses a tertiary amine group in a cyclohexyl ring with an *n*-butyl side chain (Fig. 16). Analogues with shorter or longer side chains were less successful at displacing ^{3}H-bupivacaine in both organic and aqueous MIAs [47]. The discrimination against molecules with longer side chains can be readily understood in terms of steric exclusion. The discrimination against molecules with shorter side chains, however, can only be due to van der Waals/hydrophobic interactions contributing to the binding. Interestingly, the cross-reactivity of mepivacaine, ethycaine, and ropivacaine is similar in toluene (22–42%), whereas in the aqueous MIA, the cross-reactivity decreases dramatically with the length of the chain (7% for ropivacaine, 0.4% for mepivacaine). The authors suggest this confirms the importance of the hydrophobic effect for binding in the aqueous MIA. Further evidence for this suggestion was sought via a variable temperature study—the hydrophobic contribution

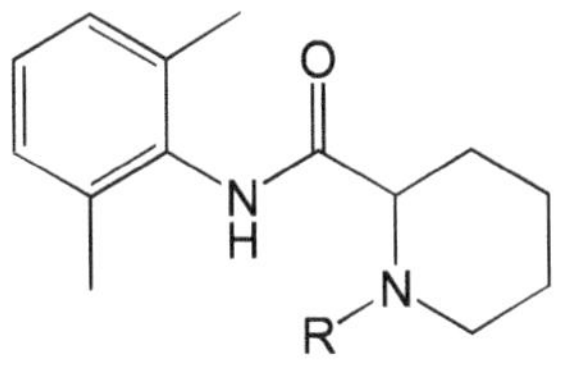

R= methyl   Mepivacaine
   ethyl   Ethycaine
   n-propyl   Ropivacaine
   n-butyl   Bupivacaine
   n-pentyl   Pentycaine

Figure 16 Bupivacaine and analogues studied by in a MIA for bupivacaine by Karlsson et al. [47].

to binding, being entropy-driven, should increase with temperature, but the results are somewhat inconclusive.

C. Binding Site Heterogeneity

For noncovalent-imprinted polymers, binding site heterogeneity (/polyclonality) arises largely from the preequilibria between template and different template–monomer complexes in the prepolymerization mixture:

$$
\begin{aligned}
T + M &\xrightarrow{\beta_1} TM \\
T + 2M &\xrightarrow{\beta_2} TM_2 \\
\vdots \quad &\vdots \quad \vdots \quad \vdots \\
T + (n-1)M &\xrightarrow{\beta_{n-1}} TM_{n-1} \\
T + nM &\xrightarrow{\beta_n} TM_n
\end{aligned}
\tag{12}
$$

where T is the template, M the monomer, the complexes TM to TM_n represent the variety of stoichiometries of complex present and β_1 to β_n represent the various compound association constants (complexes involving more than one template molecule are of course possible but are neglected for simplicity). Assuming the binding sites in the final MIP reflect this distribution of complexes, the strongest binding sites are expected to result from the complexes with the maximum number of interactions, TM_n. These sites are also expected to be the most selective, while the weakest and least selective sites will result from free monomer in the prepolymerization mixture that is not involved in complexes at all, M, and the lower-order complexes TM, etc.

For a "good" MIP, the monomer–ligand interactions are strong such that the "good" sites, arising from TM_n complexes, constitute a significant proportion of the total sites present. In such a case, the MIA solvent can be optimized such that a relatively small amount of MIP is used and yet the situation described as ideal in Fig. 8 still pertains. For instance, the S-propranolol MIA optimized by Andersson, the aqueous solvent was optimized such that only 50 μg of polymer was required per assay [38].

For the conventional radiolabel MIA, "hot" probe and "cold" analyte are chemically identical, so a binding isotherm for the analyte can be derived as in Section II.G. When this was done for Andersson's S-propranolol MIP and the data fit to a bis-Langmuir model, the strongest binding sites were found to have $n_{B1} = 0.63$ μmol g^{-1} and $K_{a1} = 2.5 \times 10^8$ M^{-1} [38], thus the strong sites were much stronger and more numerous than for instance, in the caffeine MIP described in

Section II.G, although still few in comparison to the theoretical total sites based on the amount of template used. However, this demonstrates how for a "good" MIP, the amount of MIP required per assay can be reduced by optimization of the solvent.

In cases where the template is expensive, it is clearly desirable to reduce the amount of MIP required, and this should be done both by optimizing the imprinting procedure to produce as many sites as possible arising from the optimum TM_n complexes, and by optimizing the MIA solvent.

Strategies to increase the yield and homogeneity of binding sites in MIPs (to achieve "stoichiometric imprinting" or "monoclonal MIPs") are a subject of much attention and are described elsewhere in this volume (Chapter 5). However, when noncovalent MIPs are synthesized for use in MIA, a simple design strategy is possible of using a greatly reduced quantity of template, for instance, a template:monomer ratio of 1:1000 instead of the 1:4 or 1:10 commonly employed. This strategy may yield less TM_n complexes in the prepolymerization mixture, and so less of the strongest binding sites in the MIP, but these will be a much higher proportion of the total template used, as explained below.

The concentration of TM_n is related to that of free M by

$$\frac{[TM_n]}{[T]} = \beta_n \times [M]^n \tag{13}$$

where $[T]$ is the concentration of free template. The traditional template:monomer ratio of approximately 1:4 is based on early optimization of MIPs for chromatographic stationary phases, and is a result of a trade-off, to increase $[TM_n]$, as far as possible without too high $[M]$. A consequence is that most of the template is involved in lower-order complexes $TM, TM_2, \ldots, TM_{n-1}$, which may give rise to sites of intermediate strength and selectivity, still useful for chromatography but not sufficiently good for MIA. Dramatically reducing the amount of template, say to a template:-monomer ratio of 1:1000, will increase the fraction $[TM_n]/[T]$ since $[M]$ will be present in far greater excess. TM_n will be present at lower concentration than in the 1:4 case, but T and the lower complexes $TM, TM_2, \ldots, TM_{n-1}$ will be present in *much* lower concentration. The approach will leave many more free M, hence many more weak and nonspecific binding sites in the resulting MIP. However, while the latter may be problematic for applications such as chromatography they will not interfere with the MIA because the solvent is optimized such that the probe only "sees" the strongest sites.

Thus, Mayes and Lowe showed that an MIA for morphine worked equally well when the MIP was synthesized using a morphine:MAA ratio of 1:500 as 1:5 [21], while Yilmaz et al. showed an MIA for theophylline worked equally well with a theophylline:MAA ratio of 1:1000 as 1:4 [22]: Theophylline-imprinted polymers employing only 2.5 μmol template per gram of monomers (compared with 151 μmol g^{-1} in Vlatakis et al.'s MIA [16]) were employed, and the MIA functioned well, with caffeine cross-reacting less than 0.1%. These works demonstrate that the oft-quoted drawback of MIPs, the cost of the template required for their preparation, may be overcome: MIA is applicable even to expensive templates.

Because MIA probes only the sites of the MIP with strongest affinity for the probe, significant binding site heterogeneity is not detrimental. Provided the highest affinity sites exhibit good selectivity, the features of the rest of the MIP matrix are much less important than in other applications where all sites, of even weak affinity, are probed (e.g., chromatography). Thus, at higher concentrations, the MIP employed for the caffeine MIA described in Section II.E–II.G actually binds significantly more theophylline than caffeine, owing to the greater basicity of theophylline and a large number of nonselective binding sites with only weak affinity for caffeine. However, because the sites with high affinity for caffeine are selective, the polymer can still be employed in a selective assay for caffeine and theophylline cross-reacts less than 0.6% as measured by IC_{50} (Fig. 11). Similarly Andersson showed that an MIP imprinted with racemic R,S-propranolol could be employed in MIA assays for S-propranolol [38]. Although the polymer must contain an equal number of sites of high affinity for each enantiomer, R-propranolol competes only weakly with ^{3}H-S-propranolol for the strongest S-binding sites and in the assay cross-reacted only 1.4%. This implies that impure analyte could be used as a template in imprinting, and the resulting MIP could still be used successfully in a selective MIA, provided that the probe employed in the assay is pure.

V. VARIATIONS

A. Novel Polymer Formats

Although the majority of MIA assays reported have employed "traditional" noncovalent vinyl/acrylate-based MIPs, fabricated in organic solvents as macroporous monoliths then ground and sieved into useful particles, there have been many innovations in both the composition of MIPs and their macromorphology:

Functional monomers: Methacrylic acid remains often the monomer of first choice due to its ability to interact with templates as a hydrogen bond donor or acceptor, and form ion pairs. However, the more acidic trifluoromethylacrylic acid has been shown to be superior for some templates [22,26]. Haupt et al. [23] employed 4-vinylpyridine, a basic monomer, for the imprinting of the acidic template 2,4-dichloro phenoxyacetic acid (2,4-D). They also employed an aqueous-methanolic sovent in the MIP fabrication, reasoning that 2,4-D should also interact with the monomer via hydrophobic interactions. Although there is ample evidence that MIPs, fabricated in nonpolar organic solvents, may be employed for MIA in aqueous mixtures (above), there are many templates of interest that are simply insoluble in organics. The use of hydrophobic interactions during imprinting provides a route to MIPs selective for these species.

Haupt et al.'s 2,4-D MIA was highly successful, in particular, the ability to discriminate 2,4-D methyl ester, which cross-reacts approximately 100% with most anti-2,4-D antibodies [49]. This reflects the need to couple 2,4-D to a carrier protein in order to raise antibodies, which is most readily achieved via formation of an ester, and demonstrates one advantage of using MIPs in assays for small molecules.

Idziak and Benrebouh [46] also used 4-vinylpyridine as functional monomer to imprint 17-α-ethynylestradiol—though in this case the optimal imprinting solvent (due to the hydrophobicity of the template) was toluene. An MIA was developed in toluene, and exhibited excellent selectivity ($< 0.1\%$ cross-reactivity with 17-β-estradiol). Recognition in this case is considered to be through hydrogen bonds and π-stacking interactions.

Magnetic MIP beads: Particularly for chromatographic applications, it is desirable to prepare MIPs as monosized spherical particles rather than as monoliths that must be fragmented into irregular particles. The first attempt in this direction used suspension polymerization in perfluorocarbon liquids [50]: the resulting MIP beads (~15 µm) not only gave better-shaped chromatographic peaks but also seemed to have a higher capacity. A further advantage of this polymerization route was that magnetic iron oxide could be incorporated in the synthesis, producing hybrid superparamagnetic MIPs [25]. Ansell and Mosbach showed that these magnetic methacrylic acid-based MIPs imprinted with *R,S*-propranolol could be employed in a MIA in aqueous phase for *S*-propranolol, exhibiting a similar pattern of selectivity in terms of IC_{50} values to the bulk MIP-based *S*-propranolol MIA developed by Andersson [38]. The advantage of the magnetic MIA was that polymer could be removed from solution using a magnetic field, obviating the need for centrifugation. This is particularly likely to be useful for assays involving very large sample numbers, and for enabling automation.

Fine (<1 µm) *particles*: Haupt et al. in their several assay designs for 2,4-D [23,51,52], have employed a 2,4-D-imprinted MIP prepared as a monolith and ground in the traditional manner, but utilized not the intermediate-sized (~25 µm) particles as commonly done, but the finest particles. The fine particles had the same selectivity characteristics as larger particles, but the incubation times were reduced due to shorter diffusion distances, and handling was easier since the particles remained longer in suspension so could be pipetted more accurately.

Microspheres: Mosbach's group demonstrated that uniform microsphere MIPs (<1 µm) could be produced by precipitation polymerization under very dilute conditions [30]. Microsphere MIPs have subsequently been applied in MIAs for theophylline, 17-β-estradiol, caffeine, *S*-propranolol (all methacrylic acid-based MIPs) [31,33,39], and 2,4-D (4-vinylpyridine-based) [53]. The microspheres exhibit the same handling advantages as "fine" particles. They are nonporous so that accessible recognition sites are considered to be limited to the surface of the microspheres, yet Scatchard analysis suggested the best recognition sites in the theophylline–MIP microspheres were of similar affinity to those of bulk theophylline MIPs [16], and were present at a higher number per gram. The selectivity of the theophylline MIA also appeared to be similar with the microspheres as with the ground MIP particles, although fewer competitors were tested with the microspheres and the MIA was not validated on real samples. Because the microspheres' recognition sites are at the surface they could be adapted to MIA for 2,4-D with enzyme labels (Section IV.B), and because they form stable suspensions they could be used in a proximity-scintillation-based MIA for *S*-propranolol where separation of the solid and supernatant was not required, a pseudo-homogenous MIA (Section IV.C).

Ye et al. [33] also showed that microspheres imprinted with theophylline or 17-β-estradiol but using low cross-linking ratios (MAA:EDMA ratio 2.3:1) could be employed in MIAs for each of these analytes and worked as well as MIPs with more conventional higher cross-linking ratios. A benefit of this was that further monomers could be incorporated: it was shown that the "latent monomer" nitrophenylacrylate in the theophylline MIP was not detrimental to its affinity for ^{3}H-theophylline, although it was not shown whether cross-reactivity of interferents was affected and it is unclear what benefit the additional monomer would bring to an MIA.

Immobilized templates: Yilmaz et al. [32] demonstrated an MIA for theophylline based on an MIP produced by polymerizing TFMAA/DVB in the pores of a silica gel matrix on which theophylline had been immobilized. The silica gel was dissolved and the resulting polymer processed to yield particles of low surface area with surface recognition sites for theophylline. More polymer was required for the MIA than when a conventional MIP imprinted with free theophylline molecules was used, suggesting a lower density of good recognition sites, but the selectivity of the assay was similar. The immobilized-template approach could be of use in imprinting compounds which are poorly soluble in solvents of choice, although a simpler solution in such cases might be to imprint the template in solution at very low concentrations [21,22]. The new approach could though allow the imprinting of larger molecules including proteins that are not suited to "conventional" imprinting (since they are unable to diffuse through a highly cross-linked polymer network), and moreover might be anticipated to yield more homogenous recognition sites.

Surface-imprinted silica: Despite the variations described above, all MIAs had been based on imprinted vinyl or acrylate-based copolymers until the demonstration of a MIA for *N*-acetylated μ-conotoxin GIIIB (NacGIIIB) based on surface-imprinted silica. The NacGIIIB-imprinted silica was generated by treatment of silica gel with a mixture of amino-functionalized and nonfunctionalized silanes in the presence of NacGIIIB [45]. Unlabelled NacGIIIB could displace labelled ^{14}C-Nac-GIIIB from the material more effectively than the related peptide NacGIIIA or the nonacetylated forms GIIIA or GIIIB. However, a nonimprinted polymer was not studied and since the competitors studied are all more basic than the probe, it is unclear to what extent selective recognition sites are responsible for the results. Surface imprinting of silica is a promising method for imprinting large molecules but is much less well-studied and understood than approaches based on organic polymers.

Thin polymer films on glass: Thin acrylate or silane-based films imprinted with *S*-propranolol were employed in MIAs by Marx and Liron [40]. MAA/EDMA/chloroform, MAA/TRIM/chloroform, or partially gelated silane/organo-silane/ethoxyethanol/aqueous HCl incorporating *S*-propranolol template were spin-cast onto glass cover slips, polymerized, and template extracted. The silane-based films were found to exhibit higher specific binding of ^{3}H-*S*-propranolol in phosphate buffer pH 7.6 than the acrylate-based ones, as determined by B/T (MIP) − B/T (NIP). In a buffer-based competitive MIA, *S*-propranolol displaced ^{3}H-*S*-propranolol more effectively than *R*-propranolol. Metoprolol and timolol were also discriminated but it is hard to quantify the selectivity observed as an IC_{50} value for *S*-propranolol itself is not given. Although films made in this way are of great interest in sensor applications, it is unclear whether they hold any advantages for application in MIA.

Thin layers on microtiter plate wells: The microtiter plate format is universally employed in biological immunoassays such as ELISA and enables a very high assay throughput, in particular when combined with automation. Thus, thin layers of MIPs in microtiter plate wells are of immense interest for MIAs involving colorimetric or fluorescence detection. Two approaches have been described: the entrapment of 2,4-D-imprinted vinylpyridine/TRIM microspheres in films of polyvinylalcohol hydrogel [53] and the preparation of imprinted poly-3-aminophenylboronic acid and related conjugated polymers which precipitate as films on the microwell surface [19,54]. The polyvinyl alcohol composite films have been used in combination with an enzyme-labelled probe, the conjugated polymers with fluorophore- and enzyme-labelled probes [Section IV.B).

Capillaries: Danielsson and coworkers have used 2,4-D-imprinted polymer coatings on the interior surface of capillaries in a flow-injection-based assay for 2,4-D using an enzyme-labelled probe (Section IV.B) [55,56]. The inner surface of a 0.9 mm internal diameter glass capillary was treated with 3-methacryloxypropyl trimethoxysilane, and a polymerization mixture similar to that employed to produce microspheres was introduced and polymerized at 60°C.

B. Novel Probes

Most reported MIAs have employed radiolabelled probes (Table 2), but alternatives using fluorescent, electroactive, or enzyme-labelled probes are of interest due to safety and regulatory incentives to reduce the use of radioactive substances, and an increasing number of nonradiolabel MIAs have been developed (Table 4) [29,51–63]. The great advantage of using radiolabelled probes (besides the inherent sensitivity) is that the imprint molecule, the probe, and the target analyte may be chemically identical. Thus, MIP recognition sites that bind the probe most strongly are likely to be highly selective for the analyte. When the probe and the analyte differ, there is no guarantee that the sites interrogated by the probe have the best selectivity for the analyte. Four approaches may be envisaged:

1.   The native analyte is imprinted, a labelled form of the analyte is used as the MIA probe. The most likely problem in this case is that the best recognition sites for the analyte will be unable to accommodate the larger labelled molecule. The probe molecule may be more likely to bind at less precise sites, and the selectivity of the assay will be compromised. Nevertheless, several MIAs have been developed following this principle using colorimetric [29,57,58], fluorescent [54,59,60], and enzyme labels [18,19,53–56,61] (Table 4). MIPs with recognition sites mostly at the surface, rather than within the polymer network, are expected to be most useful in this strategy.

2.   The labelled analyte is imprinted, the labelled analyte is used in the MIA as the probe. The problem in this case is that the recognition sites in the MIP are likely to be complementary not just to features of the analyte moiety, but also to those of the label moiety. Thus, competitors resembling the label may compete successfully in the MIA and appear as "false positives". No successful MIA following this approach has been reported.

3.   The native analyte is imprinted, an unrelated molecule is used as the probe. The success of this strategy depends on the probe binding at the best-imprinted sites. Otherwise the analyte will bind first to sites in the MIP that have low affinity for the probe (reducing the MIA sensitivity) and the sites interrogated by the probe will be more heterogenous (reducing selectivity). Clearly, the choice of probe is crucial: it should have similar functionalities to the analyte, and not be larger than the analyte for the reasons in 1 above. It may bind quite weakly to the well-imprinted sites, but it must bind even more weakly to the poorly imprinted sites. Thus, for instance, it may be wise to avoid a probe which is more basic than the imprint molecule in an MIA using a methacrylic acid-based MIP. With a good choice of probe, however, good results have been obtained [51,52,62].

Table 4 Reported MIAs employing nonradiolabelled probes

Template	Probe	Assay solvent	Competitors	Reference
Chromophore-labelled probes				
Chloramphenicol	Chloramphenicol-methyl red	MeCN	Chloramphenicol, chloramphenicol-diacetate, thiamphenicol	[29,57]
Biotin methyl ester	Biotin nitrophenyl ester	MeCN	Biotin methyl ester	[58]
Fluorophore-labelled probes				
Triazine	5-(4,6-dichlorotriazinyl)aminofluoresceine	EtOH	Triazine, atrazine, simazine	[59]
Chloramphenicol	Dansylated chloramphenicol	MeCN	Chloramphenicol, chloramphenicoldiacetate, thiamphenicol	[60]
Atrazine	5-(4,6-dichlorotriazinyl)aminofluoresceine	Water	Atrazine, atraton-D, metribuzin	[54]
Enzyme-labelled probes				
2,4-D	2,4-D-tobacco peroxidase	Phosphate pH 7/Triton X-100 (999:1)	2,4-D, related acids, and esters	[18,53,55,56]
Epinephrine	Norepinephrine-horse-radish peroxidase	Phosphate pH 6	Epinephrine, related catechols	[19,54]
Microcystin-LR	Microcystin-horse-radish peroxidase	Phosphate pH 7	Microcystin-LR, related peptides	[61]
Unrelated chromophore/ fluorophore probes				
2,4-D	7-carboxy-methoxy-4-methylcoumarin	Phosphate pH 7/Triton X-100 (999:1)	2,4-D, related aromatics	[51,62]
Electroactive probes				
2,4-D	Homogentisic acid	Phosphate pH 7/MeOH (9:1)	2,4-D	[52,62]
2-chloro-4-hydroxy-phenoxyacetic acid (2-C-4-H)	2-C-4-H	Phosphate pH 7.4/EtOH (9:1)	2,4-D	[63]

4. An unrelated molecule is imprinted and employed as the probe. This strategy may suffer from similar disadvantages to 2, depending on the extent to which the probe and analyte differ. In practice, only one example has been reported, in which the molecules differ only very slightly [63]. Finding an electroactive or fluorescent molecule with sufficiently high degree of structural similarity is not always expected to be trivial.

The Karube group reported the first nonradiolabel MIA [59], although it was presented as a sensor rather than an assay. 1,3,5-Triazine was imprinted in the traditional way using EDMA, MAA, and diethylaminoethyl methacrylate, and the fluorescent probe was 5-(4,6-dichlorotriazinyl)aminofluorescein (DCTAF) (Fig. 17). Triazine displaced more probe from the MIA than did the N-alkyl-amino substituted triazines simazine or atrazine, but the structure of the probe raises several questions: given the bulk of the fluorophore and of the two chlorine atoms attached to the triazine ring it seems surprising that it binds to many of the binding sites in a conventional triazine-imprinted MIP at all, in particular that it should bind to the most selective ones. The acidic groups on the fluorescein moiety might also be expected to give strong nonspecific interactions with the basic MIP. Another issue is that the chloro groups of the probe are quite reactive and would be expected to react with any nucleophile present. Given the complex structure of the probe relative to the imprint molecule it would have been useful to study a wider range of competitors.

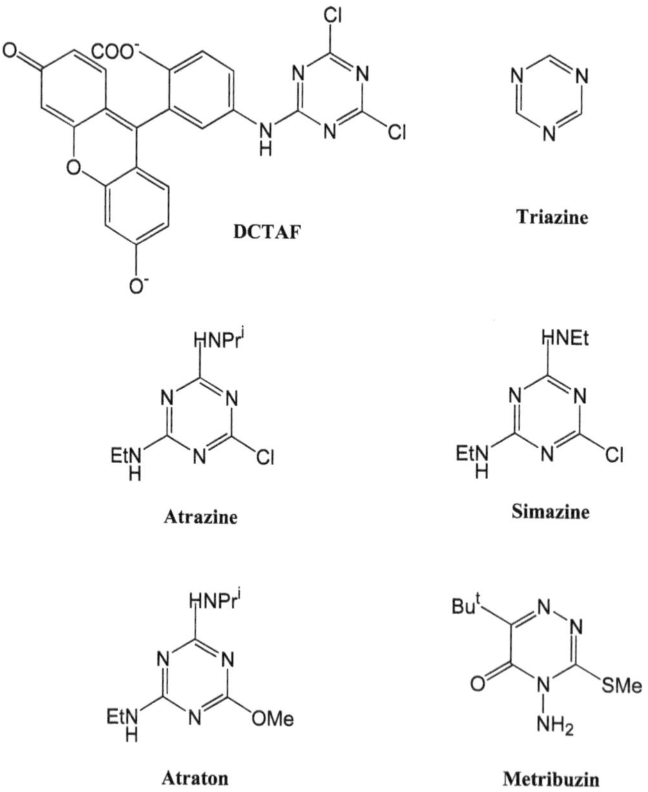

Figure 17   Molecules used in fluorescent probe MIAs for triazine and atrazine [54,59].

A chromophore-labelled probe was used for detection of chloramphenicol [29], but in this case the chloramphenicol-imprinted MIP was used in HPLC mode with methyl red-labelled chloramphenicol incorporated in the acetonitrile mobile phase (Fig. 18) (Section IV.D). The probe in this case is expected to be inert and more closely resembles the template/analyte. The approach was later refined [57] and reproduced using an alternative fluorophore-labelled probe, dansylated chloramphenicol, which should exhibit better sensitivity and selectivity since the probe may be used at lower concentrations and hence to probe fewer (better) binding sites [60] (Fig. 18). The approach was also investigated for β-estradiol, based on a β-estradiol-imprinted stationary phase and a dansylated β-estradiol probe, but in this case the probe seems to bind only weakly to the MIP and no evidence for effective competition was observed [64]. A related approach using a colorimetric label was employed by Takeuchi et al. [58] to develop a flow-through MIA for biotin methyl ester but although the analyte successfully displaced the probe no competitors were studied.

The same probe as used by Piletsky et al. [54] for the triazine MIA above, DCTAF, was employed in an MIA for atrazine based on atrazine-imprinted MIPs formed in the wells of microtitre plates. The MIPs were formed by oxidative polymerization of 3-thiophenylboronic acid and 3-aminophenylboronic acid in ethanol/ aqueous potassium dichromate in the presence of atrazine, which yields thin, colored films. The authors suggest ion-pair interactions between the boronic acids and the amine groups of atrazine in combination with π-interactions are responsible for imprinting, though it is not clear this will be sufficient to form a stable complex in the aqueous solvent, which also becomes very acidic during oxidative polymerization, or whether the resulting polymer network is really

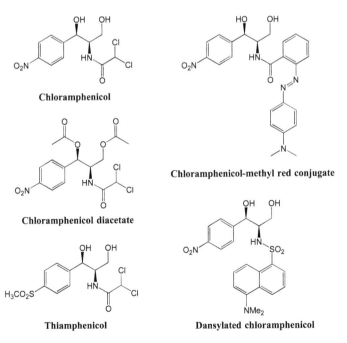

Figure 18 Molecules used in flow-through MIAs of chloramphenicol [29,57,60].

dense and cross-linked enough to retain any memory of the template after washing. The results do suggest that in the MIA, atrazine is more successful at displacing the probe than are the interferents Atraton (which differs from atrazine by just a methoxy/chlorine substitution) or Metribuzin (which is a molecule of very different structure, Fig. 17). It is unclear though how much of the probe is actually displaced, since the results are presented as the change in free probe concentration rather than change in bound: hence IC_{50} values cannot be deduced. In combination with the concerns mentioned above about the reactivity of the probe, and its structural difference to the analyte, more evidence is required to demonstrate that this system can be a useful MIA. However, given the ease and convenience of the polymer preparation method, further studies of this system are eagerly awaited.

Enzyme labels are popular in immunoassay because a small quantity of bound or displaced probe can easily be detected through amplification, so assays may achieve similar sensitivity to those using radiolabels. Enzyme MIAs have been reported by the groups of Mosbach/Haupt/Danielsson and Piletsky/Turner. Enzyme labels even more than fluophores or chromophores present the problem of the size of probe relative to the recognition sites and demand recognition sites at the surface of the MIP: the Mosbach group [18] used 2,4-D-imprinted TRIM-4-VPy microspheres, either in suspension or trapped in a thin film of hydrogel in microtiter wells [53]. The microspheres, which should have a greater proportion of surface recognition sites than conventional ground particulate MIPs, were first employed in a conventional radiolabel MIA for 2,4-D. Figure 19 illustrates some of the structural analogues investigated as competitors. Table 5 shows that relative to the fine particles used by Haupt et al. the microspheres led to a 20-fold increase in IC_{50} to about $10 \, \mu g \, mL^{-1}$ (so loss in sensitivity), but exhibited similar selectivity. Then, tobacco peroxidase-labelled 2,4-D was employed as probe in phosphate buffer, and the unbound conjugate probe was quantified using either a colorimetric assay (1,4-phenylenediamine substrate) or a chemiluminescence assay (luminol substrate).

Figure 19 2,4-D, analogues and nonrelated probes studied in MIAs [18,23,51,52,53,55,63]. 2,4-D, 2,4-dichlorophenoxyacetic acid; CPOAC, 4-chlorophenoxyacetic acid; POAC, phenoxyacetic acid; CMMC, 7-carboxy-methoxy-4-methylcoumarin; HGA, homogentisic acid (2,5-dihydroxyphenylacetic acid); 2-C-4-H, 2-chloro-4-hydroxyphenoxyacetic acid.

Table 5 Results obtained in various MIAs for the herbicide 2,4-D. IC50 values were estimated by this author from graphs of signal vs. concentration presented in each work. All MIAs were performed in phosphate buffer pH 7 with ethanol or surfactant added, except the fluorescent probe (CMMC) assay performed in acetonitrile.

MIA	IC50 (2,4-D) $\mu g\,mL^{-1}$	Cross-reactivities CPOAC	(%) POAC
Radiometric, fine particles [23]	~0.5	24	2
Radiometric, microspheres [18]	~10	25	n.d.
Enzyme-linked colorimetric, microspheres [18]	~200	<1	<1
Enzyme-linked chemiluminescence, microspheres [18]	~20	<1	<1
Enzyme-linked microplate, immobilized microspheres [53]	~10	32	3
Enzyme-linked flow-injection, coated capillary [55]	~1	<1	<1
Fluorescent probe (CMMC), fines [51]	~1000	42	9
Fluorescent probe (CMMC), fines, MeCN [51]	~400	50	14
Electroactive probe (HGA), fines [52]	~5000	n.d.	n.d.
Electroactive probe (2-C-4-H), fines [63]	~4000	n.d.	n.d.

n.d. = not determined.

Some nonspecific adsorption of the enzyme to the MIP occurred but could be reduced by addition of Triton X-100 surfactant. 2,4-D displaced probe from the MIP: with the colorimetric assay more probe was required for detection and the IC_{50} value for 2,4-D was about 200 $\mu g\,mL^{-1}$, with the chemiluminescence assay less conjugate was required and the IC_{50} value was about 20 $\mu g\,mL^{-1}$. The analogues 4-chlorophenoxyacetic acid (CPOAC) and phenoxyacetic acid (POAC) did not compete significantly in either case. Subsequently, the chemiluminescence assay was adapted to be performed in microtiter plates with luminescence detection of the bound conjugate via an imaging CCD camera [53]. The IC_{50} value for 2,4-D was about 10 $\mu g\,mL^{-1}$, CPOAC and POAC exhibited similar cross-reactivities as in the radiolabel MIA [23].

Further, the enzyme MIA was adapted to a capillary flow-injection format [55,56] as described in Section IV.C. The IC_{50} for 2,4-D using this approach was reduced to about 0.5 ng mL^{-1}, while the cross-reactivities of interferents were again <1% (Table 5). The improvement obtained with the capillary-based assay is dramatic and promising, although long-term stability of the MIP in the capillary and reproducibility are issues that remain.

Piletsky et al. employed an enzyme-labelled probe in combination with their novel imprinting strategy for formation of conjugated polymer films in microtiter plates. Oxidation–polymerization of 3-aminophenylboronic acid was conducted in the presence of the template epinephrine to yield thin films [19,54]. Horseradish peroxidase-labelled norepinephrine was employed as probe in phosphate buffer. The buffer conditions were chosen to minimize nonspecific binding of the enzyme to the MIP. Epinephrine exhibited an IC_{50} value about 10 μM with other benzenediols competing less. Although the diol group of epinephrine may form a boronate ester with the monomers, this may not be very stable under the polymer synthesis conditions and it remains unclear how recognition sites are formed in this noncross-linked polymer. The stability of the polymer is another issue (the IC_{50} was shown to increase if too many washing cycles were employed before the MIA [54]), but the system definitely merits further investigation.

Few workers in the field would anticipate a probe labelled with an enzyme such as horseradish peroxidase ($M_r = 44,000$) being able to bind to a significant proportion of the good binding sites of a conventional acrylate/vinyl-based ground MIP. Recently, however, Piletsky and coworkers [61] have published an enzyme-MIA for the cyclic heptapeptide cyanobacteria toxin microcystin-L,R, which employs conventional EDMA cross-linked MIP particles with diameter 45–63 μm. The MIA is conducted in phosphate buffer pH 7, using microcystin-L,R-horseradish peroxidase conjugate as a probe. The supernatant is separated from the polymer by filtration and the peroxidase activity in the supernatant quantified. Although the results suggest the probe is displaced by the analyte, and related cyclic peptides displace less, the data presented are incomplete. In particular, it is unclear how much of the probe is actually bound, a very narrow range of analyte/interferent concentrations are studied ($0.1–0.8 \, \mu g \, mL^{-1}$, whereas at least five decades of concentration would be appropriate) so IC_{50} values cannot be determined, and the mathematical treatment of the data is obfuscating.

An MIA using a nonrelated fluorescent probe was demonstrated by Haupt [51] and Haupt et al. [62] again for the herbicide 2,4-D. The 2,4-D MIP fine particles previously applied in a radiolabel MIA were employed with CMMC as a probe (Fig. 19). Comparing the structure of the probe with that of the template, it may be seen that many features are shared, in particular the oxyacetic acid side chain that probably interacts with pyridine groups in the imprinted recognition sites. Preliminary experiments confirmed that the probe bound with a degree of selectivity at the same sites as the imprint molecule. MIA was then performed in phosphate buffer pH 7 or in acetonitrile, with nonbound CMMC being determined via fluorimetry. Relative to the radiolabel MIA, the IC_{50} for 2,4-D increased greatly, but the selectivity was only slightly compromised (Table 5). A nonrelated electroactive probe was then studied in work aiming toward the development of a 2,4-D sensor [52]. Again the probe shares some of the features of 2,4-D but it is less similar than the fluorescent probe. In preliminary electro MIA studies, a higher concentration of 2,4-D was required to displace 50% of the probe, and selectivity was not studied (Table 5). A related study was reported by Schollhorn et al. [63] using the electroactive analogue 2-C-4-H as both imprint molecule and probe. In the electro MIA, the IC_{50} value for 2,4-D was again high and no competitors were studied, but the authors suggest their results could be improved with some optimization of their MIP composition.

Thus, the range of probes investigated with 2,4-D suggest that formats using alternative probes are often just as selective as radiolabelled 2,4-D. The sensitivity is a function not only of the successful competition between probe and analyte for good recognition sites, but also of the inherent sensitivity with which the probe may be detected. The most promising alternative format at this stage appears to be the capillary-based enzyme MIA. The formation of imprinted films by oxidative polymerization of conjugated polymers in microtiter plate wells is of great interest since the method is straightforward and the resulting materials readily adaptable to enzyme MIA but the approach needs further validation.

C. Novel Assay Formats

The flow-through MIA developed for chloramphenicol by Levi et al. [29] is illustrated in Fig. 20. The probe is present at constant concentration in the mobile phase so a constant baseline absorbance is detected. When samples containing analyte or interferent (plus the probe, added at the same concentration as in the mobile phase) were injected, chloramphenicol displaced the probe which was detected eluting from the column as a peak of increased absorbance, above the baseline. Thiamphenicol displaced less dye, with chloramphenicol diacetate cross-reacting minimally. Parameters such as the probe concentration and flow rate were investigated—the latter was found to have a significant influence on peak height, with competition being less effective at high flow rates reflecting the kinetics of the exchange process. Chloramphenicol in bovine serum was quantified by extracting into ethyl acetate, evaporating the solvent and replacing with acetonitrile before injection. The response was linear over the examined concentration range with a limit of detection about $5 \, \mu g \, mL^{-1}$ and standard error of the mean lower than 10%. A similar approach was used in the works of McNiven et al. [57], Suarez-Rodriguez et al. [60], Rachkov et al. [64] and Takeuchi et al. [58]. Unfortunately in none of these cases were the results compared with a static, equilibrium MIA using the same reagents, which would have provided a useful indicator of the relative qualities of the two approaches. The limit of detection for chloramphenicol quoted above is relatively average compared with most conventional equilibrium-based MIAs. The flow-through MIA certainly has an advantage in terms of handling, although the necessity of running samples in series rather than parallel may also be a disadvantage.

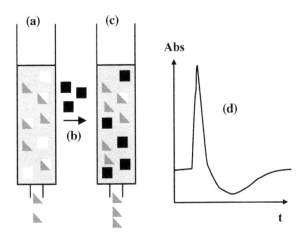

Figure 20 Schematic of the flow-through competitive MIA [29,56–58,60]. (a) There is a constant concentration of chromophore/fluorophore in the mobile phase, and a constant amount in the binding sites of the MIP. (b) When analyte is injected, it displaces the (intensely colored/fluorescent) probe from the MIP binding sites so the concentration in the mobile phase transiently increases. (c) The presence of the analyte is detected as a peak, followed by a trough as the analyte itself is eluted from the MIP and probe is readsorbed from the mobile phase. Interferents that do not bind to the MIP do not displace probe, so no signal is detected. Sensitivity is enhanced over conventional chromatography because the probe is much more readily detected via absorbance or fluorescence than is the analyte itself.

The flow-injection enzyme MIA developed by Surugiu and coworkers [55,56] differs in that the probe is not present at constant concentration but simply injected together with the analyte. 2,4-D-imprinted MIP was formed on the inner wall of a 0.9 mm internal diameter glass capillary as described in Section IV.A. After extracting template, 2,4-D-peroxidase conjugate and free analyte in phosphate buffer were passed through the capillary. Binding of the probe in this case is sufficiently strong that after the excess probe and analyte are eluted the bound probe remains on the MIP. After a washing step, the chemiluminescence substrate was injected and the bound form of the probe quantified by measuring the emitted light. The capillary could be regenerated for repeated use by flushing with pH 2.5 glycine buffer to remove the probe. The results using this approach were excellent as indicated in Table 5, though the stability of the capillary and reproducibility remain to be demonstrated.

Homogenous binding assay formats are highly popular with biological immunoassays because there is no need for physical separation of solid phase and supernatant: reagents are simply added and measurements recorded. Such assays as SPA and EMIT are therefore well suited to automation and high throughput. The recent description of soluble-imprinted microgels by the Wulff group [65] may provide a step towards homogenous MIA. Obviating the need for separation does not entirely depend on the use of soluble antibody, however, a binding matrix that stays in stable suspension may serve as well.

The scintillation proximity MIA developed by Ye and Mosbach [39] was the first MIA to obviate the need for a separation step. They utilized microsphere MIPs imprinted with S-propranolol and containing the scintillation monomer 4-hydroxymethyl-2,5-diphenyloxazole acrylate. Binding of the probe ^{3}H-S-propranolol to the microspheres enabled excitation of the scintillant and photoemission could be detected, which was suppressed when analyte was present and displaced the probe. The initial MIA was performed in toluene/AcOH (99.5:0.5) and appeared to be of

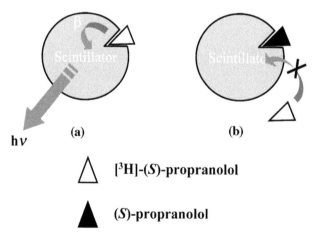

$\triangle$ [^{3}H]-(S)-propranolol

$\blacktriangle$ (S)-propranolol

Figure 21 Schematic of the scintillation proximity MIA for (S)-propranolol [39,41]. (a) The bound, tritium-labelled (S)-propranolol triggers the scintillator to generate the fluorescent light. (b) When the tritium-labelled (S)-propranolol is displaced by the unlabelled (S)-propranolol, it is too far from the antenna and the scintillator to efficiently transfer the radiation energy; therefore, no fluorescence can be generated.

similar sensitivity to the bulk MIP MIA developed by Andersson [38], with a clear signal at $10-100 \, \mathrm{ng \, mL^{-1}}$ analyte, and similar selectivity against R-propranolol, although data did not extend far enough to determine IC_{50}s. Toluene was required in order to effect energy transfer between the β-emitter and the scintillant (Fig. 21). The assay did not work in other solvents since although the probe bound, energy was not transferred effectively to the scintillant. This problem was addressed in later work by incorporating DVB as cross-linker (41): its aromatic groups fill the same relay function as the toluene. The new DVB-based MIPs were used in MIAs in acetonitrile/acetic acid (99.5:0.5 v/v) and in acetonitrile/citrate buffer pH 6 (50:50 v/v). Both MIAs employed polymer concentrations of $0.2 \, \mathrm{mg \, mL^{-1}}$ and the IC_{50} values for S-propranolol and R-propranolol were approximately 10^3 and $10^5 \, \mathrm{ng \, mL^{-1}}$ (in organic solvent), $10^3 \, \mathrm{ng \, mL^{-1}}$ and $3 \times 10^4 \, \mathrm{ng \, mL^{-1}}$ (in cosolvent mixture). The better stereoselectivity of the organic MIA corresponds to the observations of Andersson on his MIA using bulk S-propranolol-imprinted MIPs: however, the selectivity in the cosolvent mixture for the scintillation proximity assay is better than for Anderssons MIA in aqueous solution. A concern is the relative inefficiency of imprinting in microsphere MIPs in some cases as demonstrated by 2,4-D, but if an MIP can be made for an analyte in this way scintillation proximity MIA should be generally applicable and represents an exciting step forward.

VI. CONCLUSION

MIPs provide a complementary range of reagents to antibodies for application in ligand binding assays, in particular for small molecules. Many of the perceived drawbacks of MIPs do not hinder their successful application in MIAs: many aqueous-phase MIAs have been demonstrated (but the ability to perform MIA in organic solvents may sometimes be useful), the quantity of template needed in MIP syntheses may be dramatically reduced when they are to be used in MIAs, the presence of a heterogenous distribution of binding sites is not a problem provided that the fraction of sites that bind the analyte strongest are highly selective.

MIAs have been developed for an expanding range of small molecules of medical and environmental interest. Rapid advances are being made in terms of new MIP formats (for easier handling or compatibility with ubiquitous instrumentation, e.g., microplates) and new assay formats replacing radiolabels with fluorophores, electroactive groups, and enzyme labels. In many cases, studies have been limited to proof of principle and it is expected that there will be more emphasis on demonstrating complete analytical procedures based on MIA as the technique becomes more widely accepted.

REFERENCES

1. Price, C.P.; Newman, D.J. Principles and Practice of Immunoassay, 2nd Ed., Macmillan: London, 1997.
2. Hage, D.S. Immunoassays. Anal. Chem. **1999**, *71*, 294R–304R.
3. Bock, J.L. The new era of automated immunoassay. Am. J. Clin. Pathol. **2000**, *113*, 628–646.
4. Ansell, R.J.; Ramstrom, O.; Mosbach, K. Towards artificial antibodies prepared by molecular imprinting. Clin. Chem. **1996**, *42*, 1506–1512.

5. Andersson, L.I. Molecular imprinting for drug bioanalysis—A review on the application of imprinted polymers to solid-phase extraction and binding assay. J. Chromatogr. B **2000**, *739*, 163–173.

6. Sellergren, B.; Andersson, L.I. Application of imprinted synthetic polymers in binding assay development. Methods **2000**, *22*, 92–106.

7. Andersson, L.I. Application of molecularly imprinted polymers in competitive ligand binding assays for analysis of biological samples. In *Molecularly Imprinted Polymers: Man-made Mimics of Antibodies and Their Applications in Analytical Chemistry*; Sellergren, B., Ed.; Elsevier: Amsterdam, 2001; 341–354.

8. Ansell, R.J. MIP-ligand binding assays (pseudo-immunoassays). Bioseparation **2002**, *10*, 365–377.

9. Mazumdar, P.M.H. The template theory of antibody formation and the chemical synthesis of the twenties. In *Immunology 1930–1980: Essays on the History of Immunology*; Mazumdar, P.M.H., Ed.; Wall and Thompson: Toronto, 1989, 13–33.

10. Pauling, L. A theory of the structure and process of formation of antibodies. J. Am. Chem. Soc. **1940**, *62*, 2643–2657.

11. Burnet, F.M. A modification of Jerne's theory of antibody production using the concept of clonal selection. Australian J. Sci. **1957**, *20*, 67–69.

12. Forsdyke, D.R. The origins of the clonal selection theory of immunity—a case study for evaluation in science. FASEB J. **1995**, *9*, 164–166.

13. Roitt, I.; Delves, P.J. *Roitt's Essential Immunology 10th Ed.*; Blackwell Science Publishers, Oxford, UK, **2001**.

14. Köhler, G.; Milstein, C. Continuous cultures of fused cells secreting antibody of predefined specificity. Nature **1975**, *256*, 495–497.

15. Catt, K.; Niall, H.D.; Tregear, G.W. Solid phase radioimmunoassay. Nature **1967**, *213*, 825–827.

16. Vlatakis, G.; Andersson, L.I.; Muller, R.; Mosbach, K. Drug assay using antibody mimics made by molecular imprinting. Nature **1993**, *361*, 645–647.

17. Senholdt, M.; Siemann, M.; Mosbach, K.; Andersson, L.I. Determination of cyclosporin A and metabolites total concentration using a molecularly imprinted polymer based radioligand binding assay. Anal. Lett. **1997**, *30*, 1809–1821.

18. Surugiu, I.; Ye, L.; Yilmaz, E.; Dzgoev, A.; Danielsson, B.; Mosbach, K.; Haupt, K. An enzyme-linked molecularly imprinted sorbent assay. Analyst **1999**, *125*, 13–16.

19. Piletsky, S.A.; Piletska, E.V.; Chen, B.N.; Karim, K.; Weston, D.; Barrett, G.; Lowe, P.; Turner, A.P.F. Chemical grafting of molecularly imprinted homopolymers to the surface of microplates. Application of artificial adrenergic receptor in enzyme-linked assay for β-agonists determination. Anal. Chem. **2000**, *72*, 4381–4385.

20. Bengtsson, H.; Roos, U.; Andersson, L.I. Molecular imprint based radioassay for direct determination of *S*-propranolol in human plasma. Anal. Commun. **1997**, *34*, 233–235.

21. Mayes, A.G.; Lowe, C.R. Optimization of molecularly imprinted polymers for radio-ligand binding assays. In *Methodological Surveys in Bioanalysis of Drugs. Drug Development Assay Approaches Including Molecular Imprinting and Biomarkers*; Reid, E.D., Hill, H.M., Wilson, I.D., Eds.; Royal Society of Chemistry, Cambridge, UK, 1998; Vol. 25, 28–36.

22. Yilmaz, E.; Mosbach, K.; Haupt, K. Influence of functional and cross-linking monomers and the amount of template on the performance of molecularly imprinted polymers in binding assays. Anal. Commun. **1999**, *36*, 167–170.

23. Haupt, K.; Dzgoev, A.; Mosbach, K. Assay system for the herbicide 2,4-dichlorophenoxyacetic acid using a molecularly imprinted polymer as an artificial recognition element. Anal. Chem. **1998**, *70*, 628–631.

24. Ansell, R.J.; Andersson, L.I.; Mosbach, K. Unpublished.

25. Ansell, R.J.; Mosbach, K. Magnetic molecularly imprinted polymer beads for drug radioligand binding assay. Analyst **1998**, *123*, 1611–1616.

26. Ansell, R.J.; Gamlien, A.; Berglund, J.; Mosbach, K.; Haupt, K. Unpublished.

27. Umpleby, R.J.I.; Bode, M.; Shimizu, K.D. Measurement of the continuous distribution of binding sites in molecularly imprinted polymers. Analyst **2000**, *125*, 1261–1265.

28. Noerby, J.G.; Ottolenghi, P.; Jensen, J. Scatchard plot: common misinterpretation of binding experiments. Anal. Biochem. **1980**, *102*, 318–320.

29. Levi, R.; McNiven, S.; Piletsky, S.A.; Cheong, S.-H.; Yano, K.; Karube, I. Optical detection of chloramphenicol using molecularly imprinted polymers. Anal. Chem. **1997**, *69*, 2017–2021.

30. Ye, L.; Cormack, P.A.G.; Mosbach, K. Molecularly imprinted monodisperse microspheres for competitive radioassay. Anal. Commun. **1999**, *36*, 35–38.

31. Ye, L.; Weiss, R.; Mosbach, K. Synthesis and characterization of molecularly imprinted microspheres. Macromolecules **2000**, *33*, 8239–8245.

32. Yilmaz, E.; Haupt, K.; Mosbach, K. The use of immobilised templates—a new approach in molecular imprinting. Angew. Chem. Int. Ed. **2000**, *39*, 2115–2118.

33. Ye, L.; Cormack, P.A.G.; Mosbach, K. Molecular imprinting on microgel spheres. Anal. Chim. Acta **2001**, *435*, 187–196.

34. Mayes, A.G.; Andersson, L.I.; Mosbach, K. Sugar binding polymers showing high anomeric and epimeric discrimination obtained by noncovalent molecular imprinting. Anal. Biochem. **1994**, *222*, 483–488.

35. Muldoon, M.T.; Stanker, L.H. Polymer synthesis and characterization of a molecularly imprinted sorbent assay for atrazine. J. Agric. Food Chem. **1995**, *43*, 1424–1427.

36. Siemann, M.; Andersson, L.I.; Mosbach, K. Selective recognition of the herbicide atrazine by noncovalent molecularly imprinted polymers. J. Agric. Food Chem. **1996**, *44*, 141–145.

37. Andersson, L.I.; Muller, R.; Vlatakis, G.; Mosbach, K. Mimics of the binding-sites of opioid receptors obtained by molecular imprinting of enkephalin and morphine. Proc. Natl. Acad. Sci. USA **1995**, *92*, 4788–4792.

38. Andersson, L.I. Application of molecular imprinting to the development of aqueous buffer and organic solvent based radioligand binding assays for (*S*)-propranolol. Anal. Chem. **1996**, *68*, 111–117.

39. Ye, L.; Mosbach, K. Polymers recognizing biomolecules based on a combination of molecular imprinting and proximity scintillation: a new sensor concept. J. Am. Chem. Soc. **2001**, *123*, 2901–2902.

40. Marx, S.; Liron, Z. Molecular imprinting in thin films of organic–inorganic hybrid sol–gel and acrylic polymers. Chem. Mat. **2001**, *13*, 3624–3630.

41. Ye, L.; Surugiu, I.; Haupt, K. Scintillation proximity assay using molecularly imprinted microspheres. Anal. Chem. **2002**, *74*, 959–964.

42. Ramstrom, O.; Ye, L.; Mosbach, K. Artificial antibodies to corticosteroids prepared by molecular imprinting. Chem. Biol. **1996**, *3*, 471–477.

43. Berglund, J.; Nicholls, I.A.; Lindbladh, C.; Mosbach, K. Recognition in molecularly imprinted polymer α2-adrenoreceptor mimics. Bioorg. Med. Chem. Lett. **1996**, *6*, 2237–2242.

44. Svenson, J.; Nicholls, I.A. On the thermal and chemical stability of molecularly imprinted polymers. Anal. Chim. Acta **2001**, *435*, 19–24.

45. Iqbal, S.S.; Lulka, M.F.; Chambers, J.P.; Thompson, R.G.; Valdes, J.J. Artificial receptors: molecular imprints discern closely related toxins. Mater. Sci. Eng. C **2000**, *7*, 77–81.

46. Idziak, I.; Benrebouh, A. A molecularly imprinted polymer for 17 alpha-ethynylestradiol evaluated by immunoassay. Analyst **2000**, *125*, 1415–1417.

47. Karlsson, J.G.; Andersson, L.I.; Nicholls, I.A. Probing the molecular basis for ligand-selective recognition in molecularly imprinted polymers selective for the local anaesthetic bupivacaine. Anal. Chim. Acta **2001**, *435*, 57–64.

48. Janotta, M.; Weiss, R.; Mizaikoff, B.; Bruggemann, O.; Ye, L.; Mosbach, K. Molecularly imprinted polymers for nitrophenols—an advanced separation material for environmental analysis. Int. J. Environ. Anal. Chem. **2001**, *80*, 75–86.
49. Franek, M.; Kolar, V.; Granatova, M.; Nevorankova, Z. Monoclonal ELISA for 2,4-dichlorophenoxyacetic acid: characterisation of antibodies and assay optimization. J. Agric. Food Chem. **1994**, *42*, 1369–1374.
50. Mayes, A.G.; Mosbach, K. Molecularly imprinted polymer beads: suspension polymerization using a liquid perfluorocarbon as the dispersing phase. Anal. Chem. **1996**, *68*, 3769–3774.
51. Haupt, K.; Mayes, A.G.; Mosbach, K. Herbicide assay using an imprinted polymer based system analogous to competitive fluoroimmunoassays. Anal. Chem. **1998**, *70*, 3936–3939.
52. Kroeger, S.; Turner, A.P.F.; Mosbach, K.; Haupt, K. Imprinted polymer-based sensor system for herbicides using differential-pulse voltammetry on screen-printed electrodes. Anal. Chem. **1999**, *71*, 3698–3702.
53. Surugiu, I.; Danielsson, B.; Ye, L.; Mosbach, K.; Haupt, K. Chemiluminescence imaging ELISA using an imprinted polymer as the recognition element instead of an antibody. Anal. Chem. **2001**, *73*, 487–491.
54. Piletsky, S.A.; Piletska, E.V.; Bossi, A.; Karim, K.; Lowe, P.; Turner, A.P.F. Substitution of antibodies and receptors with molecularly imprinted polymers in enzyme-linked and fluorescent assays. Biosens. Bioelectron. **2001**, *16*, 701–707.
55. Surugiu, I.; Svitel, J.; Ye, L.; Haupt, K.; Danielsson, B. Development of a flow injection capillary chemiluminescent ELISA using an imprinted polymer instead of the antibody. Anal. Chem. **2001**, *73*, 4388–4392.
56. Svitel, J.; Surugiu, I.; Dzgoev, A.; Ramanathan, K.; Danielsson, B. Functionalized surfaces for optical biosensors: applications to in vitro pesticide residual analysis. J. Mater. Sci. Mater. Med. **2001**, *12*, 1075–1078.
57. McNiven, S.; Kato, M.; Levi, R.; Yano, K.; Karube, I. Chloramphenicol sensor based on an in situ imprinted polymer. **1998**, *365*, 69–74.
58. Takeuchi, T.; Dobashi, A.; Kimura, K. Molecular imprinting of biotin derivatives and its application to competitive binding assay using nonisotopic labeled ligands. Anal. Chem. **2000**, *72*, 2418–2422.
59. Piletsky, S.A.; Piletskaya, E.V.; El'Skaya, A.V.; Levi, R.; Yano, K.; Karube, I. Optical detection system for triazine based on molecularly-imprinted polymers. Anal. Lett. **1997**, *30*, 445–455.
60. Suarez-Rodriguez, J.L.; Diaz-Garcia, M.E. Fluorescent competitive flow-through assay for chloramphenicol using molecularly imprinted polymers. Biosens. Bioelectron. **2001**, *16*, 955–961.
61. Chianella, I.; Lotierzo, M.; Piletsky, S.A.; Tothill, I.E.; Chen, B.N.; Karim, K.; Turner, A.P.F. Rational design of a polymer specific for microcystin-LR using a computational approach. Anal. Chem. **2002**, *74*, 1288–1293.
62. Haupt, K. Molecularly imprinted sorbent assays and the use of non-related probes. React. Funct. Polym. **1999**, *41*, 125–131.
63. Schollhorn, B.; Maurice, C.; Flohic, G.; Limoges, B. Competitive assay of 2,4-dichlorophenoxyacetic acid using a polymer imprinted with an electrochemically active tracer closely related to the analyte. Analyst **2000**, *125*, 665–667.
64. Rachkov, A.; McNiven, S.; Elskaya, A.; Yano, K.; Karube, I. Fluorescence detection of β-estradiol using a molecularly imprinted polymer. Anal. Chim. Acta **2000**, *405*, 23–29.
65. Biffis, A.; Graham, N.B.; Siedlaczek, G.; Stalberg, S.; Wulff, G. The synthesis, characterisation and molecular recognition properties of imprinted microgels. Macromol. Chem. Phys. **2001**, *202*, 163–171.

26
Molecularly Imprinted Polymers as Recognition Elements in Sensors: Mass and Electrochemical Sensors

Karsten Haupt Compiègne University of Technology, Compiègne, France

I. INTRODUCTION

Chemical sensors and biosensors are of increasing interest within the field of modern analytical chemistry. This is essentially due to new demands particularly in clinical diagnostics, environmental analysis, food analysis, and production monitoring, as well as the detection of illicit drugs, genotoxicity, and chemical warfare agents. Another application area that has been opening up during the last few years is drug screening.

The central part of a chemical or biosensor is the recognition element, which is in close contact with an interrogative transducer. The recognition element is responsible for specifically recognizing and binding the target analyte in an often complex sample. The transducer then translates the chemical signal generated upon analyte binding or conversion into an easily quantifiable output signal. Biosensors rely on biological entities such as antibodies, enzymes, receptors, or whole cells as the recognition elements.

An alternative approach involves the use of biomimetic receptor systems capable of binding target molecules with affinities and specificities on a par with natural receptors. One technique that is being increasingly adopted is molecular imprinting. The binding sites that are generated during the imprinting process often have affinities and selectivities approaching those of antibodies [1]. Molecularly imprinted polymers, sometimes referred to as antibody mimics, display some clear advantages over real antibodies for sensor technology: due to their highly cross-linked polymeric nature, they are intrinsically stable and robust, facilitating their application in extreme environments, such as, in the presence of acids or bases, in organic solvents, or at high temperatures and pressures. Moreover, these materials can be stored in the dry state at room temperature for long periods of time. They are also relatively cheap to produce, and their integration with standard industrial fabrication processes should be comparatively easy. It is, therefore, not surprising that there is now a strong development towards the use of molecularly imprinted polymers (MIPs) as the recognition elements in sensors.

This chapter focuses on theoretical and practical aspects of the application of molecularly imprinted polymers in sensors. Thereby, special emphasis will be on electrochemical and acoustic sensors, whereas for optical sensors the reader is directed to Chapter 27.

II. GENERAL CONSIDERATIONS

In biosensors, a signal is generated upon the binding of the analyte to the recognition element. The transducer then translates this signal into a quantifiable (in most cases electrical) output signal. The same general principle applies if an MIP is used as the recognition element instead of a biomolecule (Fig. 1). The main parameters describing the performance of a sensor are selectivity, sensitivity, stability, and reusability, dynamic range, and response time. These are determined both by the recognition element and the transducer, thus the choice of the right combination between the two is important.

Early attempts to utilize the recognition properties of MIPs for chemical sensing were for example ellipsometric measurements on thin Vitamin K_1-imprinted polymer layers [2], the measurement of changes in the electrical streaming potential over a HPLC column packed with a MIP [3], or permeability studies of imprinted polymer membranes [4]. The first reported *integrated* sensor based on an MIP was a capacitance sensor. The device consisted of a field-effect capacitor containing a thin phenylalanine anilide-imprinted polymer membrane. Binding of this model analyte resulted in a change in capacitance of the device, thus allowing for the detection of the analyte in a qualitative manner (Fig. 2) [5].

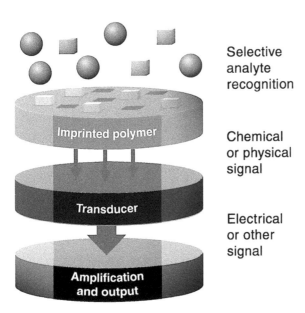

Figure 1 Schematic representation of an MIP-based biomimetic sensor.

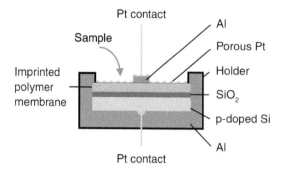

Figure 2 Capacitance sensor employing a field-effect capacitor as the transducer. (From Ref. 5.)

III. TRANSDUCERS

Scheme 1 depicts the three different possibilities of transducing the binding event, which are described in more detail in the following section. In the simplest case, a certain change in one or more physicochemical parameters of the system (such as, mass accumulation) upon analyte binding is used for detection. In order to increase sensitivity and the signal-to-noise ratio, reporter groups may be incorporated into the polymer that generate or enhance the sensor response. If the analyte possesses a specific property (such as, fluorescence or electrochemical activity) this can also be used for detection.

A. General Detection Principles

Mass-senstitive acoustic and optical transducers belong to the first group of sensors as depicted in Scheme 1. These sensors detect mass accumulation via a change in oscillation frequency, like the quartz crystal microbalance, or via a change in certain optical parameters, like the refractive index in surface plasmon resonance devices.

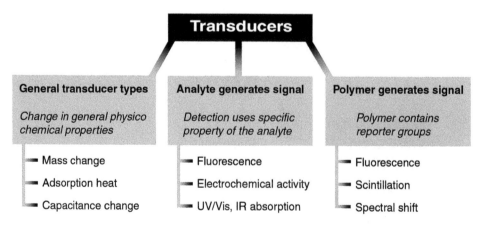

Scheme 1 Different approaches to the transduction of the binding signal in MIP-based sensors.

This principle is widely applicable and more or less independent of the nature of the analyte. Another sensor design belonging to the same group is based on conducto-metric transducers [6–8]. Here, two electrodes are separated by an imprinted polymer, often in the form of a membrane. Binding of the analyte to the polymer changes its conductivity, which is translated into an electrical signal. Acoustic and conductometric transducers are described more in detail below.

A very general means of transducing the analyte binding to the recognition element is by measuring the adsorption heat that is produced or taken up by any adsorption or desorption process. This heat, although small, can be measured using a sensitive calorimetric device. MIP-sensors with calorimetric transducers have not yet been described, but the feasibility of quantifying analyte binding to an MIP by isothermal titration microcalorimetry has recently been demonstrated [9]. The use, for example, of thermistor-based devices [10] should allow for the construction of integrated sensors.

B. The Analyte Generates the Signal

If the target analyte exhibits a special property such as fluorescence or electrochemical activity, this can be exploited for the design of MIP-based sensors. For example, an amperometric morphine sensor was developed where the analyte morphine was selectively enriched on an MIP, and subsequently quantified by electro-oxidation [11]. Optical sensors for the detection of fluorescent analytes belong to the same group. A potential problem that can arise when this detection principle is used is that traces of the imprint molecule can remain entrapped in the polymer, which may cause a high background signal resulting in decreased sensitivity. A remedy could be to imprint the polymer with a structurally related molecule similar to the analyte but not having that special property [12].

C. The Polymer Generates the Signal

An attractive design of the recognition element/transducer couple is to have the signal generated by the polymer itself. One example for such a format is a glucose-sensing polymer that works in ligand exchange-mode [13]. A complex of a polymerizable copper chelate and methylglucoside was used during preparation of the polymer. Extraction of copper and methylglucoside from the polymer and subsequent reloading with copper yielded the active form of the polymer. Addition of the analyte glucose resulted in its coordination to the metal chelate accompanied by proton release (Fig. 3), which was a function of analyte concentration and could be quantified by simple pH measurements. Since a polymer prepared in the presence of ethylene glycol instead of methylglucoside released only half as many protons upon analyte binding, the authors suggested that the templating with methylglucoside might have increased the specificity of the polymer for glucose. They also demonstrated that the polymer could be used to measure glucose in blood plasma, although with a slightly reduced response compared to the measurements in a pure saline solution.

In particular optical sensing systems belong to this group, which are described in detail in Chapter 27. For example, fluorescent reporter groups are incorporated into the MIP, the properties of which are altered upon analyte binding [14–16]. A very sensitive sensor for a hydrolysis product of the chemical warfare agent Soman has been described based on a polymer-coated fiber optic probe and a luminescent

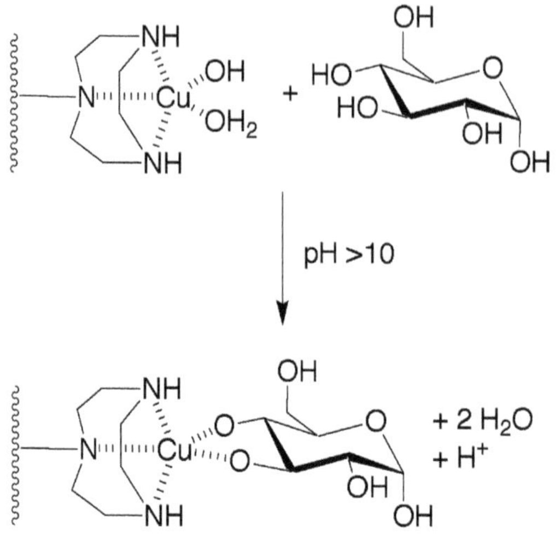

Figure 3 Analyte binding site in a glucose-sensing polymer. Upon coordination of the metal chelate by glucose, a proton is released. (From Ref. 13.)

europium complex for detection [17]. The complex of europium ligated by divinyl-methyl benzoate (ligating monomer) and by the analyte pinacoyl methylphosphonate was copolymerized with styrene, whereafter the analyte molecule was removed by washing. Rebinding of the analyte was quantified from laser-excited luminescence spectra. Although it is not clear whether imprinting has contributed to the selectivity of the sensor, this detection principle appears very promising, taking into account the very low detection limits that can be obtained (seven parts per trillion in this particular case).

IV. MASS-SENSITIVE TRANSDUCERS

Mass-sensitive transducers are either piezoelectric crystals in which the resonance frequency of the bulk material is measured (for frequenzies in the lower MHz range), or surface-acoustic wave devises (SAWs) which comprise a separate waveguide and piezoelectric transmitter and receiver (for frequenzies up to 2.5 GHz) [18]. SAWs are limited to gas-phase applications due to heavy damping of the acoustic wave in liquids. A variant of SAWs, so-called shear transverse wave resonators (STW, for frequenzies in the higher MHz range) show excellent performance even in the liquid phase [19,20]. In these devices, the oscillation frequency changes in response to mass changes at the transducer surface upon analyte binding to the recognition element. During the last years, acoustic transducers like the SAW oscillator [21,22], the Love-wave oscillator [23] or the quartz crystal microbalance (QCM), [21,22,24–28], have become increasingly used for the construction of MIP-based sensors. QCM sensors, which are bulk-acoustic wave devices, have been particularly popular because of their relatively low price, their robustness and ease of use. These sensors are based on the first group of transducers (Scheme 1). They consist of a thin quartz disk with electrode layers on both sides, which can be put into oscillation using the piezoelectric effect (Fig. 4). A thin imprinted layer is deposited on one side of the

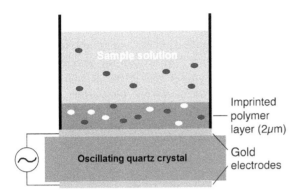

Figure 4 Schematic representation of an MIP-based acoustic sensor using a quartz crystal microbalance. (From Ref. 25.)

disk. Analyte accumulation in the MIP results in a mass change, which in turn causes a decrease in oscillation frequency that can easily be quantified by frequency counting. In addition, it is relatively easy to interface the MIP with the sensor. Measurements with QCM can be performed both in solution, [22,25,26,28] and in the gas phase [21,27]. A problem associated with this transducer type, in particular if used in the liquid phase, is that they are sensitive not only to mass changes but also to changes in viscoelasticity close to the surface. Thus, the sample matrix can cause artifacts, and the use of a reference sensor coated with a nonselective polymer of the same type as the MIP has been proposed to eliminate these. Such reference sensor can be placed of the same quartz crystal as the selective sensor [29], which eliminates at the same time temperature effects.

For example, QCM has been used to construct an imprinted polymer-based sensor for glucose [26]. The polymer, poly(o-phenylene diamine), was electrosynthesized directly at the sensor surface in the presence of 20 mM glucose. In that way, a very thin (10 nm) polymer layer was obtained that could rebind glucose with certain selectivity over other compounds such as ascorbic acid, paracetamol, cysteine and to some extent fructose. However, only millimolar concentrations of the analyte could be measured. Thin TiO_2 sol–gels have been used for imprinting of azobenzene carboxylic acid [30]. In a recent application for the detection of cells [29], imprints of whole yeast cells in polyurethane layers and in sol–gel layers have been produced at the surface of a QCM crystal using a stamping method. The sensor could be used to quantify yeast cells in suspension at concentrations between 1×10^4 and 1×10^9 cells per mL under flow conditions. Others have relied on common acrylic polymers for the design of MIP-based QCM sensors, [25,27,28,31]. One reason for that was probably the abundance of know-how available on such polymers, and their adaptability to many different template molecules due to the plethora of available functional monomers. With such polymers, it has been demonstrated that the sensor selectivities are similar to those obtained in other applications of acrylic MIPs. For example, a QCM sensor coated with an S-propranolol-imprinted polymer was able to discriminate between the R and S-enantiomers of the drug with a selectivity coefficient of 5 [25]. The general procedure for the preparation of this sensor is shown in Section IX.

V. ELECTROCHEMICAL TRANSDUCERS

Electrochemical sensors can use amperometry, potentiometry, or conductometry as transduction principle [32]. Potentiometric sensors make use of the development of an electrical potential at an electrode surface in contact with ions that exchange with the surface. The potential is measured under zero-current conditions against a reference electrode and is proportional to the logarithm of the analyte activity in the sample. Potentiometric sensors are limited to the measurement of charged species or of gases that dissociate to yield charged species in an electrolyte. Ion-selective electrodes are one example of this sensor type.

In amperometric sensors, the current flowing between a working electrode and a counter electrode is measured as a function of the applied potential and usually normalized by incorporating a reference electrode. A current peak is obtained due to the transfer of electrons to or from a species. The most popular amperometric method for chemical sensing is cyclic voltammetry, but other methods such as, differential pulse voltammetry or square-wave voltammetry are also used. Since the electrochemical behavior, such as the oxidation potential, is characteristic for a certain compound, additional information on the analyte identity can sometimes be obtained. To use this transducer type, the analyte has to be electrochemically active. If this is not the case, a competitive or displacement sensor format may be used [11,33]. A labeled analyte derivative is allowed to compete with the analyte for the binding sites in the MIP, or the labeled analyte is allowed to bind first and is subsequently displaced upon binding of the analyte. If an analyte cannot be labeled easily, the use of non-related probes for detection can sometimes be a solution [34,35]. These can be conceived as compounds that can bind to some extent to the imprinted sites in the polymer, without being functionally related to the target analyte. Such compounds may need to have at least some degree of structural similarity with the analyte. The selectivity of sensors based on such competitive formats is not jeopardized, since selectivity is determined by the specificity of analyte (the original imprint molecule) binding to the imprinted sites. However, competitive sensors normally have broader dynamic ranges than sensors that directly measure the analyte.

For example, a voltammetric sensor for the herbicide 2,4-D was constructed [33] where the electroactive compound 2,5-dihydroxyphenylacetic acid was used as a probe instead of the labeled analyte. MIP particles (0.5 mg) were coated as a thin layer onto a screen-printed carbon electrode, and incubated with the sample to which the probe was added. In the presence of the analyte, some of the probe was competed out of the imprinted sites, whereas the remaining probe was directly quantified by differential pulse-voltammetric measurements (Fig. 5). Section X resumes the steps for the preparation of this sensor.

Conductometric sensors measure the change in conductivity of a selective layer in contact with two electrodes upon its interaction with the analyte. Conductometric sensors are often based on field-effect devices. For example, capacitance sensors such as the above-mentioned field-effect capacitor [5] belong to this group. Capacitive detection was also employed in conjunction with imprinted electropolymerized polyphenol layers on gold electrodes [36]. The sensitive layer was prepared by electropolymerization of phenol on the electrode in the presence of the template phenylalanine. The insulating properties of the polymer layer were studied by electrochemical impedance spectroscopy. Electrical leakages through the polymer layer

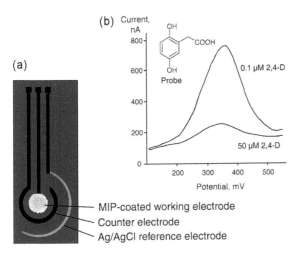

Figure 5 (a) Disposable sensor element based on a screen-printed carbon electrode [33]. The MIP was coated onto the carbon working electrode (middle) which is surrounded by a carbon counter electrode (smaller arc) and an Ag/AgCl reference electrode (larger arc); (b) differential pulse voltammetric scan after incubation of the electrode in 10 mM of the electroactive probe 2,5-dihydroxyphenylacetic acid in the presence of two different concentration of the competing analyte 2,4-D.

were suppressed by deposition of a self-assembled monolayer of mercaptophenol before polymerization and of alkanethiol after polymerization. After that, the template was removed. The multilayer system obtained displayed a decrease in electrical capacitance on addition of phenylalanine. Only a low response was observed toward other amino acids and phenol. The same authors also reported a capacitive creatinine sensor based on a photografted molecularly imprinted polymer [37]. Another example for this sensor type is a sensing device for the herbicide atrazine which is based on a freestanding atrazine-imprinted acrylic polymer membrane and conductometric measurements (Fig. 6) [38]. The authors carefully optimized the polymer recipe, in particular with respect to the kind and molar ratio of cross-linking monomers used, and the relative amount of porogenic solvent in the imprinting mixture. This turned out to be an important factor not only in obtaining flexible and stable membranes, but also because the conductometric response seemed to depend on the ability of the MIP to change its conformation upon analyte binding. Attractive features of this sensor were the comparatively short time required for one measurement (6–10 min), its rather low detection limit of 5 nM, and its good selectivity for atrazine over structurally related triazine herbicides. Section XI resumes the procedure of preparation of this sensor.

VI. INTERFACING THE MIP WITH THE TRANSDUCER

An important aspect in the design of an MIP-based sensor is to find an appropriate way of interfacing the polymer with the transducer. In most cases, the MIP has to be brought into close contact with the transducer surface. An obvious advantage would

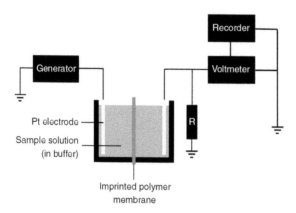

Figure 6 Conductometric sensor based on an MIP-membrane as the recognition element. (From Ref. 38.)

be to integrate this step in an automated production process. Thereby, the polymer can either be synthesized in situ at the transducer surface, or the surface can be coated with a preformed polymer.

In situ synthesis of a polymer can be done by electropolymerization on conducting surfaces like gold [26,36]. This is convenient but sometimes requires specialized polymer recipes that are, at least to date, less thoroughly studied than the common acrylic and vinyl polymers with respect to the possibility of being molecularly imprinted. Other, more general approaches are chemical grafting on a surface using polymerization initiators chemically bound [39] or physically adsorbed [40] to the surface. More generally applicable are standard surface coating techniques like spin-coating and spray coating, which can be used with the polymers commonly used for imprinting. With these two techniques, thin and even polymer layers can be produced. Spin-coating and spray-coating techniques have been employed to apply a thin film of monomer solution to acoustic transducer surfaces [21,23]. However, if radically polymerized vinyl or acrylic systems are used, the coating has to be done under oxygen-free conditions, due to the radical scavenging effect of oxygen which would inhibit polymerization. A rather simple way to synthesize a polymer layer on a flat surface in laboratory scale is to use a sandwich technique. The imprinting solution is cast between the transducer surface and another flat surface such as a glass or quartz disk, whereafter the polymerization is initiated [25,31]. The thickness of the layer can be somewhat controlled by varying the weight applied onto the sandwich assembly. Often the surface has to be activated before using the above coating techniques, to ensure stable adhesion of the MIP to the surface. For example, polymerizable double bonds can be generated at glass surfaces by silanization with methacryloyloxy trimethoxysilane.

Preformed polymers, for example in the form of nano- or micrometer-sized particles, can be interfaced with the transducer in different ways. It has been suggested to entrap MIP-particles into gels [11] or behind a membrane [33,41] for use with electrochemical transducers. Others have spin-coated, a suspension of MIP particles in a solution of an inert, soluble polymer (polyvinyl chloride), which served as glue, onto an acoustic transducer surface [28]. Among the potential problems that can arise when using these approaches, are diffusion limitations resulting in long response times of the sensor, nonspecific analyte binding, or a decrease in binding capacity.

Imprinted polymers exhibiting, at the same time, electrical conductivity facilitate their assembly with an electrochemical transducer in an integrated device. In this context, the preparation of composite particles consisting of an electrically conducting polymer (polypyrrole) and an acrylic MIP should be mentioned [42]. The polypyrrole that was grown into the preformed porous MIP did not alter its recognition properties. However, in this way MIP particles could be mechanically and electrically connected to a gold-covered silica substrate.

VII. OUTLOOK

The signals generated by most of the above-mentioned transducer types are two-dimensional and provide only limited information about the composition of the sample. Although this is normally compensated by the high selectivity of MIPs, a different strategy could conceivably be the use of "intelligent" transducer mechanisms, which generate signals with a higher inherent information content. One way to achieve that is to exploit the high molecular specificity of absorption spectra in the mid-infrared spectral region ($3500–500 \, cm^{-1}$). The combination of MIPs and FTIR spectrometry might allow analytical problems to be addressed where the selectivity of the MIP alone is not sufficient, e.g., when samples with complex matrices are to be investigated, or when structurally very similar analytes are present in the sample. A first approach towards a chemical sensor based on an imprinted polymer and infrared evanescent-wave spectroscopy has been reported recently [43].

In some biosensors, enzymes are involved as the recognition element and/or for the generation or amplification of the signal. Such sensors are often superior to sensors in which the signal is only due to the binding event itself. Analyte conversion and turnover result in an increased sensitivity and lower interference by nonspecific binding. There are some obvious parallels to MIP-based sensors, that is, with catalytic MIPs, a similar approach could conceivably be used. In an early attempt to create MIPs with catalytic activity, imprints of a substrate analogue were made, and the resulting polymer was able to hydrolyze the p-nitrophenyl ester of an amino acid [44]. Since then, several reports on catalytic MIPs using substrate, transition state or product analogues as the imprint molecule have followed [45–47]. Unfortunately, the rate enhancements which have been achieved so far are still modest compared to enzymes or even catalytic antibodies, and the application of catalytic MIPs in sensor technology will depend upon further improvement of their performance.

An important aspect in the development of sensor technology is the need for mass-produced and low-cost disposable transducers [48]. This is especially relevant for environmental and biomedical analysis. For electrochemical sensors, screen-printed electrodes fulfill this need, and the ease of preparation and low cost of MIPs make them attractive as recognition elements for such devices. A first report on this topic demonstrated that an imprinted polymer could be coated onto screen-printed carbon electrodes, and the resulting devices could be used as an amperometric sensor [33].

VIII. CONCLUSIONS

In terms of sensitivity, MIP-based biomimetic sensors are, with some honorable exceptions as outlined above, still somewhat inferior to biosensors. MIP-sensors measure only analyte binding: until now, no MIP-sensor has been described that uses an enzyme reaction for signal amplification as do certain biosensors. However, the situation will certainly improve through further optimization of the MIPs and the transducers. In particular, what one hopes to achieve is the development of MIPs that contain a more homogeneous binding site population, have a higher affinity for the target analyte, and that can be used in aqueous solvents. A considerable part of the current research efforts on MIPs already deals with these problems. In fact, some of the above-mentioned detection methods are, apart from their use in sensors, equally well suited for investigating the recognition of analytes by MIPs at the molecular level. On the other hand, the outstanding stability of MIPs, their low price, as well as the fact that they can be tailor-made for analytes for which a biological receptor cannot be found, are among the properties that make them especially suitable for sensor applications. It appears that the development of imprinted polymer-based sensors is just about to leave the proof-of-principle stage, and researchers are starting to address specific analytical problems and to measure real-world samples.

IX. EXPERIMENTAL PROTOCOL 1

A. Construction of an Acoustic Sensor Combining an Imprinted Polymer with a Quartz Crystal Microbalance Transducer [25]

The oscillating quartz crystal and its resonance frequency to be used depend on the QCM equipment available. As an example, AT-cut, 14 mm in diameter, plano-convex quartz crystals of 5 MHz resonant frequency can be used. A 5 mm diameter gold electrode film is vacuum deposited on each face of the crystal. The gold surface can be activated for better adherence of the polymer film by depositing a mixed layer of aminoethane thiol and propane thiol, followed by reacting the amino groups with methacrylic anhydride to introduce polymerizable double bonds. A copolymer of trimethylolpropane trimethacrylate (TRIM) and methacrylic acid (MAA) is used for molecular imprinting, and S-propranolol as the model template. A polymerization solution consisting of TRIM (6 mmol), MAA (10 mmol), toluene (3675 mL) and S-propranolol (1.25 mmol), with AIBN (0.28 mmol) as the polymerization initiator, is cooled on ice and purged with nitrogen for 2 min. The film is then cast by dispensing a 1-μL drop of the solution onto the surface of the gold electrode. The drop is immediately covered with a 6 mm diameter quartz disk. This sandwich assembly is placed on the flat window of a UV lamp with the cover quartz disk facing the lamp, and a weight of 0.5 kg is applied axially to the assembly. Polymerization is carried out under UV irradiation at 350 nm for 10 min. This results in a polymer layer with a thickness of about 2 μm. The sandwich is soaked in 80% ethanol in order to wash away unpolymerized monomers and facilitate removal of the cover disk from the polymer film. The S-propranolol template is then washed out by incubating the polymer layer in a solution of 20% acetic acid in acetonitrile. Measurements are ideally done using a flow cell. Incubation of the sensor in solutions of S-propranolol will result in a decrease of resonance frequency. To demonstrate the selectivity of the polymer, the opposite enantiomer, R-propranolol, can be used as a control.

X. EXPERIMENTAL PROTOCOL 2

A. Example for the Preparation of an MIP-Based Voltammetric Sensor using a Competitive Format with an Electroactive Probe [33]

1. Imprinted Polymer Preparation

An imprinted polymer block is prepared using 2,4-dichlorophenoxyacetic acid (2,4-D) as the template, 4-vinylpyridine (4-VP) as the functional monomer and ethylene-glycol dimethacrylate (EDMA) as the cross-linker. EDMA (20 mmol), 4-VP (4 mmol), of 2,4-D (1 mmol) and the polymerization initiator 2,2'-azobis(2,4-dimethylvaleronitrile (0.31 mmol) are weighed into a glass test tube and dissolved in 4 mL of methanol and 1 mL of ultrapure water. The solution is sonicated, cooled on ice, bubbled with nitrogen for 2 min, and placed in a thermostated water bath at 45°C for 4 h, followed by 2 h at 60°C. A control polymer can be prepared using the same recipe but without the addition of the template. The resultant hard bulk polymer is ground in a mechanical mortar, and the particles are then resuspended in acetone and allowed to settle for 1 h. The fine particles that remain in suspension (diameter 0.5–1 μm) are collected. If not enough fines are obtained, the grinding and sedimentation procedures are repeated. The particles are washed by incubation in methanol/acetic acid (7:3) (2×), acetonitrile/acetic acid (9:1) (2×), acetonitrile, and methanol (2×) for 2 h each time, followed by centrifugation. The solvent is then removed by centrifugation, and the particles dried in vacuo.

2. Preparation of Polymer-Coated Electrodes

The screen-printed electrodes used should be three-electrode systems, for example the model shown in Fig. 5 [49], which have carbon working and counter electrodes and an Ag/AgCl reference. An electrode is covered with a mask of adhesive tape leaving the working electrode free. Twenty-five microliters of a 20 mg/mL suspension of imprinted polymer particles in methanol are pipetted onto the working electrode and left to dry. The mask is then removed and the particle layer overlaid with 10 μL of a hot 2% low-melting agarose solution in water and covered with a thin plastic foil of a diameter slightly larger than the working electrode. Once gelling of the agarose is complete, the plastic foil is removed, and the electrode dried at 40°C. In this way, the agarose forms a thin membrane covering the working electrode.

3. Electrochemical Measurements

The electrodes are placed in solutions containing the electroactive probe 2,5-dihydroxyphenylacetic acid (10 μM) and the analyte 2,4-D in 20 mM phosphate buffer, pH 7, 10% methanol, for 1 h. Following incubation, the electrodes are rinsed briefly with ultrapure water to avoid carryover of the probe. The electrodes are then transferred into a solution of 0.1 M KCl in 20 mM phosphate buffer, pH 7, 10% methanol, and the amount of bound probe is quantified by differential-pulse voltammetry. The measurement parameters are as follows: potential window 100–550 mV, scan rate 30 mV/s, sample width 17 ms, pulse amplitude 50 mV, pulse width (modulation time) 50 ms, pulse period (interval) 200 ms, quiet time 2 s, and sensitivity 10–5 A/V. Peak currents are determined either after subtraction of a manually added baseline or as absolute peak heights above zero.

XI. EXPERIMENTAL PROTOCOL 3

A. Example for the Preparation of a MIP-Based Conductometric Sensor [38]

1. MIP Membrane Preparation

Imprinted polymer membranes are prepared using atrazine as the template, methacrylic acid as a functional monomer, and tri(ethylene glycol) dimethacrylate (TEDMA) as a cross-linker. The molar ratio of the functional monomer to the template is 5:1. This ratio has to be optimized for each template. In order to obtain thin, flexible and mechanically stable membranes, oligourethane acrylate (molecular mass 2600) is added to the monomer mixture. Preparation of the molecularly imprinted polymer membrane is done as follows. Atrazine (20 mg) is mixed with methacrylic acid (40 mg), TEDMA (289 mg), oligourethane acrylate (51 mg), AIBN (2 mg) and 30% v/v of chloroform. Then a 60–120 μm gap between two quartz slides is filled with the monomer mixture. To initiate polymerization, the slides with the mixture are exposed to UV radiation (365 nm, intensity 20 W m^{-2}) for 30 min. After polymerization, atrazine is extracted with ethanol in a Soxhlet apparatus for 2 h. This should not cause any visible changes in the MIP membrane. A membrane for control experiments can be prepared similarly except that no atrazine is added to the monomer mixture.

2. Conductivity Measurements

The membrane is then mounted into an experimental setup as shown in Fig. 6. The conductivity measurements are carried out with an electrochemical cell in which two platinum electrodes are separated by the imprinted-polymer membrane. Potassium phosphate buffer (25 mM, pH 7.5) containing 35 mM NaCl is used as the background electrolyte. To avoid electrode polarization, a probing ac signal of small amplitude (60 mV) in the frequency range 20 Hz–2 kHz, generated by a waveform generator, is used. The voltage drop on the resistor connected in series with the cell is recorded with a nanovoltmeter. Responses to the analyte are obtained by adding different amounts of atrazine to the electrochemical cell. The change in the membrane electrical conductivity is then calculated and plotted as a function of atrazine concentration.

REFERENCES

1. Vlatakis, G.; Andersson, L.I.; Müller, R.; Mosbach, K. Drug Assay Using Antibody Mimics Made by Molecular Imprinting. Nature **1993**, *361*, 645–647.
2. Andersson, L.; Mandenius, C.F.; Mosbach, K. Studies on Guest Selective Molecular Recognition on an Octadecyl Silylated Silicon Surface Using Ellipsometry. Tetrahedron Lett. **1988**, *29*, 5437–5440.
3. Andersson, L.I.; Miyabayashi, A.; O'Shannessy, D.J.; Mosbach, K. Enantiometric Resolution of Amino Acid Derivatives on Molecularly Imprinted Polymers as Monitored by Potentiometric Measurements. J. Chromatogr. **1990**, *516*, 323–331.
4. Piletsky, S.A.; Parhometz, Y.P.; Lavryk, N.V.; Panasyuk, T.L.; El'skaya, A.V. Sensors for Low-Weight Organic Molecules Based on Molecular Imprinting Technique. Sens. Actuators B **1994**, *18–19*, 629–631.

5. Hedborg, E.; Winquist, F.; Lundström, I.; Andersson, L.I.; Mosbach, K. Some Studies of Molecularly Imprinted Polymer Membranes in Combination with Field-Effect Devices. Sen. Actuators A **1993**, *36–38*, 796–799.

6. Kriz, D.; Kempe, M.; Mosbach, K. Introduction of Molecularly Imprinted Polymers as Recognition Elements in Conductometric Chemical Sensors. Sens. Actuators B **1996**, *33*, 178–181.

7. Piletsky, S.A.; Piletskaya, E.V.; Elgersma, A.V.; Yano, K.; Karube, I.; Parhometz, Y.P.; El'skaya, A.V. Atrazine Sensing by Molecularly Imprinted Membranes. Biosens. Bioelectron. **1995**, *10*, 959–964.

8. Piletsky, S.; Piletskaya, E.V.; Panasyuk, T.L.; El'skaya, A.V.; Levi, R.; Karube, I.; Wulff, G. Imprinted Membranes for Sensor Technology: Opposite Behavior of Covalently and Noncovalently Imprinted Membranes. Macromolecules **1998**, *31*, 2137–2140.

9. Weber, A.; Dettling, M.; Brunner, H.; Tovar, G.E.M. Isothermal Titration Calorimetry of Molecularly Imprinted Polymer Nanospheres. Macromol. Rapid Commun. **2002**, *23*, 824–828.

10. Xie, B.; Ramanathan, K.; Danielsson, B. Principles of Enzyme Thermistor Systems: Applications of Biomedical and Other Measurements. Adv. Biochem. Eng. Biotechnol. **1999**, *64*, 1–33.

11. Kriz, D.; Mosbach, K. Competitive Amperometric Morphine Sensor Based on an Agarose Immobilized Molecularly Imprinted Polymer. Anal. Chim. Acta **1995**, *300*, 71–75.

12. Andersson, L.I.; Paprica, A.; Arvidsson, T. A Highly Selective Solid Phase Extraction Sorbent for Pre-Concentration of Sameridine Made by Molecular Imprinting. Chromatographia **1997**, *46*, 57–62.

13. Chen, G.H.; Guan, Z.B.; Chen, C.T.; Fu, L.T.; Sundaresan, V.; Arnold, F.H. A Glucose Sensing Polymer. Nature Biotechnol. **1997**, *15*, 354–357.

14. Cooper, M.E.; Hoag, B.P.; Gin, D.L. Design and Synthesis of Novel Fluorescent Chemosensors for Biologically Active Molecules. Polym. Prepr. **1997**, *38*, 209–210.

15. Turkewitsch, P.; Wandelt, B.; Darling, G.D.; Powell, W.S. Fluorescent Recognition Sites through Molecular Imprinting. A Polymer-Based Fluorescent Chemosensor for Aqueous Camp. Anal. Chem. **1998**, *70*, 2025–2030.

16. Liao, Y.; Wang, W.; Wang, B. Building Fluorescent Sensors by Template Polymerization: The Preperation of a Fluorescent Sensor for L-Tryptophan. Bioorg. Chem. **1999**, *27*, 463–476.

17. Jenkins, A.L.; Uy, O.M.; Murray, G.M. Polymer-Based Lanthanide Luminescent Sensor for Detection of the Hydrolysis Product of the Nerve Agent Soman in Water. Anal. Chem. **1999**, *71*, 373–378.

18. Benes, E.; Groschl, M.; Burger, W.; Schmidt, M. Sensors Based on Piezoelectric Resonators. Sens. Actuators A **1995**, *48*, 1–21.

19. Tom-Moy, M.; Baer, R.L.; Spira-Solomon, D.; Doherty, T.P. Atrazine Measurements Using Surface Transverse Wave Devices. Anal. Chem. **1995**, *67*, 1510–1516.

20. Dickert, F.L.; Hayden, O.; Halikias, K.P. Synthetic Receptors as Sensor Coatings for Molecules and Living Cells. Analyst **2001**, *126*, 766–771.

21. Dickert, F.L.; Forth, P.; Lieberzeit, P.; Tortschanoff, M. Molecular Imprinting in Chemical Sensing - Detection of Aromatic and Halogenated Hydrocarbons as Well as Polar Solvent Vapors. Fresenius J. Anal. Chem. **1998**, *360*, 759–762.

22. Dickert, F.L.; Tortschanoff, M.; Bulst, W.E.; Fischerauer, G. Molecularly Imprinted Sensor Layers for the Detection of Polycyclic Aromatic Hydrocarbons in Water. Anal. Chem. **1999**, *71*, 4559–4563.

23. Jakoby, B.; Ismail, G.M.; Byfield, M.P.; Vellekoop, M.J. A Novel Molecularly Imprinted Thin Film Applied to a Love Wave Gas Sensor. Sens. Actuators A **1999**, *76*, 93–97.

24. Dickert, F.L.; Thierer, S. Molecularly Imprinted Polymers for Optochemical Sensors. Adv. Mater. **1996**, *8*, 987–990.

25. Haupt, K.; Noworyta, K.; Kutner, W. Imprinted Polymer-Based Enantioselective Acoustic Sensor Using a Quartz Crystal Microbalance. Anal. Commun. **1999**, *36*, 391–393.

26. Malitesta, C.; Losito, I.; Zambonin, P.G. Molecularly Imprinted Electrosynthesized Polymers: New Materials for Biomimetic Sensors. Anal. Chem. **1999**, *71*, 1366–1370.

27. Ji, H.-S.; McNiven, S.; Ikebukuro, K.; Karube, I. Selective Piezoelectric Odor Sensors Using Molecularly Imprinted Polymers. Anal. Chim. Acta **1999**, *390*, 93–100.

28. Liang, C.; Peng, H.; Bao, X.; Nie, L.; Yao, S. Study of Molecular Imprinting Polymer Coated Baw Bio-Mimetic Sensor and Its Application to the Determination of Caffeine in Human Serum and Urine. Analyst **1999**, *124*, 1781–1785.

29. Dickert, F.L.; Hayden, O. Bioimprinting of Polymers and Sol-Gel Phases. Selective Detection of Yeasts with Imprinted Polymers. Anal. Chem **2002**, *74*, 1302–1306.

30. Lee, S.W.; Ichinose, I.; Kunitake, T. Molecular Imprinting of Azobenzene Carboxylic Acid on a Tio2 Ultrathin Film by the Surface Sol-Gel Process. Langmuir **1998**, *14*, 2857–2863.

31. Kugimiya, A.; Takeuchi, T. Molecularly Imprinted Polymer-Coated Quartz Crystal Microbalance for Detection of Biological Hormone. Electroanalysis **1999**, *11*, 1158–1160.

32. Fabry, P.; Siebert, E. Electrochemical Sensors. In *The CRC Handbook of Solid-State Electrochemistry*; Gellings, P.J., Bouwmeester, H.J.M., Eds.; CRC: Boca Raton, 1997; pp. 329–369.

33. Kröger, S.; Turner, A.P.F.; Mosbach, K.; Haupt, K. Imprinted Polymer-Based Sensor System for Herbicides Using Differential Pulse Voltammetry on Screen-Printed Electrodes. Anal. Chem. **1999**, *71*, 3698–3702.

34. Haupt, K. Molecularly Imprinted Sorbent Assays and the Use of Non-Related Probes. React. Funct. Polym. **1999**, *41*, 125–131.

35. Piletsky, S.A.; Terpetschnik, E.; Andersson, H.S.; Nicholls, I.A.; Wolfbeis, O.S. Application of Non-Specific Fluorescent Dyes for Monitoring Enantio-Selective Ligand Binding to Molecularly Imprinted Polymers. Fresenius' J. Anal. Chem. **1999**, *364*, 512–516.

36. Panasyuk, T.L.; Mirsky, V.M.; Piletsky, S.A.; Wolfbeis, O.S. Electropolymerized Molecularly Imprinted Polymers as Receptor Layers in Capacitive Chemical Sensors. Anal. Chem. **1999**, *71*, 4609–4613.

37. Panasyuk-Delaney, T.; Mirsky, V.M.; Wolfbeis, O.S. Capacitive Creatinine Sensor Based on a Photografted Molecularly Imprinted Polymer. Electroanalysis **2002**, *14*, 221–224.

38. Sergeyeva, T.A.; Piletsky, S.A.; Brovko, A.A.; Slinchenko, E.A.; Sergeeva, L.M.; Panasyuk, T.L.; Elskaya, A.V. Conductometric Sensor for Atrazine Detection Based on Molecularly Imprinted Polymer Membranes. Analyst **1999**, *124*, 331–334.

39. Sulitzky, C.; Ruckert, B.; Hall, A.J.; Lanza, F.; Unger, K.; Sellergren, B. Grafting of Molecularly Imprinted Polymer Films on Silica Supports Containing Surface-Bound Free Radical Initiators. Macromolecules **2002**, *35*, 79–91.

40. Panasyuk-Delaney, T.; Mirsky, V.M.; Ulbricht, M.; Wolfbeis, O.S. Impedometric Herbicide Chemosensors Based on Molecularly Imprinted Polymers. Anal. Chim. Acta **2001**, *435*, 157–162.

41. Kriz, D.; Ramström, O.; Svensson, A.; Mosbach, K. Introducing Biomimetic Sensors Based on Molecularly Imprinted Polymers as Recognition Elements. Anal. Chem. **1995**, *67*, 2142–2144.

42. Kriz, D.; Andersson, L.I.; Khayyami, M.; Danielsson, B.; Larsson, P.-O.; Mosbach, K. Preparation and Characterization of Composite Polymers Exhibiting Both Selective Molecular Recognition and Electrical Conductivity. Biomimetics **1995**, *3*, 81–90.

43. Jakusch, M.; Janotta, M.; Mizaikoff, B.; Mosbach, K.; Haupt, K. Molecularly Imprinted Polymers and Infrared Evanescent Wave Spectroscopy. A Chemical Sensors Appraoch. Anal. Chem. **1999**, *71*, 4786–4791.
44. Leonhardt, A.; Mosbach, K. Enzyme-Mimicking Polymers Exhibiting Specific Substrate Binding and Catalytic Functions. React. Polym. **1987**, *6*, 285–290.
45. Sellergren, B.; Shea, K.J. Enantioselective Ester Hydrolysis Catalyzed by Imprinted Polymers. Tetrahedron: Asymmetry **1994**, *5*, 1403–1406.
46. Ohkubo, K.; Funakoshi, Y.; Urata, Y.; Hirota, S.; Usui, S.; Sagawa., T. High Esterolytic Activity of a Novel Water-Soluble Polymer Catalyst Imprinted by a Transition-State Analogue. J. Chem. Soc., Chem. Commun. **1995**, *20*, 2143–2144.
47. Wulff, G.; Gross, T.; Schönfeld, R. Enzyme Models Based on Molecularly Imprinted Polymers with String Esterase Activity. Angew. Chem. Int. Ed. **1997**, *36*, 1962–1964.
48. Alvarez-Icaza, M.; Bilitewski, U. Mass production of Biosensors. Anal. Chem. **1993**, *65*, 525A–533A.
49. Kröger, S.; Turner, A.P.F. Solvent-Resistant Carbon Electrodes Screen Printed onto Plastic for Use in Biosensors. Anal. Chim. Acta **1997**, *347*, 9–18.

27

Molecularly Imprinted Polymers as Recognition Elements in Optical Sensors

Shouhai Gao   Toronto Research Chemicals, Inc., North York, Ontario, Canada

Wei Wang   University of New Mexico, Albuquerque, New Mexico, U.S.A

Binghe Wang   Georgia State University, Atlanta, Georgia, U.S.A.

I. INTRODUCTION

Custom-made fluorescent and other optical sensors for organic molecules have a wide range of potential applications [1–4]. Traditionally, such sensors have been prepared through de novo design and synthesis. For example, many very sensitive fluorescent sensors have been designed for peptides [5], metal ions [6–10], saccharides [11–13], and others [12,14–18]. However, the de novo design approach requires a thorough knowledge of the structural features of the analyte including conformational features and functional group orientations. Molecular imprinting or template-directed polymerization, on the other hand, offers an opportunity for the construction of well-defined binding cavities for a particular analyte without the prior knowledge of the three-dimensional structure of the analyte and the de novo construction of the complementary binding sites. Several other chapters in the book deal with ways to construct binding sites with high selectivity and affinity (see Part II of the book for detail). Therefore, this will not be repeated here. However, from the construction of a binding site to the development of an optical sensor, there is a huge gap. This is because a sensing device requires the output of a detectable signal. One of the major obstacles to overcome in using molecular imprinting to construct optical sensors, in addition to the problem of constructing high affinity and high selectivity binding sites, is to find a highly sensitive way to signal the binding event.

Fluorescence is one of the most sensitive methods of detection. Therefore, it is not surprising that most of the work in optical sensing in the field of molecular imprinting has been on fluorescent sensing. Consequently, this is also the focus of this chapter. The development of fluorescent sensors requires that the imprinted polymers be constructed in such a way so that binding of the analytes would generate a fluorescence signal. Assuming high affinity and selective polymeric receptors can be constructed, the fluorescent signaling of the binding event can be achieved in one of three ways. First, if the analyte itself is fluorescent, the binding itself should increase the fluorescence intensity of the imprinted polymer. Scenario 1 in Fig. 1 depicts such a system. However, such an approach is not generally applicable since most of the analytes, which one needs to make sensors for to achieve sensitive detection, are not strongly fluorescent. Second, for analytes that are not fluorescent themselves, one can always use a fluorescently labeled analyte derivative that can be used in a competitive assay (Scenario 2, Fig. 1). This is much the same as the radioligand displacement assay commonly used in receptor and antibody binding experiments. These two scenarios require washing of the assay solution to remove nonspecific binding before the determination of the fluorescence intensity and/or spectrum in the batch mode, which is different from the following cases. The third approach is the incorporation of a fluorophore as a fluorescent reporter compound into the imprinted polymer. If binding of an analyte causes the perturbation of the absorption and/or emission of the fluorophore, the binding event can be signaled with a change in fluorescence emission intensity and/or wavelength of the incorporated reporter (Scenario 3a). For analytes that cannot interact with the incorporated fluorophore strongly enough to cause a significant change in the fluorescence intensity and/or wavelength, one can use a nonspecific external quencher or modifier to affect the fluorescence change upon analyte binding (Scenario 3b). This three-component assay can be a general strategy for the construction of sensors for different kinds of analytes. The flow of this chapter will be divided into different sections based on the above classification. However, it needs to be noted that this classification is artificial and does not apply to every situation, and there are examples that could fit into more than one category. It also needs to be noted that this chapter does not strive for comprehensiveness in terms of covering all the papers published in the field of optical sensors using imprinted polymers. Instead, it focuses on using examples to illustrate principles based on which imprinted polymeric optical sensors can be prepared. Such an approach includes two specific examples with typical experimental protocols (see Sections IV.A and IV.C).

II. IMPRINTED POLYMERS FOR THE DETECTION OF FLUORESCENT ANALYTES

If the target analyte itself is fluorescent, it becomes an easy task for the detection of the analyte with imprinted polymers, since binding is expected to increase the fluorescence of the imprinted polymer (Scenario 1, Fig. 1). In such a case, the function of the imprinted polymers is solely to selectively recognize the analyte. The reporting event (fluorescence) comes from the analyte itself. Some examples of this type of imprinted polymeric sensors and their polymerization conditions are summarized in Table 1.

A fluorescent sensor for dansyl-phenylalanine (dansyl-Phe-OH) (Fig. 2, 1) has been successfully constructed by using methacrylic acid (MAA) and 2-vinyl-pyridine (2VPy) as the functional monomers and ethylene glycol dimethacrylate (EDMA) as the cross-linker. The fluorescence of the imprinted polymers upon analyte binding was detected using fiber optics, and the signal was found to be a function of the analyte concentration. The sensor also showed a moderate degree

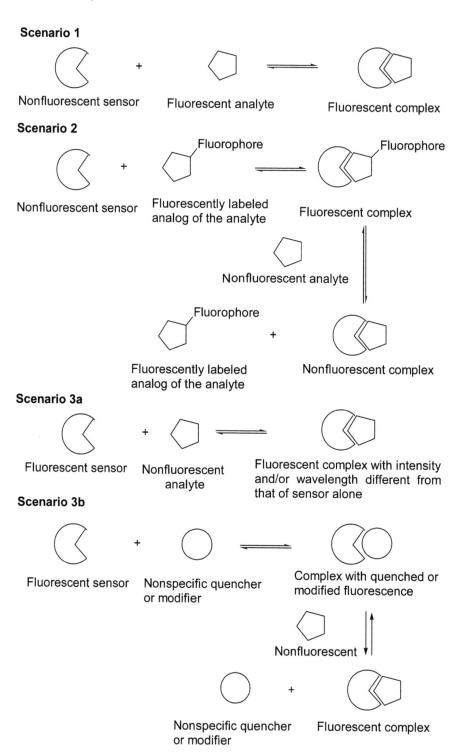

Figure 1 Different ways of achieving fluorescent sensing using imprinted polymers.

Table 1 Imprinted Polymers as the Recognition Moiety for the Detection of Analytes with Fluorescence

Analyte	Monomer	Cross-linker	Template:monomer: cross-linker	Polymerization conditions[a]	Reference
DansylPhe-OH	MAA 2VPy	EDMA	1:4:40	Initiator: ABDV Porogen: CH$_3$CN Temperature: 45°C Time: 15 h	[19]
Cinchonidine	TFMAA	EDMA	1:4:28	Initiator: AIBN Porogen: CHCl$_3$ Temperature: 4°C Time: 16 h UV irradiation	[20]
Flavonol	MAA	EDMA	1:8:37.4	Initiator: AIBN Porogen: CHCl$_3$ Temperature: 70°C Time: 14 h	[21]
β-Estradiol	MAA	EDMA	1:8:25	Initiator: AIBN Porogen: CH$_3$CN Temperature: 40°C Time: 16 h	[22]
Fluorescein NATA	TES	HAPTES	1:1:3 (v:v:v)	Initiator: none Porogen: none Temperature: 4°C Time: 8 h	[23]
PAHs	PAHs	8a,b	9a,b	Initiator: none Porogen: THF Temperature: 70°C Time: several h	[24–26]

[a]ABDV: 2,2'-azobis(2,4-dimethylvaleronitrile); AIBN: 2,2'-azobis(isobutyronitrile).

of stereoselectivity in acetonitrile for the L-form of the analyte (dansyl-L-Phe-OH) (Fig. 2, **1**), which was the original "print" molecule [19]. No binding constants were reported.

Takeuchi et al. [20] have successfully used MAA and 2-(trifluoromethyl) acrylic acid (TFMAA) for the preparation of imprinted polymers for the fluorescent cinchona alkaloids (Fig. 2, **2a,b**). The binding affinity of the imprinted polymers for the print compound was evaluated chromatographically. It was found that the imprinted polymers exhibited diastereoselectivity for the alkaloids. What is more interesting in this study is that upon binding to the imprint polymer, the fluorescence spectrum of the analyte showed a wavelength shift with the imprinted polymers prepared with TFMAA. For example, cinchonidine (Fig. 2, **2a**) binding shifted the emission λ_{max} from 390 nm in the free form to 360 nm in the bound form. With increasing concentrations of the analyte, a slow recovery of the emission λ_{max} was observed, presumably due to the increasing percentage of the analyte in the free form. The fluorescence shift offers a chance for in situ detection without the separation of the free/bound form. The fluorescence wavelength shift upon binding was thought to be due to the protonation of the analyte within the binding cavity. Since the fluorescence

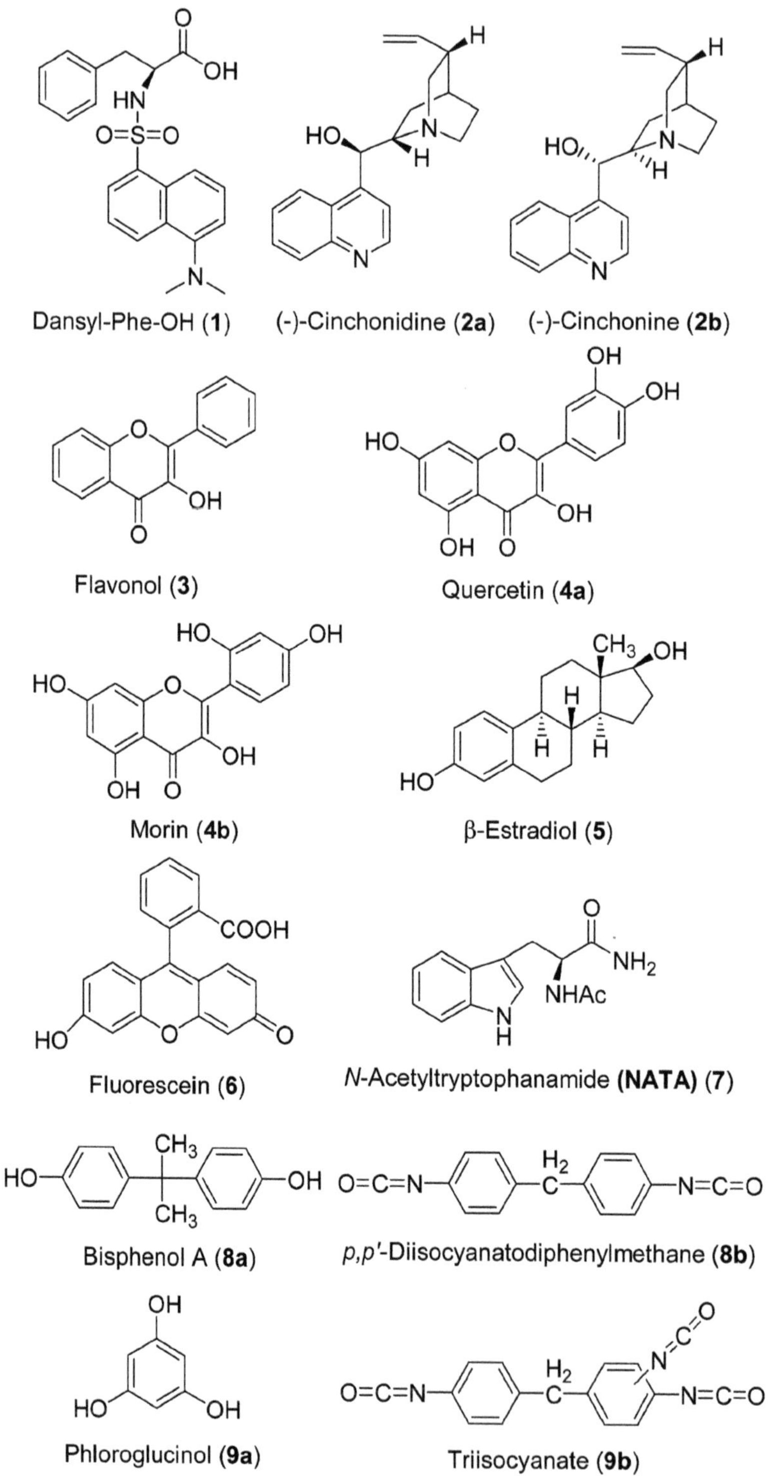

Figure 2 List of analytes.

studies were carried out in a nonaqueous solution (chloroform–acetonitrile), this seems to make sense, particularly considering that TFMAA is a strong acid (pKa 2.3). However, if protonation was indeed the reason for this shift, it would be hard to use similar systems for the development of sensors that are functional in buffered aqueous solutions since in such cases, the pH and therefore the protonation state of the analyte would be expected to remain about the same between the bound and free form.

Suarez-Rodriquez and Díaz-García [21] have developed a flow-through fluorescent sensing device for flavonol-type compounds (Fig. 2, 3, **4a,b**) using noncovalently imprinted polymers. For polymer preparation, flavonol (Fig. 2, **3**) was used as the template molecule, and MAA and EDMA were used as the functional monomer and cross-linker, respectively. The polymers prepared were able to distinguish flavonoids that differ in the placement of hydroxyl groups. Extensive experiments were carried out to examine the reproducibility, detection limit, and detection concentration range. It was found that the detection limit could reach 5×10^{-8} M. Samples in different matrices were also examined to show that the device developed could be used for the analysis of flavonoids in hydrophobic samples (foods, cosmetics, gasoline, etc.) without a separation step.

A fluorescence sensing method based on a combination of HPLC separation and fluorescence detection was developed for the steroid hormone β-estradiol (Fig. 2, **5**), which is fluorescent [22]. The imprinted polymer in this system was prepared by the co-polymerization of MAA and EDMA in the presence of the template molecule (Table 1), β-estradiol. This system was able to rapidly and effectively measure β-estradiol concentrations in the range of 0.1–4.0 μM with satisfactory reproducibility. However, this in essence is a HPLC method with the imprinted polymers used as the stationary phase.

Lulka and co-workers have imprinted two fluorescent compounds, fluorescein (Fig. 2, **6**) and N-acetyltryptophanamide (NATA) (Fig. 2, **7**) with organic silanes bis (2-hydroxyethyl) -aminopropryltriethoxysilane (HAPTES) and tetraethoxysilane (TES) [23]. The binding constants were determined with steady-state fluorescence spectroscopy. It was found that high affinity binding sites with K_d's in the range of sub- to low micromolar were achieved.

Dickert and co-workers have been working on making polyurethane-based imprinted polymers, using **8a,b** (Fig. 2) as the monomers and **9a,b** as the cross linkers, for the fluorescent detection of polyaromatic compounds (PAHs) [24–26]. Such imprinted polymers interact with the analytes through either van der Waals or hydrophobic interactions when the analysis was carried out in an aqueous solution. In their studies, the polymers were coated onto the surface of quartz or microelectronic devices. Good selectivity and high sensitivity were achieved with detection limit of up to ppt in some cases.

III. IMPRINTED POLYMERS AS THE RECOGNITION MOIETY FOR THE FLUORESCENT DETECTION OF NONFLUORESCENT ANALYTES USING A FLUORESCENTLY LABELED ANALOG IN COMPETITIVE ASSAYS

If the analyte itself is not fluorescent, a competitive or displacement system can be used for detecting the binding event (Scenario 2, Fig. 1). In this competitive sensing system, a labeled analyte derivative is allowed to compete with the analyte for the binding sites in the imprinted polymers. The relative affinity of the analyte and its labeled derivative determines the ratio of these two species bound to the imprinted polymer, which can be used for the analysis of the concentration. Table 2 offers some examples of imprinted polymers as potential sensors useful in competitive assays.

A fluorescent ligand displacement assay has been developed for the herbicide 2,4-dicholophenoxyacetic acid (2,4-D) (Fig. 3, 10) using imprinted polymers [27]. The imprinted polymer was prepared with 2,4-D as the template, 4-VPy as the functional monomer, and EDMA as the cross-linker. This assay system is analogous to a competitive fluoroimmunoassay and uses a coumarin derivative, 7-carboxy-methoxy-4-methylcoumarin (CMMC) (Fig. 3, 11) as a nonrelated fluorescent probe. This avoids the use of radiolabels commonly employed in imprinted polymer-based assays and the handling of radioactive material. The specificity and sensitivity of this assay were compared with that using radiolabeled 2,4-D, and these two methods were found to give similar results, which is remarkable. It has also been shown that this assay can be used both in aqueous buffer and in organic solvent such as acetonitrile with a detection limit of about 100 nM. However, the somewhat high cross-reactivity among analogs still makes this less practical than the corresponding antibody method, which has very low cross-reactivity. It is also interesting to note that the binding curves for 2,4-D and CMMC were very similar, particularly in acetonitrile further indicating the high cross-reactivity among analogous compounds. The same group also developed a method for the chemiluminescence imaging ELISA for 2,4-D using the same monomer and TRIM as the cross-linker [28,29]. In this case, tobacco peroxidase was used as the enzyme and the chemiluminescent reaction with luminol was used for the detection. Infrared evanescent wave spectroscopy was also used to study 2,4-D imprinted polymers prepared using 4-VPy and MMA as the

Table 2 The MIP-Based Fluorescent Sensors for the Detection of Nonfluorescent Analytes Using a Fluorescently Labeled Analog in Competitive Assays

Analyte	Monomer	Cross-linker	Template: Monomer: cross-linker	Polymerization conditions[a]	Reference
2,4-D	4-VPy	EDMA or TRIM	1:4:20	Initiator: ABDV	[27–31]
				Porogen: MeOH/ H_2O (4:1)	
				Temperature: 45°C	
				Time: 6 h	
L-Phe-NH$_2$	MAA	EDMA	1:4:16	Initiator: AIBN	[32]
				Porogen: CHCl$_3$	
				Temperature: 80°C	
				Time: 18 h	
CAP	DEAEM	EDMA	1:2:24	Initiator: AIBN	[33]
				Porogen: THF	
				Temperature: 60°C	
				Time: 48 h	

[a]ABDV: 2,2′-azobis(2,4-dimethylvaleronitrile); AIBN: 2,2′-azobis(isobutyronitrile).

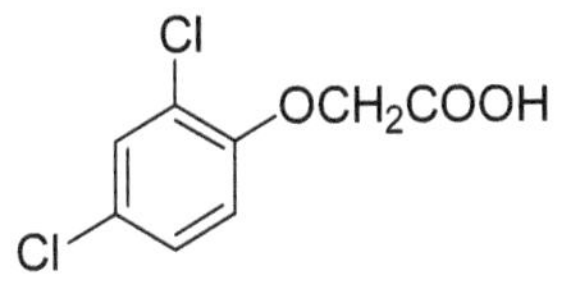

2,4-D (10)

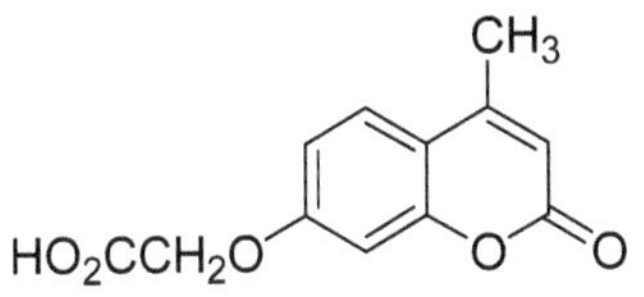

CMMC (11)

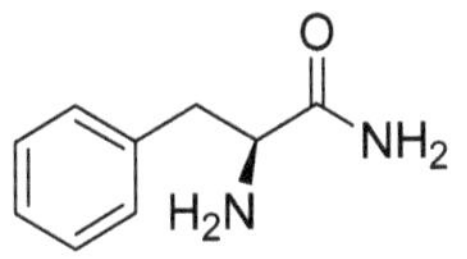

L-Phenylalaninamide (12)

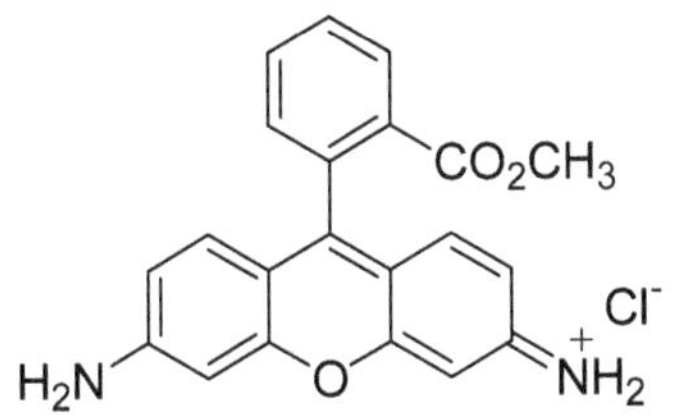

Rhodamine B (13)

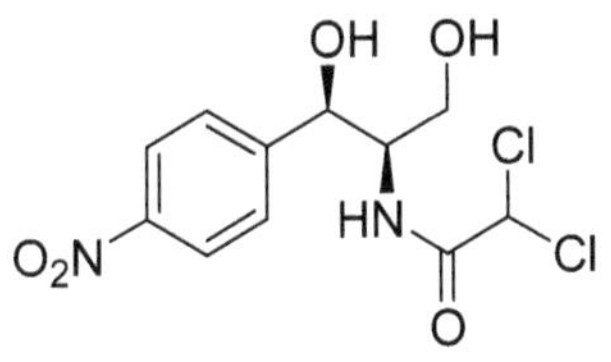

Chloramphenicol (CAP) (14)

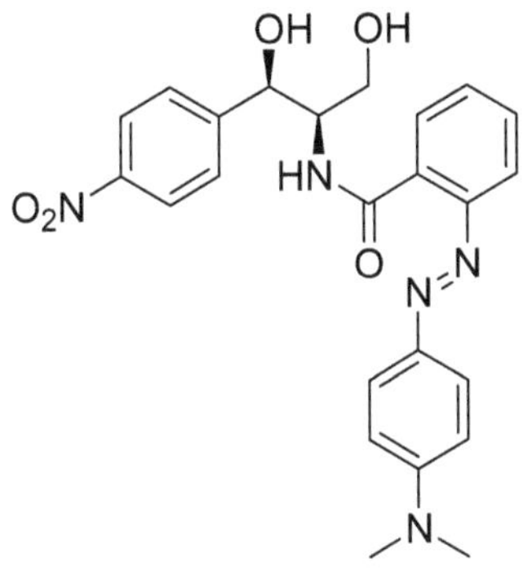

CAP-methyl red (CAP-MR) (15)

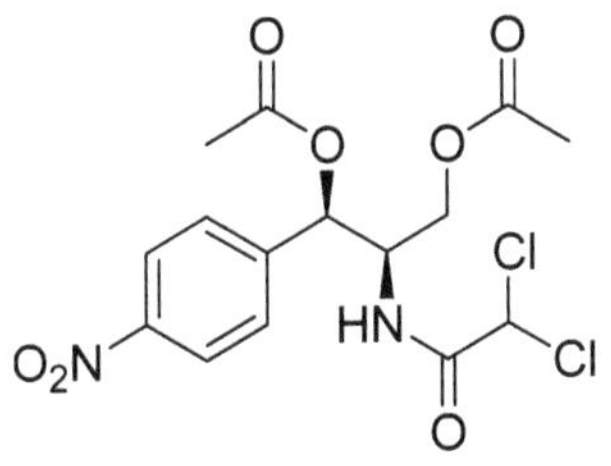

Chloramphenicol diacetate (CAP-DA) (16)

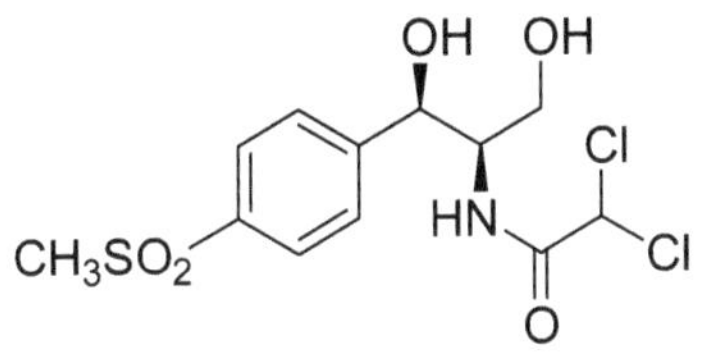

Thiamphenicol (TAM) (17)

Figure 3 List of analytes.

functional monomers and EDMA as the cross linker [30,31]. K_d's in the low micro-molar range have been obtained for these imprinted polymers.

Piletsky et al. [32] have prepared L-phenylalaninamide (L-Phe-NH$_2$) (Fig. 3, 12) imprinted polymers using MAA as the functional monomer and EDMA as the cross-linker. The polymers were evaluated for their binding to the template using HPLC through the use of nonspecific fluorescent dyes, such as rhodamine B (Fig. 3, 13). The displacement of the fluorescent dyes upon addition of the analyte was used as a way to detect the binding event. Again, this in essence is a HPLC method.

Specifically, the imprinted polymer is first loaded with the dye, and a solution of the dye in the eluent is passed through the polymer during HPLC. If analyte is injected into the dye solution in the eluent, part of the dye is competitively replaced by the analyte from the polymeric stationary phase, which causes a fluorescence signal change. The method, based on the displacement of a non-specific dye by an analyte, is highly reproducible and relatively sensitive. The advantage of this is the same as the previous example, i.e., it is generally applicable in the sense that the analytes themselves do not need to be chromophoric or fluorescent for sensitive detection. The imprinted polymers showed binding affinity for the template, L-Phe-NH$_2$, with a K_d of about 60 μM. However, the same imprinted polymer also showed a K_d of about 133 μM for the dye, rhodamine B, and the control polymer had a fairly high affinity for the template (K_d 83 μM).

An optical sensing system for the determination of chloramphenicol (CAP) (Fig. 3, 14) was developed by Karube et al. [33] using imprinted polymers. The best of these CAP-imprinted polymers was prepared using (diethylamino) ethyl methacrylate (DEAEM) as the functional monomer with a monomer:template ratio of 2:1. The method is based on competitive displacement of a chloramphenicol–methyl red (CAP-MR) (Fig. 3, 15) dye conjugate from the binding cavities in the imprinted polymer by the analyte during HPLC chromatography. The general principle of this approach is very similar to that described in the previous section and would work with any other optical tag such as a fluorophore substituting methyl red. It was reported that this system rapidly and effectively detected CAP with a linear response range of 3–100 μg/mL. The polymers also showed good selectivity. For example, chloramphenicol diacetate (CAP-DA) (Fig. 3, 16), which is structurally similar to CAP and yet lacks the appropriate functional groups, was not detected even at high concentration (1000 μg/mL). The high selectivity is understandable since this approach first involves an HPLC separation step. However, thiamphenicol (TAM) (Fig. 3, 17) did cause interference. The detection method was also tested with serum samples with high reproducibility. Overall, this approach offers a selective and rapid method for CAP detection, and is able to discriminate between similar molecules that possess a distinct arrangement of functional groups. The same group also further examined the possibility of in situ preparation of the imprinted polymer in an HPLC column [34]. The results were somewhat mixed, but comparable to that of bulk polymers packed into an HPLC column.

IV. MIP FLUORESCENT SENSORS THAT HAVE A FLUOROPHORE INCORPORATED

Although the previous two types of examples showed promising results, the imprinted polymers themselves are not fluorescent. Such polymers are essentially used only as recognition moieties. The fluorescent signal comes from something that is external to the "sensors" themselves. Such kind of design severely limits their application potentials as sensors, although they have proven to be very useful as analytical tools. Ideally, one would like to have sensors that are fluorescent, and binding of the analytes results in an intrinsic change of the fluorescence either in wavelength or intensity or both. This approach requires the incorporation of a fluorophore into the imprinted polymer as a fluorescent reporter. If binding of an analyte perturbs the fluorescent properties of the imprinted polymer, a detectable signal can be generated (Fig. 1, Scenario 3a). This approach requires careful design of a system that results in analyte–fluorophore interactions that would change the fluorescent properties of the incorporated fluorophore. If binding does not perturb the fluorescence to a sufficient degree, an external modifier, a quencher or an enhancer, can be added in a three-component assay (Fig. 1, Scenario 3b).

A. The Detection of Analytes that are Capable of Modifying the Fluorescent Properties of the Fluorescent Reporter in Imprinted Polymers

The design and synthesis of fluorescent imprinted polymers that respond to the binding event with fluorescence changes require careful consideration of the intermolecular interactions expected in the binding and careful selection of the appropriate fluorophore. There have been several examples reported in the literature that deal with the design and synthesis of special fluorescent recognition moieties that respond to the binding event with intrinsic fluorescence intensity changes. The monomers used and polymerization conditions for each such study are summarized in Table 3.

We have recently reported our efforts in developing fluorescent sensors for carbohydrates [35,36]. It has been known for a long time that boronic acid can recognize diol structures with high affinity and selectivity. This property has been extensively explored for the preparation of sugar sensors through de novo design, among other applications [37]. In the 1980s, Wulff [38] and Wulff and Schauhoff [39] also used boronic acid compound in the preparation of imprinted polymers that can recognize sugar. In an effort to make fluorescence sensors through molecular imprinting, we have designed a fluorescent boronic acid monomer using methacrylate as the polymerizable functional group. The design of the fluorescent moiety borrows from the work of Shinkai et al. [40]. In this monomer (Fig. 4, **18**), anthracene is the fluorophore. However, the presence of a benzylic amine significantly reduces the fluorescence of the anthracene moiety due to the quenching of its fluorescence by the nitrogen lone pair electrons through photoelectron transfer (PET). It is known that the Lewis acidity of the boron atom in boronic acid increase tremendously upon ester formation with diols (**18a**). In our own determination, the pKa of phenylboronic acid decreases from 8.8 to 6.8 and 4.5 upon ester formation/binding with D-glucose and D-fructose, respectively. This decreased pKa of the boron atom allows for the nitrogen lone pair electrons to be donated to the open shell of boron, and therefore abolishes or tremendously reduces the fluorescence quenching by the nitrogen lone pair electrons. Consequently, binding of sugars or other diols to a boronic acid compound analogous to the monomer (**18**) can cause a significant increase in fluorescence. Based on these known properties, we designed and synthesized a fluorescent monomer (Fig. 4, **18**) that allows for the preparation of fluorescent sensors of *cis* diols using molecular imprinting methods. This monomer has been used for the preparation of imprinted polymers as sensitive fluorescent sensors for D-fructose using both AIBN-initiated free radical polymerization and atom transfer radical polymerization (ATRP). As designed, the

Table 3 The Preparation of Imprinted Polymers that have a Fluorophore Incorporated

Analyte	Fluorophore chosen	Monomer	Cross-linker	Fluorophore: template: monomer: cross-linker	Polymerization conditions[a]	Reference
D-Fructose	Anthracene	HEMA	EDMA	1:4:50	Initiator:AIBN Porogen: THF/CH$_3$CN Temp.: 65°C Time: 45 h	[35,36]
Sialic acid	OPA	Allylamine VPBA	EDMA	1:1:1:30	Initiator: AIBN Porogen: DMF Temp.: 80°C Time: 76 h	[41]
cAMP	24	HEMA	TRIM	1:1.4:62:1028	Initiator: AIBN Porogen: CH$_3$OH Temp.: 60°C Time: 24 h	[42]
26a,b	27		TRIM	1:1:676	Initiator: AIBN Porogen: PhCH$_3$ Temp.: 50°C Time: 48 h	[43]
28	29-31		TRIA	1:1:600	Initiator: AIBN Porogen: CHCl$_3$ Temp.: 60°C Time: 24 h	[44]
9-EA 32	Porphyrin	MAA and 33	EDMA	1:1:2:48	Initiator: AIBN Porogen: CHCl$_3$ Temp.: 45°C Time: 16 h	[45]

[a]AIBN: 2,2′-azobis(isobutyronitrile).

fluorescence intensity of the imprinted polymers indeed increased upon binding with the template sugar (fructose) by several folds (Fig. 5). The imprinted polymers created with both ATRP polymerization and AIBN-initiated free radical polymerization were able to preferentially recognize the template molecule (Fig. 6), and the imprinted polymer prepared with ATRP seems to have a better selectivity than the imprinted polymer prepared using AIBN-initiated free radical polymerization. Such studies for the first time demonstrated that carefully designed monomeric units can be used for the construction of imprinted polymers that show intrinsic fluorescence intensity changes upon binding to the analyte. Such responses are specific because the mechanism of the fluorescence intensity changes depends on the specific binding event, not nonspecific intermolecular interactions. However, further fine-tuning of the system is needed before this can be used in practical applications.

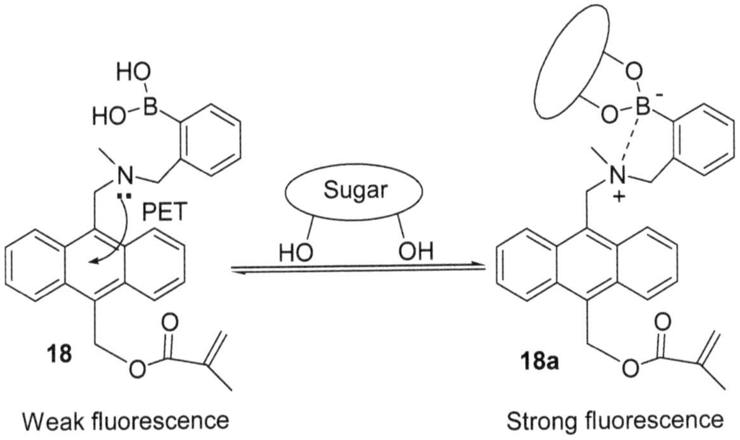

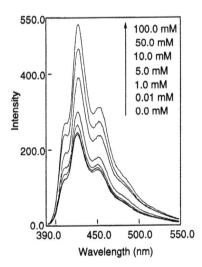

Figure 4 The boronic acid-based fluorescent monomer (**18**) used for the preparation of imprinted polymeric sensors for carbohydrates.

B. Protocol of the Preparation of the Fluorescent Sensor for D-Fructose [35,36]

1. Reagents and Equipment

D-Fructose, anhydrous dioxane, anhydrous pyridine, anhydrous THF, CH_3CN, HEMA, EDMA, AIBN, CuCl, 2,2′-bypyridine, anhydrous DMF, ethyl 2-bromoisobutyrate, MeOH, 0.1 N HCl, 50% $MeOH/H_2O$ (v/v) phosphate (0.05 M) buffer (pH 7.4). Thermolyn vortex mixer (Type 16700), INSTECT Model 1060 stirrer, quartz cuvette, spectrofluorometer.

Figure 5 A typical set of fluorescence spectra of D-fructose imprinted polymer at different concentrations of D-fructose. Reproduced with permission from Ref. 35.

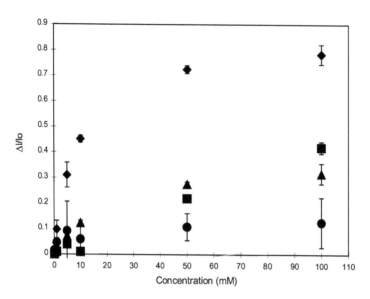

Figure 6 Fluorescence intensity changes of D-fructose imprinted polymers vs. concentration of sugars ($\blacklozenge$, D-fructose; $\blacktriangle$, D-mannose; $\blacksquare$, D-glucose; $\bullet$, control polymer) (λ_{ex}, 370 nm; λ_{em}, 426).

2. Methods

(a) Synthesis of the Fluorescent Monomer 18. The synthesis of fluorescent monomer **18** has been described in detail elsewhere [35,36] and, therefore, is only briefly described in Scheme 1.

(b) Preparation of the D-Fructose Boronate Complex. A mixture of D-fructose (18 mg, 0.1 mmol) and boronic acid monomer **18** (91 mg, 0.2 mmol) in 9.0 mL of dry dioxane and 1.0 mL of dry pyridine was stirred at 45–50°C for 26 h, then 65–70°C for 1 h. Then the solvents and water were removed under reduced pressure during 1.5 h to afford a yellow oil. The residue was further dried under an oil pump overnight. The complex was used directly without purification for the next step polymerization. FAB–MS m/z: 1015 ($M^+ + H$).

(c) Preparation of the D-Fructose Imprinted Polymer Using AIBN-Initiated Free Radical Polymerization. To a solution of the D-fructose-boronate complex (101 mg, 0.1 mmol) in 2.5 mL of dry THF and 1.5 mL of CH_3CN were added HEMA (42 mg, 0.4 mmol) and EDMA (980 mg, 5.0 mmol). The mixture was purged with argon for 5 min, then AIBN was added. The reaction flask was capped with a glass stopper, then placed in an oil bath. The polymerization was carried out at 65°C with stirring for 45 h, then 90°C for 4 h. The rigid polymer was ground and then washed with MeOH (2 × 35 mL) for 1.5–2 h, 0.1 N HCl aqueous solution with MeOH (1/1, v/v) (3 × 35 mL) for 3–5 h, and MeOH (10 × 35 mL) for 22–24 h with a fritted glass filter. The washed polymers were dried at 35°C under reduced pressure overnight to afford 880 mg (79%) of the product. The polymer particles were sieved with a 100-mesh sieve. The polymer particles smaller than 100 mesh were used for the binding studies.

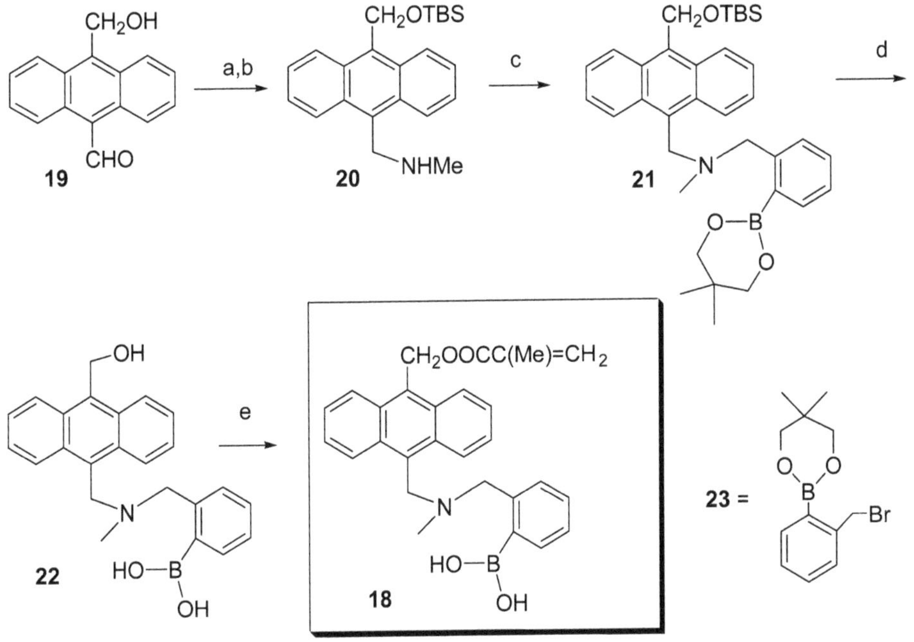

(a) TBSCl/imidazole/DMF, 91%; (b) (i) NH$_2$Me/MeOH, 6 h, (ii) NaBH$_4$/MeOH, 2 h, 89%; (c) **23**, K$_2$CO$_3$/MeCN, reflux, 42%; (d) TBAF/THF, 74%; (e) methacrylic anhydride, DMAP/TEA/CH$_2$Cl$_2$, 70%.

Scheme 1 Synthesis of the boronic acid-containing functional monomer **18**.

(d)   Preparation of the D-Fructose Imprinted Polymer Using Atom Transfer Radical Polymerization.   Copper chloride (CuCl) (12 mg, 0.12 mmol) and 2,2'-bypyridine (bpy) (39 mg, 0.24 mmol) were mixed in a round flask. It was degassed by vacuum for 1 h, and back filled with argon for three times. To this flask was added the solution of HEMA (3.12 g, 24 mmol), EDMA (4.75 g, 24 mmol) and the D-fructose–boronate complex (122 mg, 0.12 mmol) in 5 mL of anhydrous DMF, which was degassed previously by bubbling argon for 1 h. Then 72 μL of ethyl 2-bromoisobutyrate (BriB) was added, the polymerization was proceeded at 50°C for 12 h. The polymer formed was ground to fine particles and then washed with MeOH (2 × 50 mL) for 1.5–2 h, 0.1 N HCl aqueous solution with MeOH (1/1, v/v) (3 × 50 mL) for 3–5 h, and MeOH (10 × 50 mL) for 22–24 h with a fritted glass filter. The washed polymer particles were dried at 35°C under reduced pressure overnight to afford 6.52 g (82%) of the product. The polymer particles were sieved with a 100-mesh sieve. The polymer particles smaller than 100 mesh were used for the binding studies.

(e)   Preparation of the Control Polymer.   The control polymer was prepared following the same procedure for the preparation of the D-fructose imprinted polymers in the absence of the print molecule, D-fructose.

(f)   Fluorescent Binding Studies.   For each fluorescence measurement, 10 mg of the polymer particles were added to a 15 mL test tube, followed by the addition of 4.0 mL of D-fructose solution at different concentrations in 50% MeOH/H$_2$O (v/v)

phosphate (0.05 M) buffer at pH 7.4. The suspension was mixed by using a Thermolyn vortex mixer for 3 h. Then about 3.5 mL of the suspension were transferred into a cuvette for fluorescence measurement. The emission spectra were recorded from 380 to 550 nm immediately after the solution was stirred with an INSTECT stirrer for 30 s. The excitation wavelength was set at 370 nm with both the excitation and the emission slits of 1.5 nm.

Karube et al. [41] have reported a type of fluorescent receptor system specific for sialic acid. This artificial receptor was prepared by polymerization of allylamine, vinylphenylboronic acid (VPBA) and EDMA in the presence of sialic acid as the template. A boronic acid monomer was used because it is known to have high affinity recognition for structures that have diols, such as in a sugar, as described above. The general polymerization conditions are listed in Table 3. After the removal of the template molecules from the imprinted polymers, the polymer was allowed to react with an OPA (o-phtaleic aldehyde) reagent (mixture of o-phtaleic dialdehyde with β-mercaptoethanol). The fluorescence intensity of such OPA-modified polymer increased with the concentration of sialic acid. Most other sugars tested, such as glucose and mannose, showed less change in the fluorescent emission compared to sialic acid indicating that the imprinted polymer has certain selectivity for the template molecule. However, in most cases, the fluorescence intensity differences among different sugars were no more than 30%, even in the best-case scenario. It was said that the detection of sialic acid in the concentration range of 0.5–10 μM was shown to be possible. However, considering the cross-reactivity with other sugars, this detection (or more appropriately concentration determination) is only possible with samples without other sugar present. It is also interesting to note that the best results were obtained when borate buffer was used, and yet borate is known to bind to sugar in much the same fashion as boronic acid, which is the recognition moiety used in the imprinted polymer. One would expect that the competing binding would lower the affinity of the sugar to the imprinted polymer. Another unique situation with the method described in this paper is the postimprinting modification, which may negatively affect the affinity of the imprinted polymer for the analyte. It will be interesting to see what happens when similar fluorophores are used as one of the functional monomers, and therefore incorporated into the polymer through the imprinting process.

Powell et al. [42] have developed a fluorescent chemosensor for adenosine 3′:5′-cyclic monophosphate (cAMP) using imprinted polymers that contain a fluorescent dye, $trans$-4-(p-(N,N-dimethylamino)styryl)-N-vinylbenzylpyridinium chloride (Fig. 7, **24**), as an integral part of the binding cavity which serves as both a recognition element and the measuring element for the fluorescence detection of cAMP (Fig. 7, **25**) in aqueous media. The positive charge of **24** is designed to interact with the negatively charged oxygen anion, and stacking could also play an important role for their interactions. Furthermore, the fluorescence of compounds such as **24** is expected to be very sensitive to the environment due to its intramolecular charge transfer behavior, and interaction with the anion of the phosphate group perturbs such a system. Therefore, binding of the template, cAMP, to the imprinted polymer causes a drop in fluorescence intensity of the imprinted polymer due to fluorescence quenching. The fluorescent polymer was prepared by adding compound **24** to the template molecule, cAMP, and then polymerization with 2-hydroxyethyl methacrylate (HEMA) as the functional monomer and trimethylolpropane trimethacrylate (TRIM) as the cross-linker. The imprinted polymers displayed high affinity for

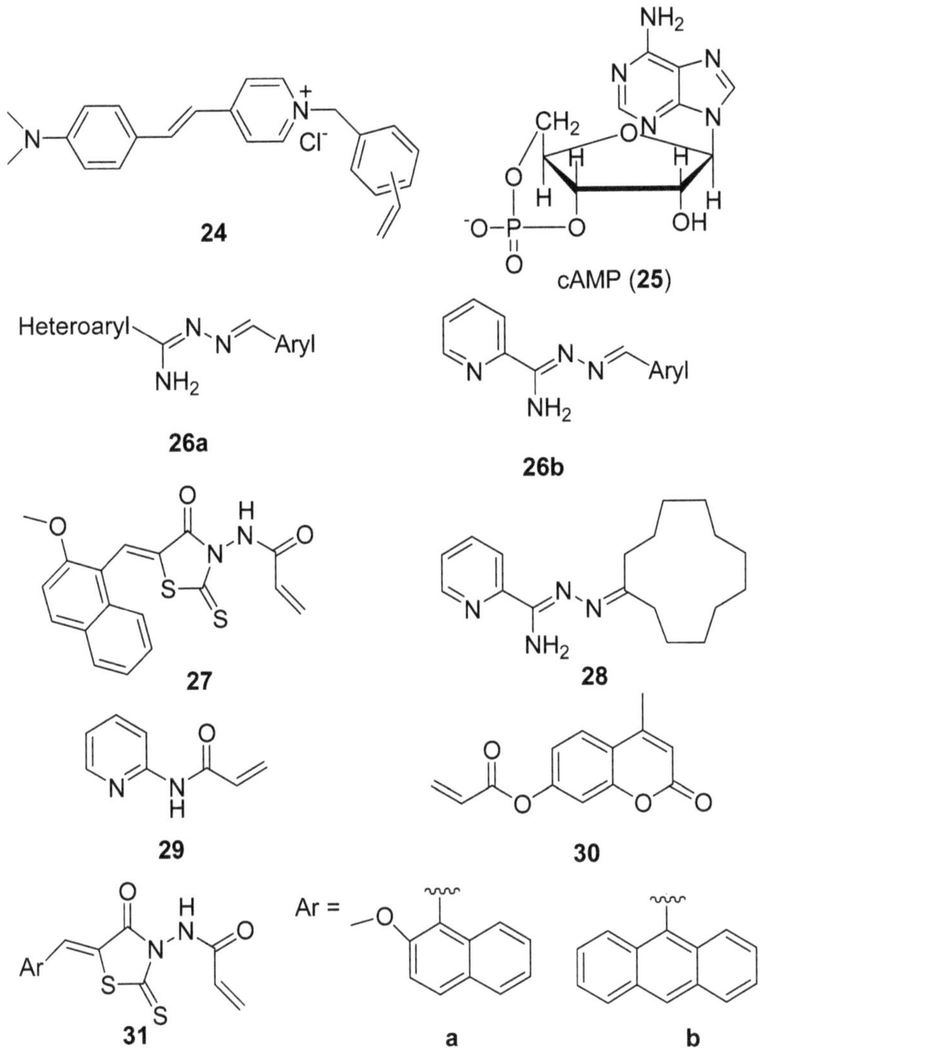

Figure 7 List of analytes.

cAMP with a K_a of about $3.5 \times 10^5 \, \text{M}^{-1}$ in aqueous solution. The imprinted polymers also showed good selectivity. For example, addition of cGMP induced almost no effect on the fluorescence intensity of the imprinted polymer.

Fluorescent polymers imprinted with various N^1-benzylidene pyridine-2-carboxamidrazones (Fig. 7, **26a,b**) were prepared by Rathbone and Ge [43] aimed at developing a high throughput screening of potential medicinal agents. Extensive and well-thought out efforts went into the design of an appropriate fluorescent monomer for the polymer synthesis. Based on computational chemistry work, it was thought that a 3-aminorhodanine-based compound **27** should be able to engage in hydrogen bond interactions with the target analytes. Imprinted polymers were prepared using the monomer designed and trimethylolpropane triacrylate as the cross-linker in toluene. As designed, the imprinted polymer indeed showed a significant fluorescence intensity change (decrease by up to 24-fold in some cases) upon re-binding with the template molecule. This was thought to be due to the quenching of the fluorescence of the monomeric unit by the template molecule through hydrogen-bond interaction. Very good selectivities were observed among analogs. It was also found that compounds with less freedom of rotation afforded better recognition. Another interesting aspect of this work is the examination of the binding and recognition events by a ratiometric method through the excitation at different wavelengths. As can be expected, concentration determination based on absolute fluorescent intensity can be affected by many factors, which could influence the accuracy. However, a ratiometric method offers an "internal standard", and offers a better chance for observing minor differences in the recognition process. In a related study, the same group also reported the preparation of imprinted polymers that recognize cyclododecylidene pyridine-2-carboxamidrazone (Fig. 7, 28) using similarly designed fluorescent monomers [44]. In this situation, 2-aminopyridine, 7-hydroxy-4-methylcoumarin, and 3-aminorhodanine were tested as potential hydrogen bond donor/acceptor scaffolds from which to construct the desired monomers (Fig. 7, 29–31), with monomer 31 giving the best results.

Takeuchi and co-workers [45] have studied the preparation of imprinted polymers using metalloporphyrin as the recognition center. In this study, 9-ethyladenine (9-EA) (Fig. 8, 32) was used as a model compound. It was found that the use of two functional monomers MAA and a Zn–porphyrin-based monomer (Fig. 8, 33) significantly increased the affinity of the resulting polymer for the template molecule. Upon addition of 9-EA to the imprinted polymer significant fluorescence intensity

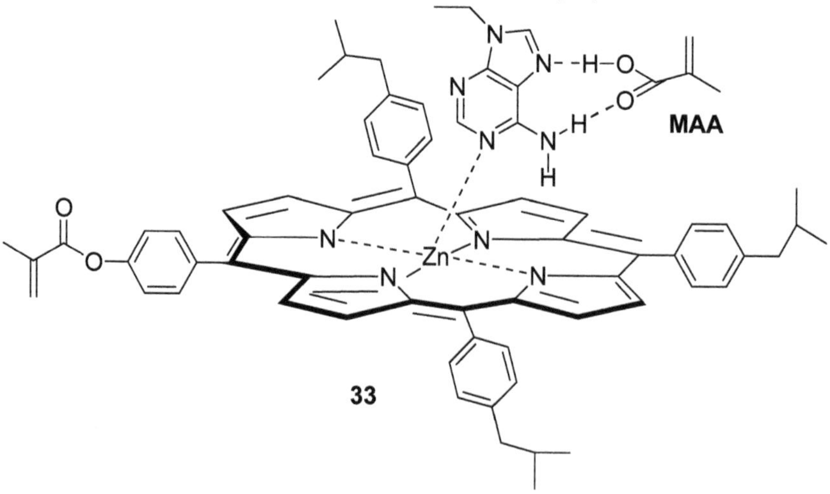

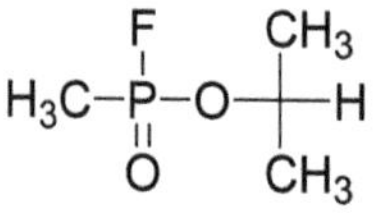

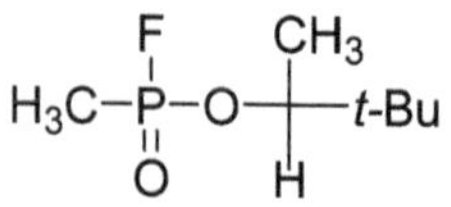

Figure 8 List of analytes.

decrease, presumably due to quenching, was observed. The sensitivity seemed to be much higher in the low concentration region, exhibiting a biphasic mode. The highest *Ka* was determined to be about $7.5 \times 10^5\ M^{-1}$. It is interesting to note that an earlier study from the same group indicated that a red shift of up to 10 nm could occur upon binding of imprinted polymers in chloroform when Zn–porphyrin was used as the monomer without using MAA [46]. It was not clear whether the same wavelength shift occurred or not with the imprinted polymers prepared using two functional monomers, and how this would affect the fluorescence reading.

Sensor devices based on lanthanide luminescence were constructed to measure and detect the nerve agents Sarin and Soman (Fig. 8, **34**, **35**). In this approach, imprinted polymers containing fluorescent Eu^{3+} were prepared. The imprinted polymers fluoresce at 610 nm upon binding with the phosphonate hydrolysis product of nerve gas agents Soman and Sarin [47]. The device developed has a remarkable detection limit of 10 ppt to 10 ppm [48]. Along the same line, the same type of recognition moieties was used for the preparation of imprinted polymers of pesticides and insecticides that have organophosphate groups with sensitivities of up to 10 ppt [49].

C. The Use of an External Fluorescence Quencher or Modifier

For analytes that cannot interact with the incorporated fluorophore strongly enough to cause a significant change in the fluorescence intensity and/or wavelength, one can use a nonspecific external quencher or modifier to affect the fluorescence change upon analyte binding (Fig. 1, Scenario 3b).

Wang et al. [50] have developed a method that could be generally applicable to the preparation of fluorescent sensors for nonfluorescent molecules by using an external fluorescence quencher in studying the binding of the imprinted fluorescent polymer. L-Tryptophan (Fig. 9, **36**) was chosen as the template molecule and the

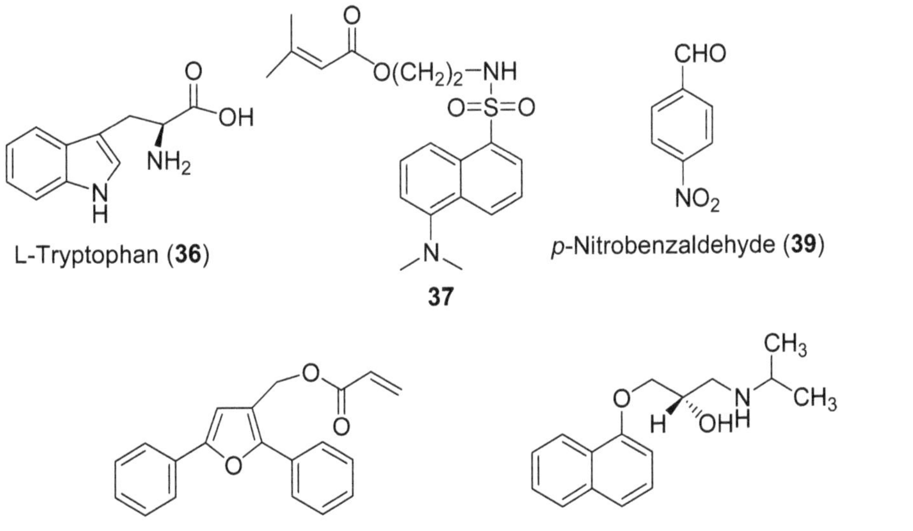

L-Tryptophan (**36**)

37

p-Nitrobenzaldehyde (**39**)

4-Hydroxymethyl-2,5-diphenyloxazolemacrylate (**40**)

(*S*)-Propranol (**41**)

Figure 9 List of analytes and monomeric materials.

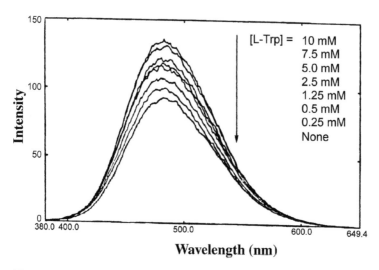

Figure 10 The effect of L-tryptophan on the fluorescence intensity of the L-tryptophan imprinted polymer in the presence of 3 mM of *p*-nitrobenzaldehyde. Reproduced with permission from Ref. 50.

imprinted polymers were prepared using 3,3-dimethylacrylic acid-based monomers. The dansyl moiety, attached to 3,3-dimethylacrylic acid via an ethanolamine linker, was used as the fluorescent tag (Fig. 9, **37**). For the fluorescence investigation, 4-nitrobenzaldehyde (Fig. 9, **39**) was used as an external fluorescence quencher. In the absence of the target molecules, the quencher resulted in quenching the fluorescence of the fluorophore. However, in the presence of the target molecules, the displacement of the quencher from the binding cavity by the template molecules increased the fluorescence intensity of the imprinted polymer in a concentration-dependent fashion (Fig. 10). The sensor exhibited selectivity among different amino acids such as phenylalanine and alanine, and enantioselectivity for the template molecule. The synthesis of the fluorescent monomer **36** follows procedures described in Scheme 2.

D. Protocols for Preparation of the Fluorescent Sensor for L-Tryptophan [50]

1. Reagents and Equipment

Acetonitrile, L-tryptophan, ethylene glycol dimethacrylate (EGMA), *p*-nitrobenzaldehyde, acetic acid, AIBN, solution of 20% acetic acid in methanol, solution of 10% acetic acid in methanol, methanol, and CHCl$_3$.

Thermolyn vortex mixer (Type 16700), INSTECT Model 1060 stirrer, quartz cuvette, Shimadzu RF-5301PC spectrofluorometer.

2. Methods

(a) Synthesis of the Dansyl-Modified Fluorescent Monomer 37 [50]. The synthesis of the dansyl-modified fluorescent monomer is briefly described in Scheme 2. More detailed experimental procedures can be found in the relevant reference [50].

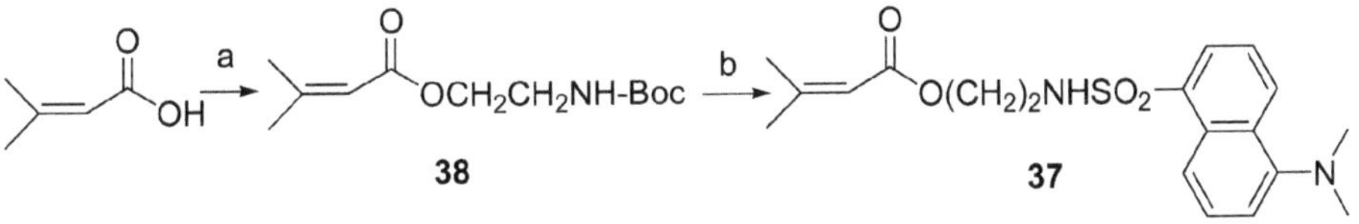

a. Boc-NH-CH$_2$CH$_2$OH, DCC, DMAP, CH$_2$Cl$_2$, 89%; b. i) 25% TFA in CH$_2$Cl$_2$; ii) Dansyl-Cl, NaHCO$_3$/EtOAc, 80%.

Scheme 2 Synthesis of dansyl-modified 3,3-dimethylacrylate fluorescent monomer **37**.

(b) Polymerization for the Preparation of the Imprinted Polymer. To a 50-mL round bottom flask was added 314 mg (0.835 mmol) of the fluorescent monomer **37** and 0.5 mL of acetonitrile at 0°C. Then 43 mg (0.211 mmol) of L-tryptophan was added to the solution with stirring. This was followed by the addition of 1.663 g (8.399 mmol) of ethylene glycol dimethacrylate. Then 126 µL of acetic acid and 10.5 µL of TFA were added to dissolve the L-tryptophan. The reaction flask was flushed with nitrogen for about 45 min. Then 20 mg (0.122 mmol) of AIBN was added. After stirring for 2 h more, the reaction temperature was increased to 45°C using an oil bath. The reaction was kept at 45°C for 48 h under a continuous flow of nitrogen.

After polymerization, the polymer was dried in a vacuum oven at 35°C overnight. Then the polymer was ground with a mortar and sieved with a 100-mesh sieve. The sieved polymer was then washed with 100 mL of 20% acetic acid in methanol, 200 mL of 10% acetic acid in methanol, and 100 mL of methanol with a fritted glass filter. Then the polymer was dried again in a vacuum oven at 35°C for 12 h. The polymerization yield was about 60%.

(c) Fluorescent Measurements of the Imprinted Polymer Without the Addition of the Quencher, p-Nitrobenzaldehyde. For each fluorescence measurement, 40 mg of the L-tryptophan imprinted polymer was added to a 25 mL tube, followed by the addition of 4 mL of CHCl$_3$. Then 4 mL of the amino acid at different concentrations in 0.03 M citric acid aqueous solution was added. The two phases were then mixed by using a Thermolyn Mixer for 4 h. After waiting for 3 min to allow for the mixture to separate into two layers, the tube was gently shaken to allow the polymer to evenly suspend in the CHCl$_3$ phase. Then 3 mL of CHCl$_3$ suspension was transferred into a cuvette for fluorescence measurement. The emission spectra were recorded from 380 to 650 nm immediately after the solution was stirred with an INSTECH stirrer for 15 s.

(d) Fluorescent Measurements of the Imprinted Polymer with the Addition of the Quencher, p-Nitrobenzaldehyde. A stock solution of 0.1 M *p*-nitrobenzaldehyde in CHCl$_3$ was used to adjust the concentration of the quencher. Otherwise, the same procedure was followed as for the experiments without the added quencher, *p*-nitrobenzaldehyde.

In an approach with an external modifier that increases the fluorescence of the polymer, an external ^{3}H-labeled template was used to generate the fluorescent signal through proximity scintillation [51]. The difference in this case is that the binding of ^{3}H-labeled template gives a fluorescent complex, and its displacement by the analyte results in a decrease in fluorescence. Therefore, a scintillation monomer, 4-hydroxymethyl-2,5-diphenyloxazole acrylate (Fig. 9, **40**), was incorporated into the imprinted polymer. Since β particles travel only a short distance, the β emission of tritium-labeled template only triggers the fluorescence of the fluor when bound. Therefore, when the imprinted polymer and the tritium-labeled template are mixed, the imprinted polymer fluoresces strongly. However, addition of the template molecule competes for the same binding sites and proportionally decreases the fluorescence intensity. This strategy was used for the preparation of polymeric sensors for propranolol (Fig. 9, 41). Comparing with the regular radioisotope displacement assay, this scintillation system avoids the separation step of the unbound radioactive ligands.

V. OTHER SENSORS

It is possible to design an optical sensor by using techniques other than fluorescence as the detection mode, if there is a change in an optical property of the imprinted polymer after the analyte interacts with the imprinted polymer. Other detection methods include surface plasmon resonance spectroscopy [52–54], Raman [55–57], IR [30,58], photo- and chemiluminescence [59,60] among others [61–63]. A few selected examples are described in the section below.

A sensor system based on the optical phenomenon of surface plasmon resonance (SPR) was developed that can be used with imprinted polymers as the sensing elements. In this particular study, MAA was used as the functional monomer and EDMA was used as the cross-linker [54]. Imprinted polymers for theophylline, caffeine, and xanthine (Fig. 11, 42–44) were prepared. These polymers were all shown to have good selectivity for the template molecule used. For detection, the polymer was layered over a silver film to serve as the analysis surface for the molecularly imprinted sorbent assay (MIA). Binding of the analyte to the imprinted polymers was examined by monitoring the shifts in the SPR angle θ_r. The linear dynamic range of the MIA was found to extend up to 6 mg/mL, with a concentration detection limit estimated at 0.4 mg/mL for theophylline in aqueous solution. In cross-reactivity studies for theophylline and caffeine imprinted polymers, eight other structurally similar compounds were examined and most showed no or very slight shifts in θ_r

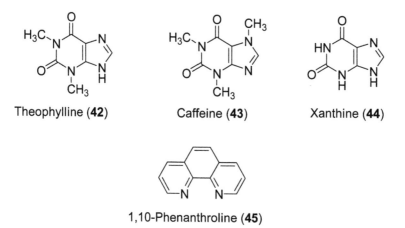

Theophylline (42)   Caffeine (43)   Xanthine (44)

1,10-Phenanthroline (45)

Figure 11 List of analytes.

as compared with the template molecule, indicating the selectivity of the imprinted polymers for the "print" molecule.

The band-edge photoluminescence (PL) of *n*-CaSe has been shown to respond to the adsorption of a variety of analytes to the semiconductor's surface. This conceivably can be used for sensor development. However, one major problem is the issue of selectivity. Although Lewis bases and acids can be readily distinguished due to their differential effect on the electronic properties of the semiconductor, analyte-specific analysis is difficult to achieve. Ellis and co-workers [64] have examined the effect of imprinted polymer coating on the surface on the response of the semiconductor. Without the coating, the bare surface of CdSe responds to the adsorption of ammonia, mono-, di-, and trimethylamine with similar PL enhancement. However, upon coating the surface with ammonia imprinted poly(arylic acid) (PAA), CdSe only responded to the presence of ammonia, but not trimethylamine. On the other hand, CdSe coated with trimethylamine-imprinted polymer does not provide this selectivity, indicating the selectivity was mostly due to the size effect.

The Lin group has developed an interesting concept of combining molecular imprinting recognition and chemiluminescence for the detection of 1,10-phenanthroline (Fig. 11, **45**) [59,60]. The idea is based on the fact that metal [Co(II) or Cu(II)] complexes with phenanthroline and pyridine can catalyze the decomposition of hydrogen peroxide. The superoxide radical ion formed in the process can in turn react with phenanthroline giving rise to chemiluminescent species. Therefore, incorporation of the ternary complex of vinylpyridine-Cu(II) -1,10-phenanthroline into a polymer through template-polymerization gives rise to polymers that are capable of recognizing 1,10-phenanthroline itself. The detection limit in a flow through system was in the low micromolar range with the limit for the Cu(II) lower than that of the Co(II).

VI. CONCLUSIONS AND PERSPECTIVES

Overall, a great deal of progress has been made during the last few years in the preparation of optical sensors using molecular imprinting techniques. Particularly important are those studies using tailor-made monomers for the specific recognition of selected analytes and structural features. Further design of functional group-specific fluorescent recognition moieties will undoubtedly help the advancement of this field. However, fundamental progress in this area will also heavily rely on the advancement of new ways to create imprinted polymers. With the most commonly used acrylic acid-based monomers, the polymerization process involves the conversion of two sp^2 carbons per monomer to sp^3 carbons. Such a gross change in configuration means that even if there are optimal analyte–monomer interactions before polymerization, the polymerization process will severely distort the binding cavities thus formed from that of the "optimal" shape and functional group arrangements. This presents a tremendous problem in the preparation of imprinted polymers with high selectivity and affinity. One can envision that a substitution-based polymerization method with properly designed monomers should not result in such kinds of gross changes in the configuration upon polymerization. Furthermore, a slow stepwise polymerization (e.g., living polymerization) will also allow the polymers to adjust to and compensate for these configurational changes through the re-alignment of new functional monomers. We also need to recognize that the currently available imprinting methods give the equivalents of "polyclone" antibody mimics. It will be very hard for these "polyclone" mimics to reach the kind of selectivity and specificity desired for highly specific sensors. Therefore, it will be important to find ways to make "monoclone" versions of imprinted polymers and have ways to select and optimize the "receptors" prepared. To address all these problems will require expertise from more than one field. Thus it is hoped that this chapter and this book will help in further communications among interested groups so that these fundamental problems can be collectively addressed.

ACKNOWLEDGMENT

Work performed in our laboratory has been supported by the National Institutes of Health (DK55062, NOI-CO-27184 and CA88343).

REFERENCES

1. de Silva, A.P.; Gunaratne, H.Q.N.; Gunnlaugsson, T.; Huxley, A.J.M.; McCoy, C.P.; Rademacher, T.; Rice, T.E. Signaling recognition events with fluorescent sensors and switches. Chem. Rev. **1997**, *97*, 1515–1566.
2. Fabbrizzi, L.; Poggi, A. Sensors and switches from supramolecular chemistry. Chem. Soc. Rev. **1995**, *24*, 197–202.
3. Scheller, F.W.; Schubert, F.; Fedrowitz, J. In *Frontiers in Biosensorics I. Fundamental Aspects*; Birkhauser Verlag: Berlin, 1997, 1–175 pp.
4. Scheller, F.W.; Schubert, F.; Fedrowitz, J. In *Frontiers in Biosensorics II. Practical Applications*. Birkhauser Verlag: Berlin, 1997, 1–287 pp.
5. Chen, C.T.; Wagner, H.; Still, W.C. Fluorescent, sequence-selective peptide detection by synthetic small molecules. Science **1998**, *279*, 851–853.
6. Torrado, A.; Walkup, G.K.; Imperiali, B. Exploiting polypeptide motifs for the design of selective Cu(II) ion chemosensors. J. Am. Chem. Soc. **1998**, *120*, 609–610.
7. Watanabe, S.; Onogawa, O.; Komatsu, Y.; Yoshida, K. Luminescent metalloreceptor with a neutral bis (acrylaminoimidazoline) binding site: optical sensing of anionic and neutral phosphodiesters. J. Am. Chem. Soc. **1998**, *120*, 229–230.
8. Marsella, M.J.; Newland, R.J.; Carroll, P.J.; Swager, T.M. Ionoresistivity as a highly sensitive sensory probe: investigation of polythiophenes functionalized with calix[4]arene-based ion receptors. J. Am. Chem. Soc. **1995**, *117*, 9842–9848.
9. Winkler, J.D.; Bowen, C.M.; Michelet, V. Photodynamic fluorescent metal ion sensors with parts per billion sensitivity. J. Am. Chem. Soc. **1998**, *120*, 3237–3242.
10. Xia, W.S.; Schmehl, R.H.; Li, C.J. A highly selective fluorescent chemosensor for K+ from a bis 15-crown-5 derivative. J. Am. Chem. Soc. **1999**, *121*, 5599–5600.
11. Marvin, J.S.; Hellinga, H.W. Engineering biosensors by introducing fluorescent allosteric signal transducers: construction of a novel glucose sensor. J. Am. Chem. Soc. **1998**, *120*, 7–11.
12. James, T.D.; Sandanayake, K.R.A.S.; Shinkai, S. Chiral discrimination of monosaccharides using a fluorescent molecular sensor. Nature (London) **1995**, *374*, 345–347.
13. Kataoka, K.; Hisamitsu, I.; Sayama, N.; Okano, T.; Sakurai, Y. Novel sensing system for glucose based on the complex formation between phenylborate and fluorescent diol compounds. J. Biochem.-Tokyo **1995**, *117*, 1145–1147.
14. Ueno, A.; Ikeda, H.; Wang, J. Signal Transduction in Chemosensors of Modified Cyclodextrins; Desvergne, J.P., Czarnik, A.W., Ed.; In *Chemosensors of Ion and Molecule Recognition*, Desvergne, J.P., Czarnik, A.W., Eds.; Kluwer Academic Publishers: The Netherlands, 1997, pp. 105–119.

15. Cooper, M.E.; Hoag, B.P.; Gin, D.L. Design and synthesis of novel fluorescent chemosensors for biologically active molecules. Polym. Prepr. **1997**, *38*, 209–210.
16. Huston, M.E.; Akkaya, E.U.; Czarnik, A.W. Chelation enhanced fluorescent detection of non-metal ions. J. Am. Chem. Soc. **1989**, *111*, 8735–8737.
17. Marsella, M.J.; Carroll, P.J.; Swager, T.M. Design of chemoresistive sensory materials: polythiophene-based pseudopolyrotaxanes. J. Am. Chem. Soc. **1995**, *117*, 9832–9841.
18. Zhou, Q.; Swager, T.M. Fluorescent chemosensors based on energy migration in conjugated polymers: the molecular wire approach to increased sensitivity. J. Am. Chem. Soc. **1995**, *117*, 12593–12602.
19. Kriz, D.; Ramström, O.; Svensson, A.; Mosbach, K. Introducing biomimetic sensors based on molecularly imprinted polymers as recognition elements. Anal. Chem. **1995**, *67*, 2142–2144.
20. Matsui, J.; Kubo, H.; Takeuchi, T. Molecualrly imprinted fluorescent-shift receptors prepared with 2-(trifluoromethyl)acrylic acid. Anal. Chem. **2000**, *72*, 3286–3290.
21. Suárez-Rodríguez, J.L.; Díaz-García, M.E. Flavonol fluorescent flow-through sensing based on a molecular imprinted polymer. Anal. Chim. Acta. **2000**, *405*, 67–76.
22. Rachkov, A.; McNiven, S.; El'skaya, A.; Yano, K.; Karube, I. Fluorescent detection of β-estradiol using a molecularly imprinted polymer. Anal. Chim. Acta. **2000**, *405*, 23–29.
23. Lulka, M.F.; Chambers, J.P.; Valdes, E.R.; Thompson, R.G.; Valdes, J.J. Molecular imprinting of small molecules with organic silanes: fluorescence detection. Anal. Lett. **1997**, *30*, 2301–2313.
24. Dickert, F.L.; Thierer, S. Molecularly imprinted polymers for optiochemical sensors. Adv. Mater. **1996**, *8*, 987–990.
25. Dickert, F.L.; Besenbock, H.; Tortschanoff, M. Molecular imprinting through van der Waals interactions: fluorescence detection of PHAs in water. Adv. Mater. **1998**, *10*, 149–151.
26. Dickert, F.L.; Tortschanoff, M. Molecularly imprinted sensor layers for the detection of polycyclic aromatic hydrocarbons in water. Anal. Chem. **1999**, *71*, 4559–4563.
27. Haupt, K.; Mayes, A.G.; Mosbach, K. Herbicide assay using an imprinted polymer-based system analogous to competitive fluoroimmunoassays. Anal. Chem. **1998**, *70*, 3936–3939.
28. Surugiu, I.; Danielsson, B.; Ye, L.; Mosbach, K.; Haupt, K. Chemiluminescence imaging ELISA using an imprinted polymer as the recognition element instead of an antibody. Anal. Chem. **2001**, *73*, 487–491.
29. Surugiu, I.; Svitel, J.; Ye, L.; Haupt, K.; Danielsson, B. Development of a flow injection capillary chemiluminescent ELISA using an imprinted polymer instead of the antibody. Anal. Chem. **2001**, *73*, 4388–4392.
30. Jakusch, M.; Janotta, M.; Mizaikoff, B.; Mosbach, K.; Haupt, K. Molecularly imprinted polymers and infrared evanescent wave spectroscopy. A chemical sensors approach. Anal. Chem. **1999**, *71*, 4786–4791.
31. Haupt, K. Molecularly imprinted sorbent assays and the use of non-related probes. Reactive Funct. Polymers **1999**, *41*, 125–131.
32. Piletsky, S.A.; Terpetschnig, E.; Andersson, H.S.; Nicholls, I.A.; Wolfbeis, O.S. Application of non-specific fluorescent dyes for monitoring enantio-selective ligand binding to molecularly imprinted polymers. Fresenius J. Anal. Chem. **1999**, *364*, 512–516.
33. Levi, R.; McNiven, S.; Piletsky, S.A.; Cheng, S.-H.; Yano, K.; Karube, I. Optical detection of chloramphenicol using molecularly imprinted polymers. Anal. Chem. **1997**, *69*, 2017–2021.
34. McNiven, S.; Kato, M.; Levi, R.; Yano, K.; Karube, I. Chloramphenicol sensor based on an in situ imprinted polymer. Anal. Chim. Acta. **1998**, *365*, 69–74.
35. Wang, W.; Gao, S.; Wang, B. Building fluorescent sensors by template polymerization: the preparation of a fluorescent sensor for D-fructose. Org. Lett. **1999**, *1*, 1209–1212.

36. Gao, S.; Wang, W.; Wang, B. Building fluorescent sensors for carbohydrates using template-directed polymerizations. Bioorg. Chem. **2001**, *29*, 308–320.

37. James, T.D.; Sandanayake, K.R.A.S.; Shinkai, S. Saccharide sensing with molecular-receptors based on boronic acid. Angew. Chem. Int. Ed. Eng. **1996**, *35*, 1910–1922.

38. Wulff, G. Selective binding to polymer via covalent bonds. The construction of chiral cavities as specific receptor sites. Pure Appl. Chem. **1982**, *54*, 2093–2102.

39. Wulff, G.; Schauhoff, S. Racemic resolution of free sugars with macroporous polymers prepared by molecular imprinting. Selective dependence on the arrangement of functional groups versus spatial requirements. J. Org. Chem. **1991**, *56*, 395–400.

40. James, T.D.; Sandanayake, K.R.A.S.; Iguchi, R.; Shinkai, S. Novel saccharide-photoinduced electron transfer sensors based on the interaction of boronic acid and amine. J. Am. Chem. Soc. **1995**, *117*, 8982–8987.

41. Piletsky, S.A.; Piletskyaya, E.V.; Yano, K.; Kugimiya, A.; Elgersma, A.V.; Levi, R.; Kahlow, U.; Takeuchi, T.; Karube, I.; Panasyuk, T.I.; El'skaya, A.V. A biomimetic receptor system for sialic acid based on molecular imprinting. Anal. Lett. **1996**, *29*, 157–170.

42. Turkewitsch, P.; Wandelt, B.; Darling, G.D.; Powell, W.S. Fluorescent functional recognition sites through molecular imprinting. A polymer-based fluorescent chemosensor for aqueous cAMP. Anal. Chem. **1998**, *70*, 2025–2030.

43. Rathbone, D.L.; Ge, Y. Selectivity of response in fluorescent polymers imprinted with N1-benzylidene pyridine-2-carboxamidrazones. Anal. Chim. Acta. **2001**, *435*, 129–136.

44. Rathbone, D.L.; Su, D.; Wang, Y.; Billington, D.C. Molecular recognition by fluorescent imprinted polymers. Tetrahedron Lett. **2000**, *41*, 123–126.

45. Matsui, J.; Higashi, M.; Takeuchi, T. Molecularly imprinted polymer as 9-ethyladenine receptor having a porphyrin-based recognition center. J. Am. Chem. Soc. **2000**, *122*, 5218–5219.

46. Matsui, J.; Tachibana, Y.; Takeuchi, T. Molecularly imprinted polymer having metalloporphyrin-based signaling binding site. Anal. Commun. **1998**, *35*, 225–227.

47. Jenkins, A.L.; Uy, O.M.; Murray, G.M. Polymer based lanthanide luminescence sensors for the detection of nerve agents. Anal. Commun. **1997**, *34*, 221–224.

48. Jenkins, A.; Uy, O.M.; Murray, G.M. Polymer-based lanthanide luminescent sensor for detection of the hydrolysis product of the nerve agent soman in water. Anal. Chem. **1999**, *71*, 373–378.

49. Jenkins, A.L.; Yin, R.; Jensen, J.L. Molecularly imprinted polymer sensors for pesticide and insecticide detection in water. Analyst. **2001**, *126*, 798–802.

50. Liao, Y.; Wang, W.; Wang, B. Building fluorescent sensors by template polymerization: the preparation of a fluorescent sensor for L-tryptophan. Bioorg. Chem. **1999**, *27*, 463–476.

51. Ye, L.; Mosbach, K. Polymers recognizing biomolecules based on a combination of molecular imprinting and proximity scintillation: a new sensor concept. J. Am. Chem. Soc. **2001**, *123*, 2901–2902.

52. Li, P.; Huang, Y.; Hu, J.; Yuan, C.; Lin, B. Surface plasmon resonance studies on molecular imprinting. Sensors **2002**, *2*, 35–40.

53. Kugimiya, A.; Takeuchi, T. Surface plasmon resonance sensor using molecularly imprinted polymer for detection of sialic acid. Biosens. Bioelectron. **2001**, *16*, 1059–1062.

54. Lai, E.P.C.; Fafara, A.; VanderNoot, V.A.; Kono, M.; Polsky, B. Surface plasmon resonance sensors using molecularly imprinted polymers for sorbent assay of theophylline, caffeine, and xanthine. Can. J. Chem. **1998**, *76*, 265–273.

55. Al-Obaidi, A.H.R; McStay, D.; Quinn, P.J. In Glaxo Group Limited: UK, United Kingdom PCT Int. Appl. (2002), 52 pp. CODEN: PIXXD2 WO 2002008735 A1 20020131 CAN 136: 131226 AN 2002: 90328.

56. McStay, D.; Quinn, P.J.; Hoskins, R.; Al-Obaidi, A.H. Raman Spectroscopy of Molecular Imprinted Polymers. In *Sensors and Their Applications XI*, Proceedings of

the Conference on Sensors and their Applications, 11th, London, United Kingdom; 2001, Grattan, K.T.V., Khan, S.H., Eds.; Institute of Physics Publishing: 397–401.

57. Rosen, R.B.; Kruger, E.F.; Katz, A.; Alfano, R.R. US Pat. 2002095257 A1 20020718 Appl. Publ. **2002**.

58. Mizaikoff, B. Proceedings of SPIE; The International Society for Optical Engineering: 1999; Vol. 3849, 7–18.

59. Lin, J.-M.; Yamada, M. Chemiluminescent flow-through sensor for 1,10-phenanthroline based on the combination of molecular imprinting and chemiluminescence. Analyst **2001**, *126*, 810–815.

60. Lin, J.-M.; Yamada, M. Chemiluminescent reaction of fluorescent organic compounds with $KHSO_5$ using Cobalt (II) as catalyst and its first application to molecular imprinting. Anal. Chem. **2000**, *72*, 1148–1155.

61. Murray, G.M.; Uy, O. Ionic sensors based on molecularly imprinted polymers. Tech. Instrum. Anal. Chem. **2002**, *23*, 441–465.

62. Haupt, K.; Mosbach, K. Molecularly imprinted polymers and their use in biomimetic sensors. Chem. Rev. **2000**, *100*, 2495–2504.

63. Mcniven, S.; Karube, I. Toward optical sensors for biologically active molecules. Tech. Instrum. Anal. Chem. **2001**, *23*, 467–501.

64. Nickel, A.-M.L.; Seker, F.; Ziemer, B.P.; Ellis, A.B. Imprinted poly(acrylic acid) films on cadmium selenide. A composite sensor structure that couples selective amine binding with semiconductor substrate photoluminescence. Chem. Mater. **2001**, *13*, 1391–1397.

Appendix: Useful Addresses and Links

SUPPLIERS

Organic Monomers, Polymers, Crosslinkers, and Initiators

PolyScience
6600 W. Touhy Ave.
Niles, IL 60714
USA
Phone: (800) 229-7569, (847) 647-0611
www.polyscience.com

Scientific Polymer Products, Inc.
6265 Dean Parkway
Ontario, NY 14519
USA
Phone: 585-265-0413
www.scientificpolymer.com

Monomer-Polymer & Dajac Laboratories, Inc.
1675 Bustleton Pike
Feasterville, PA 19053
USA
Phone: 215-364-1155

Wako Chemicals USA, Inc.
1600 Bellwood Road
Richmond, VA 23237-1326
USA
Phone: 1-800-992-9256, 804-271-7677
www.wakousa.com

Inorganic Building Blocks

Gelest, Inc.
11 East Steel Road
Morrisville, PA 19067
USA

Phone: (215)547-1015
www.gelest.com

United Chemical Technologies, Inc.
2731 Bartram Road
Bristol, PA 19007-6893
USA
Phone: 800-541-0559, (215) 781-9255
www.unitedchem.com

EVERYTHING ABOUT IMPRINTING AND MORE...!

Society for Molecular Imprinting
www.smi.tu-berlin.de

EQUIPMENT

Milling & Sieving

Retsch GmbH & Co. KG
Rheinische Straße 36
42781 Haan
Germany
Phone: +49 (0) 21 29 / 55 61-0
www.retsch.de

Crescent Dental Mfg. Co.
7750 West 47th Street
Lyons, IL 60534
USA
Phone: (800) 323-8952, (708) 447-8050

Photochemical Equipment

Southern New England Ultraviolet Company
550-29 East Main Street
Branford, CT 06405
USA
Phone: (203) 483-5810

Hanovia Ltd
145 Farnham Road
Slough
Berkshire
SL1 4XB
UK
Phone: +44 (0)1753 515300
www.hanovia.com

Index

Milton Keynes UK
Ingram Content Group UK Ltd.
UKHW051859071024
449327UK00025B/2030